CHROMATIN STRUCTURE AND DYNAMICS: STATE-OF-THE-ART

New Comprehensive Biochemistry

Volume 39

General Editor

G. BERNARDI
Paris

ELSEVIER

Amsterdam · Boston · Heidelberg · London · New York · Oxford
Paris · San Diego · San Francisco · Singapore · Sydney · Tokyo

Chromatin Structure and Dynamics: State-of-the-Art

Editors

J. Zlatanova

Polytechnic University,
Brooklyn, USA

S.H. Leuba

University of Pittsburgh
School of Medicine,
Pittsburgh, USA

2004

ELSEVIER

Amsterdam · Boston · Heidelberg · London · New York · Oxford
Paris · San Diego · San Francisco · Singapore · Sydney · Tokyo

ELSEVIER B.V.	ELSEVIER Inc.	ELSEVIER Ltd	ELSEVIER Ltd
Sara Burgerhartstraat 25	525 B Street, Suite 1900	The Boulevard, Langford Lane	84 Theobalds Road
P.O. Box 211, 1000 AE	San Diego, CA 92101-4495	Kidlington, Oxford OX5 1GB	London WC1X 8RR
Amsterdam, The Netherlands	USA	UK	UK

First edition 2004

Library of Congress Cataloging in Publication Data
A catalog record is available from the Library of Congress.

British Library Cataloguing in Publication Data
A catalogue record is available from the British Library.

ISBN: HARDBOUND: 0-444-51594-1
ISBN: PAPER BACK: 0-444-51595-X
ISBN: 0-444-80303-3 (Series)
ISSN: 0167-7306

∞ The paper used in this publication meets the requirements of ANSI/NISO Z39.48-1992 (Permanence of Paper).
Printed in The Netherlands.

To Ken van Holde, the scientist, the humanist, the person

Preface

This book comes at a time of unprecedented upheaval in chromatin research. The past decade has witnessed many new developments in the field, and many 'rediscoveries' of already forgotten or neglected observations or ideas. The challenge of understanding how genomes and genes function in the context of chromatin is even greater than before: the more we learn, the more we understand that our knowledge is much too limited, that we have only seen the tip of the iceberg, and that we need to combine efforts to not only describe new phenomena but to understand the structures underlying these phenomena. The horizon has broadened enormously; now we need to go for the depth.

The idea for this book germinated from our efforts to organize an international symposium of the same name in May of 2002 (meeting reviewed by E. M. Bradbury in *Molecular Cell* **10**, 13–19, 2002). The excitement that meeting created in us and the participants indicated that we had hit a raw nerve in bringing a field to its structural roots.

Fifteen years have passed since the Green Bible of Ken van Holde was published. The compilation of the present comprehensive in-depth chapters was motivated by the desire to fill, at least in part, the vacuum in overviewing the chromatin structure and dynamics field in a way that attempts to give a unified view of a complex and fast-moving field. Although a compilation of chapters written by different authors cannot be, by definition, as good as a monograph in terms of a unified perspective, it has its own advantages in that it provides the readers with broader, sometimes even contrasting views; having such views appearing in a single book is certainly helpful to the development of any field of science. We have selected our authors in a most careful way, so that the entire chromatin structure/dynamics field is represented in sufficient depth. Our authors are all recognized experts in their areas of research, which we believe is a major condition (and grounds) for success. The anonymous reviewers also made major contributions to the quality of each and every chapter. To all authors and reviewers, many, many thanks for their effort and endurance.

We would like, with this book, to welcome the new investigators coming to our fascinating field. Let us, the more established researchers, embrace these people and give them all the support they may need to succeed.

Thanks, and enjoy.

Jordanka Zlatanova
Brooklyn

Sanford H. Leuba
Pittsburgh
August 2003

List of contributors*

D. Wade Abbott 241
Department of Biochemistry and Microbiology, University of Victoria, P.O. Box 3055, Petch Building, 220 Victoria, British Columbia, Canada V8W 3P6

Juan Ausió 241
Department of Biochemistry and Microbiology, University of Victoria, P.O. Box 3055, Petch Building, 220 Victoria, British Columbia, Canada V8W 3P6

David P. Bazett-Jones 343
Programme in Cell Biology, The Hospital for Sick Children, 555 University Avenue, Toronto, ON M5G 1X8, Canada

E. Morton Bradbury 1
Department of Biological Chemistry, School of Medicine, U.C. Davis, Davis, CA 95616 and Biosciences Division, Los Alamos National Laboratory, Los Alamos, NM 87545, USA

Gerard J. Bunick 13
Department of Biochemistry, Cellular and Molecular Biology and Graduate School of Genome Science & Technology, The University of Tennessee, Knoxville, TN 37996, USA

Michael Bustin 135
Protein Section, National Cancer Institute, Bldg. 37, Room 3122B, NIH, Bethesda, MD 20892, USA

Paola Caiafa 309
Department of Cellular Biotechnology and Hematology, University of Rome 'La Sapienza', 00161 Rome, Italy

James R. Davie 205
Manitoba Institute of Cell Biology, University of Manitoba, 675 McDermot Avenue, Winnipeg, Manitoba, Canada R3E 0V9

Dale Edberg 155
Washington State University, Biochemistry and Biophysics, School of Molecular Biosciences, Pullman WA 99164-4660, USA

---

* Authors' names are followed by the starting page number(s) of their contribution(s).

Christopher H. Eskiw 343
Programme in Cell Biology, The Hospital for Sick Children, 555 University Avenue, Toronto, ON M5G 1X8, Canada

B. Leif Hanson 13
Department of Biochemistry, Cellular and Molecular Biology and Graduate School of Genome Science & Technology, The University of Tennessee, Knoxville, TN 37996, USA

Joel M. Harp 13
Department of Biochemistry and Macromolecular Crystallography Facility, Vanderbilt University School of Medicine, Nashville, TN 37232-8725, USA

Vaughn Jackson 467
Department of Biochemistry, Medical College of Wisconsin, Milwaukee, WI 53226, USA

A. Jerzmanowski 75
Laboratory of Plant Molecular Biology, Warsaw University, and Institute of Biochemistry and Biophysics, Polish Academy of Sciences, Pawinskiego 5A, 02-1-6 Warsaw, Poland

Jörg Langowski 397
Division Biophysics of Macromolecules (B040), Deutsches Krebsforschungszentrum, Im Neuenheimer Feld 580, D-69120 Heidelberg, Germany

Sanford H. Leuba 369
Department of Cell Biology and Physiology, University of Pittsburgh School of Medicine, Hillman Cancer Center, UPCI Research Pavilion, Pittsburgh, PA 15213, USA

Tom Owen-Hughes 421
Division of Gene Regulation and Expression, Wellcome Trust Biocentre, University of Dundee, Dundee DD1 5EH, UK

John R. Pehrson 181
Department of Animal Biology, School of Veterinary Medicine, University of Pennsylvania, Philadelphia, PA 19104, USA

Ariel Prunell 45
Institut Jacques Monod, Centre National de la Recherche Scientifique, Université Denis Diderot Paris 7, et Université P. et M. Curie Paris 6, 2 place Jussieu, 75251, Paris Cédex 05, France

Raymond Reeves 155
Washington State University, Biochemistry and Biophysics, School of Molecular Biosciences, Pullman WA 99164-4660, USA

Helmut Schiessel 397
Max Planck Institute for Polymer Research, Theory group, PO Box 3148, D-55021 Mainz, Germany

Andrei Sivolob 45
Department of General and Molecular Genetics, National Shevchenko University, 252601, Kiev, Ukraine

Irina Stancheva 309
Department of Biomedical Sciences, University of Edinburgh, Edinburgh EH8 9XD, UK

Jean O. Thomas 103
Cambridge Centre for Molecular Recognition & Department of Biochemistry, 80 Tennis Court Road, Cambridge CB2 1GA, UK

Andrew A. Travers 103, 421
MRC Laboratory of Molecular Biology, Hills Road, Cambridge CB2 2QH, UK

Bryan M. Turner 291
Chromatin and Gene Expression Group, Institute of Biomedical Research, University of Birmingham Medical School, Birmingham B15 2TT, UK

K.E. van Holde 1
Department of Biochemistry and Biophysics, Oregon State University, Corvallis, OR 97331-7305, USA

Katherine L. West 135
Protein Section, Laboratory of Metabolism, Center for Cancer Research, Bldg. 37, Room 3E24, National Cancer Institute, National Institutes of Health, Bethesda, MD 20892, USA and Division of Cancer Sciences and Molecular Pathology, Department of Pathology, University of Glasgow, Glasgow, G11 6NT, UK

Jordanka Zlatanova 309, 369
Department of Chemical and Biological Sciences and Engineering, Polytechnic University, Brooklyn, NY 11201, USA

Contents

Other volumes in the series

Volume 28. *Free Radical Damage and its Control* (1994)
C. Rice-Evans and R.H. Burdon (Eds.)

Volume 29a. *Glycoproteins* (1995)
J.Montreuil, J.F.G. Vliegenthart and H. Schachter (Eds.)

Volume 29b. *Glycoproteins II* (1997)
J. Montreuil, J.F.G. Vliegenthart and H. Schachter (Eds.)

Volume 30. *Glycoproteins and Disease* (1996)
J. Montreuil, J.F.G. Vliegenthart and H. Schachter (Eds.)

Volume 31. *Biochemistry of Lipids, Lipoproteins and Membranes* (1996)
D.E. Vance and J. Vance (Eds.)

Volume 32. *Computational Methods in Molecular Biology* (1998)
S.L. Salzberg, D.B. Searls and S. Kasif (Eds.)

Volume 33. *Biochemistry and Molecular Biology of Plant Hormones* (1999)
P.J.J. Hooykaas, M.A. Hall and K.R. Libbenga (Eds.)

Volume 34. *Biological Complexity and the Dynamics of Life Processes* (1999)
J. Ricard

Volume 35. *Brain Lipids and Disorders in Biological Psychiatry* (2002)
E.R. Skinner (Ed.)

Volume 36. *Biochemistry of Lipids, Lipoproteins and Membranes* (2003)
D.E. Vance and J. Vance (Eds.)

Volume 37. *Structural and Evolutionary Genomics: Natural Selection in Genome Evolution* (2004)
G. Bernardi

Volume 38. *Gene Transfer and Expression in Mammalian Cells* (2003)
Savvas C. Makrides (Ed.)

Volume 39. *Chromatin Structure and Dynamics: State of the Art* (2004)
J. Zlatanova and S.H. Leuba (Eds.)

Volume 40. *Emergent Collective Properties: Networks and Information in Biology* (2004)
J. Ricard

J. Zlatanova and S.H. Leuba (Eds.) *Chromatin Structure and Dynamics: State-of-the-Art*

DOI: 10.1016/S0167-7306(03)39001-5

CHAPTER 1

Chromatin structure and dynamics: a historical perspective

E. Morton Bradbury[1] and K.E. van Holde[2]

[1] *Department of Biological Chemistry, School of Medicine, U.C. Davis, Davis, CA 95616 and Biosciences Division, Los Alamos National Laboratory, Los Alamos, NM 87545, USA. E-mail: emb@lanl.gov*

[2] *Department of Biochemistry and Biophysics, Oregon State University, Corvallis, OR 97331-7305, USA. Tel.: 541-737-4155; Fax: 541-737-0481; E-mail: vanholdk@ucs.orst.edu*

1. Introduction

As an introduction to this book, which summarizes the latest advances in chromatin research, it is of interest to briefly compare our current knowledge to that of 25 years earlier—1978. A quarter of a century seems an appropriate interval over which to assess scientific progress, and 1978 seems an excellent year from which to start. By that date, the great revolution in our view of chromatin structure had been largely completed—after 1974 we thought in terms of chromatin subunits instead of "uniform supercoils" (see Ref. [1]). As we shall show in the following pages, even by 1978 a surprising amount of real information had already been accumulated about this new structure. Since then, we have accrued vast amounts of additional information, yet we may ask: How many new *fundamental* insights have been gained? How many of the major questions outstanding in 1978 remain unresolved today?

In the following sections, we shall deal with a number of topic areas in the field of chromatin research, in each case contrasting the current level of understanding to that in 1978. The chromatin field is too vast, and our own expertise too limited, to cover all areas. However, we feel that these are a representative group of important topics.

2. Advances in selected areas of chromatin research

2.1. Nucleosomes

2.1.1. The core particle

The idea that chromatin possessed some kind of repetitive particulate structure, rather than existing as a uniform, histone-coated DNA supercoil, emerged from a number of laboratories in the early 1970s (see Refs. [2–6]). At any event, by 1978, the basic concept of the nucleosomal core particle, as currently envisioned, was well established. The histone octamer, involving strong H3·H4 and H2A·H2B interactions, was recognized as the basis for the structure [7] and convincing evidence that the DNA was coiled about this core had been developed from both nuclease digestion [8] and neutron scattering [9,10]. The latter technique, along with electron microscopy [11] and low-resolution X-ray diffraction [12] had also provided approximate dimensions for the core particle very close to those known today. The length of DNA wrapped about the core particle had been quite accurately determined [13]. Reconstitution of core particles from histones and DNA had been accomplished, and the properties of these particles shown to be virtually identical to "native" core particles [14].

Of course, nothing was known in 1978 concerning the internal structure of the histone core, nor of the interactions of the histones with the DNA. That information has only been gained through a magnificent series of high-resolution X-ray diffraction studies ([15–18]; see also Harp *et al.*, this volume, p. 13). These have discovered a remarkable uniformity in core histone structure, referred to as the "histone fold" [15] which in turn appears to exist in numerous, sometimes seemingly unrelated proteins (see Ref. [19]). As a consequence of high resolution X-ray diffraction studies, we now know the exact interactions between the core histones, and their contacts with the DNA [18,20]. Unfortunately, these studies have been remarkably uninformative with respect to the biological *functions* of the histone octamer, or of the nucleosome, for that matter. We think we know why—all of the covalent modifications that seem to modulate nucleosomal function in chromatin occur on the N- and C-terminal tails of the histones (see Ref. [21]), and *it is precisely these regions that are largely unresolved in the X-ray studies conducted to date*. Thus, as to the question of how nucleosomes participate in either the formation of higher-order chromatin structure (see below) or processes like transcription, replication, or DNA repair, we are almost as ignorant today as in 1978. Indeed, Luger and Richmond [20] list these as "questions that remain to be answered" in the conclusion of their excellent paper.

2.1.2. The chromatosome, and the role of the lysine-rich histones

In 1978, the chromatosome, a particle containing 160–170 bp DNA, the core histone octamer, and one molecule of a lysine-rich histone such as H1, was characterized by Simpson [22]. Evidence existed that the binding site of the lysine-rich histone (or "linker" histone) lay near the ends of the DNA coiled about the octamer core. Although there has been an enormous amount of research dedicated to precise linker histone localization and considerable controversy (for discussion, see Wolffe [23], pp. 53–58, also Jerzmanowski, this volume, p. 75), we know virtually nothing more with certainty to this day. Indeed, over many years it would appear that all possible binding sites of H1 to the nucleosome have been proposed. Recent reports on the dynamics of H1 binding to chromatin in living cells may be relevant to this situation. Early studies reported that H1 subtypes exchange between chromatin segments *in vitro* [24] and *in vivo* [25]. Fluorescence redistribution after photobleaching (FRAP) assays of H1 dynamics have confirmed and extended these earlier findings. Using green fluorescent protein (GFP) labeling of H1 subtypes H1.1 [26] and H1.C and H1.0 [27], it has been shown that H1 subtypes are in steady exchange in the cell nucleus. The residence time of these subtypes on chromatin is between 1 and 2 min and FRAP kinetics give the time between binding events as 200 to 400 ms. Except for the core histones [28], it would appear that most nuclear proteins are in rapid exchange between their binding sites and the nucleoplasm (see Ref. [29]). These findings have relevance to our understanding of how DNA processing enzyme complexes gain access to their DNA binding sites in chromatin. It is possible therefore that several low affinity sites may be present on the nucleosome for H1 binding. Also the binding of H1 subtypes, except H5, involves largely the sidechain amino groups of lysine residues which are also in dynamic exchange in their binding to DNA phosphate groups. The binding of H1 has been shown to suppress the sliding of nucleosome cores on chromatin constructs [30]. However, reports on the effect of H1 binding on nucleosome sliding *in vivo* are conflicting [31,32]. It should be commented that mobility of nucleosome cores following the dissociation of H1 would also allow access of DNA enzyme complexes to their DNA sequence binding sites. For further discussion of current research on linker histones, see Jerzmanowski, this volume, p. 75.

In Section 2.2, we shall describe advances (or lack of same) in our understanding of higher-order structure in chromatin. Again, the role of lysine-rich histone remains unclear. Although it is evident that they are required for maximum compaction, what structural role they play therein remains elusive.

2.1.3. *Nucleosome assembly*

Even in 1978, it was realized the assembly of nucleosomes *in vivo* was likely to be a facilitated process. In fact, Laskey *et al.* [33] had discovered a factor, *nucleoplasmin* that assisted in this assembly. Very recently, X-ray diffraction studies have revealed much about this protein, and its probable role in nucleosome assembly [34] or disassembly [35]. At the same time, single-molecule studies (see Zlatanova and Leuba, this volume, p. 369) have provided an insight into the dynamics and energetics of nucleosome formation. This appears likely to be an area in which rapid progress is possible.

2.2. *Higher-order chromatin structure*

With the demise of the uniform fiber model in 1974, it became necessary to devise other models to account for the early electron micrographs of chromatin fibers and the X-ray diffraction studies (see Ref. [1], Chapter 1). Two models appeared in 1976, and were the major contenders for consideration in 1978. The "superbead" model of Franke *et al.* [36] envisioned the chromatin fiber as a compaction of multi-nucleosome "superbeads". The "solenoid" model of Finch and Klug [37] postulated a regular helical array of nucleosomes, with approximately six nucleosomes per turn and a pitch of 10 nm. Although a number of competing helical models appeared in the 1980s (see Ref. [1], Chapter 7) the solenoid model remains a serious contender to this day. Structural details of this model, such as the precise disposition of linker DNA, are still lacking.

Despite enormous effort, attempts to *experimentally* define the higher-order structure of chromatin under "physiological" conditions have met with much frustration. A variety of new techniques have been employed, including cryo-electron microscopy [38,39] (see also Bazett-Jones and Eskiw, this volume, p. 343) and atomic force microscopy ([40]; see also Zlatanova and Leuba, this volume, p. 369). A major problem has been the difficulty in clearly observing chromatin fiber internal structure in the highly condensed state found at physiological salt concentrations. In transmission EM, chromatin fibers prepared under these conditions give a knobby, irregular appearance, with an average diameter of about 30 nm [37]. Little evidence for internal structure can be seen. Quite convincing studies by cryoelectron microscopy and atomic force microscopy (see above) at lower ionic strength demonstrate an irregular "folded ribbon" in which linker DNA crosses back and forth within the fiber. However, we still do not know what happens to this structure when it condenses at physiological salt concentration. For further discussion see Langowski and Schiessel, this volume, p. 397.

In brief, 25 years of dedicated research and structural speculation have not brought us to the point where we can describe the structures of the condensed chromatin fiber *in vivo* with any degree of certainty. Indeed, the significance of the "30 nm fiber" as an *in vivo* structure has been questioned [41].

One aspect of higher-order chromatin structure that was entirely obscure in 1978 concerns the arrangement of nucleosomes on the DNA fiber. The concepts of specific positioning and phasing of nucleosomes, that we understand clearly today, had not as yet been defined. In fact, what information and speculation existed tended toward the idea that nucleosomes were randomly arranged (see, for example, Ref. [42]). We now know, of course, that there are precisely defined positions for many nucleosomes *in vivo* and the thermodynamic and structural rules for determining these positions are becoming clear (see Ref. [43] for a very complete discussion). Truly major advances have been made in understanding how the precise arrangement of nucleosomes (and their rearrangement, see Section 2.4) modulates the expression of specific genes.

The recognition of "positioning sequences" has also made possible the construction of "minichromosomes" of regular defined structure (e.g., Ref. [44]), and reconstituted nucleosomes containing a defined DNA sequence. This latter advance was essential for the high-resolution X-ray diffraction studies of nucleosomes that have been accomplished (see Section 2.1). Defined minichromosomes have proved a powerful tool in many studies; for a recent example, see Fan *et al.* [45].

2.3. *Histones*

2.3.1. *Histone sequences and variants*

In 1978, not only were all the major classes of histone recognized, but also sequences for the major variants of each had been determined. For example, all four core histones for calf had been sequenced (see Ref. [1], Chapter 4). The existence of certain minor variant forms had been established by electrophoretic analysis as early as 1966 [46]. However, the conclusive evidence for non-allelic variants came from the work of Newrock *et al.* [47] who demonstrated the existence of separate variant mRNAs in sea urchin. Needless to say, there was no clear evidence at that time for the physiological role of variants.

In the period from 1978 to the present, the catalog of histone variants has increased (see Ref. [19] and the chapters by Pehrson, p. 181, and by Ausio and Abbott, p. 241 in this volume). Unfortunately, we still do not have any clear idea as to the specific functions of most of these. Much interest has been generated recently by the emerging evidence for biological importance of a subset of histone variants called *replacement histones* (e.g., H1T, H10, H2AX, H2AZ, H3.3...) (see meeting review, Ref. [48]). Unlike the major histone subtypes that are synthesized in S-phase of the cell cycle for the packaging of the bulk of eukaryotic genomes into chromosomes, replacement histones are synthesized through the cell cycle and in terminally differentiated cells. Whereas, the genes for the major histone variants are found in clusters, those of replacement histones are found as singlets. Replacement histone H2AX contains a C-terminal extension to the major variant H2A with a conserved serine 139. Of much interest is the finding by Bonner's group [49] that this serine 139 is phosphorylated almost immediately following the induction of DNA double-strand breaks, but not other types of DNA damage. In an amplified response, about 1000 phosphorylated H2AX molecules are distributed over 1 to 2 Mb of DNA, i.e., the size of a large chromatin domain. Another H2A replacement histone H2AZ is involved in early metazoan development, but is not known how H2AZ modulates chromatin structure or its functions. Luger's group has reported only minor changes in the crystal structure of a core particle in which H2A has been replaced by H2AZ [50]. It has been proposed that such subtle changes in nucleosome structure can nevertheless have large effects on higher-order chromatin structures [45]. A third H2A variant, *macro H2A*, has been shown to be preferentially located in the inactive X chromosome, suggesting a role in transcriptional silencing [51]. In *Drosophila*, whereas the major H3 subtype is incorporated into chromatin during S-phase, the incorporation of replacement histone H3.3 is replication-independent. H3.3 has also been found to be located in particular loci, including rDNA arrays [52]. Another H3 homologue, CENP-A, is found only in centromeric chromatin [53]. Because most of the chromatin in *S. cerevisiae* is in an accessible state it is significant that *S. cerevisiae* H2A and H3 are homologues of replacement histones H2AX and H3.3 found in higher eukaryotes. The specificity of the H1 variant H5 for red cells of birds and some reptiles was recognized as early as 1977 [54], although its functional role is still not fully understood. In view of the emerging evidence for the importance of histone variants, it is surprising that Jenuwein and Allis [55] in their otherwise thoughtful discussion of the "histone code" pay virtually no attention to the possible role of variants as part of the message, but concentrate wholly on histone modifications. It is difficult to escape the conclusion that histone variants, particularly the replacement histones, are required to modulate or label nucleosomes for specialized chromosomal functions, and thus be a part of such a "code". Their importance has emerged only recently and is clearly an important advance compared to our understanding in 1978.

2.3.2. Histone modifications

By 1978, all of the kinds of histone modification we recognize today—acetylation, methylation, phosphorylation, ADP ribosylation, and ubiquitylation—had been discovered. Further, most of the specific sites on histones for such modifications had been identified. Some of this work was already old, dating back to the early sixties. It is noteworthy that over a decade earlier, Allfrey *et al.* [56] suggested a role for acetylation and methylation in transcriptional regulation. Not only were the types and most locations of modification clear by 1978; specific modifying enzymes were recognized as well. These include acetylases and de-acetylases, methylases and de-methylases, kinases and phosphatases, and the enzymes involved in ADP-ribosylation and ubiquitination. (For details on all of this early work on histone modification, see Chapter 4 in Ref. [1].)

So what have we learned about histone modification that is new? First, we have learned much more about patterns of modification, and how they relate to chromatin condensation and decondensation (see Ref. [57], for example, and the chapters by Davie, p. 205 and by Turner, p. 291 in this volume). More important, perhaps, is the new recognition that enzymes like acetylases and de-acetylases are usually found *in vivo* as part of large multi-protein complexes (see Refs. [58,59]). Such complexes, by virtue of "recognition" proteins within them, can be specifically targeted to genome regions or even specifically marked nucleosomes. The realization that modification can be spatially defined, and in turn serve for recognition by other factors, has led to the use of the term "histone code" ([57]; see also Turner, this volume, p. 291). The basic idea, however, is by no means new. The concept that a histone code could serve as the basis for epigenetic inheritance had been put forward in the early seventies, even before nucleosomal structure was recognized, by Tsanev and Sendov [60,61]. The term is a catchy one, but we must be a bit careful in how we use it, for such a "code" will almost surely turn out to be a non-linear one. For example, it is quite likely that phosphorylation at site A and acetylation at site B on the same nucleosome will mean something quite different than either modification alone. Perhaps it should be called a "Boolean" code. Furthermore, most discussion of the "histone code" specifically excludes consideration of histone variants. It would be very surprising if they did not constitute an important part of the code. In any event, this corner of the chromatin field is the center of intense activity at the present time.

2.3.3. Histone–histone interactions

It is not widely appreciated that the major aspects of core histone interactions were well understood even before the development of the "nucleosome" model. Evidence for strong H2A · H2B dimer interactions and an H3·H4 tetramer was available in the early seventies (see Ref. [1], Chapter 2). By 1978, the rigorous sedimentation equilibrium studies from Moudrianakis' laboratory had elucidated the thermodynamics of octamer formation [7]. What was missing, of course, was any structural information concerning these interactions. This was overcome by arduous X-ray diffraction studies, culminating in the elegantly detailed structures we have today [15,17,18], see also Harp *et al.*, this volume, p. 13. We now know *how* the core histones interact with one another and with the DNA. However, all of this "internal" knowledge has not helped much in explaining nucleosome function and dynamics, which appear to be expressed and controlled on the exterior, via modification of the N- and C-terminal tails and by the incorporation of histone variants.

On the other hand, recognition of the histone fold in archaeal chromatin and its implications for the formation of nucleosome-like structures, has provided important insights into the probable evolution of the eukaryotic chromatin [62].

2.4. Chromatin and transcription

Although the potential importance of chromatin structure in regulating or modifying gene transcription was clearly recognized in 1978, there existed at that point virtually no relevant experimental data. There had been pioneering studies of globin gene transcription by Weintraub and Groudine [63], and of ovalbumin transcription by Garel *et al.* [64]. Some important studies of ribosomal genes had also been done (see, for example, Ref. [65]). But there existed no overall picture of a mechanism for either activation or repression of specific genes. The huge gap in the picture in 1978 was the lack of recognition of those multitudinous proteins we now call *transcription factors*. Indeed, the first clear identification of such a substance came only in 1983 [66].

The enthusiastic search for more and more transcription factors that ensued in the following decade diverted the attention of many molecular biologists from the fundamental problem of how transcription can be initiated or proceed in a chromatin matrix. However, three lines of research continued throughout the 1980s and 1990s that have converged with transcription factor analysis to build the detailed, if still confusing picture we have today.

(a) *Isolation and characterization of the polymerase II holoenzyme complex, and associated proteins.* Outstanding in this field has been the elegant analysis of the yeast pol II holoenzyme and associated mediator complex by the Kornberg group (see, for overviews, Refs. [67,68]). Studies of this kind have shown us how complex a machine the polymerase really is, and how it can interact with general transcription factors. They have not led so far to in depth understanding of how this enormous machinery can interact with a polynucleosomal template.

(b) *Analysis of the nucleosome positioning in promoters.* The development of methods to accurately map, to the nucleotide level [69], the positions of nucleosomes *in situ* has opened the way to understanding how chromatin structure can influence the initiation of transcription. For an overview and introduction to a number of systems, see Turner (Ref. [70], Chapter 7). Most interesting, perhaps, is the new realization that chromatin in protomers can be "remodeled" in an ATP-dependent manner (see Refs. [71,72], and Travers and Owen-Hughes (this volume, p. 421) for contemporary summaries). It is clear from many examples that this remodeling can include directed, ATP driven translation of nucleosomes.

(c) *Transcription elongation in chromatin.* Just how an RNA polymerase can traverse an array of nucleosomes in chromatin was pretty much a mystery in 1978. Pioneering experiments by Williamson and Felsenfeld [73] had shown that the elongation rate for *E. coli* polymerase was decreased markedly—but not entirely—by the presence of nucleosomes on a DNA template. Over the following decades, numerous similar studies were carried out, using a variety of polymerases and both natural and synthetic nucleosomal arrays (see, for example, Ref. [74]; Wolffe [23] gives an excellent summary in Chapter 4).

The most incisive studies of the problem at the molecular level are those from the Felsenfeld laboratory (see, for example, Refs. [75,76]). They have shown that at least under some circumstances, a polymerase can "step around" a nucleosome, displacing it in *cis*, but not causing dissociation. It is not yet clear, however, if this mechanism is physiologically relevant and/or whether it is the *only* mechanism. There exist results in apparent conflict with this model (i.e., Ref. [77]). That the *in vivo* process is certainly more complex than the *in vitro* models used to date is further indicated by the discovery of *elongation factors* that markedly increase transcriptional rates and suppress pausing (see, for example, Conaway and Conaway [78]). Thus, the question as to how transcription elongation occurs in a chromatin template remains at least partially unresolved. For a further discussion, see Jackson, this volume, p. 467.

3. *Conclusions and overview*

In summary: what have we learned in 25 years? In some areas, surprisingly little—for example, we cannot say that we really understand the condensed chromatin fiber structure much better than we did in 1978. Although the significance of the great majority of histone variants remains unknown, replacement histones appear now to be involved in major chromosomal functions. There are areas in which we have accrued incredible amounts of detailed information yet still do not quite know what to do with it. Histone acetylation is a prime example. Allfrey *et al.* [56] could predict its role in a general sense in 1964. We now know a whole rogue's gallery of acetylases and deacetylases plus the specific histone sites for many. Nevertheless, authorities in the field must still write in 2000, "The mechanisms by which histone acetylation affects chromatin structure and transcription is not yet clear" [58].

On the other hand, there is no question that enormous strides have been made. It is now possible to describe in detail the chromatin structures in many specific promoters, and then show how they are remodeled for transcriptional activation. Different kinds of chromatin organization are now recognized for different levels of developmental control. Despite the remarkable advance in detailed information that the past 25 years have provided, the overall picture of transcription in chromatin remains strangely obscured. There is almost too much information, at least too much to handle in the absence of new unifying concepts. However, there exist certain lines of research that seem on the verge of merging to provide such unification. For example, there are strong indications that the interaction of histone-modifying enzymes with tissue-specific factors and cofactors can target certain nucleosomes in specific promoters for modification, and that such modification can in turn mark that chromatin region for remodeling or not. If this is generally true, we can hope to understand in one unifying concept what histone modification really signifies, what the histone tails are for, and what remodeling is all about. It may well be that understanding of some of the long-standing puzzles is finally in view.

Further, we must emphasize the potential of powerful new techniques, in particular at the single molecule level, to provide new kinds of information that have not been hitherto available (see, for example, Zlatanova and Leuba, this volume, p. 369).

At the same time, we must be careful to remember that it is in the *nucleus* that events like transcription, replication, and repair occur, and that we still know little of that environment, or how chromatin is disposed therein. It would seem likely that a next stage of development in the field, once *in vitro* mechanisms are understood, is to see how these translate to their native environment. Although the nuclear matrix was first defined in 1975 [79] only a few intrepid explorers have continued investigation of the disposition of chromatin in the nucleus. A thoughtful review of the current status of knowledge about large-scale chromatin structure and function is given by Mahy *et al.* [80] and an intriguing view of chromatin dynamics *in situ* is provided by Roix and Misteli [29]. An excellent brief overview is provided by Bazett-Jones and Eskiw (this volume, p. 343). This may well be the new frontier.

References

1. van Holde, K.E. (1988) Chromatin. Springer-Verlag, New York.
2. Hewish, D.R. and Burgoyne, L.A. (1973) Biochem. Biophys. Res. Commun. 52, 504–510.
3. Sahasrabuddhe, C.G. and van Holde, K.E. (1974) J. Biol. Chem. 249, 152–156.
4. Olins, A.L. and Olins, D.E. (1974) Science 183, 330–332.
5. Kornberg, R.D. and Thomas, J.O. (1974) Science 184, 865–868.
6. Kornberg, R.D. (1974) Science 184, 868–871.
7. Eickbush, T.H. and Moudrianakis, E.N. (1978) Biochemistry 17, 4955–4964.
8. Noll, M. (1974) Nucleic Acids Res. 1, 1573–1578.
9. Pardon, J.F., Worcester, D.L., Wooley, J.C., Cotter, R.I., Lilley, D.M., and Richards, R.M. (1977) Nucleic Acids Res. 4, 3199–3214.
10. Suau, P., Kneale, G.G., Braddock, G.W., Baldwin, J.P., and Bradbury, E.M. (1977) Nucleic Acids Res. 4, 3769–3786.
11. van Holde, K.E., Sahasrabuddhe, C.G., Shaw, B.R., van Bruggen, E.F.J., and Arnberg, A.C. (1974) Biochem. Biophys. Res. Commun. 60, 1365–1370.
12. Finch, J.T., Lutter, L.C., Rhodes, D., Brown, R.S., Rushton, B., Levitt, M., and Klug, A. (1977) Nature 269, 29–36.
13. Mirzabekov, A.D., Shick, V.V., Belyavsky, A.V., and Bavykin, S.G. (1978) Proc. Natl. Acad. Sci. USA 75, 4184–4188.

14. Tatchell, K. and van Holde, K.E. (1977) Biochemistry 16, 5295–5303.
15. Arents, G., Burlingame, R.W., Wang, B.C., Love, W.E., and Moudrianakis, E.N. (1991) Proc. Natl. Acad. Sci. USA 88, 10148–10152.
16. Arents, G. and Moudrianakis, E.N. (1995) Proc. Natl. Acad. Sci. USA 92, 11170–11174.
17. Luger, K., Mader, A.W., Richmond, R.K., Sargent, D.F., and Richmond, T.J. (1997) Nature 389, 251–260.
18. Harp, J.M., Hanson, B.L., Timm, D.E., and Bunick, G.J. (2000) Acta Crystallogr. D 56, 1513–1534.
19. Wolffe, A.P. and Pruss, D. (1996) Trends Genet. 12, 58–62.
20. Luger, K. and Richmond, T.J. (1998) Curr. Opin. Struct. Biol. 8, 33–40.
21. Bradbury, E.M. (1992) Bioessays 14, 9–16.
22. Simpson, R.T. (1978) Biochemistry 17, 5524–5531.
23. Wolffe, A.P. (1998) Chromatin: Structure and Function. Academic Press, New York.
24. Caron, F. and Thomas, J.O. (1981) J. Mol. Biol. 146, 513–537.
25. Louters, L. and Chalkley, R. (1985) Biochemistry 24, 3080–3085.
26. Lever, M.A., Th'ng, J.P., Sun, X., and Hendzel, M.J. (2000) Nature 408, 873–876.
27. Misteli, T., Gunjan, A., Hock, R., Bustin, M., and Brown, D.T. (2000) Nature 408, 877–881.
28. Kimura, H. and Cook, P.R. (2001) J. Cell Biol. 153, 1341–1353.
29. Roix, J. and Misteli, T. (2002) Histochem. Cell Biol. 118, 105–116.
30. Pennings, S., Meersseman, G., and Bradbury, E.M. (1991) J. Mol. Biol. 220, 101–110.
31. Ura, K., Hayes, J.J., and Wolffe, A.P. (1995) EMBO J. 14, 3752–3765.
32. Varga-Weisz, P.D., Blank, T.A., and Becker, P.B. (1995) EMBO J. 14, 2209–2216.
33. Laskey, R.A., Honda, B.M., Mills, A.D., and Finch, J.T. (1978) Nature 275, 416–420.
34. Dutta, S., Akey, I.V., Dingwall, C., Hartman, K.L., Laue, T., Nolte, R.T., Head, J.F., and Akey, C.W. (2001) Mol. Cell 8, 841–853.
35. Chen, H., Li, B., and Workman, J.L. (1994) EMBO J. 13, 380–390.
36. Franke, W.W., Scheer, U., Trendelenburg, M.F., Spring, H., and Zentgraf, H. (1976) Cytobiologie 13, 401–434.
37. Finch, J.T. and Klug, A. (1976) Proc. Natl. Acad. Sci. USA 73, 1897–1901.
38. Woodcock, C.L., Grigoryev, S.A., Horowitz, R.A., and Whitaker, N. (1993) Proc. Natl. Acad. Sci. USA 90, 9021–9025.
39. Horowitz, R.A., Agard, D.A., Sedat, J.W., and Woodcock, C.L. (1994) J. Cell Biol. 125, 1–10.
40. Leuba, S.H., Yang, G., Robert, C., Samori, B., van Holde, K., Zlatanova, J., and Bustamante, C. (1994) Proc. Natl. Acad. Sci. USA 91, 11621–11625.
41. van Holde, K. and Zlatanova, J. (1995) J. Biol. Chem. 270, 8373–8376.
42. Prunell, A. and Kornberg, R.D. (1978) Cold Spring Harb. Symp. Quant. Biol. 42, 103–108.
43. Widom, J. (2001) Q. Rev. Biophys. 34, 269–324.
44. Simpson, R.T., Thoma, F., and Brubaker, J.M. (1985) Cell 42, 799–808.
45. Fan, J.Y., Gordon, F., Luger, K., Hansen, J.C., and Tremethick, D.J. (2002) Nat. Struct. Biol. 9, 172–176.
46. Kinkade, J.M., Jr. and Cole, R.D. (1966) J. Biol. Chem. 241, 5798–5805.
47. Newrock, K.M., Cohen, L.H., Hendricks, M.B., Donnelly, R.J., and Weinberg, E.S. (1978) Cell 14, 327–336.
48. Bradbury, E.M. (2002) Mol. Cell 10, 13–19.
49. Rogakou, E.P., Pilch, D.R., Orr, A.H., Ivanova, V.S., and Bonner, W.M. (1998) J. Biol. Chem. 273, 5858–5868.
50. Suto, R.K., Clarkson, M.J., Tremethick, D.J., and Luger, K. (2000) Nat. Struct. Biol. 7, 1121–1124.
51. Changolkar, L.N. and Pehrson, J.R. (2002) Biochemistry 41, 179–184.
52. Ahmad, K. and Henikoff, S. (2002) Mol. Cell 9, 1191–1200.
53. Sullivan, K.F., Hechenberger, M., and Masri, K. (1994) J. Cell Biol. 127, 581–592.
54. Huang, P.C., Branes, L.P., Mura, C., Quagliarello, V., and Kropowski-Bohdan, P. (1977) In: Ts'o, P.O.P. (ed.) Molecular Biology of the Mammalian Genetic Apparatus. North Holland Pub. Co, Amsterdam, pp. 105–122.
55. Jenuwein, T. and Allis, C.D. (2001) Science 293, 1074–1080.

56. Allfrey, V.G., Faulkner, R.M., and Mirsky, A.E. (1964) Proc. Natl. Acad. Sci. USA 51, 786–794.
57. Strahl, B.D. and Allis, C.D. (2000) Nature 403, 41–45.
58. Berger, S.L., Grant, P.A., Workman, J.L., and Allis, C.D. (2000) In: Elgin, S.C.R. and Workman, J.L. (eds.) Chromatin Structure and Gene Expression. Oxford University Press, Oxford, pp. 135–155.
59. Free, A., Grunstein, M., Bird, A., and Vogelauer, M. (2000) In: Elgin, S.C.R. and Workman, J.L. (eds.) Chromatin Structure and Gene Expression. Oxford University Press, Oxford, pp. 156–181.
60. Tsanev, R. and Sendov, B. (1971) J. Theor. Biol. 30, 337–393.
61. Tsanev, R. and Sendov, B. (1971) Z. Krebsforsch Klin. Onkol. Cancer Res. Clin. Oncol. 76, 299–319.
62. Pereira, S.L., Grayling, R.A., Lurz, R., and Reeve, J.N. (1997) Proc. Natl. Acad. Sci. USA 94, 12633–12637.
63. Weintraub, H. and Groudine, M. (1976) Science 193, 848–856.
64. Garel, A., Zolan, M., and Axel, R. (1977) Proc. Natl. Acad. Sci. USA 74, 4867–4871.
65. Foe, V.E., Wilkinson, L.E., and Laird, C.D. (1976) Cell 9, 131–146.
66. Dynan, W.S. and Tjian, R. (1983) Cell 32, 669–680.
67. Kornberg, R.D. (1998) Cold Spring Harb. Symp. Quant. Biol. 63, 229–232.
68. Myers, L.C. and Kornberg, R.D. (2000) Annu. Rev. Biochem. 69, 729–749.
69. Flaus, A., Luger, K., Tan, S., and Richmond, T.J. (1996) Proc. Natl. Acad. Sci. USA 93, 1370–1375.
70. Turner, B.M. (2001) Chromatin and Gene Regulation. Blackwell Science Ltd, Oxford.
71. Becker, P.B. and Hörz, W. (2002) Annu. Rev. Biochem. 71, 247–273.
72. Flaus, A. and Owen-Hughes, T. (2001) Curr. Opin. Genet. Dev. 11, 148–154.
73. Williamson, P. and Felsenfeld, G. (1978) Biochemistry 17, 5695–5705.
74. O'Neill, T.E., Roberge, M., and Bradbury, E.M. (1992) J. Mol. Biol. 223, 67–78.
75. Studitsky, V.M., Clark, D.J., and Felsenfeld, G. (1994) Cell 76, 371–382.
76. Studitsky, V.M., Kassavetis, G.A., Geiduschek, E.P., and Felsenfeld, G. (1997) Science 278, 1960–1963.
77. Pfaffle, P., Gerlach, V., Bunzel, L., and Jackson, V. (1990) J. Biol. Chem. 265, 16830–16840.
78. Conaway, J.W. and Conaway, R.C. (1999) Annu. Rev. Biochem. 68, 301–319.
79. Berezney, R. and Coffey, D.S. (1975) In: Weber, G. (ed.) Advances in Enzyme Regulation. Pergamon Press, Oxford, pp. 63–100.
80. Mahy, N.L., Bickmore, W.A., Tumbar, T., and Belmont, A.S. (2000) In: Elgin, S.C.R. and Workman, J.L. (eds.) Chromatin Structure and Gene Expression. Oxford University Press, Oxford, pp. 300–321.

J. Zlatanova and S.H. Leuba (Eds.) *Chromatin Structure and Dynamics: State-of-the-Art*

DOI: 10.1016/S0167-7306(03)39002-7

CHAPTER 2

The core particle of the nucleosome

Joel M. Harp[1], B. Leif Hanson[2], and Gerard J. Bunick[2,3]

[1]*Department of Biochemistry and Macromolecular Crytallography Facility, Vanderbilt University School of Medicine, Nashville, TN 37232-8725, USA*
[2]*Department of Biochemistry, Cellular and Molecular Biology and Graduate School of Genome Science & Technology, The University of Tennessee, Knoxville, TN 37996, USA*
[3]*Structural Biology, Life Sciences Division, Oak Ridge National Laboratory, P.O. Box 2008, Oak Ridge, TN 37831-6480, USA. Tel.: 865-576-2685; Cell: 865-806-9631; Fax: 865-574-0004; E-mail: bunickgj@ornl.gov, gjbunick@utk.edu*

1. *Introduction*

The nucleosome is the fundamental repeating structural unit of chromatin. It is composed of two molecules of the core histones H2A, H2B, H3, H4, approximately two superhelical turns of double-stranded DNA, and linker histone H1 (H5). In addition to biochemical studies, the existence of the nucleosome was established in electron micrographs (Fig. 1a) [1,2], and the name nucleosome, coined to incorporate the concept of the spherical nu-bodies [3]. Micrococcal nuclease limit digestion of chromatin established the nucleosome core particle (NCP) as the portion of the nucleosome containing only the core histones surrounded by ~1.75 superhelical turns of double-stranded DNA [4,5].

Once it became apparent that NCP could be isolated, a major goal was to determine the crystal structure of this fundamental component of chromatin. The first report of NCP crystals is attributed to researchers in Russia [6]. The crystals were prepared from NCPs isolated from mouse Erlich ascites tumor cells and yielded X-ray powder diffraction data. The powder diffraction pattern showed a strong ring at about 51 Å with weaker data for higher *d*-spacings. A 25 Å resolution model of the NCP was constructed based on X-ray and neutron single crystal diffraction data from centric zones [7,8], electron microscopy of crystals [9], and electron micrographic image reconstruction of the histone core at 22 Å resolution [10]. Based on these findings the NCP is an approximate wedge shape bipartite particle of 110 Å diameter with a maximum thickness of 57 Å. The approximately 1.75 turns of superhelical DNA is coiled around the ramp-like surface of the histone core. These data suggest that the particle has two-fold symmetry. Improved crystals of the NCP isolated and purified from several sources provided low-resolution three-dimensional models from analysis of X-ray and neutron diffraction data [11–13]. Figure 1b shows a slice of electron density through the NCP at 16 Å resolution from Ref. [13]. The resolution was too low to distinguish and trace individual peptide chains; as a consequence, the assignments of density to individual histones is incorrect in this model and other NCP structures derived from naturally isolated bulk nucleosomes.

Nucleosomes isolated and purified from chicken erythrocytes and beef kidney crystallized and diffracted to limits of 5–6 Å, but led to structural models of 7–8 Å resolution [12,14]. In these structures, the path of the DNA around the histone core is clearly seen. With the exception of positions about 1.5 and 4.5 helical turns from the center of the nucleosomal DNA in either direction, the DNA appears uniformly bent. Significant compression of the DNA occurs where the minor grooves face

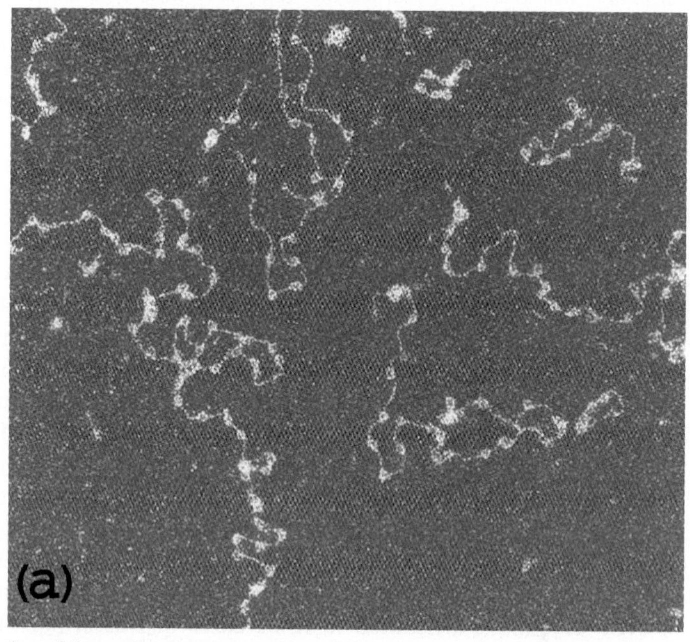

Fig. 1. Evolution of the nucleosome structure model. (a) An unstained dark field electron micrograph of soluble chromatin isolated from hypotrichous ciliate protozoan macronuclei. The soluble chromatin in low ionic strength buffer was spread onto a carbon film treated by plasma discharge in the presence of amyl amine. The beads-on-a-string appearance of the spread is the result of the sample preparation in low ionic strength buffers. The width of the nucleosome in these images is about 7 nm (70 Å) providing an effective resolution of 30–40 Å. (b) Improving the resolution of the nucleosome, this 16 Å resolution section through the electron density map shows that the DNA is wound around a histone spool. (c) Sections of the electron density through the 8 Å NCP map showing histone core and the central turn of DNA. The crystals used for the 16 Å and 8 Å X-ray structures were grown from chicken erythrocyte nucleosomes isolated from soluble chromatin by micrococcal nuclease digest. The resolution is limited because of DNA length heterogeneity and DNA base-sequence heterogeneity in the nucleosome core particles, and two-fold nucleosome packing disorder in the crystals. (d) The 2.5 Å NCP model with the DNA in red and blue shown in a wire frame representation, and the histones chains shown in different colors in a ribbon Cα presentation. The small dots represent positions of ordered water molecules and ions on the molecule. This model is based on reconstituted nucleosome core particles containing histones isolated from chicken erythrocytes and the defined palindromic DNA sequence based on human X-chromosome alpha satellite DNA repeats. The interpretation of the location of histone proteins in the progression of structures shows a decided improvement with increasing resolution.

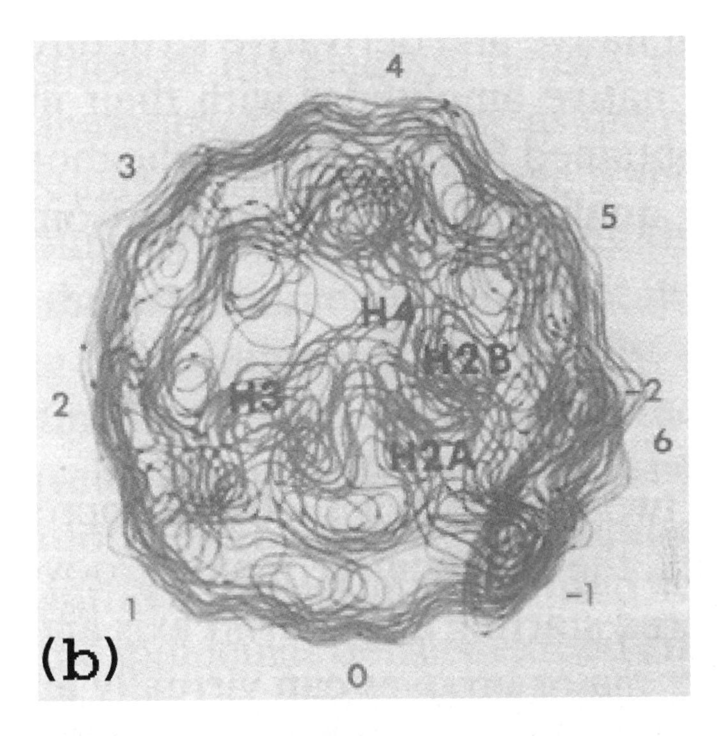

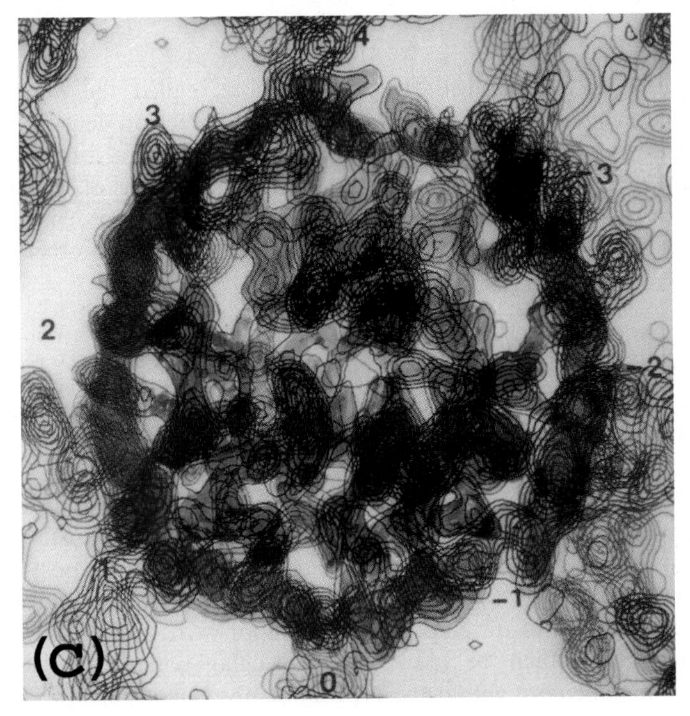

Fig. 1. Continued.

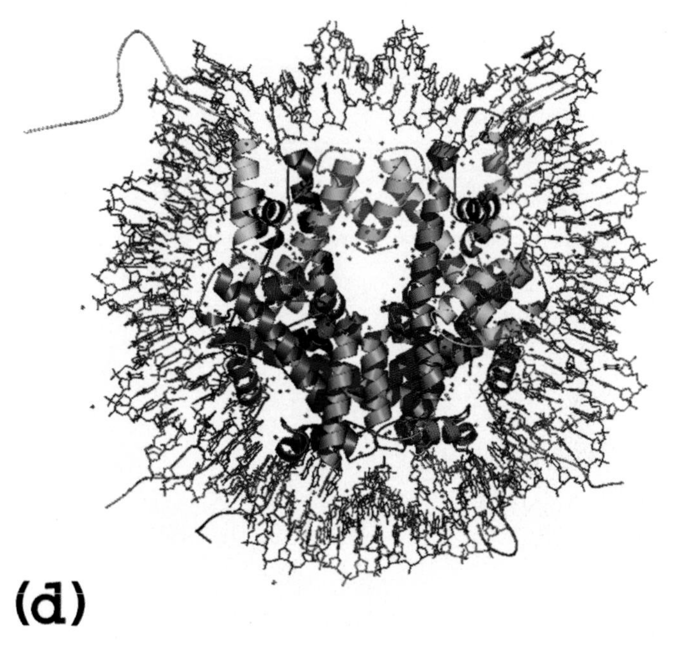

Fig. 1. Continued.

toward and are in contact with the histone core, called the minor groove-in positions. Figure 1c consists of several sections of the electron density from the 8 Å NCP map showing parts of the histone core and the central turn of DNA [14]. A prominent feature in this structure was an extension of electron density protruding between the DNA gyres which bears a similarity to density seen in the 16 Å neutron structure [11], but which was not apparent in the other model.

2. Toward higher resolution nucleosome structure

Several steps were needed to determine the structure of the core particle to higher resolution (Fig. 1d). The X-ray phases of the low-resolution models were insufficient to extend the structure to higher resolution, since the resolution of the early models of the NCP was severely limited by disorder in the crystals. The disorder was presumed to derive from both the random sequences of the DNA and from heterogeneity of the histone proteins caused by variability in post-translational modification of the native proteins. One strategy for developing an atomic position model of the NCP was to develop a high-resolution structure of the histone core. This structure could then be used with molecular replacement techniques to determine the histone core within the NCP and subsequently identify the DNA in difference Fourier electron density maps.

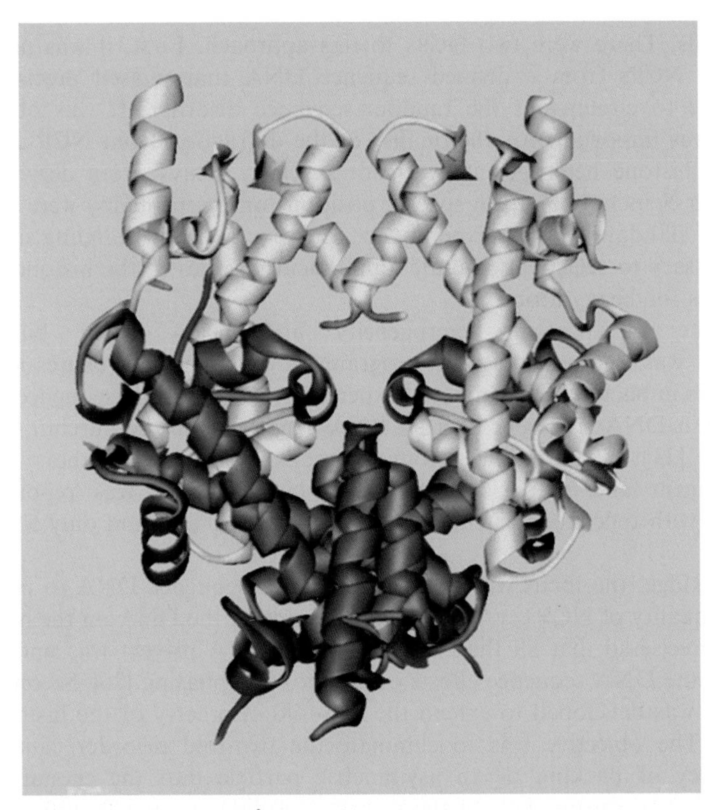

Fig. 2. The histone octamer. The 3.1 Å X-ray diffraction data model of Arents *et al.* [20] is shown in secondary structure cartoon format. The core of the histone octamer is well defined, but more than 30% of the histone sequence is in regions without secondary structure. These are unfortunately the most interesting regions in terms of epigenetic signaling. 25% of the molecule located in the N-terminal tails (and the C-termini of H2A) in the 3.1 Å octamer structure has no interpretable electron density. Despite these limitations, this structure is sufficient to use as a starting model for molecular replacement phasing of the NCP. (Image courtesy of E. Moudrianakis.)

The crystallization and structural determination of the histone octamer was first reported in 1984 [34]. However, the overall dimensions of the 3.3 Å structure [15] did not appear to fit within the known X-ray structures of the nucleosome core particle [12,13]. In an elegant analysis [16], re-examination of the original phasing of the histone octamer data revealed misplacement of the heavy atom site by 2.7 Å. The structure was resolved, after which it was possible to build molecular models of the individual histones into the 3.1 Å resolution electron density map of the histone core of the nucleosome [17]. Figure 2 shows the first atomic resolution model of the core histone octamer. Several additional publications followed in which the histone octamer structure formed the basis for constructing models of the NCP [17–21].

An alternative approach to higher resolution nucleosome structure was to solve the complete NCP structure by increasing the diffracting resolution of

NCP crystals. There were two facets to this approach. First, it was necessary to reconstitute NCPs from a defined sequence DNA that phased precisely on the histone core to circumvent the random sequence disorder. It was obvious that the DNA was important for the quality of the diffraction from NCP crystals but the role of histone heterogeneity was not so clear. Heavy atom derivatives (i.e., electron rich elements bound in specific positions on the proteins) were not readily prepared by standard soaking experiments, due to a paucity of binding sites. Hence, it was necessary to selectively mutate amino acid residues in the histones to create binding sites for heavy atoms.

The issue of protein heterogeneity and heavy atom binding site engineering was pursued with a program of cloning and expression of the core histones in bacteria [22]. In initial experiments, the random sequence DNA was replaced by a DNA fragment of a 5S RNA gene from the sea urchin, *Lytechinus variegatus* [23]. Diffraction from crystals containing the 5S RNA DNA fragment and native chicken erythrocyte histones was reported to be anisotropic with reflections extending to 3.0 Å on the *c*-axis and only 5.0 Å on the *b*-axis [24].

At Oak Ridge, the focus was to develop specific-sequence DNA to improve the diffraction quality of NCP crystals. The positioning of the DNA on the histone core has to be precise so that all the NCPs are identical. A project was undertaken to understand the DNA sequence effects on nucleosome phasing [25]. Second, a DNA palindrome was developed to extend the two-fold symmetry of the histone core to the DNA. The objective was to eliminate the two-fold disorder caused by the indeterminacy of packing of an asymmetric particle into the crystal lattice. A palindrome based on one-half of the primary candidate sequence was constructed and methods were developed to produce the palindrome fragment in large quantities for reconstitution of NCPs.

The primary candidate sequence, the non-palindromic, 145 bp native human α-satellite DNA was used in reconstitution experiments, crystallization, and diffraction experiments prior to the availability of the palindrome sequence. It was known that even the 145 bp α-satellite native sequence extended crystal diffraction well past the 7–8 Å resolution obtained from crystals of isolated core particles. At that time, the technology of PCR cloning was not readily available, so the palindrome had to be constructed using standard cloning technology. Subcloning of the palindrome made use of an *Alu* I site occurring near one of the predicted nucleosome centers. The *Alu* I cleaved half-nucleosome DNA fragment was ligated to a commercially available *Eco*R I linker. The half-palindrome fragment was then ligated to itself to form a 146 bp DNA palindrome. A significant point is that the nucleosome phasing sequence used was chosen from 12 possible phasing sites. The resulting α-satellite DNA palindrome was thus 1 of 24 possibilities and was chosen based on the existence of the *Alu* I site in an appropriate location for sub-cloning (Fig. 3).

After the α-satellite DNA palindrome was constructed and cloned, the task remained to produce it in large quantities. To do this multiple copies of the half-palindrome fragment were cloned into a vector [26]. The half-palindrome fragment

Human α-satellite sequence

ATCAATATCC ACCTGCAGAT TCTACCAAAA GTGTATTTGG
AAACTGCTCC ATCAAAAGGC ATGTTCAGCT CTGTGAGTGA
AACTCCATCA TCACAAAGAA TATTCTGAGA ACGCTTCCGT
TTGCCTTTTA TATGAACTTC CTGAT

(A)

α-satellite palindrome

ATCAATATCC ACCTGCAGAT TCTACCAAAA GTGTATTTGG
AAACTGCTCC ATCAAAAGGC ATGTTCAGCG GAA|TTCCGCT
GAACATGCCT TTTGATGGAG CAGTTTCCAA ATACACTTTT
GGTAGAATCT GCAGGTGGAT ATTGAT

(B)

Fig. 3. DNA sequence used in Oak Ridge NCP structural studies. The DNA sequence of the human α-satellite sequence (A) and the α-satellite palindrome (B) for which it was a model. The palindrome is one of 24 potential phasing sequences taken from the human α-satellite DNA repeats. The center of the palindrome sequence is marked by a vertical bar. The α-satellite sequence continues to serve as the starting point for high-resolution NCP structures.

is produced in large quantity from a plasmid containing as many as 32 copies of the half-palindrome. The half-palindrome is then ligated to provide the required quantities of α-satellite DNA palindrome. It was not possible to maintain multiple copies of the full palindrome in plasmids.

Reconstitution of NCPs using the α-satellite DNA palindrome was accomplished by the method of salt gradient dialysis [27]. Multiple phases of the DNA position on the histone core were present in some reconstitution experiments. The symmetrically phased form was identified on polyacrylamide gels as the resistant band in a time course of micrococcal nuclease cleavage. The symmetrical phasing meant that both DNA termini were protected from nuclease attack while the so-called degenerate phases possessed asymmetrically bound DNA with one terminus exposed to varying degrees. Several corrective methods were used when the reconstituted NCPs contained degenerate DNA phasing. If the NCPs were stored at 4°C for one-to-two weeks, it was found that the proportion of improperly phased molecules decreased. Incubating the NCPs at 29°C or 37°C reduced the time needed for the DNA to slide into the symmetrically phased state on the histone core. If the incubation procedure did not produce a homogeneous product by particle gel analysis, the degenerate phases could be separated from the correctly phased particles on non-denaturing polyacrylamide gels. A method for purification of the correctly phased NCPs was developed [28] using preparative polyacrylamide gel electrophoresis based on work done earlier [29].

The crystals of NCPs containing α-satellite DNA palindrome and chicken erythrocyte histones diffracted isotropically to ~3.0 Å using an in-house rotating anode X-ray source and to better than 2.5 Å at a moderate intensity synchrotron beamline [30,31]. The crystals used for structure determination were grown in the microgravity environment using a counter-diffusion apparatus [32]. Ground-based crystallization consistently resulted in crystals with some degree of anisotropy in the diffraction data so that resolution varied by the direction of the crystal orientation. Diffraction data from microgravity grown NCP crystals demonstrated greatly reduced anisotropy effectively increasing the resolution and overall quality of the data. Subsequently, a number of NCP crystal structures have been reported [31,33–37]. Human α-satellite DNA sequences, especially the palindrome pioneered at Oak Ridge, form the basis for all reported high-resolution structures of the NCP [38].

3. *Units of nucleosome structure*

A primary value of molecular models is the heuristic support they provide to biological questions. The central repository for macromolecular structures is the Protein Data Bank (PDB) [38]. At this site, coordinate files from experimentally determined molecular structures can be accessed and displayed using molecular graphics software. In addition to coordinates for macromolecules, information included in the coordinate deposition files can assist the user in determining the reliability of the experimentally determined structures. In the discussion below, the structural information is taken from the Oak Ridge NCP structure [31], PDB access code 1EQZ. The molecular graphics images were created using the programs PyMOL, XtalView, and MolScript [39–41]. Other NCP coordinate files are available from the PDB. These structures can be accessed and used to address specific structural questions that are not included in the following discussion.

The canonical nucleosome as first defined by Kornberg [5,42] is composed of a variable length, roughly 200 bp, of DNA along with two subunits each of the core histones H2A, H2B, H3, and H4. The nucleosome may also be associated with one molecule of linker histone such as H1 or H5. The protein core of the nucleosome is known as the histone octamer as it contains eight subunits, two each of the core histones. Functional subunits of the histone octamer are recognized from the behavior of the heterodimer associations. Histones H3 and H4 associate in two heterodimers; these further assemble into a tetramer. Thus, the particle formed by H3 and H4, $(H3{:}H4)_2$, is known as the histone tetramer (Fig. 4a). Histones H2A and H2B form independent heterodimers, (H2A:H2B), which are known simply as histone dimers (Fig. 4b). The core histones bound together into the histone octamer core of the nucleosome display an apparent two-fold symmetry with the symmetry axis passing through the octamer and intersecting the DNA at about the midpoint of the bound DNA sequence (Fig. 5). The symmetry axis of the histone octamer is perpendicular to the superhelical axis of the DNA on the nucleosome.

Despite the apparent symmetry of the histones within the core particle, deviations of the histones and DNA result in an asymmetric structure. In the histones, these deviations can occur throughout protein structure, not just in the relatively unstructured tails (Fig. 6). The most pronounced deviations appear in histone H3.

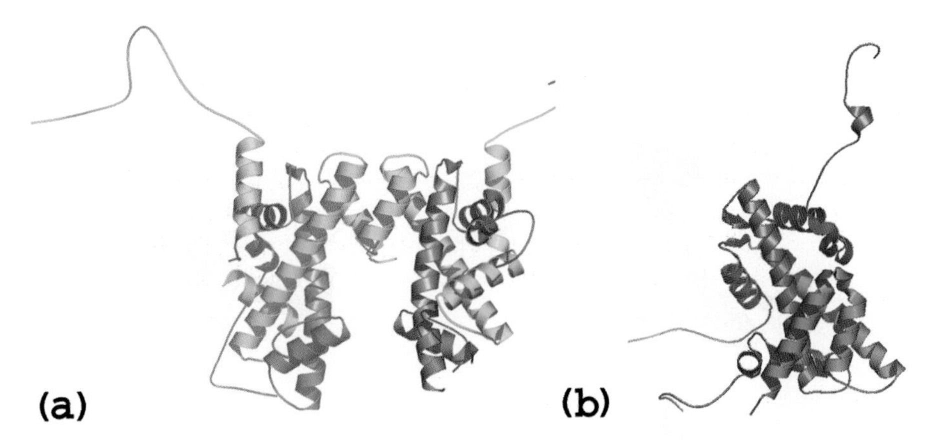

Fig. 4. The functional subsets of the histone core. (a) Ribbon Cα model of the H3 : H4 tetramer. (b) Ribbon Cα model of the H2A : H2B dimer.

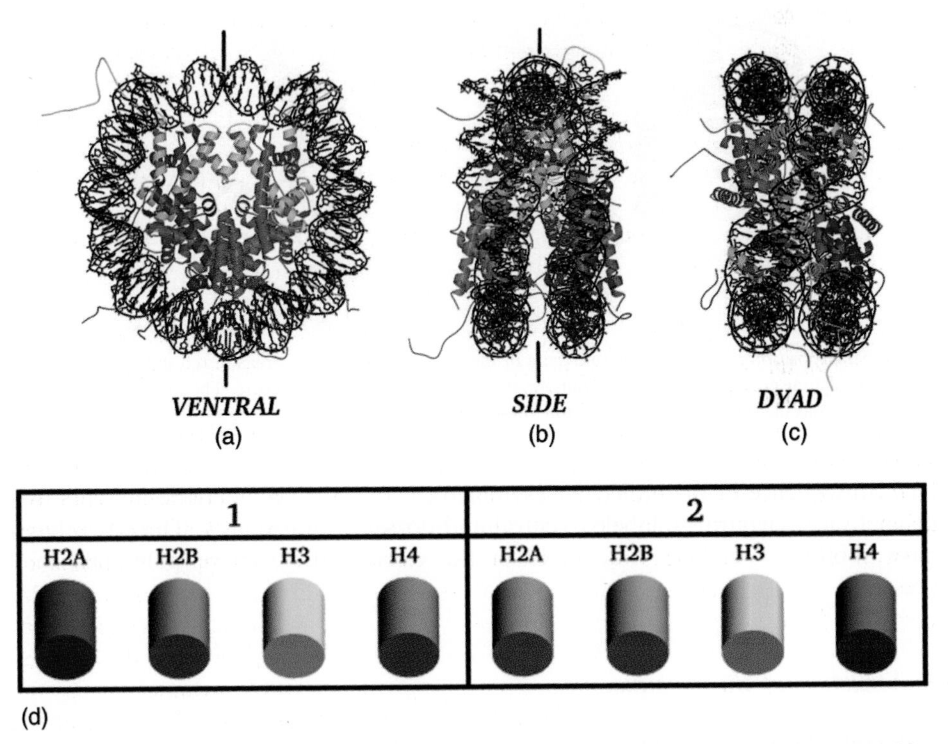

Fig. 5. Three views of the NCP from Harp *et al.* [31]. (a) Ventral surface view. (b) Side view. (c) View down the molecular pseudo-dyad axis. The histones are represented by Cα ribbon models of the secondary structure elements, and the DNA model indicates the base pairing between complementary strands. The DNA is positioned asymmetrically by one-half base pair on the NCP. This results in a two sides arbitrarily referred to a dorsal and ventral (the surface shown here). The ventral surface of the NCP is best recognized by the extended N-terminal H3 tail protruding to the right. In these images, the pseudo-dyad axis is represented by vertical bars for both the ventral and side view. The pseudo-dyad axis passes through the center of the dyad view orthogonal to the plane of the page. (d) Color code for histone chains in the figures in this chapter. Note the change in hue denoting the two sides of the histone octamer.

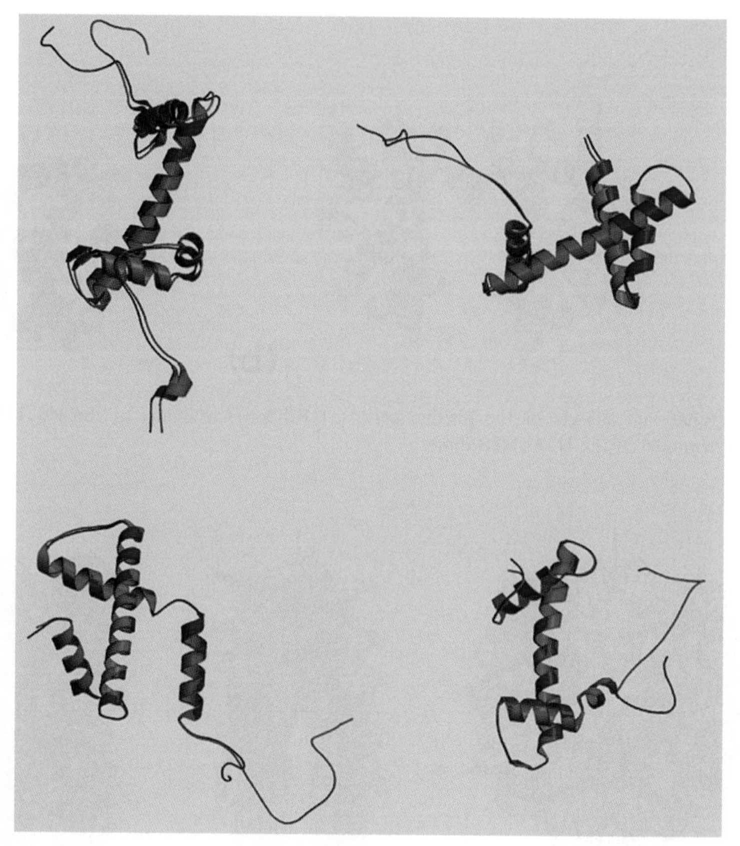

Fig. 6. Superposition of the four histone pairs in the NCP structure. The areas in red indicate deviations between a histone component and its symmetry equivalent in the NCP structure. The histone octamer is a symmetric molecule when crystallized in the absence of DNA.

One consequence of the molecular asymmetry is that the core particle presents two distinct faces, arbitrarily labeled ventral and dorsal in our images. These two faces have subtle yet distinct differences in the electrostatic surface potentials they present.

4. Structure of the core histones

Although the histone fold was first described from the structure of the histone octamer core of the nucleosome [17], the high α-helical content was predicted much earlier [43]. The core histones possess three functional domains; (1) the histone fold domain, (2) an N-terminal tail domain, and (3) various accessory helices and less structured regions. The N-terminal tail domains of the core histones are currently the focus of intense research. Covalent modifications of residues in these unstructured domains appear to modify local chromatin structure, either directly or

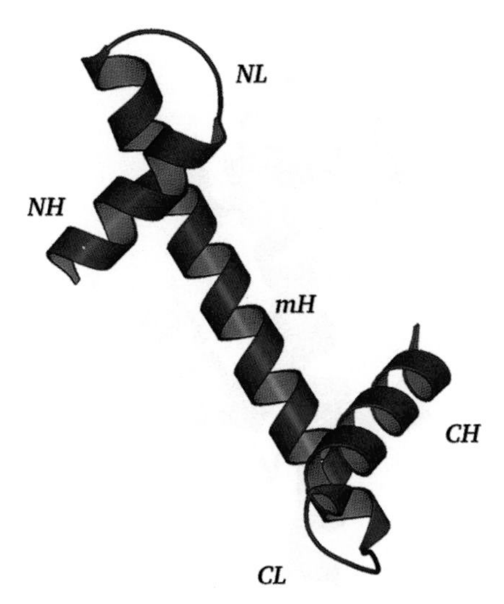

Fig. 7. Ribbon Cα model of H4 showing the folding pattern of an exemplar histone. The canonical histone fold includes one long medial α-helix (mH) with two shorter α-helices towards the N- and C-termini (NH, CH), bound by loops (NL, CL) to the primary helix.

through interactions with regulatory proteins that recognize specific covalent modifications. These modifications may involve addition of acetyl, methyl, or phosphate groups or more bulky adducts such as ubiquitination or ADP-ribosylation. Current research activities are identifying regulatory relationships that appear to be part of an epigenetic code [44] regulating gene activity through modification of higher order chromatin structure.

The histone fold appears as a symmetrical duplication of a helix-fold-helix motif [18,19] with a long median helix, the mH helix, and two shorter terminal helices, the N-terminal or NH helix and the C-terminal or CH helix (Fig. 7). The helices are joined by loops N-terminal, the NL loop, and C-terminal, the CL loop, to the mH helix. The formation of heterodimer pairs is accomplished through a "handshake" [17] pairing in which the mH helices of the handshake partners align in opposite orientations such that the NL loop of one partner aligns in a parallel orientation with the CL loop of the other (Fig. 8).

Accessory domains in the core histones include an N-terminal accessory helix in H3 that is involved in binding to and stabilizing DNA as it enters and exits the nucleosome (Fig. 9a). Histone H2B possesses a short C-terminal accessory helix, which with the rest of the C-terminal tail, forms a large portion of the protein faces of the nucleosome (Fig. 9b). The C-terminal domain of H2A is responsible in docking of the dimer to the tetramer and the C-terminus is known to also interact with linker DNA. Significantly, archael histones possess only the histone fold domain, which is the essential structural motif for bending DNA to the nucleosome structure. The structure of the histone core of the nucleosome is highly conserved and is evidently designed for compaction of average B-form DNA to the greatest extent. It is not likely to be coincidental that the persistence

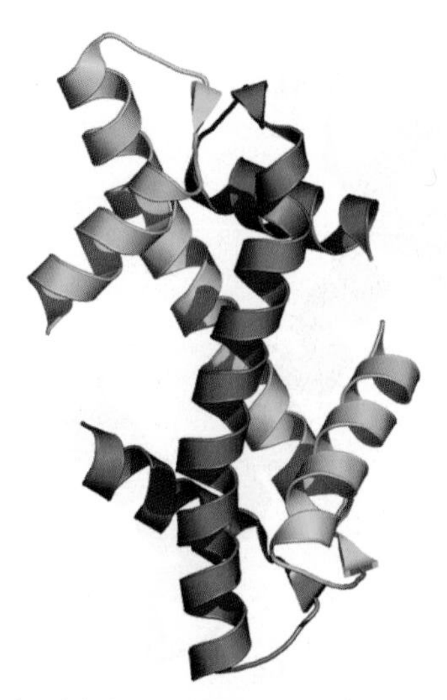

Fig. 8. The classical histone handshake motif shown in a ribbon Cα model of the interdigitating H3 : H4 heterodimer. Additional stabilization of the heterodimeric structure is provided by the formation of a short parallel β-bridge between the C-terminal loop occurring after the medial helix of H3 and the N-terminal loop occurring before the medial helix of H4. This β-bridge provides a platform for DNA binding.

length of average B-form DNA corresponds to the length of DNA in the nucleosome core particle [45]. The binding of DNA to the histone core results in conformation change in the histone octamer, primarily of the H2A and H4 medial helices (Fig. 10).

Heterodimer formation is extended to higher order interactions within the histone octamer core by interactions through the histone fold. One such interaction is the formation of a four helix bundle from the two H3 subunits in one of two homologous histone interactions within the nucleosome (Fig. 11a). The four helix bundle made by the H3 subunits forms the $(H3:H4)_2$ tetramer and represents the "hinge region" of the nucleosome. This homologous interaction is stabilized with hydrogen bonds utilizing Arg and Asp side chains as well as side chain backbone interactions. In a theme that is repeated in dimer–dimer interactions, His 113 residue also participates in H3 : H3 hinge stabilization. A four helix bundle formed by the histone fold domain of H4 and H2B serves to join the (H2A : H2B) dimers to the $(H3:H4)_2$ tetramer (Fig. 11b). This interaction is stabilized both with hydrogen bonds and with an aromatic stack (Tyr 83 of H2B and Tyr 72 of H4). In the tetramer : dimer interaction His 75 of H4 is within hydrogen bonding distance of Glu 93 of H2B. The two (H2A : H2B) dimers are joined to each other by hydrogen bonds between Asp 38 and Gln 41 residues of the different H2A chains (Fig. 11c). Connections between the His 82 of H2B and Gln 41 of H2A also contribute to this site of heterodimer interaction. All three dimer interactions have at least one histidine participating in the contact residues. This provides a protonation state dependent bonding pattern that could be manipulated in the nucleosome to stabilize or destabilize the histone core as needed.

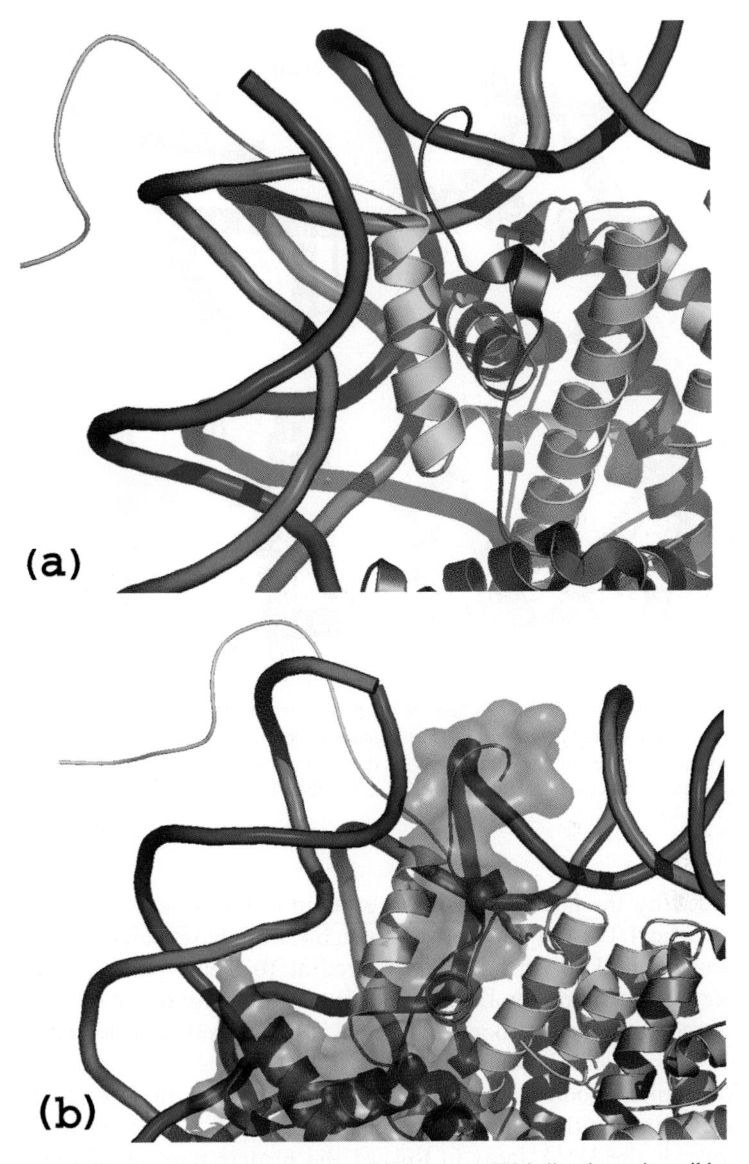

Fig. 9. Accessory helices in core histone structures. (a) Accessory H3 helix, shown in a ribbon Cα model, interacts with the DNA entering and leave the nucleosome. A short helix in the tail of H2A is seen between the accessory and medial helix of H3. (b) Solvent accessible surface representation of the C-terminal residues of H2A showing the contribution of these residues to the ventral surface of the NCP.

5. DNA structure

The ~1.75 superhelical turns of DNA on the NCP are the longest DNA sequences to date whose structure has been experimentally determined. Although the DNA is B-form, deviations from ideality exist. Wrapping of the DNA onto the

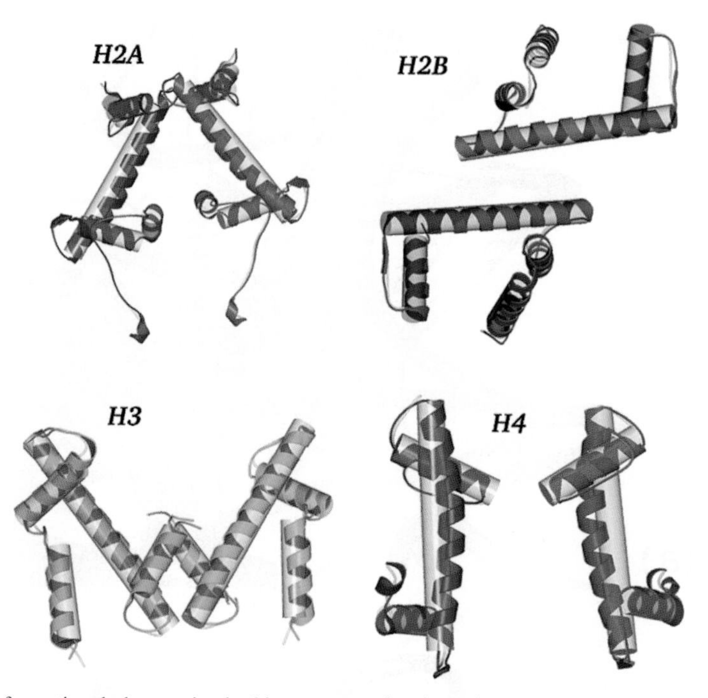

Fig. 10. Conformational changes in the histone core when bound to DNA. The ribbon Cα models of the histones in the NCP are shown with a least squares overlay of the histones from the octamer. The transparent cylinders represent the boundaries of the octamer structure. Histones H2A and H4 show the most pronounced structural shift in the medial helix when bound to DNA.

ramp provided by the histone octamer results in overwinding of the DNA and an increase in helical twist angle between neighboring base pairs. In solution, the helical repeat of the DNA has been measured at 10.5 base pairs. On the NCP the overwinding from bending the DNA results in a periodicity of 9.9 base pair per turn when measured with respect to a local helical axis or 10.33 base pairs per turn with respect to a global helical axis [46].

Tracing the phosphodiester backbone of the two gyres of DNA shows regions of asymmetry between them, most pronounced at the 10 o'clock position where the minor-groove of the DNA faces in toward the protein core of the nucleosome, a "minor-groove in" (mGI) site. Other unmatched kinks in the backbone are seen on the opposite side of the gyre. These are both regions of underwound DNA, bound by a mixture of binding interactions with the histones, described in greater detail below. These multiple types of histone–DNA binding could provide increased plasticity in the interaction between the two components of NCP resulting in a region that deviates from ideality (Fig. 12).

The center of the palindromic DNA was of necessity changed from the native α-satellite to provide for construction of the palindrome [30]. The deviation of the palindrome from perfect symmetry on the histone octamer may be a

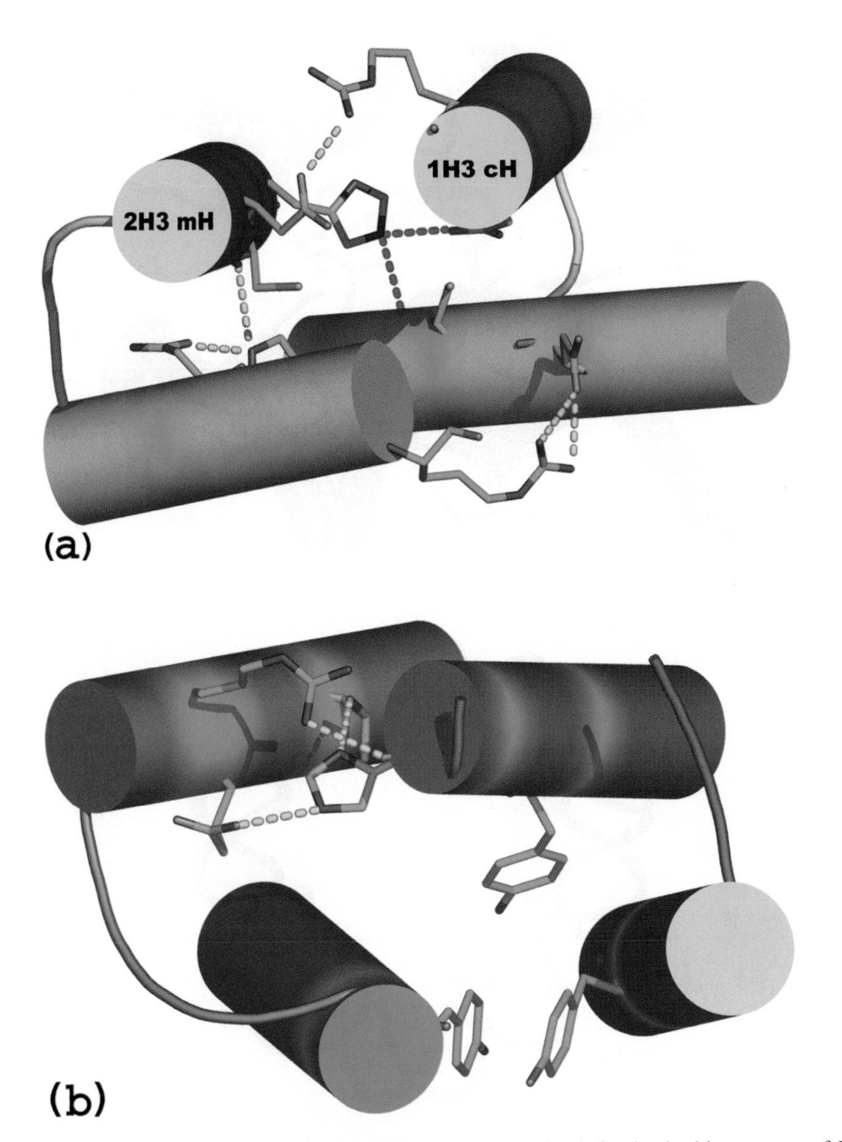

Fig. 11. Homomeric and heteromeric interactions between protein chains in the histone core of NCP. (a) The homomeric interaction between H3 chains in the "hinge" region, showing the interdigitation of the medial and C-terminal helices of each chain. There are asymmetric interactions between the two chains in this image, the two His 113–Cys 110 carbonyl oxygen and His 113–Asp 123 distances are not equal (1H3 His–2H3 Cys bond distance 2.93 Å, 2H3 His–1H3 Cys bond distance 3.23 Å; 1H3 His–2H3 Asp bond distance 2.83 Å, 2H3 His–1H3 Asp bond distance 2.57 Å) and only one hydrogen bond is present between the 1H3 Arg 129 and 2H3 Asp 106 residues. (b) The heteromeric interaction between H4 and H2B. This interaction is also mediated with a His residue (H4 His 75). More interestingly, there is a strong π-bonding ring stack formed between the co-planar Tyrs 83 (H2B) and 88 (H4) and the orthogonal Tyr 72 (H4). The stacking distances are less than 4 Å. This is the only strong hydrophobic interaction seen in the interactions between chains in the histone core. (c) The homomeric/heteromeric interactions between the H2A : H2B heterodimer. The His residues of the H2B histones that could affect the interactions are shown in green.

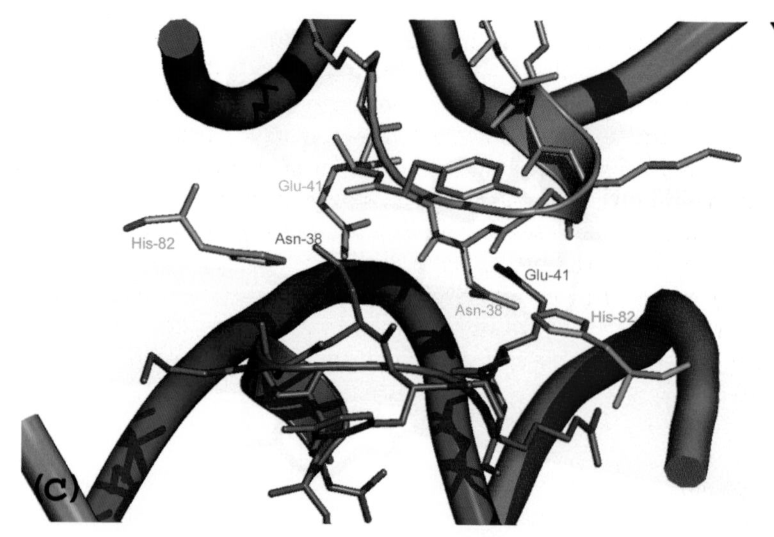

Fig. 11. Continued.

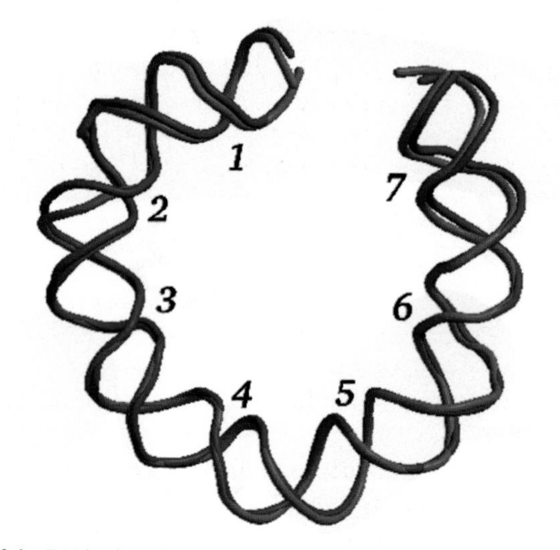

Fig. 12. Asymmetry of the DNA phosphodiester backbone trace as seen in the Oak Ridge NCP structural model (PDB access code 1EQZ). No attempt has been made to regularize the geometry of the DNA: positions of phosphates are based solely on the experimental electron density. Here the two DNA gyres are overlaid, with the 72 bp ventral gyre in red and the 73 bp dorsal gyre in blue. The minor groove positions facing the histone core are numbered sequentially from the dyad axis. The most pronounced asymmetry is seen in position 2 (10 o'clock).

consequence of the synthetic *Eco*RI site at the dyad. Energy minimization studies of this sequence with the DUPLEX program [47] clearly demonstrate a tendency for this segment of DNA to bend away from the histone surface, with the bend, centered at the dyad. Other modeling studies support this conclusion (Victor Zhurkin, personal communication).

Bonded to the DNA in the crystal model are numerous divalent manganese ions and water molecules. The Mn^{2+} ions are primarily associated with the N7 atom of the purine bases, however a number of them are also bound through water bridges in the negatively charged environment of the minor groove of the DNA. Also found associated with the N7 of purine residues were potassium ions. They are distinguishable from the Mn^{2+} ions by greater distance in the bond length. Well-defined water molecules associate in a 2:1 ratio preferentially with the phosphate backbone versus the bases. Binding of the histones to the DNA will most likely result in water displacement primarily at mGI sites throughout the molecule and can provide sites for hydrophobic interactions elsewhere in the DNA, as detailed below. Water molecules participate in a number of bridging networks between the DNA and the protein in the crystal structure.

6. DNA–histone binding

The predominant structural mechanism of DNA binding to the histone surface is mediated by the insertion of arginine side chains into the DNA minor groove. For nomenclature, the binding motifs at these minor groove-in (mGI) positions can be differentiated by the secondary structural features of the histone participating in the binding. Thus two types of binding motifs with DNA minor grooves are recognized, those with α-helices forming the basic protein architecture supporting the insertion of the arginine (α-mGI), and those with β-bridges forming the basic architecture (β-mGI).

The α-mGI motif operates primarily through the induced dipole of the α helices directed toward DNA phosphates (Fig. 13a) and was termed the paired element motif (PEM) [17], although paired ends of helices is more descriptive. The guanido group is restrained from binding deeply into the mGI by a threonine from another histone. This steric hindrance preserves the avidity of the histone core for DNA, but prevents overly strong binding that would disrupt the plasticity of the histone–DNA interaction.

In the β-mGI binding motif, short loops from different histone chains join in a short parallel β-bridge structure forming a platform for positioning the phosphate backbone of the mGI. In most cases the bonding with the DNA is restrained by a threonine on the opposite strand of the β-bridge, hydrogen bonding to the arginine (Fig. 13b). This is not invariant however, and the threonine may be present to position the arginine side-chain when DNA is not bound. Such an interaction would place the guanidino group as a "sticky" site at which initiation of the DNA–histone interaction could occur.

Major-groove interactions (MGI) are also present in the NCP structure. One of the strongest is seen in Fig. 13c, where the Arg residues on the H3 are hydrogen-bonded to phosphate groups on both sides of the major-groove. Arg 69 and Arg 72 interact with the phosphates along the sugar-phosphate backbone at the major groove and have shorter hydrogen-bonding distances than does a neighboring Arg 63 located in the minor groove. Perhaps most interestingly, Leu 65 of H3 weakly hydrogen bonds to a phosphate group oxygen of a thymine in a poly(dT) tract. In the histone octamer structure this leucine is exposed to solvent. This hydrophobic interaction in the major groove is intriguing in light of the significance of poly(dA) tracts in the DNA sequence dependent phasing of nucleosomes. The position of the leucine is in the center of a run of four thymine residues.

7. *Surface features of the NCP*

The standard presentation of NCP molecular models uses skeleton building blocks: ball and stick representations of atoms and bonds, or secondary structural cartoon features that emphasize the α-helical nature of the histone core. While these simplifications are necessary to convey some of the character of the salient

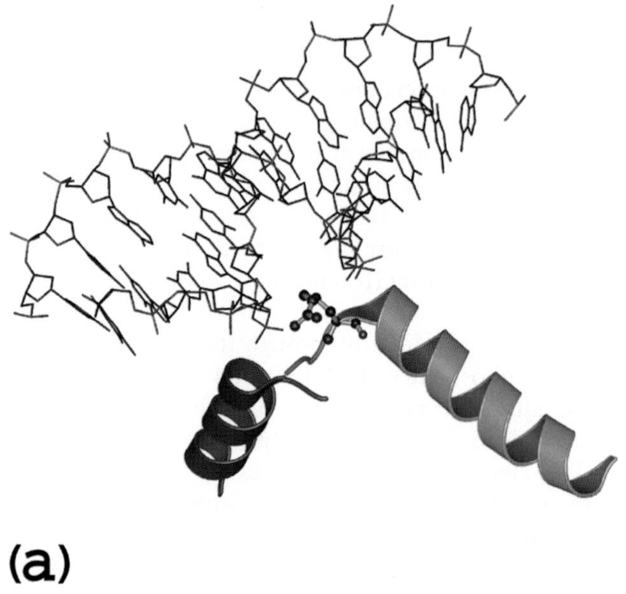

Fig. 13. Images of histone and DNA binding motifs as described in the text. Histone–DNA interactions that occur between the DNA minor-groove or the DNA major-groove and the protein are denoted mGI and MGI, respectively. There are two mGI (or minor-groove in) binding motifs characterized by an arginine side chain inserted into the DNA minor-groove. (a) The α-mGI binding motif, utilizes the induced dipole of the α-helix to facilitate placement of the arginine in the minor-groove. (b) The β-mGI binding motif consists of a short β-bridge platform for the threonine assisted positioning of the guanido side group of arginine in the DNA minor groove. (c) Protein–DNA interactions are depicted between the N–helix of H3 and the major-groove located between minor-groove positions 2 and 3 of the DNA. This interaction occurs on both the ventral (shown) and dorsal sides of the molecule. The arginine in the minor-groove (Arg 63) is the same as that shown in Fig. 15. In this representation, the arginines are bound to phosphate group oxygen atoms and a leucine forms a weak H bond with thymine (C–H–O distance 3.44 Å, C–HH–C distance 3.81 Å). The distances between Arg 69 and the phosphate group oxygens are 2.5 Å and 2.6 Å. The distances between the phosphate oxygens and Arg 72 are 2.78 Å and 2.88 Å. By contrast, the distance between Arg 63 and phosphate oxygens seen in Fig. 15 for the minor-groove interaction at mGI 2 (Fig. 14) is about 3.1 Å.

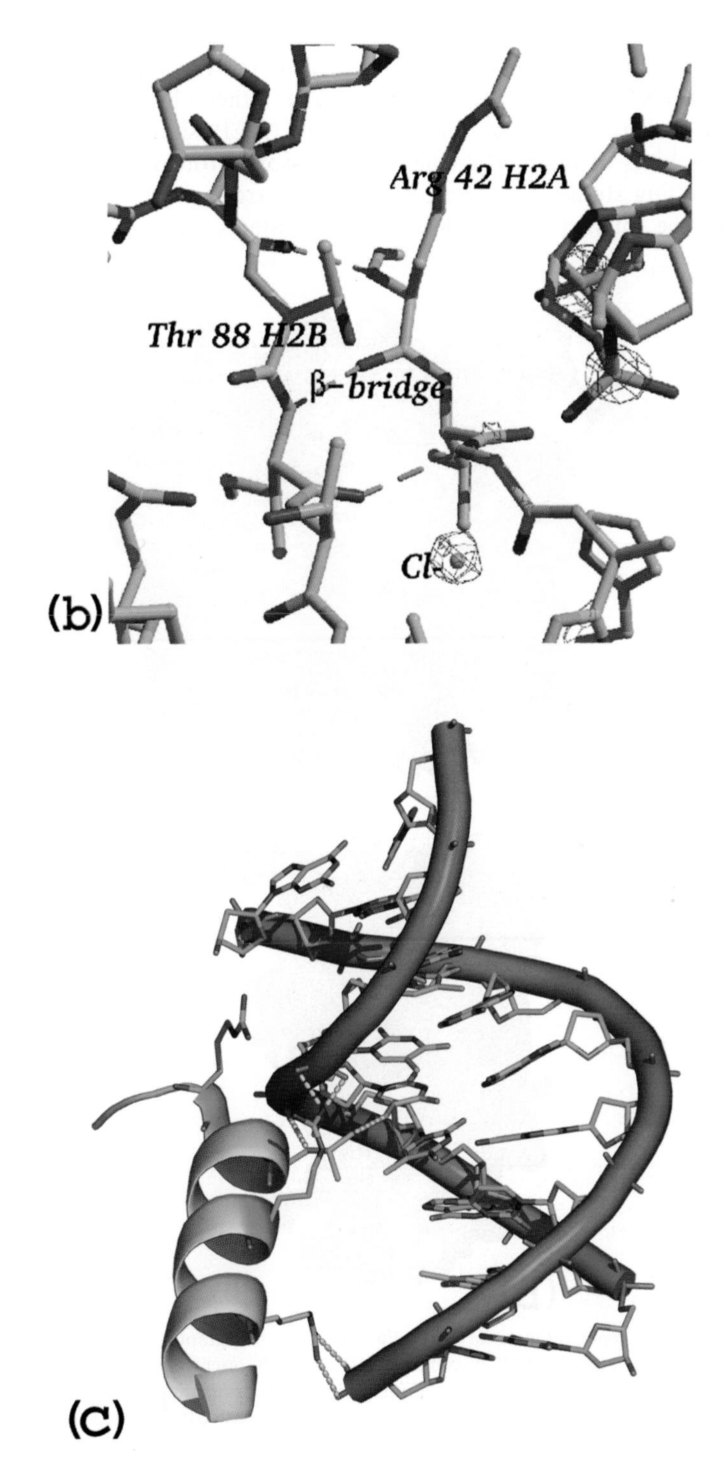

Fig. 13. Continued.

features of the NCP, they restrict the ability to convey information about the surface of the NCP where the interaction with other chromatin elements may occur. High-resolution surface features of the NCP show a complex surface that provides hints of binding cavities for accessory proteins (Fig. 14). Generally in protein binding studies, the greater the surface area covered during interaction, the stronger the interaction [48]. Regions of greater surface relief are then more

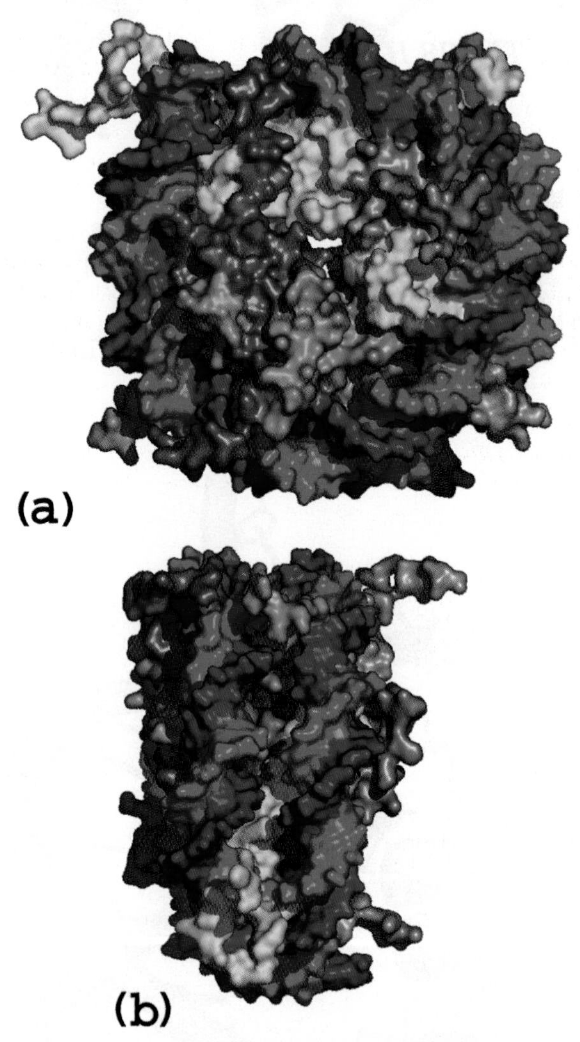

Fig. 14. Surface feature representations of NCP, illustrating the topographic features of the nucleosome core. (a) The ventral surface of the NCP; compare with similarly oriented image in Fig. 6. Regions of greatest surface relief may indicate likely sites of other chromatin protein interactions. (b) One potential site of interaction near the egress of the H2B N-terminal tail from the NCP. The deep furrow between the DNA gyres and the hole into the body of the NCP could be potential protein binding sites. This region has been implicated as the locale for HMGN1 binding.

likely to host stable interactions with chromatin elements. It is somewhat surprising that to date no single stable interaction has been identified with the central void of the nucleosome core, which stains heavily with uranyl acetate in electron micrographs.

A region of greater surface relief is seen between the DNA gyres near the emergence of the H2B N-terminal tails (Fig. 14b). This region is characterized by the lack of protein between the DNA gyres and a deep hole that intrudes to near the center of mass for the NCP. These two interesting surface features, symmetrically disposed relative to the molecular dyad axis, are possibly the binding sites of the HMGN1 proteins to the nucleosome [49,50]. If the interaction between HMGN1 proteins and the nucleosome resembles a lock and key, one would expect HMGN1 proteins to have two axes of interaction, one that lies between the DNA gyres and a process that extends into the interior of the nucleosome. The DNA atoms in these regions are characterized by high thermal factors in the crystal structure (Fig. 15).

One surface feature that we see in our model of the NCP may have biological implications, or may be an artifact of the crystallization process. This is the presence of a cacodylate ion that serves as an interparticle contact in the crystal (Fig. 16). This interaction occurs within a pocket formed by side chains of Gln 76 and Asp 77 of 2H3 and the main chain carbonyl of Leu 22 of 2H4 on the NCP and by side chains of Glu 64 of 1H2A and His 49 of 2H2B as well as the main chain carbonyl of Val 48 of 1H2B on the symmetry neighbor. The possibility of cacodylate ion mediating an interaction between chromatin elements could be a clue

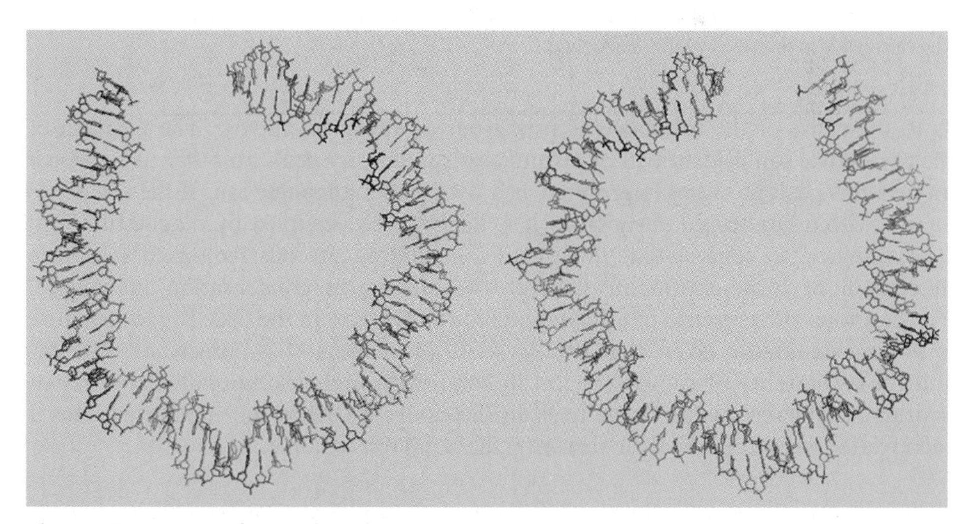

Fig. 15. Temperature factors (B values) for DNA showing areas of structure indeterminacy are represented in a pseudo-color model. Blue represents the lowest temperature factors and red the highest. Each gyre of the DNA is shown, ventral left, dorsal right. The temperature factor is one means of describing the atomic disorder. Sites indicated in red correlate with the holes shown in the previous figure and could be positions of HMGN1 binding.

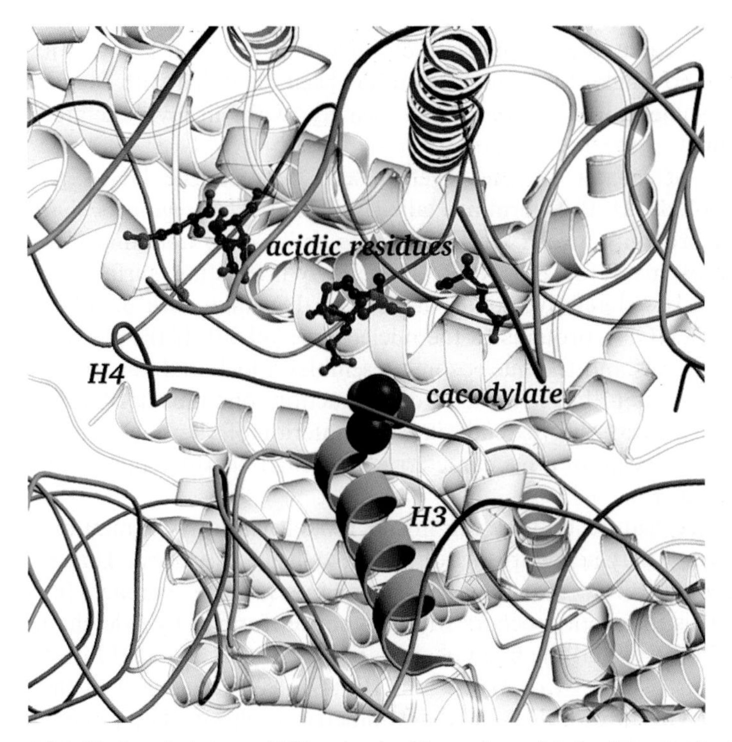

Fig. 16. Cacodylate binding site between NCP molecules. Figure shows details of the stacking interactions between two neighboring molecules involving the dorsal H3 : H4 tetramer face of one NCP and the ventral H2A : H2B dimer face an adjacent NCP. The acidic residues forming a surface patch on the dimer face are rendered in ball-and-stick. The H4 tail is in blue and the H3 region involved in the interaction is in yellow. The cacodylate is shown as a CPK rendering.

as to the cause of the carcinogenic properties of arsenicals *in vivo*. The position of the cacodylate ion is identified by strong electron density peak. In other nucleosome models this peak has been interpreted as a hydrated manganese ion. If the site exists *in vivo* within condensed chromatin, it is likely to be occupied by magnesium ion. It is tempting to suggest that this metal ion binding site has biological relevance in control of local chromatin structure or chromatin condensation in general. Furthermore, the presence of a cacodylate ion in this site in the Oak Ridge structure is not unreasonable given that the crystallization method is somewhat different and the availability of cacodylate ion in solution is high. Additionally, the size of hydrated manganese is too large to fit in the cavity we see in our model, whereas a cacodylate ion is consistent in size with the available volume.

8. Translation, libration, and screw-axis motions of NCP elements

The NCP is a dynamic structure, even within the crystal. The diffraction from an NCP crystal is affected by disorder or motions of subunits, the individual amino acids and nucleotides, and each atom within the structure. Translation, libration, and screw-axis (TLS) analysis permits the separation of motion components to better understand the sources of atomic position uncertainty in diffraction-based macromolecular structures. Translation refers to the displacement of the position of the origin of the axes of motion from the center of mass for a given analyzed element. Libration refers to the magnitude of the motions about the center of motion for the analyzed element, and screw-axis analysis gives the orientation of the libration axes.

Analysis of the Oak Ridge NCP X-ray diffraction data with the Refmac5 program of the CCP4 crystallographic data analysis suite [51,52] has provided some insight into the translation, libration and screw-axis movements (TLS) of NCP subunits [53,54]. These motions can be thought of as a group vibration contributing to the overall positional uncertainty of a particular NCP subunit. Analysis of the TLS molecular motion has provided insight into the behavior of some systems [55], making it tempting to ascribe biological importance to these motions. Other studies [56–58] have shown the TLS parameters for the correlated motions of subunits and of individual amino acids and nucleotides make the largest contribution to overall atomic displacements. In practice, without atomic resolution data, it is impossible to separate the translations and librations of the major elements of a molecule from the motions of torsion angle components of individual amino acids and nucleotides. Thus, our analysis of the NCP diffraction data contains both these elements of the total atomic motion for the NCP atoms.

Within the NCP we have defined a number of subgroups for this analysis. For the TLS refinement of the histones we have used the two H2A : H2B dimers, the H3 : H4 dimers and the $(H3:H4)_2$ tetramer as subunits (Fig. 17). We have divided the DNA at the dyad axis into separate groups for each gyre. In general, the centers of motion for each subunit undergo a minimal translation from the subunit center of mass, although the origins for the axes of motion for the DNA are not coincident. This suggests that the motion of the DNA is not mirrored about the dyad. This behavior is most likely a manifestation of the asymmetry of the NCP structure. Within the histones the libration has a large component parallel with the dyad axis of the DNA and along the direction of the major alpha-helix bundle of the H3 : H4 dimers, suggestive of a dynamic tension between the curved DNA and protein required to hold it in that position. The orientation axes of the H3 : H4 dimers cross from one surface of the NCP to the other to intersect with the DNA gyre on the opposite side of the dyad. The libration of the two H2A : H2B dimers is not co-linear with the dyad axis, rather the principle axes of libration for the dimers are co-linear with the H2A main helices of each respective dimer. The orientation axes of the H2A : H2B dimers remain on the same surface of the NCP as their respective dimer group, coplanar with one gyre of DNA. The axis of motion for H2A or H2B appears to be dictated by the medial helix. In each of H2A or H2B chains the primary axis of motion matches the axis of motion for the medial helix.

One possible interpretation of these motions is that the ends of the DNA on the NCP are exerting a force on the histones, resulting in tension in the protein in this direction. An analysis of the libration in the DNA gyres extrinsic to the dyad DNA fragment shows the predominant axis in the direction of the DNA ends, in agreement with direction of the dimer libration tension (Fig. 18). The libration axes of the histones indicate a motion that is consistent with an elliptically shaped nucleosome. Imaging and modeling of nucleosome shapes in the 11 nm chromatin fiber suggest that an elliptical form may be a favored conformation [59–61].

When TLS refinement is applied to the tetramer subunit, the resulting libration is less easy to interpret, and much less correlated with an intranucleosomal element. The center of motion for the tetramer shifts away from the dyad axis

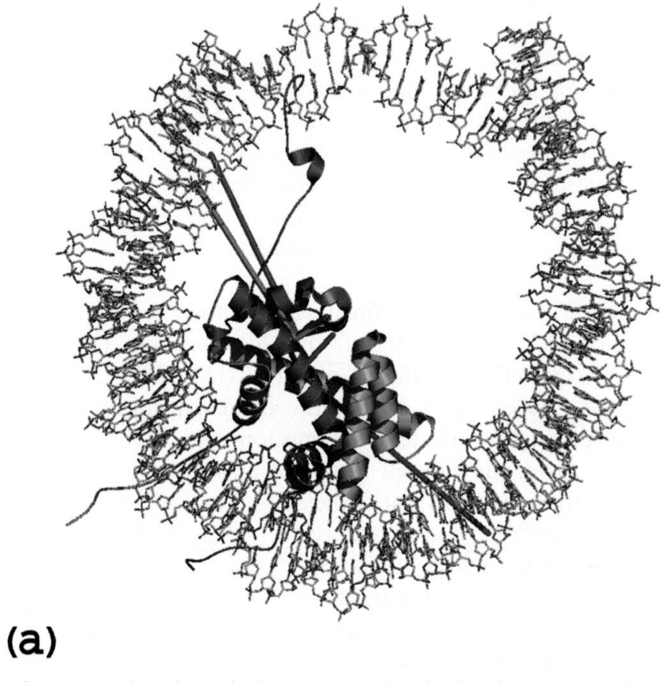

Fig. 17. Composite structural motions of subunits can be described with translation, libration, and screw-axis (TLS) analysis of the NCP. Analysis of the histone subunits are shown here. (a) Composite motion of histones H2A (blue) and H2B (blue) considered as individual elements and combined as H2A : H2B dimer (red). Note for the individual histones that the axis of motion is parallel with the medial α-helix of the histone. The origin of the TLS axes are within the structural positions of the histone. The composite motion for the H2A : H2B dimer is dominated by the motion of H2A, as is seen in the similarity of orientation and position of the two axes. (b) The orientation and motion of the two H2A : H2B dimers appear symmetric across the dyad axis of the NCP. (c) H3 : H4 composite motions when considered as dimers (blue) and as the tetramer (red). Interpretation is more complex because of the asymmetric magnitude of motion for the two dimers, and the different position in the axis of primary motion for the tetramer. These motions are most likely the consequence of packing interactions, described in greater detail in the text.

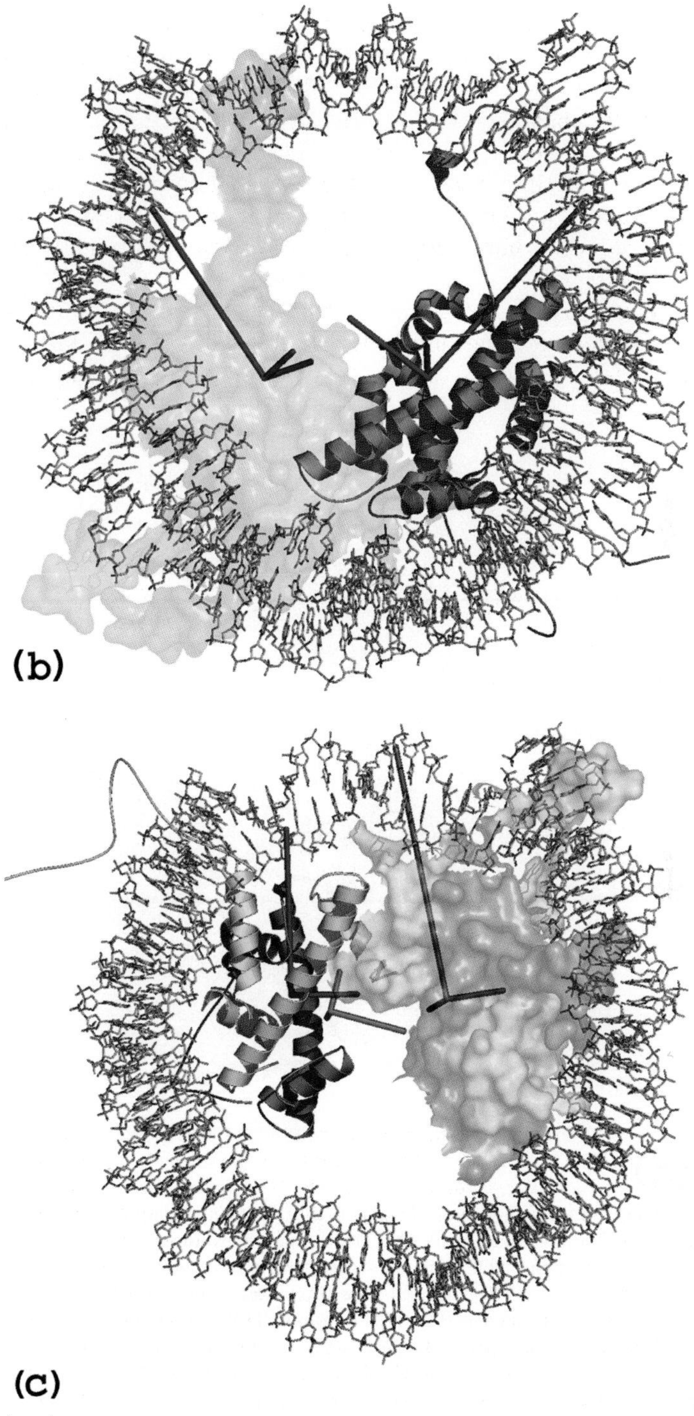

Fig. 17. Continued.

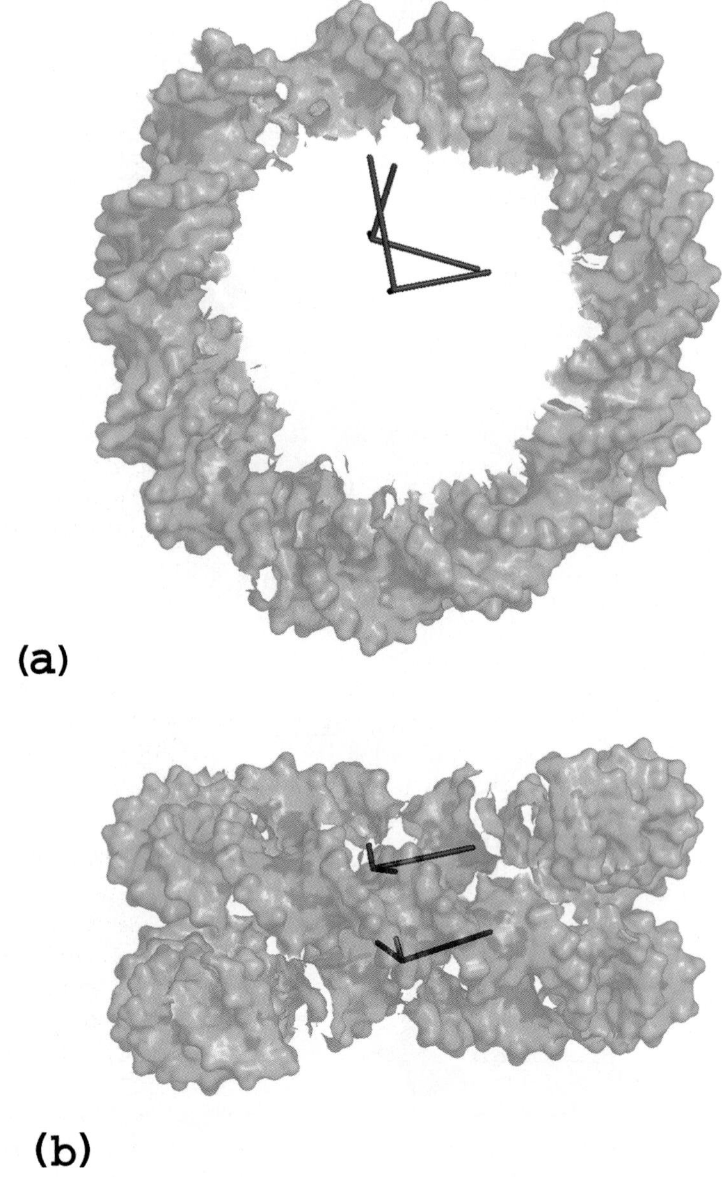

Fig. 18. TLS analysis of the palindromic DNA on the NCP. (a) Composite motions of the DNA gyres looking at the ventral surface with the DNA colored by atom type. The two gyres reflect the structural asymmetry of the NCP with non-coincident axes of motion and different orientations for the primary axis of motion. The ventral gyre TLS axes more closely resemble the composite motions of the individual H3 : H4 dimers (Fig. 17c), the dorsal TLS axes resemble the composite motion of the tetramer. (b) The composite motions of the DNA gyres are shown in a view down the dyad axis. The DNA is shown in a surface representation colored by atom type. Note that axes of motion appear parallel and in plane with the pitch of the DNA. In this view the ventral surface is on the bottom of the image.

and displays a lateral component. It would appear that the composite libration for the H3 : H4 dimers cancel each other in the direction of the dyad axis. Instead, the primary axis of motion runs towards the 146–147 bp end of the DNA. What results in the cancellation of motion in the direction of the dyad is unclear. One possible explanation is that the two H3 : H4 dimers synchronously oscillate in opposite directions so that the net motion for the paired dimers is cancelled. The lateral component of motion in the tetramer is also seen in the 40 bp DNA segment that includes the dyad position. Theoretical modeling studies of bending of the dyad portion of the DNA, which was formed from a commercially available linker and was not part of the native alpha-satellite sequence, shows that it prefers to bend in the opposite direction from the histone core of the NCP. Thus, this portion of the DNA may be providing the tension responsible for the libration of the H3 : H4 tetramer. In the absence of other possible interactions, the lateral libration of the tetramer and the dyad DNA is possibly a consequence of the NCP crystal packing.

Another way to look at the composite motions of libration in the core particle is to recognize that the major motions are perpendicular to the DNA superhelical axis. Although the directions of librations from the H2A : H2B and H3 : H4 tetramer are not strictly radial, there is a radial component when all histones are considered as single element. From these motions, it is possible to picture the dynamic tension between the DNA and the histone core as one in which the DNA is trying to straighten out and fly off the histone core with the α-helical regions of the histone core flexing in response, not unlike an oscillating spring (Fig. 19). The overall effect is that of a molecule in dynamic equilibrium, pulsating in a radial direction perpendicular to the superhelical axis of the DNA. Comparison of histone core structural changes between the octamer and NCP show that binding DNA results in contraction and displacement of the α-helices of the histones, especially in histones H2A and H4. This displacement in the central helix mirrors the primary TLS axis for these histones. It is interesting that none of the libration axes for any component of the NCP appear oriented with the zones of greatest thermal motion in the DNA.

9. Crystal packing features of NCP and implications for higher order chromatin structure

One salient question about crystallographic studies of the NCP is whether these structural models can provide us with useful information regarding higher order chromatin structure. Perhaps the most important unanswered question of the experimentally determined NCP model is the inability to definitively place the binding site of histone H1. The longevity of this controversy is remarkable in light of the information, such as positioning of histone tails, which can be gleaned from a high quality NCP crystal structure. The answer to H1 positioning will await an atomic resolution structure of the chromatosome, or perhaps an oligonucleosome.

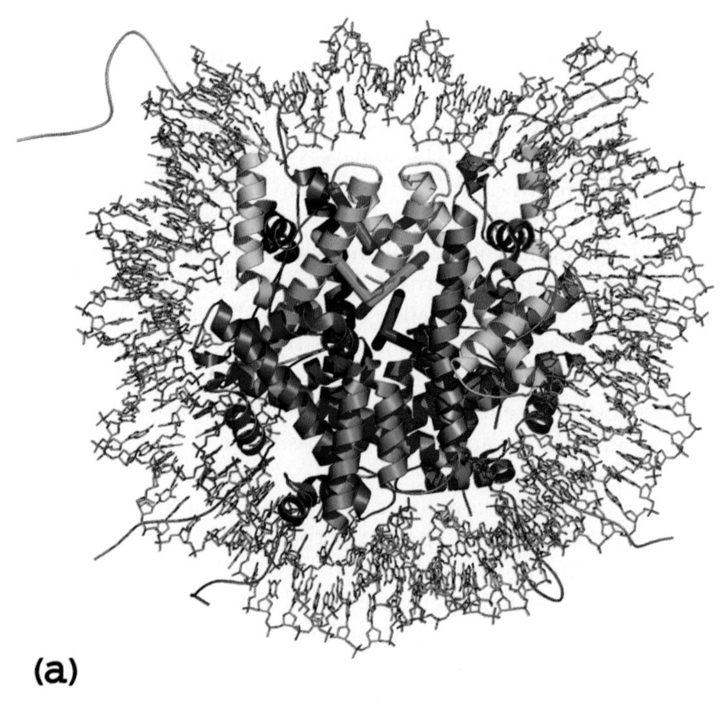

Fig. 19. TLS analysis of the NCP, DNA, and histone core. In these ventral and dorsal views of the NCP model, the composite motion axes of the DNA, histones, and the NCP are shown in red, blue, and green, respectively. The center of motion axes for the DNA and the histones are non-coincident, the TLS axis for the DNA is furthest from the center of mass of the NCP. This may reflect the dominance of the DNA ends in the overall displacement of the DNA. The TLS analysis shows that DNA regions with high B-values, seen in Fig. 15, have little contribution to the overall motion of the DNA on the NCP. The overall motion of the NCP appears dominated by the DNA motion, with the TLS origin shifted in the direction and appearing congruent with the DNA. Overall, the primary axes of motion are in plane with the DNA, hence the interpretation that the composite motions are dominated by dynamic tension between the DNA and the histones, with deviation from these general motions the consequence of packing interactions.

Questions concerning the structure or existence of a 30 nm chromatin fibril seem to be beyond the scope of crystallographic analysis at this time.

Two features in NCP crystal packing do have implications for higher order chromatin structure. First, the interactions between histones of separate NCPs can occur within the core histone region or between N-terminal tails (Fig. 20a). These interactions can be direct between oppositely charged amino acid residues or mediated with metal ions or water molecules. The histone tails can be profoundly asymmetric in positioning, resulting in the two faces of the core particle presenting chemically different environments. The different faces could in turn provide a favored orientation for binding other chromatin elements. Modification of the amino acid residues in the tail will affect internucleosomal interactions and modify the polarity of the nucleosome faces. The level of interaction complexity

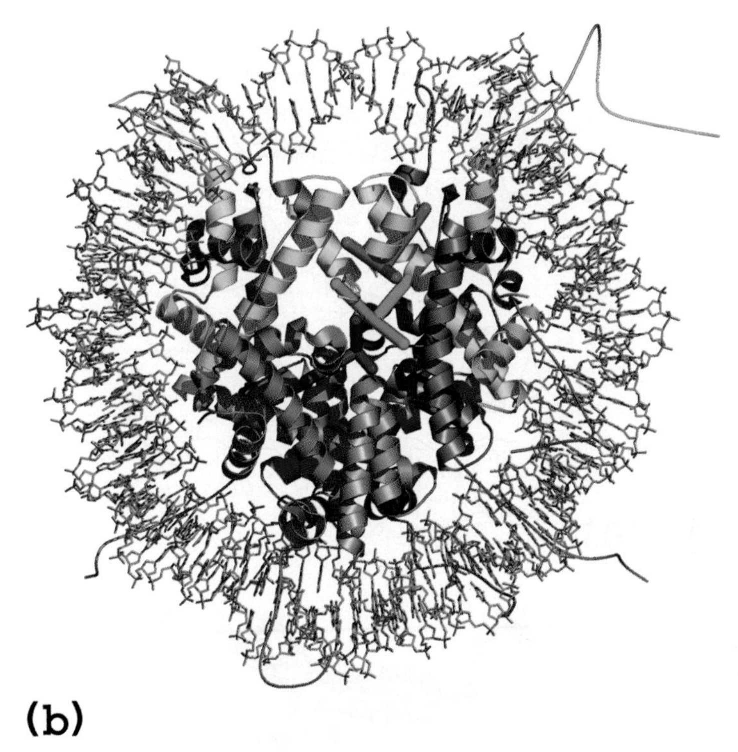

Fig. 19. Continued.

can quickly proliferate with a small number of modifications to the nucleosome structure.

The other feature with implications for higher order chromatin structure is the base stacking interactions between the DNA ends of adjoining molecules (Fig. 20b). The preservation of the DNA base stacking in the crystal packing is so important energetically to cause the displacement of the terminal phosphate groups and 5′ to 5′ interactions between DNA ends of different NCPs. The DNA strongly stabilizes the NCP arrangement in the crystal. Thus, the two features described here have somewhat divergent roles. The histones provide relatively non-specific interactions between nucleosomes and other chromatin proteins. However, despite the wealth of potential charge interaction between the histones of one NCP and the DNA ends of another, the base stacking is preserved and the DNA termini abut each other.

The two previously described interactions are sufficient to provide a two-dimensional or sheeted chromatin structure. The intermolecular interaction that provides the third dimensional packing of chromatin sheets is most likely the metal–ion interaction described previously. While it is tempting to describe this feature as a general chromatin interaction, it is absent in the histone octamer crystal interaction (PDB access number 1HQ3). Instead the contact residues (the side chains of Gln 76 and Asp 77 of 2H3 and the main chain carbonyl of Leu 22 of 2H4

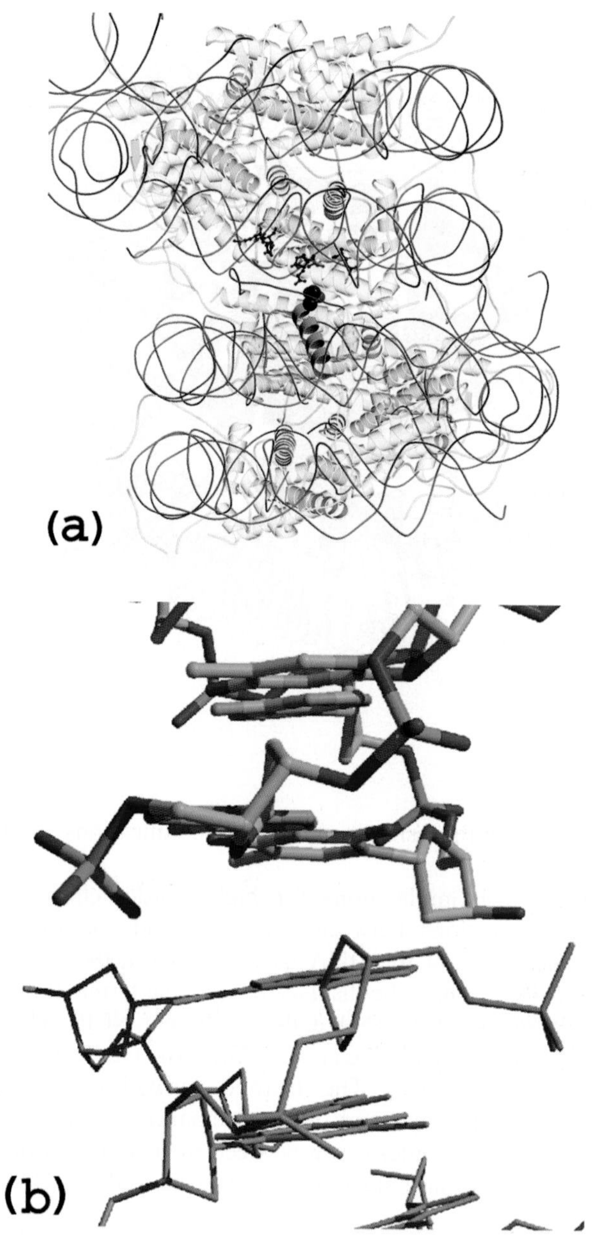

Fig. 20. Packing interactions between NCP molecules, which are a consequence of crystallization, nevertheless provide hints for higher order chromatin structure assembly. (a) Histone–histone interactions shown at the site of the cacodylate ion. In addition to binding interactions with the cacodylate ion, the N-terminal tail is involved in significant interactions with the patch of acidic residues on the dimer face of the neighbor NCP. The orientation of the dyad alternates between the two NCP molecules. (b) DNA base stacking is continuous between neighboring NCP molecules in the crystal lattice as the DNA exits one NCP and enters the next. The stacking interaction is strong enough to force a shift in the terminal phosphates for adjoining 5′ termini.

on the NCP and by side chains of Glu 64 of 1H2A and His 49 of 2H2B as well as the main chain carbonyl of Val 48 of 1H2B on the symmetry neighbor) in the octamer structure are freely accessible to the solvent. The generality of this NCP packing interaction could be investigated by structural analysis of a monoclinic NCP crystal form. These crystals (space group $P2_1$) are not grown in the presence of cacodylic acid. The structure might suggest whether this site has a preference for metal–ion interactions that mediate nucleosome associations in chromatin.

References

1. Olins, A.L. and Olins, D.E. (1973) J. Cell Biol. 59, 252a.
2. Woodcock, C.F.L. (1973) J. Cell Biol. 59, 368a.
3. Oudet, P., Gross-Bellard, M., and Chambon, P. (1975) Cell 4, 281–300.
4. Hewish, D.R. and Burgoyne, L.A. (1973) Biochem. Biophy. Res. Commun. 52, 504–510.
5. Kornberg, R. (1974) Science 184, 868–871.
6. Bakayev, V.V., Melnikov, A.A., Osicka, V.D., and Varshavsky, A.J. (1975) Nucleic Acids Res. 2, 1401.
7. Finch, J.T., Brown, R.S., Rhodes, D., Richmond, T., Ruston, B., Lutter, L.C., and Klug, A. (1981) J. Mol. Biol. 145, 757–769.
8. Bentley, G.A., Finch, J.T., and Lewit-Bentley, A. (1981) J. Mol. Biol. 145, 771.
9. Finch, J.T., Lutter, L.C., Rhodes, D., Brown, R.S., Rushton, B., Levitt, M., and Klug, A. (1977) Nature 269, 29–36.
10. Klug, A., Rhodes, D., Smith, J., Finch, J.T., and Thomas, J.O. (1980) Nature 287, 509.
11. Bentley, G.A., Lewit-Bentley, A., Finch, J.T., Podjarny, A.J., and Roth, M. (1984) J. Mol. Biol. 176, 55–75.
12. Richmond, T.J., Finch, J.T., Rushton, B., Rhodes, D., and Klug, A. (1984) Nature 311, 532–537.
13. Uberbacher, E.C. and Bunick, G.J. (1985) J. Biomol. Struct. Dyn. 2, 1033–1055.
14. Uberbacher, E.C. and Bunick, G.J. (1989) J. Biomol. Struct. Dyn. 7, 1–18.
15. Burlingame, R.W., Love, W.E., Wang, B.-C., Hamlin, R., Xuong, N.-H., and Moudrianakis, E.N. (1985) Science 228, 546–553.
16. Wang, B.-C., Rose, J., Arents, G., and Moudrianakis, E.N. (1994) J. Mol. Biol. 236, 179–188.
17. Arents, G., Burlingame, R.W., Wang, B.-C., Love, W.E., and Moudrianakis, E.N. (1991) Proc. Natl. Acad. Sci. USA 88, 10148–10152.
18. Arents, G. and Moudrianakis, E.N. (1993) Proc. Natl. Acad. Sci. USA 90, 10489–10493.
19. Moudrianakis, E.N. and Arents, G. (1993) Cold Spring Harb. Symp. Quant. Biol. 58, 273–279.
20. Arents, G. and Moudrianakis, E.N. (1995) Proc. Natl. Acad. Sci. USA 92, 11170–11174.
21. Baxevanis, A.D., Arents, G, Moudrianakis, E.N., and Landsman, D. (1995) Nucleic Acids Res. 23, 2685.
22. Luger, K., Rechsteiner, T.J., Flaus, A.J., Waye, M.M.Y., and Richmond, T.J. (1997) J. Mol. Biol. 272, 301–311.
23. Simpson, R.T and Stafford, D.W. (1983) Proc. Natl. Acad. Sci. USA 80, 51–55.
24. Richmond, T.J., Searles, M.A., and Simpson, R.T. (1988) J. Mol. Biol. 199, 161–170.
25. Uberbacher, E.C., Wilkinson-Singley, E., Harp, J.M., and Bunick, G.J. (1988) In: Olson, W.K., Sarma, M.H., Sarma, R.H., and Sundaralingam, M. (eds.) DNA Bending and Curvature. Adenine Press, Schenectady, NY.
26. Palmer, E.L., Geweiss, A., Harp, J.M., York, M.H., and Bunick, G.J. (1996) Anal. Biochem. 231, 109–114.
27. Thomas, J.O. and Butler, P.J.G. (1977) J. Mol. Biol. 116, 769–781.
28. Harp, J.M., Palmer, E.L., York, M.H., Geweiss, A., Davis, M., and Bunick, G.J. (1995) Electrophoresis 16, 1861–1864.

29. Uberbacher, E.C. and Bunick, G.J. (1986) Purification of nucleoprotein particles by elution preparative gel electrophoresis ORNL/TM-10230, Oak Ridge National Laboratory, No. DE87001853 (available from the National Technical Information Service, 5285 Port Royal Rd., Springfield, VA 22161, USA).
30. Harp, J.M., Uberbacher, E.C., Roberson, A., and Bunick, G.J. (1996) Acta Crystallogr. D 52, 283–288.
31. Harp, J.M., Hanson, B.L., Timm, D.E., and Bunick, G.J. (2000) Acta Crystallogr. D 56, 1513–1534.
32. Carter, D.C., Wright, B., Miller, T., Chapman, J., Twigg, P., Keeling, K., Moody, K., White, M., Click, J., Ruble, J., Ho, J.X., Adcock-Downey, L., Bunick, G.J., and Harp, J.M. (1999) J. Crystal Growth 196, 602–609.
33. Luger, K., Mader, A.W., Richmond, R.K., Sargent, D.F., and Richmond, T.J. (1997) Nature 389, 251–260.
34. Suto, R.K., Clarkson, M.J., Tremethick, D.J., and Luger, K. (2000) Nat. Struct. Biol. 7, 1121–1124.
35. White, C.L., Suto, R.K., and Luger, K. (2001) EMBO J. 20, 5207–5218.
36. Davey, C.A., Sargent, D.F., Luger, K., Mader, A.W., and Richmond, T.J. (2002) J. Mol. Biol. 319, 1097–1113.
37. Suto, R.K., Edayathumangalam, R.S., White, C.L., Melander, C., Gottesfeld, J.M., Dervan, P.B., and Luger, K. (2003) J. Mol. Biol. 326, 371–380.
38. Berman, H.M., Westbrook, J., Feng, A., Gilliland, G., Bhat, T.N., Weissig, H., Shindyalov, I.N., and Bourne, P.E. (2000) Nucleic Acid Res. 28, 235–242.
39. DeLano, W.L. (2002) The PyMOL Molecular Graphics System. DeLano Scientific, San Carlos, CA, USA. http://www.pymol.org.
40. McRee, D.E. (1993) Practical Protein Crystallography. Academic Press, San Diego, CA, USA. http://www.scripps.edu/pub/dem-web/toc.html.
41. Kraulis, P.J. (1991) J. App. Crystal. 24, 946–950.
42. Kornberg, R.D. and Lorch, Y. (1999) Cell 98, 285–294.
43. Thomas, G.J., Jr., Prescott, B., and Olins, D.E. (1977) Science 197, 385–388.
44. Strahl, B.D. and Allis, C.D. (2000) Nature 403, 41–45.
45. Smith, S.B., Cui, Y., and Bustamante, C. (1996) Science 271, 795–799.
46. Lu, X., Shakked, Z., and Olson, W.K. (2000) J. Mol. Biol. 300, 819–840.
47. Hingerty, B.E., Figueroa, S., Hayden, T.L., and Broyde, S. (1989) Biopolymers 28, 1195–1222.
48. Edmundson, A.B., Harris, D.L., Fan, Z.-C., Guddat, L.W., Schley, B.T., Hanson, B.L., Tribbick, G., and Geysen, H.M. (1993) Proteins: Struct. Funct. Genet. 16, 246–267.
49. Uberbacher, E.C., Mardian, J.K.W., Rossi, R.M., Olins, D.E., and Bunick, G.J. (1982) Proc. Natl. Acad. Sci. USA 79, 5258–5262.
50. Postnikov, Y.Y. and Bustin, M. (1999) Methods Enzymol. 304, 133–155.
51. Collaborative Computational Project, Number 4. (1994). Acta Crystallogr. D 50, 760–763.
52. Winn, M.D., Isupov, M.N., and Murshudov, G.N. (2001) Acta Crystallogr. D 57, 122–133.
53. Schomaker, V. and Trueblood, K.N. (1968) Acta Crystallogr. B 24, 63–76.
54. Trueblood, K.N., Buergi, H.-B., Burzlaff, H., Dunitz, J.D., Gamaccioli, C.M., Schultz, H.H., Shmueli, U., and Abrahams, S.C. (1996) Acta Crystallogr. A 52, 770–781.
55. Wilson, M.A. and Brunger, A.T. (2000) J. Mol. Biol. 301, 1237–1256.
56. Diamond, R. (1990) Acta Crystallogr. A 46, 425–435.
57. Kidera, A. and Go, N. (1990) Proc. Natl. Acad. Sci. USA 87, 3718–3722.
58. Kidera, A. and Go, N. (1992) J. Mol. Biol. 225, 457–475.
59. Fritzsche, W. and Henderson, E. (1996) Biophy. J. 71, 2222–2226.
60. Bishop, T.C. and Hearst, J.E. (1998) J. Phy. Chem. B 102, 6433–6439.
61. Bishop, T.C. and Zhmudsky, O.O. (2002) J. Biomol. Struct. Dyn. 19, 1–11.

J. Zlatanova and S.H. Leuba (Eds.) *Chromatin Structure and Dynamics: State-of-the-Art*

DOI: 10.1016/S0167-7306(03)39003-9

CHAPTER 3

Paradox lost: nucleosome structure and dynamics by the DNA minicircle approach

Ariel Prunell[1] and Andrei Sivolob[2]

[1]*Institut Jacques Monod, Centre National de la Recherche Scientifique, Université Denis Diderot Paris 7, et Université P. et M. Curie Paris 6, 2 place Jussieu, 75251, Paris Cédex 05, France. E-mail: prunell@ccr.jussieu.fr*
[2]*Department of General and Molecular Genetics, National Shevchenko University, 252601, Kiev, Ukraine*

1. Introduction: the linking number paradox and DNA local helical periodicity on the histone surface

In the past few years, new concepts have emerged in the understanding of the role of chromatin to mediate gene activity through regulated access of transcriptional factors to their target sites. Although restriction site exposure essays have shown that nucleosomal DNA can be spontaneously accessible to DNA binding proteins [1,2], there is an obvious need for a large enhancement of that exposure as well as for its modulation. The central mechanism appears to be chromatin remodeling, both chemical, through histone covalent modifications (in particular acetylation) and physical, in which the energy of ATP hydrolysis is used to mobilize and structurally alter nucleosomes (see Ref. [3] for a review), the latter possibly utilizing above mentioned nucleosomal DNA inherent dynamic character. Chemical and physical remodelings may cooperate with each other (see Ref. [4] for a review), in particular through co-recruitment to the same chromatin sites by matrix/scaffold attachment region binding proteins [5].

In spite of these progresses, and perhaps because of them, some early basic problems have not received all the attention they deserved, even if their solutions are likely to fill crucial gaps in our understanding of chromatin function. Among them, the so-called "linking number paradox" [6,7], previously reviewed by one of the authors [8], has raised heated debate in the past [9–11]. As will be shown below, this paradox is at the heart of nucleosome conformational dynamics within the context of chromatin superstructure.

This problem first emerged from the necessity to reconcile topological and structural data of nucleosomes and chromatin. As soon as a minichromosome could be reconstituted from pure DNA and histones, the total reduction of the DNA linking number (Lk) was found to be equal to the number of nucleosomes, which was also true for the native H1-bearing SV40 minichromosome [12]. On the other hand, the first low-resolution crystal of the core particle showed that DNA was wrapped with 1 3/4 turns of a left-handed superhelix. Assuming linker DNAs continued the trajectories so defined, two complete turns of the superhelix would be formed which should have instead reduced Lk by two turns per nucleosome (one-turn reduction per negative crossing) [13].

The proposed solution to that paradox was contained in the classical equation [14–16]:

$$\Delta \mathrm{Lk} = \Delta \mathrm{Tw} + \mathrm{Wr} \tag{1}$$

formally valid for the whole minichromosome, but which also applies to individual nucleosomes if the minichromosome was equivalent to a random juxtaposition of independent topological domains with no net contribution of linker DNAs to ΔTw and Wr. This equation shows that the linking number change (−1 per nucleosome) has two contributions: not only the writhe (Wr), but also a possible change in the twist of DNA upon wrapping (ΔTw). With $\mathrm{Wr} \sim -1.7$ (see below; Wr would be exactly equal to the number of crossings only if DNA diameter and superhelix pitch were equal to zero), the proposal was that a $\Delta\mathrm{Tw} = 0.7$ turn overtwisting, which would satisfy Eq. (1), occurred over the $N_n \sim 160$ bp wrapped in the 2-turn nucleosome. This overtwisting would change the helical periodicity of the wrapped DNA according to the equation [17]:

$$\Delta \mathrm{Tw} = \mathrm{N_n}(1/h - 1/h_o) \tag{2}$$

in which $h_o \sim 10.5$ bp/turn is the helical periodicity of DNA free in solution. It comes $h \sim 10.0$ bp/turn. Such an integral value had merits of its own. It would optimize the interactions of the double helix with the histone surface [6], somehow reminiscent of the situation prevailing in wet DNA fibers, where the helical periodicity was also integral. Moreover, 10 bp/turn was a sufficiently large decrease below the solution value to be testable by DNA cleavage *in situ*.

DNase I cuts only the exposed regions of each strand, as opposed to the histone-bound regions which are protected. A limited digestion then gives rise to a ladder of fragments of size multiple of the exposure (or protection) periodicity which could be resolved through denaturing gel electrophoresis. Early results, obtained with native chromatin from rat liver nuclei, pointed to h values rather comprised between 10.3 and 10.4 bp/turn [18,19], an increase of 0.3–0.4 bp/turn over the first estimate [20]. It was soon realized that this discrepancy did not necessarily disprove the overtwisting hypothesis. The histone surface could be convex, and, with the DNase cutting along the normal to that surface at each point, the cleavage periodicity would be larger than the real DNA periodicity [8,18]. The problem was given a new twist when it was later realized that the cleavage periodicity could not be equated to h in Eq. (2), even if the histone surface was perfectly cylindrical. Indeed, wrapping of DNA into a superhelix, even in the absence of torsional constraint, was not only bending it, but also twisting it, altering the exposure periodicity, while keeping $\Delta\mathrm{Tw} = 0$ in Eq. (2). This led to the notion of the DNA *local* helical periodicity, h_{loc}, most conveniently visualized by the periodicity of the DNA contact points with the surface [21]. h_{loc}, therefore, depended on the wrapping surface. In contrast, h in Eq. (2) does not, and was termed intrinsic (h_{int}) when referring to DNA on a surface. For simple geometrical surfaces, h_{int} can be easily deduced from h_{loc}. For a perfect cylinder, and with the superhelix parameters of that time [13], the conversion formula was [21]

$$h_{int} = h_{loc} + 0.15 \tag{3}$$

resulting in $h_{int} = 10.3–10.4 + 0.15 = 10.45$ to 10.55 bp/turn. (0.15 in Eq. (3) decreases to 0.13 with the most recent parameters in Ref. [22].) This value, similar to h_o, suggested (but did not prove, in the absence of a precise knowledge of the shape of the histone surface) that DNA in these mixed-sequence nucleosomes was wrapped without overtwisting.

Other cleavage agents were subsequently used, the most convenient of them probably being hydroxyl radicals generated by the Fenton reaction [23]. Their advantage over DNase I was their virtual lack of DNA sequence specificity, which allowed the cleavage periodicity to be obtained with nucleosomes reconstituted at specific locations of unique DNA fragments, in contrast with DNase I which would cleave only some of the potential sites. An average cleavage periodicity of 10.2 (± 0.05) bp/turn was found with a "5S" nucleosome, a significant decrease below the above 10.3–10.4 bp/turn figure of DNase I, with a larger 10.7 bp/turn periodicity in the ~ 30 bp central region [24]. Because the same average cleavage periodicity and discrepancy in the center were found with mixed-sequence nucleosomes, it was proposed that OH radicals cleaved all nucleosomes in the same way [25]. But we have subsequently found, from our limited collection of unique nucleosomes, that average OH radicals cleavage periodicity ranges between ~ 10.2 and 10.5 bp/turn, with a mean value potentially similar to that of DNase I, 10.3–10.4 bp/turn. Moreover, the discrepancy in the center region is rather in both directions, when it exists at all (C. Lavelle, A. Sivolob and A. Prunell, unpublished results).

The high-resolution crystal structure of a unique 146 bp nucleosome core particle reconstituted on an inverted repeat of human α-satellite [22] offered the first opportunity to directly estimate h_{loc} [8]. This was conveniently done by measuring the spacing between arginine lateral chains inserted at regular intervals into the narrowed small groove of the double helix. A value of 10.23 bp/turn was obtained between the most distal histone–DNA binding sites (superhelix locations (SHL) ± 6.5 and 133 bp of DNA; see Fig. 1). This periodicity was expected to increase to 10.31 bp/turn in the symmetrical, more stable, 147 bp particle whose structure has recently been reported [26]. These two figures are consistent with above cleavage periodicity data. In contrast, h_{loc} was 10.0 bp/turn between the distal $(H3–H4)_2$ tetramer binding sites (SHL ± 2.5 and 50 bp) and 10.43 bp/turn (10.57 bp/turn in the 147 bp particle) between proximal H2A–H2B dimer binding sites (SHL ± 3.5). This periodicity jump reflects an untwisting of the two DNA stretches (12 and 11 bp, respectively) in between SHL 2.5 and 3.5 and SHL −2.5

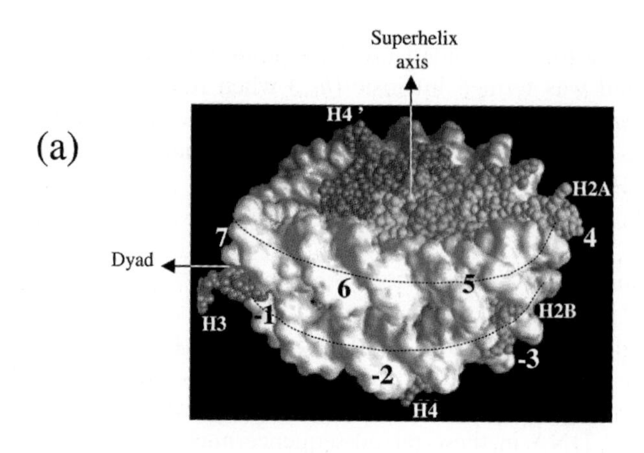

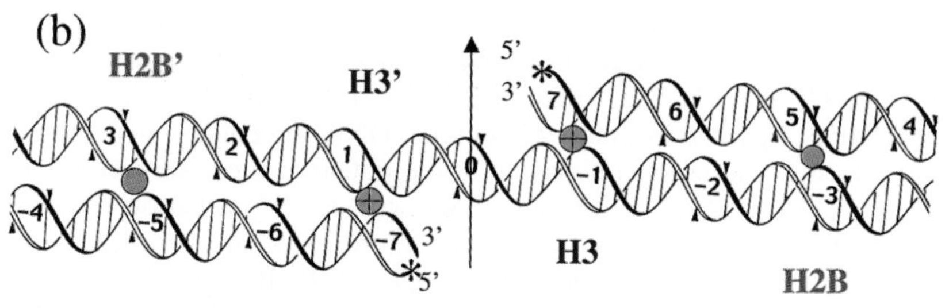

Fig. 1. The core particle, the DNA superhelix and H2B and H3 N-terminal tails. (a) Space-filling representation of the 2.8 Å crystal structure of the 146 bp human α-satellite nucleosome core particle [22]. The dyad is in the plane of the paper and the superhelix axis slightly off that plane. Positive and negative numbers mark the superhelix locations (SHL) in the upper and lower gyres, respectively, and the dotted curve follows the path of the double helix axis. (b) Ribbon representation of the DNA superhelix slit along a line parallel to its axis, opened out and laid flat on the paper surface. SHL are also indicated, together with H2B and H3 tails passage points between the gyres. (From Fig. 5 in Ref. [29].)

and −3.5, presumably at SHL 3 and −3 where H2B N-terminal tails pass between the two gyres through the channel formed by the aligned minor grooves (Fig. 1).

In conclusion, nucleosome cleavage periodicity data, together with direct measurements of a defined nucleosome crystal structure, suggest that wrapped DNA may be overtwisted to different extents, but in no case do these overtwistings appear sufficient to explain the paradox ($\sim$0.5 bp/turn overtwisting required, against $\sim$0.2 bp/turn observed at the most).

2. *Early topological studies: a single nucleosome on a DNA minicircle*

The advantage of such a simple system was twice: first, to suppress a potential linker contribution to the mean ΔLk per nucleosome, termed $\langle \Delta Lk_n \rangle$ from now on, which could result from nucleosome/nucleosome interactions in a minichromosome;

and second, to permit $\langle \Delta Lk_n \rangle$ to be directly correlated with DNA path on the histone surface and in the entry–exit region. $\langle \Delta Lk_n \rangle$ was measured by relaxation with DNA topoisomerase I (see Eq. (5) below), and DNA path visualized by electron microscopy. At the same time, there was a need for accurate measurements of DNA potentially sequence-dependent helical periodicity under relaxation conditions (h_o). This was achieved by using first an original method developed in this laboratory (the V-curve method; Ref. [27]), and later a more accurate procedure [17] involving the determination of topoisomer ratios in relaxation of DNA minicircles of neighboring sizes [28,29].

Early studies mainly focused on nucleosomes reconstituted on a $N = 359$ bp DNA minicircle originating from plasmid pBR322, which relaxed into a single product, corresponding to a topoisomer of $\Delta Lk = -1$ (topoisomer -1), also leading to $\langle \Delta Lk_n \rangle = -1$ [30]. ΔLk is the topoisomer constraint, defined by

$$\Delta Lk = Lk - Lk_o \tag{4}$$

where $Lk_o = Tw_o = N/h_o$ is the most probable twist [17]. $\langle \Delta Lk_n \rangle$ is the mean ΔLk of the topoisomer equilibrium distribution

$$\langle \Delta Lk_n \rangle = \sum_i n_i \Delta Lk_i \tag{5}$$

where n_i is the relative proportion of topoisomer i. With $h_o = 10.56$ (± 0.01) bp/turn [27], 359 bp was an integral multiple of h_o and Lk_o an integer. Nucleosomes on topoisomer -2 ($\Delta Lk = -2$) of that minicircle adopted two distinct "open" and "closed" conformations under electron microscopy (Fig. 2). The "closed" conformation ((c) in Fig. 2) had the expected 2-turn wrapping and two negative crossings (the first one is on the histone surface), whereas the "open" $\sim$1.4-turn conformation ((b) in Fig. 2) showed a breaking of the most distal histone–DNA binding sites, at SHL ± 6.5 (see Fig. 1), and no crossing of the entering and exiting DNAs [31]. Interestingly, the two conformations were equally represented under the low salt conditions of the electron micrograph in Fig. 2 (see legend), whereas the "open" conformation became rare upon addition of 50 mM NaCl (not shown). With topoisomer -1, in contrast, only the "open" conformation was observed, regardless of the salt concentration. This result implied that the unit linking number reduction was associated with the "open" conformation, i.e., with one negative crossing, as intuitively expected in the absence of DNA overtwisting upon wrapping [ΔTw (ΔTw_n) $= 0$ in Eq. (1)]. A theoretical model of a 1.4-turn nucleosome on a DNA minicircle confirmed this intuition: computation by a Monte Carlo method showed Wr of such a nucleosome with its nicked loop in the most probable conformation to be very close to -1.0 [32]. At this point, the conclusion could have been that there was no paradox in the first place.

However, this less-than-two-turn conformation clearly resulted from the system attempt to minimize the energy of loop bending within the topological constraints

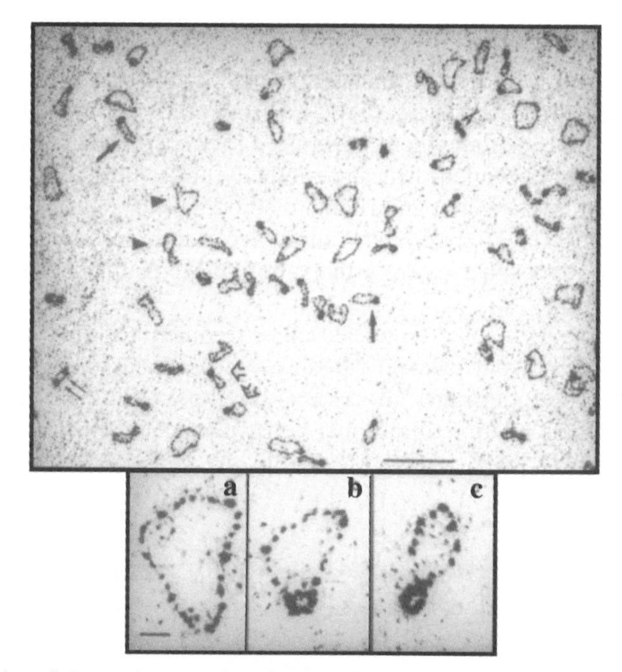

Fig. 2. Visualization of chromatin reconstituted on 359 bp $\Delta Lk = -2$ topoisomer by dark field electron microscopy. A negative is shown. The fragment originates from an Sau 3A-digest of plasmid pBR322. Samples were diluted in TE buffer (10 mM Tris–HCl, pH 7.5, and 1 mM EDTA) before adsorption to the grids. Unreacted naked minicircles (arrowheads), nucleosome monomers (single arrows) and dimers (arrow doublets) are clearly recognizable. The gallery shows enlargements of a naked minicircle (a) and mononucleosomes in the "open" (b) and "closed negative" (c) conformations. Note the open conformation of the naked minicircle, which results from enhanced intramolecular electrostatic repulsion in low salt. Bars: 100 nm and ~300 bp (field); 10 nm and 30 bp (gallery). (Adapted from Fig. 5 in Ref. [31].)

of the DNA minicircle. For this reason, it was not clear whether the $\Delta Tw_n = 0$ conclusion would apply to constraint-free nucleosomes on linear fragments or nucleosomes in a minichromosome. And yet subsequent work with mononucleosomes reconstituted on a linear 5S 256 bp fragment supported this single-crossing explanation of the paradox: their visualization in 3 dimensions under cryo-electron microscopy showed a wrapping of ~ 1.6 (± 0.15) turns but no crossing of the DNA arms, which bended away from the histone surface, presumably as a result of DNA–DNA electrostatic repulsion [33]. Subsequent classical electron microscopy [34] confirmed that result (see representative nucleosomes and scheme of entry/exit DNA conformation in Fig. 3(a)) and further showed that it also applied to GH5-containing nucleosomes (GH5 is H5 globular domain). Although the wrapping was now close to two turns, an accentuation of the bends still prevented the arms to cross (Fig. 3(b)).

Further reconstitution with native H5 showed that its long C-terminal tail could overcome the repulsion between the DNA arms and bridge them together into a

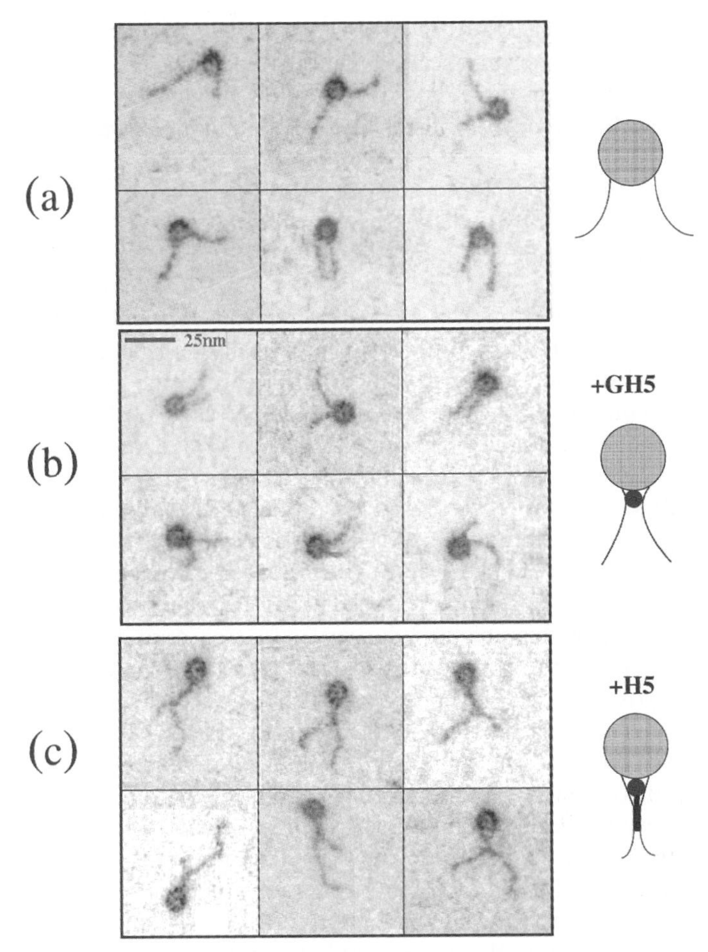

Fig. 3. Gallery of representative nucleosomes reconstituted in the absence (a) or presence of GH5 (b) or H5 (c), and visualized by scanning transmission electron microscopy. (a) and (b): 256 bp 5S rDNA fragment [65]. (c): 357 bp fragment from the "5S" series (see text). Samples were diluted in TE buffer supplemented with 50 mM NaCl and 5 mM $MgCl_2$ before adsorption to the grids. Note the nucleosome different positions relative to the DNA ends. Bars: 25 nm and ~75 bp. (Adapted from Fig. 7,9, and 10 in Ref. [34].) Schemes of the corresponding DNA conformations are shown.

two-duplex stem structure extending on an average 30 bp from the nucleosome [34] (Fig. 3(c)). Stem DNA topology was investigated by relaxation of H5-containing mononucleosomes on 359 bp minicircles. Two species were now observed in the equilibrium, corresponding to topoisomers −1 and −2, in proportions leading to $\langle \Delta Lk_n \rangle = -1.6$ to -1.7 [35]. Strikingly, this figure coincided with the Wr value calculated from the above theoretical model extended to a 2-turn nucleosome with a negative crossing in the loop [32]. This coincidence, in removing the qualifications attached to a 1.4-turn nucleosome, supported the conclusion of an absence of significant DNA overtwisting ($\Delta Tw_n \sim 0$) in these nucleosomes. However, in showing the occurrence of an effective negative crossing of the loop as a result of stem formation, these data did not allow the single-crossing explanation of the paradox to be extended to linker histone-containing chromatin.

3. Polymerase-induced positive supercoiling and linker positive crossing in nucleosomes

If the initiation of transcription may mostly rely on chromatin remodeling (see Introduction), its elongation has not lost much of its mystery since early experiments showed that nucleosomes were virtually transparent to the polymerase [36–38]. Soon after the concept of the positive supercoiling wave pushed by polymerase tracking along the DNA double helix emerged [39], this wave was proposed to play a major role in elongation by destabilizing/releasing H2A–H2B dimers. The polymerase would then transcribe DNA on the tetramer without releasing it, and this tetramer would subsequently serve as a nucleation point for nucleosome regeneration, a process facilitated by the negative supercoiling wave in the wake of the polymerase [40,41]. That release indeed fitted the two observations of a dimer exchange with the histone pool during *in vivo* transcription [42,43,40], and of a dimer deficit in core particles originating from transcriptionally active chromatin [44]. This proposal was also consistent with physico-chemical essays which revealed the lability of the dimers in the tripartite histone octamer (a tetramer flanked by two dimers; Ref. [45]). But transcription experiments with phage polymerases rather showed a random redistribution of the whole octamer [46] or, depending on experimental conditions, its translocation behind the polymerase [47,48]. A dimer release by RNA polymerase II was reported [49], although this release appears to have little to do with positive torsional stress, which was free to escape from the ends of the untethered DNA fragment, but rather with some peculiarity of polymerase II, as opposed to polymerase III [50] or phage polymerases.

In the absence of conclusive data on the role of a positive supercoiling wave, static positive supercoiling elicited by nucleosome reconstitution on relaxed or slightly positively-supercoiled plasmids [51] or by ethidium bromide intercalation in the loop of mononucleosomes on DNA minicircles [52] did not succeed either in releasing dimers. Moreover, circular dichroism, histone chemical modification and H3-thiol accessibility failed to detect an even slight alteration in the structure of such torsionally-stressed nucleosomes [51]. The reason was later found to lie in the ability of nucleosome entry/exit DNAs to form a positive crossing [52].

Linker positive crossing provided an attractive explanation to the linking number paradox. Positive crossing would indeed partially or totally compensate for the internal negative crossing, so that the $\langle \Delta Lk_n \rangle$ value associated with such "closed positive" nucleosomes would be close to zero. This explanation further required nucleosomes to spontaneously fluctuate to that conformation, rather than being forced only by applied positive supercoiling. Appropriate steady-state proportions between the two "closed negative" and "positive" nucleosome species in chromatin would insure $\langle \Delta Lk_n \rangle = -1$, as observed, taking into account that nucleosomes in the "open" conformation (of $\langle \Delta Lk_n \rangle = -1$; see above) would not impinge on that result.

4. A nucleosome on an homologous series of DNA minicircles: a dynamic equilibrium between three distinct DNA conformational states

4.1. Qualitative analysis

Results of electron microscopy in Fig. 2 already pointed to nucleosome thermal fluctuations between "open" and "closed negative" conformations. And yet these nucleosomes relaxed into topoisomer −1, i.e., to the "open" conformation (see above), with no trace of neighboring topoisomers −2 and 0 as expected products of the "closed negative" and "positive" conformations. Thermodynamics teaches that there could be only one reason for this: nucleosomes on topoisomers −2 and 0 (nucleosomes could indeed form on topoisomer 0 alone; Ref. [31]) had significantly higher free energies than nucleosomes on topoisomer −1. In other words, these nucleosomes, even in "closed negative" and "positive" conformations, had a significant residual constraint in their loop, in contrast to nucleosomes on topoisomer −1 whose loop was relaxed (see above). This in turn suggested that ΔLk_n of these conformations were significantly different from −2 and 0. Simulation has shown that Wr (ΔLk_n in the absence of overtwisting; see Eq. (1)) of the "closed negative" conformation was indeed −1.6 to −1.7. As for the "closed positive" conformation, its ΔLk_n could have been closer to −0.5 than to 0 if the positive crossing was not complete (given the left-handed wrapping around the histones, a complete positive crossing may have required too much of DNA bending).

Because topoisomers with fractional ΔLk values (such as −1.6 and −1.7, or −0.5) could only be obtained from minicircles of sizes smaller or larger than 359 bp (see Eq. (4)), nucleosomes were relaxed on a homologous series of DNA minicircles of unique sequence differing by 1–2 bp (the 351–366 bp "pBR" series, originating from a fragment of plasmid pBR322; Ref. [28]). The particular example of the 356 bp minicircle from that series is shown in Fig. 4. Figure 4 (a) shows the "chromatin" gel, with starting nucleosomes on topoisomer −2.9 in lane 1, and relaxed products in lane 2, which migrate as a doublet. The lower band in this doublet corresponds to nucleosomes on topoisomer −1.9 which slightly predominate (see topoisomer composition in Fig. 4(b) and (c)) and whose conformation should be "closed negative". The higher band in the doublet corresponds to nucleosomes on topoisomer −0.9, expectedly in the "open" conformation. Strikingly, relative amounts of these two nucleosomes are strongly altered when histones are acetylated and relaxation performed in phosphate buffer (lane 5 in Fig. 4(a); phosphate destabilizes the histone tails from their DNA interactions, somehow enhancing

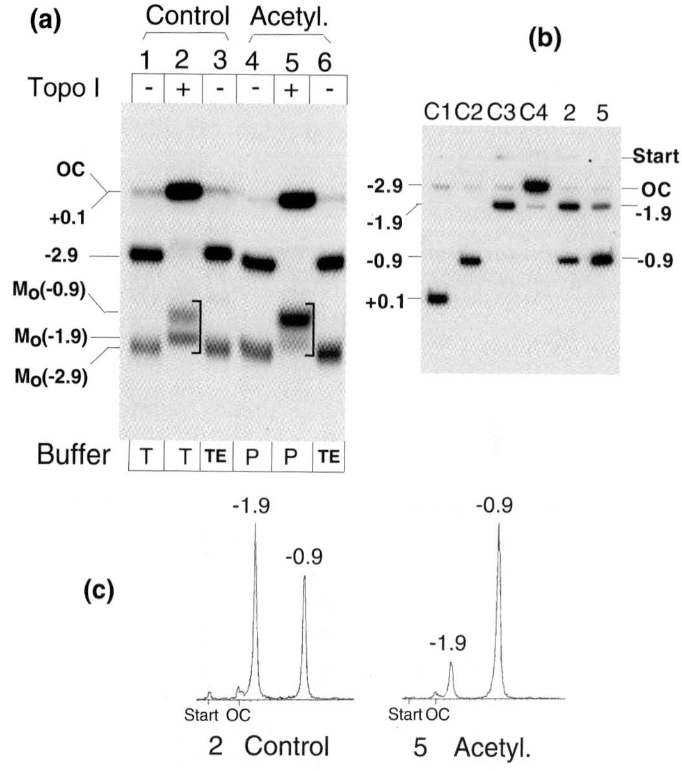

Fig. 4. Nucleosome relaxation, and influence of histone N-terminal tails. Example of nucleosomes on 356 bp $\Delta Lk = -2.9$ topoisomer from the "pBR" DNA minicircle series [28]. (a) Mononucleosomes (M_O) were reconstituted with control (Control) or acetylated (Acetyl) histones, incubated at 37 °C in Tris buffer [T: 50 mM Tris–HCl (pH 7.5), 0.1 mM EDTA, 50 mM KCl, 5 mM $MgCl_2$, and 0.5 mM dithiothreitol] or phosphate buffer [P: same as Tris buffer with 50 mM potassium phosphate (pH 7.5) instead of 50 mM Tris–HCl] in the absence (Topo I −) or presence (Topo I +) of topoisomerase I, and electrophoresed in a native polyacrylamide gel at room temperature. Note the splitting of nucleosome relaxation products in two bands. TE: starting chromatin in TE buffer. (b) Gel slices (brackets) were cut out, and eluted DNAs were electrophoresed in a chloroquine-containing native polyacrylamide gel, together with control naked topoisomers (C1–C4). Lanes were numbered as in the (a) gel. Autoradiograms are shown. (c) Radioactivity profiles of lanes 2 and 5 in the (b) gel. Topoisomers are indicated by their ΔLk values. (Adapted from Fig. 2 in Ref. [28].)

the effects of acetylation; Ref. [53]). Nucleosomes on topoisomer −0.9 now strongly predominate in the relaxation equilibrium (Fig. 4(b) and (c)), which indicates an hindrance of the "closed negative" conformation as a consequence of tail destabilization, to the benefit of the "open" conformation.

Results of the 11 DNA minicircles of the series [28] are shown in Fig. 5(a) and (b) in the form of plots of the relative amounts of the two, sometimes three, adjacent topoisomers in the relaxation equilibrium distributions as functions of their ΔLk [29]. The pBR "control nucleosome" plot (Fig. 5(a)) shows shoulders or peaks centered at ΔLk values around −1.7, −1, and −0.5, which correspond to the "closed

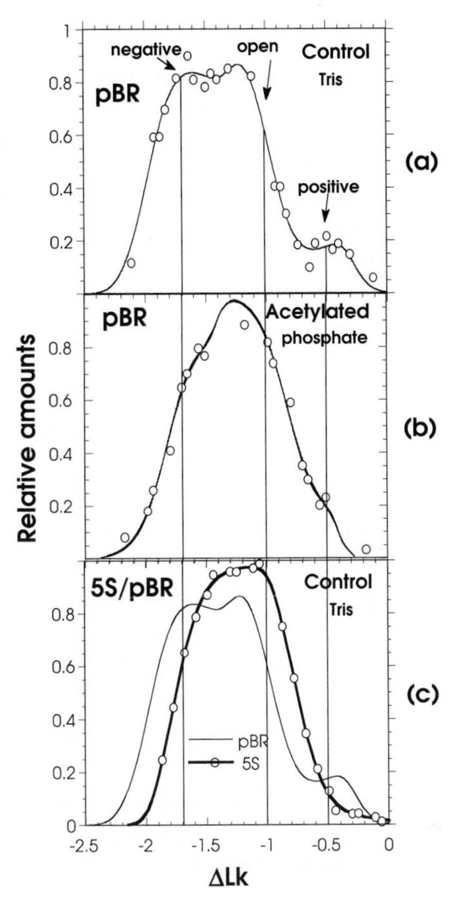

Fig. 5. Nucleosome relaxation data on the two DNA minicircle series. (a)–(c) Nucleosomes were reconstituted with control (Control) or acetylated histones (Acetylated) on $\Delta Lk = -2.4$ to -3.3 topoisomers of "pBR" 351–366 bp eleven DNA minicircles series or "5S" 349–363 bp ten DNA minicircles series, and relaxed in Tris (Tris) or phosphate buffer (phosphate), as described in legend to Fig. 4(a). Topoisomer relative amounts in the equilibrium distributions (see examples in Fig. 4(c)) were plotted as functions of their ΔLk, calculated from Eq. (4) using $h_o = 10.49_4$ (± 0.003) and 10.47_5 (± 0.003) bp/turn for pBR DNA in Tris and phosphate buffers [28], respectively, and $h_o = 10.53_8$ (± 0.006) bp/turn for 5S DNA in Tris buffer [29]. Smooth curves were calculated as described in the text. [Drawn from data in Ref. [28] (Acetylated/phosphate) and adapted from Fig. 3 in Ref. [29] (Control/Tris).]

negative", "open", and "closed positive" DNA conformations referred to above. Following the above discussion, these peaks or shoulders must result from the relative energy benefit of relaxing into these particular topoisomers of $\Delta Lk = \Delta Lk_n$, because only these topoisomers can provide a relaxed loop to the nucleosomes in these particular conformations. The "acetylated nucleosome" plot (Fig. 5(b)) shows a much smaller "$\Delta Lk = -1.7$" peak, consistent with results of 356 bp nucleosomes in Fig. 4, and virtually no "$\Delta Lk = -0.5$" peak, which demonstrates that tail destabilization hinders both "closed" states. Knowing that entry/exit DNAs are closer and must repulse each other more in “closed” than in “open” conformations, such a tail-modulated hindrance implies that the unacetylated tails interacted with these DNAs in the control (note that H3 N-terminal tails are longer than those of the other histones and most favorably located for these interactions; Fig. 1(a)). Such tail interactions with entry–exit DNAs were confirmed by UV laser-induced cross-linking of long mononucleosomes, which further showed that the tails rearrange to a nucleosomal location upon trimming down to core particles [54].

4.2. *Quantitative analysis*

4.2.1. *Topology: general equations*

For a nucleosome on a topoisomer of constraint ΔLk (see Eq. (4)) and in conformation i (i = 1–3 for “closed negative”, “open”, and “closed positive”), one has

$$\Delta \mathrm{Lk} = \Delta \mathrm{Lk}_{\mathrm{n}}(i) + \Delta \mathrm{Lk}_{\mathrm{l}}(i) \tag{6}$$

This equation provides an explicit definition of the linking difference associated with conformation i, $\Delta \mathrm{Lk}_{\mathrm{n}}(i)$, as being the total ΔLk of the topoisomer when the loop is relaxed in that conformation ($\Delta \mathrm{Lk}_{\mathrm{l}}(i) = 0$). One also has

$$\Delta \mathrm{Lk} = \Delta \mathrm{Tw}_{\mathrm{n}}(i) + \Delta \mathrm{Tw}_{\mathrm{l}}(i) + \mathrm{Wr}(i) \tag{7}$$

in which $\Delta \mathrm{Tw}_{\mathrm{n}}(i)$ and $\Delta \mathrm{Tw}_{\mathrm{l}}(i)$ are the twist changes on the histone surface and in the loop, and $\mathrm{Wr}(i)$ the total writhe of the partially-wrapped minicircle. When the loop is relaxed, $\Delta \mathrm{Tw}_{\mathrm{l}} = 0$, $\mathrm{Wr}(i) = \mathrm{Wr}_{\mathrm{o}}(i)$, and Eqs. (6) and (7) combine into

$$\Delta \mathrm{Lk}_{\mathrm{n}}(i) = \Delta \mathrm{Tw}_{\mathrm{n}}(i) + \mathrm{Wr}_{\mathrm{o}}(i) \tag{8}$$

which is Eq. (1) applied to each conformation. $\Delta \mathrm{Tw}_{\mathrm{n}}$ allows h, i.e., h_{int}, of the wrapped DNA to be derived from Eq. (2), in which N_{n} was taken equal to 113 and 132 bp in nucleosome “open” 1.45- and “closed” 1.7-turn conformations, respectively.

4.2.2. *Energetics*

The free energy of the particle (DNA plus protein; in kT) in state i on topoisomer Lk (of size N) is

$$G(i, N, \mathrm{Lk}) = G_{\mathrm{sc}}(i) + \mathrm{G}_{\mathrm{n}}(i) \tag{9}$$

with

$$G_{\mathrm{sc}}(i) = \{K_{\mathrm{sc}}(i)/N_{\mathrm{l}}\}\{\Delta \mathrm{Lk}_{\mathrm{l}}(i)\}^2$$

which gives, by replacing $\Delta Lk_l(i)$ in Eq. (6)

$$G_{sc}(i) = \{K_{sc}(i)/N_l\}\{Lk - Lk_o - \Delta Lk_n(i)\}^2 \tag{10}$$

$G_{sc}(i)$ is the supercoiling free energy of the loop for the nucleosome in state i, $N_l = N - N_n$ is the loop size, and $K_{sc}(i)/N_l$ the reduced loop supercoiling force constant (K in Ref. [17]). It is noteworthy that K_{sc}/N_l somehow measures the loop conformational rigidity, and depends not only on the DNA intrinsic rigidity parameters (its bending and twisting constants; see below), but also on the loop conformation. $G_n(i)$ in Eq. (9) represents the bending energy of the torsionally relaxed loop plus all the other contributions to the free energy (originating in particular from the protein; see below).

The probability of a nucleosome in state i on topoisomer Lk in the relaxation equilibrium is proportional to

$$Z(i, N, Lk) = \exp[-G(i, N, Lk)] \tag{11}$$

which gives for the probability of all nucleosomes on topoisomer Lk, regardless of their state

$$\rho(N, Lk) = \sum_i Z(i, N, Lk) \Big/ \sum_{Lk}\sum_i Z(i, N, Lk) \tag{12}$$

From this, the mean linking number of the topoisomer equilibrium distribution, at size N, is

$$\langle Lk_n \rangle = \sum_{Lk} Lk \cdot \rho(N, Lk) \tag{13}$$

whereas the mean linking difference is given by the equation

$$\langle \Delta Lk_n \rangle = \langle Lk_n \rangle - Lk_o \tag{14}$$

equivalent to Eq. (5). Neglecting the $\pm 2\%$ variation in N between 351 and 366 bp, N could be replaced by its mean, 359 bp, in Eqs. (9) and (11). ρ in Eq. (12) can then be expressed as a function of ΔLk by

$$\rho(\Delta Lk) = \sum_i Z(i, \Delta Lk) \Big/ \Big(\cdots + \sum_i Z(i, \Delta Lk - 1) + \sum_i Z(i, \Delta Lk) + \sum_i Z(i, \Delta Lk + 1) + \cdots\Big) \tag{15}$$

Table 1
Nucleosome fitted and calculated parameters

	pBR		K_{sc}/N_l [b]	5S		State	Wr_o [b]
	ΔLk_n [a]	ΔG_n (kT) [a]		ΔLk_n [a]	ΔG_n (kT) [a]		
Control tris	−1.69	−0.8	12	−1.40	−1.7	"negative"	-1.6_5
	−1.04	0	12	−0.72	0	"open"	−1.0
	−0.56	1.2	12	−0.41	2.2_5	"positive"	−0.3
Acetylated phosphate	−1.73	0.8	12			"negative"	-1.6_5
	−1.02	0	12			"open"	−1.0
	−0.61	3.6	12			"positive"	−0.3

Errors were ±0.02, ±1, and ±0.1 for ΔLk_n, K_{sc}/N_l and ΔG_n, respectively.
[a]Linking number (ΔLk_n) and free energy (ΔG_n) differences were derived from fitting of the three-state model to topoisomer relative amounts-versus-ΔLk data in Fig. 5.
[b]Writhes and reduced supercoiling force constant (Wr and K_{sc}/N_l; with N_l the loop size) were calculated from Fig. 6(b) and (d), and (a) and (c), respectively, using the theory as described in the text. K_{sc}/N_l values were 12, 11.5 and 11 in "closed negative", "open", and "closed positive" states, respectively, but a unique value of 12 was used to fit experimental data points in Fig. 5.

Equation (15) was fitted to the topoisomer relative amounts-versus-ΔLk plots in Fig. 5, using a unique K_{sc}/N_l value of 12 for all states (see below), to find the values of $\Delta Lk_n(i)$ and $G_n(i)$. $G_n(i)$ is actually measured by reference to the "open" state, and noted as $\Delta G_n(i)$ in Table 1. All values were then used in reverse to recalculate the topoisomer amounts-versus-ΔLk plots (smooth curves in Fig. 5) [29]. As shown in Table 1, the energetically most favored state shifted from "closed negative" in pBR control nucleosomes to "open" in the same acetylated nucleosomes in phosphate, confirming above qualitative conclusions. At the same time, the energy difference between the "closed positive" state and the most favored state almost doubled (2 to 3.6 kT). This latter value, ~4 kT, may therefore be considered as an upper limit for that state to be accessible under equilibrium conditions.

The relative steady-state occupancy of state i, $f_n(i)$, by a nucleosome with a nicked loop, i.e., free from torsional constraint ($G_{sc}(i) = 0$), could then be calculated from $\Delta G_n(i)$ in Table 1 and the equation

$$f_n(i) = \exp[-\Delta G_n(i)] \Big/ \sum_i \exp[-\Delta G_n(i)] \tag{16}$$

One obtains 63% of pBR control nucleosomes in the "closed negative" state against 28% in the "open" state, with approximately reverse figures for the same acetylated nucleosomes in phosphate: respectively 32% against 65% (complements to 100% are "closed positive" nucleosomes).

4.2.3. Loop most probable conformations and elastic supercoiling free energies

The theory. Obtaining K_{sc}/N_l from Fig. 5 plots would require a fitting with six parameters (assuming K_{sc}/N_l is identical in the three states) instead of five, which is not experimentally feasible. Instead, K_{sc}/N_l was calculated for each state using the explicit solutions to the equations of the equilibrium in the theory of the elastic rod model for DNA derived by Coleman and co-workers [55–57]. These authors considered the loop as a segment with specified conditions at the end points where it contacts the protein surface. The end-conditions define the unit vectors tangent to the double helix axis at these points. They depend on the geometry of the histone-bound DNA (considered as an ideal superhelix of pitch p and radius r), and were obtained as described [57]. The DNA segment was then treated as an inextensible, homogeneous body obeying the rod theory of Kirchhoff. The solutions to the equations lead to the most probable configuration of the loop which minimizes the elastic free energy, an objective which was previously achieved using the Monte-Carlo method referred to above, but owing to more tedious calculations. Given the pair of p and r parameters, the most probable configuration of the loop can then be calculated for any given topoisomer size and ΔLk.

A prerequisite to these calculations is the knowledge of bending and torsional rigidity coefficients (A and C, respectively) of the naked DNA. A was calculated from a persistence length $a = A/\mathrm{kT} = 50$ nm [58], and C through the following equation, valid only for a naked DNA minicircle near relaxation [59]

$$K_{sc} = 2\pi^2(C/A)a/0.34 - (3/4)N(C/A)^2 \tag{17}$$

which relates K_{sc} to A and C under these conditions. K_{sc} (together with h_o; see legend to Fig. 5) was measured through relaxation of minicircles from the series at different temperatures [29]. This resulted in $K_{sc} = 4000$ (± 300), i.e., $K_{sc}/N = 11$ (± 1), and in $C/A = 1.6$ (± 0.1) and $C = 3.3$ (± 0.2) 10^{-19} erg cm at 300°K. This C value is in good agreement with the value derived from topoisomer distribution analysis [60,61] and fluorescence measurements [62].

Until recently, the theory did not allow the configuration of the "positive" state to be described, due to entry/exit DNAs interpenetration upon application of the positive constraint to the loop. A recent development [63] takes the DNA impenetrability into account and deals with the resulting DNA self-contacts, which were allowed to slide freely, following the needs of the energy minimization process.

The results. Once the loop most probable conformations are known for any given ΔLk of the topoisomer, the bending and twisting contributions to the supercoiling free energy of the loop could be derived. The sum of these contributions, G_{sc}, and associated Wr and ΔTw_l (see Eq. (7)), were plotted as functions of ΔLk in Fig. 6(a) and (c), and (b) and (d), respectively, for the normal ("straight", as opposed to "curved"; see below) DNA superhelix. G_{sc} shows (local) minima corresponding to the different states, whose conformations are displayed in Fig. 6(e). $\Delta Tw_l = 0$ at these minima, i.e., the twisting contribution

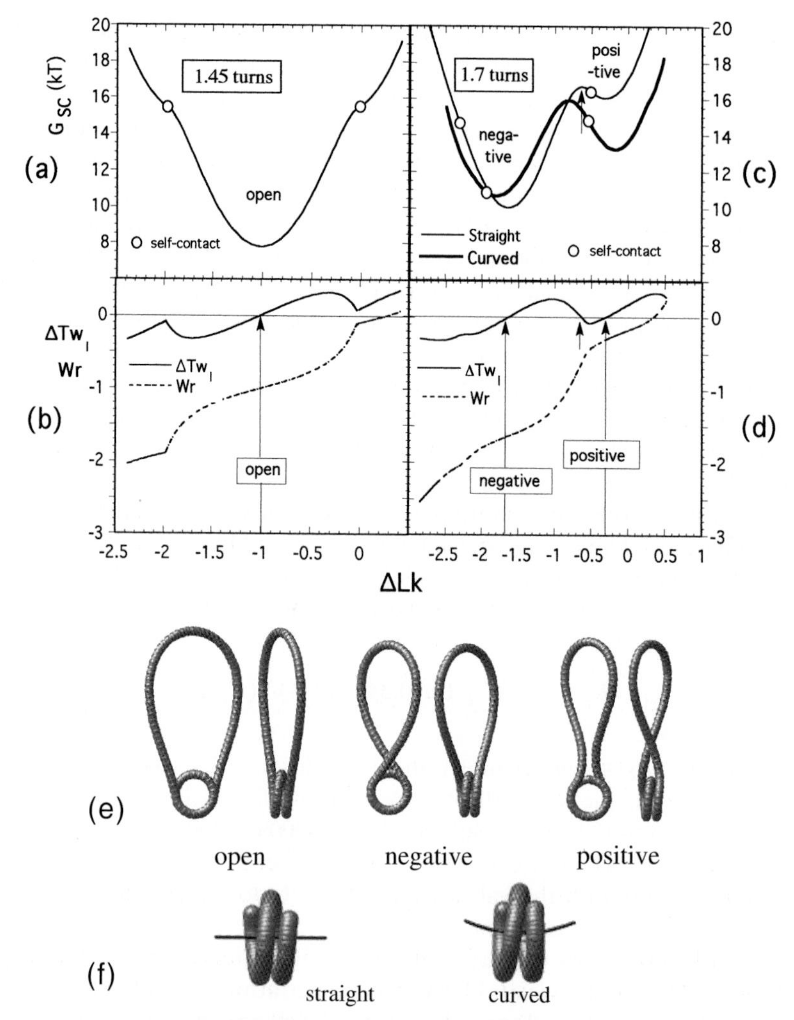

Fig. 6. Theoretical ΔLk dependence of the loop elastic energy (G_{sc}) and twist (ΔTw_l), and total writhe (Wr) in the three states, and corresponding representations of relaxed DNA conformations. The calculation was performed with no change in twist of the wrapped DNA, for regular (straight; 1.45- and 1.7-turns nucleosomes) and curved (curved; 1.7-turns nucleosome) DNA superhelices (see text). One G_{sc} minimum in (a) and two in (c) at $\Delta Lk \sim -1$, -1.7, and -0.3 (and $\Delta Tw_l = 0$ in (b) and (d)) correspond to "open", "closed negative", and "positive" states, respectively. A local energy maximum around $\Delta Lk = -0.6$ in (c) (small arrow) also has $\Delta Tw_l = 0$, but does not correspond to a stable conformation. Starting from the mid-region of the energy profiles in (a) and (c), the points at which a DNA self-contact first occurs in the loop are indicated by empty circles. The energy calculation used $A/kT = 50$ nm as the persistence length, and $C/A = 1.6$ (C is the twisting constant). Variations of C/A from 1.4 to 2, while keeping A constant, have no effect on ΔG_{sc}. Variations in A have more influence. However, decreasing A/kT to an unrealistic 30 nm low would decrease ΔG_{sc} between "closed positive" and "negative" states from 6 kT to 3.6 kT, a value still too large to account for the existence of the "closed positive" state (see text). Scale representations of DNA equilibrium conformations in the different states (e) and of straight and curved (radius of axis curvature = 15 nm) DNA superhelices (f) were obtained with the "MacMolecule2" software (Molecular Ventures Inc.). (From Fig. 6 in Ref. [29].)

is cancelled, and the entire loop elastic energy is in bending. Because G_{sc} varies with ΔLk approximately according to a second degree polynomial in the vicinity of those minima, K_{sc}/N_l could be calculated through identification with G_{sc} in Eq. (10). K_{sc}/N_l figures were found to be only slightly different from one state to another (see legend of Table 1), justifying the use of a unique value for fitting.

Wr_o of the states, i.e., Wr values at $\Delta Tw_l = 0$, were obtained from Fig. 6(b) and (d). Wr_o of "closed negative" and "open" states, -1.65 and -1.0 (Table 1; similar values were derived by Monte-Carlo calculation in Ref. [32]), are virtually identical to the corresponding fitted pBR ΔLk_n values, which confirms the absence of DNA overtwisting upon wrapping in these nucleosomes ($\Delta Tw_n(i) = 0$ in Eq. (8)). In contrast, Wr_o and ΔLk_n values of the "closed positive" state, ~ -0.3 against ~ -0.6, differ. The reason is not in the DNA twist, but rather in DNA/DNA electrostatic repulsion which was not taken into account in the theory. G_{sc} profile in Fig. 6(c) indeed shows that a DNA self-contact (empty circle) occurs in the loop before the "positive" state is reached. The effect of that repulsion will be to increase the apparent diameter of the double helix (2 nm was taken for calculation) so that the self-contact and the minimum will occur at a smaller ΔLk and higher energy than shown in the figure (see below for more details). For the "closed negative" state and even more so for the "open" state, in contrast, the self-contacts did not interfere because they occur far beyond those states (see Fig. 6(c)).

As far as loop bending energy is concerned, and consistent with conformations in Fig. 6(e), the "open" state should be favored over the "closed negative" state (by ~ 2 kT; compare minima in Figs. 6(a) and (c)). This is true for acetylated nucleosomes in phosphate, but not for control nucleosomes which rather favor the "closed negative" state (Table 1). The reason is in the unaccounted for contributions to $G_n(i)$ in Eq. (9). Three such contributions can be identified. The first two favor the "open" state: (i) the above mentioned tail-modulated electrostatic repulsion between entering and exiting DNAs, which is lower in that state; and (ii) the straightening of the unwrapped DNA at the edges. The third contribution originates from the protein and instead favors the "closed negative" state: histone–DNA binding sites at SHL ± 6.5 (Fig. 1) are engaged in the latter state, but break in the "open" state at a significant energy cost.

In contrast, there is no expected change in the protein contribution between "closed negative" and "positive" states, potentially leaving the bending energy (ΔG_{sc}) and the electrostatic repulsion as the sole contributors to $G_n(i)$. The latter contribution can only be larger in the "closed positive" state, due to the early-occurring DNA self-contact (see above). As a consequence, $\Delta G_{sc} \sim 6$ kT in Fig. 6(c) should be a minor estimate for the free energy difference between these two states. In contrast with this prediction, the actual ΔG_n of pBR nucleosome is much smaller (2–3 kT; see Table 1). Bending and torsional rigidity coefficients are not responsible for this discrepancy because even extreme values of these coefficients fail to decrease ΔG_{sc} significantly below the ~ 4 kT limit beyond which the "closed positive" state does not exist (see legend of Fig. 6, and above). The potential solution to that dilemma will be given below.

4.3. DNA sequence-dependent nucleosome structural and dynamic polymorphism. A role for H2B N-terminal tail proximal domain

Above results refer to an "average" nucleosome, which may be different from any of the 15 or so unique nucleosomes located at alternative positions on the DNA minicircles [64,34]. Previously mentioned differential cleavage periodicities of individual nucleosomes with hydroxyl radicals support this view. To investigate this question further, a new 349–363 bp homologous series of DNA minicircles was constructed from the 5S rDNA nucleosome positioning sequence [65]. The main peak in the corresponding control plot of topoisomer relative amounts-versus-ΔLk (Fig. 5(c)) appears displaced toward larger ΔLk values, relative to the pBR profile. Fitting of the three-state model resulted in a 0.3 increase in ΔLk_n of both "closed negative" and "open" states, relative to pBR values (Table 1) [29]. This reflected a $\Delta Tw_n \sim 0.3$ overtwisting relative to naked 5S DNA (see Eq. (8)) and a $\Delta Tw_n \sim 0.2_5$, i.e., $\sim$2 bp, overtwisting relative to pBR nucleosome, taking into account the h_o difference, in the opposite direction, between the two naked DNAs (see legend of Fig. 5). The fact that ΔTw_n was the same in both conformations further implied that the overtwisting did not take place on the nucleosome edges, but rather within SHL $\pm$5.5, knowing that histone/DNA binding sites at SHL $\pm$6.5 were broken in the "open" state (see Fig. 1).

Further DNase I footprints of the two nucleosomes trimmed to core particles revealed that they were untwisted by 0.5–1 bp in between both SHL 2 and 3 and SHL -2 and -3, very similar to the human α-satellite nucleosome, and presumably for the same reason: the passage of H2B N-terminal tails between the two gyres at SHL $\pm$3 (see Fig. 1(b)). In contrast, and consistent with topological data, only the pBR nucleosome was untwisted by $\sim$1 bp at each of the two positions of the opposite gyre, in between SHL 4 and 5 and SHL -4 and -5 [29]. Therefore, H2B tails may untwist DNA only at the dyad-proximal SHL $\pm$3 sites (both pBR and 5S nucleosomes, and the human α-satellite nucleosome), or also at the dyad-distal SHL $\pm$5 sites on the opposite gyre (pBR nucleosome). This discrepancy presumably depends on local flexibility and relative width of the two aligned small grooves. It is interesting that these results point to a large overtwisting of nucleosomal DNA, if it were not for H2B tail untwistings ($h_{loc} = 10.0$ bp/turn in the crystal structure after removal of the two SHL -2.5 to -3.5 and 2.5 to 3.5 segments; see Ref. [8]). This somehow gives an *a posteriori* credence to the initial proposal of an overtwisting as the explanation of the paradox (see Introduction).

Dynamics of 5S and pBR nucleosomes were also different: as shown by the virtual disappearance of the $\Delta Lk \sim -0.5$ shoulder in the 5S topoisomer amounts-versus-ΔLk profile in Fig. 5(c), 5S nucleosomes have difficulty to get access to the "closed positive" state. At the same time, the "closed negative" state was twice as favorable (relative to the "open" state) as in pBR control nucleosome (Table 1). As a result, the free energy difference between "positive" and "negative" states, $\sim$4 kT, is now more consistent with the model (see Fig. 6(c)). Such a reverse change in "negative" and "positive" state free energies suggests an alteration in the relative orientation of entry/exit DNAs from one nucleosome to the other. If these DNAs were slightly less divergent (when viewed from the superhelix axis) in pBR nucleosome, the positive crossing could indeed be expected to become easier and the negative crossing harder. In a modification of the model, the DNA superhelix axis was curved so as to bring the two DNA gyres in contact at the entry–exit points (Fig. 6(f)). As shown by the "curved" energy profile in Fig. 6(c), the state energies were shifted as predicted, with their difference, 2 kT, being now consistent with that of pBR nucleosome.

In conclusion, the pBR nucleosome differs from the 5S nucleosome in two main respects: its 1-bp untwisting at each of SHL 5 and −5, and its easy access to the "positive" conformation, which may result from a slight reorientation of entry/exit DNAs (see above). Importantly, these two features are necessarily related, at least statistically, because, as previously mentioned, each nucleosome represents an average of individual nucleosomes which differ from one another in a number of criteria, in addition to their twist (see below). Our model [29] proposed that reorientation is mediated by these local untwistings, which act at a distance to modify the interactions of H3 N-terminal tails with entry/exit DNAs at SHL $\pm$7 (Fig. 1(b)) and, perhaps, also further along them, through the rotation they inflict to the distal DNA stretches between SHL 5 to 7 and −5 to −7.

H2B N-terminal tail intercalation into the small groove may also have a role in locking the octamer into position on the DNA. The other intercalating H3 N-terminal tails may not be as efficient in this respect because of the relative weakness of DNA/histone interactions at SHL $\pm$6.5 [66], whose breakage conditions the "open" conformation of both circular (Fig. 2(b)) and linear nucleosomes (Fig. 3(a)). Consistently, only H2B N-terminal tails were found to be crucial in preventing sliding of linear nucleosomes during native gel electrophoresis [67]. This observation concurs with our own results which showed, from our limited collection of unique 5S nucleosomes, an inverse correlation between H2B-tail untwisting activity near SHL $\pm$5 and nucleosome ability to slide at elevated temperatures (C. Lavelle and A. Prunell, unpublished results).

5. *Nucleosomes in chromatin: a dynamic equilibrium*

These results raise the prospect of dynamics of nucleosomes in linker histone-free chromatin, that is, of a thermal fluctuation of nucleosomes between "closed negative", "open", and "closed positive" states identified in the minicircle system. If this equilibrium exists, an extra supercoiling constraint applied to the fiber should displace it in one direction or the other depending on the sign of that constraint, and this displacement should be reversible upon its removal.

5.1. *A displaceable equilibrium*

5.1.1. *Supercoiling constraints*

A puzzling observation in the above mentioned investigation of the effect of positive supercoiling on nucleosome structure [51] was that the number of nucleosomes being formed on the plasmid seemed limited only by the physical space available rather than by the requirement to dissipate the increasing positive torsional stress elicited by their formation, for example through DNA rotation on the histone surface. As already mentioned, ΔLk_n of positively-crossed nucleosomes would be close to zero if a full positive crossing was achieved, in which case they would be topologically neutral, and the elicited positive torsional stress indeed minimal. It was further observed that the positively constrained chromatin went back to normal ($\langle \Delta Lk_n \rangle = -1$) upon incubation with topoisomerase I, demonstrating the reversibility of that displacement.

In another experiment, negative supercoiling was applied to reconstituted minichromosomes by treatment with DNA gyrase [68]. The maximal supercoiling generated, corresponding to a supercoiling density $\sigma \sim -0.1$, was found to be nearly identical for the naked plasmid and for the reconstituted complex after deproteinization, as if nucleosomes were transparent to the enzyme. Again, there was no need for DNA untwisting on the histone surface or any kind of nucleosome disruption if the equilibrium was displaced so that all nucleosomes became negatively crossed. Assuming linker DNA took up the same $\sigma = -0.1$ as naked DNA, it was indeed calculated that an overall $\sigma = -0.1$ could be achieved with $\langle \Delta Lk_n \rangle \sim -1.6$ [30], equal to ΔLk_n of the "closed negative" state (Table 1). Moreover, the process was also reversible by subsequent treatment with topoisomerase I.

5.1.2. Histone acetylation. Toward an invariant of chromatin dynamics: the ΔLk-per-nucleosome parameter

Nucleosome dynamic equilibrium in a minichromosome may be displaced by histone hyperacetylation, as suggested by the correlative drop observed in $\langle \Delta Lk_n \rangle$: -1.04 (± 0.08) to -0.82 (± 0.05) [69]. Equilibrium displacement of pBR nucleosome can also be calculated in the minicircle system from Eq. (16), using $\Delta Lk_n(i)$ values in Table 1. Interestingly, a similar 0.2 drop is found between control, $\langle \Delta Lk_n \rangle = -1.41$, and acetylated nucleosomes in phosphate, $\langle \Delta Lk_n \rangle = -1.23$. This drop is due to the "open" state being favored as a result of tail destabilization or release (see above). There is every reason to believe that the shift in minichromosomes has the same origin: acetylated nucleosomes tend to adopt the "open" conformation. In the minicircle system, $\langle \Delta Lk_n \rangle = -1.23$ was obtained with 65% of "open" nucleosomes, and increasing that proportion to 100% would not increase $\langle \Delta Lk_n \rangle$ above the -1.0 limit. For this, "closed positive" nucleosomes should exist in significant proportions, which is unlikely because tail release also hinders that state (see Table 1). Alternatively, nucleosomes in that minichromosome may overtwist DNA as do 5S nucleosomes. In that case, with an "open" state ΔLk_n of -0.7, $\sim 15\%$ of "closed negative" nucleosomes in the equilibrium would even be required in order to decrease $\langle \Delta Lk_n \rangle$ from -0.7 to the actual -0.82 value. This overtwisting is supported by that minichromosome being made of tandem repeats of the same 5S rDNA sequence [70].

But what should happen upon histone hyperacetylation if nucleosomes, on average, do not overtwist DNA? Then no change in $\langle \Delta Lk_n \rangle$ should be observed.

This was precisely the case of SV40 minichromosomes and a transfected plasmid assembled *in vivo* with control or hyperacetylated histones [71]. A condition for that conclusion, however, is that $\langle \Delta Lk_n \rangle$ is equal to -1.0 also in these minichromosomes. This, although likely, in particular for SV40 minichromosome, has not been rigorously demonstrated in this [71] or other studies [12], due to lack of accurate measurements of both DNA supercoiling and nucleosome number, such as those performed with the above 5S-repeat minichromosome [70,69].

Minichromosome data, therefore, suggest a unit value of $\langle \Delta Lk_n \rangle$, independent of the twist of nucleosomal DNA. The strength of this invariance could be tested by constructing minichromosomes from repeats of various nucleosome-positioning sequences with different overtwisting potentials (minichromosomes on complex sequences, such as viruses or plasmids, are unlikely to provide a sufficient range of mean overtwistings). It is interesting that the minicircle system shows a trend toward a similar $\langle \Delta Lk_n \rangle$ invariance. $\langle \Delta Lk_n \rangle$ indeed varies from ~ -1.4 for "pBR" control nucleosomes (see above) to ~ -1.3 for "5S" nucleosomes (similarly calculated from Eq. (16) and $\Delta Lk_n(i)$ values in Table 1), a small increase compared to the $\Delta Tw_n = 0.3$ overtwisting of 5S nucleosome. The reason is known, and is presumably the same in minichromosomes: untwisted nucleosomes (e.g., pBR), with otherwise large negative $\langle \Delta Lk_n \rangle$ value, get access to the "closed positive" conformation, which draws $\langle \Delta Lk_n \rangle$ closer to -1. In contrast, overtwisted nucleosomes (e.g., 5S) have a $\langle \Delta Lk_n \rangle$ value already closer to -1, and are not so much in need of the "closed positive" conformation.

5.2. Superstructural context of nucleosome dynamics in chromatin

A common view for the superstructure of the H1/H5-containing 30 nm chromatin fiber has emerged in the last few years. Microscopic techniques have shown an irregular 3D zig-zag of nucleosomes with straight linkers projecting towards the fiber interior [72–75] (Fig. 7(a)). Such a cross-linker model was further supported by the demonstration of an internal location of H1/H5 [76–78], and by the observation of a bridging together of entry/exit DNAs into a stem through interactions with H1/H5 C-terminal tail (see above; Ref. [34]). This stem has been observed also in native chromatin fragments [79] and chromatin reconstituted on tandem repeats of the 5S rDNA sequence [80]. Interestingly, comparative hydrodynamic studies of normal or tail-less H5-free and H5-containing nucleosome arrays suggested that linker DNA-interacting core histone N-terminal tails were dislodged upon stem formation, making them free to interact with neighboring nucleosomes to mediate fiber folding and oligomerization [81,82]. It is noteworthy that the diameter of such a fiber may not depend much on the nucleosome repeat length if the stem size varies in direct proportion to that length. This would explain another paradox of those times, that is, the 30 nm filaments formed with extreme and intermediate repeat length chromatins of sea urchin sperm ($\sim$240 bp), chicken erythrocytes ($\sim$200 bp), and yeast ($\sim$165 bp) were indistinguishable under electron microscopy and X-ray diffraction [83–85], which was later confirmed *in situ* by electron microscopy of sperm cells and chicken erythrocytes frozen hydrated sections [86]. Very likely, the

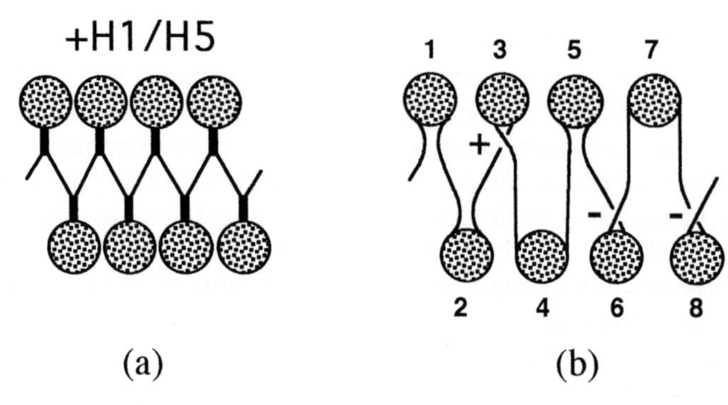

Fig. 7. Cross-linker model for nucleosome arrangement in the chromatin fiber superstructure in the presence (a) or absence (b) of H1/H5, based on data in the literature (see text) and H5-containing mononucleosome stem structure in Fig. 3(c). In 3D, the plane of the nucleosomes is expected to rotate more or less regularly around the fiber axis, forming a solenoid-like superstructure. Nucleosomes 1, 2 and 5 are in the "open" conformation of Fig. 3(a), nucleosomes 4 and 7 in the "open" conformation of Fig. 2(b), and other nucleosomes in the "closed negative" (Fig. 2(c)) or "positive" conformations. Nucleosomes are expected to thermally fluctuate between the different conformations, within an overall dynamic equilibrium of $\langle \Delta Lk_n \rangle \sim -1$ (see text). + and − refer to node polarities. (From Fig. 5 in Ref. [28].)

stem could not form in the yeast filament, where most nucleosomes may adopt the "open" conformation (nucleosomes 4 and 7 in Fig. 7(b)). This in turn would be consistent with yeast chromatin high transcriptional activity [87] and DNA thermal flexibility [88,89].

In the absence of H1/H5, an extended bead-on-a-string structure occurs [90], which is still able to condense at physiological ionic strength [91–93,81], and resembles the zig-zag arrangement of H1/H5-containing chromatin, even if less compact (Fig. 7(b)). Clearly, such a zig-zag structure, which is also predicted by theoretical modeling [94–96], lends strong support to the possibility for nucleosomes to rotate around their dyad axis relative to one another, and fluctuate between the three conformations (Fig. 7(b)). This is in contrast with a stacked nucleosome arrangement (see Ref. [97] for an example), where such a rotation may be prevented. As previously mentioned, the solution to the linking number paradox then appears to be in the equilibrium steady-state proportions of nucleosomes in the different conformations, which result into $\langle \Delta Lk_n \rangle = -1$. But this formally applies only to linker histone-free chromatin.

5.3. Topology and dynamics of linker histone-containing nucleosomes in chromatin

Although the $\langle \Delta Lk_n \rangle$ measurement [12,98,99] was not as accurate as for H1/H5-free chromatin, it is generally accepted that DNA overall topology is not significantly altered in H1/H5-containing chromatin. A simple explanation for this could be that H1/H5 does not interfere with the equilibrium before "freezing" it through nucleosome/nucleosome interactions (Fig. 7(a)). In this case, the stem formed with entering and exiting DNAs should be able to accommodate (i) negative and positive crossings of the linkers, and (ii) no crossing at all. Recent relaxation experiments with H5-containing "pBR" and "5S" mononucleosomes [99b] showed the first proposition to be true, but not the second. H5 indeed suppressed not only the "open" conformation but, more surprisingly, also the barrier against the "positive" conformation in "5S" nucleosomes. This was accompanied by a large loop flexibility in both remaining states ($K_{sc}/N_l \sim 5$, against 12 for H5-free nucleosomes in Table 1), and by opposite shifts in their ΔLk_n: ΔLk_n was closer to zero for the "closed positive" state and more negative for the "closed negative" state, both by 0.2–0.3. This indicated that the two DNA duplexes in the stem are wound around each other into a superhelix, which is, in the particular case of the "closed negative" conformation, slightly longer than necessary to make an effective crossing. Finally, it is interesting to note that thermal fluctuations between negative and positive crossings may not only switch the handedness of that stem superhelix, but also the conformation of DNA at the nucleosome very entry–exit where linker histone globular domain is bound [100]. Such a DNA fluctuation should participate in linker histone lability (see below).

Such remarkable properties of the stem may appear as a waste for nucleosomes destined to be immobilized in a superstructure. However, there are at least two identifiable opportunities for these properties to be useful: (1) in having a large rotation capability, nucleosomes in the fiber could optimally orient for a tight compaction. With above $\Delta Lk_n(i)$ values of H5-containing nucleosomes, and $\langle \Delta Lk_n \rangle = -1$, the proportion of positively-crossed nucleosomes in the fiber could be as large as 50%. (2) Linker histone may also act at the nucleosome level in a locally-decondensed environment [82]. H1 exchange *in vitro* [101,102] and *in vivo* [42,103,104], and HMGN proteins binding competition with H1 *in vivo* [105], do support a view of dynamic nucleosomes fluctuating around their stem in moderately-compacted H1/H5-containing chromatin. Moreover, a growing bulk of evidence suggests that the overall structure of *in vivo* transcribing chromatin remains relatively condensed, at the 30 nm fiber level and beyond [106,107].

6. *Conclusions*

This review has focused on an old problem, the so-called "linking number paradox", which, because its solution was believed to reside in an overtwisting of nucleosomal DNA (see Introduction), was an incentive to early chromatin studies aiming at measuring the local helical periodicity of nucleosomal DNA *in situ*. In addition to above mentioned DNase and chemical cleavages, the distribution of specific di- and tri-nucleotides was also found to reflect the local helical periodicity, due to the sequence-dependent properties of these oligonucleotides in conferring on the double helix an ability to curve, or be curved, in the same direction every 10 bp or so on the histone surface [108–110]. The scope of these studies broadened considerably when the existence of naturally-bent DNA was experimentally proved [111,112] and new rules of DNA–protein recognition established [113,114]. The limit of the "h_{loc} approach" was reached when it was finally realized that if nucleosomal DNA could indeed show some overtwisting, its extent was not large enough to explain the paradox [24]. Remarkably enough, there was an early attempt to explain the paradox by a specific arrangement of nucleosomes relative to each other in the fiber [115], but, with the understanding of chromatin superstructure lagging behind, there was no way to bring that explanation to a test at that time. More recently, some consensus emerged toward a superstructural cross-linker model, but linker path resolution has remained largely insufficient for that model to have any direct relevance to the paradox. The DNA minicircle approach, detailed in this review, found the solution to the paradox to lie not in nucleosomal DNA structure, neither in nucleosome relative arrangement, but in the ability of nucleosome immediate entry–exit DNA to cross positively, a counter-intuitive notion in view of the DNA left-handed wrapping around the histones. More recently, this ability was found to depend on the DNA sequence wrapped around the histones. In the mechanism involved, the local untwistings, when generated at the dyad-distal sites by H2B N-terminal tails passing between the two gyres, appear to act at a distance on H3-tail interactions to reorient the nucleosome entry/exit DNAs [29]. Interestingly, these two manifestations of nucleosome polymorphism, differential twist and positive crossing ability, counteract each other in their topological effects, with the apparent consequence of insuring a twist-independent unit value to the ΔLk-per-nucleosome parameter, which may then be regarded as an invariant of chromatin dynamics. The "purpose" of this potential invariance, even if intriguing, is not known. In the presence of linker histones, all nucleosomes, regardless of their sequence, can cross positively, but loose access to the "open" conformation. This "open" nucleosome, in which the distal, and also weakest, histone–DNA binding sites are broken (consistent with the conclusion from site exposure essays that DNA near the edges is more exposed [1,2]), is likely to be physiologically relevant. This is suggested by the large increase in its probability upon histone N-terminal tail acetylation, a landmark of transcriptional activation *in vivo*. Consistent with H2A–H2B dimers enhanced lability in the "open" state (C. Lavelle and A. Prunell, unpublished results), one of its roles may be to prepare the way to the tetrasome, which its chiral transition [116–118] may place at the culminating position in nucleosome dynamics.

Acknowledgement

The authors wish to thank NATO for a Collaborative Linkage Grant.

References

1. Polach, K.J. and Widom, J. (1995) Mechanism of protein access to specific DNA sequences in chromatin: a dynamic equilibrium model for gene regulation. J. Mol. Biol. 254, 130–149.
2. Anderson, J.D., Thastrom, A., and Widom, J. (2002) Spontaneous access of proteins to buried nucleosomal DNA target sites occurs via a mechanism that is distinct from nucleosome translocation. Mol. Cell. Biol. 20, 7147–7157.

3. Tsukiyama, T. (2002) The in vivo functions of ATP-dependent chromatin-remodelling factors. Nat. Rev. Mol. Cell Biol. 3, 422–429.
4. Narlikar, G.J., Fan, H.Y., and Kingston, R.E. (2002) Cooperation between complexes that regulate chromatin structure and transcription. Cell 108, 475–487.
5. Yasul, D., Mlyano, M., Cal, S., Varga-Weisz, P., and Kohwi-Shigematsu, T. (2002) SATB1 targets chromatin remodelling to regulate genes over long distances. Nature 419, 641–645.
6. Klug, A. and Lutter, L. (1981) The helical periodicity of DNA on the nucleosome. Nucl. Acids Res. 9, 4267–4283.
7. Wang, J.C. (1982) The path of DNA in the nucleosome. Cell 29, 724–726.
8. Prunell, A. (1998) A topological approach to nucleosome structure and dynamics. The linking number paradox and other issues. Biophys. J. 74, 2531–2544.
9. Morse, R.H. and Simpson, R.T. (1988) DNA in the nucleosome. Cell 54, 285–287.
10. Klug, A. and Travers, A.A. (1989) The helical repeat of nucleosome-wrapped DNA. Cell 56, 10–11.
11. White, J.H. and Bauer, W.R. (1989) The helical repeat of nucleosome-wrapped DNA. Cell 56, 9–10.
12. Germond, J.E., Hirt, B., Oudet, P., Gross-Bellard, M., and Chambon, P. (1975) Folding of the DNA double helix in chromatin-like structures from simian virus 40. Proc. Natl. Acad. Sci. USA 72, 1843–1847.
13. Finch, J.T., Lutter, L.C., Rhodes, D., Brown, R.S., Rushton, B., Levitt, M., and Klug, A. (1977) Structure of nucleosome core particles of chromatin. Nature 269, 29–36.
14. White, J.H. (1969) Self-linking and the Gauss integral in higher dimensions. Am. J. Math. 91, 693–728.
15. Fuller, F.B. (1971) The writhing number of a space curve. Proc. Natl. Acad. Sci. USA 68, 815–819.
16. Crick, F.H.C. (1976) Linking numbers and nucleosomes. Proc. Natl. Acad. Sci. USA 73, 2639–2643.
17. Horowitz, D.S. and Wang, J.C. (1984) Torsional rigidity of DNA and length dependence of the free energy of DNA supercoiling. J. Mol. Biol. 173, 75–91.
18. Prunell, A., Kornberg, R.D., Lutter, L.C., Klug, A., Levitt, M., and Crick, F.H.C. (1979) Periodicity of deoxyribonuclease I digestion of chromatin. Science 204, 855–858.
19. Lutter, L.C. (1979) Precise location of DNase I cutting sites in the nucleosome core determined by high resolution gel electrophoresis. Nucl. Acids Res. 6, 41–56.
20. Noll, M. (1974) Internal structure of the chromatin subunit. Nucl. Acids Res. 1, 1573–1579.
21. Ulanovsky, L.E. and Trifonov, E.N. (1983) Superhelicity of nucleosomal DNA changes its double-helical repeat. Cell Biophys. 5, 281–283.
22. Luger, K., Mäder, A.W., Richmond, R.K., Sargent, D.F., and Richmond, T.J. (1997) Crystal structure of the nucleosome core particle at 2.8 Å resolution. Nature 389, 251–260.
23. Tullius, T.D. and Dombroski, B.A. (1985) Iron(II) EDTA used to measure the helical twist along any DNA molecule. Science 230, 679–681.
24. Hayes, J.J., Tullius, T.D., and Wolffe, A.P. (1990) The structure of DNA in a nucleosome. Proc. Natl. Acad. Sci. USA 87, 7405–7409.
25. Hayes, J.J., Clark, D.J., and Wolffe, A.P. (1991) Histone contributions to the structure of DNA in the nucleosome. Proc. Natl. Acad. Sci. USA 88, 6829–6833.
26. Davey, C.A., Sargent, D.F., Luger, K., Maeder, A.W., and Richmond, T.J. (2002) Solvent mediated interactions in the structure of the nucleosome core particle at 1.9 Å resolution. J. Mol. Biol. 319, 1097–1113.
27. Goulet, I., Zivanovic, Y., and Prunell, A. (1987) Helical repeat of DNA in solution. The V curve method. Nucleic Acids Res. 15, 2803–2823.
28. De Lucia, F., Alilat, M., Sivolob, A., and Prunell, A. (1999) Nucleosome dynamics. III. Histone-tail dependent fluctuation of nucleosomes between open and closed DNA conformations: implications for chromatin dynamics and the linking number paradox. A relaxation study of mononucleosomes on DNA minicircles. J. Mol. Biol. 285, 1101–1119.
29. Sivolob, A., Lavelle, C., and Prunell, A. (2003) Sequence-dependent nucleosome structural and dynamic polymorphism. Potential involvement of histone H2B N-terminal tail proximal domain. J. Mol. Biol. 326, 49–63.

30. Zivanovic, Y., Goulet, I., Revet, B., Le Bret, M., and Prunell, A. (1988) Chromatin reconstitution on small DNA rings. II. DNA supercoiling on the nucleosome. J. Mol. Biol. 200, 267–285.
31. Goulet, I., Zivanovic, Y., Prunell, A., and Révet, B. (1988) Chromatin reconstitution on small DNA rings. I. J. Mol. Biol. 200, 253–266.
32. Le Bret, M. (1988) Computation of the helical twist of nucleosomal DNA. J. Mol. Biol. 200, 285–290.
33. Furrer, P., Bednar, J., Dubochet, J., Hamiche, A., and Prunell, A. (1995) DNA at the entry-exit of the nucleosome observed by cryoelectron microscopy. J. Struct. Biol. 114, 177–183.
34. Hamiche, A., Schultz, P., Ramakrishnan, V., Oudet, P., and Prunell, A. (1996) Linker histone-dependent DNA structure in linear mononucleosomes. J. Mol. Biol. 257, 30–42.
35. Zivanovic, Y., Duband-Goulet, I., Schultz, P., Stofer, E., Oudet, P., and Prunell, A. (1990) Chromatin reconstitution on small DNA rings. III. Histone H5 dependence of DNA supercoiling in the nucleosome. J. Mol. Biol. 214, 479–495.
36. Williamson, P. and Felsenfeld, G. (1978) Transcription of histone-covered T7 DNA by Escherichia coli RNA polymerase. Biochemistry 17, 5695–5705.
37. Wasylyk, B., Thevenin, G., Oudet, P., and Chambon, P. (1979) Transcription of in vitro assembled chromatin by Escherichia coli RNA polymerase. J. Mol. Biol. 128, 411–440.
38. Wasylyk, B. and Chambon, P. (1980) Studies on the mechanism of transcription of nucleosomal complexes. Eur. J. Biochem. 103, 219–226.
39. Liu, L.F. and Wang, J.C. (1987) Supercoiling of the DNA template during transcription. Proc. Natl. Acad. Sci. USA 84, 7024–7027.
40. Jackson, V. (1990) In vivo studies on the dynamics of histone-DNA interaction: evidence for nucleosome dissolution during replication and transcription and a low level of dissolution independent of both. Biochemistry. 29, 719–731.
41. Jackson, V. (1993) Influence of positive stress on nucleosome assembly. Biochemistry 32, 5901–5912.
42. Louters, L. and Chalkley, R. (1985) Exchange of histones H1, H2A, and H2B in vivo. Biochemistry 24, 3080–3085.
43. Schwager, S., Retief, J.D., de Groot, P., and von Holt, C. (1985) Rapid exchange of histones H2A and H2B in sea urchin embryo chromatin. FEBS Lett. 189, 305–309.
44. Baer, B.W. and Rhodes, D. (1983) Eukaryotic RNA polymerase II binds to nucleosome cores from transcribed genes. Nature 301, 482–488.
45. Eickbush, T.H. and Moudrianakis, E.N. (1978) The histone core complex: an octamer assembled by two sets of protein-protein interactions. Biochemistry 17, 4955–4964.
46. O'Donohue, M.F., Duband-Goulet, I., Hamiche, A., and Prunell, A. (1994) Octamer displacement and redistribution in transcription of single nucleosomes. Nucleic Acids Res. 22, 937–945.
47. Studitsky, V.M., Clark, D.J., and Felsenfeld, G. (1994) A histone octamer can step around a transcribing polymerase without leaving the template. Cell 76, 371–382.
48. Felsenfeld, G., Clark, D., and Studitsky, V. (2000) Transcription through nucleosomes. Biophys. Chem. 86, 231–237.
49. Kireeva, M.L., Walter, W., Tchernajenko, V., Bondarenko, V., Kashlev, M., and Studitsky, V.M. (2002) Nucleosome remodeling induced by RNA polymerase II: loss of the H2A/H2B dimer during transcription. Mol. Cell 9, 541–552.
50. Studitsky, V.M., Kassavetis, G.A., Geiduschek, E.P., and Felsenfeld, G. (1997) Mechanism of transcription through the nucleosome by eukaryotic RNA polymerase. Science 278, 1899–1901.
51. Clark, D.J. and Felsenfeld, G. (1991) Formation of nucleosomes on positively supercoiled DNA. EMBO J. 10, 387–395.
52. Sivolob, A., De Lucia, F., Révet, B., and Prunell, A. (1999) Nucleosome dynamics II. High flexibility of nucleosome entering and exiting DNAs to positive crossing. An ethidium bromide fluorescence study of mononucleosomes on DNA minicircles. J. Mol. Biol. 285, 1081–1099.
53. Hong, L., Schroth, G.P., Matthews, H.R., Yau, P., and Bradbury, E.M. (1993) Studies of the DNA binding properties of histone H4 amino terminus. Thermal denaturation studies reveal that acetylation markedly reduces the binding constant of the H4 "tail" to DNA. J. Biol. Chem. 268, 305–314.

54. Angelov, D., Vitolo, J.M., Mutskov, V., Dimitrov, S., and Hayes, J.J. (2001) Preferential interaction of the core histone tail domains with linker DNA. Proc. Natl. Acad. Sci. USA 98, 6599–6604.
55. Tobias, I., Coleman, B.D., and Olson, W. (1994) The dependence of DNA tertiary structure on end conditions: theory and implications for topological transitions. J. Chem. Phys. 101, 10990–10996.
56. Coleman, B.D., Tobias, I., and Swigon, D. (1995) Theory of the influence of end conditions on self-contact in DNA loops. J. Chem. Phys. 103, 9101–9109.
57. Swigon, D., Coleman, B.D., and Tobias, I. (1998) The elastic rod model for DNA and its application to the tertiary structure of DNA minicircles in mononucleosomes. Biophys. J. 74, 2515–2530.
58. Hagerman, P.J. (1988) Flexibility of DNA. Annu. Rev. Biophys. Biophys. Chem. 17, 265–286.
59. Tobias, I. (1998) A theory of thermal fluctuations in DNA miniplasmids. Biophys. J. 74, 2545–2553.
60. Shimada, J. and Yamakawa, H. (1985) Statistical mechanics of DNA topoisomers. The helical warmlike chain. J. Mol. Biol. 184, 319–329.
61. Frank-Kamenetskii, M.D., Lukashin, A.V., Anshelevich, A.V., and Vologodskii, A.V. (1985) Torsional and bending rigidity of double helix from data on small DNA rings. J. Biomol. Struct. Dynam. 2, 1005–1012.
62. Heath, P.J., Clendenning, J.B., Fujimoto, B.S., and Shurr, J.M. (1996) Effect of bending strain on the torsion elastic constant of DNA. J. Mol. Biol. 260, 718–730.
63. Tobias, I., Swigon, D., and Coleman, B.D. (2000) Elastic stability of DNA configurations. I. General theory. Physical Review E 61, 747–758.
64. Duband-Goulet, I., Carot, V., Ulyanov, A.V., Douc-Rasy, S., and Prunell, A. (1992) Chromatin reconstitution on small DNA rings. IV. DNA supercoiling and nucleosome sequence preference. J. Mol. Biol. 224, 981–1001.
65. Simpson, R.T. and Stafford, D.W. (1983) Structural features of a phased nucleosome core particle. Proc. Natl. Acad. Sci. USA 80, 51–55.
66. Luger, K. and Richmond, T.J. (1998) DNA binding within the nucleosome core. Curr. Opin. Struct. Biol. 8, 33–40.
67. Hamiche, A., Kang, J.-G., Dennis, C., Xiao, H., and Wu, C. (2001) Histone tails modulate nucleosome mobility and regulate ATP-dependent nucleosome sliding by NURF. Proc. Natl. Acad. Sci. USA 98, 14316–14321.
68. Garner, M.M., Felsenfeld, G., O'Dea, M.H., and Gellert, M. (1987) Effects of DNA supercoiling on the topological properties of nucleosomes. Proc. Natl. Acad. Sci. USA 84, 2620–2623.
69. Norton, V.G., Imai, B.S., Yau, P., and Bradbury, E.M. (1989) Histone acetylation reduces nucleosome core particle linking number change. Cell 57, 449–457.
70. Simpson, R.T., Thoma, F., and Brubaker, J.M. (1985) Chromatin reconstituted from tandemly repeated cloned DNA fragments and core histones. A model system for study of higher order structure. Cell 42, 799–808.
71. Lutter, L.C., Judis, L., and Paretti, R.F. (1992) Effects of histone acetylation on chromatin topology in vivo. Mol. Cell. Biol. 12, 5004–5014.
72. Woodcock, C.L., Grigoryev, S.A., Horowitz, R.A., and Whitaker, N. (1993) A chromatin folding model that incorporates linker variability generates fibers resembling the native structures. Proc. Natl. Acad. Sci. USA 90, 9021–9025.
73. Horowitz, R.A., Agard, D.A., Sedat, J.W., and Woodcock, C.L. (1994) The three-dimensional architecture of chromatin *in situ*: electron tomography reveals fibers composed of a continuously variable zig-zag nucleosomal ribbon. J. Cell Biol. 125, 1–10.
74. Leuba, S.H., Yang, G., Robert, C., Samori, B., van Holde, K., Zlatanova, J., and Bustamante, C. (1994) Three-dimensional structure of extended chromatin fibers as revealed by tapping-mode scanning force microscopy. Proc. Nat. Acad. Sci. USA 91, 11621–11625.
75. van Holde, K.E. and Zlatanova, J. (1996) What determines the folding of the chromatin fiber? Proc. Natl. Acad. Sci. USA 93, 10548–10555.
76. Leuba, S.H., Zlatanova, J., and van Holde, K. (1993) On the location of histones H1 anf H5 in the chromatin fiber. Studies with immobilized trypsin and chymotrypsin. J. Mol. Biol. 229, 917–929.
77. Graziano, V., Gerchman, S.E., Schneider, D.K., and Ramakrishnan, V. (1994) Histone H1 is located in the interior of the chromatin 30-nm filament. Nature 368, 351–353.

78. Zlatanova, J., Leuba, S.H., Yang, G., Bustamente, C., and van Holde, K.E. (1994) Linker DNA accessibility in chromatin fibers of different conformations: a reevaluation. Proc. Natl. Acad. Sci. USA 91, 5277–5280.
79. Bednar, J., Horowitz, R.A., Dubochet, J., and Woodcock, C.L. (1995) Chromatin conformation and salt-induced compaction: three-dimensional structural information from cryoelectron microscopy. J. Cell Biol. 131, 1365–1376.
80. Bednar, J., Horowitz, R.A., Grigoryev, S.A., Carruthers, L.M., Hansen, J.C., Koster, A.J., and Woodcock, C.L. (1998) Nucleosomes, linker DNA, and linker histone form a unique structural motif that directs the higher-order folding and compaction of chromatin. Proc. Natl. Acad. Sci. USA 95, 14173–14178.
81. Carruthers, L.M., Bednar, J., Woodcock, C.L., and Hansen, J.C. (1998) Linker histones stabilize the intrinsic salt-dependent folding of nucleosomal arrays: mechanistic ramifications for higher-order chromatin folding. Biochemistry 37, 14776–14787.
82. Carruthers, L.M. and Hansen, J.C. (2000) The core histone N termini function independently of linker histones during chromatin condensation. J. Biol. Chem. 275, 37285–37290.
83. Widom, J. and Klug, A. (1985) Structure of the 300A chromatin filament: X-ray diffraction from oriented samples. Cell 43, 207–213.
84. Widom, J, Finch, J.T., and Thomas, J.O. (1985) Higher-order structure of long repeat chromatin. EMBO J. 4, 3189–3194.
85. Lowary, P.T. and Widom, J. (1989) Higher-order structure of Saccharomyces cerevisiae chromatin. Proc. Natl. Acad. Sci. USA 86, 8266–8270.
86. Woodcock, C.L. (1994) Chromatin fibers observed in situ in frozen hydrated sections. Native fiber diameter is not correlated with nucleosome repeat length. J. Cell. Biol. 125, 11–19.
87. Grunstein, M. (1990) Histone function in transcription. Annu. Rev. Cell Biol. 6, 643–678.
88. Saavedra, R.A. and Huberman, J.A. (1986) Both DNA topoisomerases I and II relax 2 micron plasmid DNA in living yeast cells. Cell 45, 65–70.
89. Morse, R.H., Pederson, D.S., Dean, A., and Simpson, R.T. (1987) Yeast nucleosomes allow thermal untwisting of DNA. Nucleic Acids Res. 15, 10311–10330.
90. Bustamante, C., Zuccheri, G., Leuba, S.H., Yang, G., and Samori, B. (1997) Visualization and analysis of chromatin by scanning force microscopy. Methods 12, 73–83.
91. Thoma, F., Koller, Th., and Klug, A. (1979) Involvement of histone H1 in the organization of the nucleosome and of the salt-dependent superstructures of chromatin. J. Cell Biol. 83, 403–427.
92. Schwarz, P.M. and Hansen, J.C. (1994) Formation and stability of higher order chromatin structures-contributions of the histone octamer. J. Biol. Chem. 269, 16284–16289.
93. Moore, S.C. and Ausio, J. (1997) Major role of the histones H3–H4 in the folding of the chromatin fiber. Biochem. Biophys. Res. Comm. 230, 136–139.
94. Katritch, V., Bustamante, C., and Olson, W.K. (2000) Pulling chromatin fibers: computer simulations of direct physical micromanipulations. J. Mol. Biol. 295, 29–40.
95. Beard, D.A. and Schlick, T. (2001) Computational modeling predicts the structure and dynamics of chromatin fiber. Structure (Camb) 9, 105–114.
96. Wedemann, G. and Langowski, J. (2002) Computer simulation of the 30-nanometer chromatin fiber. Biophys. J. 82, 2847–2859.
97. McGhee, J.D., Nickol, J.D., Felsenfeld, G., and Rau, D.C. (1983) Higher order structure of chromatin: orientation of nucleosomes within the 30 nm chromatin solenoid is independent of species and linker length. Cell 33, 831–841.
98. Stein, A. (1980) DNA wrapping in nucleosomes. The linking number problem re-examined. Nucl. Acids Res. 8, 4803–4820.
99. Morse, R.H. and Cantor, C.R. (1986) Effect of trypsinization and histone H5 addition on DNA twist and topology in reconstituted minichromosomes. Nucl. Acids Res. 14, 3293–3310.
99b. Sivolob, A. and Prunell, A. (2003) Linker histone-dependent organization and dynamics of nucleosome entry/exit DNAs. J. Mol. Biol. 331, 1025-1040.

100. Zhou, Y.B., Gerchman, S.E., Ramakrishnan, V., Travers, A., and Muyldermans, S. (1998) Position and orientation of the globular domain of linker histone H5 on the nucleosome. Nature 395, 402–405.
101. Caron, F. and Thomas, J.O. (1981) Exchange of histone H1 between segments of chromatin. J. Mol. Biol. 146, 513–537.
102. Jin, Y.J. and Cole, R.D. (1986) H1 histone exchange is limited to particular regions of chromatin that differ in aggregation properties. J. Biol. Chem. 261, 3420–3427.
103. Lever, M.A., Th'ng, J.P., Sun, X., and Hendzel, M.J. (2000) Rapid exchange of histone H1.1 on chromatin in living human cells. Nature 408, 873–876.
104. Misteli, T., Gunjan, A., Hock, R., Bustin, M., and Brown, D.T. (2000) Dynamic binding of histone H1 to chromatin in living cells. Nature 408, 877–881.
105. Catez, F., Brown, D., Misteli, T., and Bustin, M. (2002) Competition between histone H1 and HMGN proteins for chromatin binding sites. EMBO Rep. 3, 760–766.
106. Tumbar, T., Sudlow, G., and Belmont, A.S. (1999) Large-scale chromatin unfolding and remodeling induced by VP16 acidic activation domain. J. Cell Biol. 145, 1341–1354.
107. Muller, W.G., Walker, D., Hager, G.L., and McNally, J.G. (2001) Large-scale chromatin decondensation and recondensation regulated by transcription from a natural promoter. J. Cell Biol. 154, 33–48.
108. Trifonov, E.N. and Sussman, J.L. (1980) The pitch of chromatin DNA is reflected in its nucleotide sequence. Proc. Natl. Acad. Sci. USA 77, 3816–3820.
109. Drew, H.R. and Travers, A.A. (1985) DNA bending and its relation to nucleosome positioning. J. Mol. Biol. 186, 773–790.
110. Satchwell, S.C., Drew, H.R., and Travers, A.A. (1986) Sequence periodicities in chicken nucleosome core DNA. J. Mol. Biol. 191, 285–290.
111. Marini, J.C., Levene, S.D., Crothers, D.M., and Englund, P.T. (1982) Bent helical structure in kinetoplast DNA. Proc. Natl. Acad. Sci. USA 79, 7664–7668.
112. Trifonov, E.N. (1985) Curved DNA. CRC Crit. Rev. Biochem. 19, 89–106.
113. Trifonov, E.N. (1983) Sequence-dependent variations of B-DNA structure and protein-DNA recognition. Cold Spring Harb. Symp. Quant. Biol. 47, 271–278.
114. Travers, A. and Drew, H. (1997) DNA recognition and nucleosome organization. Biopolymers 44, 423–433.
115. Worcel, A., Strogatz, S., and Riley, D. (1981) Structure of chromatin and the linking number of DNA. Proc. Natl. Acad. Sci. USA 78, 1461–1465.
116. Hamiche, A., Carot, V., Alilat, M., De Lucia, F., O'Donohue, M.F., Révet, B., and Prunell, A. (1996) Interaction of the histone $(H3–H4)_2$ tetramer of the nucleosome with positively supercoiled DNA minicircles. Potential flipping of the protein from a left- to a right-handed superhelical form. PNAS (USA) 93, 7588–7593.
117. Alilat, M., Sivolob, A., Révet, B., and Prunell, A. (1999) Nucleosome dynamics. IV. Protein and DNA contributions in the chiral transition of the tetrasome, the histone $(H3–H4)_2$ tetramer-DNA particle. J. Mol. Biol. 291, 815–841.
118. Sivolob, A., De Lucia, F., Alilat, M., and Prunell, A. (2000) Nucleosome dynamics. VI. Histone tail regulation of tetrasome chiral transition. A relaxation study of tetrasomes on DNA minicircles. J. Mol. Biol. 295, 55–69.

J. Zlatanova and S.H. Leuba (Eds.) *Chromatin Structure and Dynamics: State-of-the-Art*

DOI: 10.1016/S0167-7306(03)39004-0

CHAPTER 4

The linker histones

A. Jerzmanowski

Laboratory of Plant Molecular Biology, Warsaw University and Institute of Biochemistry and Biophysics of the Polish Academy of Sciences, Pawinskiego 5A, 02-1-6 Warsaw, Poland. E-mail: andyj@ibb.waw.pl

1. Introduction

The proteins known today as linker or H1 histones were initially described as the abundant lysine-rich nuclear proteins that could be separated by chromatography on ion exchange resin from other major basic nuclear proteins known today as core histones (for review see Refs. [1,2]). During gel electrophoresis of histones the H1 fraction migrated as the slowest and most heterogeneous band. Upon the discovery of nucleosomal organization of chromatin in the mid 1970s it turned out that linker histones are not involved in the assembly of the nucleosomal protein core, but bind to DNA between nucleosomes (hence their name).

A major goal of the current phase of linker histone research is to reconcile the results of experiments on structural and biochemical properties of these proteins with the results of works aimed to establish their function *in vivo*. The data from sophisticated experiments probing the position of linker histones in nucleosomes and in the compact forms of nucleosomal filaments as well as from studies on the distribution and properties of different H1 variants were invariably interpreted as indicative of a critical function in chromatin architecture and dynamics and consequently in the regulation of such fundamental processes as transcription and chromosome condensation. At the same time complete knockouts of genes encoding H1 in unicellular Protista (*Tetrahymena*) and fungi (yeast, *Aspergillus*, *Ascobolus*) demonstrated that these simple Eukaryotes survive and prosper without linker histones (but see Section 5.1. for the criticism of the oversimplified conclusions based on these data). Similarly, the knockouts and overexpression of genes encoding different H1 variants in mouse produced no evidence so far that the variability of H1 in this complex multicellular organism is indeed functionally important. How can these paradoxes be explained?

2. Core and linker histones: a common name for different proteins

Despite many biochemical similarities between linker and core histones the proteins of these two groups differ in architecture, evolutionary origin, and function. Each of the four core histones has a characteristic "histone fold" domain. The latter is an old and ubiquitous structural motif used in DNA compaction and protein dimerization [3]. Linker histones do not have a histone fold. The canonical metazoan linker histone molecule contains three domains: two highly basic and unstructured tails (the shorter N-terminal and the longer C-terminal) and a non-polar central globular domain [4]. The globular domain of H1 (abbreviated GH1) is approximately 80 amino acids long and belongs to the "winged helix" family of DNA-binding proteins. This is surmised from crystallographic studies on the globular domain of the closely related linker histone H5 (a variant of H1 occurring in avian erythrocytes) [5]. It adopts a mixed α/β fold built of three α-helices (helices I–III) and three β-strands (S1–S3) (Fig. 1a). The bundle made of the three helices forms the core of the domain. Helix I is joined to helix II by a β strand, S1. The characteristic wing structure (from which this family of DNA binding proteins takes its name) lies within the domain located C-terminally to helix III. The latter is followed by two antiparallel β strands (S2 and S3) which together with S1 form a three-stranded β-sheet. The "wing" (W) is an extended loop, which joins S2 and S3. The comparison of GH5 and chicken histone GH1 in two-dimensional NMR studies demonstrated that the 3D structure of the globular domain is conserved among linker histones [6].

Analysis of the sequence and 3D structure of canonical linker histone proteins suggests that they could be the product of an ancient fusion of a DNA-condensing lysine rich-type protein already present in Eubacteria (the part corresponding to the C-terminal tail) and an architectural protein, probably of eukaryotic origin, with the property of binding to the four-way junction-like structures occurring at the site of entry and exit of DNA helices in the nucleosome (the part corresponding to GH1) [7,8]. While the core histone octamer provides a universal spool for compacting and restricting access to the DNA, the linker histones, although probably involved in stabilization of some forms of higher order chromatin structures, do not seem to be absolutely required for DNA compaction (see Section 3.2).

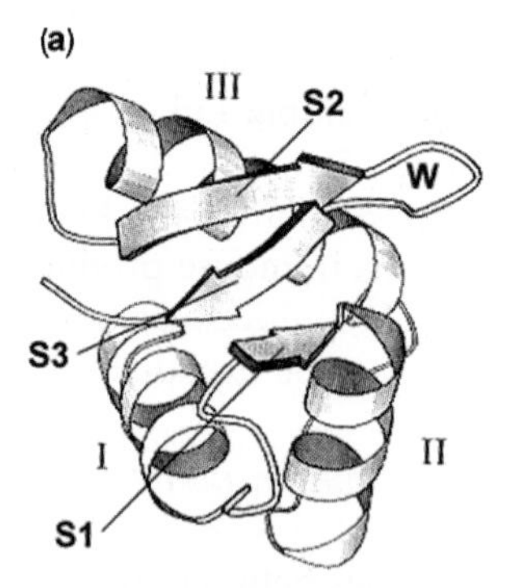

Fig. 1. Globular domain of chicken histone H5 (GH5) and its mode of binding to DNA. (a) Schematic view of GH5 according to the crystal structure by Ramakrishnan *et al.* [5] Reproduced by permission. Letters marking the structural elements of the GH5 correspond to the description in the text; (b) A model of GH5 binding to DNA based on its similarity to CAP and HNF-3 proteins, showing the conserved Lys/Arg and His residues, according to [28] (b). Reproduced by permission.

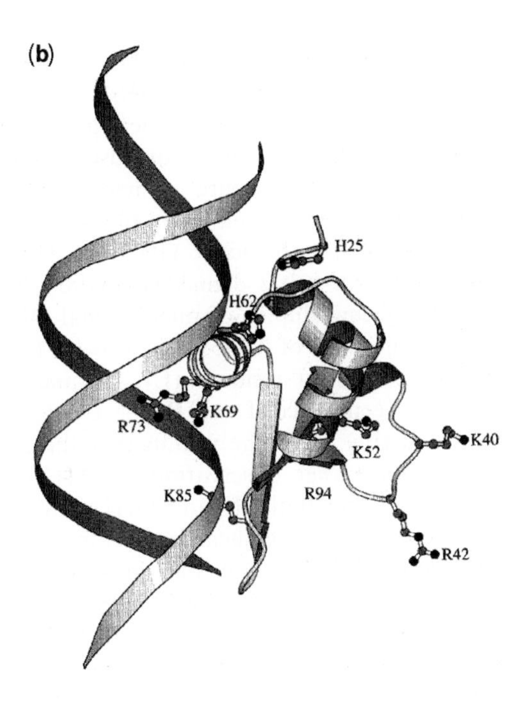

Fig. 1. Continued.

3. *Linker histones and chromatin structure*

3.1. *Mode of binding and location of histone H1 in the nucleosome*

In bulk chromatin, on an average histone H1 occurs at a frequency of one molecule per nucleosome. The peripheral location of H1 with respect to the nucleosome core was deduced from the observation that its presence causes a pause in the digestion of chromatin DNA by micrococcal nuclease, providing transient protection to a small fragment of DNA (~20 bp) in addition to the 146 bp stretch protected by interaction with the core histone octamer. Upon digestion of this additional fragment, H1 is lost from the nucleosome [9]. A nucleosome containing H1 plus ~166 bp of protected DNA is called a chromatosome. Based on experiments which showed that the isolated globular domain of H1 was as effective as the intact H1 molecule in protecting the extra 20 bp of chromatosomal DNA against degradation, a model was proposed in which the GH1 domain was placed on the dyad-axis of the nucleosome, contacting symmetrically both the entering and the exiting DNA duplexes [10]. The model required that GH1 have atleast two DNA-binding sites.

Because of the similarity between the X-ray crystal structure of GH5 (the globular domain of H5) and that of the helix-turn-helix protein CAP (bacterial Catabolite Activator Protein), for which the structure of the complex with DNA has been solved [11], it has been suggested that the primary binding site on GH1 (site I) is helix III. This helix is referred to as the DNA recognition helix and is predicted to interact with the major groove of DNA via Lys 69 and Arg 73, with Lys 85 located at the base of the "wing" interacting with the DNA backbone at an adjacent minor groove [5] (Fig. 1b). These conclusions gained strong support following solution of the co-crystal structure of hepatocyte nuclear factor 3γ (HNF-3) with DNA. The DNA-recognition domain of HNF-3 is a typical winged-helix motif with strong structural resemblance to GH5 [12]. The crystal structure of GH5 also suggests the presence of a less defined secondary binding site (site II) which can interact with an adjacent duplex of nucleosomal DNA [5]. This second site is located on the opposite side of GH5, 25 Å away from the recognition helix III and involves two conserved residues (Lys 40 and Arg 42) which are part of a disordered loop between helices I and II, the Lys 52 in helix II and Arg 94 placed in the S3 β strand (Fig. 1b). The notion that there could be a second binding site is consistent with earlier observations of H1/GH1 (and H5/GH5) DNA complexes which showed that both H1 and GH1 could bind co-operatively to two molecules of DNA and assemble them into 'tramline' complexes [13,14]. Is there any firm supporting evidence that the conserved residues mentioned above are indeed critical for the binding of GH5? Lys 85 located at the beginning of the wing is efficiently protected by association of H5 with the nucleosome [15] and its replacement with glutamine or glutamic acid results in loss of protection of an additional 20 bp of nucleosomal DNA [16]. The importance for correct binding of GH5 of Lys 69 and Arg 73 located in helix III and of all four residues proposed to form DNA contacts in site II was confirmed by mutagenesis [17].

The location of the globular domain of H1 on the nucleosome has recently been the subject of intense debate. The older symmetric model of Allan *et al.* [10] assumed that GH1 was positioned centrally on the "pseudo-dyad" axis of symmetry of the core particle and thereby interacted with equal lengths (about 10 bp) of both the "entering" and "exiting" DNA duplexes (Fig. 2a). However, the more recent data of Zhou *et al.* [18] who used GH5 reconstituted on mixed-sequence chicken mononucleosomes have contradicted this model. The authors, using a site-specific protein–DNA photocrosslinking method, have concluded that the globular domain spans DNA gyres between one terminus of the chromatosomal DNA which it binds via site I and the DNA close to the "pseudo-dyad" which

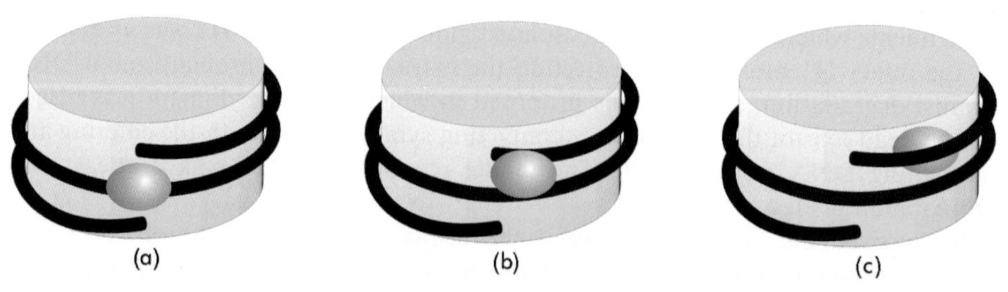

Fig. 2. Schematic view of the location of GH1 in the nucleosome. (a) Symmetric model of Allan *et al.* [9]; (b) Asymmetric model of Zhou *et al.* [17]; (c) Model of Pruss *et al.* [23].

it binds via site II. The asymmetric protection afforded by H1 with about 20 bp of DNA protected on one side of the core particle was also demonstrated for reconstituted chromatosomes by An *et al.* [19]. The asymmetric model of GH1 binding to the nucleosome (Fig. 2b) is supported by the finding that the winged helix transcription factor HNF-3, which is structurally similar to GH5, also binds to nucleosomal DNA in an asymmetric manner [20]. While the model accounts for the observation that the protection of an additional 20 bp of DNA in positioned chromatosomes is asymmetric and concerns only one end of the DNA wrapped around the core particle [19,21], it does not answer the question of where the two binding sites of GH1 are located in relation to the core histone octamer. It also does not provide an explanation of what determines the selection of one of two locations available for GH1 on the surface of the core particle. Both the symmetric and the asymmetric location of GH1 permit the C-terminal tail to bind the entering and/or the exiting DNA duplexes. This is in full agreement with the finding that the lengths of stems of linker DNA protruding from mononucleosomes correlate with the length of the engineered H5 C-terminal tails [22]. The model of the asymmetric location of a naturally asymmetric H1 molecule on the core particle adds another argument in favor of the concept of directionality in the folding of the basic nucleosomal filament (see discussion in the following section). The asymmetric mode of binding may also have important implications for the function of H1. Certain DNA sequences may stabilize the position of H1 on the nucleosome [19] and could influence the selection of one of two possible asymmetric locations. A strong H1 "positioning" sequence present on a single nucleosome might start a wave of polar H1 alignment along an entire nucleosomal array. An example of a similar effect has been demonstrated in an *in vitro* system consisting of nucleosomes reconstituted on a 4-kb plasmid that contains the *Xenopus laevis* gene encoding the oocyte-type 5S RNA. When the 120 bp gene was flanked by natural *Xenopus* sequences rich in AT, the addition of H1 to the reconstituted core particles resulted in large scale chromatin rearrangements encompassing the whole "minichromosome" [23].

In addition to the models which place GH1 close to or within the dyad and outside the DNA gyres, a third model of GH1 location has been proposed by Pruss *et al.* [24] (Fig. 2c). It was based on the results of crosslinking and mapping studies performed on chromatosomes reconstituted on a *Xenopus borealis* 5S-rDNA sequence. The third model postulates the asymmetric location of GH1, which is similar to the model of Zhou *et al.* [18]. However, in further details it departs radically from the two models discussed so far. It proposes that GH1 binds to only one DNA gyre at a single internal site 65 bp away from the dyad. Most importantly, it locates GH1 not outside but inside the DNA gyre. The model of Pruss *et al.* has been criticized mostly because of the incorrect assumption by the authors that chromatosomes reconstituted on the *X. borealis* 5S rDNA sequence occupy a single translational position. It has been shown in later experiments that the reconstitution procedure of Pruss *et al.* results in a mixture of chromatosomes with several different translational settings and therefore the GH1-DNA contacts deduced from crosslinking and DNA cleavage data cannot be unambiguously interpreted [19,25]. Thus, the most plausible model remains that of Zhou *et al.* [18] postulating the asymmetric location of GH1 close to the dyad and the bridging by the globular domain of two DNA gyres, to which it binds, via site I on helix III and site II, located 25 Å away.

It should be remembered however, that so long as the results of the X-ray crystallographic studies on the chromatosome are not available we are still dealing with models based on numerous assumptions, not all of which may be correct. One of these assumptions was that GH5 binds to DNA in a very similar way as CAP [5] and the monomeric HNF-3 [12], i.e., via helix III inserted into the major groove, with the "wing" lying along the DNA. A recent analysis of a 1.5 Å resolution structure of the human RFX1 protein by Gajiwala *et al.* [26] has demonstrated that the presence of the "wing" can be much more important for interaction of linker histones with DNA than was previously assumed. RFX1 is a member of the winged-helix family of DNA binding proteins. While the DNA-binding domains of RFX1 and HNF-3 reveal only 8% sequence identity, the overall topology of RFX1 strongly resembles that of HNF-3 (the overall root mean square deviation for Cα atoms of the two superimposed structures is 3.5 Å). Despite this similarity RFX1 binds DNA in a completely different way from that shown for HNF-3 and other winged-helix proteins [26,27]. Helix III of RFX1 is not located in the major grove but instead inserts its N-terminal end into the minor groove and makes a single base contact there. Because the two antiparallel β strands S2 and S3 run along the adjacent major grove it is the wing (W) that forms a key recognition element of the whole DNA binding domain (Fig. 3A). Since in HNF-3 and RFX1 proteins the same structural motifs interact with DNA it is impossibile to distinguish between the two modes of binding based only on the mutagenesis of the conserved residues located in helix III and the wing.

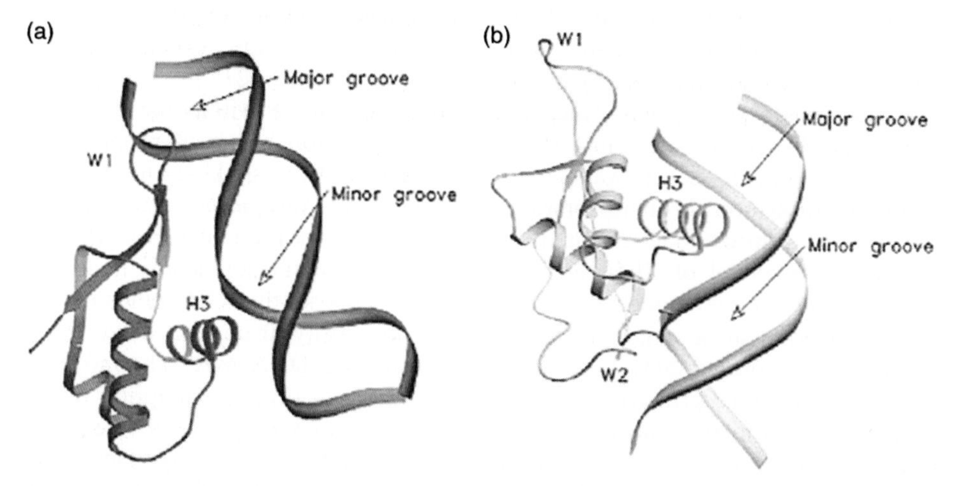

Fig. 3. Comparison of DNA binding by two types of winged-helix proteins. (a) RFX1 and (b) HNF-3. Reproduced by permission from Ref. [26], W1 and H3 correspond to "Wing" domain and helix III, respectively, of the GH5 on Fig. 1.

The reason for this entirely different mode of binding of HNF-3 and RFX1 (compare Figs. 3A and B) lies in part in the difference in the electrostatic surface potential of the two proteins. Proteins binding to highly electronegative DNA typically have strongly electropositive surface potential in the area which directly interacts with DNA. In HNF-3 (and in other members of the family, like DP2 and E2F4) the region of helix III is significantly more electropositive than the wing. The opposite is true for RFX1. Gajiwala *et al.* [26] showed that the electrostatic properties of GH5 are much more similar to those of RFX1 than to those of HNF-3 and thus place it in the same sub-class of winged-helix proteins as human RFX1. It is likely that previous studies have underestimated the importance for GH5 binding to DNA of the residues located more centrally in the wing domain. Interestingly, the evolutionary analysis of linker histone sequences reveals considerable differences in the size of the wing between major taxa (see Section 4.1). If linker histones do indeed interact with DNA in the same way as RFX1, these differences could have important functional consequences.

3.2. Linker histones and higher order chromatin structures

Under *in vitro* conditions, an increase in the concentration of mono- or divalent cations causes loosely organized extended nucleosomal filaments to condense into more compact structures. The first well defined level of compaction is reached at >60 mM Na^+ or >0.3 mM Mg^{2+} [2] and has the appearance of a rod-like compact fiber with a diameter of about 30 nm (but see the remarks below). The detailed architecture of this fiber and in particular whether or not it contains substantial amounts of regular helical structure has been the subject of numerous debates over the last 20 years. The models best substantiated by experimental evidence fall into two classes: the solenoid- and the zig-zag-type models (for review see Refs. [28,29]). According to the supporters of the solenoid-type models [30,31] the 30-nm compact fiber results from winding the nucleosomal filament into a one-start helical structure with about six nucleosomes per helical turn. The nucleosomes lie with their long axes parallel and their dyad axes perpendicular to the fiber axis, while the linker DNA entry/exit site on each nucleosome faces the center of the solenoid. The linker DNA is bent and continues the superhelical path induced by the nucleosome core which allows the consecutive nucleosomes to come into face-to-face contact with one another. The zig-zag-type models [32] are based on the observation that at very low ionic strength the nucleosomal filament has a zig-zag appearance. Moreover, it has been suggested (based on results obtained with scanning force microscopy) that under these conditions the loose filament already has some sort of 3D organization with an average diameter of 30 nm [33,34]. It is proposed that condensation of the initial zig-zag of nucleosomes results in a ribbon built of two parallel rows of nucleosomes. This ribbon then coils into a compact 30-nm fiber. In the zig-zag models the linker DNA is not bent and coiled but straight and the consecutive nucleosomes are not in close contact. It is generally agreed that the occurrence of the compact 30-nm fiber is the result of reaching a critical level of neutralization of the repulsive negative phosphate charges along the DNA backbone. In addition to free cations, the presence of the positively charged N-terminal domains of the core histones is critical for the formation of the fiber [35,36]. The authors of both the solenoid- and of the zig-zag-type models propose that histone H1 has an important function in promoting folding into the compact 30-nm fiber. Indeed, the available data show that the globular and C-terminal domains of H1 are involved in the induction of higher order chromatin structures [30,37] when correctly positioning H1, the globular domain on the nucleosome and the C-terminal domain on the linker DNA. The solenoid model implies that folding into the 30-nm fiber can be mediated by interactions between H1 molecules, perhaps in a directional manner assuming the asymmetric location of GH1. It is possible that these interactions are co-operative like the interactions of H1 with naked DNA [38]. The function of H1 and linker DNA in the zig-zag models is to induce and maintain a rigid nucleosomal architecture. An increased concentration of free cations causes a decrease in the linker entry/exit angle and an accordion-like compaction of the filament [39,40]. The observation that H1's C-terminal domain bridges two duplexes of linker DNA into a stem-like structure [22] creates a problem for both types of models. Indeed, it is difficult to reconcile with the idea of a smoothly bent superhelical DNA path postulated for the solenoid-type fiber. On the other hand, the stem structure is also a major obstacle for any alteration of the angle between the entering and exiting duplexes of linker DNA, unless one makes the additional assumption that the angle change concerns linker fragments extending beyond the stem [40]. The results of Graziano *et al.* [41] who used small-angle neutron scattering to measure the mass per unit length and cross-sectional radius of gyration of the 30-nm compact fiber are consistent with a model of the fiber in which H1 is placed internally at about the same radial location as the inner face of the nucleosome. However, this is in agreement with the predictions of both models. Is it possible to assess how important H1 really is for the formation of the 30-nm compact fiber, independently of which of the models will be proved correct?

Following the studies of Thoma and Koller [42] and Thoma *et al.* [30] it has been generally believed that H1-depleted chromatin is unable to form stable compact 30-nm fibers upon increasing monovalent cations but instead forms disordered clumps. Leuba *et al.* [43] reported that the 3D organization of a zig-zag nucleosomal filament observed at very low ionic strength requires the globular domain of H1 and either the tails of H1 or the N-terminal portion of histone H3. However, H1-depleted chromatin is still able to undergo folding upon increasing the salt concentration [44,45]. Yao *et al.* [46] showed that H1-depleted dinucleosomes fold similarly to those with H1, except that the folding transitions are shifted to a slightly higher concentration of monovalent cations (see also Ref. [47]). Although not much is known about the detailed structure of H1-depleted chromatin compacted by increased salt, early neutron scattering experiments on H1-depleted nucleosomal arrays studied at high chromatin concentration suggested that they were organized into 30-nm fibers [48]. A more recent study showed that folding of H1-depleted physiologically spaced 12 nucleosomal arrays induced by

Mg^{2+} (but not Na^{+}) results in a structure that is equivalent to the 30-nm compact fiber in the extent of condensation [49]. Finally, the independent and critical function of core histone N-termini in chromatin condensation was demonstrated by showing that nucleosomal filaments reconstituted from core histones lacking their N-terminal domains are unable to condense into folded structures upon an increase of Mg^{2+}, despite the presence of properly bound histone H5 ([50,51], see also Ref. [52] for the discussion of the special role of H3 and H4 tails). Thus, the presence of H1 is not a *sine-qua-non* condition for salt-induced chromatin folding, which can proceed in H1's absence and is an intrinsic property of filaments consisting of spaced core particles. A key question is how many of the features of the native 30-nm compact fiber are due to the presence of histone H1? From the available data it seems that H1 may influence the intrinsic folding pathway of the chromatin filament by stabilizing a single ordered conformation. This property can have much to do with the cooperativity of H1 interactions within chromatin but also with the way H1 is bound to the nucleosome and with the effect it exerts on the path of linker DNA.

3.3. Dynamic character of H1 binding to chromatin

As early as 1981 *in vitro* experiments showed that H1 could be readily exchanged between chromatin fragments at an ionic strength at which chromatin is folded into the compact 30-nm fiber [53]. This was consistent with the older results of Appels *et al.* who demonstrated that H1 migrates from HeLa nuclei to inactive chicken erythrocyte nuclei during cell fusion experiments [54,55]. Later, in an analysis of H1 exchange between different chromosomes in interspecies cell hybrids, Wu *et al.* [56] demonstrated that the displacement and migration of H1 occurs *in vivo* and takes place specifically and exclusively during mitosis. The recent kinetic studies using fluorescence recovery after photobleaching (FRAP) demonstrated that histone H1 exchanges even more rapidly than was estimated, in both condensed and decondensed chromatin. The exchange does not require fiber–fiber interactions, and so must occur through a soluble intermediate [57,58]. It has been proposed that *in vivo* H1 binds chromatin for about 1–2 min then dissociates and diffuses freely through the nucleoplasm for a short time before it reassociates with another available binding site on chromatin (see Ref. [59] for extended discussion). Such behavior of H1 contrasts markedly with that of core histones [57] but fits well into a current dynamic model of nuclear organization in which many proteins, even those with mostly architectural function in different nuclear compartments, are maintained in a flux-equilibrium allowing rapid local and global reorganizations of the structures in the nucleus [60]. However, it should be remembered that most immunolocalization experiments show that H1 is strictly co-localized with chromosomes. Perhaps these earlier studies were unable to differentiate between the majority of chromatin bound H1 and a considerably smaller pool of free H1, which is believed to be constantly present in the vicinity of chromosomes. An important result of the FRAP experiments of Misteli *et al.* [57] was the demonstration that a larger fraction of H1 is stably bound to heterochromatin than to euchromatin. The demonstration of the oscillations of H1 between bound and free forms indicates that the chromatosome is a highly dynamic structure. Moreover, it suggests that the occurrence of the compact 30-nm fiber (in which most of the nuclear chromatin is maintained) cannot depend on the static binding of H1 as compaction itself is clearly no obstacle for H1 mobility. The steady-state occupancy of the chromatin-binding site by H1 is more reminiscent of the mode of transcription factor binding than that of the core histones. However, by analogy with transcription factors, such status does not diminish the importance of specific protein–DNA interactions, which in the case of linker histones is the binding of GH1 to the nucleosome. An important unanswered question is to what extent H1 exchange is able to homogenize the distribution of different H1 variants over the chromatin surface.

The finding that H1 migrates between free and bound states lead on to the question of which factors can shift the equilibrium between the two states? The FRAP analysis showed that H1 exchanges in both condensed (mitotic) and diffused (interphase) chromatin with similar kinetics [58]. However, the data of Wu *et al.* [56] indicate that the interchromosomal translocation of H1 only occurs during a period of full mitotic condensation of chromatin. Is there something unique to that stage that allows long distance translocations? H1 is massively phosphorylated during mitosis and during the HeLa cell cycle the highest density of extrachromosomal H1 was observed in the M-phase by using an antibody against phosphorylated H1 (however, it cannot be completely ruled out that the antibody also recognized some other phosphoproteins) [61]. The reversible mitotic phosphorylation of H1 is unique in that it concerns the whole H1 pool in the nucleus and results in phosphorylation of multiple sites in H1 tails [62]. However, while often described as a prerequisite for loosening the H1–DNA contacts [63] phosphorylation may in fact have nothing to do with H1 release. First, the rapid exchange of H1 in the nucleus was shown not to be immediately affected by depletion of cellular ATP [58]. Second, it was shown in an *in vitro* system that phosphorylation of chromatin-bound H1 by a mitotic H1 kinase requires prior dissociation of H1 from the chromatin [64]. The mitotic phosphorylation could thus be simply a consequence of the availability of free H1 as a result of its dynamic behavior in chromatin and the transient increase of H1 kinase activity occurring during mitosis. However, there is some indication that mitotic chromatin condensation itself affects H1 binding. H1 without the C-terminal domain, known to bind to its target sites on nucleosomes *in vitro*, was shown by FRAP analysis to be completely displaced from chromosomes during mitosis [58]. HMG1 protein which binds to nucleosome *in vitro* in much the same way as H1, in *Xenopus* and *Drosophila* replaces H1 during early embryogenesis [65–68], was shown to be completely detached from condensed metaphase chromosomes in NIH 3T3 cells [69]. In view of the above facts it seems plausible that mitotic condensation distorts or partially abolishes at least some of the GH1 binding sites on nucleosomes. One of the reasons could be a several-fold increase in the amount of divalent cations (Mg^{2+} and Ca^{2+}) bound to condensed as compared to decondensed chromatin, as recently shown in elegant experiments employing secondary ion mass spectrometry by Strick *et al.* [70]. The several-fold increase in the neutralization of DNA phosphates by divalent cations may lead to distortions in local DNA structures, for example bending towards the neutralized region, which could influence the site of H1 binding. That H1 is not absolutely required for chromatin transitions during mitosis was shown by Ohsumi *et al.* [71] who demonstrated that sperm chromatin naturally deprived of H1 undergoes normal mitotic condensation when incubated with H1-depleted *Xenopus* mitotic extract. If mitotic condensation is likely to create less favorable conditions for H1 binding in chromatin, the massive mitotic phosphorylation of H1's C-terminal domain is a sure candidate to shift the equilibrium towards the unbound form of H1. In *Tetrahymena*, the substitution of phosphorylation sites in H1 for glutamic acids, a change mimicking a fully phosphorylated state of histone H1, has the same effect on gene expression as the loss of H1 from chromatin [72]. If the normal equilibrium between chromatin bound and free H1 was shifted towards the free form during mitosis, a distinct gradient of free H1 could occur in close vicinity to the condensed chromosomes. An important question to be answered is whether such a gradient could have functional significance?

4. Variability of linker histones

4.1. Evolutionary perspective

In discussing the complex matter of the variability of linker histones and its possible biological significance it is useful to start from a general picture of evolutionary relationships emerging from the analysis of currently available H1 sequences. The first issue that needs clarification is the relationship between the typical or "canonical" linker histones characterized in animals, plants, fungi, and certain groups of protists (the green algae Chlorophytes) and represented by proteins with tripartite structural organization (a "winged-helix"-type globular domain flanked by N- and C-terminal basic tails), and the small histone H1-like proteins which compositionally and structurally resemble the C-terminal basic tail of canonical H1 and have been isolated from organisms representing certain other groups of protists, like *Trypanosoma cruzi*, *Trypanosoma brucei* (Kinetoplastids), and *Tetrahymena thermophila* (Ciliates) [73–75]. The argument in favor of an idea that the H1-like proteins are functional counterparts of the canonical H1 comes from the finding that they bind the to linker DNA of the nucleosome arrays [74,76]. However, it should be remembered that much of the linker DNA is effectively accessible and any DNA binding protein will non-specifically bind to it. Are the H1's C-terminal tail and the H1-like proteins of *Protista* real homologues, i.e., with a common evolutionary origin? Assuming that selection pressure acted to preserve the characteristic amino acid composition rather than a particular sequence, it is not impossible. However, an evaluation with statistical methods indicates that such homology is very uncertain [77]. Although the H1-like proteins are certainly capable of charge neutralization of the DNA phosphates similar to that provided by the

C-terminal tail of a canonical H1 (see Ref. 7 for an in depth review) they are unable to bind to nucleosomes in a GH1-specific manner. Therefore, the conclusions of the genetic analyses obtained for *Tetrahymena* H1-like proteins (see Section 5.1) may be extrapolated to the C-terminal tail but not necessarily to the complete molecule of the canonical H1.

The low complexity of H1 terminal tail sequences precludes their use for detailed evolutionary analysis. Therefore, conclusions concerning the relationships among canonical linker histones must be based on comparisons of their globular domains (GH1s). Alignment of the sequences of known H1 histones shows a general conservation of the length of the fragment comprising the GH1 (70–81 amino acids). This, together with the highly conservative distribution of amino acids with characteristic properties, is indicative of a conserved three-dimensional structure. The alignments of known GH1s shows an insertion of a five amino acid fragment in GH1 sequences in the branch of animals. This is in agreement with the earlier demonstration that the H1 sequences of plants lack a highly conserved pentapeptide motif GXGAX located in the "wing" (see Fig. 1a) of GH1 in all animal H1 histones [78]. Identification of a rare molecular event that unites taxa, such as the insertion of a conserved pentapeptide that occurred in the GH1 of animals, enables independent verification of phylogenies established by other methods. An analysis with parsimony performed with the whole set of GH1 sequences to reveal the phylogenetic relations with respect to the size of the insert yielded a four-branch

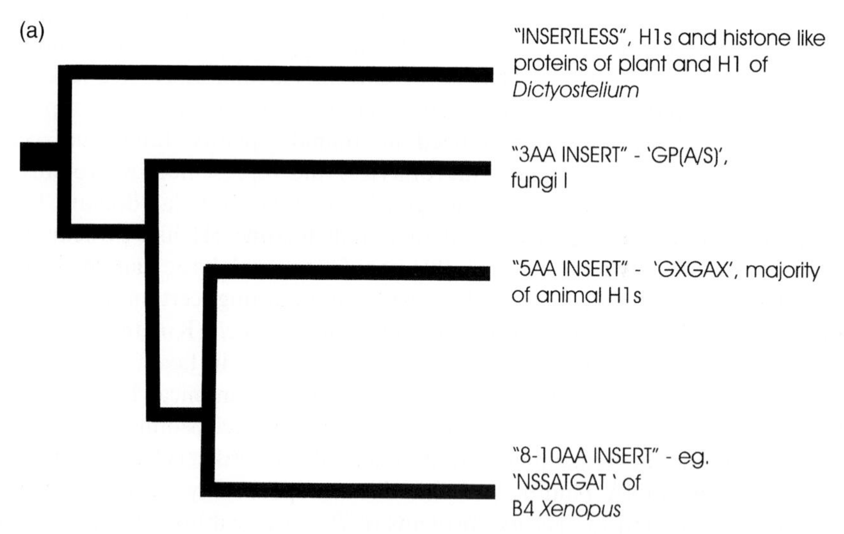

Fig. 4. The evolution of the globular domain of histone H1. (a) The parsimony tree for the structure of the wing subdomain of GH1. The length of the insert was calculated from the alignment of the GH1 sequences. (b) The parsimony tree for histone H1 based on sequences of GH1 domains of known H1 histones. Percentage values on each branch represent the corresponding bootstrap probabilities. Branches for which the relationships are unclear (supported by less than 50% probability) are marked by thin lines. Reproduced with permission from Ref. [72].

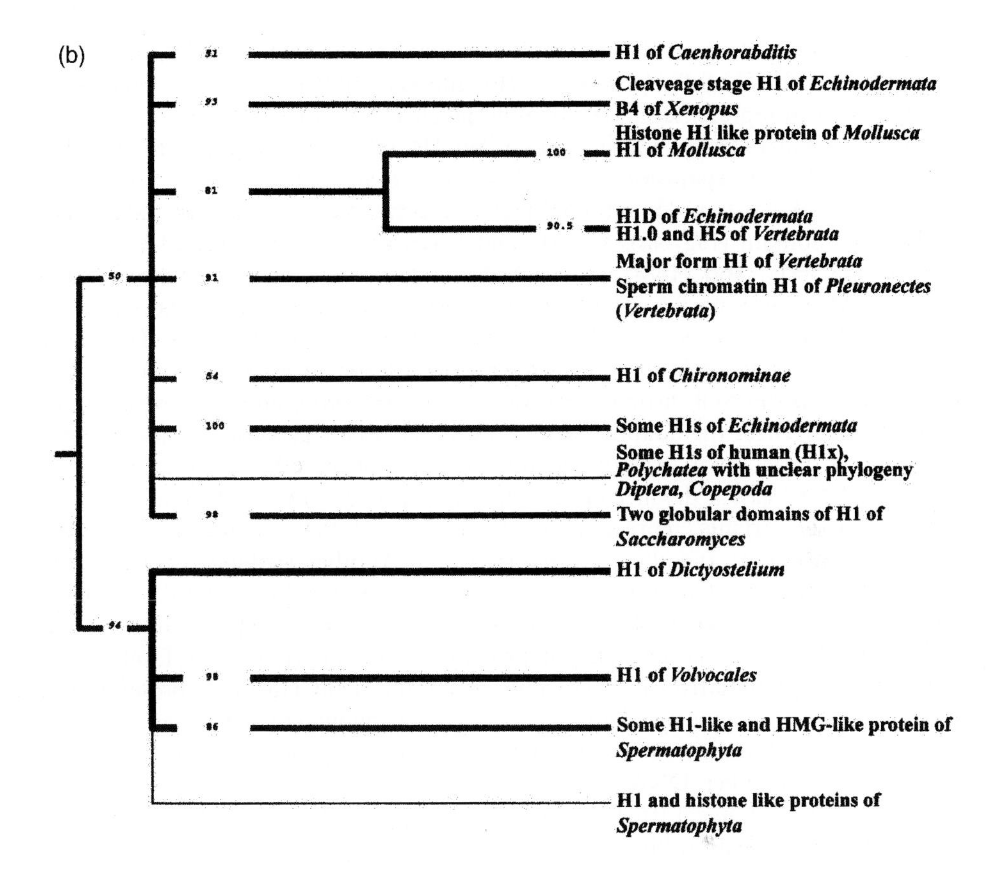

Fig. 4. Continued.

tree [77] (Fig. 4a). Based on this tree, H1 histones can be classified into four major groups: (1) H1 without an insert, including all known plant H1s (including green algae) and *Dictyostelium* H1; (2) H1 with a 3 amino acid insert including yeast and other fungi H1s; (3) H1 with a 5 amino acid insert including the majority of animal H1s; and (4) H1 with a 8–10 amino acid insert including some special "cleavage-stage" animal H1 variants.

The validity of the above classification is emphasized by the fact that the parsimonious tree based on the full length GH1 (Fig. 4b) has the same topology as the tree based on comparison of the size of the insert in the wing subdomain [77]. Both trees separate H1 into two main branches: the branch of *Dictiostelium* and plants and the branch of fungi and animals.

The relationships between linker histones within different kingdoms can be deduced from a detailed analysis of the main branches of the tree in Fig. 4a. Within animal H1s there are several well separated isoforms. One is the major form of H1 in vertebrates. The other easily distinguished isoform is represented by the cell cycle-independent forms of H1 characteristic of the whole subkingdom of deuterostomes: H1D of *Echinodermata*, H5, and H1° of *Vertebrata*. It is interesting that the isoform represented in the tree by the cleavage stage H1 of *Echinodermata* and B4 (H1M) of *Xenopus* also contains an H1 of *Mollusca* and thus probably evolved before the separation of *Mollusca* and *Deuterostomes*. It is noted that this isoform belongs to a separate group characterized by an 8–10 amino acid insert in the wing domain (Fig. 4a). Schulze and Schulze [79] proposed that vertebrate histones H5, H1°, and H1M are direct descendants of predecessor genes which retained the evolutionarily old sequence information by acquiring the status of an isolated "orphan" genes at the time when the main group of H1 genes underwent concerted evolution due to the process of sequence homogenization. The occurrence of novel well-distinguished isoforms of H1 throughout the evolution of animals could reflect the adaptive evolution of H1. Plant H1s evolved in a similar way (Fig. 5). Apart from a clearly separate grouping of the GH1s of *Volvocales*, the tree in Fig. 5 reveals a distinct branch of the so-called "stress-inducible" variants of H1. These variants represent an old isoform, which occurred before the separation of mono- and dicotyledonous plants. Another striking feature of the phylogenetic tree of plant H1s is the branch representing the "hybrid" proteins. This branch groups sequences derived from proteins in which a typical GH1 domain is fused to domains characteristic of proteins from outside the H1 family. The most typical representatives of such proteins are the plant HMGI/Y proteins [80].

In summary, the sequence comparisons offer the following general picture: there are two major families of partly similar abundant proteins interacting with internucleosomal linker DNA: these are the H1-like proteins occurring in some *Protista* and the canonical tripartite H1s of animals, plants, fungi, and green algae. The evolutionary relationship between these two families is not certain. Based on the analysis of the GH1 domain, the major differences between the canonical linker histones concern the length of the highly conserved "wing" region. Plants and *Dictyostelium* have GH1s with a shorter wing than that present in GH1 of animals and fungi. A further divergence of the globular domain of animal linker histones is best seen in vertebrates in which the GH1 of the major form of H1 is distinctively different from that of the cell cycle-independent forms H5, H1°, and H1M. The diversification of GH1 into separate isoforms can also be seen during the evolution of plants.

The above general scheme based on comparisons of the most evolutionarily conserved GH1 domain does not reveal the rich microheterogeneity of linker histones which stems from differences between the less well conserved basic tails. Such variants, often referred to as somatic subtypes, occur in plants (for review see Ref. [80]) and animals, both invertebrates and vertebrates (for review see Refs. [81–83]). For example in mammals five somatic subtypes represent the major form of H1: $H1^s$-1, $H1^s$-2, $H1^s$-3, $H1^s$-4, and H1a, according to the nomenclature proposed in Ref. [82]. The N- and C-terminal tails of the subtypes differ in length, the amino acid composition and the frequency and distribution of phosphorylation sites. The testis specific H1t can be considered the most diverged subtype of the major form.

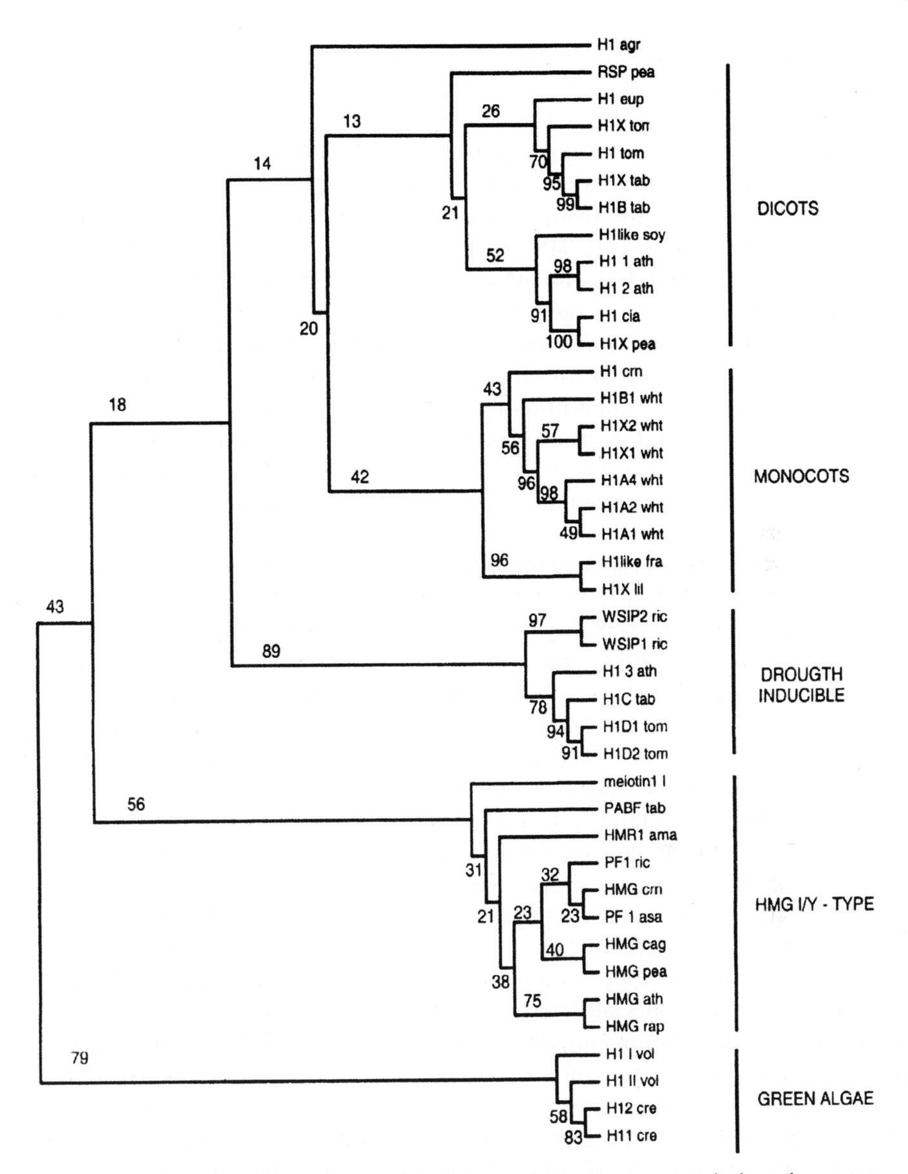

Fig. 5. The evolution of plant histone H1 proteins. Percentages on each branch represent the corresponding bootstrap probability values. Abbreviations: agr, Celery (*Apium graveolens*); ama, Snapdragon (*Antirrhinum majus*); asa, Oat (*Avena sativa*); ath, Thale cress (*Arabidopsis thaliana*); cag, Sword bean (*Canavalia gladiata*); cia, Chickpea (*Cicer arietinum*); cre, *Chlamydomonas reinhardtii*; crn, Corn (*Zea mays*); eup, Leafy spurge (*Euphorbia esula*); fra, *Fritillaria agrestis*; lil, Lilly (*Lilium longiflorum*); pea, Pea (*Pisum sativum*); rap, Rape (*Brassica napus*); ric, Rice (*Oryza sativa*); soy, Soybean (*Glycine max*); tab, Common tobacco (*Nicotiana tabacum*); tom, Tomato (common) (*Lycopersicon esculentum*); vol, Volvox (*Volvox carteri*); wht, Wheat (*Triticum aestivum*). Reproduced with permission from Ref. [75].

4.2. Biological significance of H1 diversity—evidence from biochemical and molecular studies

There is ample evidence pointing to the biological significance of linker histone diversity. It comes from several areas of research. The comparative analysis of the sequences of H1 variants from different animal species, belonging both to characteristic old isoforms (like H5 and H1°) and different somatic subtypes of the major vertebrate H1 (like H1t and H1s-1 to $H1^s$-4 in mammals) shows a variety of non-synonymous substitution rates [84], which can be taken as an indication of a differentiated function. Further evidence comes from studies on the distribution of variants in tissues and in chromatin. In *Echinodermata* and vertebrates the cleavage stage-specific variants (homologues of H1M or B4 in *Xenopus*) are found exclusively in oocytes and in early cleavage stages of the embryo [85–89]. H5, an avian-specific variant is expressed only in erythrocytes [90], and the H1° is found only in terminally differentiated cells of mammals [91,92]. A testis-specific H1t is detected only in spermatogenic cells and in developing spermatids of vertebrates and invertebrates [93,94]. Of the other subtypes of the major H1 variant in mammals the H1a seems to be restricted to thymus, testis, and spleen [95], while $H1^s$-1 to $H1^s$-4 subtypes occur ubiquitously [82,83]. There is also evidence of a non-random distribution of atleast some of the H1 variants in chromatin. The first report came from Mohr *et al.* [96] who by using monoclonal antibodies against variants of insect (*C. thummi*) H1 showed the specific intrachromosomal localization with preference for heterochromatin regions of the variant H1 I-1. This variant differed from other *C. thummi* H1s by a characteristic amino acid motif in the N-terminal tail. Similar motifs were identified in H1s from *C. elegans* and from *Volvox* [97]. Differences in the distribution of human somatic H1 subtypes with respect to DNA sequences representing active and inactive genes was recently reported based on statistical analysis of the results of immunofractionation of chromatin using subtype-specific antibodies [82,98]. Also in plants, differences in chromatin distribution between major somatic H1s and a "stress-inducible" H1 variant have been reported [99].

Some of the most compelling evidence of the functional importance of H1 diversity comes from studies of H1 expression patterns in animal development. Differential patterns of expression of H1 variants have been convincingly shown for both invertebrates and vertebrates (for review see Ref. [100]). In the dipteran insect *C. thummi*, the expression of specific H1 variant H1 I-1 dominates over the expression of other variants in early development and this situation is reversed in later stages [101]. H1 I-1 contains in the N-terminal tail an inserted sequence KAPKAPKSPKAE that is lacking in other H1 variants of *C. thummi*. Homologous motifs occur in specific H1 variants from different insect species as well as in H1 of *Volvox* and in sea urchin sperm H1 [102].

It is striking that a similar pattern of H1 variant expression during early development has been maintained in the whole subkingdom of Deuterostomes, as illustrated by studies in sea urchin and *X. laevis*. This possibly dates back to an earlier stage before the separation of Deuterostomes and *Mollusca* (see the discussion above). In both the sea urchin and the frog a specific cleavage stage H1 variant that belongs to a distinct class, with 8–10 amino acids inserts in a wing domain (see Fig. 4a), is the major or only form of H1 during initial divisions after fertilization of the oocyte. Beginning from the blastula–gastrula transition the other somatic-type variants take over with the H1°-type variant becoming more abundant concomitantly with the progression of differentiation and lengthening of the cell cycle [85, 103–108].

Given the variability of the expression patterns of different H1 variants it was not surprising to find that different *cis*-regulatory elements are involved in the control of their transcription. Detailed studies of vertebrate H1 genes resulted in the identification of characteristic assemblies of *cis*-regulatory DNA motifs responsible for differential control of three major types of variants: cleavage stage H1 [87], major-type vertebrate H1 (or replication-dependent H1) [109,110], and differentiation-specific H1 (H1° and H5) [107,111]. Additional specific *cis*-regulatory sequences have been shown to control the characteristic spatial pattern of H1t expression [112,113].

Another regulatory element of potential importance in differentiating the expression pattern of H1 variants resides in the structure of mRNA. All major-type (replication-dependent) somatic variants of vertebrate H1 have a polyA$^-$ (non-polyadenylated) mRNA, whereas the cleavage stage and differentiation variants (replication-independent) have a polyA$^+$ (polyadenylated) mRNA. The unique 3′ hairpin structure in the 3′ UTR of the polyA$^-$ mRNA is involved in the regulation of mRNA stability in the cytoplasm [114] and is thought to play a key role in coordinating histone protein synthesis with the DNA replication phase of the cell cycle. However, the requirement for this structure in the coordination of H1 and DNA synthesis throughout the cell cycle is not absolute, since no polyA$^-$ H1 mRNA occurs in angiosperm plants or nematodes [77].

5. *Functional analysis of the role of linker histones in cells and organisms*

With the availability of genetic manipulation techniques, gene knockouts and overexpression became a common approach in studies of protein function. Because the linker histones of multicellular plants and animals are encoded by separate genes representing different variants, the complete elimination of H1 expression in these organisms is difficult to achieve. Therefore, the natural choice for studies on the effect of complete elimination of H1 were the simpler Eukaryotes, in which there is no variability of H1 and only a single H1 gene. However, in the absence of analogous experiments with complete elimination of linker histones during growth and development of plants or animals, the general conclusions from studies using simple Eukaryotes cannot be simply extrapolated to Eukaryotes with a more complex organization. In contrast to plants and animals, unicellular or undifferentiated Eukaryotes, like *Tetrahymena* or yeast, express the majority of their genes under normal growth conditions. This means that chromatin of most genes is open and transcription is controlled by mechanisms which do not involve switching portions of the genome between inactive and active states or securing the inaccessibility of chromatin regions for a long period of time. Whenever more stringent repression is required, for example in the silent mating type loci in yeast, specialized chromatin structures can be induced, like those mediated by SIR3/4 proteins [115].

5.1. Function of linker histones in simple eukaryotes

The complete elimination of H1 by knocking out the corresponding gene (or genes) has so far been accomplished in four organisms: the ciliate protist *Tetrahymena* [116] and three species of fungi, the yeast *Saccharomyces cerevisiae* [117–120], *Aspergillus nidulans* [121], and *Ascobolus immersus* [122]. Except for *Tetrahymena*, which has an H1 lacking the globular domain and yeast where the H1 has an unusual organization with two globular domains, the remaining two organisms have canonical tripartite linker histones which localize in chromatin. In *Aspergillus* the amount of H1 in chromatin was shown to be of the same order as that of H4 [123]. All four organisms were able to live without H1 and did not show any immediate phenotypic effects caused by its complete elimination. The more detailed studies revealed that in *Tetrahymena* the loss of H1 did not affect nucleosome positioning [124] and resulted in a change in transcriptional levels of only a few genes, some of which were up- and some downregulated [125]. In yeast without H1 there was a slight decrease in the transcription of the majority of genes (over 6000), although only 27 genes showed a decrease of two-fold or more. A small fraction of genes showed a slight increase in transcription [120]. No data concerning the effects of H1 loss on transcription are available for *Aspergillus* or *Ascobolus*. In *Aspergillus* neither the nucleosomal repeat length nor the DNA susceptibility to MNase were affected by the loss of H1. In *Ascobolus*, the loss of H1 was correlated with increased susceptibility to MNase, but did not cause a change in the nucleosomal repeat. Interestingly, the loss of H1 in *Ascobolus*, while not affecting methylation—mediated gene silencing was correlated with hypermethylation of cytosines in the non-CpG sites. *Ascobolus* was also the only one of the four organisms with completely eliminated H1 to show a delayed phenotypic change. Although the loss of H1 had no effect on the early vegetative life or on sexual reproduction of this fungus it consistently caused a short-life-span phenotype. While the wild-type strains grew normally for over 20 days, strains without H1 suddenly stopped growing after 6–13 days. To account for this effect Barra *et al.* [122] proposed that the loss of H1 induces chromatin changes which repress genes essential for growth.

The consequences of histone H1 elimination in *Tetrahymena* and in the three species of fungi are consistent with the interpretation that linker histones do not have a role which is essential for the normal life-cycle of these organisms including sexual reproduction. Indeed, it has been reported that in yeast the H1 occurs with a stoichiometry of about one molecule per 37 nucleosomes showing a preferential association with rDNA sequences [126]. The analyses of the effects of H1 loss on transcription, particularly in *Tetrahymena* and yeast, provide strong evidence that in the four studied species linker histones do not act as general repressors of genes. Therefore, what is their function? In *Tetrahymena*, yeast, and *Aspergillus* it may be limited to that of specialized regulators for a selected set of genes. However, the effect of H1 elimination on the life span in *Ascobolus* might be indicative of a role in a long-term chromatin state necessary for maintaining particular types of cellular differentiation.

5.2. Function of linker histones in complex multicellular eukaryotes

5.2.1. Experiments employing cell lines

In studies aimed at establishing the role of six different H1 variants occurring in B-cell chicken lymphoma line, Takami and Nakayama, using mutant DT40 cell lines, showed that the expression of only one copy of a single H1 variant was enough for normal proliferation of the cells [127]. The lack of the other variants was compensated by the upregulation of the remaining single variant, a finding consistent with the earlier demonstration of an inherent ability of histone gene families of the DT40 cells to compensate for the disruption of an entire single allele bearing a major histone gene cluster including all six variants of H1 [128]. In DT40 cells the disruption of any of the six H1-encoding genes had no influence on the growth rate of the cells [129]. In the mutant lines the genes encoding the remaining H1 variants were transcribed at higher levels, maintaining a constant steady-state level of total H1 mRNA, However, analysis by 2D electrophoresis of the total proteins extracted from mutant cell lines revealed that they differed in characteristic subsets of proteins, indicating that each H1 variant plays a role in the transcriptional control of specific genes. The above studies demonstrate that the variability of linker histones is not essential for proliferation of DT40 cells and point to the well developed ability of these cells to control the physiological stoichiometry of H1. However, the conclusions regarding the role of H1 in multicellular organisms cannot be established based on the observations of a tumor cell line growing *in vitro*. These cells have numerous aberrations (for example, the DT40 cells expressing only single copy of H1 gene, also had 17 core histone allels removed) and hardly resemble a typical cell within a multicellular environment.

A different approach was taken by the group of Brown *et al.* [130] who developed a system for the inducible overexpression of individual H1 variants and their mutagenized versions in mouse 3T3 fibroblast cells. These researchers found noticeable differences in the effects caused by the overdose of the differentiation-specific $H1^{\circ}$ and one of the somatic subtypes, H1c ($H1^{s}$-1 according to nomenclature in Ref. [82]). Overexpression of $H1^{\circ}$ caused a transient delay in cell cycle progression and a decrease in steady-state levels of all the RNA polII transcripts studied. Overexpression of H1c did not affect the cell cycle and caused either no change or a considerable increase in the steady-state levels of the studied transcripts. The differential effects of overexpressing these two H1 variants were shown to be due to differences in the globular domain [131]. This is surprising, given the relatively small difference between the structures of these two globules.

Interestingly, upon extended observation, the changes induced by overexpression of either of the two variants of H1 become neutralized by a compensatory mechanism which adjusted all cellular parameters to a normal level [132].

5.2.2. Experiments employing whole organisms

The developmental regulation of 5S RNA gene transcription in *Xenopus laevis* has been a most thoroughly documented case of histone H1 involvement in modulation of transcription of a defined set of genes *in vivo*. Two 5S RNA gene families are transcribed in the early stages of embryonic development in *Xenopus*: the major oocyte-type, occurring at 20,000 copies per haploid genome, and the somatic-type, occurring at 400 copies per haploid genome. From the late gastrulation stage, transcription of the oocyte-type 5S RNA genes becomes largely repressed whereas that of the somatic-type 5S RNA genes continues unaffected throughout consecutive developmental stages and during the adult life of the frog. Studies on the *in vitro* transcription of *Xenopus* somatic cell chromatin were the first to establish that somatic-type histone H1 is necessary to maintain the repression of oocyte-type 5S RNA genes in somatic cells [133]. This finding correlated well with the later observation that normal somatic H1 protein was absent in oocytes [66]. Instead, the oocyte chromatin contains the "cleavage-stage" H1 variant, B4 [134] (see the discussion in 4b). During early embryogenesis the B4 variant is gradually replaced by the somatic-type variant of H1, its level in chromatin correlating with the repression of the oocyte-type 5S RNA genes [123,79]. A direct cause-and-effect relationship between the accumulation of somatic histone H1 and the repression of oocyte-type 5S RNA synthesis was demonstrated by elimination of somatic H1 using ribozyme strategies during early *Xenopus* embryogenesis. The elimination of somatic H1 *in vivo* led to continued expression of oocyte-type 5S RNA genes in embryos that would have normally had these genes switched-off [135,136]. These studies were complemented by experiments carried out by several groups, using chromatin and nucleosomes reconstituted *in vitro* on oocyte- and somatic-type *Xenopus* 5S RNA genes. Based on the results of these experiments it seems that the sequences of the 5S genes themselves and of their surrounding DNA are critical for the H1-mediated differential effect on transcription of somatic- and oocyte-type 5S RNA genes [23,25,137–140]. The discrepancies concern the proposed molecular mechanisms by which H1 affects the transcription of the oocyte-type genes. It is agreed that in the oocyte-type genes, the chromatosome is positioned directly over the 5S gene, so that the binding site for a critical transcription factor, TFIIIA, is occluded by the positioned nucleosome. On the somatic-type gene, the nucleosome lies about 65 bp further upstream leaving the TFIIIA site partly free. According to Panetta *et al.* [25] the TFIIIA and the H1 are competing for their respective binding sites, with H1 preferentially binding the oocyte nucleosome and TFIIIA preferentially binding the somatic nucleosome. According to Sera and Wolffe [139] and Howe *et al.* [140] histone H1 interacts with nucleosomes assembled on oocyte and somatic 5S genes with equal affinities but it is the H1 that determines the nucleosome positioning over the 5S oocyte gene and the occlusion of the TFIIIA site. It should be remembered that the 5S oocyte gene is flanked by the AT-rich sequences which interact with H1 with high affinity whereas the 5S somatic gene is surrounded by the GC rich flanks which show no specific preference for H1 [137]. The transcriptionally repressive state resulting from H1 binding to the oocyte nucleosome is additionally stabilized due to H1's preferential and strong interaction with the AT-rich flank of the oocyte 5S gene ([23,138] for review see [141]). In a later experiment Vermak *et al.* [142] demonstrated that the globular domain of somatic H1 alone was capable of repressing the transcription of oocyte-type 5S RNA genes *in vivo*. This again confirms that the structural differences in the globular domains of H1 variants are critical for their differential effects on chromatin.

The specific regulatory role of H1 documented for the *Xenopus* 5S RNA gene points to the importance of AT-rich tracts in the genome as potential sites of strong and preferential H1 binding. Indeed, it has been shown that histone H1 is probably involved in the repression of the murine β-interferon promoter by binding to its upstream AT-rich region [143].

In *Xenopus*, in addition to the role in 5S RNA gene regulation, somatic H1 has been shown to be involved in further stages of differentiation. The restriction of *myoD* expression, a marker for the loss of the ability by ectodermal cells to differentiate into mesoderm, requires the presence of somatic histone H1 [144]. Again, the globular domain alone and not the whole H1 molecule is required to confer this effect [142].

Gene knockout mice have been extensively used to study the effects of the elimination of different histone H1 variants on mammalian growth and development. The results obtained independently by three different laboratories showed that mice with targeted disruption of a gene encoding testis-specific subtype H1t develop normally and are fertile [145–147]. Mice lacking either H1a, another variant which accumulates at specific stages of spermatogenesis, or differentiation-specific variant $H1^{o}$ also developed normally [148,149]. The analysis of chromatin from $H1^{o}$-null mice showed that other H1 variants, especially $H1^{s}$-2, $H1^{s}$-3, and $H1^{s}$-4 compensate for the loss of $H1^{o}$ to maintain a normal H1-to-nucleosome ratio, even in tissues that normally contain abundant amounts of $H1^{o}$ [149]. In recently described experiments using double knockout strains Fan *et al.* [150] showed that mice lacking $H1^{o}$ and any of the three somatic subtypes $H1^{s}$-2, $H1^{s}$-3, or $H1^{s}$-4 were normal and maintained the physiological ratio of total H1 to nucleosomes. The main conclusion from knockout mice studies is that any individual variant including $H1^{o}$ and H1t is dispensable for mouse development. Moreover, the individual somatic subtypes $H1^{s}$-2 to $H1^{s}$-4 are dispensable even in mice lacking the $H1^{o}$ variant. However, the highly efficient compensation for the loss of individual variants by other variants indicates that there is a strong pressure to maintain normal H1-to-nucleosome stoichiometry in chromatin.

Given the lack of any clear phenotypic effects of changed proportions of linker histone variants in mice, it is particularly interesting that such effects were documented in invertebrates and in angiosperm plants. In *Caeanorhabditis elegans* the dsRNA-mediated decrease in H1.1, one of the eight variants of H1 occurring in this invertebrate, led to the desilencing of a normally repressed reporter transgene in germ line cells without affecting the expression of this reporter in somatic cells. Correlated with this desilencing were defects in germline proliferation and differentiation, and in some cases even sterility [151]. H1.1 is the most abundant of H1 variants in *C. elegans* yet the desilencing probably occurred following even a slight decrease in its content. Interestingly, the specific depletion of any of the other H1 variants did not cause desilencing. Most importantly, the defects in germline development caused by the decrease in H1.1 were highly reminiscent of those associated with mutation of the *mes* genes involved in germline-specific chromatin repression in *C. elegans*. Some of the *mes* genes are *C. elegans* homologues of the *Drosophila* Polycomb Group (Pc-G) genes involved in gene repression during development. In tobacco, the reversal of the normal ratio of major (H1A and H1B) to minor (H1C to H1E) H1 variants achieved by antisense silencing of H1A and H1B resulted in a compensatory increase of H1C to H1E. Although this had no effect on the growth and development of the plant, it did lead to disturbances in male gametogenesis and eventually to a male sterility phenotype. The underlying cause of this phenotype could be the observed aberrant meiotic divisions of the male gamete mother cells, although specific effects on the transcription profile during male gametogenesis cannot be excluded [152].

6. Conclusions and perspectives

There are several generalizations and suggestions concerning linker histones and their biological functions that can be made based on the discussed experimental results. The conserved "winged-helix" structure and characteristic binding to nucleosomes by the globular domain of canonical H1, reminiscent of the structure and DNA recognition properties of winged-helix transcription factors, suggest a decisive role for GH1 in determining the effects of H1 on chromatin. The contribution of GH1 could be due to its role in mediating the polarity and cooperativity of H1 binding and its preference for specific sequences or methylated DNA [153]. A recent demonstration of the direct interaction between H1 and chromodomain-containing non-histone chromatin proteins HP1 and Polycomb [154] suggests that linker histones can facilitate and/or mediate the functions of these proteins. The role of H1's strongly charged tails can be compared to that of divalent cations, except that neutralization of the DNA phosphates by these tails is more directed due to a specific localization of H1 in the nucleosome and can therefore influence the geometry of linker DNA in a more regular way. Given the intrinsic property of filaments of spaced nucleosome core particles to fold upon DNA charge neutralization, the overall role of H1 in generating higher-order chromatin structures is most consistent with that of a selector and stabilizer of a favorable folding pathway. It is however important to note that in some studies, for example in the experiments with H1 overexpression in *Xenopus* and in mammalian cell lines, the observed effects of H1 only required the presence of the GH1 domain.

Based on the available data there is no ground for the assumption that linker histones have a single essential function that is universal to all Eukaryotes. The absence of phenotypic effects caused by the elimination of H1 in *Tetrahymena* and some fungi (yeast and *Aspergillus*) fits better into a scenario where the gradual acquisition of functions by linker histones occurs in parallel with the appearance of novel organismal strategies, such as multicellularity and the need for efficient mechanisms of epigenetic inheritance. While it is possible that the evolution of novel H1 functions is reflected by structural changes in the globular domain, one has to take into account that H1's role was probably also shaped by the evolution of other proteins which could use its unique position in nucleosomes to exert influence on the chromatin state. In any case, it is risky to extrapolate from conclusions based on the data obtained for *Tetrahymena* H1-like proteins which lack the globular domain or for the H1 of rather primitively differentiated fungi to the situation in highly complex multicellular plants and animals.

As regards the H1 structure in organisms with tripartite linker histone, the main division line is between plants and animals with an insert extending the length of the "wing" subdomain of GH1 occurring in the latter. Although there are no data yet on how the length of the wing affects DNA binding or other interactions of linker histones, this is by far the most distinctive difference detected by phylogenetic analysis of known linker histone globular domain sequences. Given the critical role of GH1 in the *in vivo* effects of H1, it is tempting to speculate that such a major difference in a region probably defining the principal DNA binding site can have profound and general consequences for chromatin structure. For example, it may influence the strength and stability of the repression resulting from additive action of H1 and some non-histone proteins. The validity of this suggestion is open for experimental testing.

Only in two studied cases involving complex multicellular Eukaryotes (*C. elegans* and tobacco) have changes in native proportions of H1 variants resulted in clear phenotypic aberrations, both concerning the development of gametes. In animals and plants the elimination of any particular H1 variant is not lethal and does not affect the somatic growth and development. In mice such changes do not affect the germline development, either. However, in all studied plants and animals there is an efficient compensation mechanism for the eliminated variant by remaining variants, indicating the importance of the overall stoichiometry of H1-to-nucleosome. Although, one cannot completely rule out the possibility that certain other non-H1 proteins could, at least over a short-range, compensate for the lack of H1, the above results strongly suggest that a considerable lowering of the H1-to-nucleosome ratio will prove lethal in both animals and plants. However, given the distinct character and evolutionary stability of the main H1 variants in animals and plants, the occurrence of the variant-specific *cis*-regulatory elements upstream of their respective genes and the characteristic profiles of their expression during development, how can one explain the lack of any immediate consequences of the elimination of any one of them for the somatic growth and development of a complex organism? The answer may lie in a pleiotropy and partial redundancy of the variants. Pleiotropy means that each variant is involved in more than one specific function. The widespread occurrence of the pleiotropy of genes has been recently shown in a convincing way for yeast [155]. Partial redundancy implies that between any two variants only some of the functions are overlapping while others are unique. Each variant performs distinct non-essential functions which together add up to a common essential function. Such overlapping functions have been recently shown for the yeast oxysterol-binding protein homologues [156]. Theoretical considerations show that a system consisting of pleiotropic and partially redundant genes will be evolutionarily stable [157]. Such a system can be extremely tolerant with respect to fluctuations in the variant proportions, especially in the short term (during few generations) and under laboratory conditions. On the other hand, the limits of tolerance can be set at different levels in different organisms as documented for *C. elegans* and tobacco.

The spectrum of pleiotropic functions of different H1 variants in different taxa has yet to be established. With respect to interactions within chromatin, the elucidation of the role of linker histones in transmitting or enhancing the effects of other non-histone proteins will be of particular interest. Given the dynamic behavior of H1 in the nucleus, the possibility of auxiliary extrachromosomal functions performed by the free H1 pool should be explored. An example of such a function is the recent finding that free H1 is involved in the import of adenovirus DNA into the nucleus [158]. Other potential functions, especially at the period of maximum chromatin condensation, could depend on the gradient of free H1 around the chromosomes.

Acknowledgements

I thank M. H. Parseghian, U. Grossbach, E. Schulze, J. Gittins, S. Kaczanowski, and M. R. Przewloka for a critical reading and comments on the manuscript and S. Świezewski for the technical help. This work was supported by a Howard Hughes Medical Institute grant No. 55000312, a KBN grant No. 6PO4A 00320, and a Foundation for Polish Science grant No. 2/2000. The author is supported by Center of Excellence in Molecular Biotechnology.

References

1. Cole, R.D. (1987) Int. J. Pept. Protein Res. 30, 433–449.
2. Van Holde, K.E. (1989) Chromatin. Springer-Verlag, New York, pp. 69–180.
3. Arents, G. and Moudrianakis, E.N. (1995) Proc. Natl. Acad. Sci. USA 92, 11170–11174.
4. Hartman, P.G., Chapman, G.E., Moss, T., and Bradbury, E.M. (1977) Eur. J. Biochem. 77, 45–51.
5. Ramakrishnan, V., Fich, J.T., Graziano, V., Lee, P.L., and Sweet, R.M. (1993) Nature 362, 219–223.
6. Cerf, C., Lippens, G., Ramakrishnan, V., Muyldermans, S., Segers, A., Wyns, L., Wodak, S.J., and Hallenga, K. (1994) Biochemistry 33, 11079–11086.
7. Kasinsky, H.E., Lewis, J.D., Dacks, J.B., and Ausio, J. (2001) FASEB J. 15, 34–42.
8. Zlatanova, J. and van Holde, K. (1998) FASEB J. 12, 421–431.
9. Noll, M. and Kornberg, R.D. (1977) J. Mol. Biol. 109, 393–404.

10. Allan, J., Hartman, P.G., Crane-Robinson, C., and Aviles, F.X. (1980) Nature 288, 675–679.
11. Schultz, S.C., Shields, G.C., and Steitz, T.A. (1991) Science 253, 1001–1007.
12. Clark, K.L., Halay, E.D., Lai, E., and Burley, S.K. (1993) Nature 364, 412–420.
13. Draves, P.H., Lowary, P.T., and Widom, J. (1992) J. Mol. Biol. 225, 1105–1121.
14. Thomas, J.O., Rees, C., and Finch, J.T. (1992) Nucleic Acids Res. 20, 187–194.
15. Thomas, J.O. and Wilson, C.M. (1986) EMBO J. 5, 3531–3537.
16. Buckle, R.S., Maman, J.D., and Allan, J. (1992) J. Mol. Biol. 223, 651–659.
17. Goytisolo, F.A., Gerchman, S.-E., Yu, X., Rees, C., Graziano, V., Ramakrishnan, V., and Thomas, J.O. (1996) EMBO J. 15, 3421–3429.
18. Zhou, Y.-B., Gerchman, S.E., Ramakrishnan, V., Travers, A., and Muyldermans, G. (1998) Nature 395, 402–405.
19. An, W., Leuba, S.H., van Holde, K., and Zlatanova, J. (1998) Proc. Natl. Acad. Sci. USA 95, 3396–3401.
20. Cirillo, L.A., McPherson, C.E., Bossard, P., Stevens, K., Cherian, S., Shim, E.Y., Clark, K.L., and Zaret, K.S. (1998) EMBO J. 17, 244–254.
21. Wong, J.M., Li, Q, Levi, B.Z., Shi, Y.B., and Wolffe, A.P. (1997) EMBO J. 16, 7130–7145.
22. Hamiche, A., Schultz, P., Ramakrishnan, V., Oudet, P., and Prunell, A. (1996) J. Mol. Biol. 257, 30–42.
23. Tomaszewski, R. and Jerzmanowski, A. (1997) Nucleic Acids Res. 25, 458–465.
24. Pruss, D., Bartholomew, B., Persinger, J., Hayes, J., Arents, G., Moudrianakis, E.N., and Wolffe, A.P. (1996) Science 274, 614–617.
25. Panetta, G., Buttinelli, M., Flaus, A., Richmond, T.J., and Rhodes, D. (1998) J. Mol. Biol. 282, 683–697.
26. Gajiwala, K.S., Chen, H., Cornille, F., Roques, B.P., Reith, W., Mach, B., and Burley, S.K. (2000) Nature 403, 916–921.
27. Wolberg, C. and Campbell, R. (2000) Nature Struct. Biol. 7, 261–262.
28. van Holde, K. and Zlatanova, J. (1995) J. Biol. Chem. 270, 8373–8376.
29. Ramakrishnan, V. (1997) Crit. Rev. Eukaryot. Gene Expr. 7, 215–230.
30. Thoma, F., Koller, T., and Klug, A. (1979) J. Cell Biol. 83, 403–427.
31. McGhee, J.D., Nickol, J.M., Felsenfeld, G., and Rau, D.C. (1983) Cell 33, 831–841.
32. Woodcock, C.L., Frado, L.L., and Rattner, J.B. (1984) J. Cell Biol. 99, 42–52.
33. Leuba, S.H., Yang, G., Robert, C., Samori, B., van Holde, K., Zlatanova, J., and Bustamante, C. (1994) Proc. Natl. Acad. Sci. USA 91, 11621–11625.
34. Yang, G., Leuba, S.H., Bustamante, C., Zlatanova, J., and van Holde, K. (1994) Nature Struct. Biol. 1, 761–763.
35. Garcia-Ramirez, M., Dong, F., and Ausio, J. (1992) J. Biol. Chem. 267, 19587–19595.
36. Luger, K., Mäder, A.W., Richmond, R.K., Sargent, D.F., and Richmond, T.J. (1997) Nature 389, 251–260.
37. Allan, J., Mitchell, T., Harborne, N., Böhm, L., and Crane-Robinson, C. (1986) J. Mol. Biol. 187, 591–601.
38. Clark, D.J. and Thomas, J.O. (1986) J. Mol. Biol. 187, 569–580.
39. Bednar, J., Horovitz, R.A., Grigoryev, S.A., Carruthers, L.M., Hansen, J.C., Koster, A.J., and Woodcock, C.L. (1998) Proc. Natl. Acad. Sci. USA 95, 14173–14178.
40. Zlatanova, J., Leuba, S.H., and van Holde, K. (1999) Critic. Rev. Eukaryot. Gene Expr. 9, 245–255.
41. Graziano, V., Gerchman, S.E., Schneider, D.K., and Ramakrishnan, V. (1994) Nature 368, 351–354.
42. Thoma, F. and Koller, T. (1977) Cell 12, 101–107.
43. Zlatanova, J., Leuba, S.H., and van Holde, K. (1998) Biophys. J. 74, 2554–2566.
44. Clark, D.J. and Kimura, T. (1990) J. Mol. Biol. 211, 883–896.
45. Hansen, J.C., Ausio, J., Stanik, V.H., and van Holde, K.E. (1989) Biochemistry 28, 9129–9136.
46. Yao, J., Lowary, P.T., and Widom, J. (1991) Biochemistry 30, 8408–8414.
47. Butler, P.J.G. and Thomas, J.O. (1998) 281, 401–407.
48. Carpenter, B.G., Baldwin, J.P., Bradbury, E.M., and Ibel, K. (1976) Nucleic Acids Res. 3, 1739–1746.

49. Schwarz, P.M. and Hansen, J.C. (1994) J. Biol. Chem. 269, 16284–16289.
50. Fletcher, T.M. and Hansen, J.C. (1995) J. Biol. Chem. 270, 25359–25362.
51. Carruthers, L.M. and Hansen, J.C. (2000) J. Biol. Chem. 275, 37285–37290.
52. Schwarz, P.M., Felthauser, A., Fletcher, T.M., and Hansen, J.F. (1996) Biochemistry 35, 4009–4015.
53. Caron, F. and Thomas, J.O. (1981) J. Mol. Biol. 146, 513–537.
54. Appels, R., Bolund, L., and Ringertz, N.R. (1974) J. Mol. Biol. 87, 339–355.
55. Appels, R., Bolund, L., Goto, S., and Ringertz, N.R. (1974) Exp. Cell Res. 85, 182–190.
56. Wu, L.H., Kuehl, I., and Rechsteiner, M. (1986) J. Biol. Chem. 103, 465–474.
57. Misteli, T., Gunjan, A., Hock, R., Bustin, M., and Brown, D.T. (2000) Nature 408, 877–881.
58. Lever, M.A., Th'ng, P.H., Sun, X., and Hendzel, M.J. (2000) Nature 408, 873–876.
59. Misteli, T. (2001) Science 291, 843–847.
60. Heun, P., Taddei, A., and Gasser, S.M. (2001) Trends Cell Biol. 11, 519–525.
61. Bleher, R. and Martin, R. (1999) Chromosoma 108, 308–316.
62. Lennox, R.W. and Cohen, L.H. (1988) In: Adolph, K.W. (ed.) Chromosomes and Chromatin. CRC Press, Boca Raton, FL, pp. 33–56.
63. Roth, S.Y. and Allis, C.D. (1992) Trends Biochem. Sci. 17, 93–98.
64. Jerzmanowski, A. and Cole, R.D. (1992) J. Biol. Chem. 267, 8514–8520.
65. Nightingale, K., Dimitrov, S., Reevs, R., and Wolffe, A.P. (1996) EMBO J. 15, 548–561.
66. Dimitrov, S., Almouzni, G., Dasso, M.C., and Wolffe, A.P. (1993) Dev. Biol. 160, 214–227.
67. Dimitrov, S., Dasso, M.C., and Wolffe, A.P. (1994) J. Cell Biol. 126, 591–601.
68. Ner, S.S. and Travers, A.A. (1994) EMBO J. 13, 1817–1822.
69. Falciola, L., Spada, F., Calogero, S., Langst, G., Voit, R., Grummt, I., and Bianchi, M.E. (1997) J. Cell Biol. 137, 19–26.
70. Strick, R., Strissel, P.L., Gavrilov, K., and Levi-Setti, R. (2001) J. Cell Biol. 155, 899–910.
71. Ohsumi, K., Katagiri, C., and Kishimoto, T. (1993) Science 262, 2033–2035.
72. Dou, Y., Mizzen, C.A., Abrams, M., Allis, C.D., and Gorovsky, M.A. (1999) Mol. Cell 4, 641–647.
73. Toro, G.C. and Galanti, N. (1988) Exp. Cell Res. 174, 16–24.
74. Burri, M., Schlimme, W., Betschart, B., Kampfer, U., Schaller, J., and Hecker, H.C. (1993) Parasitol. Res. 79, 649–659.
75. Hayashi, T., Hayashi, H., and Iwai, K. (1987) J. Biochem. 102, 369–376.
76. Wu, M., Allis, C.D., Sweet, M.T., Cook, R.G., Thatcher, T.H., and Gorovsky, M.A. (1994) Mol. Cell Biol. 14, 10–20.
77. Kaczanowski, S. and Jerzmanowski, A. (2001) J. Mol. Evol. 53, 19–30.
78. Lindauer, A., Muller, K., and Schmitt, R. (1993) Gene 129, 59–68.
79. Schulze, E. and Schulze, B. (1995) J. Mol. Evol. 41, 833–840.
80. Jerzmanowski, A., Przewloka, M., and Grasser, K.D. (2000) Plant Biol. 2, 586–597.
81. Wiśniewski, J.R. and Grossbach, U. (1996) Int. J. Dev. Biol. 40, 177–187.
82. Parseghian, M.H. and Hamkalo, B.A. (2001) Biochem. Cell Biol. 79, 289–304.
83. Parseghian, M.H., Newcomb, R.L., and Hamkalo, B.A. (2001) J. Cell. Biochem. 83, 643–659.
84. Ponte, I., Vidal-Taboada, J.M., and Suau, P. (1998) Mol. Biol. Evol. 15, 702–708.
85. Smith, R.C., Dworkin-Rastl, E., and Dworkin, M.B. (1988) Genes Dev. 2, 1284–1295.
86. Ohsumi, K. and Katagiri, C. (1991) Dev. Biol. 147, 110–120.
87. Cho, H. and Wolffe, A.P. (1994) Gene 143, 233–238.
88. Mandl, B., Brandt, W.F., Superti-Furga, G., Graninger, P.G., Birnstiel, M.L., and Busslinger, M. (1997) Mol. Cell Biol. 17, 1189–1200.
89. Tanaka, M., Hennebold, J.D., Macfarlane, J., and Adashi, E.Y. (2001) Development 128, 655–664.
90. Moss, B.A. (1974) Biochim. Biophys. Acta 395, 205–209.
91. Smith, B.J., Walker, J.M., and Johns, E.W. (1980) FEBS Lett. 112, 42–44.
92. Doenecke, D. and Tonjes, R. (1986) J. Mol. Biol. 187, 461–464.
93. Seyedin, S.M., Cole, R.D., and Kistler, W.S. (1981) Exp. Cell Res. 136, 399–405.
94. Doenecke, D., Drabent, B., Bode, C., Bramlage, B., Franke, K., Gavenis, K., Kosciessa, U., and Witt, O. (1997) Adv. Exp. Med. Biol. 424, 37–48.
95. Franke, K., Drabent, B., and Doenecke, D. (1998) Biochim. Biophys. Acta 1398, 232–242.

96. Mohr, E., Trieschmann, L., and Grossbach, U. (1989) Proc. Natl. Acad. Sci. USA 86, 9308–9312.
97. Schulze, E., Trieschmann, L., Schulze, B., Schmidt, E.R., Pitzel, S.A., Zechel, K., and Grossbach, U. (1993) Proc. Natl. Acad. Sci. USA 90, 2481–2485.
98. Parseghian, M.H., Newcomb, R.L., Winokur, S.T., and Hamkalo, B.A. (2000) Chromosome Res. 8, 405–424.
99. Ascenzi, R. and Gantt, J.S. (1999) Chromosoma 108, 345–355.
100. Khochbin, S. (2001) Gene 271, 1–12.
101. Trieschmann, L., Schulze, E., Schulze, B., and Grossbach, U. (1997) Eur. J. Biochem. 250, 184–196.
102. Schulze, E., Nagel, S., Gavenis, K., and Grossbach, U. (1994) J. Cell Biol. 127, 1789–1798.
103. Knowles, J.A. and Childs, G.J. (1986) Nucleic Acids Res. 14, 8121–8133.
104. Knowles, J.A., Lai, Z.C., and Childs, G.J. (1987) Mol. Cell Biol. 7, 478–485.
105. Lieber, T., Angerer, L.M., Angerer, R.C., and Childs, G.J. (1988) Proc. Natl. Acad. Sci. USA 85, 4123–4127.
106. Moorman, A.F., de Boer, P.A., Charles, R., and Lamers, W.H. (1987) Differentiation 35, 100–107.
107. Khochbin, S. and Wolffe, A.P. (1993) Gene 128, 173–180.
108. Grunwald, D., Lawrence, J.J., and Khochbin, S. (1995) Exp. Cell Res. 218, 586–595.
109. Duncliffe, K.N., Rondahl, M.E., and Wells, J.R. (1995) Gene 163, 227–232.
110. Meergans, T., Albig, W., and Doenecke, D. (1998) Eur. J. Biochem. 256, 436–446.
111. Zlatanova, J. and Doenecke, D. (1994) FASEB J. 8, 1260–1268.
112. Clare, S.E., Fantz, D.A., Kistler, W.S., and Kistler, M.K. (1997) J. Biol. Chem. 272, 33028–33036.
113. Wolfe, S.A., Mottram, P.J., vanWert, J.M., and Grimes, S.R. (1999) Biol. Reprod. 61, 1005–1011.
114. Birnstiel, M.L., Busslinger, M., and Strub, K. (1985) Cell 41, 349–359.
115. Hecht, A., Laroche, T., Strahl-Boilsinger, S., Gasser, S.M., and Grunstein, M. (1995) Cell 80, 583–592.
116. Shen, X., Yu, L., Weir, J.W., and Gorovsky, M.A. (1995) Cell 82, 47–56.
117. Ushinsky, S.C., Bussey, H., Ahmed, A.A., Wang, Y., Friesen, J., Williams, B.A., and Storms, R.K. (1997) Yeast 13, 151–161.
118. Patterton, H.G., Landel, C.C., Landsman, D., Peterson, C.L., and Simpson, R.T. (1998) J. Biol. Chem. 273, 7268–7276.
119. Escher, D. and Schaffner, W. (1997) Mol. Gen. Genetics 256, 456–461.
120. Hellauer, K., Sirard, E., and Turcotte, B. (2001) J. Biol. Chem. 276, 13587–13592.
121. Ramon, A., Muro-Pastor, M.I., Scazzocchio, C., and Gonzalez, R. (2000) Mol. Microbiol. 35, 223–233.
122. Barra, J.L., Rhounim, L., Rossignol, J.L., and Faugeron, G. (2000) Mol. Cell. Biol. 20, 61–69.
123. Felden, R.A., Sanders, M.M., and Morris, R. (1976) J. Cell Biol. 68, 430–439.
124. Karrer, K.M. and VanNuland, T.A. (1999) J. Biol. Chem. 274, 33020–33024.
125. Shen, X. and Gorovsky, M.A. (1996) Cell 86, 475–483.
126. Freidkin, I. and Katcoff, D.J. (2001) Nucleic Acids Res. 29, 4043–4051.
127. Takami, Y. and Nakayama, T. (1997) Genes Cells 2, 711–723.
128. Takami, Y. and Nakayama, T. (1997) Biochim. Biophys. Acta 1354, 105–115.
129. Takami, Y., Nishi, R., and Nakayama, T. (2000) Biochem. Biophys. Rec. Commun. 268, 501–508.
130. Brown, D.T., Alexander, B.T., and Sittman, D.B. (1996) Nucleic Acids Res. 24, 486–493.
131. Brown, D.T., Gunjan, A., Alexander, B.T., and Sittman, D.B. (1997) Nucleic Acids Res. 25, 5003–5009.
132. Brown, D.T. (2001) Genome Biol. reviews 0006.1–0006.6.
133. Schlissel, M.S. and Brown, D.D. (1984) Cell 37, 903–911.
134. Dworkin-Rastl, E., Kandolf, H., and Smith, R.C. (1994) Dev. Biol. 161, 423–439.
135. Bouvet, p., Dimitrov, S., and Wolffe, A.P. (1994) Genes Dev. 8, 1147–1159.
136. Kandolf, H. (1994) Proc. Natl. Acad. Sci USA 91, 7257–7261.
137. Jerzmanowski, A. and Cole, R.D. (1990) J. Biol. Chem. 265, 10726–10732.
138. Tomaszewski, R., Mogielnicka, E., and Jerzmanowski, A. (1998) Nucleic Acids Res. 26, 5596–5601.
139. Sera, T. and Wolffe, A.P. (1998) Mol. Cell. Biol. 18, 3668–3680.
140. Howe, L., Itoh, T., Katagiri, C., and Ausio, J. (1998) Biochemistry 37, 7077–7082.

141. Crane-Robinson, C. (1999) BioEssays 21, 367–371.
142. Vermaak, D., Steinbach, O.C., Dimitrov, S., Rupp, R.A.W., and Wolffe, A.P. (1998) Curr. Biol. 8, 533–536.
143. Bonnefoy, E., Bandu, M.-T., and Doly, J. (1999) Mol. Cell. Biol. 19, 2803–2816.
144. Steinbach, O.C., Wolffe, A.P., and Rupp, R.A. (1997) Nature 389, 395–399.
145. Fantz, D.A., Hatfield, W.R., Horvath, G., Kistler, M.K., and Kistler, W.S. (2001) Biol. Reprod. 64, 425–431.
146. Drabent, B., Saftig, P., Bode, C., and Doenecke, D. (2000) Histochem. Cell. Biol. 113, 433–442.
147. Lin, Q., Sirotkin, A., and Skoultchi, A.I. (2000) Mol. Cell. Biol. 20, 2122–2128.
148. Rabini, S., Franke, K., Saftig, P., Bode, C., Doenecke, D., and Drabent, B. (2000) Exp. Cell Res. 255, 114–124.
149. Sirotkin, A.M., Edelmann, W., Cheng, G., Klein-Szanto, A., Kucherlapati, R., and Skoultchi, A.I. (1995) Proc. Natl. Acad. Sci. USA 92, 6434–6438.
150. Fan, Y., Sirotkin, A., Russell, R.G., Ayala, J., and Skoultchi, A.I. (2001) Mol. Cell. Biol. 21, 7933–7943.
151. Jedrusik, M.A. and Schulze, E. (2001) Development 128, 1069–1080.
152. Prymakowska-Bosak, M., Przewloka, M.R., Slusarczyk, J., Kuras, M., Lichota, J., Kilianczyk, B., and Jerzmanowski, A. (1999) Plant Cell 11, 2317–2329.
153. Karymov, M.A., Tomschik, M., Leuba, S.H., Caiafa, P., and Zlatanova, J. (2001) FASEB J. 15, 2631–2641.
154. Nielsen, A.L., Oulad-Abdelghani, M., Ortiz, J.A., Remboutsika, E., Chambon, P., and Losson, R. (2001) Mol. Cell 7, 729–739.
155. Gavin, A.-C., Bosche, M., Krause, R., Grandi, P., Marzioch, M., Bauer, A., Schultz, J., Rick, J.M. et al. (2002) Nature 415, 141–147.
156. Beh, C.T., Cool, L., Phillips, J., and Rine, J. (2001) Genetics 157, 1117–1140.
157. Nowak, M.A., Boerlijst, M.C., Cooke, J., and Maynard Smith, J. (1997) Nature 388, 167–171.
158. Trotman, L.C., Mosberger, N., Fornerod, M., Stidwill, R.P., and Greber, U.F. (2001) Nature Cell Biol. 3, 1092–1100.

J. Zlatanova and S.H. Leuba (Eds.) *Chromatin Structure and Dynamics: State-of-the-Art*

DOI: 10.1016/S0167-7306(03)39005-2

CHAPTER 5

Chromosomal HMG-box proteins

Andrew A. Travers[1] and Jean O. Thomas[2]

[1]*MRC Laboratory of Molecular Biology, Hills Road, Cambridge CB2 2QH, UK. Tel.: +44-1223-402419; Fax: +44-1223-412142; E-mail: aat@mrc-lmb.cam.ac.uk*

[2]*Cambridge Centre for Molecular Recognition & Department of Biochemistry, 80 Tennis Court Road, Cambridge CB2 1GA, UK. Tel.: +44-1223-333670; Fax: +44-1223-333743 (or 766002); E-mail: jot1@bioc.cam.ac.uk*

1. Introduction

The HMG-box (HMGB) chromosomal proteins are one of three classes of abundant chromatin associated proteins–the High Mobility Group proteins [1]. HMGB proteins are characterised by the HMG-box, a DNA-binding domain specific to eukaryotes. This domain binds preferentially to a variety of distorted DNA structures, including negatively supercoiled DNA, small DNA circles, cruciforms, DNA bulges and cisplatin modified DNA [2–9]. The HMGB proteins appear to act primarily as architectural facilitators in the manipulation of nucleoprotein complexes, for example, in the assembly of complexes involved in recombination and the initiation of transcription, as well as in the assembly and organisation of chromatin (for recent reviews see Refs. [10–14]).

The HMG-box domain is found in several related types of protein [15], all of which are predominantly nuclear. The archetypal HMGB proteins are highly abundant (~10–20 copies per nucleosome in the mammalian nucleus [16]) and often occur as two major species, HMGB1 and HMGB2, originally termed HMG1 and HMG2, in vertebrates [17] (for recent nomenclature changes see Ref. [18]). The two distinguishing features of these highly homologous proteins are to similar, but distinct, tandem HMG-box domains (A and B), and a long acidic C-terminal 'tail', consisting of ~30 (HMGB1) or 20 (HMGB2) acidic (aspartic and glutamic acid) residues, linked to the boxes by a short, predominantly basic linker. Putative counterparts of HMGB1 and HMGB2, Hmo1p and Hmo2p, also with two boxes but lacking a distinct acidic tail, are found in *Saccharomyces cerevisiae* [19]. However the most abundant HMG-box proteins in *S. cerevisiae*, Nhp6ap, and Nhp6bp (non-histone proteins 6A and 6B, respectively), contain only a single HMG box, and again lack an acidic tail (Fig. 1a). Likewise the two major HMG-box proteins in *Drosophila melanogaster*, HMG-D and HMG-Z, have only a single HMG box but, unlike the yeast proteins, contain a short C-terminal acidic tail in addition to a basic region. These abundant proteins in yeast and *Drosophila* may be the general functional counterparts of HMGB1 and 2 in vertebrates. Characterised single HMG-box proteins from plants contain basic extensions N-terminal, and sometimes C-terminal, to the HMG box, and (often relatively short) C-terminal

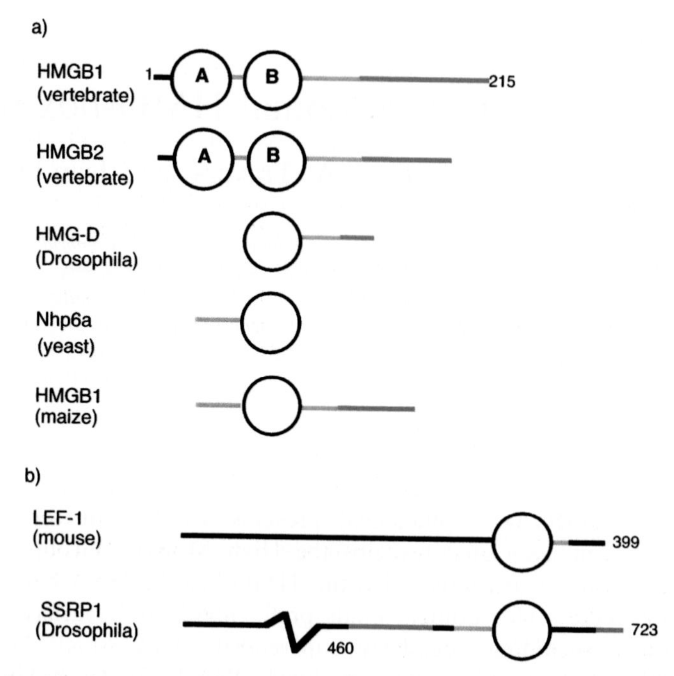

Fig. 1. HMG boxes in the abundant chromosomal HMGB proteins and in other DNA-binding proteins. (a) The "archetypal" vertebrate HMGB1 and HGMB2 and abundant *single* HMG-box proteins in other organisms that are presumed to be the functional counterparts of HMGB1 and 2. HMG-D is from *Drosophila*, Nhp6a is from yeast (*Saccharomyces cerevisiae*) and the plant protein, HMGB1, is from maize; the family of plant proteins contains a basic N-terminal extension and acidic tails of variable length [20]: the latter are separated from the HMG box by a region of variable length and positive charge. (b) Organisation of LEF-1, a typical sequence-specific HMG-box transcription factor, and of *Drosophila* SSRP1. Circles: HMG boxes; blue lines: basic regions; red lines: acidic tails; black lines: other amino acid sequence. The lengths of the lines are roughly proportional to the length of amino acid sequence in (a), but are not to scale in (b). Adapted from Ref. [13] with permission.

acidic tails [20]. Single HMG-box domains are also found in a number of disparate proteins (e.g., SSRP1, BAP111, and BAF57) associated with chromatin remodeling complexes (see Section 3.3), and some of them also contain acidic regions. In addition, single HMG boxes, with no accompanying acidic tail, occur in a class of sequence-specific transcription factors, such as SRY (sex reversal on the Y chromosome) and LEF-1 (lymphocyte enhancer factor 1).

2. *Structure and DNA binding*

2.1. *The HMG-box domain*

The global fold of the HMG box is well conserved and consists of three α-helices arranged in an L-shape (Fig. 2) [21–28]. The structures of the A and B domain

HMG boxes of HMGB1 are generally very similar [13,14,27]. At a detailed level, the B domain is highly similar to the HMG-box domains of *S. cerevisiae* Nhp6ap and *Drosophila* HMG-D [23,28] and rather less similar to the sequence-specific SRY and LEF-1 HMG-boxes [24,25], whereas the A domain differs significantly in the relative disposition of helices I and II and in the trajectory of the helix I–II loop [27] (Fig. 2c). These structural differences are accompanied by functional differences, at least *in vitro*. Both the A and B domains of HMGB1 bind in the minor groove of DNA. However the A domain has a much higher preference for distorted DNA structures such as cruciforms, four-way junctions, and also cisplatin-modified DNA [29–31] (see below). In contrast, the B domain binds less selectively to distorted DNA structures but on binding linear DNA both the B domain and the related B-type domains [32] untwist the double helix and introduce an approximately right-angled bend [32–34] (whereas the A domain does not bend DNA as effectively [29,35]). In addition, HMG-D, and presumably other HMGB proteins, can bind double-stranded RNA *in vitro* with high affinity [36] but there is no evidence for such a role *in vivo*.

The DNA bend induced by non-sequence-specific B-type domains is substantial, different determinations, using a variety of methods, yielding values varying from $\sim 80^\circ$ for the B domain of HMGB1 [34] to $\sim 111^\circ$ in the crystal structure of HMG-D bound to a DNA decamer [32]. However, comparison of the bends induced by different HMG-box domains, using the same method, indicates that the bend angle is conserved [33,37], suggesting that the reported range of bend angles is at least partly due to the particular method used. Surprisingly high values of ~ 130–150° have been reported for the DNA bend induced by insect HMGB proteins [37,38]; it seems likely that more than one HMG domain might be bound to the DNA in this case.

2.2. *DNA binding*

2.2.1. *Structural basis for DNA binding and bending*

How do the B-type HMG-box domains bend DNA by over 90° within a single turn of the DNA duplex? In B-type domain complexes, as exemplified by the sequence-specific HMG boxes (e.g., SRY and LEF-1), the DNA-binding face of the domain presents a hydrophobic surface that conforms to a wide, shallow minor groove. In the center of this surface, a hydrophobic wedge, usually consisting of four spatially close residues, is inserted deep into the minor groove. One of the residues (residue 15, numbering normalised to the A domain of HMGB1 (Fig. 2)) located towards the N-terminal end of helix I partially intercalates between two base pairs [24,25,39]. This residue (corresponding to residue 16 in the A domain of HMGB1 (Fig. 2a)) is methionine in LEF-1 [25], Nhp6a [28], and HMG-D [32], and isoleucine in SRY [24,39]; the inferred residue in the B domain of HMGB1, for which there are no structural data (and no detailed model) for the protein–DNA complex, is phenylalanine. This partial intercalation introduces a kink into the bound DNA, enhancing the more uniform bend associated with the widening of the

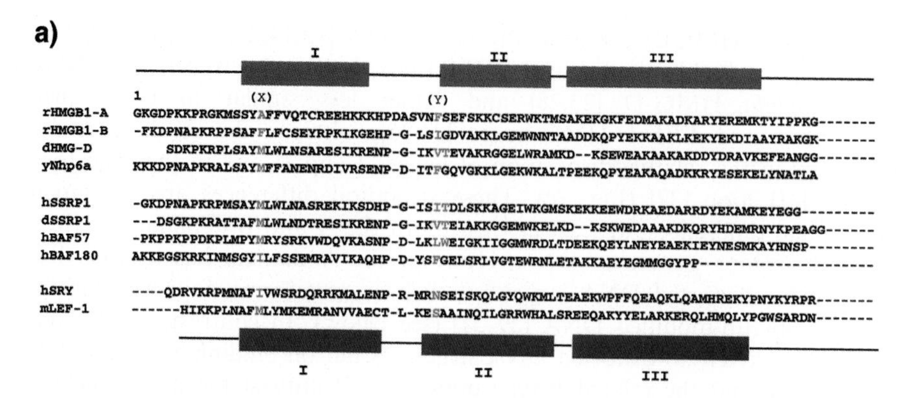

Fig. 2. Differences in intercalating residues and folds between HMG boxes. (a) Residues at the two "intercalating positions" in sequence-specific and non-sequence-specific HMG boxes. Location in the amino acid sequence (residues marked in red, indicated as X and Y) and with respect to the α-helices determined by NMR spectroscopy for the non-sequence-specific A domain [27] (shown in green above the sequence) and the sequence-specific LEF-1 HMG box [25] (shown in blue below the sequence); the exact positions of the α-helices differ slightly in the other members of the set shown. Note that in HMG-D both residue Y (Val) and the following amino acid (Thr) intercalate in the same base step. (b) Location of residues at X and Y in the 3D structures of the HMGB1 A domain (single intercalating residue at position Y only), HMGB1 B domain (inferred intercalations at X and Y) and the HMG-box domain of HMG-D (one intercalating residue at X and two adjacent residues at Y). (c) Solution structures of the A and B HMG boxes of HMGB1 and the HMG box of HMG-D, determined by NMR spectroscopy [21,23,27]. The structures are oriented to show the differences in the relative dispositions of helices I and II between the A type HMG-box (A domain from HMGB1) and the B type HMG-box (B domain from HMGB1 and HMG box from HMG-D). In the A-type box, helix I is essentially straight whereas in the B-type box it is bent, and the loop between helices I and II is longer in the A-type box. The panel is adapted from Fig. 4 in Ref. [27]. The figure is adapted from Ref. [13] with permission.

minor groove. In the A domain of HMGB1 (discussed further below) the residue corresponding to the intercalating residue is alanine, whose methyl side-chain is too small to intercalate in isolation (Table 1). Other residues within the hydrophobic wedge can also contribute to the overall bend angle. For example a naturally occurring mutation, changing Met11 (numbering corresponding to that for the A domain in Fig. 2a) of the SRY HMG box to isoleucine, reduces the observed DNA bend angle by 13° [40].

In addition to the kink induced by the "primary" intercalating residue, a second kink two base steps away from the primary kink is revealed in the crystal structure of a complex of the HMG-D box with linear DNA [32]. The second kink arises from partial "secondary" intercalation in the minor groove of two adjacent residues, valine and threonine, immediately before the N-terminal end of helix II [32,42]. In the HMG boxes of HMGB1 and 2, and other non-sequence-specific HMG proteins, a hydrophobic (and therefore potentially intercalating) residue is almost always found in the position corresponding to the valine in HMG-D [32,42] (residue Y in Table 1 and Fig. 2). In contrast, in the sequence-specific HMG

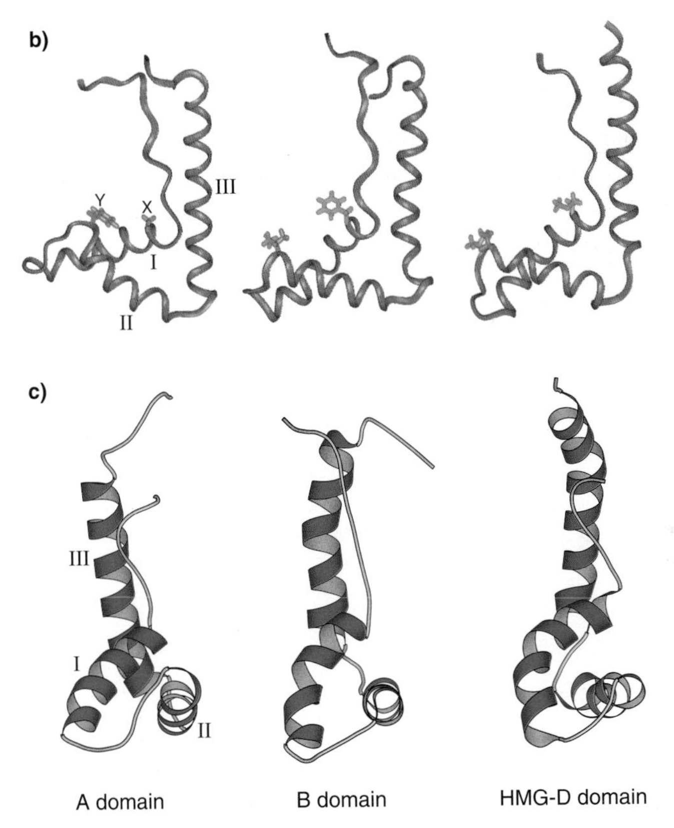

Fig. 2. Continued.

domains the residue at the potential secondary site (Y) is polar. Mutation of the secondary intercalating residues (Phe in Nhp6ap, and Val and Thr individually in HMG-D) substantially reduces both the DNA affinity and the bending capability of the proteins, as judged by the resulting decrease in circularisation efficiency [28,42]. However, the context of these residues is also important since a Val37Phe substitution in HMG-D also results in a reduction in circularisation efficiency [42].

2.2.2. *Binding to distorted DNA structures*

The basis of the preference of the A domain, relative to the B domain, of HMGB1 for binding to distorted DNA structures has become apparent from numerous structural and biophysical studies. In contrast to the B domain, the A domain has only the secondary potentially intercalating residue, namely Phe at position Y (Table 1 and Fig. 2). This presumably accounts for the smaller bend angle in the A domain/cisplatin-modified DNA complex (61°) than in the B domain complex (80–95°) [34,43]. As yet, there is no structure of a complex between the A domain and linear DNA. However, in the crystal structure of the A domain complexed with cisplatin-modified DNA the domain binds to one side of the *cis*-platinum adduct,

Table 1
DNA bending and intercalation by HMGB domains

HMG box	DNA	Reference	Res. X	Intercalation?	Res. Y	Intercalation?	Bend angle
SRY	linear 14-mer	40	Ile	yes	Asn	no	54° (NMR)
LEF-1	linear 12-mer	25	Met	yes	Ser	no	117 ± 10° (NMR)
Nhp6a	linear 15-mer	28,44	Met	yes	Phe	yes	~ 70° (NMR)
HMG-D	linear 11-mer	32	Met	yes	Val[b]	yes	111° [c] (crystallog.)
HMG-D[a]	2 base bulge 18-mer	33					95 ± 5° (FRET)[d]
	16-mer	41	Met	yes	Val[b]	yes	(NMR)
HMG1-A	*cis*-Pt 20-mer	43	Ala	no	Phe	yes	61° (crystallog.)
HMG1-A[a]	4-way junction	31	Ala	no	Phe	*stacks* against base in center[e]	n/a
HMG1-B[f]	*cis*-Pt 20-mer	34	Phe	inferred	Ile	likely	80–95° (FRET)[d]

Red: HMG boxes from sequence-specific single HMG box proteins; black: HMG boxes from abundant single box yeast and *Drosophila* HMG box proteins; blue: A box from HMGB1; green: B box from HMGB1. Adapted from Ref. [13] with permission.
[a]Evidence-based models rather than defined structures.
[b]Adjacent threonine also partially intercalates.
[c]Value may be high due to crystal packing.
[d]Fluorescence resonance energy transfer. No direct evidence.
[e]Stacks against a base at the proximal end of a junction arm (i.e. in center of junction where bases not stacked).
[f]No structure or model of protein-DNA complex; intercalation inferred.
Adapted from Ref. [13] with permission.

with Phe at position Y partially intercalating within the already kinked, platinated GpG site [43]. This intercalation involves a change in the local side-chain conformation of Phe37. In the free A domain this residue is packed against helix II, whereas in the complex with cisplatin-modified DNA the primary interaction of the aromatic ring is with the planar faces of the guanine residues (Figs. 2b and 3). A similar conformational change occurs in the Phe residue in position Y of Nhp6ap on binding to linear DNA [44]. A similar (but distinct) role for a correponding Phe residue was also deduced [31] for the A component of the AB didomain of HMGB1, which binds preferentially to the crossover of the extended (square planar) form of a four-way junction [45]; in contrast, the neighbouring B domain in AB binds to one of the arms of the junction, although the isolated B domain again binds to the crossover, like the isolated A domain or A in AB [31]. Moreover, mutation of the primary intercalating residue, Phe, in the B domain [46,47] to alanine, and of the corresponding residue, Ileu, in the SRY HMG domain to threonine [48], has no effect on binding to four-way junctions, which would be consistent with a major role for the secondary intercalating residues in this interaction. The selectivity of the A domain was attributed to two features: (1) the greater positive charge on the distal end of the short arm of the A-domain HMG box (containing helices I–II and the intervening loop), relative to the B domain (and possibly also the disposition of charges); (2) stacking of the Phe at position Y at the N-terminus of helix II against an exposed base-pair at the proximal end of one arm. The charge on the short arm would facilitate the insertion of the A domain into a central "hole" on the minor groove side of the junction, thereby stabilizing the resulting complex [31]. These differences could explain why the isolated A domain has a stronger preference than the B domain for four-way junctions [29].

B-type HMG boxes also bind to other distorted DNA structures in preference to linear DNA and here again the secondary intercalating residue(s) is important. The Phe at position Y in NHP6A provides the selectivity for binding to cisplatin-modified DNA [49]. Similarly, in the interaction of HMG-D with DNA containing a two-base bulge, the secondary intercalating residues, Val and Thr, are accommodated by the somewhat wider minor groove immediately to one side of the distorted structure but not within the DNA bulge itself [41]. (However, mutation of the secondary intercalating residue in the B domain of HMGB1 does not substantially reduce the affinity of the domain for cisplatin-modified DNA, indicating that the primary intercalating residue can also recognise the distortion in this case [50].) One common feature of the recognition of distorted DNA by wild-type A and B domains thus appears to be the involvement of the secondary intercalating residues.

2.2.3. Role of the basic region in DNA binding

A distinctive feature of many HMG-box proteins is a basic extension either C- or N-terminal to the HMG box. In the case of both the sequence-specific LEF-1 HMG box [25] and the non-sequence-specific Nhp6ap and HMG-D boxes

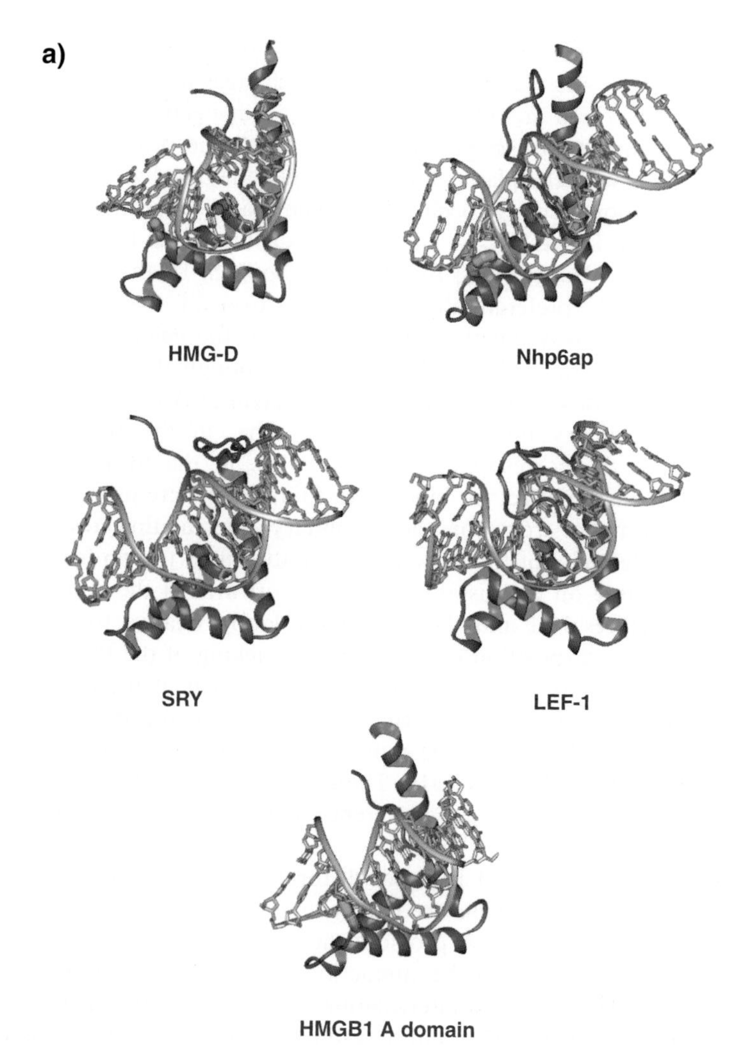

Fig. 3. (a) Structures of HMG-box/DNA complexes in which intercalation occurs by both X and Y residues (HMG-D and Nhp6ap), or by X only (SRY and LEF-1) or by Y only (HMGB1 A-domain). Further details are given in Table 1. The panels are successively: HMG-D HMG box complexed with 10 bp duplex DNA (X-ray crystal structure [32]; Nhp6ap with basic N-terminal extension complexed with 15 bp duplex DNA (NMR structure [44]); SRY HMG domain complexed with 14 bp duplex DNA (NMR structure [40]; LEF-1 HMG box with basic C-terminal extension complexed with duplex DNA (NMR structure [25]); the HMGB1 A-domain/*cis*-platinated DNA complex (X-ray crystal structure [43]); note that in this structure the orientation of the intercalating phenylalanine group differs from that in the solution structure of the free domain shown in Fig. 2. In both the Nhp6ap and SRY complexes the basic extension interacts with the major groove opposite the compressed minor groove (see text). Adapted from Ref. [44] with permission; (b) Model for the interaction of the HMGB1 A-domain with a four-way junction. The short arm of the box (helices I and II and the connecting loop) inserts into the central "hole". The isolated B domain binds in a similar manner, but in the AB-didomain the A domain preferentially binds in this mode while the B domain binds along an arm. Adapted from Ref. [31] with permission.

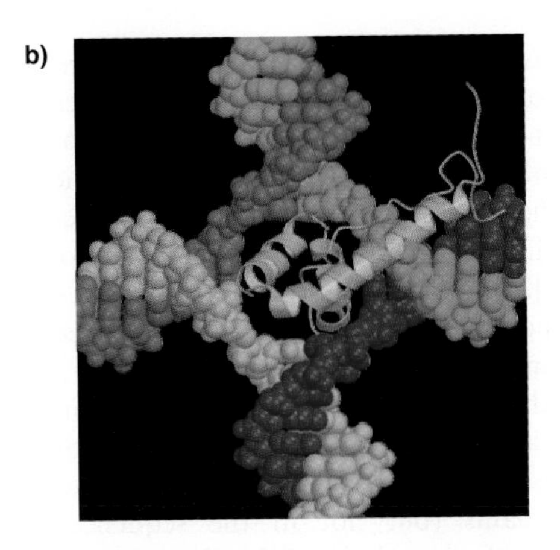

Fig. 3. Continued.

[23,28,51], this region binds in the compressed major groove on the face of the double helix opposite to the widened minor groove (Fig. 3). This stabilises the bend induced by the HMG box [42,52], presumably by charge neutralisation, and consequently facilitates circularisation by HMG-box proteins of short DNA fragments [53,54] that are well below the persistence length of DNA [55]. An analogous function may be performed by the short lysine-rich sequence between the A and B domains of HMGB1. This sequence is necessary for preferential binding of the B domain to supercoiled DNA and is also essential for a high-affinity binding of the B domain to DNA containing a site-specific 1,2-d(GpG) intrastrand DNA adduct of cisplatin [56].

2.2.4. Role of the acidic region

DNA binding by HMGB proteins can also be modulated by an acidic region. This is a characteristic C-terminal feature of HMGB1, HMGB2, and related proteins, including *Drosophila* DSP1 (dorsal switch protein) and UBF (upstream binding factor), and of some abundant single-HMG-box proteins (see above). *In vitro* the acidic tail lowers the affinity of the HMG boxes for most DNA substrates [5,6,57–59], the longer tail of HMGB1 having a larger effect than the shorter tails of HMGB2 and the closely related HMGB2a [59]. The tail might mediate this reduction in affinity by contacting one or both boxes [60,61], or simply through charge repulsion of the DNA. In contrast, the tail has little effect on the affinity of HMG-box proteins for DNA minicircles [7,59], resulting in a marked selectivity of HMGB1 and 2 for minicircles over linear DNA in a competition assay [59]. (Indeed, in the case of HMG-D, the short acidic tail with only nine acidic residues slightly increases the affinity for distorted DNA structures but nonetheless reduces the affinity for linear DNA [6].) As well as affecting DNA binding, the tail may also act as a protein chaperone (see below).

2.2.5. *Structural basis for non-sequence-specific DNA recognition*

Within the HMG-box domain itself, several differences distinguish the sequence-specific from the non-sequence-specific domains (Fig. 2a). In particular, position 13 (HMGB A-box numbering) is asparagine in all specific HMG-boxes and usually serine in non-specific boxes [15,32,62]. In the sequence-specific HMG-domain of LEF-1, the Asn directs a hydrogen-bond network to three of the four bases of a CA dinucleotide [25] whereas in the non-sequence-specific HMG domain of HMG-D, the corresponding Ser residue forms water-mediated hydrogen bonds to a ribose O4′ position and to an adenine N3 [32]. However, because any base could be bound by serine (either directly for cytosine or thymine, or indirectly through a water molecule for adenine or guanine) hydrogen bonding could occur to any sequence at this position. Other positions that are well conserved in the non-sequence-specific domains (but not in the sequence-specific domains) include (numbering based on A domain of HMGB1 (Fig. 2a)) position 8, invariably occupied by a proline, which forms part of the hydrophobic zipper between the N-terminal extended strand and helix III, and position 46 where the residue is nearly always negatively charged (glutamate or aspartate). Other positions showing preferential conservation are position 23 (usually arginine) in helix I and position 60 (usually glutamate) in helix III.

3. *HMGB function*

3.1. *DNA bending as a major feature*

The high-mobility group proteins were initially characterised as prominent components of chromatin [1,17]; each mammalian nucleus contains, on average, approximately 10^5–10^6 HMGB1 molecules [16,63]. This abundance implies a major chromatin-related function but the precise biological role of the proteins has proved elusive.

In vitro, HMGB1 and 2 and their counterparts in other organisms increase both the axial and torsional flexibility of DNA, by concomitantly untwisting and introducing a kink. This property, coupled with their abundance and lack of DNA-binding specificity, have led to the suggestion that an important role of the proteins is to act as DNA chaperones to promote the formation of protein–protein contacts in nucleoprotein complexes, or to facilitate the manipulation of DNA in other nuclear processes [55,64,65]. A common feature of these functions would be the transient introduction of DNA bends, possibly stabilising a tight loop in contrast to the much larger loops observed when HMGB1 binds to crossovers in supercoiled DNA [5]. A similar role has been postulated for the bacterial protein HU [35,66], which also bends DNA substantially, but, unlike HMGB proteins, binds to DNA on the *inside* of the induced bend. Despite the differences in the mode of DNA binding, HU and HMGB proteins behave, to a certain extent, as functional counterparts. Thus HMGB1 can substitute for HU, not only in promoting site-specific DNA inversion *in vitro* [35] but also when expressed in an *E. coli* mutant lacking the HU and FIS DNA-bending proteins; and yeast NHP6A can rescue lesions in *E. coli* nucleoid morphology [67]. Similarly, expression of the yeast mitochondrial HMGB protein HM alleviates the phenotypes of *E. coli* cells lacking HU protein and, conversely, HU can functionally complement the respiration deficiency associated with yeast strains lacking HM [68].

There are two established cases in which the assembly of nucleoprotein complexes containing sequence-specific DNA-binding proteins is promoted by the DNA-bending properties of HMGB1 and 2. First, in V(D)J recombination the lymphocyte-specific proteins RAG1 and RAG2 (human recombination activating genes 1 and 2) appear to recruit HMGB1 and 2 to the appropriate sites in chromatin, presumably by protein–protein contacts with the RAG1 homeodomain [69–75]. Here they ensure the "12/23 rule". This requires that V(D)J recombination occurs only between specific recombination signal sequences (RSS). Each RSS is made up of a conserved heptamer and nonamer sequence separated by a non-conserved spacer of either 12 or 23 base pairs. HMGB1 (in concert with RAG1,2) facilitates recombination probably by bending the DNA between the two conserved sequences spaced by 23 bp and stabilising a nucleoprotein complex. HMGB1 plays the dual role of bringing critical elements of the 23-RSS heptamer into the same phase as the 12-RSS to promote RAG binding and of assisting in the catalysis of 23-RSS cleavage. Recent footprinting experiments indicate that HMGB1 (or HMGB2) is positioned 5′ of the nonamer in 23-RSS complexes, interacting largely with the side of the duplex opposite to the one contacting the RAG proteins [76]. A second instance in which an HMGB protein may facilitate nucleoprotein complex assembly is in the formation of an enhanceosome containing the Epstein-Barr virus replication activator protein ZEBRA and HMGB1 [77]; the two proteins bind cooperatively, HMGB1 binding to, and presumably bending, a specific DNA sequence between two ZEBRA recognition sites. HMGB1 also promotes the formation of complexes between the human adeno-associated virus Rep and DNA [78], although in this case a role for DNA bending has not been established. Bending of DNA by HMGB1 and 2 has also been invoked to explain the essential role of these proteins in initiating DNA replication by loop formation at the MVM (minute virus of mice) parvovirus origin of replication [79].

3.2. *HMGB proteins and chromatin structure*

HMGB1 was first described as a "non-histone chromosomal protein" [1,80] and has been implicated in the maintenance and establishment of chromatin structure [10] as well as having more recently identified roles.

The interaction of the abundant HMGB proteins with both nucleosomal particles lacking linker histones and with linker histones themselves *in vitro* is well documented. A soluble chromatin fraction released from mouse myeloma nuclei contained mononucleosomes associated with almost stoichiometric amounts of HMGB1,2 but lacking histone H1 [81]. Nhp6ap binds directly to nucleosomal particles reconstituted with chicken histones (i.e., a single histone octamer associated with >160 bp DNA) [82] as does *Xenopus* HMGB1 [83]. Similarly all the maize (single box) HMGB variants, HMGB1–5 (not to be confused with the two-box vertebrate HMGB1 and HMGB2), as well as SSRP1, bind to nucleosomal particles [84]. Restriction of this binding to one molecule of plant HMGB1 per particle required both the N-terminal basic and C-terminal acidic domains.

Early studies showed that HMGB1/2 bound to nucleosomal particles containing 180 but not 140 bp DNA [85]. Subsequently, footprinting studies revealed that the HMG protein bound immediately proximal to a DNA exit/entry site on the core particle [83,86]. Consistent with this view HMG-D increases the nucleosome repeat length from ~165 to ~185 bp in a cell-free assembly system, an effect that is independent of the acidic tail [87]. Histone H1 and HMGN1, and HMGN2 (formerly HMG14 and HMG17 [18]) are the only other proteins so far reported to cause such an increase [88,89]. In *S. cerevisiae*, loss of the two-box protein Hmo1p results in a general increase in the sensitivity of chromatin to digestion by micrococcal nuclease [19]. This same phenotype is observed in strains that lack both Nhp6ap and Nhp6bp (M. Buttinelli and A. A. Travers, unpublished). Taken together, all these observations point to a role, or roles, of HMGB proteins in the maintenance of some elements of chromatin structure (see below for further discussion). Another observation consistent with such a role is the enhanced micrococcal nuclease sensitivity of chromatin isolated from mouse cells over-expressing HMGB1, although (perhaps surprisingly) not in cells over-expressing HMGB2, whose acidic tail is ~10 residues shorter [90]. This enhanced sensitivity is associated with a diminished linker histone content, suggesting that the binding of the HMGB proteins and linker histones may be mutually exclusive.

A transition in chromatin composition from a state containing predominantly an HMGB protein to one containing somatic H1 has been invoked to explain changes in chromatin structure during the early development of both *Drosophila* and *Xenopus*. In *Drosophila* the early pre-blastoderm syncytial nuclei are much larger than later nuclei and contain chromatin that is less compacted [91]. HMG-D, which is highly abundant (although not all molecules are necessarily available/competent for DNA binding), is deposited in the egg by the mother and with each nuclear division in the embryo the average number of HMG-D molecules per nucleus falls. After nuclear cycle 6, the amount of H1 increases substantially [91], correlating with a progressive increase in compaction of both nuclei and metaphase chromosomes until the mid-blastula transition (MBT). A similar situation has been described for *Xenopus* [92] where HMGB1 and the linker histone variant B4 (which is much less basic than the somatic H1) are believed to organise chromatin prior to the MBT and are progressively replaced by the somatic linker histone after it. In mammalian development, no comparable variations in the levels of somatic H1 have been observed during the first few nuclear divisions of embryogenesis [93]. However, disruption of 11 of the 12 H1 diploid gene copies in chicken DT40 cells resulted in an approximately two-fold increase in the levels of HMGB1 and 2 (based on gel electrophoresis) [94]. The remaining H1 gene was also up-regulated to attain half-normal levels of the protein, with no accompanying increase in the levels of the core histones. These changes appeared not to affect the global chromatin structure but did affect the protein expression profile. Whether in this case HMGB1 and HMGB2 can assume the functional, although not structural (see below), role of H1 is not clear.

Two factors at work in the replacement of HMGB proteins with H1 during *Drosophila* and *Xenopus* development are probably, first, the reversible dissociation of HMG-D and HMGB1, respectively; and second, the law of mass action operating on an increasing concentration of H1, which will occupy transiently vacant sites. Photobleaching techniques show that HMGB1 is indeed highly mobile in the nucleus, much more so than H1 [95]. *In vitro*, histone H1 can displace HMG-D from *Drosophila* chromatin in a concentration-dependent manner [87]. However, although the binding sites of H1 and HMGB1 may be sterically mutually exclusive, they are nonetheless distinct, consistent with distinct modes of binding—H1 to juxtaposed duplexes and HMGB1 potentially to a bend/kink in the DNA (or to linear DNA where incorporation of such a bend is possible).

HMGB proteins including HMGB1 ([96,97] A. Simpson and J. O. Thomas, unpublished) and HMG-D [87] can also interact directly *in vitro* with histone H1. The interaction of HMG-D is independent of the acidic tail (suggesting that, in this case at least, non-specific charge interactions involving the tail can be excluded) but requires aminoacids 75–100, which comprise a short linker and basic region between the HMG box and the tail [87].

3.3. Nucleosome assembly and remodeling

An early report showed that rat HMGB1 mixed with core histones could facilitate the formation of histone H3/H4 tetramers and to a lesser extent histone octamers [98]. The resulting complexes could be incorporated into nucleosomal particles. These observations suggest a protein chaperone function and imply that HMGB1 can interact directly with histones, although the biological relevance of this function is unclear. In contrast, however, it was subsequently reported that HMGB1 and HMGB2 inhibited nucleosome assembly *in vitro* [99]. More recently HMGB2 has been reported to be a component of the SET nucleosome assembly complex [100]. This complex is found both in the nucleus and, in a larger form, in the endoplasmic reticulum and contains, in addition to HMGB2, the nucleosome assembly protein SET (also known as TAF-1α), the tumour suppressor protein pp32 and the base excision repair enzyme APE. Little is known of the exact biological function of this complex but it has been shown that SET binds to the transcriptional co-activators CBP/p300 [101,102] and APE is the rate-limiting enzyme in the repair of apurinic/apyrimidinic sites [103]. An interesting possibility is that HMGB2 recognises distorted DNA and CBP/p300 mediates local protein acetylation, which then allows recruitment of APE to these sites [104].

In addition to a potential role of HMGB proteins in nucleosome assembly, it has recently become apparent that many chromatin remodeling complexes either contain, or can associate with, a polypeptide containing an HMG-box homologous to the HMGB1 B domain (Table 2). Examples of such complexes include the BAF (a mammalian SWI/SNF related complex [106]) and the *Drosophila* BRM (*brahma*)

Table 2
Chromatin remodeling complexes containing HMG-box proteins

Complex	No. of components	HMG domain component(s)	Reference	Comments
SET (human)	4	HMGB2	100	DNA repair function, possibly more components in active complex
SPN (yeast)	3	Nhp6a/bp	82,116	
FACT (human)	2	SSRP1	114	
SWI/SNF1 (BAF) (human)	10	BAF57	106,108	
SWI/SNF2 (PBAF) (human)	9	BAF180 + BAF57	106,108	
BRM (*Drosophila*)	≥7	BAP111	107	BAP111 has homologies with BAF57 outside the HMG box
CHD1 (*Drosophila*)	?	SSRP1	113	Colocalisation of CHD1 and SSRP1 on polytene chromosomes

complex [107], which contain respectively the HMG-domain proteins BAF57 and BAP111. None of these proteins appears to be essential for remodeling, but loss of BAP111 results in a significant reduction of function [107]. However mutations in the BAF57 subunit do impair the function of the BAF complex *in vivo* in both the silencing of the CD4 locus and the activation of the CD8 locus [108]. In addition to BAF57, the vertebrate remodeling complex SWI/SNF2 contains a polypeptide, BAF180 (also known as polybromo), with a C-terminal truncated HMG-box that retains all the residues considered essential for the folded structure (Fig. 2a) [109,110]. Likewise the related RSC remodeling complex requires Nhp6ap for the repression of the *CHA1* locus encoding the catabolic L-serine (L-threonine) dehydratase. [111,112]. Yet another instance of the association of an HMG-domain protein with a remodeling polypeptide is suggested by the cellular colocalisation of SSRP1 with the *Drosophila* ATP-dependent remodeling protein CHD1 [113].

In addition to these complexes containing an ATPase/helicase domain subunit, an HMG-domain protein is also a component of the smaller mammalian FACT remodeling complex [114] and the yeast SPN complex [82]. These related complexes facilitate nucleosome remodeling during transcription initiation and elongation [115]. They both contain Spt16p (in FACT the mammalian homologue) and an HMG-domain protein, Nhp6ap in SPN [82,114] and SSRP1 in FACT [114]. They differ in the SPN complex that contains one additional subunit, Pob3p, which has homology to SSRP1 but lacks an HMG-box domain [82,116]. The mammalian FACT complex is also a component of the DNA-damage-induced p53 Ser392 kinase assembly [117]. This contains another component, casein kinase II, leading to the suggestion that association of the kinase with FACT during transcriptional elongation may be the trigger for the kinase complex to phosphorylate p53 at DNA lesions [117].

The association of HMGB proteins with chromatin remodeling complexes suggests that their DNA bending function may be utilised in this process. The juxtaposition of positively and negatively charged regions in the proteins could also be important. This is a feature of HMGB1 and 2 (as well as of the AT-hook protein HMGA, which interacts with the nucleosome core) and also of SSRP1 in the FACT complex. *S. cerevisiae* Nhp6ap and Nhp6bp have an HMG domain and a positively charged region, but no acidic region, and it may be significant that in the SPN remodeling complex (see above) both the Spt16p and Pob3p subunits possess such a region, which could potentially compensate for the lack of a comparable region in NHP6. Juxtaposition of oppositely charged regions is also found in parts of certain chromatin remodeling motor proteins, for example dMi-2 [118]. Although these similarities could be fortuitous, it raises the possibility that these charged regions could facilitate the binding of HMGB proteins to nucleosomes by interacting with both the positive charges on the histones and the negative charges on the DNA. The ability of the HMG domain to introduce a sharp bend would allow the constraint of a short DNA loop at one exit/entry point of the wrapped nucleosomal DNA (Fig. 4) [119]. Such a loop, coupled with the untwisting in the HMG-bound DNA, could serve to increase the accessibility of wrapped nucleosomal DNA at both an exit/entry point and at internal sites. This conformational change would serve to lower the activation energy for directed remodeling. Consistent with this scenario HMGB1 facilitates histone octamer migration mediated by the ACF remodeling complex [120]. This function requires the acidic tail of HMGB1 [120].

The postulated interaction of the acidic tail of abundant HMGB proteins with core histones might suggest possible functional differences between the two major

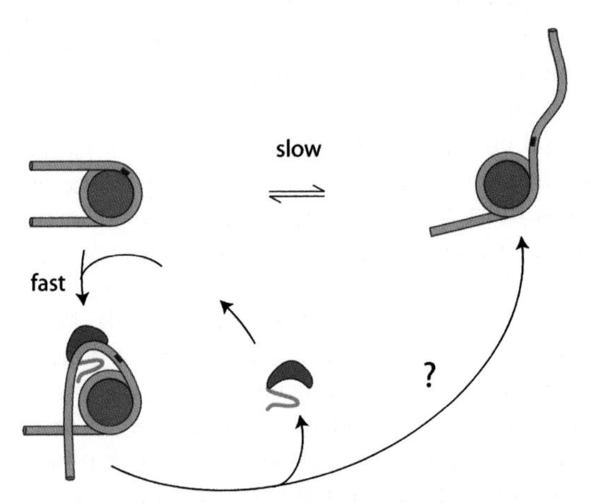

Fig. 4. Possible mechanism for HMGB proteins in facilitating the unwrapping of nucleosomal DNA. The model postulates that the binding of an HMGB protein at an exit/entry point for the wrapped DNA increases the accessibility of the wrapped DNA both at the exit/entry points adjacent to the bound HMGB protein and also at internal positions. Such a conformational change would facilitate the binding both of sequence-specific regulatory proteins and of remodeling complexes. Adapted from Ref. [119] with permission.

forms of these proteins in vertebrate and insect cells. Vertebrate HMGB1 and HMGB2 differ in both the length and amino acid composition of their acidic tails. The tail of HMGB1 (30 residues) is significantly longer than that of either HMGB2a or HMGB2b (~20 residues). In all three proteins there are more glutamate than aspartate residues but whereas in HMGB1 and HMGB2b the Glu:Asp ratio is 2, in HMGB2a it is 9.5. Similarly the much shorter acidic tails of the *Drosophila* counterparts, HMG-D and HMG-Z, although differing by only one in net charge, differ in amino acid composition [121]. The tail of HMG-D consists of a run of aspartate residues flanked by glutamate residues whereas the charged residues in the HMG-Z acidic tail are predominantly glutamate. These differences could conceivably reflect significant differences in the interactions of the HMGB proteins with nucleosomes.

3.4. *HMGB proteins and transcription*

Early studies established that HMGB1 and HMGB2 could stimulate transcription *in vitro* by RNA polymerases II and III [122–124]. Subsequently these proteins, together with their counterparts in other organisms, have also been implicated in the repression of transcription.

3.4.1. *Effects at the level of chromatin*

These effects have been most extensively studied in *S. cerevisiae* where deletion of both *NHP6A* and *NHP6B* results in changes in gene expression coupled to alterations in chromatin organization. This deletion was initially reported to abolish or decrease the activation of four genes, *CUP1*, *CYC1*, *GAL1*, and *DDR2* [125] and subsequently expression of both the *HO* gene [126,127] and the pol III *SNR6* gene [128,129]. A microarray study revealed that the *NHP6* deletion resulted in >3-fold changes in transcription of 197 genes, corresponding to 3.2% of the yeast genome [112]. Of these, 114 are up-regulated and 83 are down-regulated. These data indicate that the transcriptional effects of Nhp6P in yeast are not general but gene-specific. At the *CHA1* locus, loss of Nhp6P results both in an increase in the basal level of transcription and in a substantial decrease in the induced level [112]. This suggests an effect at the level of chromatin. The *CHA1* regulatory region contains a positioned nucleosome which occludes the TATA box under non-inducing conditions. On induction the TATA region becomes accessible [130]. However in the mutant strain, consistent with the increased basal level transcription, the chromatin structure of the TATA region in the uninduced state is similar to that in the induced wild-type strain. Nhp6P thus appears to be required for establishment of the organised chromatin structure characteristic of the uninduced state. The RSC remodeling complex is also required for this process [111] suggesting that RSC and Nhp6p may cooperate to remodel chromatin.

Further insights into how Nhp6ap and Nhp6bp function were provided by studies on the *HO* gene [127]. Loss of *NHP6* can be suppressed by mutations that increase nucleosome accessibility and mobility, and enhanced by those with the opposite effect. Mutations both in the *SIN3* and *RPD3* genes, encoding components of a histone deacetylase complex, and in *SIN4*, partially restore wild-type function in cells lacking both Nhp6a and Nhp6b, while loss of the histone acetylase Gcn5p (also a component of the SAGA histone acetylase complex) in the same cells results in a more severe phenotype. Rpd3p and Gcn5p contribute to the dynamic balance between histone acetylation and deacetylation [131]. Both histone acetylation and the sin (SWI/SNF independence) phenotype are correlated with chromatin unfolding [132,133] and/or enhanced nucleosome accessibility [134] while histone deacetylation would be expected to favour folding. On this argument one role of the Nhp6 proteins would be to antagonise folding and possibly promote nucleosome accessibility. A similar role has also been postulated for the HMGN (nucleosome-binding) proteins, although there is no indication that this class of proteins promotes nucleosome mobility [135]. Loss of either Sin4p [136] or Nhp6a/bp results in a similar general increase in accessibility of yeast chromatin to micrococcal nuclease. However, the functional consequences differ, suggesting that enhanced nucleosome accessibility and mobility both contribute to transcriptional regulation.

3.4.2. Interactions with transcription factors

In vitro HMGB proteins can enhance the binding of various transcription factors (e.g., adenovirus MLTF [137], Oct-1 and 2 [138], HoxD9 [139], p53 [140,141], steroid hormone receptors [142–144], Rel proteins [145,146], p73 [147], Dof2 [148], and the Epstein-Barr activator Rta [149]) to their cognate DNA binding sites (Table 3). Similarly rat SSRP1 has been shown to facilitate the DNA binding of serum response factor [150] and human SSRP1 is associated with the γ isoform of p63 *in vivo* at the endogenous MDM2 and p21$^{waf1/cip1}$ promoters [151]. In most of these cases, the interaction of the HMG protein with the transcription factor has been detected *in vitro* and could, in principle, serve as the mechanism for recruitment of HMGB1 or 2 to particular DNA sites (Fig. 5). In some cases transfection experiments indicate functional interactions *in vivo* (Table 3). Direct interactions between Nhp6p and the Gal4p and Tup1p transcription factors have also been inferred *in vivo* by a split-ubiquitin screen and confirmed by a pull-down assay [152]. Although the demonstrated interactions *in vitro* so far involve an HMGB protein and a single transcription factor, it is entirely possible that *in vivo*, in a natural regulatory context, the bending of DNA by HMGB1 and 2 could potentially allow the recruitment of a second transcription factor to the complex (Fig. 5), in an analogous manner to the action of sequence-specific HMG-box transcription factors [13] such as LEF-1 in the enhanceosome at the T cell receptor alpha (TCRα) [153]. HMGB1 may play a catalytic, chaperone role, since it does not appear to be stably incorporated into the final complex (Fig. 5; Ref. [13]). Although a role for HMGB1- and HMGB2-induced DNA bending in the facilitated binding of transcription factors, while being entirely plausible, has not been directly established, it is strongly suggested by the ability of HU to substitute for the HMGB1-stimulated binding of the Epstein Barr virus trans-activator Rta to its cognate binding sites [149]. A possible role for an HMGB1-induced change in DNA conformation in facilitation of transcription factor

Table 3

HMG-box proteins stimulate the binding of certain sequence-specific DNA-binding proteins to their DNA targets

Sequence-specific protein	Reference	Protein–protein interactions *in vitro*?	Effect on function?
MLTF	137	Not determined	Increased transcription *in vitro*
Oct-1 and 2	138	Yes. POU homeodomain interacts with single A or B HMG domains of HMGB2	Increased transcription *in vivo* (transfection assay)
Steroid hormone receptors	142–144	Yes. Interact with single A and B HMG boxes of HMGB1	Increased transcription *in vivo* (transfection assay)
p53	140,141,147	Yes	Increased transcription *in vivo* (transfection assay) [140,141] but can inhibit transcription from certain promoters 147
p63	151	Yes	Increased transcription *in vivo* (transfection assay)
p73	147	Yes. Interacts with single A and B HMG boxes of HMGB1	Can either increase or decrease transcription *in vivo* (transfection assay)
Hox proteins	139	Yes. Hox homeodomain interacts with single A and B HMG domains of HMGB1	Increased transcription *in vivo* (transfection assay)
Dof2	148	Yes, interacts with maize HMGB5 and to a lesser extent with HMGB1-4	
Rta	149	Tested but not observed	HMGB1 stimulates binding of factor to cognate sites
Rel proteins (Dorsal, NFκB)	145	Yes. Dorsal interaction with HMGB1 requires sequences N-terminal to A and B boxes	Repression *in vivo* (transfection assay)
	146	*Drosophila* HMGB1 related protein DSP1 has similar requirements	*In vitro* DSP1 binds to SP100B which in turn recruits Heterochromatin Protein 1 (HP1)
SRF	150	Direct interaction not tested	Stimulation of SRF binding by SSRP1. Activation (transfection assay) and association (co-immunoprecipation) *in vivo*. Activation requires basic region of SSRP1
RAG1,2	72	Yes	Increased cleavage *in vitro*
	75	RAG1 homeodomain interacts with tandem HMG boxes of HMGB1,2	Increased V(D)J recombination (transfection assay)
Rep	78	Not tested	Increased DNA cleavage by Rep *in vitro*

Adapted from Ref. [13] with permission.

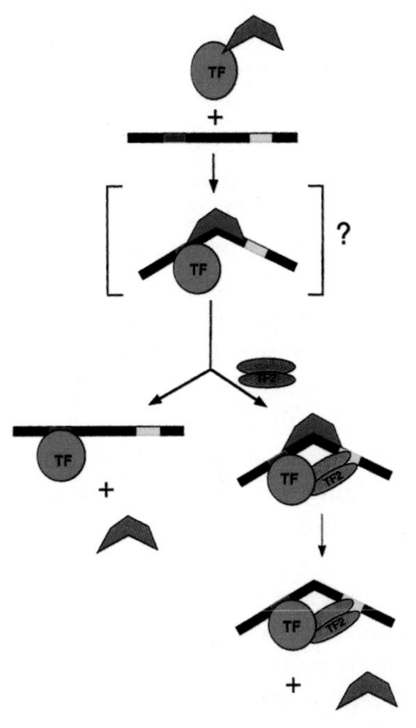

Fig. 5. A chaperone role for HMGB1 in facilitation of transcription factor binding. A similar mechanism could apply to the role of HMGB1 in V(D)J recombination (see text). The major features are: (i) Recruitment of HMGB1 to specific sites through interactions with sequence-specific proteins (these have been detected in several cases. (ii) Formation of a ternary complex, likely to be unstable but detected/inferred in some cases; in this ternary complex HMGB1 may bend the DNA, thus providing the potential for the recruitment of additional DNA-binding proteins to the complex. (iii) If no additional protein is recruited, HMGB1 may dissociate from the complex leaving the transcription factor stably bound: in this schematic, the final DNA is shown straight (as in the case of Oct-1 binding), but this is not necessarily the case (for example the progesterone receptor bends its DNA in the binary complex); see Table 3 for references. Alternatively, a second protein may be recruited, followed optionally by HMGB1 dissociation. (Note that only a single HMG box is shown, since the separate A and B HMG boxes of HMGB1 have been shown to facilitate binding of various transcription factors *in vitro* (Table 3). However, in other cases (e.g., V(D)J recombination) both HMG boxes are required; in this case, one box (e.g., box A which does not bend DNA well) may be involved in protein–protein interactions, the other (e.g., box B) in bending the DNA; see text.) Adapted from Ref. [13] with permission.

binding is also suggested by the observation that HMGB1 promotes binding of p53 to linear DNA but not to 66 bp DNA circles [141]. However, in this case the data do not distinguish between possible effects of DNA bending and untwisting.

In vitro interactions between HMG proteins and the basal transcription machinery have also been reported. Human HMGB1 binds to the TATA-box binding protein (TBP) and interferes with the normal binding of TFIIB in the pre-initiation complex [154,155], thereby inhibiting TBP function; both HMGB1 and TFIIB independently enhance binding of TBP to TATA-box DNA [154]. Similarly, Nhp6ap promotes the formation of a complex with TBP and TFIIA at the TATA box that has enhanced affinity for TFIIB [125] and also enhances the binding of TFIIIC at the *SNR6* promoter [128]. In contrast, HMGB2 has been reported to stabilise and activate the TBP-containing TFIID–TFIIA complex bound to promoter DNA [156].

Impaired glucocorticoid receptor binding would provide an explanation for the pleiotropic effects on glucose metabolism observed in mice null for HMGB1 [157]. Neither this mutation nor the loss of HMGB2 (which has a role in mouse spermatogenesis) [158] is lethal, suggesting a functional redundancy between members of the HMGB1 and 2 family. A similar situation occurs with Nhp6ap and Nhp6bp in yeast [159]. However, the different phenotypes of the HMGB1 and HMGB2 null mice probably reflect specific roles for the two proteins in different tissues [157,158]. The apparent qualitative interchangeability of HMGB1 and 2 in transfection assays suggests that acidic tail length is not a dominant feature in promoting the DNA binding of sequence-specific proteins [138]. Indeed, the tail is not essential for facilitated DNA binding *in vitro* [139]. Moreover, in some cases, individual boxes appear to be functional, at least in assays *in vitro* (Table 3).

The general observation that HMGB proteins can interact with a substantial number of other proteins raises the question of selectivity. A phage display assay revealed that HMGB1 can recognise several short peptide motifs, among which the sequences PXXPXP and WXXW were well represented [160]. However, no clear pattern was apparent, suggesting that HMGB proteins may have the potential to interact with many diverse protein sites.

Although the recruitment of transcription factors to their binding sites by HMGB proteins has been studied *in vitro* with naked DNA, *in vivo* these interactions occur in the context of DNA organised into chromatin. Here the binding sites might be located within DNase 1 hypersensitive regions, where they would be expected to be relatively accessible, or alternatively they might be sequestered in the wrapped DNA in nucleosomes. At the more accessible sites, facilitation of transcription factor binding, and of the activation and repression of transcription *in vivo* by these proteins, could operate by the mechanism proposed for the "sequence-specific" HMGB proteins such as SRY and TCF-1. In the latter case it is inferred that the bend introduced by these proteins provides a DNA scaffold, allowing the assembly of a complex nucleoprotein structure [153]. This may indeed happen in the assembly of the EBV enhanceosome [77]. However, when the binding site is incorporated into a nucleosome core, an HMGB protein could facilitate the binding of a transcriptional activator or repressor to the internal site by binding close to an exit/entry point and loosening the wrapped DNA (Fig. 4). It has been proposed that limited natural unwrapping of histone-bound DNA would render such sites transiently accessible to sequence-specific transcription factors or restriction enzymes [161], site exposure being dependent simply on stochastic random fluctuations in the extent of wrapping. Mediation of targeted binding of a factor by an HMGB protein would be expected to overcome this dependence. Alternatively, the HMGB-facilitated binding of a transcription factor to an accessible site could promote chromatin remodeling to establish either an activated or repressed configuration of chromatin.

3.5. HMGB proteins and DNA repair

Do HMGB1 and functionally related HMG-box proteins (Fig. 1) bind preferentially to structural aberrations in DNA *in vivo*? Such aberrations clearly are targets *in vivo* because both expression of the two-HMG-domain protein Ixr1p (intrastrand cross-link recognition protein) in yeast [162], and oestrogen-induced elevation (approximately two-fold) of HMGB1 in MCF-7 breast cancer cells [163,164], increase the cytotoxicity of cisplatin, presumably by shielding the DNA lesions from the nucleotide excision repair machinery [165]. High levels of expression of HMGB2 in testis have been linked to the particular efficacy of cisplatin against testicular cancer [158]. Nevertheless mouse embryonic fibroblasts lacking HMGB1 do not differ from wild-type cells in their sensitivity to cisplatin [165] suggesting that other proteins, for example HMGB2 or SSRP1 [166], may also influence the effects of the drug.

However, neither cisplatin-DNA adducts nor DNA bulges occur frequently in natural DNA. Among other possible *in vivo* ligands are UV-induced pyrimidine dimers to which HMGB1 binds preferentially *in vitro* [58]. Yet loss of Nhp6ap and Nhp6bp in yeast does not protect against the loss of viability on UV irradiation [49]. HMGB1 might also have a direct role in DNA repair. *In vitro* it acts synergistically with the DNA-binding subunit of DNA-dependent protein kinase, which is involved in double-strand break repair, and enhances kinase activity by approximately three-fold [167]. HMGB1 and 2 also promote blunt-end joining by T4 DNA ligase [168], and by mammalian DNA ligase IV [169] which is involved in both double-strand break repair and V(D)J recombination. In all these examples the binding of distorted DNA by HMGB proteins could constitute the primary recognition event.

3.6. Post-translational modifications of HMGB proteins

Post-translational modifications of HMGB proteins have been known for many years but their biological function is unclear. HMGB1 is acetylated at Lys 2 and Lys11 [170,171] and embryonic HMG-D is singly acetylated at an unknown position [87]. Recent studies have reported that acetylation of Lys 2 in HMGB1 in some way increases the affinity of the protein for UV-irradiated DNA *in vitro* by approximately 10-fold, but only by 2-fold for non-irradiated DNA [172]. *In vivo* the effects of acetylation might be mediated in a more complex manner (e.g., through interaction with a bromodomain [173] on another protein(s)). Insect HMGB proteins are also phosphorylated *in vivo* at multiple serine residues located within their short acidic tails [87,174,175]. This modification reduces the affinity of the proteins for 4-way junctions by $\sim$100-fold *in vitro* but has no effect on their binding to linear DNA [174] and could conceivably be involved in differential recognition of different DNA substrates.

4. Other functions for HMGB proteins

There are persistent reports that HMGB1, although predominantly nuclear, can also occur in the cytoplasm [176] as first observed over 20 years ago [177]. It also occurs on the external surface of the plasma membrane [178,179], in which context it has been previously studied as amphoterin. (The name recognises its bipolar (basic/acidic) nature.) It can be released from cells either in a controlled manner or passively as a result of cell damage. It signals to neighbouring cells, with important biological consequences (see below). The only receptor for HMGB1 so far identified is the cell-surface RAGE receptor (*r*eceptor for *a*dvanced *g*lycation *e*ndproducts, a multiligand member of the immunoglobulin superfamily) [180] present on many cell types, which binds a number of ligands other than end-products of glycation and activates several different intracellular signaling systems [181].

In the absence of cell damage, HMGB1 is released by an unidentified mechanism that apparently does not involve the endoplasmic reticulum [182]. Early observations showed that HMGB1 (amphoterin) secreted by neurons, where it is associated with the plasma membrane of the advancing filopodia, promotes the outgrowth of neurites [178,179], suggesting a role in neural development and cell motility. Extracellular HMGB1 has also been reported to stimulate the differentiation of murine erythroleukaemia cells, an effect that appears to be independent of the RAGE receptor [182]. More recently HMGB1 (and its individual DNA-binding domains) were shown to stimulate cell migration [183,184]. HMGB1 is also secreted by activated monocytes and plays important roles in inflammation and tumour metastasis ([185,186], reviewed in Refs. [187,188]). Extracellular amphoterin (HMGB1) stimulates pro-inflammatory cytokine synthesis in monocytes [189], and its release therefore needs to be tightly regulated. This regulation breaks down in necrotic cells, which leak HMGB1 (and presumably HMGB2) in an uncontrolled manner, leading to an inflammatory response in target cells [95]. Necrotic HMGB1 knockout cells have a greatly reduced ability to cause inflammation [95].

In contrast to necrotic cells, where HMGB1 release is effectively a marker for cell death, apoptotic cells do not release HMGB1 and do not generate an inflammatory response. Release can however be triggered by the deacetylase inhibitor trichostatin A, leading to the suggestion that in apoptotic cells chromatin hypoacetylation is responsible for HMGB1 retention [95]. Another cell death pathway, induced by the cytotoxic T-lymphocyte protease granzyme A, is caspase-independent and is characterised by DNA single-stranded nicking rather than oligonucleosomal fragmentation. Granzyme A is nuclear and cleaves HMGB2, but not HMGB1, within the A-domain [100]. In mammalian K562 and HeLa cells, HMGB2 is found associated with the SET nucleosome assembly and DNA repair complex in the endoplasmic reticulum [100]. Since both HMGB2 and APE, another component of the SET complex, are substrates for granzyme A this physical association has led to the suggestion that on transport to the nucleus SET is inactivated by cleavage of HMGB2 and also APE [100].

It seems likely that the extracellular effects are independent of the roles of HMGB1,2 in organising chromatin structure and function, although in some cases their release is evidently a marker for cell death and chromatin degradation, although not through the apoptotic route. Release into the extracellular space for "productive" signaling purposes must be very finely controlled to achieve local effects (e.g., leading to cell migration) without the damage triggered by indiscriminate release through necrotic cell death.

Acknowledgements

We thank Drs. Juli Feigon and Reid Johnson for their generous help with Fig. 3.

References

1. Johns, E.W. and Forrester, S. (1969) Studies on nuclear proteins. The binding of extra acidic proteins to deoxyribonucleoprotein during the preparation of nuclear proteins. Eur. J. Biochem. 8, 547–551.
2. Bonne, C., Duguet, M., and de Recondo, A.M. (1980) Single-strand DNA binding protein from rat liver: interactions with supercoiled DNA. Nucleic Acids Res. 8, 4955–4968.
3. Javaherian, K., Liu, J.F., and Wang, J.C. (1978) Nonhistone proteins HMG1 and HMG2 change the DNA helical structure. Science. 199, 1345–1346.
4. Sheflin, L.G. and Spaulding, S.W. (1989) High mobility group protein 1 preferentially conserves torsion in negatively supercoiled DNA. Biochemistry 28, 5658–5664.
5. Strŏs, M., Stokrová, J., and Thomas, J.O. (1994) DNA looping by the HMG-box domains of HMG1 and modulation of DNA binding by the acidic C-terminal domain. Nucleic Acids Res. 22, 1044–1051.
6. Payet, D. and Travers, A.A. (1997) The acidic tail of the high mobility group protein HMG-D modulates the structural selectivity of DNA binding. J. Mol. Biol. 266, 66–75.
7. Webb, M., Payet, D., Lee, K.-B., Travers, A.A., and Thomas, J.O. (2001) Structural requirements for cooperative binding of HMG1 to DNA minicircles. J. Mol. Biol. 309, 79–88.
8. Bianchi, M.E. (1988) Interaction of a protein from rat liver nuclei with cruciform DNA. EMBO J. 7, 843–849.
9. Pil, P.M. and Lippard, S.J. (1992) Specific binding of chromosomal protein HMG1 to DNA damaged by the anticancer drug cisplatin. Science 256, 234–237.
10. Bustin, M. and Reeves, R. (1996) High mobility group chromosomal proteins: architectural components that facilitate chromatin function. Progr. Nucl. Acid Res. Mol. Biol. 54, 35–100.
11. Bustin, M. (1999) Regulation of DNA-dependent activities by the functional motifs of the high-mobility-group chromosomal proteins. Mol. Cell. Biol. 19, 5237–5246.
12. Bianchi, M.E. and Beltrame, M. (2000) Upwardly mobile proteins. EMBO Rep. 1, 109–119.
13. Thomas, J.O. and Travers, A.A. (2001) HMG1 and 2, and related 'architectural' DNA-binding proteins. Trends Biochem. Sci. 26, 167–174.
14. Thomas, J.O. (2001) HMG1 and 2: architectural DNA-binding proteins. Biochem. Soc. Trans. 29, 395–401.
15. Baxevanis, A.D. and Landsman, D. (1995) The HMG-1 box protein family: classification and functional relationships. Nucleic Acids Res. 23, 1604–1613.
16. Duguet, M. and de Recondo, A.M. (1978) A deoxyribonucleic acid unwinding protein isolated from regenerating rat liver. Physical and functional properties. J. Biol. Chem. 253, 1660–1666.
17. Goodwin, G.H., Sanders, C., and Johns, E.W. (1973) A new group of chromatin-associated proteins with a high content of acidic and basic amino acids. Eur. J. Biochem. 38, 14–19.

18. Bustin, M. (2001) Revised nomenclature for high mobility group (HMG) chromosomal proteins. Trends Biochem. Sci. 26, 152–153.
19. Lu, J., Kobayashi, R., and Brill, S.J. (1996) Characterization of a high mobility group 1/2 homolog in yeast. J. Biol. Chem. 271, 33678–33685.
20. Grasser, K.D. (1998) HMG1 and HU proteins: architectural elements in plant chromatin. Trends Plant Sci. 3, 260–265.
21. Weir, H.M., Kraulis, P.J., Hill, C.S., Raine, A.R., Laue, E.D., and Thomas, J.O. (1993) Structure of the HMG box motif in the B-domain of HMG1. EMBO J. 12, 1311–1319.
22. Read, C.M., Cary, P.D., Crane-Robinson, C., Driscoll, P.C., and Norman, D.G. (1993) Solution structure of a DNA-binding domain from HMG1. Nucleic Acids Res. 21, 3427–3436.
23. Jones, D.N.M, Searles, M.A., Shaw, G.L., Churchill, M.E.A., Ner, S.S., Keeler, J., Travers, A.A., and Neuhaus, D. (1994) The solution structure and dynamics of the DNA-binding domain of HMG-D from *Drosophila melanogaster*. Structure 2, 609–627.
24. Werner, M.H., Huth, J.R., Gronenborn, A.M., and Clore, G.M. (1995) Molecular-basis of human 46X,Y sex reversal revealed from the 3-dimensional solution structure of the human SRY-DNA complex. Cell 81, 705–714.
25. Love, J.J., Li, X., Case, D.A., Giese, K., Grosschedl, R., and Wright, P.E. (1995) Structural basis for DNA bending by the architectural transcription factor LEF-1. Nature 376, 791–795.
26. van Houte, L.P., Chuprina, V.P., van der Wetering, M., Boelens, R., Kaptein, R., and Clevers, H. (1995) Solution structure of the sequence-specific HMG box of the lymphocyte transcriptional activator Sox-4. J. Biol. Chem. 270, 30516–30524.
27. Hardman, C.H., Broadhurst, R.W., Raine, A.R., Grasser, K.D., Thomas, J.O., and Laue, E.D. (1995) Structure of the A domain of HMGB1 and its interaction with DNA as studied by heteronuclear three- and four-dimensional NMR spectroscopy. Biochemistry 34, 16596–16607.
28. Allain, F.H.-T., Yen, Y.-M., Masse, J.E., Schultze, P., Dieckmann, T., Johnson, R.C., and Feigon, J. (1999) Solution structure of the HMG protein NHP6A and its interaction with DNA reveals the structural determinants for non-sequence-specific binding. EMBO J. 18, 2563–2579.
29. Teo, S.-H., Grasser, K.D., and Thomas, J.O. (1995) Differences in the DNA-binding properties of the HMG-box domains of HMG1 and the sex-determining factor SRY. Eur. J. Biochem. 230, 943–950.
30. Dunham, S.U. and Lippard, S.J. (1997) DNA sequence context and protein composition modulate HMG-domain protein recognition of cisplatin-modified DNA. Biochemistry 36, 11428–11436.
31. Webb, M. and Thomas, J.O. (1999) Structure-specific-binding of the HMG1 didomain to four-way junction DNA is mediated by the A domain. J. Mol. Biol. 294, 373–387.
32. Murphy, F.V. 4th, Sweet, R.M., and Churchill, M.E.A. (1999) The structure of a chromosomal high mobility group protein–DNA complex reveals sequence-neutral mechanisms important for non-sequence specific DNA recognition EMBO J. 18, 6610–6618.
33. Lorenz, M., Hillisch, A., Payet, D., Buttinelli, M., Travers, A., and Diekmann, S. (1999) DNA bending induced by high mobility group proteins studied by fluorescence resonance energy transfer. Biochemistry 38, 12150–12158.
34. Jamieson, E.R., Jacobson, M.P., Barnes, C.M., Chow, C.S., and Lippard, S.J. (1999) Structural and kinetic studies of a cisplatin-modified DNA icosamer binding to HMGB1 domain B. J. Biol. Chem. 274, 12346–12354.
35. Paull, T.T., Haykinson, M.J., and Johnson, R.C. (1993) The nonspecific DNA-binding and -bending proteins HMG1 and HMG2 promote the assembly of complex nucleoprotein structures. Genes Dev. 7, 1521–1534.
36. Arimondo, P.B., Gelus, N., Hamy, F., Payet, D., Travers, A., and Bailly, C. (2000) The chromosomal protein HMG-D binds to the TAR and RBE RNA of HIV-1. FEBS Lett. 485, 47–52.
37. Wisniewski, J.R., Krohn, N.M., Heyduk, E., Grasser, K.D., and Heyduk, T. (1999) HMG1 proteins from evolutionary distant organisms distort B-DNA conformation in similar way. Biochim. Biophys. Acta 1447, 25–34.

38. Heyduk, E., Heyduk, T., Claus, P., and Wisniewski, J.R. (1997) Conformational changes of DNA induced by binding of Chironomus high mobility group protein 1a (cHMGB1a). Regions flanking an HMGB1 box domain do not influence the bend angle of the DNA. J. Biol. Chem. 272, 19763–19770.
39. King, C.-Y. and Weiss, M.A. (1993) The SRY high-mobility-group box recognizes DNA by partial intercalation in the minor groove: a topological mechanism of sequence specificity. Proc. Natl. Acad. Sci. USA 90, 11990–11994.
40. Murphy, E.C., Zhurkin, V.B., Louis, J.M., Cornilescu, G., and Clore, G.M. (2001) Structural basis for SRY-dependent 46-X,Y sex reversal: modulation of DNA bending by a naturally occurring point mutation. J. Mol. Biol. 312, 481–499.
41. Cerdan, R., Payet, D., Yang, J.-C., Travers, A.A., and Neuhaus, D. (2001) HMG-D complexed to a bulge DNA: an NMR model. Protein Sci. 10, 504–518.
42. Payet, D., Hillisch, A., Lowe, N., Diekmann, S., and Travers, A. (1999) The recognition of distorted DNA structures by HMG-D: a footprinting and molecular modelling study. J. Mol. Biol. 294, 79–91.
43. Ohndorf, U.-M., Rould, M.A., He, Q., Pabo, C.O., and Lippard, S.J. (1999) Basis for recognition of cisplatin-modified DNA by high-mobility group proteins. Nature 399, 708–712.
44. Masse, J.E., Wong, B., Yen, Y.-M., Allain, F.H.-T., Johnson, R.C., and Feigon, J. (2002) The *S. cerevisiae* architectural HMGB protein NHP6A complexed with DNA: DNA and protein conformational changes on binding. J. Mol. Biol. 323, 263–284.
45. Pohler, J.R., Norman, D.G., Bramham, J., Bianchi, M.E., and Lilley, D.M.J. (1998) HMG box proteins bind to four-way DNA junctions in their open conformation. EMBO J. 17, 817–826.
46. Taudte, S., Xin, H., and Kallenbach, N.R. (2000) Alanine mutagenesis of high-mobility-group-protein-1 box B (HMG1-B). Biochem. J. 347, 807–814.
47. Strŏs, M. and Muselíková, E. (2000) A role of basic residues and the putative intercalating phenylalanine of the HMGB1-box B in DNA supercoiling and binding to four-way DNA junctions. J. Biol. Chem. 275, 35699–35707.
48. Haqq, C.M., King, C.Y., Ukiyama, E., Falsafi, S., Haqq, T.N., Donahoe, P.K., and Weiss, M.A. (1994) Molecular basis of mammalian sexual determination: activation of Mullerian inhibiting substance gene expression by SRY. Science 266, 1494–1500.
49. Wong, B., Masse, J.E., Yen, Y.-M., Feigon, J., and Johnson, R.C. (2002) Binding to cisplatin-modified DNA by the *Saccharomyces cerevisiae* HMGB protein NHP6A. Biochemistry 41, 5404–5414.
50. He, Q., Ohndorf, U.M., and Lippard, S.J. (2000) Intercalating residues determine the mode of HMG1 domains A and B binding to cisplatin-modified DNA. Biochemistry 39, 14426–14435.
51. Dow, L.K., Jones, D.N., Wolfe, S.A., Verdine, G.L., and Churchill, M.E.A. (2000) Structural studies of the high mobility group globular domain and basic tail of HMG-D bound to disulfide cross-linked DNA. Biochemistry 39, 9725–9736.
52. Lnenicek-Allen, M., Read, C.M., and Crane-Robinson, C. (1996) The DNA bend angle and binding affinity of an HMG box is increased by the presence of short terminal arms. Nucleic Acids Res. 24, 1047–1051.
53. Grasser, K.D., Teo, S.-H., Lee, K.-B., Broadhurst, R.W., Rees, C., Hardman, C.H., and Thomas, J.O. (1998) DNA-binding properties of the tandem HMG boxes of HMG1. Eur. J. Biochem. 253, 787–795.
54. Yen, Y.-M., Wong, B., and Johnson, R.C. (1998) Determinants of DNA binding and bending by the *Saccharomyces cerevisiae* high mobility group protein NHP6A that are important for its biological activities. Role of the unique N terminus and putative intercalating methionine. J. Biol. Chem. 273, 4424–4435.
55. Crothers, D.M. (1993) Architectural elements in nucleoprotein structures. Curr. Biol. 3, 675–676.
56. Strŏs, M. (2001) Two mutations of basic residues within the N-terminus of HMG-1 B domain with different effects on DNA supercoiling and binding to bent DNA. Biochemistry 40, 4769–4779.

57. Sheflin, L.G., Fucile, N.W., and Spaulding, S.W. (1993) The specific interactions of HMG1 and 2 with negatively supercoiled DNA are modulated by their acidic C-terminal domains and involve cysteine residues in their HMG1/2 boxes. Biochemistry 32, 3238–3248.
58. Pasheva, E.A., Pashev, I.G., and Favre, A. (1998) Preferential binding of high mobility group 1 protein to UV-damaged DNA. Role of the COOH-terminal domain. J. Biol. Chem. 273, 24730–24736.
59. Lee, K.-B. and Thomas, J.O. (2000) The effect of the acidic tail on the DNA-binding properties of the HMG1,2 class of proteins: insights from tail switching and tail removal. J. Mol. Biol. 304, 135–149.
60. Ramstein, J., Locker, D., Bianchi, M.E., and Leng, M. (1999) Domain–domain interactions in high mobility group 1 protein (HMGB1). Eur. J. Biochem. 260, 692–700.
61. Petry, I., Wisniewski, J.R., and Szewczuk, Z. (2001) Conformational stability of six truncated cHMG1a proteins studied in their mixture by H/D exchange and electrospray ionization mass spectrometry. Acta Biochim. Pol. 48, 1131–1136.
62. Travers, A.A. (1995) Reading the minor groove. Nat. Struct. Biol. 2, 615–618.
63. Bonne, C., Sautière, P., Duguet, M., and de Recondo, A.M. (1982) Identification of a single-stranded DNA binding protein from rat liver with high mobility group protein 1. J. Biol. Chem. 257, 2722–2725.
64. Travers, A.A., Ner, S.S., and Churchill, M.E.A. (1994) DNA chaperones: a solution to a persistence problem? Cell 77, 167–169.
65. Ross, E.D., Hardwidge, P.R., and Maher, L.J., 3rd. (2001) HMG proteins and DNA flexibility in transcription activation Mol. Cell. Biol. 21, 6598–6605.
66. Bianchi, M.E. (1994) Prokaryotic HU and eukaryotic HMG1: a kinked relationship. Mol. Microbiol. 14, 1–5.
67. Paull, T.T. and Johnson, R.C. (1995) DNA looping by *Saccharomyces cerevisiae* high mobility group proteins NHP6A/B. Consequences for nucleoprotein complex assembly and chromatin condensation. J. Biol. Chem. 270, 8744–8754.
68. Megraw, T.L. and Chae, C.B. (1993) Functional complementarity between the HMG1-like yeast mitochondrial histone HM and the bacterial histone-like protein HU. J. Biol. Chem. 268, 12758–12763.
69. Shirakata, M., Huppi, K., Usuda, S., Okazaki, K., Yoshida, K., and Sakano, H. (1991) HMG1-related DNA-binding protein isolated with V-(D)-J recombination signal probes. Mol. Cell. Biol. 11, 4528–4536.
70. Agrawal, A. and Schatz, D.G. (1997) RAG1 and RAG2 form a stable postcleavage synaptic complex with DNA containing signal ends in V(D)J recombination. Cell 89, 43–53.
71. Sawchuk, D.J., Weis-Garcia, F., Malik, S., Besmer, E., Bustin, M., Nussenzweig, M.C., and Cortes, P. (1997) V(D)J recombination: modulation of RAG1 and RAG2 cleavage activity on 12/23 substrates by whole cell extract and DNA-bending proteins. J. Exp. Med. 185, 2025–2032.
72. van Gent, D.C., Hiom, K., Paull, T.T., and Gellert, M. (1997) Stimulation of V(D)J cleavage by high mobility group proteins. EMBO J. 16, 2665–2670.
73. Kwon, J., Imbalzano, A.N., Matthews, A., and Oettinger, M.A. (1998) Accessibility of nucleosomal DNA to V(D)J cleavage is modulated by RSS positioning and HMG1. Mol. Cell 2, 829–839.
74. West, R.B. and Lieber, M.R. (1998) The RAG-HMG1 complex enforces the 12/23 rule of V(D)J recombination specifically at the double-hairpin formation step. Mol. Cell. Biol. 18, 6408–6415.
75. Aidinis, V., Bonaldi, T., Beltrame, M., Santagata, S., Bianchi, M.E., and Spanopoulou, E. (1999) The RAG1 homeodomain recruits HMG1 and HMG2 to facilitate recombination signal sequence binding and to enhance the intrinsic DNA-bending activity of RAG1–RAG2. Mol. Cell. Biol. 19, 6532–6542.
76. Swanson, P.C. (2002) Fine structure and activity of discrete RAG–HMG complexes on V(D)J recombination signals. Mol. Cell. Biol. 22, 1340–1351.
77. Ellwood, K.B., Yen, Y.M., Johnson, R.C., and Carey, M. (2000) Mechanism for specificity by HMG-1 in enhanceosome assembly. Mol. Cell. Biol. 20, 4359–4370.
78. Costello, E., Saudan, P., Winocour, E., Pizer, L., and Beard, P. (1997) High mobility group chromosomal protein 1 binds to the adeno-associated virus replication protein (Rep) and promotes

Rep-mediated site-specific cleavage of DNA, ATPase activity and transcriptional repression. EMBO J. 16, 5943–5954.
79. Cotmore, S.F., Christensen, J., and Tattersall, P. (2000) Two widely spaced initiator binding sites create an HMG1-dependent parvovirus rolling-hairpin replication origin. J. Virol. 74, 1332–1341.
80. Johns, E.W. (ed.) The HMG Chromosomal Proteins. Academic Press, New York.
81. Jackson, J.B., Pollock, J.M., Jr., and Rill, R.L. (1979) Chromatin fractionation procedure that yields nucleosomes containing near-stoichiometric amounts of high mobility group nonhistone chromosomal proteins. Biochemistry 18, 3739–3748.
82. Formosa, T., Eriksson, P., Wittmeyer, J., Ginn, J., Yu, Y., and Stillman, D.J. (2001) Spt16-Pob3 and the HMG protein Nhp6 combine to form the nucleosome-binding factor SPN. EMBO J. 20, 3506–3517.
83. Nightingale, K., Dimitrov, S., Reeves, R., and Wolffe, A.P. (1996) Evidence for a shared structural role for HMG1 and linker histones B4 and H1 in organizing chromatin. EMBO J. 15, 548–561.
84. Lichota, J. and Grasser, K.D. (2001) Differential association and nucleosome binding of the maize HMGA, HMGB, and SSRP1 proteins. Biochemistry 40, 7860–7867.
85. Schroter, H. and Bode, J. (1982) The binding sites for large and small high-mobility-group (HMG) proteins. Studies on HMG-nucleosome interactions in vitro. Eur. J. Biochem. 127, 429–436.
86. An, W., van Holde, K., and Zlatanova, J. (1998) The non-histone chromatin protein HMG1 protects linker DNA on the side opposite to that protected by linker histones. J. Biol. Chem. 273, 26289–26291.
87. Ner, S.S., Blank, T., Pérez-Parallé, M.L., Grigliatti, T.A., Becker, P.B., and Travers, A.A. (2001) HMG-D and histone H1 interplay during chromatin assembly and early embryogenesis. J. Biol. Chem. 276, 37569–37576.
88. Blank, T.A. and Becker, P.B. (1995) Electrostatic mechanism of nucleosome spacing. J. Mol. Biol. 252, 305–313.
89. Tremethick, D.J. and Drew, H.R. (1993) High mobility group proteins 14 and 17 can space nucleosomes in vitro. J Biol. Chem. 268, 11389–11393.
90. Ogawa, Y., Aizawa, S., Shirakawa, H., and Yoshida, M. (1995) Stimulation of transcription accompanying relaxation of chromatin structure in cells overexpressing high mobility group 1 protein. J. Biol. Chem. 270, 9272–9280.
91. Ner, S.S. and Travers, A.A. (1994) HMG-D, the Drosophila melanogaster homologue of HMG 1 protein, is associated with early embryonic chromatin in the absence of histone H1. EMBO J. 13, 1817–1822.
92. Ura, K., Nightingale, K., and Wolffe, A.P. (1996) Differential association of HMG1 and linker histones B4 and H1 with dinucleosomal DNA: structural transitions and transcriptional repression. EMBO J. 15, 4959–4969.
93. Adenot, P.G., Campion, E., Legouy, E., Allis, C.D., Dimitrov, S., Renard, J., and Thompson, E.M. (2000) Somatic linker histone H1 is present throughout mouse embryogenesis and is not replaced by variant H1°. J. Cell Sci. 113, 2897–2907.
94. Takami, Y. and Nakayama, T. (1997) A single copy of linker H1 genes is enough for proliferation of the DT40 chicken B cell line, and linker H1 variants participate in regulation of gene expression. Genes Cells 2, 711–723.
95. Scaffidi, P., Misteli, T., and Bianchi, M.E. (2002) Release of chromatin protein HMGB1 by necrotic cells triggers inflammation. Nature 418, 191–195.
96. Carballo, M., Puigdomenech, P., and Palau, J. (1983) DNA and histone H1 interact with different domains of HMG 1 and 2 proteins. EMBO J. 2, 1759–1764.
97. Kohlstaedt, L.A. and Cole, R.D. (1994) Specific interaction between H1 histone and high mobility protein HMG1. Biochemistry 33, 570–575.
98. Bonne-Andrea, C., Harper, F., Sobczak, J., and de Recondo, A.M. (1984) Rat liver HMG1: a physiological nucleosome assembly factor. EMBO J. 3, 1193–1199.
99. Waga, S., Mizuno, S., and Yoshida, M. (1989) Nonhistone proteins HMG1 and HMG2 suppress the nucleosome assembly at physiological ionic strength. Biochim. Biophys. Acta 1007, 209–214.

100. Fan, Z., Beresford, P.J., Zhang, D., and Lieberman, J. (2002) HMG2 interacts with the nucleosome assembly protein SET and is a target of the cytotoxic T-lymphocyte protease granzyme A. Mol. Cell Biol. 22, 2810–2820.
101. Seo, S.B., McNamara, P., Heo, S., Turner, A., Lane, W.S., and Chakravarti, D. (2001) Regulation of histone acetylation and transcription by INHAT, a human cellular complex containing the set oncoprotein. Cell 104, 119–130.
102. Shikama, N., Chan, H.M., Krstic-Demonacos, M., Smith, L., Lee, C.W., Cairns, W., and La Thangue, N.B. (2000) Functional interaction between nucleosome assembly proteins and p300/CREB-binding protein family coactivators. Mol. Cell. Biol. 20, 8933–8943.
103. Bennett, R.A., Wilson, D.M., 3rd, Wong, D., and Demple, B. (1997) Interaction of human apurinic endonuclease and DNA polymerase beta in the base excision repair pathway. Proc. Natl. Acad. Sci. USA 94, 7166–7169.
104. Tini, M., Benecke, A., Um, S.J., Torchia, J., Evans, R.M., and Chambon, P. (2002) Association of CBP/p300 acetylase and thymine DNA glycosylase links DNA repair and transcription. Mol. Cell. 9, 265–277.
105. Deleted.
106. Wang, W., Chi, T., Xue, Y., Zhou, S., Kuo, A., and Crabtree, G.R. (1998) Architectural DNA binding by a high-mobility-group/kinesin-like subunit in mammalian SWI/SNF-related complexes. Proc. Natl. Acad. Sci. USA 95, 492–498.
107. Papoulas, O., Daubresse, G., Armstrong, J.A., Jin, J., Scott, M.P., and Tamkun, J. (2001) The HMG-domain protein BAP111 is important for the function of the BRM chromatin-remodeling complex *in vivo*. Proc. Natl. Acad. Sci. USA 98, 5728–5733.
108. Chi, T.H., Wan, M., Zhao, K., Taniuchi, I., Chen, L., Littman, D.R., and Crabtree, G.R. (2002) Reciprocal regulation of CD4/CD8 expression by SWI/SNF-like BAF complexes. Nature 418, 195–199.
109. Nicolas, R.H. and Goodwin, G.H. (1996) Molecular cloning of polybromo, a nuclear protein containing multiple domains including five bromodomains, a truncated HMG-box, and two repeats of a novel domain. Gene 175, 233–240.
110. Xue, Y., Canman, J.C., Lee, C.S., Nie, Z., Yang, D., Moreno, G.T., Young, M.K., Salmon, E.D., and Wang, W. (2000) The human SWI/SNF-B chromatin-remodeling complex is related to yeast Rsc and localizes at kinetochores of mitotic chromosomes. Proc. Natl. Acad. Sci. USA 97, 13015–13020.
111. Moreira, J.M. and Holmberg, S. (1999) Transcriptional repression of the yeast *CHA1* gene requires the chromatin-remodeling complex RSC. EMBO J. 18, 2836–2844.
112. Moreira, J.M.A. and Holmberg, S. (2000) Chromatin-mediated transcriptional regulation by the yeast architectural factors NHP6A and NHP6B. EMBO J. 19, 6804–6813.
113. Kelley, D.E., Stikes, D.G., and Perry, R.P. (1999) CHD1 interacts with SSRP1 and depends on both its chromodomain and its ATPase/helicase-like domain for proper association with chromatin. Chromosoma 108, 10–25.
114. Orphanides, G., Wu, W.-H., Lane, W.S., Hampsey, M., and Reinberg, D. (1999) The chromatin-specific transcription factor FACT comprises human SPT16 and SSRP1 proteins. Nature 400, 284–288.
115. Wada, T., Orphanides, G., Hasegawa, J., Kim, D.K., Shima, D., Yamaguchi, Y., Fukuda, A., Hisatake, K., Oh, S., Reinberg, D., and Handa, H. (2000) FACT relieves DSIF/NELF-mediated inhibition of transcriptional elongation and reveals functional differences between P-TEFb and TFIIH. Mol. Cell 5, 1067–1072.
116. Brewster, N.K., Johnston, G.C., and Singer, R.A. (2001) A bipartite yeast SSRP1 analog comprised of Pob3 and Nhp6 proteins modulates transcription. Mol. Cell. Biol. 21, 3491–3502.
117. Keller, D.M., Zeng, X., Wang, Y., Zhang, Q.H., Kapoor, M., Shu, H., Goodman, R., Lozano, G., Zhao, Y., and Lu, H. (2001) A DNA damage-induced p53 serine 392 kinase complex contains CK2, hSpt16, and SSRP1. Mol. Cell 7, 283–292.

118. Kehle, J., Beuchle, D., Treuheit, S., Christen, B., Kennison, J.A., Bienz, M., and Müller, J. (1998) dMi-2, a Hunchback-interacting protein that functions in Polycomb repression. Science 282, 1897–1900.
119. Travers, A.A. (2003) Priming the nucleosome—a role for HMGB proteins? EMBO Rep. 4, 131–136.
120. Bonaldi, T., Längst, G., Strohner, R., Becker, P.B., and Bianchi, M.E. (2002) The DNA chaperone HMGB1 facilitates ACF/CHRAC-dependent nucleosome sliding. EMBO J. 21, 6865–6873.
121. Ner, S.S., Churchill, M.E., Searles, M.A., and Travers, A.A. (1993) dHMG-Z, a second HMG-1-related protein in *Drosophila melanogaster*. Nucleic Acids Res. 21, 4369–4371.
122. Stoute, J.A. and Marzluff, W.F. (1982) HMG-proteins 1 and 2 are required for transcription of chromatin by endogenous RNA polymerase. Biochem. Biophys. Res. Comm. 107, 1279–1284.
123. Tremethick, D.J. and Molloy, P.L. (1986) High mobility group proteins 1 and 2 stimulate transcription in vitro by RNA polymerases II and III. J. Biol. Chem. 261, 6986–6992.
124. Singh, J. and Dixon, G.H. (1990) High mobility group proteins 1 and 2 function as general class II transcription factors. Biochemistry 29, 6295–6302.
125. Paull, T.T., Carey, M., and Johnson, R.C. (1996) Yeast HMG proteins NHP6A/B potentiate promoter-specific transcriptional activation in vivo and assembly of preinitiation complexes in vitro. Genes Dev. 10, 2769–2781.
126. Sidorova, J. and Breeden, L. (1999) The *MSN* and *NHP6A* genes suppress *SWI6* defects in *Saccharomyces cerevisiae*. Genetics 151, 45–55.
127. Yu, Y., Eriksson, P., and Stillman, D.J. (2000) Architectural factors and the SAGA complex function in parallel pathways to activate transcription. Mol. Cell. Biol. 20, 2350–2357.
128. Kruppa, M., Moir, R.D., Kolodrubetz, D., and Willis, I.M. (2001) Nhp6, an HMG1 protein, functions in SNR6 transcription by RNA polymerase III in *S. cerevisiae*. Mol. Cell 7, 309–318.
129. Lopez, S., Livingstone-Zatchej, M., Jourdain, S., Thoma, F., Sentenac, A., and Marsolier, M.-C. (2001) High-mobility-group proteins NHP6A and NHP6B participate in activation of the RNA polymerase III SNR6 Gene. Mol. Cell. Biol. 21, 3096–3104.
130. Moreira, J.M. and Holmberg, S. (1998) Nucleosome structure of the yeast CHA1 promoter: analysis of activation-dependent chromatin remodeling of an RNA-polymerase-II-transcribed gene in TBP and RNA pol II mutants defective *in vivo* in response to acidic activators. EMBO J. 17, 6028–6038.
131. Verdone, L., Wu, J., van Riper, K., Kacherovsky, N., Vogelauer, M., Young, E.T., Grunstein, M., Di Mauro, E., and Caserta, M. (2002) Hyperacetylation of chromatin at the ADH2 promoter allows Adr1 to bind in repressed conditions. EMBO J. 21, 1101–1111.
132. Tse, C., Sera, T., Wolffe, A.P., and Hansen, J.C. (1998) Disruption of higher-order folding by core histone acetylation dramatically enhances transcription of nucleosomal arrays by RNA polymerase III. Mol. Cell. Biol. 18, 4629–4638.
133. Horn, P.J., Crowley, K.A., Carruthers, L.M., Hansen, J.C., and Peterson, C.L. (2002) The SIN domain of the histone octamer is essential for intramolecular folding of nucleosomal arrays. Nat. Struct. Biol. 9, 167–171.
134. Anderson, J.D., Lowary, P.T., and Widom, J. (2001) Effects of histone acetylation on the equilibrium accessibility of nucleosomal DNA target sites. J. Mol. Biol. 307, 977–985.
135. Bustin, M. (2001) Chromatin unfolding and activation by HMGN chromosomal proteins. Trends Biochem. Sci. 26, 431–437.
136. Macatee, T., Jiang, Y.W., Stillman, D.J., and Roth, S.Y. (1997) Global alterations in chromatin accessibility associated with loss of *SIN4* function. Nucleic Acids Res. 25, 1240–1247.
137. Watt, F. and Molloy, P.L. (1988) High mobility group proteins 1 and 2 stimulate binding of a specific transcription factor to the adenovirus major late promoter. Nucleic Acids Res. 16, 1471–1486.
138. Zwilling, S., Konig, H., and Wirth, T. (1995) High mobility group protein 2 funtionally interacts with the POU domains of octamer transcription factors. EMBO J. 14, 1198–1208.
139. Zappavigna, V., Falciola, L., Helmer-Citterich, M., Mavilio, F., and Bianchi, M.E. (1996) HMG1 interacts with Hox proteins and enhances their DNA-binding and transcriptional activation. EMBO J. 15, 4981–4991.

140. Jayaraman, L., Moorthy, N.C., Murthy, K.G., Manley, J.L., Bustin, M., and Prives, C. (1998) High mobility group protein-1 (HMG-1) is a unique activator of p53. Genes Dev. 12, 462–472.
141. McKinney, K. and Prives, C. (2002) Efficient specific DNA binding by p53 requires both its central and C-terminal domains as revealed by studies with High-Mobility Group 1 protein. Mol. Cell. Biol. 22, 6797–6808.
142. Boonyaratanakornkit, V., Melvin, V., Prendergast, P., Altmann, M., Ronfani, L., Bianchi, M.E., Taraseviciene, L., Nordeen, S.K., Allegretto, E.A., and Edwards, D.P. (1998) High-mobility group chromatin proteins 1 and 2 functionally interact with steroid hormone receptors to enhance their DNA binding in vitro and transcriptional activity in mammalian cells. Mol. Cell. Biol. 18, 4471–4487.
143. Verrijdt, G., Haelens, A., Schoenmakers, E., Rombauts, W., and Claessens, F. (2002) Comparative analysis of the influence of the high-mobility group box 1 protein on DNA binding and transcriptional activation by the androgen, glucocorticoid, progesterone and mineralocorticoid receptors. Biochem. J. 361, 97–103.
144. Melvin, V.S., Roemer, S.C., Churchill, M.E.A., and Edwards, D.P. (2002) The C-terminal extension (CTE) of the nuclear hormone receptor DNA binding domain determines interactions and functional response to the HMGB-1/-2 co-regulatory proteins. J. Biol. Chem. 277, 25115–25124.
145. Brickman, J.M., Adam, M., and Ptashne, M. (1999) Interactions between an HMG-1 protein and members of the Rel family. Proc. Natl. Acad. Sci. USA 96, 10679–10683.
146. Decoville, M., Giraud-Panis, M.J., Mosrin-Huaman, C., Leng, M., and Locker, D. (2000) HMG boxes of DSP1 protein interact with the rel homology domain of transcription factors. Nucleic Acids Res. 28, 454–462.
147. Strŏs, M., Ozaki, T., Bacíková, A., Kageyama, H., and Nakagawara, A. (2002) HMGB1 and HMGB2 cell-specifically down-regulate the p53- and p73-dependent sequence-specific transactivation from the human Bax gene promoter. J. Biol. Chem. 277, 7157–7164.
148. Krohn, N.M., Yanagisawa, S., and Grasser, K.D. (2002) Specificity of the stimulatory interaction between chromosomal HMGB proteins and the transcription factor Dof2 and its negative regulation by protein kinase CK2-mediated phosphorylation. J. Biol. Chem. 277, 32438–32444.
149. Mitsouras, M., Wong, B., Arayata, C., Johnson, R.C., and Carey, M. (2002) The DNA architectural protein HMGB1 displays two distinct modes of action that promote enhanceosome assembly. Mol. Cell. Biol. 22, 4390–4401.
150. Spencer, J.A., Baron, M.H., and Olson, E.N. (1999) Cooperative transcriptional activation by serum response factor and the high mobility group protein SSRP1. J. Biol. Chem. 274, 15686–15693.
151. Zeng, S.X., Dai, M.S., Keller, D.M., and Lu, H. (2002) SSRP1 functions as a co-activator of the transcriptional activator p63. EMBO J. 21, 5487–5497.
152. Laser, H., Bongards, C., Schüller, J., Heck, S., Johnsson, N., and Lehming, N. (2000) A new screen for protein interactions reveals that the *Saccharomyces cerevisiae* high mobility group proteins Nhp6A/B are involved in the regulation of the GAL1 promoter. Proc. Natl. Acad. Sci. USA 97, 13732–13737.
153. Giese, K., Kingsley, C., Kirshner, J.R., and Grosschedl, R. (1995) Assembly and function of a TCRα enhancer complex is dependent on LEF-1-induced DNA bending and multiple protein–protein interactions. Genes Dev. 9, 995–1008.
154. Ge, H. and Roeder, R.G. (1994) The high mobility group protein HMG1 can reversibly inhibit class II gene transcription by interaction with the TATA-binding protein. J. Biol. Chem. 269, 17136–17140.
155. Sutrias-Grau, M., Bianchi, M.E., and Bernués, J. (1999) High mobility group protein 1 interacts specifically with the core domain of human TATA box-binding protein and interferes with transcription factor IIB within the pre-initiation complex. J. Biol. Chem. 274, 1628–1634.
156. Shykind, B.M., Kim, J., and Sharp, P.A. (1995) Activation of the TFIID-TFIIA complex with HMG-2. Genes Dev. 9, 1354–1365.

157. Calogero, S., Grassi, F., Aguzzi, A., Voigtlander, T., Ferrier, P., Ferrari, S., and Bianchi, M.E. (1999) The lack of chromosomal protein HMGB1 does not disrupt cell growth but causes hypoglycaemia in newborn mice. Nat. Genet. 22, 276–279.
158. Ronfani, L., Ferraguti, M., Croci, L., Ovitt, C.E., Schöler, H.R., Consalez, G.G., and Bianchi, M.E. (2001) Reduced fertility and spermatogenesis defects in mice lacking chromosomal protein Hmgb2. Development 128, 1265–1273.
159. Costigan, C., Kolodrubetz, D., and Snyder, M. (1994) NHP6A and NHP6B, which encode HMG1-like proteins, are candidates for downstream components of the yeast SLT2 mitogen-activated protein kinase pathway. Mol. Cell. Biol. 14, 2391–2403.
160. Dintilhac, A. and Bernués, J. (2002) HMGB1 interacts with many apparently unrelated proteins by recognizing short amino acid sequences. J. Biol. Chem. 277, 7021–7028.
161. Anderson, J.D. and Widom, J. (2000) Sequence and position-dependence of the equilibrium accessibility of nucleosomal DNA target sites. J. Mol. Biol. 296, 979–987.
162. Brown, S.J, Kellett, P.J., and Lippard, S.J. (1993) Ixr1, a yeast protein that binds to platinated DNA and confers sensitivity to cisplatin. Science 261, 603–605.
163. Chau, K.Y., Lam, H.Y., and Lee, K.L. (1998) Estrogen treatment induces elevated expression of HMG1 in MCF-7 cells. Exp. Cell Res. 241, 269–272.
164. He, Q., Liang, C.H., and Lippard, S.J. (2000) Steroid hormones induce HMG1 overexpression and sensitize breast cancer cells to cisplatin and carboplatin. Proc. Natl. Acad. Sci. USA 97, 5768–5772.
165. Wei, M., Burenkova, O., and Lippard, S.J. (2003) Cisplatin sensitivity in Hmgb1−/− and Hmgb1+/+ mouse cells. J. Biol. Chem., 278, 1769–1773.
166. Bruhn, S.L., Pil, P.M., Essigmann, J.M., Housman, D.E., and Lippard, S.J. (1992) Isolation and characterization of human cDNA clones encoding a high mobility group box protein that recognizes structural distortions to DNA caused by binding of the anticancer agent cisplatin. Proc. Natl. Acad. Sci. USA 89, 2307–2311.
167. Yumoto, Y., Shirakawa, H., Yoshida, M., Suwa, A., Watanabe, F, and Teraoka, H. (1998) High mobility group proteins 1 and 2 can function as DNA-binding regulatory components for DNA-dependent protein kinase in vitro. J. Biochem. 124, 519–527.
168. Stros, M., Cherny, D., and Jovin, T.M. (2000) HMG1 protein stimulates DNA end joining by promoting association of DNA molecules via their ends. Eur. J. Biochem. 267, 4088–4097.
169. Nagaki, S., Yamamoto, M., Yumoto, Y., Shirakawa, H., Yoshida, M., and Teraoka, H. (1998) Non-histone chromosomal proteins HMG1 and 2 enhance ligation reaction of DNA double-strand breaks. Biochem. Biophys. Res. Commun. 246, 137–141.
170. Sterner, R., Vidali, G., Heinrikson, R.L., and Allfrey, V.G. (1978) Postsynthetic modification of high mobility group proteins. Evidence that high mobility group proteins are acetylated. J. Biol. Chem. 253, 7601–7604.
171. Sterner, R., Vidali, G., and Allfrey, V.G. (1979) Studies of acetylation and deacetylation in high mobility group proteins. Identification of the sites of acetylation in HMG-1. J. Biol. Chem. 254, 11577–11583.
172. Ugrinova, I., Pasheva, E.A., Armengaud, J., and Pashev, I.G. (2001) In vivo acetylation of HMG1 protein enhances its binding affinity to distorted DNA. Biochemistry 40, 14655–14660.
173. Marmorstein, R. and Berger, S.L. (2001) Structure and function of bromodomains in chromatin-regulating complexes. Gene 272, 1–9.
174. Wisniewski, J.R., Szewczuk, Z., Petry, I., Schwanbeck, R., and Renner, U. (1999) Constitutive phosphorylation of the acidic tails of the high mobility group 1 proteins by casein kinase II alters their conformation, stability, and DNA binding specificity. J Biol. Chem. 274, 20116–20122.
175. Renner, U., Ghidelli, S., Schafer, M.A., and Wisniewski, J.R. (2000) Alterations in titer and distribution of high mobility group proteins during embryonic development of Drosophila melanogaster. Biochim. Biophys. Acta 1475, 99–108.
176. Falciola, L., Spada, F., Calogero, S., Längst, G., Voit, R., Grummt, I., and Bianchi, M.E. (1997) High mobility group 1 protein is not stably associated with the chromosomes of somatic cells. J. Cell Biol. 137, 19–26.

177. Bustin, M. and Neihart, N.K. (1979) Antibodies against chromosomal HMG proteins stain the cytoplasm of mammalian cells. Cell 16, 181–189.
178. Merenmies, J., Pihlaskari, R., Laitinen, J., Wartiovaara, J., and Rauvala, H. (1991) 30-kDa heparin-binding protein of brain (amphoterin) involved in neurite outgrowth. Amino acid sequence and localization in the filopodia of the advancing plasma membrane. J. Biol. Chem. 266, 16722–16729.
179. Parkkinen, J., Raulo, E., Merenmies, J., Nolo, R., Kajander, E.O., Baumann, M., and Rauvala, H. (1993) Amphoterin, the 30-kDa protein in a family of HMG1-type polypeptides. Enhanced expression in transformed cells, leading edge localization, and interactions with plasminogen activation. J. Biol. Chem. 268, 19726–19738.
180. Hori, O., Brett, J., Slattery, T., Cao, R., Zhang, J., Chen, J.X., Nagashima, M., Lundh, E.R., Vijay, S., Nitecki, D., Morser, J., Stern, D., and Schmidt, A.M. (1995) The receptor for advanced glycation end products (RAGE) is a cellular binding site for amphoterin. J. Biol. Chem. 270, 25752–25761.
181. Schmidt, A.M. and Stern, D.M. (2001) Receptor for AGE (RAGE) is a gene within the major histocompatibility class III region: implications for host response mechanisms in homeostasis and chronic disease. Front. Biosci. 6, D1151–D1160.
182. Sparatore, B., Pedrazzi, M., Passalacqua, M., Gaggero, D., Patrone, M., Pontremoli, S., and Melloni, E. (2002) Stimulation of erythroleukaemia cell differentiation by extracellular high-mobility group-box protein 1 is independent of the receptor for advanced glycation end-products. Biochem. J. 363, 529–535.
183. Fages, C., Nolo, R., Huttunen, H.J., Eskelinen, E, and Rauvala, H. (2000) Regulation of cell migration by amphoterin. J Cell. Sci. 113, 611–620.
184. Degryse, B., Bonaldi, T., Scaffidi, P., Muller, S., Resnati, M., Sanvito, F., Arrigoni, G., and Bianchi, M.E. (2001) The high mobility group (HMG) boxes of the nuclear protein HMG1 induce chemotaxis and cytoskeleton reorganization in rat smooth muscle cells. J. Cell. Biol. 152, 1197–1206.
185. Wang, H., Bloom, O., Zhang, M., Vishnubhakat, J.M., Ombrellino, M., Che, J., Frazier, A., Yang, H., Ivanova, S., Borovikova, L., Manogue, K.R., Faist, E., Abraham, E., Andersson, J., Andersson, U., Molina, P.E., Abumrad, N.N., Sama, A., and Tracey, K.J. (1999) HMG-1 as a late mediator of endotoxin lethality in mice. Science 285, 248–251.
186. Taguchi, A., Blood, D.C., del Toro, G., Canet, A., Lee, D.C., Qu, W., Tanji, N., Lu, Y., Lalla, E., Fu, C., Hofmann, M.A., Kislinger, T., Ingram, M., Lu, A., Tanaka, H., Hori, O., Ogawa, S., Stern, D.M., and Schmidt, A.M. (2000) Blockage of RAGE-amphoterin signalling suppresses tumor growth and metastases. Nature 405, 354–360.
187. Müller, S., Scaffidi, P., Degryse, B., Bonaldi, T., Ronfani, L., Agresti, A., Beltrame, M. and Bianchi, M.E. (2001) The double life of HMGB1 chromatin protein: architectural factor and extracellular signal. EMBO J. 20, 4337–4340.
188. Bustin, M. (2002) At the crossroads of necrosis and apoptosis: signaling to multiple cellular targets by HMGB1. Science's STKE.
189. Andersson, U., Wang, H., Palmblad, K., Aveberger, A.C., Bloom, O., Erlandsson-Harris, H., Janson, A., Kokkola, R., Zhang, M., Yang, H., and Tracey, K.J. (2000) High mobility group 1 protein (HMG-1) stimulates proinflammatory cytokine synthesis in human monocytes. J. Exp. Med. 192, 565–570.

J. Zlatanova and S.H. Leuba (Eds.) *Chromatin Structure and Dynamics: State-of-the-Art*

DOI: 10.1016/S0167-7306(03)39006-4

CHAPTER 6

The role of HMGN proteins in chromatin function

Katherine L. West[1,2] and Michael Bustin[1]

[1]*Protein Section, Laboratory of Metabolism, Center for Cancer Research, Bldg. 37, National Cancer Institute, National Institutes of Health, Bethesda, MD 20892, USA. Tel.: 301-496-5235; Fax: 301-496-8419; E-mail: bustin@helix.nih.gov*
[2]*Division of Cancer Sciences and Molecular Pathology, Department of Pathology, University of Glasgow, Glasgow, G11 6NT, UK*

1. *Introduction*

The high mobility group (HMG) proteins are ubiquitous nuclear proteins that regulate and facilitate various DNA-related activities such as transcription, replication, recombination, and repair [1–6]. They bind to DNA and chromatin and act as "architectural elements" that induce both short- and long-range changes in the structure of their binding sites. HMG proteins were originally isolated from mammalian cells and arbitrarily classed as a specific type of nonhistone based on the observation that they are present in all mammalian and many vertebrate cells, that they share certain physical properties, and that they are associated with isolated chromatin. Numerous proteins have since been identified that contain one of the three types of HMG DNA binding motif, and they are subdivided into three superfamilies accordingly. HMGB (formerly HMG-1/-2) family members contain a HMG-box, HMGA (formerly HMG-I/Y/C) proteins contain an AT-hook motif, and HMGN (formerly HMG-14/-17) proteins have a nucleosome binding domain [3] (and see http://www.informatics.jax.org/mgihome/nomen/genefamilies/hmgfamily.shtml). This review is concerned solely with the HMGN family members, which specifically bind to nucleosomes, the building blocks of the chromatin fiber. These small, basic proteins have been shown to promote chromatin unfolding and enhance transcription from chromatin templates *in vitro*, and there is increasing evidence that they play important roles in the regulation of transcription *in vivo*, both in cells in culture and in whole organisms.

2. *Members of the HMGN family*

The founding members of the HMGN family are HMGN1 (formerly HMG-14) and HMGN2 (formerly HMG-17). Four additional HMGN proteins have been recently identified: HMGN3a and HMGN3b [7,8], HMGN4 [9], and NSBP1 [10,11]. HMGN1, N2, N3a, and N3b are ubiquitous in higher eukaryotes, but have not been detected in yeast, flies or other lower eukaryotes. They are well conserved from frogs to man, and there are related proteins in fish. Birds have two types of HMGN1 protein [12–14]. The main component, HMGN1a, has a higher molecular weight than most HMGN1/N2 proteins, whereas the minor component, HMGN1b, is a homologue of mammalian HMGN1. The evolutionary conservation of HMGN1, N2, and N3 implies that these proteins may play important and distinct roles within vertebrates. HMGN4 and NSBP1 have only been identified in human and mouse to date.

2.1. Conservation between HMGN family members

The HMGN proteins are classed as a distinct family because they contain three distinct function domains: a highly conserved nucleosome binding domain, a bipartite nuclear localization signal, and a negatively charged chromatin unfolding domain (Fig. 1). Outside the three functional domains there is little sequence conservation, and HMGN1 and N2 share less than 52% identity overall at the amino acid level. HMGN4 is highly similar to HMGN2, being 88% identical at the amino acid level [9]. The amino acid sequence of HMGN3a is 41% and 54% identical to those of HMGN1 and HMGN2, respectively [7,8]. HMGN3b is a splice variant of HMGN3a, and lacks the C-terminal chromatin unfolding domain [8]. NSBP1 is less closely related to the other HMGN proteins [10,11]. It contains the nucleosome binding domain, part of the nuclear localization signal and part of the chromatin unfolding domain. However, it has an additional highly acidic C-terminal domain of over 200 amino acids which is unusually rich in glutamic acid residues.

2.2. Genomic organization of HMGN family members

The genes for human *HMGN1*, *N2*, *N3*, *N4*, and *NSBP1* lie at chromosomal positions 21q22.3 [15], 1p36.1 [16,17], 6q27 [8], 6p21.3 [9], and Xq13 [11], respectively. The genomic region of chromosome 21 containing *HMGN1* is associated with the etiology of Down syndrome, and chromosomal abnormalities of the distal region of chromosome 1 where *HMGN2* is located are associated with certain types of malignancies and neoplasia. The organization of all the *HMGN* genes is highly conserved. The genes for chicken, mouse and human *HMGN1* and *N2*, and for human *HMGN3* and *NSBP1* all have six exons, and the intron/exon boundaries are highly conserved [8,11,15,17,18], so it is likely they all evolved from a common ancestor. The shorter splice variant *HMGN3b* arises due to a truncation of exon V. The structure of the *HMGN4* gene differs from the other *HMGN* family members as it contains no introns, and is thus likely to have originated from a fortuitous insertion of an *HMGN2* retropseudogene 3′ to an active promoter [9]. Mammalian genomes contain multiple retropseudogenes for *HMGN1*

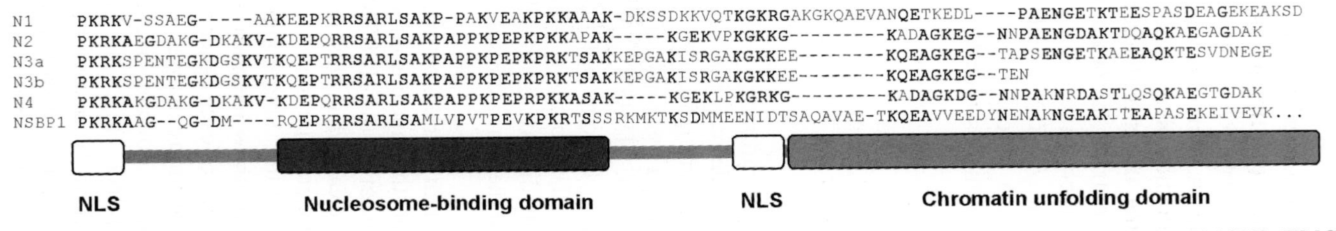

Fig. 1. HMGN proteins contain three highly conserved domains. An alignment of the amino acid sequences of human HMGN1, HMGN2, HMGN3a, HMGN3b, HMGN4, and NSBP1 is shown. Conserved amino acids are in bold. The conserved bipartite nucleosome localization signal, the nucleosome binding domain and the chromatin unfolding domain are indicated below the alignment.

and *N2*, and these are among the largest known retropseudogene families in mice and humans [19].

The genes for *HMGN1* and *HMGN2* are like most housekeeping genes in that they have CpG islands that span the entire promoter region and terminate at the start of exon II [15,17,20]. The gene for *HMGN3* also has a CpG island but it is less prominent than those of *HMGN1* and *N2* [8]. The promoter regions of the three genes share little homology, and the only major element common to all three is a CCAAT box that fits the consensus sequence for the transcription factor NF-Y [21,22]. In contrast, the promoter of the *NSBP1* gene is AT rich, and does not have a CCAAT box transcription factor binding site [11]. The lack of conserved elements in the *HMGN* promoters indicates that these genes are differentially regulated, which is consistent with the distinct tissue-specific expression patterns that have been observed for each family member (see Section 9).

3. Interaction of HMGN proteins with the nucleosome core particle

The minimal binding target of HMGN in chromatin is the 147 bp nucleosome core particle [8,10,23–25]. Two HMGN molecules can bind each nucleosome core, and in buffers close to physiological ionic strength (180 mM Tris-borate) the binding of the two molecules is cooperative, whereas at low ionic strengths (9 mM Tris-borate) the binding is non-cooperative [24,26,27]. HMGN1, N2, N3a, and NSBP1 all have similar dissociation constants for nucleosome binding of around 10^{-7} M^{-1} at physiological ionic strength [8,10,28]. The binding affinity is about 100 fold higher at low ionic strength, presumably due to additional non-specific ionic interactions between the HMGN protein and the charged residues in the nucleosome core particle [26,28].

The concentration of HMGN protein varies between tissues, but is never sufficient to bind more than 10% of the nucleosomes within the nucleus. It seems likely that a mechanism exists to target the HMGN proteins to a particular subset of nucleosomes, but it is not clear what this mechanism is, since there appear to be no DNA sequence elements that serve as HMGN targets [8,29], nor does HMGN have a preference for ubiquitinated H2A [30]. HMGN proteins and linker histone can coexist on the same nucleosome [23,31], although it is possible that they modify the interaction of each other with the nucleosome core.

Immunochemical assays [32] and deletion analysis [33,34] revealed that the minimal HMGN nucleosome binding domain is 24 amino acids long, spanning residues 17 to 40 in HMGN2 (Fig. 1). The nucleosome binding domain is positively charged and is rich in lysine and proline residues. This data confirmed earlier observations using NMR spectroscopy that HMGN proteins interact with nucleosome cores via their central positively charged domain [35,36]. The interaction with nucleosome cores can be greatly reduced by single point mutations within the nucleosome binding domain [28]. In contrast, mutations outside the nucleosome binding domain have minimal effects on the affinity for nucleosome cores.

HMGN1 and N2 bind to nucleosome cores to form, exclusively, complexes containing homodimers of either HMGN1 or HMGN2 [37]. At ionic strengths which are close to physiological conditions, C-terminal deletion mutants and the isolated nucleosome binding domain bind as homodimers, suggesting that the underlying mechanism for homodimer formation does not depend on interactions between the HMGN molecules themselves. It is likely that the binding of the first HMGN protein induces a specific allosteric change within the nucleosome core particle to create a second binding site with greater affinity for a second, identical molecule. The allosteric changes are highly specific, as heterodimers between C-terminal deletion mutants and their wild type equivalents are not observed [37]. The structural transitions must also be small, as HMGN proteins bind cooperatively to formaldehyde crosslinked nucleosome cores, in which larger structural changes would be prevented [27]. HMGN-induced conformational changes are also consistent with the increase in the radius of gyration of nucleosome cores observed upon HMGN binding [38].

The binding of HMGN1/N2 to the nucleosome core stabilizes its structure as measured by thermal denaturation [27,39]. The melting temperature profile of a core particle has two transition points. The first transition occurs at 62°C and corresponds to the melting of the DNA at the entry/exit point of the nucleosome core. The second transition, at around 76°C, corresponds to the melting of the rest of the DNA in the core, leading to the disaggregation of the particle. Both structural transitions are shifted to a higher temperature when HMGN2 or the minimal nucleosome binding domain are bound, suggesting a stabilization of the core structure [27,33,39].

Crosslinking and footprinting experiments have provided several clues as to how HMGN proteins interact with nucleosome cores. Despite their sequence differences, HMGN1 and N2 interact with the nucleosome core in an identical manner [40]. The evidence points to two major sites of interaction for each HMGN molecule: a region 25 bp from the ends of the DNA, and a region near the dyad axis (Fig. 2). DNase I footprinting [24,25,33], hydroxyl radical footprinting [40] and protein–DNA crosslinking [31] all reveal HMGN–DNA interactions 25 and 125 bp from the end of the DNA. The isolated nucleosome binding domain also protects the 25 bp site from DNase I [33]. The 25 and 125 bp positions lie across the minor groove from each other on the side of the DNA helix that does not contact the nucleosome. Although initial analysis of the protein–DNA crosslinking data suggested that HMGN was binding on the nucleosomal side of the DNA double helix, and thus may loop underneath the DNA, comparison with the more recent high resolution crystal structure of the nucleosome [41] does not support this conclusion. Adjacent to this region of DNA is part of the H2B histone fold domain which can be crosslinked to the N-terminal region of HMGN1 [42].

The second major area of interaction is near the dyad axis. Hydroxyl radical footprinting revealed protection at 65 and 75 bp from the DNA ends [40]. On the

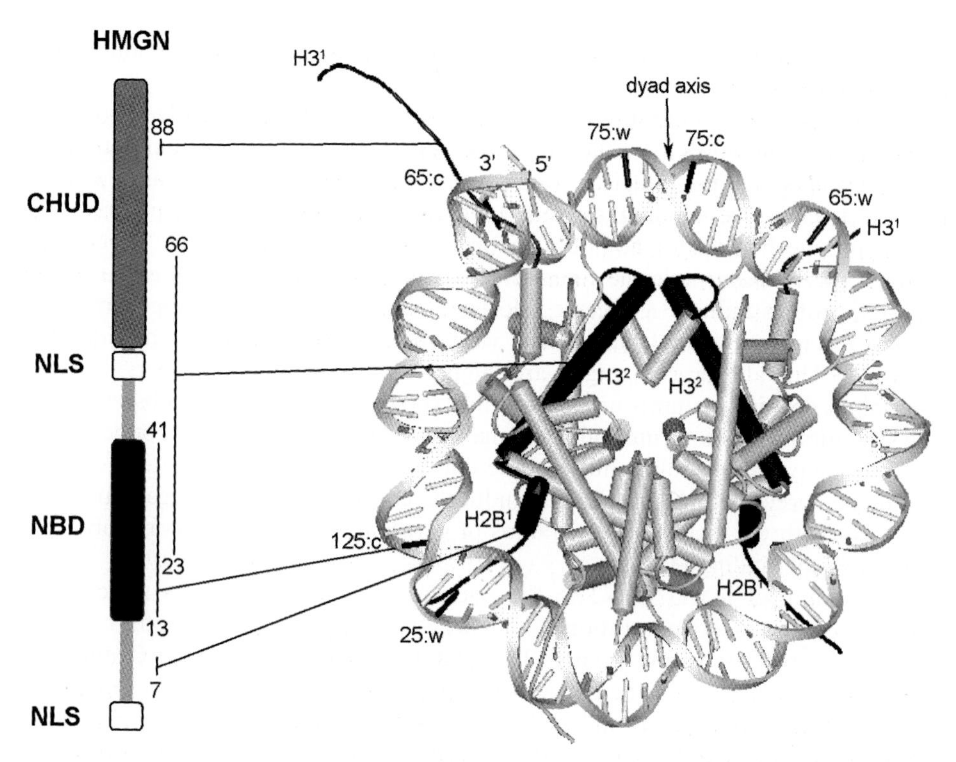

Fig. 2. Interactions of HMGN proteins with the nucleosome core particle. The interactions of HMGN1 with various regions of the nucleosome core particle (shown in black) as revealed by footprinting and crosslinking studies are indicated [41]. A single superhelical turn of the core particle DNA is shown for clarity. Footprinting experiments revealed that full length HMGN protects bases 65 and 75 on each DNA strand (63:w, 75:w, 65:c and 75:c) [40]. Both full length HMGN and the isolated nucleosome binding domain (NBD; residues 13–41 of HMGN1) protect positions 25 and 125 on both DNA strands (25:w and 125:c; 25:c and 125:w not shown) [24,25,31,33,40]. Protein–protein crosslinking has shown that residue 7 of HMGN1 can be crosslinked to peptide 25–66 of H2B (indicated as $H2B^1$), residue 88 of HMGN1 can be crosslinked to peptide 20–50 of H3 (indicated as $H3^1$) [42], and peptide 23–66 of HMGN1 can be crosslinked to peptide 91–120 of H3 (indicated by $H3^2$) [43].

complementary DNA strands, these positions face each other across the major groove next to the dyad axis. This region of the nucleosome is near a portion of H3 that can be crosslinked to the central domain of HMGN [43]. Furthermore, the C-terminal domain of HMGN1 can be crosslinked to part of the H3 N-terminal tail, which emerges from the nucleosome very close to the dyad axis [41,42]. The interaction of the C-terminal domain of HMGN with the H3 tail is supported by data showing that while HMGN1 and N2 inhibit the acetylation of the H3 N-terminal tail by PCAF, a HMGN2 mutant lacking the C-terminal domain inhibits acetylation much less efficiently [44]. The importance of the histone tails for HMGN binding is also suggested by the observation that the interaction of HMGN with trypsinized nucleosome cores is much less specific, resulting in a smear in gel mobility shift assays [33]. These results are consistent with studies showing that the C-terminal domain of HMGN is necessary for chromatin unfolding (see Section 6) [34,45].

Taken together, these data are consistent with a model in which two identical HMGN molecules bind cooperatively to one nucleosome core, causing small allosteric changes in the nucleosome structure. The N-terminal region of the protein, including the start of the nucleosome binding domain, interacts with the nucleosome core near base 25 of the DNA. The remainder of the nucleosome binding domain then bends around the nucleosome to the dyad axis, where it binds in the major groove. The C-terminal domain, which interacts less strongly with the nucleosome core [36], is then positioned to interact with the N-terminal tail of histone H3. The HMGN protein thus bridges the two adjacent DNA strands, stabilizing the nucleosome structure.

4. Interaction of HMGN proteins with nucleosome arrays

The nucleosome core particle is a relatively stable and homogenous structure that is easily prepared, and as such has formed the basis for numerous studies into chromatin structure and function. However, several recent studies have suggested that what is true for the nucleosome core may not always be true for nucleosome arrays, nor even for nucleosomes containing linker DNA. For example, the core histone tails preferentially interact with linker DNA when is it present, whereas they are constrained to bind intranucleosomal DNA in core particles [46–48]. Consequently, the activities of proteins that require access to the tails or the DNA may be affected, and it has been shown that both DNA ligase and P/CAF are less active on nucleosome core particles than other chromatin substrates [49,50]. Similar concerns apply to the interaction of HMGN proteins with nucleosome core particles, and results from studies of these complexes must be considered in the wider context of how these proteins may interact with nucleosome arrays.

Early studies suggested that HMGN1 or N2 confer enhanced accessibility on the chromatin of transcriptionally competent genes [51–53], although some of the findings are still controversial [54,55]. A more detailed study using sedimentation analysis and electric dichroism showed no effect of HMGN proteins on chromatin structure [56], although a more recent analysis using neutron scattering did show that HMGN decreased the compaction of linker histone-containing chromatin [57].

In contrast to earlier work where HMGN proteins were added to pre-assembled chromatin, significant effects of HMGN proteins on chromatin structure have been observed when HMGN proteins are incorporated into the chromatin during its assembly *in vitro* [58–61]. Nucleosomal arrays assembled by *Xenopus* egg extracts in the presence of HMGN proteins are less condensed than those without HMGN, as shown by electron microscopy [62], rates of digestion by micrococcal nuclease and restriction enzymes, and by slower sedimentation through a sucrose gradient [58,59]. The rate of nucleosome assembly and the number of nucleosomes incorporated into the template are unaffected [58–60,62] It is notable that addition of HMGN proteins after nucleosome assembly is complete has no effect on the chromatin structure [58,59]. Chromatin unfolding was also observed when HMGN proteins were added to isolated H1-containing SV40 minichromosomes [45]. Based on micrococcal nuclease digestions, it has been suggested that incorporation of HMGN proteins during nucleosome assembly increases the spacing between nucleosomes [63], especially nucleosomes deficient in H2A/H2B dimers [64]. However, it seems more likely that the HMGN does not affect nucleosome spacing but rather minimizes nucleosome sliding during the period of nuclease digestion, thus decreasing the exonucleolytic digestion which would result in a time-dependent decrease in the length of the mononucleosomal DNA. This is consistent with the observations that HMGN proteins bind near the entry/exit point of the nucleosomal DNA and that they stabilize the nucleosome core (see Section 3). Taken together, the data support a model whereby HMGN proteins decompact chromatin by increasing the separation between adjacent nucleosomes, possibly by altering the entry/exit angle or the trajectory of the linker DNA, rather than by increasing the number of base pairs separating each nucleosome.

5. *Post-transcriptional modification of HMGN proteins*

5.1. *Phosphorylation*

HMGN proteins are modified by acetylation and phosphorylation *in vivo*. HMGN1 and N2 are both phosphorylated on two serines within the nucleosome binding domain, ser 20 and ser 24 in HMGN1 and ser 24 and ser 28 in HMGN2 [65–69]. HMGN1 is also phosphorylated on ser 6, but HMGN2 does not have this serine residue [65,67,68]. Protein kinase A, Msk1, and RSK2 can modify ser 6 of HMGN1 *in vitro* [69–71], whereas the serines in the nucleosome binding domain of HMGN1 and N2 are targeted by PKC [69,72]. Casein kinase II can also phosphorylate HMGN1 at ser 89 and ser 99 *in vitro*, but these modifications have not been observed *in vivo* [73].

Phosphorylation of HMGN proteins affects their interaction with chromatin and also their nuclear localization [67,69,74,75]. During mitosis, nearly all HMGN1 and N2 molecules are phosphorylated in their nucleosome binding domain and HMGN1 is also phosphorylated on serine 6, events which coincide with the loss of HMGN from mitotic chromosomes [69,74]. *In vitro*, phosphorylation of the nucleosome binding domain prevents HMGN1 or N2 from binding to nucleosomes, as does the HMGN1 double mutation ser20glu/ser24glu, which mimics phosphorylation [69]. Furthermore, analysis of the mobility of HMGN1 within living cells by FLIP (Fluorescence Loss In Photobleaching) showed that the double point mutant has a lower residence time on chromatin than the native protein, providing *in vivo* evidence that the negative charge prevents HMGN binding to chromatin [69]. Phosphorylation of serine 6 of HMGN1 reduces, but does not abolish, binding to nucleosomes *in vitro*. It is likely, therefore, that phosphorylation of the nucleosome binding domain is responsible for the delocalization of HMGN proteins from chromatin during mitosis. The kinase which modifies the nucleosome binding domain at mitosis is unknown.

Phosphorylation of HMGN proteins also affects their nuclear localization [67] (Bustin, unpublished). Treatment of cells with protein phosphatase inhibitors increases the level of HMGN phosphorylation and also increases the proportion of HMGN protein in the cytoplasm as opposed to the nucleus [67]. Immunofluorescence and *in vitro* nuclear import studies have shown that mitotic phosphorylation of the nucleosome binding domain of HMGN1 prevents its re-entry to the newly formed nucleus in late telophase [74,115]. This effect is a specific consequence of the phosphate modification and is not due to the presence of negative charges. The phosphorylated protein interacts with specific isotypes of 14.3.3, and it is likely that this interaction impedes the active transport of HMGN1 into the nucleus [74,115].

Serine 6 of HMGN1 is also phosphorylated as part of the immediate early gene response [65]. Diverse immediate-early gene inducing agents including EGF, tetradecanoyl phorbol acetate (TPA) and anisomycin induce rapid phosphorylation of both histone H3 and HMGN1, but not HMGN2 [65]. Virtually all of the cellular HMGN1 becomes phosphorylated [65], in contrast to the very small proportion of H3 that is modified under these conditions [76]. The phosphorylation of HMGN1 and H3 is independent of transcription, and is thus a primary response to mitogen stimulation [65]. The kinase MSK1 was identified as a strong candidate for the phosphorylation of both HMGN1 and H3 as it is activated by both the ERK and p38 signal transduction pathways, and phosphorylates H3 and HMGN1 efficiently *in vitro* [70].

5.2. Acetylation

Both HMGN1 and N2 are acetylated *in vivo* [77,78]. In duck erythrocytes, both HMGN1 and N2 are acetylated on lys 2 and 4, and HMGN2 is also acetylated on lys10 [77]. In HeLa cells, HMGN1 is acetylated on lys 2 and also on three sites within the nucleosome binding domain and three sites in the second nuclear localization signal [78]. These sites correspond to those modified by p300 *in vitro*, and this acetylation significantly reduces the interaction of HMGN1 with nucleosomes *in vitro* [78]. P/CAF has more limited activity on HMGN proteins, as it specifically acetylates lys 2 in HMGN2, but does not act on HMGN1 [44]. HMGN2 that has been acetylated by P/CAF has a reduced affinity for nucleosomes [44]. The differential acetylation of HMGN1 and HMGN2 is the first experimental evidence that the two proteins are not equivalent, and this may be indicative of functional specificity *in vivo*.

The ability of HMGN proteins to bind nucleosomes can be thus regulated by both phosphorylation and acetylation. This provides a mechanism for regulating HMGN activity throughout the whole nucleus when large scale changes in gene expression are required, for example in response to mitogen stimulation, differentiation signals (see Section 9), or the G2/M transition. It could also enable a local population of HMGN molecules to be regulated, for example by a histone acetyltransferase recruited to an adjacent region of chromatin.

6. *Activation of transcription by HMGN proteins in vitro*

Several studies have shown that HMGN proteins can enhance transcription when incorporated into chromatin during nucleosome assembly. Incorporation of HMGN proteins into minichromosomes containing the 5S rDNA gene has been shown to increase RNA polymerase III transcription by 2–8 fold [34,59,60]. No transcription enhancement was observed on naked DNA templates, and the HMGN must be added during nucleosome assembly, not afterwards. In the presence of heparin, which prevents re-initiation of transcription, HMGN proteins have a minimal effect on RNA production, indicating that they increase the utilization of transcriptionally active templates rather than increasing their number [59,60]. HMGN2 also increases the efficiency of replication of minichromosomes in this system [62]. Both the initiation of replication and the rate of replication fork movement were stimulated when HMGN2 was incorporated into the minichromosome during nucleosome assembly, but not when HMGN2 was added afterwards. Stimulation of transcription by HMGN2 has also been observed on chromatin assembled using *Drosophila* embryo extracts [61]. In this study, HMGN2 stimulated RNA polymerase II transcription by 7–40 fold when incorporated into chromatin along with the sequence-specific activator GAL4-VP16. HMGN2 increased the efficiency of transcription initiation in this system, possibly by increasing the binding efficiency of GAL4-VP16, but did not increase the rate of elongation.

Studies on isolated SV40 minichromosomes also showed that HMGN proteins can increase RNA polymerase II transcription by 2–5 fold, and that this results from an increase in the rate of transcription elongation [79]. As with minichromosomes assembled in *Xenopus* egg extracts, the minichromosome was shown to be less tightly folded after the addition of HMGN [45] (see Section 4). In the SV40 system, however, HMGN proteins can be added to the minichromosome after it has been assembled, and furthermore, transcriptional activation by HMGN was dependent on the presence of the linker histone H1. HMGN was, in effect, counteracting the chromatin compaction and transcription repression due to H1, even though H1 remained bound to the minichromosome in the presence of HMGN [45]. The difference in activity of HMGN on SV40 minichromosomes compared to *in vitro*-reconstituted templates can be attributed to differences in the structure of the chromatin templates. SV40 minichromosomes are transcriptionally competent cellular chromatin containing many chromatin-bound proteins, and the promoter region is essentially nucleosome-free. In contrast, *Xenopus* egg extract and *Drosophila* embryo extract assembly systems normally produce a transcriptionally repressed chromatin structure [80].

The C-terminal domain of HMGN was required for transcriptional enhancement in both SV40 minichromosomes and in minichromosomes assembled using

Xenopus egg extracts. In the latter system, the C-terminal 26 or 22 amino acids of HMGN1 or HMGN2 respectively were required for transcription activation, and mutation of a conserved acidic residue in this region of HMGN1, Glu76Gln, abolishes activation [34]. In SV40 minichromosomes, the final 26 residues of HMGN1 were essential for both chromatin unfolding and transcription stimulation, thus demonstrating a direct link between chromatin unfolding and transcription activation by HMGN proteins. It is interesting that this domain could be replaced with the acidic activation domains of either GAL4 or HMGB2 without loss of activity [45]. The C-terminal domain of NSBP1 is also required for transcription activation, although this protein differs from other HMGN family members in that it can activate transcription *in vivo* from transiently transfected reporter genes [Bustin, unpublished] [10]. NSBP1 thus has the potential to function as a general transcriptional activator, possibly by recruiting other transcriptional factors to the template, or by modifying the chromatin structure of the transfected plasmid.

HMGN proteins are thus able to decompact chromatin and activate transcription from chromatin templates. Both activities are dependent on the C-terminal domain, and it is likely that the transcriptional enhancement is a direct result of the ability of these proteins to unfold chromatin.

7. *Models for chromatin unfolding by HMGN proteins*

7.1. *Interaction with core histone tails*

The mechanism by which HMGN proteins unfold chromatin is still unclear. One possible mechanism is based on the observation that the chromatin unfolding domain of HMGN1 can be crosslinked to the N-terminal tail of histone H3 [42]. The chromatin unfolding domain is also required for HMGN to inhibit the acetylation of the H3 tail [81], and the presence of core histone tails stabilizes the interaction of HMGN with nucleosome core particles [33]. The N-terminal tails of the core histones are necessary for chromatin folding, and their acetylation or removal prevents the salt-induced compaction of the chromatin fiber [82–84]. It is thought that the tails may interact with linker DNA [48] or adjacent nucleosomes [83,85,86] to promote chromatin compaction, and that their acetylation disrupts this interaction. HMGN proteins could compete for interaction with the core histone tails and promote their rearrangement, leading to decompaction of the nucleosome array.

7.2. *Counteracting linker histone compaction*

A second model for HMGN action is that it counteracts chromatin compaction by linker histones. The ability of HMGN proteins to unfold SV40 minichromosomes and stimulate transcription from them is dependent on the presence of linker histones, and the data is consistent with HMGN counteracting the repressive effects of H1. In support of this hypothesis, microinjection of HMGN protein into the nuclei of HeLa cells reduces the residence time of H1 on chromatin [116]. HMGN and linker histones apparently bind close to each other near the dyad axis of the nucleosome core (see Section 3) [87–89], although a different, asymmetrical, binding site has been proposed for its binding to nucleosomes containing the *Xenopus* 5S gene [90,91]. The binding of HMGN and linker histones is not mutually exclusive, however, as both proteins can coexist on the same nucleosome with no major changes in the sites of protein–DNA crosslinking for either protein [23,31]. Linker histones interact with nucleosomes through their central globular domain, and are thought to stabilize chromatin folding by electrostatic neutralization of linker DNA with their positively charged C-terminal domain [92]. The apparent close proximity of the HMGN molecules and the linker histone when bound to the same nucleosome raises the possibility that they influence each other, potentially altering the nature of their interactions with the nucleosome core, linker DNA or adjacent nucleosomes. Taken together, the data indicate that HMGN proteins may promote chromatin unfolding by modulating the interaction of linker histone with the chromatin fiber, possibly by reducing the affinity of its globular domain for the nucleosome, or by preventing its C-terminal domain from interacting with linker DNA and thus compacting chromatin.

The two models for HMGN action are not mutually exclusive, as HMGN may use both mechanisms to alter chromatin structure. Several observations point to an interplay between linker histones and the core histone tails. For example, linker histone inhibits the acetylation of H3 by P/CAF [93]. The inhibition was shown to result from steric hindrance of the tail by linker histones, rather than tail inaccessibility due to chromatin folding. It has also been shown that the removal of core histone tails may reduce the affinity of linker histone H1 for the nucleosome [94], and that the interaction of the C-terminal tail of core histone H2A with linker DNA is rearranged in the presence of H1 [47]. It is conceivable, therefore, that HMGN modulates the activity both of the core histone tails and of linker histones within the same nucleosome.

8. *Association of HMGN proteins with transcription in vivo*

Nucleosomes and oligonucleosomes which bind to HMGN1 and N2 affinity columns are enriched in acetylated histones and DNase-1 sensitive chromatin [51,95], and several studies have claimed to show that HMGN proteins are preferentially associated with transcribed sequences [96–98]. However, the chromatin isolation procedures used in these analyses eliminated particular subfractions of chromatin, including those containing linker histones, and did not account for possible HMGN redistribution [99], and thus the data is open to question. A more recent study, in which protein–DNA complexes were crosslinked to prevent rearrangement, prior to sonication and immunoprecipitation from the whole chromatin population, showed that HMGN1/N2 are present at 1.5–2.5

fold higher density on transcribed sequences (β globin) than on non-transcribed genes (ovalbumin and lysozyme) [100].

The involvement of HMGN proteins in transcription *in vivo* was initially demonstrated by Einck [101], who showed that transcription is inhibited when antibodies against HMGN2 are injected into cells. It was subsequently shown that incubation of permeabilised cells with a peptide corresponding to nucleosome binding domain of HMGN, which competes with HMGN binding to nucleosomes, also inhibits transcription [102].

Confocal immunofluorescence microscopy has shown that HMGN1 and N2 are clustered into foci which are scattered throughout the nucleus and not subcompartmentalized into any specific nuclear region. These foci contain either HMGN1 or HMGN2, indicating that the two proteins segregate into distinct nuclear domains [103]. This is consistent with immunofractionation data showing that HMGN2-containing nucleosomes are clustered into domains that, on average, consist of six adjacent HMGN2-containing nucleosomes, and no HMGN1-containing nucleosomes [103]. The HMGN foci colocalise with sites of active transcription, as marked by BrUTP-labeled nascent transcripts or RNA pol II [102]. Inhibition of pol II transcription with α-amanitin resulted in a significant reduction in the HMGN2 fluorescence signal in permeabilised cells, suggesting that the intranuclear distribution of HMGN2 is linked to pol II transcription [102]. In addition to co-localization with sites of transcription, some of the HMGN foci co-localize with speckles containing the splicing factor SC35, or with interchromatin granules. On inhibition of transcription in living cells with actinomycin D, HMGN2 is gradually redistributed and accumulates within the interchromatin granules. In contrast, the localization of linker histone H1 was unaffected by the inhibition of transcription [102]. These results indicate that the distribution of HMGN proteins within the nucleus is dynamic and is functionally related to the transcriptional state of the cells.

The dynamic nature of HMGN proteins has been clearly demonstrated using FRAP (Fluorescence Recovery After Photobleaching) technology [104]. In this procedure, a small area of the nucleus of a cell containing GFP-labeled HMGN2 is bleached with a laser. The recovery of fluorescent signal, as bleached HMGN molecules migrate out of the area and fluorescent molecules move in, is measured as a function of time, and reflects the mobility of the labeled protein within the cell [104]. These results showed that HMGN2 moves rapidly about the nucleus in an ATP-independent manner, with a mobility comparable to the pre-mRNA splicing factor SF2/ASF and the nucleolar protein fibrillarin. The mean residence time for HMGN2 on chromatin was less than 30 s, and it can be calculated that each HMGN molecule would encounter a nucleosome every 0.2 s, and that a specific nucleosome would come into contact with an HMGN molecule every 20 s. In contrast, histone H2B is virtually immobile over the experimental period. FLIP (Fluorescence Loss In Photobleaching) experiments showed that repeated photobleaching of a particular area resulted in the loss of over 95% of HMGN2-GFP fluorescence from the whole nucleus, indicating that the HMGN in any one focus is continuously and rapidly exchanged with the whole nucleoplasmic pool of HMGN [104].

In summary, it is known that HMGN proteins are localized into foci containing only proteins of their own type, that the foci are associated with regions of transcription, that HMGN proteins have a small preference for binding to active genes, and that they move rapidly throughout the nucleus. However, there is no information as to how they are targeted to particular regions since they have no DNA sequence specificity for binding to nucleosomes. One possibility is that they bind preferentially to unfolded chromatin, as is found in transcribed genes, and thus help to maintain an open chromatin conformation for subsequent transcriptional events, rather than participating in the initial chromatin unfolding event [105]. Alternatively, HMGN proteins may be targeted to active genes through interactions with other proteins, for example transcription factors, chromatin remodellers or components of the polymerase machinery. In support of this hypothesis, it has been recently demonstrated that up to 50% of HMGN molecules are organized into several distinct macromolecular complexes [29,106]. The complexes are relatively unstable, and the inclusion of HMGN varies with the metabolic state of the cell. It is notable that HMGN proteins within one of the isolated complexes had a higher affinity for nucleosomes than free HMGN [106]. If the complex is recruited to certain active genes by virtue of its other components recognizing specific DNA sequence elements and/or specifically modified histone tails, then HMGN proteins would be targeted to these same active regions and could be efficiently transferred from the complex to the nucleosomes.

9. Tissue-specific expression of HMGN family members in vivo

Northern and RNA dot blot analyses have shown that each *HMGN* mRNA is expressed in most tissues, although there is often significant variation in expression levels between different tissues, and the tissue-specific expression pattern varies considerably between family members. For example, human *HMGN2* is highly expressed in the thymus and bone marrow, whereas *HMGN4* is most highly expressed in the thyroid gland [9]. There are also differences between the expression patterns in human and mouse. For example, human *HMGN3* is most highly expressed in the pancreas, whereas its highest expression in mouse is in the eye and brain, with low expression in liver and skeletal muscle [8]. Human *NSBP1* is most highly expressed in the liver and kidney [11] whereas mouse *Nsbp1* is particularly highly expressed in 7 day embryos [10]. The distinct tissue-specific expression patterns raise the possibility that these proteins function as co-activators in tissue-specific gene expression.

The expression of *Hmgn2* during embryogenesis has been studied more systematically by *in situ* hybridization of 14- and 17-day mouse embryos [107]. At day 14, when most organs are undergoing active differentiation, *Hmgn2* was widely expressed throughout the embryo, although the levels of mRNA varied between and within different organs. By day 17, when both mature and differentiated structures are present, *Hmgn2* expression was generally restricted to cells undergoing active differentiation, whereas it was much less abundant in mature structures.

This was particularly apparent in organs where differentiating and mature regions can be distinguished, such as the kidney, lung, and intestine. *Hmgn2* expression did not correlate well with regions of proliferating cells [107,108]. Further studies on kidney cell differentiation *in vitro* showed that culturing mesenchyme cells under induction conditions led to increased expression of *Hmgn2*, followed by mesenchyme-to-epithelium differentiation. The expression of *Hmgn2* in the mesenchyme is thus a consequence of induction to differentiate [108]. These results suggest that HMGN2 may play a role in the activation of genes necessary for tissue specific differentiation.

Studies of erythropoeisis and myogenesis have also shown that HMGN proteins are downregulated after differentiation [109,110], In chicken embryo erythroid cells, the level of *HMGN* mRNA decreases by 2.5–10 fold as cells differentiate from primitive erythrocytes at day 5 to definitive mature erythrocytes at day 14 [110]. Furthermore, the balance between the different HMGN proteins changes over this period. At days 3–5, the major HMGN protein is HMGN1a, with barely detectable levels of HMGN2 or HMGN1b. By day 9, the level of HMGN1a protein has decreased significantly, whereas the levels of HMGN1b and HMGN2 protein have increased slightly so that HMGN1b is now the predominant HMGN present. The expression of HMGN proteins is thus developmentally regulated and may be important for the proper differentiation of the erythroid lineage [110]. The levels of *Hmgn1/n2* mRNA and protein are also downregulated during myogenesis [109]. It was demonstrated that this downregulation is dependent on the terminal differentiation of myoblast cells into myotubes in that it was not observed in either growth-inhibited cells [109] or non-fusing variants [111]. The functional importance of downregulating HMGN proteins was demonstrated by stably transfecting myoblasts with an inducible expression vector for *HMGN1*. Induction of *HMGN1* expression prevented the differentiation of the myoblasts, and this correlated with the downregulation of myogenic determination factors such as MyoD and myogenin [109].

The regulated expression of HMGN1 and N2 thus appears to be particularly important for differentiation, and in general both proteins are downregulated in mature tissues and organs. Within the mature organism, however, there is considerable variation in the expression of the different HMGN family members, and it is possible that each protein plays a distinct role in tissue-specific gene expression.

10. Role of HMGN proteins in vivo

The role of HMGN proteins *in vivo* has been investigated by knockout and antisense technologies in several systems, but there are still many unanswered questions about their functions. HMGN1 and N2 were shown to be important for the timing of early embryonic development by injecting one-cell mouse embryos with antisense oligonucleotides targeting *Hmgn1* and *Hmgn2* [112]. HMGN1 and N2 are constitutive components of mouse oocyte and embryonic chromatin, and their depletion in antisense-injected embryos lead to a transient reduction in both RNA and protein synthesis. A subsequent delay in development was observed, with embryos taking longer to complete each round of cell division. The embryos ultimately developed to the blastocyst stage with the same frequency as the controls, indicating that the development was delayed but not arrested. Depletion of both *Hmgn1* and *Hmgn2* was required to delay development, as targeting only one of the proteins had no effect [112]. The observation that the embryos continued to develop, albeit more slowly, in the absence of HMGN1 and N2 is consistent with an earlier study showing that chicken DT40 cells are viable when both the *HMGN1a* and *HMGN2* genes are knocked out [113].

The role of HMGN1 in whole organisms has been studied using transgenic mice in which HMGN1 content was either increased [114] or knocked out [117]. The gene for *HMGN1* lies in the region of chromosome 21 that is associated with Down syndrome when it is present in the triploid state. Increasing the HMGN1 content by three fold, in order to mimic the situation in Down syndrome, resulted in a higher incidence of thymic cysts, but did not result in other phenotypic abnormalities [114]. It can be concluded that overexpression of HMGN1 does not cause the etiology of Down syndrome, although it may contribute to the phenotype when other genes in this region of chromosome 21 are also overexpressed. In contrast, *Hmgn1* −/− mice have several phenotypic abnormalities, including reduced fertility [117].

11. Conclusions

HMGN proteins are small, ubiquitous, chromatin architectural elements that bind to the 147 bp nucleosome core via their highly conserved nucleosome binding domain. They bind as homodimers, interacting near base 25 of the nucleosome DNA, and near the dyad axis. When incorporated into minichromosomes, HMGN proteins decrease chromatin folding in a manner that is dependent on their C-terminal domain, with a concomitant increase in transcription.

The mechanism of chromatin unfolding by HMGN proteins is unclear. One model is based on the observation that HMGN1 interacts with the N-terminal tail of histone H3. HMGN proteins could prevent core histone tails from interacting with linker DNA or adjacent nucleosomes, and thus reduce their ability to compact chromatin. Another model for HMGN action is that it counteracts the chromatin compaction of linker histones, based on studies of the effect of HMGN on linker histone containing-SV40 minichromosomes. HMGN and linker histones can bind one nucleosome simultaneously, and their locations on the nucleosome are thought to be close to each other, so it is possible that HMGN modulates linker histone activity by reducing its binding affinity or preventing it from compacting chromatin.

It has been shown that HMGN proteins stabilize the nucleosome core and prevent the dissociation of the DNA ends, observations which appear to be at odds with the ability of these proteins to promote transcription. However, it is likely that the major effect of HMGN on chromatin is that of decompaction, and that the nucleosome core stabilization has a minimal impact within the context of the chromatin fiber. It is still possible that nucleosome stabilization is important *in vivo*, however; for example, it could serve to maintain a remodeled nucleosome state after the action of chromatin remodeling machinery, or perhaps the maintenance of a regularly spaced nucleosome array is important for efficient polymerase elongation.

HMGN proteins are generally found at sites of active transcription within the nucleus, and are relocated to interchromatin granules upon inhibition of transcription. Within the nucleus, they are highly dynamic, binding at any one place within the nucleus for only a short period of time, and interacting with other proteins to form metastable macromolecular complexes. The HMGN molecules are thus constantly sampling different chromatin environments in a "touch and go" manner. They may reside longer at some sites compared to others due to a particular chromatin structure, the presence of certain histone modifications or other cofactors, and at these sites HMGN could participate in chromatin decompaction and potentiate transcription.

The different HMGN family members are well conserved in most vertebrates, and have distinct patterns of tissue specific expression, suggesting that they have individual roles to play in promoting tissue-specific gene expression. Expression of *HMGN2* is broad during early embryogenesis, particularly in differentiating tissues, and both *HMGN1* and *N2* tend to be downregulated after differentiation. Mice lacking *HMGN1* show multiple pleiotropic effects. The question of whether HMGN proteins promote the transcription of specific genes rather than the whole transcriptome, and whether each HMGN family member acts on a different subset of genes are crucial to understanding the roles of HMGN proteins *in vivo*. Studies with knockout mice will provide insights into these questions

Abbreviations

HMG high mobility group

References

1. Reeves, R. (2001) Gene 277(1–2), 63–81.
2. Thomas, J.O. and Travers, A.A. (2001) Trends Biochem. Sci. 26(3), 167–174.
3. Bustin, M. (2001) Trends Biochem. Sci. 26(7), 431–437.
4. Bustin, M. and Reeves, R. (1996) Cohn, W.E. and Moldave, K. (eds.) In: Progress in Nucleic Acid Research and Molecular Biology, Vol. 54. Academic Press, San Diego, pp. 35–100.
5. Bustin, M. (1999) Mol. Cell. Biol. 19(8), 5237–5246.
6. Bianchi, M.E. and Beltrame, M. (2000) EMBO Rep. 1(2), 109–114.
7. Lee, J.W., Choi, H.S., Gyuris, J., Brent, R., and Moore, D.D. (1995) Mol. Endocrinol. 9, 243–254.

8. West, K.L., Ito, Y., Birger, Y., Postnikov, Y., Shirakawa, H., and Bustin, M. (2001) J. Biol. Chem. 276(28), 25959–25969.
9. Birger, Y., Ito, Y., West, K.L., Landsman, D., and Bustin, M. (2001) DNA and Cell Biology, 20(5), 257–264.
10. Shirakawa, H., Landsman, D., Postnikov, Y.V., and Bustin, M. (2000) J. Biol. Chem. 275(9), 6368–6374.
11. King, L.M. and Francomano, C.A. (2001) Genomics 71(2), 163–173.
12. Landsman, D. and Bustin, M. (1987) Nucleic Acids Res. 15(16), 6750.
13. Dodgson, J.B., Browne, D.L., and Black, A.J. (1988) Gene 63(2), 287–295.
14. Srikantha, T., Landsman, D., and Bustin, M. (1988) J. Biol. Chem. 263(27), 13500–13503.
15. Landsman, D., McBride, O.W., Soares, N., Crippa, M.P., Srikantha, T., and Bustin, M. (1989) J. Biol. Chem. 264(6), 3421–3427.
16. Popescu, N., Landsman, D., and Bustin, M. (1990) Hum. Genet. 85(3), 376–378.
17. Landsman, D., McBride, O.W., and Bustin, M. (1989) Nucleic Acids Res. 17(6), 2301–2314.
18. Landsman, D., Srikantha, T., and Bustin, M. (1988) J. Biol. Chem. 263(8), 3917–3923.
19. Srikantha, T., Landsman, D., and Bustin, M. (1987) J. Mol. Biol. 197, 405–413.
20. Gardiner-Garden, M. and Frommer, M. (1987) J. Mol. Biol. 196(2), 261–282.
21. Mantovani, R. (1998) Nucleic Acids Res. 26(5), 1135–1143.
22. Mantovani, R. (1999) Gene 239(1), 15–27.
23. Albright, S.C., Wiseman, J.M., Lange, R.A., and Garrard, W.T. (1980) J. Biol. Chem. 255(8), 3673–3684.
24. Sandeen, G., Wood, W.I., and Felsenfeld, G. (1980) Nucleic Acids Res. 8(17), 3757–3778.
25. Mardian, J.K., Paton, A.E., Bunick, G.J., and Olins, D.E. (1980) Science 209(4464), 1534–1536.
26. Schroter, H. and Bode, J. (1982) Europ. J. Biochem. 127, 429–436.
27. Paton, A.E., Wilkinson, S.E., and Olins, D.E. (1983) J. Biol. Chem. 258(21), 13221–13229.
28. Postnikov, Y.V., Lehn, D.A., Robinson, R.C., Friedman, F.K., Shiloach, J., and Bustin, M. (1994) Nucleic Acid Res. 22, 4520–4526.
29. Shirakawa, H., Herrera, J.E., Bustin, M., and Postnikov, Y. (2000) J. Biol. Chem. 275(48), 37937–37944.
30. Swerdlow, P.S. and Varshavsky, A. (1983) Nucleic Acids Res. 11(2), 387–401.
31. Shick, V.V., Belyavsky, A.V., and Mirzabekov, A.D. (1985) J. Mol. Biol. 185, 329–339.
32. Bustin, M., Crippa, M.P., and Pash, J.M. (1990) J. Biol. Chem. 265(33), 20077–20080.
33. Crippa, M.P., Alfonso, P.J., and Bustin, M. (1992) J. Mol. Biol. 228(2), 442–449.
34. Trieschmann, L., Postnikov, Y., Rickers, A., and Bustin, M. (1995) Mol. Cell Biol. 15, 6663–6669.
35. Abercrombie, B.D., Kneale, G.G., Crane-Robinson, C., Bradbury, E.M., Goodwin, G.H., Walker, J.M., and Johns, E.W. (1978) Eur. J. Biochem. 84(1), 173–177.
36. Cook, G.R., Minch, M., Schroth, G.P., and Bradbury, E.M. (1989) J. Biol. Chem. 264(3), 1799–1803.
37. Postnikov, Y.V., Trieschmann, L., Rickers, A., and Bustin, M. (1995) J. Mol. Biol. 252, 423–432.
38. Uberbacher, E.C., Mardian, J.K., Rossi, R.M., Olins, D.E., and Bunick, G.J. (1982) Proc. Natl. Acad. Sci. USA 79(17), 5258–5262.
39. Sasi, R., Huvos, P.E., and Fasman, G.D. (1982) J. Biol. Chem. 257(19), 11448–11454.
40. Alfonso, P.J., Crippa, M.P., Hayes, J.J., and Bustin, M. (1994) J. Mol. Biol. 236, 189–198.
41. Luger, K., Mader, A.W., Richmond, R.K., Sargent, D.F., and Richmond, T.J. (1997) Nature 389, 251–260.
42. Trieschmann, L., Martin, B., and Bustin, M. (1998) Proc. Natl. Acad. Sci. 95, 5468–5473.
43. Brawley, J.V. and Martinson, H.G. (1992) Biochemistry 31(2), 364–370.
44. Herrera, J., Sakaguchi, K., Bergel, M., Trieschmann, L., Nakatani, Y., and Bustin, M. (1999) Mol. Cell. Biol. 19, 3466–3473.
45. Ding, H.F., Bustin, M., and Hansen, U. (1997) Mol. Cell. Biol. 17, 5843–5855.
46. Usachenko, S.I., Bavykin, S.G., Gavin, I.M., and Bradbury, E.M. (1994) Proc. Natl. Acad. Sci. USA 91(15), 6845–6849.
47. Lee, K.M. and Hayes, J.J. (1998) Biochemistry 37(24), 8622–8628.

48. Angelov, D., Vitolo, J.M., Mutskov, V., Dimitrov, S., and Hayes, J.J. (2001) Proc. Natl. Acad. Sci. USA 98(12), 6599–6604.
49. Chafin, D.R., Vitolo, J.M., Henricksen, L.A., Bambara, R.A., and Hayes, J.J. (2000) EMBO J. 19(20), 5492–5501.
50. Herrera, J.E., Schiltz, R.L., and Bustin, M. (2000) J. Biol. Chem. 275(17), 12994–12999.
51. Weisbrod, S. and Weintraub, H. (1981) Cell 23(2), 391–400.
52. Weisbrod, S. and Weintraub, H. (1979) Proc. Natl. Acad. Sci. 76, 630–635.
53. Weisbrod, S., Groudine, M., and Weintraub, H. (1980) Cell 19(1), 289–301.
54. Goodwin, G.H., Nicolas, R.H., Cockerill, P.N., Zavou, S., and Wright, C.A. (1985) Nucleic Acids Res. 13(10), 3561–3579.
55. Nicolas, R.H., Wright, C.A., Cockerill, P.N., Wyke, J.A., and Goodwin, G.H. (1983) Nucleic Acids Res. 11(3), 753–772.
56. McGhee, J.D., Rau, D.C., and Felsenfeld, G. (1982) Nucleic Acids Res. 10(6), 2007–2016.
57. Graziano, V. and Ramakrishnan, V. (1990) J. Mol. Biol. 214(4), 897–910.
58. Crippa, M.P., Trieschmann, L., Alfonso, P.J., Wolffe, A.P., and Bustin, M. (1993) EMBO J. 12(10), 3855–3864.
59. Trieschmann, L., Alfonso, P.J., Crippa, M.P., Wolffe, A.P., and Bustin, M. (1995) EMBO J. 14, 1478–1489.
60. Weigmann, N., Trieschmann, L., and Bustin, M. (1997) DNA Cell Biol. 16, 1207–1216.
61. Paranjape, S.M., Krumm, A., and Kadonaga, J.T. (1995) Genes Dev. 9(16), 1978–1991.
62. Vestner, B., Bustin, M., and Gruss, C. (1998) J. Biol. Chem. 273, 9409–9414.
63. Paranjape, S.M., Kamakaka, R.T., and Kadonaga, J.T. (1994) Annu. Rev. Biochem. 63, 265–297.
64. Tremethick, D.J. (1994) J. Biol. Chem. 269, 28436–28442.
65. Barratt, M.J., Hazzalin, C.A., Zhelev, N., and Mahadevan, L.C. (1994) EMBO J. 13, 4524–4535.
66. Clayton, A.L., Rose, S., Barratt, M.J., and Mahadevan, L.C. (2000) EMBO J. 19(14), 3714–3726.
67. Louie, D.F., Gloor, K.K., Galasinski, S.C., Resing, K.A., and Ahn, N.G. (2000) Protein Sci. 9(1), 170–179.
68. Spaulding, S.W., Fucile, N.W., Bofinger, D.P., and Sheflin, L.G. (1991) Mol. Endocrinol. 5(1), 42–50.
69. Prymakowska-Bosak, M., Misteli, T., Herrera, J.E., Shirakawa, H., Birger, Y., Garfield, S., and Bustin, M. (2001) Mol. Cell. Biol. 21(15), 5169–5178.
70. Thomson, S., Clayton, A.L., Hazzalin, C.A., Rose, S., Barratt, M.J., and Mahadevan, L.C. (1999) EMBO J. 18(17), 4779–4793.
71. Walton, G.M., Spiess, J., and Gill, G.N. (1982) J. Biol. Chem. 257(8), 4661–4668.
72. Palvimo, J., Mahonen, A., and Maenpaa, P.H. (1987) Biochim. Biophys. Acta 931(3), 376–383.
73. Walton, G.M., Spiess, J., and Gill, G.N. (1985) J. Biol. Chem. 260(8), 4745–4750.
74. Hock, R., Scheer, U., and Bustin, M. (1998) J. Cell. Biol. 143(6), 1427–1436.
75. Lund, T. and Berg, K. (1991) Febs Lett. 289(1), 113–116.
76. Barratt, J.M., Hazzalin, C.A., Cano, E., and Mahadevan, L.C. (1994) Proc. Natl. Acad. Sci. USA 91, 4781–4785.
77. Sterner, R., Vidali, G., and Allfrey, V.G. (1981) J. Biol. Chem. 256, 8892–8895.
78. Bergel, M., Herrera, J.E., Thatcher, B.J., Prymakowska-Bosak, M., Vassilev, A., Nakatani, Y., Martin, B., and Bustin, M. (2000) J. Biol. Chem. 275(15), 11514–11520.
79. Ding, H.F., Rimsky, S., Batson, S.C., Bustin, M., and Hansen, U. (1994) Science 265(5173), 796–799.
80. Kamakaka, R.T., Bulger, M., and Kadonaga, J.T. (1993) Genes Dev. 7, 1779–1795.
81. Herrera, J.E., Sakaguchi, K., Bergel, M., Trieschmann, L., Nakatani, Y., and Bustin, M. (1999) Mol. Cell. Biol. 19(5), 3466–3473.
82. Hansen, J.C., Tse, C., and Wolffe, A.P. (1998) Biochemistry 37(51), 17637–17641.
83. Fletcher, T.M. and Hansen, J.C. (1995) J. Biol. Chem. 270(43), 25359–25362.
84. Tse, C., Sera, T., Wolffe, A.P., and Hansen, J.C. (1998) Mol. Cell. Biol. 18(8), 4629–4638.
85. Allan, J., Harborne, N., Rau, D.C., and Gould, H. (1982) J. Cell. Biol. 93(2), 285–297.
86. Garcia-Ramirez, M., Dong, F., and Ausio, J. (1992) J. Biol. Chem. 267, 19587–19595.
87. Thomas, J.O. (1999) Curr. Opin. Cell. Biol. 11(3), 312–317.

88. Staynov, D.Z. and Crane-Robinson, C. (1988) EMBO J. 7(12), 3685–3691.
89. Allan, J., Hartman, P.G., Crane-Robinson, C., and Aviles, F.X. (1980) Nature 288(5792), 675–679.
90. Pruss, D., Bartholomew, B., Persinger, J., Hayes, J., Arents, G., Moudrianakis, E.N., and Wolffe, A.P. (1996) Science 274(5287), 614–617.
91. Hayes, J.J. and Wolffe, A.P. (1993) Proc. Natl. Acad. Sci. USA 90(14), 6415–6419.
92. Wolffe, A.P. and Hayes, J.J. (1999) Nucleic Acids Res. 27(3), 711–720.
93. Herrera, J.E., West, K.L., Schiltz, R.L., Nakatani, Y., and Bustin, M. (2000) Mol. Cell. Biol. 20(2), 523–529.
94. Juan, L.J., Utley, R.T., Adams, C.C., Vettese-Dadey, M., and Workman, J.L. (1994) EMBO J. 13(24), 6031–6040.
95. Malik, N., Smulson, M., and Bustin, M. (1984) J. Biol. Chem. 259, 699–702.
96. Goodwin, G.H., Mathew, C.G., Wright, C.A., Venkov, C.D., and Johns, E.W. (1979) Nucleic Acids Res. 7(7), 1815–1835.
97. Dorbic, T. and Wittig, B. (1986) Nucleic Acids Res. 14(8), 3363–3376.
98. Dorbic, T. and Wittig, B. (1987) EMBO J. 6(8), 2393–2399.
99. Landsman, D., Mendelson, E., Druckmann, S., and Bustin, M. (1986) Exp. Cell. Res. 163(1), 95–102.
100. Postnikov, Y.V., Shick, V.V., Belyavsky, A.V., Khrapko, K.R., Brodolin, K.L., Nikolskaya, T.A., and Mirzabekov, A.D. (1991) Nucleic Acids Res. 19(4), 717–725.
101. Einck, L. and Bustin, M. (1983) Proc. Natl. Acad. Sci. 80, 6735–6739.
102. Hock, R., Wilde, F., Scheer, U., and Bustin, M. (1998) EMBO J. 17(23), 6992–7001.
103. Postnikov, Y.V., Herrera, J.E., Hock, R., Scheer, U., and Bustin, M. (1997) J. Mol. Biol. 274, 454–465.
104. Phair, R.D. and Misteli, T. (2000) Nature 404, 604–609.
105. Levinger, L., Barsoum, J., and Varshavsky, A. (1981) J. Mol. Biol. 146(3), 287–304.
106. Lim, J.H., Bustin, M., Ogryzko, V.V., and Postnikov, Y.V. (2002) J. Biol. Chem., 277, 20774–20782.
107. Lehtonen, S., Olkkonen, V.M., Stapleton, M., Zerial, M., and Lehtonen, E. (1998) Int. J. Dev. Biol. 42(6), 775–782.
108. Lehtonen, S. and Lehtonen, E. (2001) Differentiation 67(4–5), 154–163.
109. Pash, J.M., Alfonso, P.J., and Bustin, M. (1993) J. Biol. Chem. 268(18), 13632–13638.
110. Crippa, M.P., Nickol, J.M., and Bustin, M. (1991) J. Biol. Chem. 266(5), 2712–2714.
111. Begum, N., Pash, J.M., and Bhorjee, J.S. (1990) J. Biol. Chem. 265(20), 11936–11941.
112. Mohamed, O.A., Bustin, M., and Clarke, H.J. (2001) Dev. Biol. 229(1), 237–249.
113. Li, Y., Strahler, J.R., and Dodgson, J.B. (1997) Nucleic Acids Res. 25(2), 283–288.
114. Bustin, M., Alfonso, P.J., Pash, J.M., Ward, J.M., Gearhart, J.D., and Reeves, R.H. (1995) DNA Cell Biol. 14(12), 997–1005.
115. Prymakowska-Bosak, M., Hock, R., Catez, F., Lim, J.-H., Birger, Y., Shirakawa, H., Lee, K., and Bustin, M. (2002) Mol. Cell. Biol. 22, 6809–6819.
116. Catez, F., Brown, D.T., Mistell, T., and Bustin, M. (2002) EMBO Rep. 2002, 3, 760–766.
117. Birger, Y., West, K.L., Postnikov, Y.V., Lim, J.H., Furusawa, T., Wagner, J.P., Laufer, C.S., Kraemer, K.H., and Bustin, M. (2002) EMBO J. 22, 1665–1675.

J. Zlatanova and S.H. Leuba (Eds.) *Chromatin Structure and Dynamics: State-of-the-Art*

DOI: 10.1016/S0167-7306(03)39007-6

CHAPTER 7

HMGA proteins: multifaceted players in nuclear function

Raymond Reeves[1] and Dale Edberg

Washington State University, Biochemistry and Biophysics, School of Molecular Biosciences, Pullman, WA 99164-4660, USA

[1]*Tel.: 509-335-1948; E-mail: reevesr@wsu.edu*

1. *Introduction*

In contrast to the well established biological functions of the histone proteins, until recently our understanding of the roles played by the "high mobility group" (HMG) of nonhistone chromatin proteins in nuclear processes has been meager but tantalizing. Fortunately for one group of these proteins, the HMGA family, this situation has now dramatically changed. The reasons for this recent tidal shift in perception are many and reflect the realization by many workers that the nucleus consists of more than just DNA, histones and various enzymes. It also contains several classes of nonhistone proteins that participate in multiple functions ranging from serving as structural components of nuclear architecture to participating as ancillary players in such processes as transcription, replication, and DNA repair. The HMG proteins are the most abundant of these chromatin proteins and the HMGA subfamily is perhaps the best understood in terms of the multiple roles these proteins play in the nucleus. Members of the HMGA group of proteins, and the genes that code for them, possess a unique constellation of biochemical, biophysical, and biological attributes that enables them to participate in a diverse variety of activities not normally accessible to more specialized components of the nucleus. Foremost among these distinguishing characteristics are their remarkable degree of intrinsic flexibility and their ability to undergo extensive and complex patterns of *in vivo* biochemical modifications in response to external and internal stimuli. Although much is now known about these remarkable proteins, only future research will reveal the full extent, and nature, of involvement of the HMGA proteins in nuclear and cellular functions.

2. *Biological functions of HMGA proteins*

Attesting to the current state of interest, the HMGA family of proteins (formerly known as the HMGI (Y) family [1]), and the genes coding for them, has been the subject of numerous recent reviews [2–13]. The HMGA proteins are coded for by two different genes: the *HMGA1* gene, whose alternatively spliced mRNA transcripts give rise primarily to the HMGA1a (a.k.a., HMG-I) and HMGA1b (a.k.a., HMG-Y) proteins, and the *HMGA2* gene whose primary product is the HMGA2 (a.k.a., HMGI-C) protein. The HMGA proteins participate in, or are targets of, a wide variety of normal and pathological biological events. For example, HMGA proteins are the down-stream targets of a number of external and internal signal transductions pathways that affect both the types and levels of secondary biochemical modifications on the proteins and, as a consequence, regulate their substrate binding characteristics and their biological functions. Furthermore, by acting as "architectural transcription factors" the HMGA proteins participate in both the positive and negative regulation of a large number of eukaryotic and viral genes and are also thought to participate in such processes as DNA replication, amplification, and repair. Evidence further suggests that they are also involved in regulating cell proliferation, differentiation and apoptotic cell death.

The HMGA genes are *bona fide* oncogenes and their induced over-expression in cells promotes both cancerous transformation and metastatic progression. Elevated levels of HMGA proteins are among the most consistent biochemical features of naturally occurring human tumors with the protein concentrations being a diagnostic marker for increasingly malignant and metastatic cancers. A likely explanation for why over-expression of HMGA proteins is found in so many different types of human cancers comes from recent experiments that demonstrate that expression of the HMGA1 gene is under negative transcriptional regulation by certain tumor-suppressing proteins and is also exquisitely sensitive to positive regulation by exposure of cells to numerous oncogenic growth factors as well as tumor promoting chemicals.

The ability to participate in such varied biological processes, and to respond to so many different external and internal signaling events, has led to the HMGA genes and proteins being referred to as "hubs" of nuclear function [12]. A cardinal position of the HMGA proteins in normal nuclear activity is supported by the fact that homozygous knockouts of the *Hmga1* gene in mice results in embryonic lethality [13] and homozygous knockouts of the *Hmga2* gene results in the diminutive *pygmy* (or "mini-mouse") phenotype in mice [14].

The *HMGA* genes and proteins possess a number of distinguishing features that contribute to their ability to play vital roles in nuclear metabolism. For example, the *HMGA1* gene has multiple promoters [15] that are regulated by different signal transduction pathways [16–18] and are responsive to a wide variety of external stimuli (reviewed in Ref. [12]). Furthermore, both the *HMGA1* [15,19,20] and *HMGA2* [21,22] genes produce a number of different isoform HMGA proteins as a result of alternative splicing of precursor mRNAs. As free molecules, the HMGA proteins are quite flexible with little intrinsic structure [12,23–26] but undergo disordered-to-ordered transitions upon binding DNA substrates [23,25]. The highly conserved DNA-binding regions of the HMGA proteins, the so-called AT-hook motifs, not only preferentially bind to the minor groove of AT-rich sequences [23], but also interact with non-B-form DNA structures such as four-way-junctions [27] and the distorted forms of DNA located at specific regions on the surface of isolated nucleosome core particles [28]. The HMGA proteins also possess the ability to specifically interact with many other protein partners. They have, for example, been demonstrated to physically associate with at least 20 different transcription factors via localized peptide regions that specifically interact with restricted areas of the transcription factors (reviewed in Ref. [12]). This combination of characteristics enables the HMGA proteins to choreograph the transcriptional activation of a number of inducible genes by participating in the formation of stereo-specific multiprotein–DNA complexes called "enhanceosomes" [29] on their promoter/ enhancer regions.

Another distinctive feature of the HMGA proteins is that they are among the most highly modified proteins in the cell nucleus being subject to complex patterns of *in vivo* phosphorylation, acetylation, and methylation. These secondary biochemical modifications are generally reversible and, in some cases, cell cycle dependent. In other cases the modifications are a consequence of the HMGA proteins being the down-stream targets of signal transduction pathways that are activated by extra-cellular stimuli (reviewed in Refs. [10,12,30,31]). The intricate patterns of *in vivo* modification found on the HMGA proteins have been demonstrated to not only affect their interactions with various molecular substrates but also to influence their biological functions [32–34]. Analogous to the modifications found on histone proteins *in vivo* [35], it has recently been suggested that the patterns of modifications found on the HMGA proteins function as a biochemical "code" that regulates, or coordinates, the many different biological activities of the HMGA proteins in living cells [10].

3. *HMGA proteins: flexible players in a structured world*

Unlike most other proteins, members of the HMGA protein family are characterized by having little, if any, detectable secondary structure while free in solution [12,23–26] with individual proteins exhibiting greater than 75% random coil characteristics when analyzed by either circular dichroism or nuclear magnetic resonance (NMR) spectrometry [10,12]. Nevertheless, when bound to substrates such as DNA or other proteins, subdomains of the HMGA proteins undergo disordered-to-ordered transitions assuming defined conformations [23,25]. This structural transition has been most convincingly demonstrated by NMR studies of co-complexes of the HMGA1 protein with an AT-rich synthetic oligonucleotide substrate [25].

As illustrated in Fig. 1, the highly conserved palindromic core peptide sequence of the DNA-binding domain of the HMGA proteins, Pro–Arg–Gly–Arg–Pro, is disordered prior to substrate binding (panel A) but assumes a planar, crescent-shaped configuration (the AT-hook motif [23]; panels B and C) which, when bound to the minor groove of AT-rich sequences (panels D and E), associates with about 5 bp of DNA, or about half a turn of B-form DNA (panel D). The peptide backbone on either side of an AT-hook bound to DNA retains a great deal of plasticity (Fig. 1E). Each HMGA protein has three independent AT-hook motifs

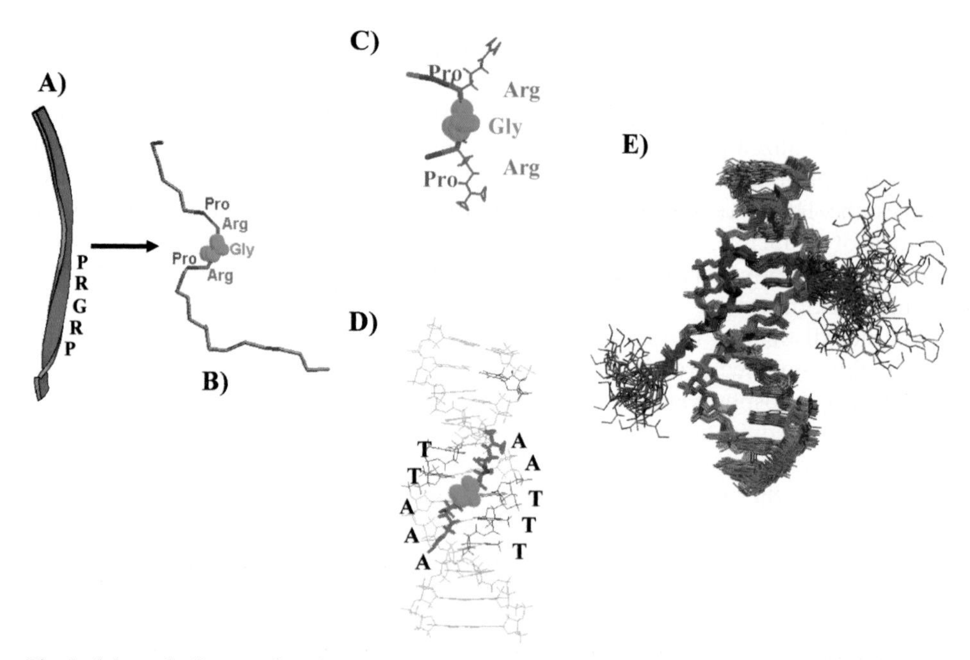

Fig. 1. Schematic diagrams based on the solution NMR structure of a complex of the second AT-hook motif of the human HMGA1a protein bound to the minor groove of an AT-rich synthetic duplex DNA fragment [25]. Various projected views of either the AT-hook peptide, or a co-complex of the peptide with DNA are shown (see text for details). Modified from Ref. [7].

that are separated by highly flexible peptide sequences. This arrangement and flexibility allows the AT-hooks of an individual protein to associate with the minor groove of either long, contiguous stretches of AT-rich sequence ($> \sim 15$ bp) or bind to three shorter stretches (4–7 bp) of sequence that are separated from each other by variable distances [23,36].

The intrinsic flexibility of HMGA proteins enables the AT-hooks of a single protein to not only bind simultaneously to AT-rich stretches on different DNA molecules, thereby forming a peptide bridge between separate DNA substrates [25], but to also bind to quite variable arrangements of AT-rich binding sites on a single DNA molecule. These unique binding capabilities facilitate a variety of biological functions including the regulation of gene transcription initiation. As the diagram in Fig. 2 illustrates, the promoter regions of most genes regulated by HMGA proteins contain variably arranged stretches of AT-rich sequence that have been postulated to represent a sort of "bar code" that is "read" by the binding of the AT-hooks of the HMGA proteins during the process of transcription activation [10]. In the majority of cases, as shown by the proximal promoter regions of the human IL-2, IL-2Rα, IL-4, iNOS, IFN-β, and E-selectin genes, the unique patterns of AT-rich binding sites occur within about 300 bp 5′-upstream of the transcription start site. In some instances, however, as illustrated by the mouse TNF-β promoter in Fig. 2, the HMGA binding sites can occur at much greater

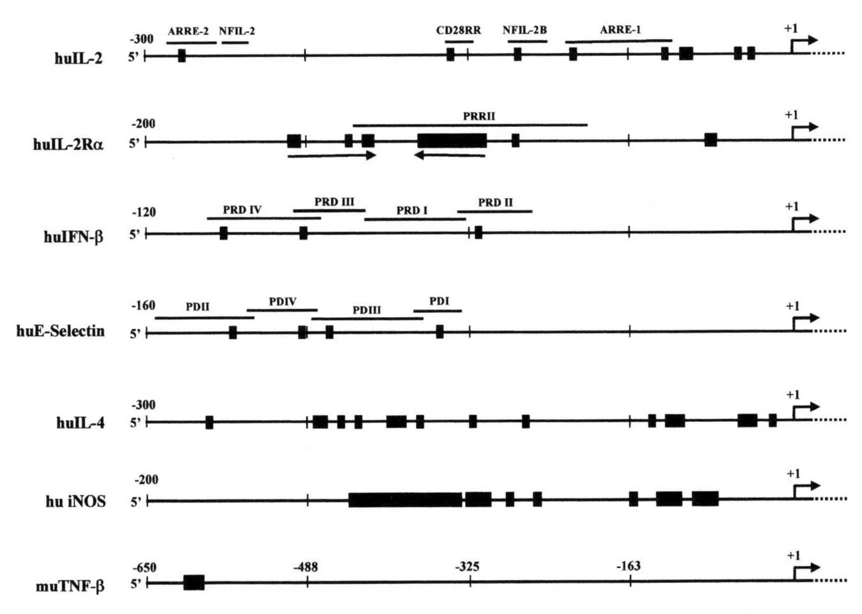

Fig. 2. The pattern of AT-rich, HMGA protein bindings sites (shown as black boxes; not to scale) in gene promoter regions form a unique "bar code" potentially involved with gene-specific transcriptional regulation ([10]; see text for details). Gene promoter sequence references: huIL-2, human interleukin-2 [38]; huIL-2Rα, human IL-2 receptor alpha subunit [60]; huIFN-β, human interferon beta [120]; huE-Selectin, human E-selectin [120]; huIL-4, human IL-4 ([121]; GenBank Accession No. M23442); hu iNOS, human inducible nitric oxide synthase ([122]; GenBank Accession No. AF045478); muTNF-β, murine tumor necrosis factor beta (a.k.a., lymphotoxin) [123]. Abbreviations: ARRE-1, -2, antigen regulated response elements-1 and -2; NFIL-2, -2B, nuclear factor interleukin-2 and -2B; CD28RE, CD28 response element; PRRII, positive regulatory region-II; PRDI-IV, positive regulatory domains I–IV.

distances 5′ upstream of the start site or can even be located 3′ downstream of the start site within intronic sequences [37].

Two additional features also contribute further combinatorial complexity to the postulated gene-specific "bar code" recognized by HMGA proteins. First, as illustrated by the arrows located beneath the human IL-2Rα promoter in Fig. 2, the HMGA proteins have been demonstrated to bind to AT-rich DNA sequences in an orientation-, or direction-specific manner [25,60]. And, second, the minor groove binding of the HMG proteins on gene promoters usually over-lap, or are near quite to, the major groove binding sites for transcription factors that interact physically with HMGA proteins during induction of gene transcription [10,12]. Together, the pattern and directionality of substrate binding, combined with specific interactions of the HMGA proteins with DNA, chromatin and other protein substrates, constitutes a collection of "determinants" that potentially allows these proteins to uniquely recognize and regulate individual gene promoter/enhancer regions among the immense number of other AT-rich binding sites present in the eukaryotic genome [60].

Even though the AT-hook peptide has no detectable secondary structure prior to substrate binding, there are inherent features of its conserved palindromic amino acid sequence that allow it to undergo a distinctive type of disordered-to-ordered transition. NMR studies have demonstrated that, as originally predicted [23], the proline residues of all three of the AT-hook peptides exist in a *trans*-configuration while the protein is free in solution [24,25]. The *trans*-configuration of the prolines restricts the flexibility of the peptide back bone on either side of a freely mobile central glycine residue (Fig. 1, panels A–C) and, thereby, predisposes the AT-hook peptide to adopt dynamic, turn-like configurations in solution [24]. When these flexible, but somewhat restrained, peptides turns encounter AT-rich stretches they are apparently "trapped" after assuming an energetically favorable planar, convex configuration that makes optimal molecular contacts with both the sides and bottom of the narrow minor groove and stabilizing ionic contacts with the phosphodiester backbones of the DNA. Major contributors to the specificity of AT-DNA binding are the side chains of the arginine residues which orient parallel to the minor groove and extend toward the central axis of the DNA (Fig. 1C), thereby allowing their guanidino groups to make hydrogen bond contacts with the O_2 position of thymines (Fig. 1D). Owing to the hydrophobic interactions between the inward projecting arginine side chains and the adenine bases, the AT-hook binds in only one direction in the minor groove. The critical importance of the prolines in the conserved AT-hook motif in facilitating both the initial DNA contacts and the subsequent conformational changes in the peptide backbone is attested to by the fact that if these residues are either changed to other amino acids, or if their position in the peptide is altered, the resulting mutant peptides will no longer preferentially bind to AT-rich DNA sequences [23]. Proteins containing such mutations act as *in vivo* dominant negative competitors for HMGA function when introduced into mammalian cells [38].

Their extreme degree of intrinsic flexibility, combined with their ability to undergo substrate-induced conformational changes, sets the HMGA proteins apart from most other highly structured nuclear proteins and plays a critical role in enabling them to participate in a wide variety of biological processes. However, the importance of intrinsically disordered regions in proteins, and transitions from disordered-to-ordered structures, is now becoming widely recognized as a significant and general feature of many different biological systems [10,39–41]. Labile transitions between disordered and ordered configurations of the HMGA proteins, most likely mediated by reversible secondary biochemical modifications (see below), are likely to regulate the formation of functional HMGA complexes in cells and, thereby, control the biological activity of these proteins *in vivo*.

4. *HMGA biochemical modifications: a labile regulatory code*

Over the last several years a substantial body of evidence has accumulated indicating the types and patterns of secondary biochemical modifications present on histones [42,43], transcriptional co-activators [44] and the HMGA proteins [33] modulate their binding to DNA, to other proteins and to protein–DNA complexes. These modifications are often reversible and are employed by cells to precisely regulate the biological activity of proteins.

The HMGA proteins are among the most highly modified proteins in the mammalian nucleus exhibiting complex patterns of phosphorylations, acetylations, methylations and possibly other covalent adductions [33,45]. These secondary biochemical modifications are both cell cycle-dependent and responsive to various environmental stimuli that activate specific signal transductions pathways (reviewed in Refs. [10,12]). Many years ago it was demonstrated that the HMGA proteins undergo cell cycle-dependent phosphorylations as a result of cdc 2 kinase activity in the G2/M phase of the cycle and that such modifications markedly reduce the affinity of binding of the proteins to AT-rich DNA *in vitro* [32]. More recent studies have shown that HMGA proteins are also the downstream targets of a number of signal transduction pathways whose activation results in phosphorylation of specific amino acid residues distributed throughout the length of the proteins. In mammalian cells, the *in vivo* signaling pathways that activate casein kinase 2 (CK-2; [46–49]) and protein kinase C (PKC; [33]) result in phosphorylation of HMGA proteins within 15–30 min of their stimulation. The HMGA1 homolog protein of the insect *Chironomous* has also been shown to be phosphoryated *in vivo* by stimulation of the mitogen-activated protein (MAP) kinase signaling pathway [50].

Interestingly, agents that activate signaling pathways leading to programmed cell death (apoptosis) also affect the phosphorylation state of HMGA proteins. Sgarra *et al.* [51] demonstrated, for example, that treatment of cells with drugs (etoposide, camptothecin) or viruses (herpes simplex virus type-I) that induce apoptosis also induce hyper-phosphorylation and mono-methylation of the HMGA1a protein soon after exposure to these agents followed a few hours later by a marked de-phosphorylation of the proteins. Since these hyper- and de-phosphorylation events occurred on the majority of the HMGA1a proteins in the cell, the authors propose that the modifications are causally connected to the global changes in chromatin structure that occur during the early and later phases of apoptotic cell death.

Recent advances in mass spectrometry (MS) technology have provided researchers with an unparalleled ability to identify the types and patterns of secondary biochemical modifications found on proteins in living cells. Matrix-assisted laser desorption/ionization-MS (MALDI-MS) analyses have shown, for example, that HMGA proteins *in vivo* are simultaneously subject to complex patterns of phosphorylation, acetylation and methylation and that, within the same cell type, different isoforms of these proteins can exhibit quite different modification patterns [33]. Furthermore, these *in vivo* modifications have been demonstrated to markedly alter the binding affinity of HMGA proteins for both DNA and chromatin substrates *in vitro* [33]. Nevertheless, due to their number and complexity, it has been difficult to determine the actual biological function(s) played by these biochemical modifications in living cells.

The use of MALDI-MS analysis alone to study *in vivo* protein modifications has several limitations, especially when it comes to identifying the specific amino acid residues in the HMGA proteins that are modified. To overcome these shortcomings, we employed a strategy in which MALDI-MS is combined with tandem mass spectrometry (MS/MS) analysis to specifically identify both the types and sites of modifications found on HMGA proteins *in vivo* [52]. This experimental approach is outlined in Fig. 3. The HMGA and other acid-soluble proteins are first isolated from cells and purified to >90% homogeneity by reverse-phase high performance liquid chromatography (RP-HPLC) employing standard techniques [53]. Enzymatic digests of the RP-HPLC purified proteins are then assessed by

Strategy for Identification of Sites of *In vivo* Post-Translational Modifications of HMGA1 Proteins by Mass Spectrometry

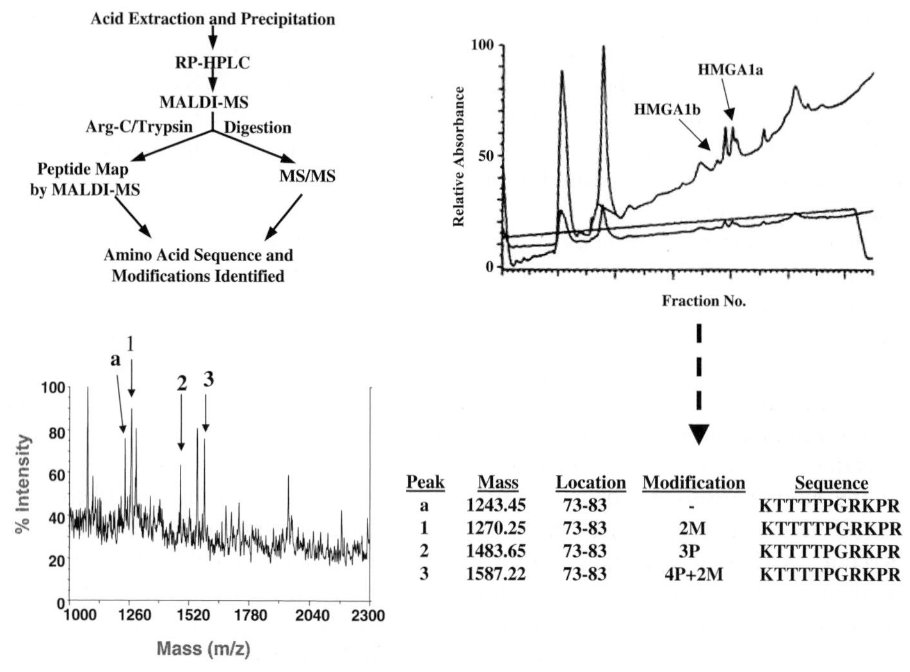

Peak	Mass	Location	Modification	Sequence
a	1243.45	73-83	-	KTTTTPGRKPR
1	1270.25	73-83	2M	KTTTTPGRKPR
2	1483.65	73-83	3P	KTTTTPGRKPR
3	1587.22	73-83	4P+2M	KTTTTPGRKPR

Fig. 3. Strategy for analyzing the patterns of native secondary biochemical modifications found on HMGA proteins in living cells using mass spectrometry techniques. The upper left side of the figure shows steps of a standard protocol for determining both the amino acid sequence and sites of biochemical modification of native HMGA proteins isolated from cells. The upper right side of the figure shows the profile of a reverse-phase HPLC chromatogram of acid soluble proteins isolated from living cells, the initial fractionation step for isolating *in vivo* modified HMGA proteins. The lower left side of the figure shows an example of a restricted region of a MALDI/MS spectrum of a HMGA1 peptide digest containing the *same* peptide fragment with varying degrees of *in vivo* secondary biochemical modifications. Peaks: a, unmodified peptide; 1, the di-methylated peptide; 2, the tri-phosphorylated peptide; and, 3, the tetra-phosphorylated plus di-methylated peptide. The table on the lower right hand side shows the sequence and types of modifications present on the peptides shown in the chromatograph. See text for details.

MALDI-MS to determine the types and extent of modifications found on different peptides fragments. These digests are also analyzed by ion trap MS/MS to directly obtain the sequence, types and sites of specific amino acid modifications present on individual peptides. These MS analytical techniques are very rapid, extremely accurate and require only small amounts of protein to obtain peptide sequences and amino acid modification information [54]. As an example, the restricted region of a MALDI/MS chromatograph illustrated in the lower left side of Fig. 3 shows peaks corresponding to the *same peptide fragment* with either no modifications (labeled “a”) or containing the various types of secondary modifications (peaks 1–3) listed in the table on the lower right side of the figure.

Figure 4 shows some of the sites and types of *in vivo* modifications found by MALDI/MS on the HMGA1a protein isolated from human MCF7 mammary epithelial cells [33]. The sequence of the HMGA1a protein is shown in the center of the figure and shaded boxes indicate the three AT-hook motifs (I, II, and III) in the HMGA1a protein and the clear box indicates the 11 internal amino acid residues that are deleted from the HMGA1b protein as a consequence of alternative mRNA splicing [15]. The types of secondary modifications found on the various amino acid residues are shown above the diagram and, where known, the enzymes thought to be responsible for these modifications (e.g., cdc-2, PKC, CK-2, etc.) are indicated above the sequence [10,33]. The lines below the sequence indicate the regions of the HMGA1 proteins that have been demonstrated to interact physically with other transcription factors [12]. It is important to note that *in vivo* the most highly modified part of the HMGA1 proteins is located between the second (II) and third (III) AT-hooks and corresponds to the region of the protein that has the most identified interacting protein partners [12]. The concurrence of numerous *in vivo* sites of reversible biochemical modifications with those of direct physical association with other proteins suggests that this region of the HMGA1 proteins is important for regulation of their biological function(s) in cells.

Support for the suggestion that biochemical modifications regulate the biological function of HMGA1 proteins in cells comes from experiments that demonstrate isolated native HMGA proteins exhibit markedly different affinities and specificities, compared to unmodified recombinant proteins, for binding to various DNA and nucleosomal substrates *in vitro* [33]. This point is illustrated by the results of the *in vitro* electrophoretic mobility shift assays (EMSAs) shown in Fig. 5. Panel A shows the profile of bands obtained when increasing concentrations of unmodified recombinant HMGA1a protein are bound to a radio-labeled DNA substrate containing multiple (>7) AT-rich binding sites for the protein whereas panel B illustrates the results obtained when identical concentrations of *in vivo* modified protein are added to the DNA probe. It is obvious from these results that the modified HMGA1a proteins bind to the DNA probe with considerably less affinity than does the unmodified recombinant protein. Likewise, as shown in panels C and D, a similar marked reduction is observed in the affinity of binding of *in vivo* modified HMGA1b proteins, compared to unmodified proteins, to isolated nucleosome core particles. Additional evidence that the biochemical modifications

Sites of *In vivo* Modification of Human HMGA1 Proteins Identified by Mass Spectrometry

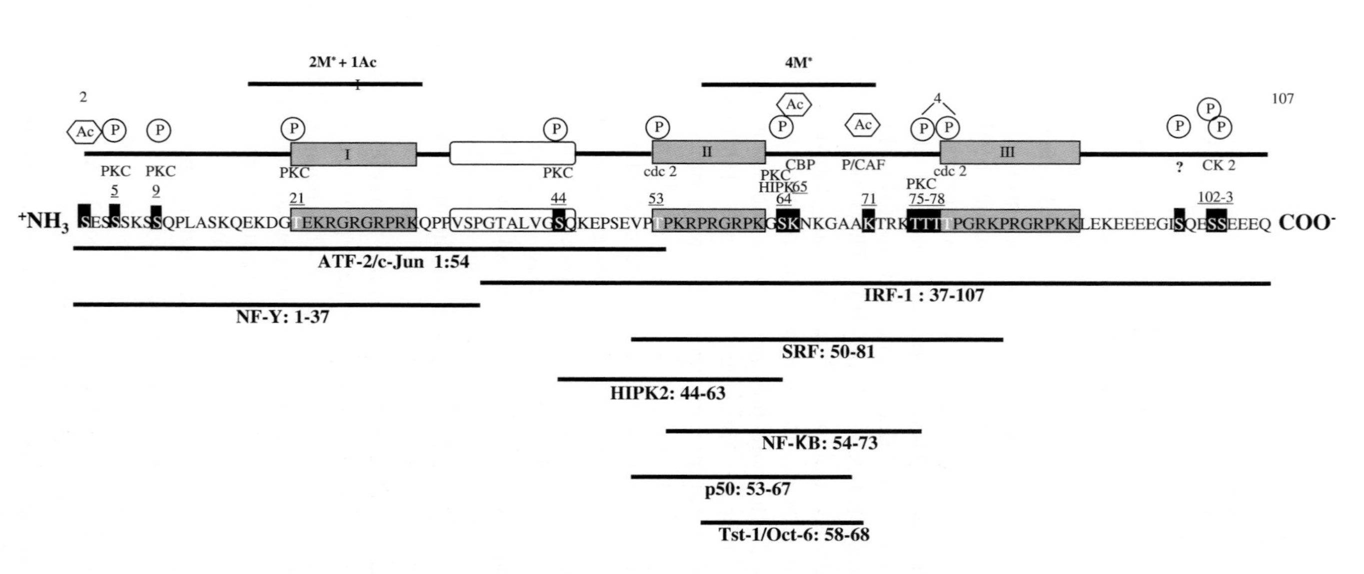

Regions of the HMGA1 Proteins that Physically Interact with Other Transcription Factors

Fig. 4. Diagram of the human HMGA1 protein showing, previously identified sites of *in vivo* biochemical modifications [10], sites of modification confirmed by mass spectrometry and the regions of the proteins identified as minimal areas required for specific interactions with other proteins. The upper line illustrates the full-length HMGA1a protein and the rectangular boxes (shaded) indicate the positions of the three AT-hook DNA-binding domains (I, II, III). The elliptical box (clear) indicates the position where an 11 amino acid residue deletion occurs that gives rise to the HMGA1b isoform protein as a result of alternative mRNA splicing. The amino acid sequence and numbering of the HMGA1a protein are shown in the middle of the diagram. The sites of *in vivo* modifications of the HMGA1 protein that have been confirmed by MALDI-MS and MS/MS are depicted by black boxes with white lettering or white lettering within the three AT-hooks. Serine and threonine residues are the sites of labile phosphorylation whereas variable acetylation occurs exclusively on lysine residues. The existence of methyl groups on HMGA1 proteins has previously been reported in the literature; however, the specific sites of such modifications have not yet been identified. Enzymes that are known to modify HMGA1 are indicated between the sequence and the diagram of the protein. The lines below the amino acid sequence show the areas of the protein that have been identified as the minimal required for specific protein–protein interactions with other transcription factors. The amino acid residues involved in these protein–protein interactions are indicated by numbers following the colons. The original sources demonstrating these physical protein interactions are: NF-κB p50/p65 heterodimer, ATF-2/c-Jun heterodimer and IRF-1, Yie *et al.* [113]; NF-Y, Currie [114]; SRF (serum response factor), Chin *et al.* [115]; NF-κB p50 homodimer, Zhang and Verdine [116]; Tst-1/Oct-6, Leger *et al.* [117]; HIPK2, Pierantoni *et al.* [118]. Figure modified and updated from Ref. [12].

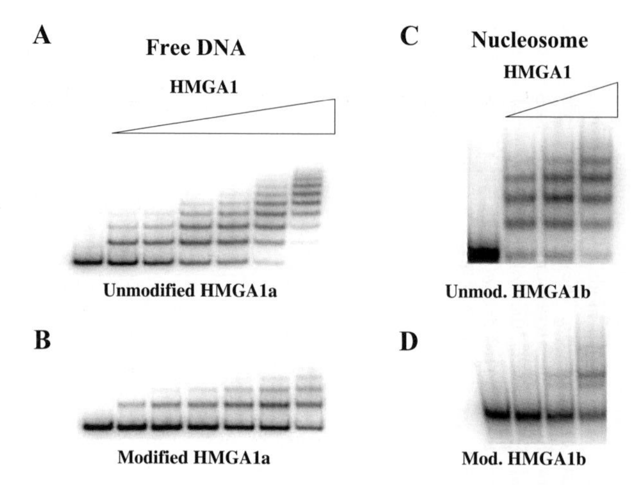

Fig. 5. Secondary *in vivo* biochemical modifications of HMGA proteins reduce the binding affinity of HMGA proteins for both free AT-rich DNA substrates (shown on left side of figure) and random sequence nucleosome core particles (right hand side of figure). Electrophoretic mobility shift assays (EMSAs) using radio-labeled free DNA or isolated nucleosome substrates were reacted with either unmodified recombinant human HMGA1 proteins (upper half of figure) or with native HMGA proteins isolated from cells containing complex patterns of secondary biochemical modifications (lower half of figure). See text for further details.

found on HMGA proteins may serve specific regulatory functions in cells comes from experiments by Munshi *et al.* [34] who investigated the role of acetylation of specific amino acid residues on the inducible regulation of the human interferon-β (IFN-β) gene. These workers demonstrated that acetylation of the HMGA1a protein at residue Lys^{71} by the P/CAF acetyltransferase facilitates transcriptional activation of the IFN-β by promoting formation of an enhanceosome on the gene's promoter region in cells infected with viruses. In contrast, acetylation of the nearby residue Lys^{65} by a different acetyltransferase enzyme, CBP, was shown to turn off transcription of the IFN-β gene by promoting destabilization and disassembly of a previously assembled enhanceosome. The cumulative data, therefore, support not only a role for secondary modifications in regulating the biological function of HMGA proteins but also suggest that the complex patterns of such modifications provide the cell with mechanisms for exerting exquisitely fine control over their *in vivo* activities.

5. *HMGA proteins, AT-hooks and chromatin remodeling*

As well as being accessory regulators of gene transcription, HMGA proteins are also integral components of chromatin and are thought to be involved with controlling the mechanics of chromosome structure, function, and dynamics (reviewed in Refs. 2,10,30). In contrast to these global influences on chromosome architecture, a second, much more restricted, effect of HMGA proteins on altering

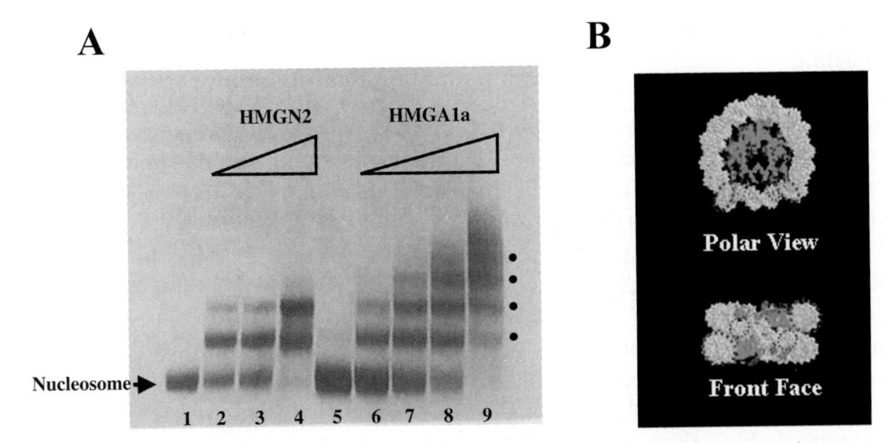

Fig. 6. Purified recombinant HMGA proteins bind to four regions of DNA on random sequence nucleosome core particles. Panel A: The results of EMSA gel assays in which increasing concentrations of either purified nonhistone HMGN2 (a.k.a., HMG-17, which binds to two sites on nucleosome core particles) or recombinant human HMGA1a protein were bound nucleosome core particles isolated from chicken erythrocytes [57]. Panel B: Two different views of the nucleosome taken from the X-ray structure of Luger *et al.* [119] showing the sites of binding of HMGA proteins (dashed circles) determined by DNA foot-printing analyses and other techniques (see text for details).

localized nucleosome structure and function has also been proposed. Indeed, one of the first biological activities suggested for the HMGA proteins (originally referred to as α proteins) was to induce positioning of nucleosomes on the AT-rich α-satellite DNA sequences of chromosomes in monkey cells [55,56]. It was later discovered, however, that the highly repetitive α-satellite DNA sequences are capable of positioning nucleosomes *in vitro* independent of HMGA proteins. Nevertheless, as shown in Fig. 6, HMGA proteins are among only a few known transcription factors that can bind directly to DNA on the surface of nucleosome core particles [57]. Panel A (see also Fig. 5C) shows the results of EMSA analyses that indicate HMGA proteins form four discrete complexes when bound to random sequence core particles isolated from chicken erythrocytes [57], whereas another well characterized nuclear protein, HMGN2 (a.k.a., HMG-17), forms only two complexes [58]. As illustrated by the schematic diagram in panel B, DNA foot-printing and other analyses have demonstrated that these four sites are located at the entrance and exits of DNA from the nucleosome and at the junctions of the over- and under-wound regions of DNA flanking either side of the dyad axis of the core particle [57,59]. In addition to these four sites, HMGA proteins are also able to bind to AT-rich stretches located on the surface of nucleosomes that have been reconstituted *in vitro* using core histones and cloned fragments of DNA of defined nucleotide sequence [59,60]. Protein domain-swap experiments have, furthermore, demonstrated that it is the AT-hook regions of the HMGA proteins that are responsible for nucleosome core particle binding [28]. Importantly, HMGA binding to either random sequence or defined sequence core particles has been shown to induce localized changes in the rotational setting of DNA on the surface of nucleosome [59], thus mediating a restricted form of chromatin remodeling.

The HMGA proteins have been proposed to participate in the localized chromatin remodeling events that accompany transcriptional activation of the promoters of certain inducible gene such as those coding for the human cytokine interleukin-2 (IL-2) [38,61] and the regulatory α subunit of its receptor, IL-2Rα [60,62]. For example, it has been demonstrated that a nucleosome is positioned over the important PRRII regulatory sequence in the promoter of the human *IL-2R*α gene (see Fig. 2) in unstimulated lymphoid cells and that this core particle undergoes a "remodeling" process during transcriptional activation of the gene in stimulated cells [60]. Importantly, additional experiments demonstrated that it is possible to reconstituted a positioned nucleosome at this same position over the PRRII element on an isolated fragment of the *IL-2R*α promoter DNA *in vitro* and, most remarkably, that the HMGA1 protein binds to this reconstituted core particle with a direction-specific polarity. This directional binding of the HMGA1 protein has been proposed to impart a stereo-specificity to the positioned nucleosome and thus "tag" or uniquely identify it for subsequent disruption by ATP-dependent chromatin remodeling complexes during the process of transcriptional activation of the *IL-2R*α promoter [60].

The ability of AT-hook peptide motifs to bind to, and induce localized changes in the structure of, nucleosome core particles is not restricted to HMGA proteins. Significantly, it has also recently been discovered that other proteins that contain AT-hook motifs are essential components of the multi-protein, ATP-dependent chromatin remodeling complexes or "machines" (CRMs) found in yeast, *Drosophila* and mammalian cells. For example, the Swi2p/Snf2p protein, which is the ATPase component of the SWI chromatin remodeling complex in yeast, contains two AT-hook motifs [63] and the ISWI ATPase component of the *Drosophila* chromatin remodeling complex NURF (nucleosome remodeling factor) contains a single AT-hook peptide [64]. Likewise, studies have demonstrated that the AT-hooks present in chromatin remodeling proteins are critically important for the biological activity of CRM complexes. For instance, the Rsc1 and Rsc2 subunits of the RSC (remodeling the structure of chromatin) complex in the yeast *S. cerevisiae* each contain a single AT-hook motif that, when mutated or deleted, destroys the chromatin remodeling activity of the RSC complex and results in cell lethality [65]. Similarly, mammalian SWI/SNF-like CRM complexes contain one or the other of two essential and closely related ATPases, known as brm/SNF2α (also called BAF) and BRG-1/SNF2β (also called PBAF) [66,67], each of which contains a single AT-hook motif in their C-terminal region. Yaniv and his colleagues [68] have demonstrated that when the AT-hook is deleted from brm/SNFα, the CRM complex looses its *in vivo* functional chromatin remodeling activity and also can no longer bind to chromatin substrates. And, finally, Xiao *et al.* [69] discovered that the N-terminal end of the largest subunit of the *Drosophila* NURF complex, NURF301, contains two AT-hook peptide motifs and an acidic region that have high sequence similarity to the mammalian HMGA proteins. Intriguingly, the amino acid sequence of the N-terminal end of NURF301 more closely resembles that of HMGA proteins than do the AT-hook containing domains of Rsc1, Rsc2, brm/SNFα or any of the other CRM proteins. These workers also demonstrated that the *only subunits* of the NURF complex required for the induction of nucleosome sliding (i.e., remodeling) in an *in vitro* model system are NURF301 and the ISWI ATPase protein that also contains a single AT-hook peptide motif. Quite importantly, Xiao *et al.* [69] went on to show that the N-terminal end of the NURF301 is the region of the protein responsible for binding to nucleosome core particles *in vitro* and that when the two AT-hooks of this region are deleted, the ability of the truncated protein to both bind ("tether") to core particles and induce sliding is inhibited [69]. This cumulative *in vivo* and *in vitro* evidence thus strongly supports an active role for the AT-hook motifs found in various CRMs in both nucleosome binding and ATP-dependent sliding/remodeling. Several mechanistic explanations have been advanced for explaining how AT-hook peptides in CRMs might be involved with nucleosome remodeling processes including the attractive suggestion that by selectively binding to distorted regions of DNA on core particles they induce, via ATP-driven reactions, dynamic localized rotational changes in DNA structure that are propagated in a screw-like manner to induce nucleosome translational sliding [12,69].

When considered in the context of activation of the human *IL-2Rα* gene discussed above, this information has also led to a proposal that the directional binding of HMGA1 proteins (via their AT-hook motifs) to the nucleosome positioned on the PRRII promoter region in unstimulated T-cells likely acts as a "marker" or "placeholder" for binding by the AT-hooks of CRM proteins during the subsequent ATP-dependent disruption of the core particle that occurs during transcriptional activation of the gene *in vivo* [10,60]. Considerable experimental support for this model has recently been obtained employing chromatin immunoprecipitation (ChIP) assays that demonstrated that *in vivo* the HMGA protein is bound to the positioned nucleosome on the *IL-2Rα* promoter in resting T lymphocytes and that, within 30 min of following cell stimulation, the HMGA1 protein dissociates from the nucleosome. In contrast, parallel ChIP assays showed that BRG-1, a subunit of the human SWI/SNF complex, is not bound to the positioned nucleosome in unstimulated lymphocytes but becomes associated with the core particle immediately after cell stimulation (within 5–15 min) at about the same moment that the HMGA1 protein is dissociating from the nucleosome, a time during which chromatin remodeling and transcriptional activation of the *IL-2Rα* gene is occurring [70]. Further experimental support for a functional cooperation, or active interplay, between the HMGA proteins and the AT-hook-containing CRM proteins during the process of chromatin remodeling comes from the work of Lomvardas and Thanos [71]. These workers demonstrated that an *in vitro* system composed of only purified recombinant HMGA1, SWI/SNF and TBP (TATA-binding protein) proteins is capable of efficiently inducing ATP-dependent nucleosome sliding/remodeling. It is likely that such cooperative interactions between different AT-hook containing proteins are common and that many more examples of HMGA proteins being intimately, and actively, involved in chromatin remodeling processes mediated by ATP-requiring CRM activities will be found.

6. *HMGA proteins as potential drug targets*

Given their central role in such a variety of normal and pathological processes [12], the *HMGA* genes and proteins are attractive potential targets for the development of therapeutic drugs. The experimental strategies for development of such drugs fall into several categories. Some of the more promising approaches are to develop: (i) drugs that lower the effective concentration of HMGA proteins in cells; (ii) drugs that non-specifically compete with the AT-hooks of the proteins for binding to substrates; (iii) drugs that block specific binding of HMGA proteins to gene promoter regions; (iv) drugs that either specifically inactivate HMGA proteins or selective cross-link them to DNA *in vivo*. Many of these strategies have, with differing degrees of success, already been investigated while others remain to be explored.

6.1. *Methods to lower the cellular concentrations of HMGA proteins*

There are several reports demonstrating that lowering the endogenous levels of HMGA proteins in cells by the introduction of either anti-sense or dominant-negative expression vector constructs results in the reversal or amelioration of certain pathologic conditions. These include the demonstration that anti-sense eukaryotic cell expression vectors can: (i) inhibit neoplastic transformation of normal rat thyroid cells infected with retroviruses [72]; (ii) suppress the growth rate of cancerous cells and decrease their ability for anchorage-independent growth [73]; and (iii) preferentially induce apoptotic cell death in anaplastic human thyroid carcinoma cells but not in normal thyroid cells [74]. On the other hand, there are also reports demonstrating that reduction of cellular levels of HMGA proteins interferes with normal cellular processes such as inhibition of the inducible expression of the unique-sequence genes coding for the human interferon-β [75] and interleukin-2 [76,77] proteins. These examples give some encouragement to the notion that modulation of endogenous HMGA protein levels might be a fruitful target for drug development.

However, the anti-sense and dominant negative studies reported so far have involved delivery of expression vectors to cells, either as transfected DNAs or via viral infections, processes that are inherently inefficient and therefore of limited general use. Alternative approaches of employing synthetic oligonucleotide-based strategies [80,81], such as the use of stable, anti-sense synthetic oligonucleotides [78] or anti-sense peptide-nucleic acids [79], in transfection experiments to inhibit HMGA protein expression seem feasible in principal but have not yet reported in the literature. Nevertheless, these procedures also suffer from similar limitations to those outlined for eukaryotic cell expression vectors. In order to overcome these shortcomings, there are a variety of other promising strategies (reviewed in

Refs. [82,83]) that can potentially be used to develop drugs that modulate the levels or functions of HMGA proteins in cells.

6.2. Drugs that non-specifically compete with AT-hooks peptides for DNA-binding

As illustrated in Fig. 1, the highly conserved Pro–Arg–Gly–Arg–Pro peptide core region of the AT-hook DNA-binding domains assumes, following DNA-binding, a planar, crescent-shaped structure similar in conformation to the pyrrole antibiotics netropsin and distamycin A and to the fluorescent dye Hoechst 33258, all of which also reversibly bind to the minor groove of AT-rich sequences with high selectivity. In fact, the structural similarity between the AT-hook DNA-binding peptide motif and Hoechst 33258 is the basis for an extremely sensitive fluorescence competition assay employed to quantitative determine the affinity of binding of HMGA proteins to AT-rich DNA substrates *in vitro* [23].

Distamycin A, and its chemical derivatives, are highly cytotoxic and exhibit anti-viral [84] and anti-cancer [85–87] activities. Distamycin A, and other minor groove binding drugs appear to exert their biological effects by interfering with cellular gene expression patterns through the alteration or disruption of DNA-binding by transcription factors [86] and, as a consequence, inhibiting the initiation step of transcription [87]. Interestingly, early *in vitro* protein–drug–DNA-binding studies by Wegner and his colleagues [88] led to the proposal that the cytotoxic effects of drugs that selectively bind to the minor groove of AT-rich stretches of DNA (such as distamycin A, netropsin and berenil) are likely to be a consequence of competitive displacement of HMGA proteins from their *in vivo* DNA-binding sites. *In vivo* support for displacement of HMGA proteins by such minor groove binding drugs also comes from studies by Radic *et al.* [89] who demonstrated that both Hoechst 33258 and distamycin A directly compete with HMGA proteins for binding to AT-rich satellite DNA sequences in mouse cells causing chromosome decondensation, particularly in heterochromatic regions. Similar cytological effects have recently been observed in human cells treated with these drugs [90].

The observed *in vivo* effects on chromosome structure of drugs that selectively bind to the minor groove of AT-rich sequences agree quite well with the predictions of models of global gene activation originally advanced by Laemmli and his colleagues [91,92]. These workers proposed that, during the early stages of cellular differentiation when developmental changes in chromatin domains are occurring, HMGA proteins out-compete inhibitory proteins (such as histone H1) for binding to AT-rich DNA sequences, called "scaffold attachment sites" (or SARs), that are distributed along the backbone of metaphase chromosomes and, as a consequence, "open up" selected regions of chromatin for active gene transcription [91,92]. Similarly, the results of minor groove-binding drug studies are consistent with the observation that artificially created "multiple AT-hook" (MATH) proteins that contain numerous AT-hook motifs separated by flexible peptide linkers have the ability to both condense chromatin and inhibit chromosome assembly when added to *in vitro* extracts of oocytes of the amphibian *Xenopus laevis* [93].

MATH proteins have also been shown to regulate the *in vivo* transcription of endogenous host cell genes in differentiated adult tissues when transgenes coding for these proteins were introduced into the laraval stages of the insect *Drosophila* [94].

A major problem with using drugs such as netropsin, berenil or derivatives of distamycin as potential therapeutic agents, however, is their generalized binding to the minor groove of most AT-rich sequences. These promiscuous interactions result in non-specific toxicity of the drugs for all types of cells thus greatly limiting their use as selective anti-viral or anti-tumor agents.

6.3. Drugs that block binding of HMGA proteins to specific gene promoters

One alternative that is beginning to be explored to overcome such generalized drug toxicity problems is to create membrane-permeable synthetic molecules that target only specific gene promoters that are naturally regulated by the HMGA proteins. Based on the known DNA-binding characteristics of the HMGA proteins and assuming, as previously discussed, that the promoter region of each HMGA-regulated gene has a unique "bar code" of AT-rich binding sites (Fig. 2), it should be possible to create synthetic drugs that will selectively recognize portions of this "bar code". Such promoter-specific drugs would be expected to competitively inhibiting the binding of endogenous HMGA proteins to their natural target sites in this promoter and, thereby, reduce or eliminate the expression of only this particular gene in living cells.

A number of experimental advances support the feasibility of this "designer" drug-based approach to inhibiting expression of specific HMGA-regulated genes. Principal among these has been the successful design and synthesis of small, minor groove binding molecules, called lexitropsins, that contain polymers of N-methylimidazole (Im) and N-methylpyrrole (Py) amino acids and which bind with high affinity to predetermined DNA sequences (reviewed in Refs. [83,95–97]). Importantly, the structural basis and pairing rules have been developed to design rationally pyrrole–imidazole (Py–Im) polyamide dimers [98] that bind specifically to the minor groove of either AT- [99] or GC-sequences [100] with affinities and specificities comparable to native DNA-binding proteins. Such Py–Im polyamide molecules have been demonstrated to specifically inhibit the transcription of viral and genomic genes in living cells [101] by selectively binding to only certain promoter sequences and, thereby, interfering with both the binding of TATA-box binding protein (TBP) and basal transcription by RNA polymerase II [102].

The synthesis of a Py–Im polyamide molecule that specifically recognizes a short stretch of AT-rich sequence that constitutes part of the distinctive "bar code" of an individual gene promoter is certainly possible. However, targeting of a lexitropsin molecule to only a single site in a complex promoter is unlikely to provide the necessary degree sequence recognition specificity required for gene specific regulation in eukaryotic cells. Thus, at a minimum, the challenge for specific gene regulation will be to create chimeric molecules with Py–Im moieties that bind to specific (but different) stretches of AT-DNA and which are separated from each other by a peptide linker of such length that the dimeric drug binds to only the appropriately spaced and directionally oriented AT-stretches present in a given target promoter. Encouragingly, experiments attesting to the feasibility of using such bipartite drugs as reagents for modulating gene expression in living organisms have recently been performed in the insect *Drosophila*.

In an elegant series of experiments, Janssen *et al.* [103] synthesized a series of dimeric oligopyrrole drugs containing internal flexible peptide linkers of varying length that exhibited high binding affinity for large and bipartite AT-rich DNA tracks with the various drugs in the series specifically binding to different AT-rich satellite DNA sequences in *Drosophila*. When these drugs were fed to larvae they significantly modulated normal developmental gene expression patterns and caused both gain- and loss-of-function of phenotypes in adult flies. For example, one of the polyamide drugs suppressed position-effect variegation (PEV) of the *white-mottled* locus (a consequence of increased gene expression) whereas another drug mediated homeotic transformations (caused by loss of gene function) exclusively in the *brown-dominant* locus [103]. These are remarkable biological results and if analogous phenotypic effects can be obtained by the targeting of bipartite Py–Im polymide drugs to the promoters of specific HMGA-regulated genes in mammalian cells, the path could be opened for new types of therapeutic treatments for a wide range of pathological conditions ranging from viral infections and immune disorders to the formation of atherosclerotic plaques. Of course, the success of such a drug development strategy is dependent on identifying the promoters of those genes that are the direct targets of HMGA protein regulation in cells exhibiting various pathological conditions, a task that has already been initiated [73,104].

6.4. Drugs that specifically inactivate or cross-link HMGA proteins in vivo

Drugs that recognize, and selectively interact with structural or other features of the HMGA protein themselves could potentially be used as therapeutic reagents to eliminate, for example, aberrant tumor cells that are constitutively over-express high levels of HMGA proteins. The power of combinatorial chemistry (reviewed in Ref. [105]) could, in principal, be employed to select for synthetic drugs that specifically interact with the unbound, unstructured forms of the HMGA proteins present in cells and prevent them from performing their endogenous biological functions. Likewise, cell permeable reagents that selectively react with structural features of the AT-hook peptide motif when it is directly bound to the minor groove of DNA (Fig. 1) might also prove to be an efficacious way of inactivating the HMGA proteins *in vivo*. In this connection, the drug FR900482 (4-formyl-6,9-dohydroxy-14-oxa-1,11-diazatetracyclo [7.4.1.$0^{2,7}$,$0^{10,12}$]-tetradeca-2,4,5-triene-8-yl methyl carbamate [106]), and its chemical derivative FK317 (11-acetyl-8-carbamoyloxymethyl-4-formyl-6-methoxy-14-oxa-1,11-diazatetracyclo [7.4.1.$0^{2,7}$,$0^{10,2}$]-tetradeca-2,4,6-trien-9-yl acetate [107]) are two examples of new anti-cancer drugs

Structure and Reductive Activation of the FR and FK Classes of Pro-Drugs

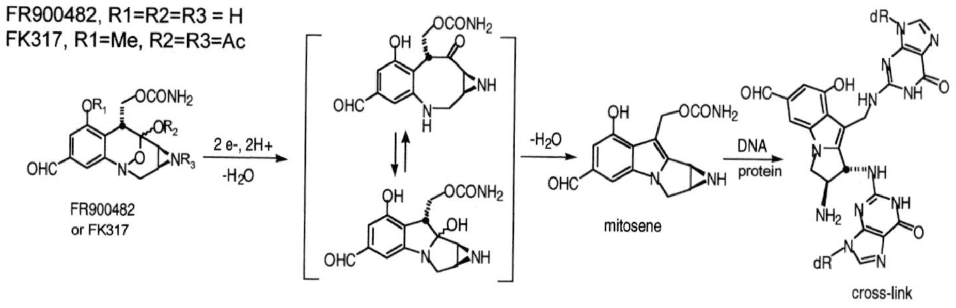

Fig. 7. Diagram of the structure of the FR900482 and FK317 pro-drugs and the proposed scheme of their reductive activation via a miosene-like intermediate inside living cells (see the text for further details). Figure modified from Ref. [108].

that have recently been demonstrated to cross-link HMGA proteins to the minor groove of DNA in living cells [108,109].

As illustrated in Fig. 7, both FR900482 and FK317 are reductively activated inside cells through a scheme that involves the thiol-mediated two-electron reduction of the N–O bond in the presence of trace Fe(II) generating a transient ketone which rapidly cyclizes to a carbinolamine derivative, followed by expulsion of water to produce the reactive electrophilic mitosene derivative [110,111]. This requirement for reductive activation is thought to be the mechanism responsible for the preferential targeting of these pro-drugs to tumorgenic cells that, in general, exhibit a more anaerobic metabolism than normal cells. The presence, or lack, of reductive activation leads to different active compounds derived from the FR900482 and FK317 drugs in normal and cancerous cells. In tumorgenic cells both drugs are deacetylated, oxidized and then reduced to create a 4-alcohol derivative that has cytotoxic properties. In contrast, in normal cells FK90082 and FK317 are deacetylated and oxidized to form a 4-carboxylic acid, which is not cytotoxic [112].

While both the FR900482 and FK317 drugs, which differ by only a single methyl group in cells after they have been reductively activated, are highly toxic to tumor cells, the two drugs have other quite different secondary biological effects. One of the major differences between the drugs is that FR900482 induces a pathological condition in treated patients known as vascular leak syndrome (VLS) where as the K317 drug does not. As a consequence, FR900482 has recently been withdrawn from clinical studies whereas FK317 is currently in phase II clinical trials for cancer treatment in Japan. One possible explanation for the difference in the ability of the two drugs to induce VLS could be differences in their abilities to up-regulate expression of the *IL-2* and *IL-2Rα* genes in lymphoid cells [109].

Although FR900482 and FK317 are not specific cross-linkers of HMGA proteins to DNA in living cells (i.e., they also cross-link other minor groove binding proteins *in vivo*), they are the first drugs demonstrated to specifically cross-link HMGA proteins to the minor groove of DNA *in vivo* and therefore present a major technical advance for studies aimed at examining the cellular dynamics of binding of these proteins to specific sequences of DNA in living cells [70]. And, importantly, the FK317 drug also potentially serves as a model for designing future therapeutic hybrid FK317-polyamide lexitropsin drugs that target specific AT-rich HMGA binding for selective killing of particular cell types. These future hybrid drugs are envisioned to consist of a cell permeable, reductively activated pro-drug (e.g., FK317) that is connected by a flexible linker to a bipartite AT-sequence recognizing Py–Im polyamide lexitropsin that specifically recognizes the promoters of specific HMGA-regulated genes. In principal, these hybrid drugs could be used to specifically treat individual pathological conditions in ways that are not currently possible. Additionally, such drugs could find use as new reagents to investigation the biological function(s) of HMGA proteins in normal processes such as embryogenesis and cell differentiation.

7. *Conclusions*

A constellation of at least three characteristics distinguishes the HMGA proteins from almost all other cellular proteins: (i) the AT-hook DNA-binding motif that recognizes the structure, rather than the nucleotide sequence, of substrates; (ii) an unusually high degree of intrinsic flexibility that allows the proteins to undergo disordered-to-ordered structural transitions as part of their biological function; and (iii) complex patterns of secondary modifications that appear to function as a biochemical "code" to precisely regulate the biological activity of the protein *in vivo*. Structural simplicity and flexibility, combined with very sophisticated regulatory control mechanisms, are the biophysical and biochemical traits that allow the HMGA proteins to function as either "generalists" or as "specialists" in so many different nuclear activities. In this review we have briefly discussed how these features enable the proteins to participate in such diverse processes as chromosome dynamics during the cell cycle, transcriptional regulation of genes and both global and localized chromatin remodeling events. Their central role in nuclear metabolism makes the HMGA proteins attractive targets for therapeutic interventions to treat several different types of pathologies. A number of current and potential approaches appear to be promising in this area. The key problem for developing such useful therapeutics, however, is to create drugs with the requisite specificity to target only certain functions of the HMGA proteins inside cells. Unfortunately, the same characteristics that allow the HMGA proteins to perform multifaceted tasks within the nucleus are precisely those that present the most challenge in creating effective and specific anti-HMGA drugs. Nevertheless, important strides have been made in our understanding of the structure and function of the HMGA proteins and thus a solid foundation has been laid for future advances in this area.

Abbreviations

ARRE-1, -2	antigen regulated response elements-1 and -2
AT-hook	DNA-binding domain peptide of HMGA proteins
bp	base pair
CD28RE	CD28 response element
CK-2	casein kinase 2
CD	circular dichroism
ChIP	chromatin immunoprecipitation assay
CK2	casein kinase 2
CRM	chromatin remodeling machine
EMSA	electrophoretic mobility shift assay
HMG	'high mobility group' nonhistone chromatin proteins
HMGA (a.k.a., HMGI/Y)	the 'AT-hook' containing family of HMG proteins
HMGB (a.k.a., HMG-1 and -2)	the "B box" containing family of HMG proteins
HMGN (a.k.a., HMG-14 and -17)	the 'Nucleosome-binding' family of HMG proteins
hu	human
IFN-β	interferon beta
IL-2	interleukin-2
IL-2Rα	alpha subunit of the IL-2 receptor
IL-4	interleukin-4
iNOS	inducible nitric oxide synthase
ISWI	imitation SWI remodeling complex
MALDI/MS	matrix-assisted laser desorption ionization/mass spectrometry
MAP kinase	mitogen activated protein kinase
MATH	synthetic multi-AT hook proteins
mRNA	messenger ribonucleic acid
mu	murine
MS	mass spectrometry
MS/MS	tandem mass spectrometry–mass spectrometry
NFIL-2	nuclear factor interleukin-2
NMR	nuclear magnetic resonance spectrometry
NURF	nucleosome remodeling factor
PEV	position effect variegation
PKC	protein kinase C
PRD-I to -IV	positive regulatory domains I to IV
PRRII	positive regulatory region II
Py–Im	pyrrole–imidazole polyamides

SRF	serum response factor
SWI/SNF	chromatin remodeling complexes
TCRα	the T cell receptor α-chain
TNF-β	tumor necrosis factor-β
SAR	scaffold attachment region
TPA	phorbol ester 12-*O*-tetradecanoylphorol-13-acetate

References

1. Bustin, M. (2001) Trends Biochem. Sci. 26, 152–153.
2. Bustin, M. and Reeves, R. (1996) Prog. Nucleic Acid Res. Mol. Biol. 54, 35–100.
3. Goodwin, G. (1998) Int. J. Biochem. Cell Biol. 30, 761–766.
4. Bustin, M. (1999) Mol. Cell Biol. 19, 5237–5246.
5. Jansen, E., Petit, M.R., Schoenmakers, E.F., Ayoubi, T.A., and van de Ven, W.J. (1999) Gene Ther. Molec. Biol. 3, 387–395.
6. Tallini, G. and Dal Cin, P. (1999) Adv. Anat. Pathol. 6, 237–246.
7. Reeves, R. (2000) Environ. Health Perspect. 108, 803–809.
8. Bianchi, M.E. and Beltrame, M. (2000) EMBO Rep. 1, 109–114.
9. Wisniewski, J.R. and Schwanbeck, R. (2000) Int. J. Mol. Med. 6, 409–419.
10. Reeves, R. and Beckerbauer, L. (2001) Biochim. Biophys. Acta 1519, 13–29.
11. Liu, F., Chau, K.Y., Arlotta, P., and Ono, S.J. (2001) Immunol. Res. 24, 13–29.
12. Reeves, R. (2001) Gene 277, 63–81.
13. Fedele, M., Battista, S., Manfioletti, G., Croce, C.M., Giancotti, V., and Fusco, A. (2001) Carcinogenesis 22, 1583–1591.
14. Ashar, H.R., Fejzo, M.S., Tkachenko, A., Zhou, X., Fletcher, J.A., Weremowicz, S., Morton, C.C., and Chada, K. (1995) Cell 82, 57–65.
15. Friedmann, M., Holth, L.T., Zoghbi, H.Y., and Reeves, R. (1993) Nucleic Acids Res. 21, 4259–4267.
16. Ogram, S.A. and Reeves, R. (1995) J. Biol. Chem. 270, 14235–14242.
17. Holth, L.T., Thorlacius, A.E., and Reeves, R. (1997) DNA Cell Biol. 16, 1299–1309.
18. Cmarik, J.L., Li, Y., Ogram, S.A., Min, H., Reeves, R., and Colburn, N.H. (1998) Oncogene 16, 3387–3396.
19. Johnson, K.R., Lehn, D.A., and Reeves, R. (1989) Mol. Cell Biol. 9, 2114–2123.
20. Nagpal, S., Ghosn, C., DiSepio, D., Molina, Y., Sutter, M., Klein, E.S., and Chandraratna, R.A. (1999) J. Biol. Chem. 274, 22563–22568.
21. Hauke, S., Rippe, V., and Bullerdiek, J. (2001) Genes Chromosomes Cancer 30, 302–304.
22. Kurose, K., Mine, N., Iida, A., Nagai, H., Harada, H., Araki, T., and Emi, M. (2001) Genes Chromosomes Cancer 30, 212–217.
23. Reeves, R. and Nissen, M.S. (1990) J. Biol. Chem. 265, 8573–8582.
24. Evans, J.N., Zajicek, J., Nissen, M.S., Munske, G., Smith, V., and Reeves, R. (1995) Int. J. Pept. Protein Res. 45, 554–560.
25. Huth, J.R., Bewley, C.A., Nissen, M.S., Evans, J.N., Reeves, R., Gronenborn, A.M., and Clore, G.M. (1997) Nat. Struct. Biol. 4, 657–665.
26. Schwanbeck, R., Gymnopoulos, M., Petry, I., Piekielko, A., Szewczuk, Z., Heyduk, T., Zechel, K., and Wisniewski, J.R. (2001) J. Biol. Chem. 267, 26012–26021.
27. Hill, D.A., Pedulla, M.L., and Reeves, R. (1999) Nucleic Acids Res. 27, 2135–2144.
28. Banks, G.C., Mohr, B., and Reeves, R. (1999) J. Biol. Chem. 274, 16536–16544.
29. Merika, M. and Thanos, D. (2001) Curr. Opin. Genet. Dev. 11, 205–208.
30. Reeves, R. (1992) Curr. Opin. Cell Biol. 4, 413–423.
31. Reeves, R. and Beckerbauer, L. (2002) Prog. Cell Cycle Res. 5, 279–286.
32. Reeves, R., Langan, T.A., and Nissen, M.S. (1991) Proc. Natl. Acad. Sci. USA 88, 1671–1675.

33. Banks, G.C., Li, Y., and Reeves, R. (2000) Biochemistry 39, 8333–8346.
34. Munshi, N., Merika, M., Yie, J., Senger, K., Chen, G., and Thanos, D. (1998) Mol. Cell 2, 457–467.
35. Strahl, B.D. and Allis, C.D. (2000) Nature 403, 41–45.
36. Maher, J.F. and Nathans, D. (1996) Proc. Natl. Acad. Sci. USA 93, 6716–6720.
37. Kim, H.-P., Kelly, J., and Leonard, W.J. (2001) Immunity 15, 159–172.
38. Himes, S.R., Reeves, R., Attema, J., Nissen, M., Li, Y., and Shannon, M.F. (2000) J. Immunol. 164, 3157–3168.
39. Plaxco, K.W. and Gross, M. (1997) Nature 386, 657, 659.
40. Wright, P.E. and Dyson, H.J. (1999) J. Mol. Biol. 293, 321–331.
41. Dunker, A.K., Lawson, D., Brown, C.J., Romero, P., Oh, J., Oldfield, C.J., Campen, A.M., Ratliff, C.M., Hipps, K.W., Ausio, J., Nissen, M.S., Reeves, R., Kang, C.H., Kissinger, C.R., Bailey, R.W., Griswold, M.D., Chiu, W., Garner, E.C., and Obradovic, Z. (2001) J. Mol. Graph. Model. 19, 1–65.
42. Cheung, P., Allis, C.D., and Sassone-Corsi, P. (2000) Cell 103, 263–271.
43. Jenuwein, T. and Allis, C.D. (2001) Science 293, 1074–1080.
44. Gamble, M.J. and Freedman, L.P. (2002) Trends Biochem. Sci. 27, 165–167.
45. Elton, T.S. (1986) Purification and Characterization of the High Mobility Group Nonhistone Chromatin Proteins. Ph.D. Thesis, pp. 1–134. Washington State University, Pullman, WA 99164.
46. Palvimo, J. and Linnala-Kankkunen, A. (1989) FEBS Lett. 257, 101–104.
47. Ferranti, P., Malorni, A., Marino, G., Pucci, P., Goodwin, G.H., Manfioletti, G., and Giancotti, V. (1992) J. Biol. Chem. 267, 22486–22489.
48. Wang, D.Z., Ray, P., and Boothby, M. (1995) J. Biol. Chem. 270, 22924–22932.
49. Wang, D., Zamorano, J., Keegan, A.D., and Boothby, M. (1997) J. Biol. Chem. 272, 25083–25090.
50. Schwanbeck, R. and Wisniewski, J.R. (1997) J. Biol. Chem. 272, 27476–27483.
51. Diana, F., Sgarra, R., Manfioletti, G., Rustighi, A., Poletto, D., Sciortino, M.T., Mastino, A., and Giancotti, V. (2001) J. Biol. Chem. 276, 11354–11361.
52. Edberg, D.D., Adkins, J., Springer, D., and Reeves, R. (2002) Unpublished data.
53. Reeves, R. and Nissen, M.S. (1999) Methods Enzymol. 304, 155–187.
54. Kinter, M. and Sherman, N.E. (2000) Protein Sequenceing and Identification Using Tandem Mass Spectrometry. Wiley-Interscience, New York.
55. Wu, K., Strauss, F., and Varshavsky, A. (1983) J. Mol. Biol. 170, 93–117.
56. Strauss, F. and Varshavsky, A. (1984) Cell 37, 889–901.
57. Reeves, R. and Nissen, M.S. (1993) J. Biol. Chem. 268, 21137–21146.
58. Paton, A.E., Wilkinson-Singley, W., and Olins, D.E. (1983) J. Biol. Chem. 258, 13221–13229.
59. Reeves, R. and Wolffe, A.P. (1996) Biochemistry 35, 5063–5074.
60. Reeves, R., Leonard, W.J., and Nissen, M.S. (2000) Mol. Cell. Biol. 20, 4666–4679.
61. Himes, S.R., Coles, L.S., Reeves, R., and Shannon, M.F. (1996) Immunity 5, 479–489.
62. John, S., Reeves, R., Lin, J.X., Child, R., Leiden, J.M., Thompson, C.B., and Leoni, L. (1995) Mol. Cell. Biol. 15, 1786–1796.
63. Laurent, B.C., Treich, I., and Carlson, M. (1993) Genes Dev. 7, 583–591.
64. Tsukiyama, T., Daniel, C., Tamkun, J., and Wu, C. (1995) Cell 83, 1021–1026.
65. Cairns, B.R., Schlichter, A., Erdjument-Bromage, H., Tempst, P., Kornberg, R.D., and Winston, F. (1999) Mol. Cell 4, 715–723.
66. Wang, W., Cot'e, J., Xue, Y., Zhou, S., Khavari, P.A., Biggar, S.R., Muchardt, C., Kalpana, G.V., Goff, S.P., Yaniv, M., Workman, J.L., and Crabtree, G.R. (1996) EMBO J. 15, 5370–5382.
67. Vignali, M., Hassan, A.H., Neely, K.E., and Workman, J.L. (2000) Mol. Cell. Biol. 20, 1899–1910.
68. Bourachot, B., Yaniv, M., and Muchardt, C. (1999) Mol. Cell. Biol. 19, 3931–3939.
69. Xiao, H., Sandaltzopoulos, R., Wang, H., Hamiche, A., Ranallo, R., Lee, K., Fu, D., and Wu, C. (2001) Mol. Cell 8, 531–543.
70. Beckerbauer, L. (2002) Role of the HMGA1 protein in transcriptional activation of the IL-2Rα gene. Ph.D. Thesis, pp. 1–187. Washington State University, Pullman, WA, 99164.
71. Lomvardas, S. and Thanos, D. (2001) Cell 106, 685–696.
72. Berlingieri, M.T., Manfioletti, G., Santoro, M., Bandiera, A., Visconti, R., Giancotti, V., and Fusco, A. (1995) Mol. Cell. Biol. 15, 1545–1553.

73. Reeves, R., Edberg, D.D., and Li, Y. (2001) Mol. Cell. Biol. 21, 575–594.
74. Scala, S., Portella, G., Fedele, M., Chiappetta, G., and Fusco, A. (2000) Proc. Natl. Acad. Sci. USA 97, 4256–4261.
75. Thanos, D. and Maniatis, T. (1992) Cell 71, 777–789.
76. Himes, S.R., Coles, L.S., Reeves, R., and Shannon, M.F. (1996) Immunity 5, 479–489.
77. Himes, S.R., Reeves, R., Attema, J., Nissen, M., Li, Y., and Shannon, M.F. (2000) J. Immunol. 164, 3157–3168.
78. Toulmen, J.J., Di Primo, C., and Moreau, S. (2001) Prog. Nucl. Acid Res. Mol. Biol. 69, 1–46.
79. Pooga, M., Land, T., Bartfai, T., and Langel, U. (2001) Biomol. Eng. 17, 183–192.
80. Giovannangeli, C. and Helene, C. (1997) Antisense Nucleic Acid Drug Dev. 7, 413–421.
81. Helene, C., Giovannangeli, C., Guieysse-Peugeot, A.L., and Praseuth, D. (1997) Ciba Found. Symp. 209, 94–102.
82. Jen, K.Y. and Gewirtz, A.M. (2000) Stem Cells 18, 307–319.
83. Gottesfeld, J.M., Turner, J.M., and Dervan, P.B. (2000) Gene Expr. 9, 77–91.
84. Howard, O.M., Oppenheim, J.J., Hollingshead, M.G., Covey, J.M., Bigelow, J., McCormack, J.J., Buckheit, R.W.J., Clanton, D.J., Turpin, J.A., and Rice, W.G. (1998) J. Med. Chem. 41, 2184–2193.
85. Cozzi, P. and Mongelli, N. (1998) Curr. Pharm. Des. 4, 181–201.
86. Broggini, M. and D'Incalci, M. (1994) Anticancer Drug Des. 9, 373–387.
87. Possati, L., Campioni, D., Sola, F., Leone, L., Ferrante, L., Trabanelli, C., Ciomei, M., Montesi, M., Rocchetti, R., Talevi, S., Bompadre, S., Caputo, A., Barbanti-Brodano, G., and Corallini, A. (1999) Clin. Exp. Metastasis 17, 575–582.
88. Wegner, M. and Grummt, F. (1990) Biochem. Biophys. Res. Comm. 166, 1110–1117.
89. Radic, M.Z., Saghbini, M., Elton, T.S., Reeves, R., and Hamkalo, B.A. (1992) Chromosoma 101, 602–608.
90. Turner, P.R. and Denny, W.A. (1996) Mutat. Res. 355, 141–169.
91. Zhao, K., Kas, E., Gonzalez, E., and Laemmli, U.K. (1993) EMBO J. 12, 3237–3247.
92. Kas, E., Poljak, L., Adachi, Y., and Laemmli, U.K. (1993) EMBO J. 12, 115–126.
93. Strick, R. and Laemmli, U.K. (1995) Cell 83, 1137–1148.
94. Girard, F., Bello, B., Laemmli, U.K., and Gehring, W.J. (1998) EMBO J. 17, 2079–2085.
95. Lown, J.W. (1994) J. Mol. Recognit. 7, 79–88.
96. Walker, W.L., Kopka, M.L., and Goodsell, D.S. (1997) Biopolymers 44, 323–334.
97. Wemmer, D.E. and Dervan, P.B. (1997) Curr. Opin. Struct. Biol. 7, 355–361.
98. Dervan, P.B. and Burli, R.W. (1999) Curr. Opin. Chem. Biol. 3, 688–693.
99. Kielkopf, C.L., White, S., Szewczyk, J.W., Turner, J.M., Baird, E.E., Dervan, P.B., and Rees, D.C. (1998) Science 282, 111–116.
100. Geierstanger, B.H., Mrksich, M., Dervan, P.B., and Wemmer, D.E. (1994) Science, 646–650.
101. Gottesfeld, J.M., Neely, L., Trauger, J.W., Baird, E.E., and Dervan, P.B. (1997) Nature 387, 202–205.
102. Ehley, J.A., Melander, C., Herman, D., Baird, E.E., Ferguson, H.A., Goodrich, J.A., Dervan, P.B., and Gottesfeld, J.M. (2002) Mol. Cell. Biol. 22, 1723–1733.
103. Janssen, S., Cuvier, O., Muller, M., and Laemmli, U.K. (2000) Mol. Cell 6, 1013–1024.
104. Henderson, A., Bunce, M., Siddon, N., Reeves, R., and Tremethick, D.J. (2000) J. Virol. 74, 10523–10534.
105. Hall, D.G., Manku, S., and Wang, F. (2001) J. Comb. Chem. 3, 125–150.
106. Iwami, M., Kiyoto, S., Terano, H., Kohsaka, M., Aoki, H., and Imanaka, H. (1987) J. Antibiot. (Tokyo) 40, 589–593.
107. Kiyoto, S., Shibata, T., Yamashita, M., Komori, T., Okuhara, M., Terano, H., Kohsaka, M., Aoki, H., and Imanaka, H. (1987) J. Antibiot. (Tokyo) 40, 594–599.
108. Beckerbauer, L., Tepe, J.J., Cullison, J., Reeves, R., and Williams, R.M. (2000) Chem. Biol. 7, 805–812.
109. Beckerbauer, L., Tepe, J.J., Eastman, R.A., Mixter, P., Williams, R.M., and Reeves, R. (2002) Chem. Biol. 9, 1–20.

110. Paz, M.M. and Hopkins, P.B. (1997) Tetrahedron Lett. 38, 343–346.
111. Paz, M.M. and Hopkins, P.B. (1997) J. Am. Chem. Soc. 119, 5999–6005.
112. Naoe, Y., Inami, M., Kawamura, I., Nishigaki, F., Tsujimoto, S., Matsumoto, S., Manda, T., and Shimomura, K. (1998, June) Jpn. J. Cancer Res., 666–672.
113. Yie, J., Liang, S., Merika, M., and Thanos, D. (1997) Mol. Cell Biol. 17, 3649–3662.
114. Currie, R.A. (1997) J. Biol. Chem. 272, 30880–30888.
115. Chin, M.T., Pellacani, A., Wang, H., Lin, S.S., Jain, M.K., Perrella, M.A., and Lee, M.E. (1998) J. Biol. Chem. 273, 9755–9760.
116. Zhang, X.M. and Verdine, G.L. (1999) J. Biol. Chem. 274, 20235–20243.
117. Leger, H., Sock, E., Renner, K., Grummt, F., and Wegner, M. (1995) Mol. Cell. Biol. 15, 3738–3747.
118. Pierantoni, G.M., Fedele, M., Pentimalli, F., Benvenuto, G., Pero, R., Viglietto, G., Santoro, M., Chiariotti, L., and Fusco, A. (2001) Oncogene 20, 6132–6141.
119. Luger, K., Mader, A.W., Richmond, R.K., Sargent, D.F., and Richmond, T.J. (1997) Nature 389, 251–260.
120. Whitley, M.Z., Thanos, D., Read, M.A., Maniatis, T., and Collins, T. (1994) Mol. Cell Biol. 14, 6464–6475.
121. Chuvpilo, S., Schomberg, C., Gerwig, R., Heinfling, A., Reeves, R., Grummt, F., and Serfling, E. (1993) Nucl. Acids Res. 21, 5694–5704.
122. Perrella, M.A., Pellacani, A., Wiesel, P., Chin, M.T., Foster, L.C., Ibanez, M., Hsieh, C.M., Reeves, R., Yet, S.F., and Lee, M.E. (1999) J. Biol. Chem. 274, 9045–9052.
123. Fashena, S.J., Reeves, R., and Ruddle, N.H. (1992) Mol. Cell Biol. 12, 894–903.

J. Zlatanova and S.H. Leuba (Eds.) *Chromatin Structure and Dynamics: State-of-the-Art*

DOI: 10.1016/S0167-7306(03)39008-8

CHAPTER 8

Core histone variants

John R. Pehrson

Department of Animal Biology, School of Veterinary Medicine, University of Pennsylvania, Philadelphia, PA 19104, USA. Tel.: 215-898-0454; Fax: 215-573-5189; E-mail: pehrson@vet.upenn.edu

The structure of the nucleosome can be adapted to specialized functions by altering its histone composition. In higher eukaryotes all of the core histones except H4 exist in multiple variant forms that have different primary structures [1]. Even H4 exists in multiple variants in at least one organism [2]. Some variants are only expressed in specific tissues, for example in sea urchin sperm and mammalian testis. In sea urchin embryos there are major changes in variant composition during development. In mammals the variant composition of many somatic tissues changes as the tissue becomes mitotically quiescent and replication variants are replaced by replacement variants. For a description of these phenomena and a discussion of possible functional roles for these variants see Ref. [3]. Some recent developments in these areas are described at the end of this chapter. The rest of this chapter focuses on four core histone variants: the H3 variant CENP-A and its homologues, which are found only in centromeric nucleosomes; H2A.Z, which appears to be involved in transcriptional regulation; H2A.X, which is involved in the repair of DNA double-stranded breaks; macroH2A, which is enriched in the inactive X chromosome of female mammals, but is also present in males and non-mammalian vertebrates.

1. CENP-A

CENP-A is an H3 variant that was discovered by its reaction with autoantibodies from patients with the CREST variant of scleroderma [4–6]. CENP-A copurifies with nucleosome core particles [7] and is localized to chromatin in the inner plate of the kinetochore [8]. CENP-A-like proteins have been found in many organisms including mammals, *Drosophila*, *C. elegans*, and yeast. They all have a C-terminal domain that is homologous to the C-terminal region of H3 (Fig. 1). They also contain an N-terminal domain that is highly variable in length and sequence. CENP-A type histones are essential in yeast [9,10], *C. elegans* [11], *Drosophila* [12], and mice [13], with the disruption of their expression leading to severe mitotic defects including non-dysjunction and mitotic arrest.

```
N-Terminal Domain
Human CENP-A   GPRRRSRKPEAPRRRSPSPTPTPGPSRRGPSLGASSHQHSRRR- 43
Mouse CENP-A   GPRR---KPQTPRRR-PS-SPAPGPSRQSSSVGSQTLRRRQK-- 37
Human H3.2     ARTKQTARKSTGGKAPRKQLATKAARKSAPATGGVKKPHRYRPG 44


C-Terminal Domain
                   6.5           1.5                    2.5
                   ↓             ↓                      ↓
Human CENP-A   QGWLKEIRKLQKSTHLLIRKLPFSRLAREICVKFTRG-VDFNWQAQALLA 92
Mouse CENP-A   FMWLKEIKTLQKSTDLLFRKKPFSMVVREICEKFSRG-VDFWWQAQALLA 86
Drosophila Cid KRMDREIRRLQHHPGTLIPKLPFSRLVREFIVKYSDD-EPLRVTEGALLA 178
Yeast Cse4p    ELALYEIRKYQRSTDLLISKIPFARLVKEVTDEFTTKDQDLRWQSMAIMA 184
Human H3.2     TVALREIRRYQKSTELLIRKLPFQRLVREIAQDFKT---DLRFQSSAVMA 91
               hhhhhhhhhhhh       hhhhhhhhhhhhhhh             hhhhhh
                    N                    1          L1         2


Human CENP-A   LQEAAEAFLVHLFEDAYLLTLHAGRVTLFPKDVQLARRIRGLEEGLG    139
Mouse CENP-A   LQEAAEAFLIHLFEDAYLLSLHAGRVTLFPKDIQLTRRIRGFEGGLP    133
Drosophila Cid MQESCEMYLTQRLADSYMLTKHRNRVTLEVRDMALMAYICDRGRQF     224
Yeast Cse4p    LQEASEAYLVGLLEHTNLLALHAKRITIMKKDMQLARRIRGQFI       228
Mouse H3.2     LQEASEAYLVGLFEDTNLCAIHAKRVTIMPKDIQLARRIRGERA       135
               hhhhhhhhhhhhhhhhhhhhhh       hhhhhhhhhhh
                          2             L2       3
```

Fig. 1. Sequence alignment of CENP-A homologues and mouse H3.2, a conventional H3. Cid is a CENP-A homologue of *Drosophila* and Cse4p of *Saccharomyces cerevisiae*. The N-terminal regions of Cse4p and Cid are not shown; these regions are much longer than the N-terminus of H3 and show little obvious homology to H3 or human CENP-A. Gaps in the alignment are indicated by "-". Numbered arrows indicate arginine residues of conventional H3 that insert into the minor groove of nucleosomal DNA [14]; the numbers indicate the location of these DNA contacts in helical turns of DNA from the central base pair of the core particle, see Ref. [14]. The underlined h's indicate the α-helical regions of conventional H3 including α-helices 1, 2, and 3 of the histone fold domain; L1 and L2 are the loop regions of the histone fold domain [14]. Sequences: human CENP-A, GI: 602414; mouse CENP-A, GI: 2465203; *S. cerevisiae* Cse4p, GI: 694027; *D. melanogaster* Cid, GI: 8101773; human H3.2, GI: 18545937.

1.1. Sequence comparisons

The C-terminal region of CENP-A proteins shows obvious homology to conventional H3 throughout the histone fold domain and most of the N-helix (Fig. 1). Over this region of homology (amino acids 48–132 of H3) human CENP-A is $\sim 63\%$ identical to a conventional human H3 variant. The only gap in the alignment occurs in L1 of the histone fold domain (the loop between helices 1 and 2) which is 2 to 3 amino acids longer in CENP-A proteins (Fig. 1). In conventional H3, L1 is involved in an interaction with nucleosomal DNA at superhelix location 2.5 [14] and the substantial sequence differences in this region of CENP-A could significantly alter the structure of this region of CENP-A nucleosomes. The N-terminal regions of CENP-A and conventional H3 show little obvious sequence homology (Fig. 1).

The primary structures of CENP-A proteins are surprisingly variable in evolution in comparison to conventional H3. The C-terminal region shows the greatest conservation. Mouse and human CENP-A are 83% identical over the 87 amino acid segment that shows obvious homology to H3 (Fig. 1), and more distantly related CENP-A proteins show substantially greater divergence. In contrast, conventional H3 is highly conserved in this region, e.g., sea urchin H3 is 99% identical to mouse H3.2 (a conventional mouse H3 subtype) [15]. The N-terminal region of CENP-A proteins varies in length from 27 to 196 amino acids and has relatively low evolutionary conservation in the amino acid sequence (Fig. 1). The N-terminal regions of human and mouse CENP-A are only ~48% identical with four gaps in the alignment (Fig. 1) and more distantly related CENP-A proteins show little obvious homology in this region.

The lower evolutionary constraint on the sequences of CENP-A-type variants in comparison to most other core histone variants could in part reflect the very limited context in which CENP-A functions, i.e., only in centromeric chromatin associated with the kinetochore. Most other variants appear to be rather widely distributed in the chromatin, and therefore, may be involved in a greater variety of interactions and modifications that could constrain their structure during evolution.

1.2. Nucleosomes

Several lines of evidence indicate that CENP-A replaces conventional H3 in the nucleosome. Biochemical studies showed that CENP-A co-sediments with nucleosome core particles [7] and a genetic analysis indicates an interaction between Cse4p, the CENP-A of *Saccharomyces cerevisiae*, and H4 [16,17]. A recent study with CENP-A purified from HeLa cells or expressed in bacteria showed that it can substitute for conventional H3 in nucleosome reconstitution [18]. Reconstituted CENP-A-containing nucleosomes appear to contain the other core histones in appropriate stoichiometry. However, they did not strongly protect 146 bp of core DNA from micrococcal nuclease, suggesting that CENP-A may significantly alter some aspects of the core nucleosome structure.

CENP-A may have a significant effect on the nucleosomal structure of the centromere. Studies in *Schizosaccharomyces pompe* indicate that centromeric chromatin has an irregular or disrupted nucleosomal organization as indicated by a smeared micrococcal nuclease digestion pattern. This smeared pattern is replaced by a normal nucleosome ladder in CENP-A knockout cells [10]. The smeared pattern could reflect a direct effect of CENP-A on nucleosome structure, as suggested by the micrococcal nuclease digestion pattern of CENP-A reconstituted nucleosomes discussed above [18], irregular spacing of CENP-A nucleosomes, or the effects of centromeric non-histone proteins.

1.3. Centromeric localization

Domain swapping studies between human CENP-A and H3 showed that the sequences that target CENP-A to the centromere are located in the histone fold region, residues 51–139 [19]. An epitope tagged chimeric protein containing this region linked to the N-terminal region of H3 co-localized with endogenous

CENP-A proteins, while a protein containing the histone fold domain of H3 and the N-terminus of CENP-A was found throughout the nucleus. Swapping experiments with subdomains of CENP-A identified two regions of the histone fold domain that are important for targeting, L1 and helix 2. In conventional nucleosomes L1 is involved in a DNA contact and helix 2 is involved in contacts with H4 and the other H3. Swapping the N-helix had a small effect on targeting and no effect on targeting was observed when helix 1 or L2 were swapped.

While DNA sequence must play some role in establishing the location of centromeres [20], it does not appear to be directly involved in targeting CENP-A to specific sites. For instance, human CENP-A can localize to non-human mammalian centromeres that have very different satellite sequences from humans [19]. Human CENP-A is also efficiently targeted to neocentromeres that have formed at sites that appear to lack the alpha-satellite sequences normally present at human centromeres [21]. Cell cycle specific regulation of CENP-A assembly appears to be important. Human CENP-A is incorporated into centromeric chromatin in the G2 phase of the cell cycle, well after the replication of CENP-A containing DNA [22]. A model has been proposed in which pre-existing CENP-A helps direct the assembly of new CENP-A nucleosomes at the centromere [23]. Other centromere proteins may also participate in the targeting of CENP-A. In the fission yeast *Schizosaccharomyces pompe*, the assembly of CENP-A in the centromere is dependent on Mis6 an inner centromere protein [10]. Thus, the location of the centromere, and CENP-A, could be propagated by an epigenetic mark established by nucleoprotein complexes that contain CENP-A and other centromeric proteins [24].

In metaphase the CENP-A containing chromatin occurs in a domain that appears to lack H3. However, a recent study indicates that CENP-A/CID-containing nucleosomes occur in linear blocks that are interspersed with blocks of H3-containing nucleosomes [25]. Thus the CENP-A chromatin domain of the inner kinetochore appears to be composed of discontinuous blocks of CENP-A containing chromatin that are grouped together.

1.4. Function

One critical function of CENP-A appears to be the recruitment of other centromeric proteins. In mammalian cells the overexpression of CENP-A throughout the cell cycle leads to its deposition to numerous non-centromeric regions [19]. This mislocalized CENP-A is able to recruit CENP-C, an inner centromeric protein, and two kinetochore proteins to these non-centromeric sites [26]. The recruitment of these proteins requires the N-terminal region of CENP-A. Consistent with these findings, the disruption of CENP-A expression leads to a loss of the localization of CENP-C to the centromere and also disrupts the structure and function of the kinetochore [12,13,27,28]. In contrast, the localization of CENP-A is not lost when CENP-C expression is disrupted [27,28]. Similar observations have been made in yeast, where genetic and biochemical studies indicate that the N-terminus of Cse4p, the CENP-A homologue of *Saccharomyces cerevisiae*, interacts with a complex of kinetochore proteins [30,60]. Taken together these studies indicate that CENP-A is a critical element in organizing the structure of the kinetochore. In this role, CENP-A can help form a stable connection between the outer kinetochore and the underlining centromeric chromatin.

One unresolved question is whether CENP-A effects the stability of centromeric chromatin. One interesting possibility is that CENP-A promotes an unusually stable chromatin structure that is adapted to withstanding the forces exerted on the centromere during chromosome segregation [23]. Reconstitution of nucleosome arrays with CENP-A may provide insights into this question.

2. *H2A.Z*

H2A.Z was first observed as a minor histone species, designated M1, that differed substantially from the major histone species in amino acid composition [31]. Peptide mapping identified it as an H2A, which was given the name H2A.Z [32]. The first H2A.Z sequence was from a chicken cDNA and the encoded protein was given the name H2A.F [33]. H2A.Z/F is similar in size to a conventional H2A and is approximately 60% identical in amino acid sequence [33–35] (Fig. 2). H2A.Z-type histones have been described in many eukaryotic organisms including mammals [35], sea urchins [36], *Drosophila* [37], yeast [38], and *Tetrahymena thermophila* [39].

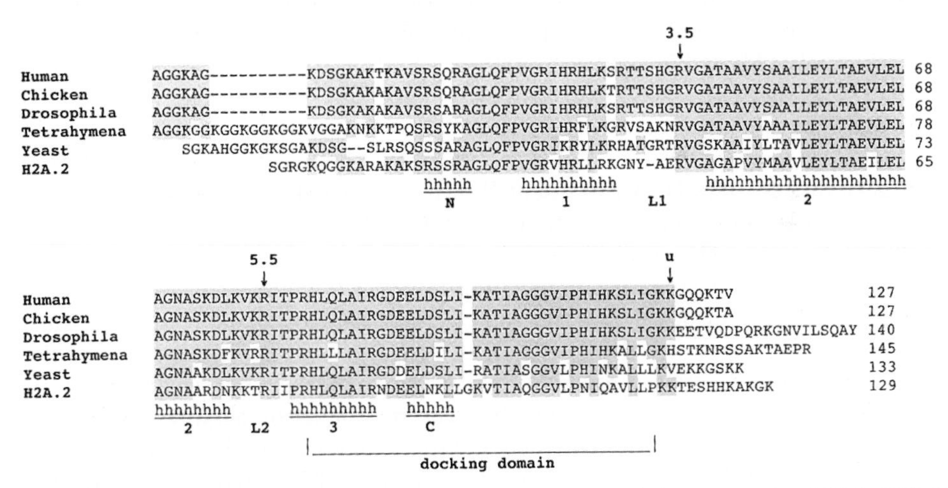

Fig. 2. Sequence alignment of H2A.Z homologues with human H2A.2, a conventional H2A. H2A.Z sequences are from human, chicken, *Drosophila Tetrahymena thermophila*, and *Saccharomyces cerevisiae*. Gaps in the alignment are indicated by "-". Numbered arrows indicate arginine residues of conventional H2A that insert into the minor groove of nucleosomal DNA [14]; numbers indicate the location of these DNA contacts in helical turns of DNA from the central base pair of the core particle, see Ref. [14]. The "u" indicates the lysine residue that is ubiquitinated [130]. The underlined h's indicate the α-helical regions of conventional H2A including α-helices 1, 2, and 3 of the histone fold domain; L1 and L2 are the loop regions of the histone fold domain [14]. The docking domain of H2A is a region of extensive interaction with the $(H3–H4)_2$ tetramer [41]. Sequences: human H2A.Z, GI: 3649600; chicken H2A.F, GI: 121988; *Drosophila* H2AvD, GI: 121989; *Tetrahymena* hv1, GI: 121991; *S. cerevisiae* Htz1p, GI: 6324562; human H2A.2 GI: 121970.

In most cells that have been examined, H2A.Z constitutes about 5–10% of the total H2A [32,40].

2.1. Nucleosomes

X-ray crystallography was used to compare the structure of nucleosome core particles reconstituted with mouse H2A.Z to those reconstituted with conventional H2A [41]. This study shows that the basic structure of the core particle is not grossly altered by H2A.Z, although several localized differences were present. One difference involves the docking domain of H2A (residues 81–119 of conventional H2A), which is involved in a major interaction with the H3–H4 tetramer. Subtle differences in this interaction appear to destablize the interaction of the H2A.Z–H2B dimer with the H3–H4 tetramer. Sequence differences in this region of H2A.Z also change the structure of the nucleosome surface and create a metal ion-binding site on the surface. Changes in the surface of the nucleosome could effect higher order nucleosomal structures such as the chromatin fiber by altering the interaction of H2A.Z-containing nucleosomes with neighboring nucleosomes. Such changes could also alter interactions between H2A.Z-containing nucleosomes and non-histone proteins. Another difference involves the interaction of the two H2As within a nucleosome. This interaction involves L1 of the histone fold domain and structural differences in this region indicate that an H2A.Z–H2A interaction would be unfavorable. This could inhibit the intermixing of H2A.Z and conventional H2A within a nucleosome.

2.2. Function

The gene that encodes H2A.Z is essential in *Drosophila* [42], *Tetrahymena* [43], and mice [44]. Deletion of the gene that encodes H2A.Z in yeast results in slow growth and chromosome segregation defects in *Schizosaccharomyces pompe* [38], and in slow growth and defects in gene regulation in *Saccharomyces cerevisiae* [45–48].

Several lines of evidence indicate that H2A.Z is involved in regulating transcription. Early studies of hv1, the H2A.Z of *Tetrahymena thermophila*, showed that it is present in the transcriptionally active macronucleus, but is absent in the transcriptionally silent micronucleus [49]. The micronucleus acquires hv1 shortly before it becomes transcriptionally active during conjugation [50]. Deletion of the gene that encodes H2A.Z in *Saccharomyces cerevisiae*, *HTZ1*, reduces the induction of *Gal1* and *Gal10* [45,48]. When the *HTZ1* deletion was combined with mutations in genes for chromatin remodeling factors such SNF2, a synergistic effect on the induction of some genes was observed [48]. Thus the absence of H2A.Z appears to make the induction of some genes more dependent on chromatin remodeling and modification complexes. Chromatin immunoprecipitation assays indicated that H2A.Z is preferentially located to the intergenic regions of the affected genes, suggesting that H2A.Z may have a direct role in the transcriptional activation of these genes [48]. The transcribed region of one affected gene had a low level of H2A.Z in comparison to the nearby intergenic region, indicating that the H2A.Z content of a individual gene can vary significantly over a relatively short distance of about 1000 bp.

The role of H2A.Z does not appear to be confined to transcriptional activation. Deletion of the gene for H2A.Z in *S. cerevisiae* also leads to defects in transcriptional silencing at the silent mating locust and telomeres [46]. Thus, H2A.Z may participate in the formation of specialized chromatin structures that can be used for either activation or silencing.

Studies of the distribution of H2AvD, the H2A.Z of *Drosophila*, indicated a complex distribution pattern in which it was associated with every sequence examined including transcribed and non-transcribed genes, and non-coding sequences such as satellite DNA [51]. The highest concentration of H2AvD was found in the heat shock gene *hsp70* and the lowest was on a satellite sequence. However, no consistent pattern was observed when transcribed and non-transcribed genes were compared, indicating that there is not a simple direct relationship between H2AvD content and transcriptional activity. The absence of such a relationship seems consistent with the yeast studies described above that indicate that H2A.Z may have a role in both activation and repression, and that the H2A.Z content of a single gene can vary significantly over a relatively short distance.

In order to identify the regions of H2A.Z that are required for its essential functions, eight different regions of H2A were swapped for the corresponding regions of H2AvD, the H2A.Z of *Drosophila* [52]. These chimeric H2AvD derivatives were tested for their ability to rescue the development of flies that had a homozygous deletion of the gene for H2AvD. One region in the C-terminal part of H2AvD (residues 92–102) was essential for rescuing fly development and a neighboring region (residues 106–119) was also very important. These two regions correspond to the docking domain of H2A discussed above. Swapping of two other regions, one in the N-terminus and one in helix 2, had lesser effects on the ability of the protein to rescue development. A similar experiment done in *Saccharomyces cerevisiae* also indicated the importance of the C-terminal region of H2A.Z [45].

Because the docking domain of H2A.Z alters the surface of the nucleosome, it could effect the interaction of H2A.Z-containing nucleosomes with other nuclear proteins [41]. Such interactions could involve proteins that activate transcription, such as factors associated with RNA polymerase II [45], or with silencing factors such as the Sir proteins [46]. H2A.Z could also alter important interactions within or between nucleosomes [41]. There is conflicting data on the effect of H2A.Z on the affinity of the H2A.Z–H2B dimer for the nucleosome and on the folding of nucleosome arrays. Abbott et al. [53] concluded that H2A.Z–H2B dimers dissociated more readily from the nucleosome than conventional H2A–H2B dimers and that H2A.Z-containing nucleosome arrays have reduced NaCl-dependent folding compared to H2A-containing arrays. In contrast, Fan et al. [54] found that H2A.Z–H2B dimers are more stably associated with nucleosomes and that H2A.Z increased the folding of nucleosome arrays in the presence of divalent cations. It is unclear whether these differences are due to differences in the preparations of histones and nucleosomes, both studies used nucleosomes reconstituted with the same DNA template, or in the conditions used to assay the nucleosomal structures, Abbott et al. [53] used NaCl to promote the folding of their arrays while Fan et al. [54] used $MgCl_2$ or $ZnCl_2$. Studies of the salt elution profile of H2A.Z from chicken erythrocyte chromatin bound to hydroxyapatite support the idea that H2A.Z is bound more tightly than conventional H2A [55].

3. H2A.X

H2A.X was first observed as a minor H2A-like histone designated M2 [31]. Peptide mapping showed that this protein, which was named H2A.X, is similar but not identical to conventional H2A subtypes [32]. Sequencing of a cDNA for human H2A.X showed that it is nearly identical to conventional H2A for its first 120 residues, but is 13 residues longer than conventional H2A and contains a 22 residue C-terminal region that does not show homology to any other known vertebrate H2A [56] (Fig. 3). H2A.X has a conserved 4 amino acid motif on its C-terminus characterized as SQ(E,D)(I,L,F,Y). This SQ motif is present on the C-terminus of H2As from many invertebrates, including the major H2A variants present in yeast [40] (Fig. 3). Thus, in some species H2A.X-like H2As appear to be the major H2A variant, whereas the H2A.X content of mammalian tissues has been reported to be 2–10% of the total H2A [57]. In *Drosophila* the H2A.X SQ motif is present on the end of the H2A.Z homologue, H2AvD (Fig. 2), indicating that H2A.Z and H2A.X functions can be combined in the same molecule.

3.1. DNA double strand breaks and H2A.X phosphorylation

The serine of the SQ motif of H2A.X is phosphorylated in response to double strand DNA breaks [57]. The threonine of a TQ sequence located just before the SQ motif in human H2A.X is also phosphorylated, but to a lesser extent. Phosphorylation of either the SQ or TQ sites alters the mobility of H2A.X in two dimensional gels leading to the formation of a species referred to as γ-H2A.X [57].

Human	NIQAVLLPKKTSATVGPKAPSGGKKATQASQEY	142
Mouse	NIQAVLLPKKSSATVGPKAPAVGKKASQASQEY	142
S. cerevisiae	NIHQNLLPKKSA------------KATKASQEL	131
S.pombe	NINAHLLPKTSG-----------RTGKPSQEL	131
H2A.2	NIQAVLLPKKTESHHKAKGK	129

Fig. 3. Sequence alignment of C-terminal regions of H2A.X homologues from human, mouse, *S. cerevisiae* and *S. pombi*, and human H2A.2. In yeast, the major H2A variants contain the SQ motif (underlined) that is characteristic of H2A.X. Mammalian H2A.X and H2A.2 are nearly identical in sequence in the regions not shown. Sequences: human H2A.X, GI: 106266; mouse H2A.X, GI: 51142; *S. cerevisiae* H2A.1, GI: 6320431; *S. pombi* H2Aα, GI: 19075680; human H2A.2, GI: 121970.

The formation of γ-H2A.X occurs rapidly after the production of double strand breaks by ionizing radiation with detectable amounts present in seconds and maximal levels reached in about 10 min [57]. γ-H2A.X disappears with kinetics consistent with the repair of the double stranded break. Western blots with antibodies that specifically recognize the phosphorylated SQ motif indicate that phosphorylation of this motif occurs in response to double breaks in many organisms, including *Drosophila* and *S. cerevisiae* [58].

One double stranded break leads to the formation of several hundred to several thousand γ-H2A.X molecules, depending on the H2A.X content of the cell [57]. Immunofluorescence studies indicate that the γ-H2A.X is present in a megabase domain of chromatin around the site of the double stranded break [58]. In addition to ionizing radiation, the formation of γ-H2A.X is induced in a variety of other circumstances that involve double strand breaks including apoptosis [59], VDJ recombination of immunoglobulin genes [29], immunoglobulin class switching [61] and meiotic recombination [62].

Members of the phosphoinositide (PI)-3 kinase family appear to be involved in the phosphorylation of H2A.X. The SQ motif matches a common target site for these kinases and the formation of γ-H2A.X in response to double stranded breaks is inhibited by wortmannin, an inhibitor of PI-3 kinases [63]. Examination of cell lines deficient in the PI-3 kinase ATM indicated that it has a major role in phosphorylating H2A.X in response to double strand breaks [64]. ATM can phosphorylate H2A.X *in vitro* suggesting that it may directly phosphorylate H2A.X *in vivo* [64]. Another PI-3 kinase ATR appears to be involved in phosphorylating H2A.X in response to replicational stress induced by treatment of dividing cells with hydroxyurea or by irradiating them with ultraviolet light [65]. It has been hypothesized that PI-3 kinases such as ATM are recruited to, or activated at, the site of the double stranded break and then phosphorylate H2A.X molecules around the break point [40,64,66].

3.2. Function

Genetic studies indicate that H2A.X has an important role in the repair of double strand breaks. Yeast cells that carry a mutation in the SQ motif of H2A are hypersensitive to DNA damaging agents that induce double strand breaks [67]. Similar to the situation with mammalian H2A.X, phosphorylation of the yeast H2A SQ motif in response to DNA damage is dependent on a specific PI-3 kinase called Mec1 [67]. H2A.X knockout mice exhibit multiple defects including reduced immunoglobulin class switching, increased sensitivity to ionizing radiation, increased chromosomal abnormalities, decreased homologous recombination and male infertility [61,68]. Most and possibly all of these defects could be related to a reduced efficiency in the repair of double strand breaks.

One important role for γ-H2A.X appears to be the recruitment of repair factors to chromatin domains that contain the DNA break. Formation of γ-H2A.X domains precedes the recruitment of repair factors such as Brca1, Rad50, and

Rad51 to the domain [63] and the recruitment of Nbs1, 53bp1, and Brca1 to such domains was absent or severely impaired in irradiated H2A.X knockout B cells [68]. In contrast, irradiation-induced Rad51 foci were present in both wild type and H2A.X knockout cells, indicating that the recruitment of this factor is not dependent on γ-H2A.X. The recruitment of repair factors could involve a direct interaction between these factors and γ-H2A.X or an indirect mechanism where γ-H2A.X produces a chromatin conformation that facilitates the recruitment of other factors to the damaged domain [58]. In order to investigate whether phosphorylation of the SQ motif of yeast H2A affects chromatin structure, the serine was mutated to a glutamate to mimic phosphorylation of this site [67]. Strains that containing this mutation were not hypersensitive DNA double strand breaks, indicating that this mutation mimics phosphorylation. Chromatin prepared from this strain was digested more rapidly by micrococcal nuclease than chromatin from wild-type cells, suggesting that phosphorylation of the SQ motif may decondense chromatin and increase the accessibility of repair factors to the DNA [67].

4. MacroH2A

MacroH2As were discovered on the basis of their copurification with rat liver mononucleosomes [69]. There are three known subtypes: macroH2A1.1 and 1.2 formed by alternate splicing of transcripts from *macroH2A1*; and macroH2A2 which is encoded by a separate gene, *macroH2A2* [69–73]. All three subtypes are nearly identical in size and have the same basic structure. They are almost three times the size of a conventional H2A and have a hybrid structure consisting of a large non-histone-region and a region that closely resembles a full length H2A (Fig. 4). The macroH2A content and subtype composition of different cell types and tissues is different and changes during development [71,72]. It was estimated that in adult rat liver there is one macroH2A molecule for every 30 nucleosomes [69].

4.1. Sequence comparisons

The H2A-regions of macroH2As are ~65% identical to a conventional H2A, and the H2A-regions of macroH2A1 and macroH2A2 subtypes are ~84% identical to one another (Fig. 5). Most of the non-histone-region appears to be derived from a gene of unknown function that has an evolutionary history that predates eukaryotes [74]. The sequences of macroH2A proteins are highly conserved among vertebrates; the H2A-region of macroH2A1.2 is completely conserved between rats and chickens and the non-histone region is ~95% identical [74]. On the basis of expressed sequence and genomic databases, macroH2A genes do not appear to be present in *Saccharomyces cerevisiae*, *Drosophila*, or *C. elegans*, but macroH2A is present in zebra fish. This suggests that macroH2As arose in the Deuterosomia branch of the animal kingdom.

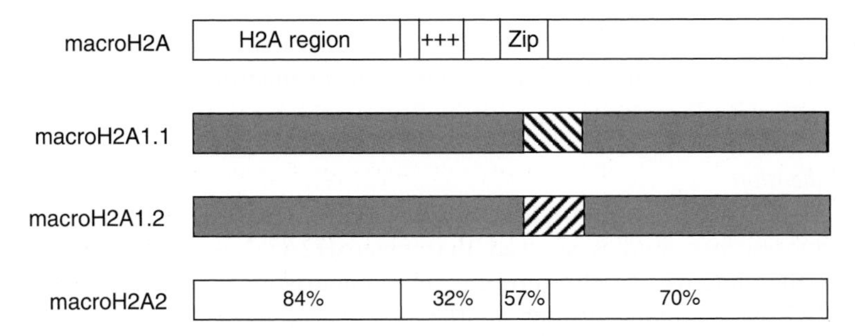

Fig. 4. Comparisons of macroH2A1.1, macroH2A1.2, and macroH2A. A diagram of macroH2A structural motifs is shown on top; "+++" indicates a lysine rich region, and "zip" indicates a region that resembles a leucine zipper [69]. The sequence relationship of macroH2A1.1 and macroH2A1.2 is shown in the middle; gray indicates regions of identical sequence, hatching indicates regions that are only 21% identical between human macroH2A1.1 and macroH2A1.2. Human macroH2A2 is 68% identical to human macroH2A1.2; the percentage identity of specific regions of human macroH2A2 compared to human macroH2A1.2 is indicated at the bottom.

Fig. 5. Sequence alignment of the H2A-region of macroH2A1 and macroH2A2, and human H2A.2, a conventional H2A. Notations are the same as in Fig. 2. Sequences: human macroH2A1.2, GI: 20336748; human macroH2A2, GI: 8923920; human H2A.2, GI: 121970.

4.2. Nucleosomes

MacroH2As are an integral part of the nucleosome, requiring the same or slightly higher salt concentration for their removal as conventional H2A [69]. MacroH2A1.2 containing nucleosomes have been reconstituted *in vitro* using macroH2A expressed in bacteria [75]. Both H2As of the nucleosome core can be replaced with macroH2A1.2 without grossly disrupting the structure of the nucleosome as judged by micrococcal nuclease digestion, DNase I digestion and sedimentation of core particles in sucrose gradients. The DNase I digestion pattern of macroH2A-containing nucleosome core particles showed a reduction in cleavage at the nucleosomal dyad in comparison to nucleosomes reconstituted with conventional H2A. This could reflect an indirect steric hindrance of this area by the non-histone region or a direct interaction of macroH2A with this region.

4.3. Localization

Immunofluorescence studies showed that macroH2A1.2 is preferentially concentrated in the inactive X chromosome in comparison to bulk chromatin [76]. MacroH2A2 is also concentrated in the inactive X [70,71]. The localization of macroH2A proteins to the inactive X is dependent on a gene called *Xist*. *Xist* is transcribed only from the inactive X and is required for initiation of X inactivation [78–82]. It produces a large non-coding nuclear RNA that coats the inactive X chromosome *in cis* [83–85]. The localization of macroH2A to the inactive X is lost when *Xist* is disrupted by Cre mediated recombination [86], or when *Xist* RNA is disrupted by peptide nucleic acids [87]. Immunoprecipitation of macroH2A-containing chromatin fragments precipitates *Xist* RNA, indicating that they are in close proximity [88]. These results suggest that the localization of macroH2A to the inactive X could involve a direct interaction with *Xist* RNA or with a factor associated with *Xist* RNA.

MacroH2A1.2 is also concentrated in the X chromosome during X-inactivation in spermatogenesis [89,90]. The X and Y chromosomes become condensed and transcriptionally inactive in the meiotic prophase of male germ cells, possibly as a mechanism to prevent damaging recombination events [91]. *Xist* is expressed in male germ cells [92–94], but its role in X inactivation and macroH2A localization in male germ cells is not known.

In vitro mutagenesis was used to identify the regions of macroH2A that are required for targeting to the inactive X [95]. Using GFP or epitope tags to trace the fate of the protein, the variant H2A-regions of macroH2A1 and macroH2A2 were concentrated in the inactive X. In contrast, neither conventional H2A nor H2B showed any such preference. Short deletions of the N- or C-terminal tails of the H2A-region did not prevent targeting, indicating that the targeting sequences are part of the central region of this domain. The non-histone region could also be targeted to the inactive X, but linkage to the nucleosome through fusion to conventional H2A or H2B was required. Linkage of the non-histone domain to chromatin proteins HP-1 or HMG-14 was not sufficient. Mutagenesis of non-histone region showed that the C-terminal 190 amino acids are important for the targeting to the inactive X.

How might these two domains of macroH2A function in localizing macroH2A to the inactive X? The non-histone region has homology to a conserved domain present in viral proteins that are bound to viral RNA and are involved in RNA synthesis [74]. This suggests that the non-histone region might bind RNA and promote localization to the inactive X by interacting with *Xist* RNA [74]. One possibility for the histone domain is self-association, either within a nucleosome by interactions involving L1 of the histone fold domain [41,95], or between nucleosomes involving interactions on the surface of the nucleosome.

4.4. Function

In mouse embryos, the preferential localization of macroH2A1.2 to the inactive X of trophectoderm cells appears to occur in the early stages of transcriptional silencing [96]. However, initiation of X inactivation in embryonic stem cells occurs before macroH2A is concentrated in the inactive X chromosome, indicating that macroH2A is not required for initiation of X inactivation [97]. The most likely role of macroH2A in X inactivation is to promote the long-term maintenance of silencing. Maintenance of inactivation appears to be mediated by multiple independent mechanisms, including histone hypoacetylation, DNA methylation and asynchronous replication [98].

The function of macroH2A proteins is not limited to X inactivation. X inactivation occurs only in female mammals, but macroH2A proteins are present in similar concentrations in male mammals [76] and are highly conserved in non-mammalian vertebrates that do not have X-inactivation [74]. Therefore, functions other than X-inactivation must underlie the conservation of macroH2As in vertebrates. The high concentration of macroH2A in the inactive X suggests that at least some of these non-inactive X functions may involve transcriptional repression or silencing.

5. H3 replacement variants

While the synthesis of histones increases dramatically during DNA replication (see Ref. [99]), some variants are synthesized and incorporated into chromatin in absence of replication [100]; these have been referred to as replication independent, basal, or replacement variants. Some variants show a mixed pattern with increased synthesis during replication, but continued expression in non-dividing cells (see Ref. [3], pp. 103–109). Several of the variants discussed above can be considered specialized replacement variants, including CENP-A [22], H2A.Z [35,100], H2A.X [56,100], and probably macroH2A. This section focuses on H3.3, an H3 replacement variant found in animals.

5.1. Sequences

H3.3 is distinguished from H3 replication coupled variants by four sequence differences at amino acids 31, 87, 89, and 90 [101,102]. These differences are conserved between mammals and *Drosophila*, indicating that they are functionally significant [103].

5.2. Localization

A recent study in *Drosophila* cells used H3-GFP fusion proteins to compare the deposition patterns of H3.3 and a replication dependent H3 [104]. The deposition of the replication dependent H3-GFP occurred at sites that co-localized with BrdU labeling, indicating that it occurs at sites of new DNA synthesis. No incorporation of H3-GFP was observed in the G2 phase of the cell cycle or when DNA synthesis was inhibited by aphidicolin. H3.3-GFP was also deposited by a replication dependent mechanism, but in addition was incorporated in G2 phase cells and when replication was blocked by aphidicolin. This difference in deposition patterns appears to be due to the sequence differences between the proteins rather than their patterns of expression. Differences in the transcriptional patterns of the constructs were minimized by expressing both constructs from the same heat shock promoter. In addition, the H3-GFP replication coupled pattern could be changed to a pattern similar to that of H3.3-GFP by point mutations that converted three of the four distinguishing amino acid residues to the H3.3 sequence. On the basis of these and other results, Ahmed and Henikoff [104] hypothesized that there are distinct chromatin assembly complexes involved in replication coupled and replication independent assembly and that these complexes recognize different H3 variants.

The replication independent deposition of H3.3 appears to occur more rapidly in transcriptionally active chromatin. Pulse labeling of quiescent chicken erythrocytes showed a preferential incorporated of H3.3 into a chromatin fraction that is highly enriched for transcriptionally active sequences [105]. However, its incorporation into this chromatin was not reduced by inhibiting transcription with actinomycin D, indicating that transcription is not required for H3 turnover. This suggests that its preferential exchange into transcriptionally active chromatin may be due to structural differences between active and inactive chromatin, rather than the displacement of histones or nucleosomes by transcription. The replication independent incorporation of a H3.3-GFP fusion protein in *Drosophila* cells had a large preference for active ribosomal genes, indicating that it preferentially exchanges into transcriptionally active chromatin [104]. H3.3 also exchanges into bulk chromatin, although apparently more slowly. In mouse hepatocytes, for instance, H3.3 replaces most of the H3.1 and H3.2 over a period of several months [106] (also see Ref. [3], pp. 103–109).

5.3. Function

An insertional mutation in the H3.3A gene, one of the two H3.3 genes in mice, indicates that this gene has important functions in development and differentiation [107]. This mutation leads to a large reduction in expression of this gene and ~50% of homozygous mutants die shortly after birth. The surviving mutants have a reduced growth rate, neuromuscular deficits and reduced male fertility.

One possible function for H3.3 and some of the other replacement variants is to maintain the integrity of chromatin structure by replacing histones that are damaged or lost during normal cellular metabolism. Ahmad and Henikoff [104] suggested that replication independent incorporation of H3.3 could provide a mechanism to switch patterns of histone modification by removing H3 molecules that may be irreversibly modified by methylation. They also suggested that H3.3 could serve as a mark of transcriptionally active chromatin. One complication for the latter idea is that this mark would be slowly erased as H3.3 replaces replication variants in bulk chromatin in long lived non-dividing cells present in many vertebrate tissues.

5.4. *Other H3 replacement variants*

H3 replacement variants are also present in plants [108] and *Tetrahymena* [109]. Phylogenetic analyses indicates that these H3 replacement variants arose independently of animal H3.3 [109]. Like H3.3, the plant H3 replacement variant also appears to be preferentially deposited into transcriptionally active chromatin [110]. The H3 replacement variant in *Tetrahymena,* hv2, is found only in the transcriptionally active macronucleus [49]. In contrast to H3.3, the amino acid differences between hv2 and the replication dependent H3 of *Tetrahymena* do not appear to be essential for replication independent incorporation into chromatin [111]. In this case constitutive expression appears to be the dominant factor that can drive replication independent deposition of hv2 or an H3 variant that is normally replication dependent. This difference between hv2 and H3.3 is not necessarily surprising because hv2 appears to have arisen independently of H3.3 and does not have the structural features characteristic of H3.3 [109].

6. *H2A.Bbd*

H2A.Bbd was discovered as an H2A-like protein encoded by human ESTs [77]. H2A.Bbd is only 42% identical to conventional H2A and is smaller due to a shortened C-terminus that lacks the ubiquitination site that is present in most other H2As (Fig. 6). H2A.Bbd appears to be associated with nucleosomes as indicated by the co-purification of an epitope-tagged H2A.Bbd with nucleosomal fragments in a sucrose gradient run at high ionic strength [77].

Messenger RNAs for H2A.Bbd have been detected in human testis, fibroblasts and lymphocytes [77], but little is known about the amount of H2A.Bbd protein in these or other human cell types or about its presence in other species. The distribution of H2A.Bbd in chromatin was examined by expressing epitope-tagged or GFP-tagged H2A.Bbd in cultured cells. These studies revealed a striking deficiency of H2A.Bbd in the inactive X chromosome, leading to its name which stands for Histone H2A Barr-body deficient [77]. The distribution of H2A.Bbd overlapped extensively with that of H4 acetylated on lysine 12, suggesting that H2A.Bbd may preferentially associate with transcriptionally active chromatin.

7. *Spermatogenesis*

Testis specific variants of H2A, H2B, and H3 have been identified in the rat and mouse (see for example, Refs. [112] and [113]). Two testis specific H2Bs have been

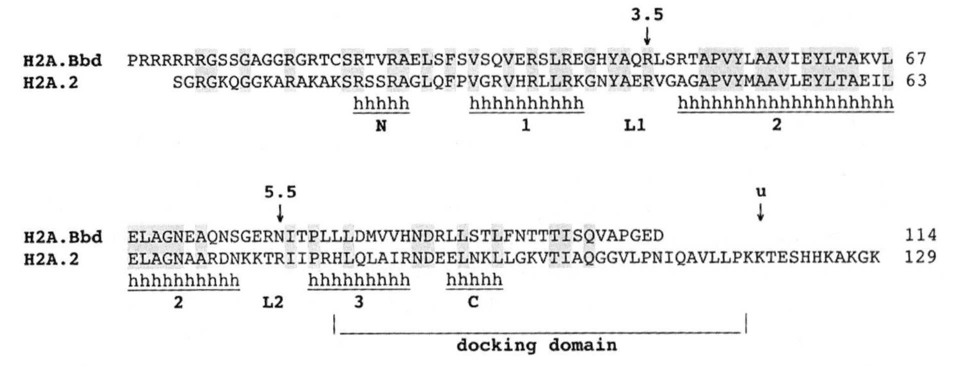

Fig. 6. Sequence alignment of H2A.Bbd and H2A.2. Notations are the same as in Fig. 2. Sequences: human H2A.Bbd, GI: 15553137; human H2A.2, GI: 121970.

sequenced, TH2B the major H2B in spermatocytes [114,115] and ssH2B a minor variant expressed in spermatids [116,117]. TH2B is similar to somatic H2Bs in the histone fold domain, but has substantial differences in the N-terminal domain. The distinguishing feature of ssH2B is a twelve amino acid extension on its C-terminus that contains seven hydrophobic residues [116]. It appears to be uniformly distributed in the nucleus, except for exclusion from the nucleolus [118] and is largely removed from spermatids prior to chromatin condensation [117]. It is unclear how TH2B, ssH2B and the other testis specific variants differ functionally from the conventional somatic variants.

In mammals the histones are removed and replaced by transition basic proteins in mid-spermatids and then the transition basic proteins are replaced by protamines in late spermitids and sperm [119]. In mouse and rat sperm, histones removal is complete or nearly so [119,120], but in humans $\sim$15% of the DNA remains associated with histones [121]. In bovine sperm more than 99% of the histones are removed, but CENP-A is quantitatively retained [122]. This retained CENP-A could be part of an epigenetic mark that allows the positions of the centromeres to be retained in sperm and on the paternal chromosomes of the zygote.

Recently, a human sperm H2B was identified that is part of protein complex that specifically binds the telomere DNA repeat [123]. This telomere-binding complex is extracted by conditions (a solution containing 0.5% Triton X-100 and 100 mM NaCl) that do not extract nucleosomal H2B, suggesting that this H2B is not extracted from a typical nucleosomal structure. The structure of this sperm H2B has not been determined.

8. Cleavage stage variants

During the first few cell divisions of the sea urchin embryo the major histones incorporated into the chromatin are the cleavage stage or CS variants [124]. CS histones are stored in the egg and are translated from stored mRNAs [124–126] . CS H2A, CS H2B, and CS H3 have distinct sequences, while the sequence of the H4 used during the cleavage stage is identical to other sea urchin H4s [127]. The distinct structures of CS histones may be related to their cytoplasmic storage and their roles in remodeling sperm chromatin following fertilization [124,127,128].

The distinguishing feature of CS H2A is a nine amino acid extension on its C-terminus. The possibility that CS H2A is the sea urchin H2A.X has been raised on the basis of its structure and its similarity to a large H2A stored in *Xenopus* eggs that was identified as H2A.X [127]; this *Xenopus* protein has been identified as H2.X on the basis of its position in two dimensional gels and its peptide map [129], but its sequence has not been determined. The C-terminal sequence of CS H2A (SMEY) resembles the SQ(ED)(ILFY) consensus of H2AX. The *Xenopus* and sea urchin proteins also show similar phosphorylation patterns during chromatin assembly [128,129]. Additional studies will be required to determine the relationship, if any, between CS H2A and H2A.X.

9. *Concluding remarks*

Studies over the last 10 years clearly demonstrate the importance of histone variants in modifying nucleosome structure and function. They play specialized roles in diverse chromatin functions including transcriptional regulation, DNA repair, chromosome segregation, spermatogenesis, and histone replacement.

The details of how the functional properties of nucleosomes and chromatin are altered by histone variants are only beginning to emerge. Some variants have unusual tails or extensions that may be involved in interactions with other proteins, DNA, RNA, or other nuclear components. These tails can also be a target for specific modifications. Identification and characterization of these interactions and modifications will be crucial for understanding these variants. Regions that are involved in the structure of the nucleosome core, including the histone fold domain, are also clearly important for some core histone variants. Structural differences in these regions may change important interactions between histones, between histones and DNA, between nucleosomes and with non-histone proteins. Progress in understanding the unusual histone variants will be linked to continued progress in understanding the functions of the conventional variants. Examining the structure and properties of nucleosomes and nucleosomal arrays reconstituted with purified histone variants should give some insights into how these proteins function. Developing approaches to examine the effects of variants on native chromatin is an important goal for the future.

References

1. Franklin, S.G. and Zweidler, A. (1977) Non-allelic variants of histones 2a, 2b and 3 in mammals. Nature 266, 273–275.
2. Ehinger, A., Denison, S.H., and May, G.S. (1990) Sequence, organization and expression of the core histone genes of Aspergillus nidulans. Mol. Gen. Genet. 222, 416–424.

3. van Holde, K.E. (1988) Chromatin. Springer-Verlag, New York, Berlin, Heidelberg, London, Paris, Tokyo, pp. 91–111.
4. Earnshaw, W.C. and Rothfield, N. (1985) Identification of a family of human centomere proteins using autoimmune sera from patients with scleroderma. Chromosoma 91, 313–321.
5. Moroi, Y., Peebles, C., Fritzler, M.J., Steigerwald, J., and Tan, E.M. (1980) Autoantibody to centromere (kinetochore) in scleroderma sera. Proc. Natl. Acad. Sci. USA 77, 1627–1631.
6. Palmer, D.K., O'Day, K., Trong, H.L., Charbonneau, H., and Margolis, R.L. (1991) Purification of the centromere-specific protein CENP-A and demonstration that is a distinctive histone. Proc. Natl. Acad. Sci. USA 88, 3734–3738.
7. Palmer, D.K., O'Day, K., Wener, M.H., Andrews, B.S., and Margolis, R.L. (1987) A 17 kD centromere protein (CENP-A) copurifies with nucleosome core particles and with histones. J. Cell Biol. 104, 805–815.
8. Warburton, P.E., Cooke, C.A., Bourassa, S., Vafa, O., Sullivan, B.A., Stetten, G., Gimelli, G., Warburton, D., Tyler-Smith, C., Sullivan, K.F., Poirier, G.G., and Earnshaw, W.C. (1997) Immunolocalization of CENP-A suggests a distinct nucleosome structure at the inner kinetochore plate of active centromeres. Current Biol. 7, 901–904.
9. Stoler, S., Keith, K.C., Curnick, K.E., and Fitzgerald-Hayes, M. (1995) A mutation in CSE4, an essential gene encoding a novel chromatin-associated protein in yeast, causes chromosome non-disjunction and cell cycle arrest at mitosis. Gene Dev. 9, 573–586.
10. Takahashi, K., Chen, E.S., and Yanagida, M. (2000) Requirement of Mis6 centromere connector for localizing a CENP-A-like protein in fission yeast. Science 288, 2215–2219.
11. Buchwitz, B.J., Ahmad, K., Moore, L.L., Roth, M.B., and Henikoff, S. (1999) A histone-H3-like protein in C. elegans. Nature 401, 547–548.
12. Blower, M.D. and Karpen, G.H. (2001) The role of Drosophila CID in kinetochore formation, cell cycle progression and heterochromatin interactions. Nature Cell Biol. 3, 730–739.
13. Howman, E.V., Fowler, K.J., Newson, A.J., Redward, S., MacDonald, A.C., Kalitsis, P., and Choo, K.H.A. (2000) Early disruption of centromeric chromatin organization in centromere protein A (*Cenpa*) null mice. Proc. Natl. Acad. Sci. USA 97, 1148–1153.
14. Luger, K., Mader, A.W., Richmond, R.K., Sargent, D.F., and Richmond, T.J. (1997) Crystal structure of the nucleosome core particle at 2.8 A resolution. Nature 389, 251–260.
15. Wells, D. and McBride, C. (1989) A comprehensive compilation and alignment of histones and histone genes, Nucleic Acids Res. 17 Suppl, 311–346.
16. Glowczewski, L., Yang, P., Kalashnikova, T., Santisteban, M.S., and Smith, M.M. (2000) Histone–histone interactions and centromere function. Mol. Cell. Biol. 20, 5700–5711.
17. Meluh, P.B., Yang, P., Glowczewski, L., Koshland, D., and Smith, M.M. (1998) Cse4p is a component of the core centromere of *Saccharomyces cerevisiae*. Cell 94, 607–613.
18. Yoda, K., Ando, S., Morishita, S., Houmura, K., Hashimoto, K., Takeyasu, K., and Okazaki, T. (2000) Human centromere protein A (CENP-A) can replace histone H3 in nucleosome reconstitution *in vitro*. Proc. Natl. Acad. Sci. USA 97, 7266–7271.
19. Sullivan, K.F., Hechenberger, M., and Masri, K. (1994) Human CENP-A contains a histone H3 related histone fold domain that is required for targeting to the centromere. J. Cell Biol. 127, 581–592.
20. Choo, K.H. (2000) Centromerization. Trends Cell Biol. 10, 182–188.
21. Lo, A.W., Magliano, D.J., Sibson, M.C., Kalitsis, P., Craig, J.M., and Choo, K.H. (2001) A novel chromatin immunoprecipitation and array (CIA) analysis identifies a 460-kb CENP-A-binding neocentromere DNA. Genome Res. 11, 448–457.
22. Shelby, R.D., Monier, K., and Sullivan, K.F. (2000) Chromatin assembly at kinetochores is uncoupled from DNA replication. J. Cell Biol. 151, 1113–1118.
23. Sullivan, K.F. (2001) A solid foundation: functional specialization of centromeric chromatin. Curr. Opin. Genet. Develop. 11, 182–188.
24. Murphy, T.D. and Karpen, G.H. (1998) Centromeres take flight: alpha satellite and the quest for the human centromere. Cell 93, 317–320.
25. Blower, M.D., Sullivan, B.A., and Karpen, G.H. (2002) Conserved organization of centromeric chromatin in flies and humans. Dev. Cell 2, 319–330.

26. Van Hooser, A.A., Ouspenski, I.I., gregson, H.C., Starr, D.A., Yen, T.J., Goldberg, M.L., Yokomori, K., Earnshaw, W.C., Sullivan, K.F., and Brinkley, B.R. (2001) Specification of kinetochore-forming chromatin by the histone H3 variant CENP-A. J. Cell Sci. 114, 3529–3542.
27. Moore, L.L. and Roth, M.B. (2001) HCP-4, a CENP-C-like protein in Caenorhabditis elegans, is required for resolution of sister centromeres. J. Cell Biol. 153, 1199–1208.
28. Oegema, K., Desai, A., Rybina, S., Kirkham, M., and Hyman, A.A. (2001) Functional analysis of kinetochore assembly in Caenorhabditis elegans. J. Cell Biol. 153, 1209–1226.
29. Chen, H.T., Bhandoola, A., Difilippantonio, M.J., Zhu, J., Brown, M.J., Tai, X., Rogakou, E.P., Brotz, T.M., Bonner, W.M., Ried, T., and Nussenzweig, A. (2000) Response of RAG-mediated VDJ cleavage by NBS1 and gamma-H2AX. Science 290, 1962–1965.
30. Ortiz, J., Stemmann, O., Rank, S., and Lechner, J. (1999) A putative protein complex consisting of Ctf19, Mcm21, and Okp1 represents a missing link in the budding yeast kinetochore. Genes Dev. 13, 1140–1155.
31. Urban, M.K., Franklin, S.G., and Zweidler, A. (1979) Isolation and characterization of the histone variants in chicken erythrocytes. Biochemistry 18, 3952–3960.
32. West, M.H.P. and Bonner, W.M. (1980) Histone 2A, a heteromorphous family of eight protein species. Biochemistry 19, 3238–3245.
33. Harvey, R.P., Whiting, J.A., Coles, L.S., Krieg, P.A., and Wells, J.R.E. (1983) H2A.F: An extremely variant histone H2A sequence expressed in the chicken embryo. Proc. Natl. Acad. Sci. USA 80, 2819–2823.
34. Ernst, S.G., Miller, H., Brenner, C.A., Nocente-McGrath, C., Francis, S., and McIsaac, R. (1987) Characterization of a cDNA clone coding for a sea urchin histone H2A variant related to the H2A.F/Z histone protein in vertebrates. Nucleic Acids Res. 15, 4629–4644.
35. Hatch, C.L. and Bonner, W.M. (1988) Sequence of cDNAs for mammalian H2A.Z, an evolutionarily diverged but conserved basal histone H2A isoprotein species. Nucleic Acids Res. 16, 1113–1124.
36. Wu, R.S., Nishioka, D., and Bonner, W.M. (1982) Differential conservation of histone 2A variants between mammals and sea urchins. J. Cell Biol. 93, 426–431.
37. van Daal, A., White, E.M., Gorovsky, M.A., and Elgin, S.C.R. (1988) *Drosophila* has a single copy of the gene encoding a highly conserved histone H2A variant of the H2A.F/Z type. Nucleic Acids Res. 16, 7487–7497.
38. Carr, A.M., Dorrington, S.M., Hindley, J., Phear, G.A., Aves, S.J., and Nurse, P. (1994) Analysis of a histone H2A variant from fission yeast: evidence for a role in chromosome stability. Mol. Gen. Genet. 245, 628–635.
39. White, E.M., Shapiro, D.L., Allis, C.D., and Gorovsky, M.A. (1988) Sequence and properties of the message encoding Tetrahymena hv1, a highly evolutionarily conserved histone H2A variant that is associated with active genes. Nucleic Acids Res. 16, 179–198.
40. Redon, C., Pilch, D., Rogakou, E., Sedelnikova, O., Newrock, K., and Bonner, W. (2002) Histone H2A variants H2AX and H2AZ. Curr. Opin. Genet. Devel. 12, 162–169.
41. Suto, R.K., Clarkson, M.J., Tremethick, D.J., and Luger, K. (2000) Crystal structure of a nucleosome core particle containing the variant histone H2A.Z. Nat. Struct. Biol. 7, 1121–1124.
42. van Daal, A. and Elgin, S.C.R. (1992) A histone variant, H2AvD, is essential in *Drosophila melanogaster*. Mol. Biol. of the Cell 3, 593–602.
43. Liu, X., Li, B., and Gorovsky, M.A. (1996) Essential and non-essential histone H2A variants in *Tetrahymena thermophila*. Mol. Cell. Biol. 16, 4305–4311.
44. Faast, R., Thonglairoam, V., Schulz, T.C., Beall, J., Wells, J.R.E., Taylor, H., Matthaei, K., Rathjen, P.D., Tremethick, D.J., and Lyons, I. (2001) Histone variant H2A.Z is required for early mammalian development. Curr. Biol. 11, 1183–1187.
45. Adam, M., Robert, F., Larochelle, M., and Gaudreau, L. (2001) H2A.Z is required for global chromatin integrity and for recruitment of RNA polymerase II under specific conditions. Mol. Cell. Biol. 21, 6270–6279.
46. Dhillon, N. and Kamakaka, R.T. (2000) A histone variant Htz1p, and a Sir1p-like protein, Esc2p, mediate silencing at *HMR*. Mol. Cell 6, 769–780.

47. Jackson, J.D. and Gorovsky, M.A. (2000) Histone H2A.Z has a conserved function that is distinct from that of the major H2A sequence variants. Nucleic Acids Res. 28, 3811–3816.
48. Santisteban, M.S., Kalashnikova, T., and Smith, M.M. (2000) Histone H2A.Z regulates transcription and is partially redundant with nucleosome remodeling complexes. Cell 103, 411–422.
49. Allis, C.D., Glover, C.V.C., Bowen, J.K., and Gorovsky, M.A. (1980) Histone variants specific to the transcriptionally active amitotically dividing macronucleus of the unicellular eukaryote, *Tetrahymena thermophila*. Cell 20, 609–617.
50. Stargell, L.A., Bowen, J., Dadd, C.A., Dedon, P.C., Davis, M., Cook, R.G., Allis, C.D., and Gorovsky, M.A. (1993) Temporal and spatial association of histone H2A variant hv1 with transcriptionally competent chromatin during nuclear development in *Tetrahymena thermophila*. Genes Dev. 7, 2641–2651.
51. Leach, T.J., Mazzeo, M., Chotkowski, H.L., Madigan, J.P., Wotring, M.G., and Glaser, R.L. (2000) Histone H2A.Z is widely but non-randomly distributed in chromosomes of *Drosophila melanogaster*. J. Biol. Chem. 275, 23267–23272.
52. Clarkson, M.J., Wells, J.R.E., Gibson, F., Saint, R., and Tremethick, D.J. (1999) Regions of variant histone His2AvD required for *Drosophila* development. Nature 399, 694–697.
53. Abbott, D.W., Ivanova, V.S., Wang, X., Bonner, W.M., and Ausio, J. (2001) Characterization of the stability and folding of H2A.Z chromatin particles. J. Biol. Chem. 276, 41945–41949.
54. Fan, J.Y., Gordon, F., Luger, K., Hansen, J.C., and Tremethick, D.J. (2002) The essential histone variant H2A.Z regulates the equilibrium between different chromatin conformational states. Nat. Struct. Biol. 9, 172–176.
55. Li, W., Nagaraja, S., Delcuve, G.P., Hendzel, M.J., and Davie, J.R. (1993) Effects of histone acetylation, ubiquitination and variants on nucleosome stability. Biochem. J. 296, 737–744.
56. Mannironi, C., Bonner, W.M., and Hatch, C.L. (1989) H2A.X a histone isoprotein with a conserved C-terminal sequence, is encoded by a novel mRNA with both DNA replication type and poly A 3′ processing signals. Nucleic Acids Res. 17, 9113–9125.
57. Rogakou, E.P., Pilch, D.R., Orr, A.H., Ivanova, V.S., and Bonner, W.M. (1998) DNA double-stranded breaks induce histone H2AX phosphorylation on serine 139. J. Biol. Chem. 273, 5858–5868.
58. Rogakou, E.P., Boon, C., Redon, C., and Bonner, W.M. (1999) Megabase chromatin domains involved in DNA double-strand breaks in vivo. J. Cell Biol. 146, 905–916.
59. Rogakou, E.P., Nieves-Neira, W., Boon, C., Pommier, Y., and Bonner, W.M. (2000) Initiation of DNA fragmentation during apoptosis induces phosphorylation of H2AX histone at serine 139. J. Biol. Chem. 275, 9390–9395.
60. Chen, Y., Baker, R.E., Keith, K.C., Harris, K., Stoler, S., and Fitzgerald-Hayes, M. (2000) The N terminus of the centromere H3-like protein Cse4p performs an essential function distinct from that of the histone fold domain. Mol. Cell. Biol. 20, 7037–7048.
61. Peterson, S., Casellas, R., Reina-San-Martin, B., Chen, H.T., Difilippantonio, M.J., Wilson, P.C., Hanitsch, L., Celeste, A., Muramatsu, M., Pilch, D.R., Redon, C., Ried, T., Bonner, W.M., Honjo, T., Nussenzweig, M.C., and Nussenzweig, A. (2001) AID is required to initiate Nbs1/gamma-H2AX focus formation and mutations at sites of class switching. Nature 414, 660–665.
62. Mahadevaiah, S.K., Turner, J.M.A., Baudat, F., Rogakou, E.P., Boer, P.d., Blanco-Rodriguez, J., Jasin, M., Keeney, S., Bonner, W.M., and Burgoyne, P.S. (2001) Recombinational DNA double-strand breaks in mice precede synapsis. Nature Genet. 27, 271–276.
63. Paull, T.T., Rogakou, E.P., Yamazaki, V., Kirchgessner, C.U., Gellert, M., and Bonner, W.M. (2000) A critical role for histone H2AX in recruitment of repair factors to nuclear foci after DNA damage. Curr. Biol. 10, 886–895.
64. Burma, S., Chen, B.P., Murphy, M., and Kurimasa, A. (2001) ATM phosphorylates histone H2AX in response to DNA double-strand breaks. J. Biol. Chem. 276, 42462–42467.
65. Ward, I.M. and Chen, J. (2001) Histone H2A.X is phosphorylated in an ATR-dependent manner in response to replicational stress. J. Biol. Chem. 276, 47759–47762.

66. Andegeko, Y., Moyal, L., Mittelman, L., Tsarfaty, I., Shiloh, Y., and Rotman, G. (2001) Nuclear retention of ATM at sites of DNA double strand breaks. J. Biol. Chem. 276, 38224–38230.
67. Downs, J.A., Lowndes, N.F., and Jackson, S.P. (2000) A role for *Saccharomyces cerevisiae* histone H2A in DNA repair. Nature 408, 1001–1004.
68. Celeste, A., Petersen, S., Romanienko, P.J., Fernandez-Capetillo, O., Chen, H.T., Sedelnikova, O.A., Reina-San-Martin, B., Coppola, V., Meffre, E., Difilippantonio, M.J., Redon, C., Pilch, D.R., Olaru, A., Eckhaus, M., Camerini-Otero, R.D., Tessarollo, L., Livak, F., Manova, K., Bonner, W.M., Nussenzweig, M.C., and Nussenzweig, A. (2002) Genomic instability in mice lacking histone H2AX. Science 296, 922–927.
69. Pehrson, J.R. and Fried, V.A. (1992) MacroH2A, a core histone containing a large nonhistone region. Science 257, 1398–1400.
70. Chadwick, B.P. and Willard, H.F. (2001) Histone H2A variants and the inactive X chromosome: identification of a second macroH2A variant. Hum. Mol. Genet. 10, 1101–1113.
71. Costanzi, C. and Pehrson, J.R. (2001) MACROH2A2, a new member of the MACROH2A core histone family. J. Biol. Chem. 276, 21776–21784.
72. Pehrson, J.R., Costanzi, C., and Dharia, C. (1997) Developmental and tissue expression patterns of histone macroH2A1 subtypes. J. Cell. Biochem. 65, 107–113.
73. Rasmussen, T.P., Huang, T., Mastrangelo, M.A., Loring, J., Panning, B., and Jaenisch, R. (1999) Messenger RNAs encoding mouse histone macroH2A1 isoforms are expressed at similar levels in male and female cells and result from alternative splicing. Nucleic Acids Res. 27, 3685–3689.
74. Pehrson, J.R. and Fuji, R.N. (1998) Evolutionary conservation of macroH2A subtypes and domains. Nucleic Acids Res. 26, 2837–2842.
75. Changolkar, L.N. and Pehrson, J.R. (2002) Reconstitution of nucleosomes with histone macro H2A1.2. Biochemistry 41, 179–184.
76. Costanzi, C. and Pehrson, J.R. (1998) Histone macroH2A1 is concentrated in the inactive X chromosome of female mammals. Nature 393, 599–601.
77. Chadwick, B.P. and Willard, H.F. (2001) A novel chromatin protein, distantly related to histone H2A, is largely excluded from the inactive X chromosome. J. Cell Biol. 152, 375–384.
78. Borsani, G., Tonlorenzi, R., Simmler, M.C., Dandolo, L., Arnaud, D., Capra, V., Grompe, M., Pizzuti, A., Muzny, D., Lawrence, C., Willard, H.F., Avner, P., and Ballabio, A. (1991) Characterization of a murine gene expressed from the inactive X chromosome. Nature 351, 325–329.
79. Brockdorff, N., Ashworth, A., Kay, G.F., Cooper, P., Smith, S., McCabe, V.M., Norris, D.P., Penny, G.D., Patel, D., and Rastan, S. (1991) Conservation of position and exclusive expression of mouse Xist from the inactive X chromosome. Nature 351, 329–331.
80. Brown, C.J., Ballabio, A., Rupert, J.L., Lafreniere, R.G., Grompe, M., Tonlorenzi, R., and Willard, H.F. (1991) A gene from the region of the human X inactivation centre is expressed exclusively from the inactive X chromosome. Nature 349, 38–44.
81. Marahrens, Y., Panning, B., Dausman, J., Strauss, W., and Jaenisch, R. (1997) Xist-deficient mice are defective in dosage compensation but not spermatogenesis. Genes Dev. 11, 156–166.
82. Penny, G.D., Kay, G.F., Sheardown, S.A., Rastan, S., and Brockdorff, N. (1996) Requirement for Xist in X chromosome inactivation. Nature 379, 131–137.
83. Brockdorff, N., Ashworth, A., Kay, G.F., McCabe, V.M., Norris, D.P., Cooper, P.J., Swift, S., and Rastan, S. (1992) The product of the mouse Xist gene is a 15 kb inactive X-specific transcript containing no conserved ORF and is located in the nucleus. Cell 71, 515–527.
84. Brown, C.J., Hendrich, B.D., Rupert, J.L., Lafreniere, R.G., Xing, Y., Lawrence, J., and Willard, H.F. (1992) The human XIST gene: analysis of a 17 kb inactive X-specific RNA that contains conserved repeats and is highly localized with the nucleus. Cell 71, 527–542.
85. Clemson, C.M., McNeil, J.A., Willard, H.F., and Lawrence, J.B. (1996) XIST RNA paints the inactive X chromosome at interphase: evidence for a novel RNA involved in nuclear/chromosome structure. J. Cell Biol. 132, 259–275.

86. Csankovszki, G., Panning, B., Bates, B., Pehrson, J.R., and Jaenisch, R. (1999) Conditional deletion of *Xist* disrupts histone macroH2A localization but not maintenance of X inactivation. Nat. Genet. 22, 323–324.
87. Beletskii, A., Hong, Y.-K., Pehrson, J., Egholm, M., and Strauss, W.M. (2001) PNA interference mapping demonstrates functional domains in the non-coding RNA Xist. Proc. Natl. Acad. Sci. USA 98, 9215–9220.
88. Gilbert, S.L., Pehrson, J.R., and Sharp, P.A. (2000) XIST RNA associates with specific regions of the inactive X chromatin. J. Biol. Chem. 275, 36491–36494.
89. Hoyer-Fender, S., Costanzi, C., and Pehrson, J.R. (2000) Histone macroH2A1.2 is concentrated in the XY-body by the early pachytene stage of spermatogenesis. Exp. Cell Res. 258, 254–260.
90. Richler, C., Dhara, S.K., and Wahrman, J. (2000) Histone macroH2A1.2 is concentrated in the XY compartment of mammalian male meiotic nuclei. Cytogene. Cell Genet. 89, 118–120.
91. McKee, B.D. and Handel, M.A. (1993) Sex chromosomes, recombination, and chromatin conformation. Chromosoma 102, 71–80.
92. McCarrey, J.R. and Dilworth, C.D. (1992) Expression of Xist in mouse germ cells correlates with X-chromosome inactivation. Nat. Genet. 2, 200–203.
93. Richler, C., Soreq, H., and Wahrman, J. (1992) X inactivation in mammalian testis is correlated with inactive X-specific transcription. Nat. Genet. 2, 192–195.
94. Salido, E.C., Yen, P.H., Mohandas, T.K., and Shapiro, L.J. (1992) Expression of the X-inactivation-associated gene XIST during spermatogenesis. Nat. Genet. 2, 196–199.
95. Chadwick, B.P., Valley, C.M., and Willard, H.F. (2001) Histone variant macroH2A contains two distinct macrochromatin domains capable of directing macroH2A to the inactive X chromosome. Nucleic Acids Res. 29, 2699–2705.
96. Costanzi, C., Stein, P., Worrad, D.M., Schultz, R.M., and Pehrson, J.R. (2000) Histone macroH2A1 is concentrated in the inactive X chromosome of female preimplantation embryos. Development 127, 2283–2289.
97. Mermoud, J.E., Costanzi, C., Pehrson, J.R., and Brockdorff, N. (1999) Histone macroH2A1.2 relocates to the inactive X chromosome after initiation and propagation of X-inactivation. J. Cell Biol., 1399–1408.
98. Csankovszki, G., Nagy, A., and Jaenisch, R. (2001) Synergism of Xist RNA, DNA methylation, and histone hypoacetylation in maintaining X chromosome inactivation. J. Cell Biol. 153, 773–784.
99. Elgin, S.C.R. and Weintraub, H. (1975) Chromosomal proteins and chromatin structure. Ann. Rev. Biochem. 44, 725–774.
100. Wu, R.S. and Bonner, W.M. (1981) Separation of basal histone synthesis from S-phase histone synthesis in dividing cells. Cell 27, 321–330.
101. Brush, D., Dodgson, J.B., Choi, O.R., Stevens, P.W., and Engel, J.D. (1985) Replacement variant histone genes contain intervening sequences. Mol. Cell. Biol. 5, 1307–1317.
102. Wells, D. and Kedes, L. (1985) Structure of a human histone cDNA: Evidence that basally expressed histone genes have intervening sequences and encode polyadenylylated mRNAs. Proc. Natl. Acad. Sci. USA 82, 2834–2838.
103. Fretzin, S., Allan, B.D., van Daal, A., and Elgin, S.C.R. (1991) A *Drosophila melanogaster* H3.3 cDNA encodes a histone variant identical with the vertebrate H3.3. Gene 107, 341–342.
104. Ahmad, K. and Henikoff, S. (2002) The histone variant H3.3 marks active chromatin by replication-independent nucleosome assembly. Mol. Cell 9, 1191–1200.
105. Hendzel, M.J. and Davie, J.R. (1990) Nucleosomal histones of transcriptionally active/competent chromatin preferentially exchange with newly synthesized histones in quiescent chicken erythrocytes. Biochem. J. 271, 67–73.
106. Zweidler, A. (1984) In: Stein, G.S., Stein, J.L., and Marzluff, W.F. (eds.) Core Histone Variants of the Mouse: Primary Structure and Differential Expression in Histone Genes. Wiley, New York, pp. 339–371.

107. Couldrey, C., Carlton, M.B.L., Nolan, P.M., Colledge, W.H., and Evans, M.J. (1999) A retroviral gene trap insertion into the histone 3.3A gene causes partial neonatal lethality, stunted growth, neuromuscular deficits and male sub-fertility in transgenic mice. Hum. Mol. Genet. 8, 2489–2495.
108. Chaubet, N., Clement, B., and Gigot, C. (1992) Genes encoding a histone H3.3-like variant in Arabidopsis contain intervening sequences. J. Mol. Biol. 225, 569–574.
109. Thatcher, T.H., MacGaffey, J., Bowen, J., Horowitz, S., Shapiro, D.L., and Gorovsky, M.A. (1994) Independent evolutionary origin of histone H3.3-like variants of animals and *Tetrahymena*. Nucleic Acids Res. 22, 180–186.
110. Waterborg, J.H. (1993) Histone synthesis and turnover in alfalfa. Fast loss of highly acetylated replacement histone variant H3.2. J. Biol. Chem. 268, 4912–4917.
111. Yu, L. and Gorovsky, M.A. (1997) Constitutive expression, not a particular primary sequence, is the important feature of the H3 replacement variant hv2 in *Tetrahymena thermophila*. Mol. Cell. Biol. 17, 6303–6310.
112. Meistrich, M.L., Bucci, L.R., Trostle-Weige, P.K., and Brock, W.A. (1985) Histone variants in rat spermatogonia and primary spermatocytes. Dev. Biol. 112, 230–240.
113. Trostle-Weige, P.K., Meistrich, M.L., Brock, W.A., Nishioka, K., and Bremer, J.W. (1982) Isolation and characterization of TH2A, a germ cell specific variant of histone 2A in rat testis. J. Biol. Chem. 257, 5560–5567.
114. Choi, Y.C., Gu, W., Hecht, N.B., Feinberg, A.P., and Chae, C.B. (1996) Molecular cloning of mouse somatic and testis-specific H2B histone genes containing a methylated CpG island. DNA Cell Biol. 15, 495–504.
115. Kim, Y.-J., Hwang, I., Tres, L.L., Kierszenbaum, A.L., and Chae, C.-B. (1987) Molecular cloning and differential expression of somatic and testis-specific H2A histone genes during rat spermatogenesis. Dev. Biol. 124, 23–34.
116. Moss, S.B., Challoner, P.B., and Groudine, M. (1989) Expression of a novel histone 2B during mouse spermiogenesis. Dev. Biol. 133, 83–92.
117. Unni, E., Zhang, Y., Kangasniemi, M., Saperstein, W., Moss, S.B., and Meistrich, M.L. (1995) Stage-specific distribution of the spermatid-specific histone 2B in the rat testis. Biol. Reprod. 53, 820–826.
118. Moss, S.B. and Orth, J.M. (1993) Localization of a spermatid-specific histone 2B protein in mouse spermiogenic cells. Biol. Reprod. 48, 1047–1056.
119. Grimes, S.R. (1986) Nuclear proteins in spermatogenesism. Comp. Biochem. Physiol. 83B, 495–500.
120. Grimes, S.R., Meistrich, M.L., Platz, R.D., and Hnilica, L.S. (1977) Nuclear protein transitions in rat testis spermatids. Exp. Cell Res. 110, 31–39.
121. Tanphaichitr, N., Sobhon, P., Taluppeth, N., and Chalermisarachai, P. (1978) Basic nuclear proteins in testicular cells and ejaculated spermatozoa in man. Exp. Cell Res. 117, 347–356.
122. Palmer, D.K., O'Day, K., and Margolis, R.L. (1990) The centromere specific histone CENP-A is selectively retained in discrete foci in mammalian sperm nuclei. Chromosoma 100, 32–36.
123. Gineitis, A.A., Zalenskaya, I.A., Yau, P.M., Bradbury, E.M., and Zalensky, A.O. (2000) Human sperm telomere-binding complex involves histone H2B and secures telomere membrane attachment. J. Cell Biol. 151, 1591–1598.
124. Newrock, K.M., Alfageme, C.R., Nardi, R.V., and Cohen, L.H. (1978) Histone changes during chromatin remodeling in embryogenesis. Cold Spring Harbor Symp. Quant. Biol. 42, 421–431.
125. Newrock, K.M., Cohen, L.H., Hendricks, M.B., Donnelly, R.J., and Weinberg, E.S. (1978) Stage-specific mRNAs coding for subtypes of H2A and H2B histones in the sea urchin embryo. Cell 14, 327–336.
126. Salik, J., Herlands, L., Hoffmann, H.P., and Poccia, D. (1981) Electrophoretic analysis of the stored histone pool in unfertilized sea urchin eggs: quantification and identification by antibody binding. J. Cell Biol. 90, 385–395.
127. Mandl, B., Brandt, W.F., Superti-Furga, G., Graninger, P.G., Birnstiel, M.L., and Busslinger, M. (1997) The five cleavage-stage (CS) histones of the sea urchin are encoded by a maternally expressed

family of replacement histone genes: functional equivalence of the CS H1 and frog H1M (B4) proteins. Mol. Cell. Biol. 17, 1189–1200.
128. Green, G.R. and Poccia, D.L. (1989) Phosphorylation of sea urchin histone CS H2A. Dev. Biol. 134, 413–419.
129. Kleinschmidt, J.A. and Steinbeisser, H. (1991) DNA-dependent phosphorylation of histone H2A.X during nucleosome assembly in Xenopus laevis oocytes: involvement of protein phosphorylation in nucleosome spacing. EMBO J. 10, 3043–3050.
130. Goldknopf, I.L. and Busch, H. (1977) Isopeptide linkage between nonhistone and histone 2A polypeptides of chromosomal conjugate-protein A24. Proc. Natl. Acad. Sci. USA 74, 864–868.

J. Zlatanova and S.H. Leuba (Eds.) *Chromatin Structure and Dynamics: State-of-the-Art*

DOI: 10.1016/S0167-7306(03)39009-X

CHAPTER 9

Histone modifications

James R. Davie

Manitoba Institute of Cell Biology, University of Manitoba, 675 McDermot Avenue, Winnipeg, Manitoba, Canada, R3E 0V9. Tel.: 204-787-2391; Fax: 204-787-2190; E-mail: davie@cc.umanitoba.ca

1. Introduction

The four core histones, H2A, H2B, H3, H4 and their variants, and the linker histone H1 subtypes are susceptible to a wide range of post-synthetic modifications, including acetylation, phosphorylation, methylation, ubiquitination, and ADP-ribosylation (Figs. 1 and 2). In this chapter, the four latter modifications and their functions in chromatin structure and function are presented.

2. Histone phosphorylation

Histone phosphorylation was first reported in 1966 [1,2]. The four core histones, histone variants, and H1 histones are phosphorylated, with the sites of phosphorylation being found in both the amino-terminal and carboxy-terminal portions of the histones [3] (Figs. 1 and 2). Phosphorylation of the core histones has been implicated in transcription, replication, chromosome condensation, and DNA repair.

Cell cycle studies of histone phosphorylation using synchronized Chinese hamster ovary cells and HeLa S-3 cells demonstrated that H1 and H3 are phosphorylated at different times during the cell cycle, while H2A and H4 are phosphorylated at uniform rates throughout the cell cycle [4–6]. Kinetic studies of the phosphorylation of H2A and H4 in trout testis indicate that these histones are phosphorylated shortly after synthesis [7]. Phosphorylation of H4 did not occur appreciably until after a series of acetylation and deacetylation events, while H2A was phosphorylated shortly after synthesis followed by dephosphorylation.

H2A.1, H2A.2, and H2A.X are phosphorylated at serine residue 1 [8,9]. H2A.Z is not phosphorylated. *In vitro* protein kinase C phosphorylates H2A at serine residue 1 [10]. *Tetrahymena* H2A is phosphorylated in the C-terminal sequence [11]. *Tetrahymena* H2A.1 is phosphorylated at serine residues 122, 124, and 129, while H2A.2 is modified at serine residues 122 and 128 (Fig. 3). Phosphorylation of H2A occurs in the transcriptionally active macronucleus of *Tetrahymena thermophila*, but not in the transcriptionally inert micronucleus [12]. *Tetrahymena* H2A variant hv1 is phosphorylated [13]. H4, like H2A, is phosphorylated at N-terminal serine

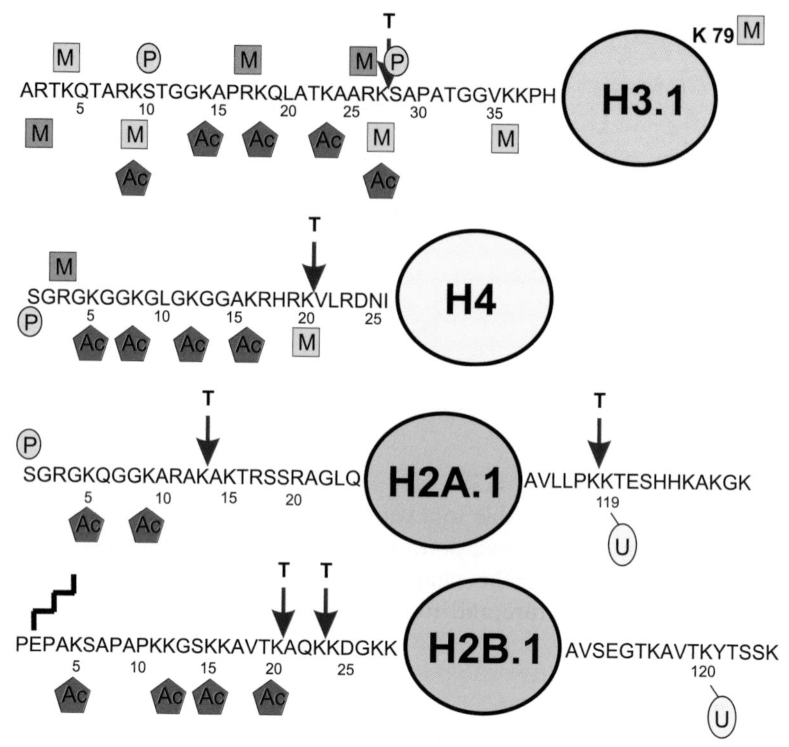

Fig. 1. Core histone modifications. Human histone N-terminal and in some cases C-terminal amino acid sequences are shown. The modifications include methylation (M), acetylation (Ac), phosphorylation (P), ubiquitination (U), and ADP ribosylation (step ladder). The sites of trypsin digestion of histones in nucleosomes are indicated (T).

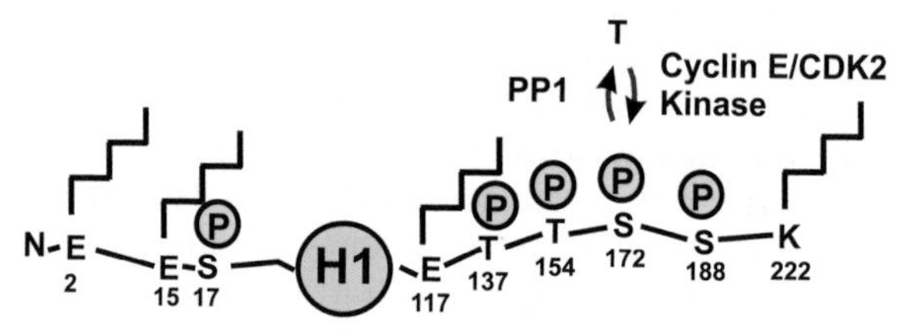

Fig. 2. Histone H1 modifications. Sites of phosphorylation (P) and ADP ribosylation (step ladder) on mouse H1^{S}-3 are shown.

residue (Fig. 1). *In vitro* nuclear enzymes, cAMP-dependent and cGMP-dependent protein kinases and protein kinase C, phosphorylate H2B at serine residues 32 and 36 [14–17]. A cAMP independent protein kinase phosphorylated preferentially Ser-32 of H2B [18]. These residues, however, are located in the histone fold part of the molecule, and in the nucleosome it would be unlikely that these serine residues would be accessible to the kinase [19]. Sea urchin sperm-specific H2B variants, H2B.1, and H2B.2, are phosphorylated at several sites in the N-terminal

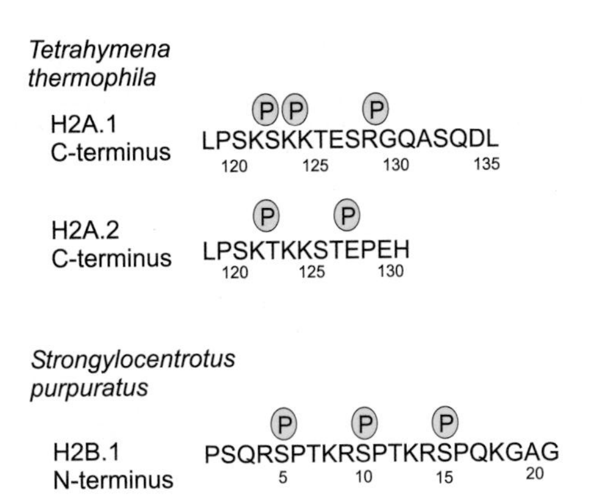

Fig. 3. Sites of phosphorylation in *Tetrahymena* H2A variants and sea urchin H2B.

domain (H2B.1, two or three sites; H2B.2, four sites) (Fig. 3). The consensus sequence for the phosphorylation sites is –Ser–Pro–X–Lys/Arg– (X is Thr, Gln, Lys or Arg). This sequence is recognized by the histone H1 growth-associated kinase or $p34^{cdc2}$ kinase [20].

2.1. Histone phosphorylation and mitosis

The core histones and H1 undergo phosphorylation on specific serine and threonine residues. H1 can be phosphorylated on Ser/Thr residues on the N terminal and C terminal domains of the molecule (Fig. 2), and H3 can be phosphorylated on Ser residues on its N terminal domain (Fig. 1). The phosphorylation of both H1 and H3 is cell cycle dependent with the highest level of phosphorylation of both histones occurring in M-phase (Fig. 4). In G1 phase of the cell cycle, the lowest number of H1 sites is phosphorylated, and there is a gradual increase in the number of sites phosphorylated throughout S and G2 phases of the cell cycle. In M phase, when chromatin is highly condensed, the maximum number of sites is phosphorylated [6,21,22]. The strong correlation between highly phosphorylated H1 and chromatin condensation at mitosis lead to the assumption that H1 phosphorylation drives mitotic chromatin condensation; however, chromatin condensation can occur in the absence of H1 phosphorylation [23]. H1 phosphorylation destabilises chromatin structure and weakens its binding to DNA. Therefore, H1 phosphorylation may lead to decondensation of chromatin and access of the DNA to factors involved in transcription and replication in G1 and S and to condensing factors present in mitosis [24].

Phosphorylation of H1 during mitosis is catalyzed by the cyclin B-Cdc2 kinase, a tightly regulated enzyme [25,26]. In *Tetrahymena* mitotically dividing nuclei, cyclic-AMP dependent kinase (PKA) or PKA-like kinase phosphorylates H1 [27,28]. Protein phosphatase 1 dephosphorylates the phosphorylated H1 [29].

H3 differs from the other core histones in that it is phosphorylated to a greater extent during mitosis than during other parts of the cell cycle [4,30] (Fig. 4). During

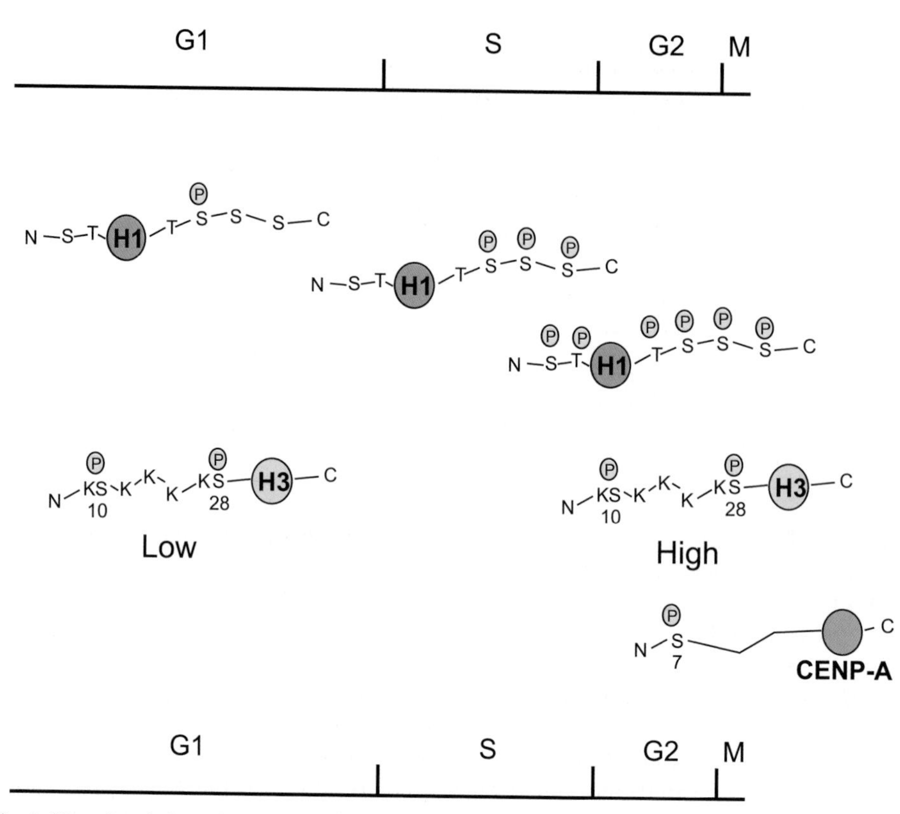

Fig. 4. Phosphorylation of histone H1, H3, and CENP-A throughout the cell cycle.

the cell cycle of Chinese hamster cells, H3 is phosphorylated during interphase but the levels of phosphorylated H3 increased dramatically during mitosis. The phosphorylated H3 was dephosphorylated when the cells left anaphase. Studies on H3 phosphorylation during mitosis have revealed that Ser-10 phosphorylation of H3 is correlated with both mitotic and meiotic divisions in *Tetrahymena* micronuclei [31], and that phosphorylation at this site is required for proper chromosome condensation and segregation [32]. In mammalian cells, mitosis-specific phosphorylation of H3 on Ser-10 initiates primarily within pericentromeric heterochromatin during late G2 and spreads in an ordered fashion throughout the condensing chromatin and is complete just prior to the formation of the prophase chromosomes [33]. Phosphorylation of H3 at Ser-10 weakens the association of the H3 tail to DNA, which may promote the binding of factors that drive chromatin condensation as cells enter mitosis [34]. Phosphorylation of H3 at Ser-28 also occurs in mammalian cells during early mitosis, suggesting that H3 phosphorylation at sites Ser-10 and Ser-28 are involved in events leading to mitotic chromosome condensation [35].

The centromeric-specific histone H3-like variant, CENP-A, that is present in all eukaryotes is phosphorylated by aurora B at Ser-7 and dephosphosphorylated by

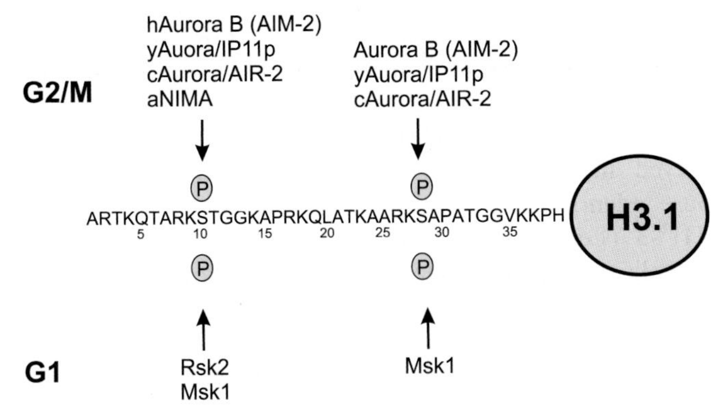

Fig. 5. Kinases involved in phosphorylating H3 at Ser-10 and Ser-28 in G2/M and G1 phases of the cell cycle. H, human; y, yeast (*S. cerevisiae)*; a, *Aspergillus nidulans*; c, *C. elegans*.

protein phosphatase 1 γ1 (PP1γ1) [36,37]. The centromeric nucleosomes appear to consist of a tetramer of CENP-A and H4 and two H2A–H2B dimers [36]. CENP-A phosphorylation initiates later than that of H3 in G2/M and commences following pericentric initiation and genome-wide stages of H3 phosphorylation (Fig. 4). Phosphorylation of CENP-A, which starts in prophase, is reversed in anaphase [38]. Dominant negative CENP-A phosphorylation mutants delayed the terminal stages of cytokinesis in HeLa cells and interfered with the localizations of inner centromere protein (INCENP), aurora B and PP1γ1 [37].

S. cerevisiae aurora/IP11p, *C. elegans* aurora/AIR-2, and mammalian aurora B (also called AIM-1) phosphorylate H3 at Ser-10 and Ser-28 during mitosis [39–41] (Fig. 5). Protein phosphatase 1 removes the phosphate at these sites [39,42]. INCENP is bound to aurora B and is essential for the proper targeting of aurora B on the chromosomes [43–45]. INCENP and aurora B (AIM-1) are overexpressed in a variety of human cancers, including colorectal cancer [43,46]. Overexpression of aurora B (AIM-1) in CHE diploid fibroblasts leads to chromosomal instability, suggesting that aurora B overexpression may play a role in carcinogenesis [46]. INCENP/aurora B and H3 phosphorylation appear to be involved in assembly of mitotic chromosomes, but not mitotic chromosome compaction [44,47].

In the filamentous fungus, *Aspergillus nidulans*, NIMA (never in mitosis, gene A) kinase phosphorylates H3 at Ser-10 [48]. At mitosis NIMA kinase association with chromatin increases and following metaphase NIMA locates to the mitotic spindle and spindle pole bodies. A human NIMA-related kinase (Nek6) was identified as a putative mitotic H1 and H3 kinase [49].

2.2. Histone H1 phosphorylation, transcription, and signal transduction

H1 phosphorylation has a role in gene transcription. Inactivation of the MMTV promoter is associated with dephosphorylation of H1, and reactivation of the promoter is associated with rephosphorylation of H1 [50]. Inhibition of the H1 kinase, Cdk2, prevented the glucocorticoid receptor (GR) from remodelling the promoter, inhibiting GR dependent transcriptional activation of MMTV [51]. Applying the H1 nomenclature proposed by Paraseghian *et al.* [25,52], Archer and colleagues provided evidence that $H1^S$-2 (Doenecke's H1-3) was the phosphorylated H1 subtype involved in modulating the chromatin structure of the MMTV promoter in mouse adenocarinoma cells [53]. In mouse $10T\frac{1}{2}$ fibroblasts phosphorylation of $H1^S$-3 (Lennox and Cohen's H1b) is dependent upon ongoing transcription and replication processes [54]. It has been proposed that inhibition of transcription and/or replication alters accessibility of $H1^S$-3 to Cdk2, which would result in decreased levels of phosphorylated $H1^S$-3 [54]. The modification of this mouse histone is unique in this regard. No other histone modification has been shown to be dependent upon these processes.

Mutation of the clustered phosphorylation sites in the N-terminal region of *Tetrahymena* H1, which lacks a globular domain, to alanines did not affect global gene expression [55,56]. However, the mutations did have a specific effect on some genes, resulting in either repression or activation [56]. This observation suggests that H1 phosphorylation may support gene expression and repression. Further mutations of this region of the molecule revealed that it was not phosphorylation recognition or a site specific charge that was influencing these transcriptional responses but the negative charge of the region referred to as the charge patch [57,58]. It is of interest to note that phosphorylation sites in the H1 subtypes are commonly clustered [25,58].

Oncogene-transformed mouse fibroblasts have a more decondensed chromatin structure than parental cell lines [59]. Phosphorylated $H1^S$-3 levels are elevated in oncogene (*ras*, *raf*, *fes*, *mos*, *myc*) and aberrantly expressed MAPKK (MEK) transformed mouse fibroblasts, which have elevated activities of MAPK (ERK1 and 2) [59] (Fig. 6). Further, *Rb*-deficient human fibroblasts have increased levels of phosphorylated H1 and a relaxed chromatin structure [60]. Cyclin E-Cdk2 was directly involved in increasing the levels of phosphorylated H1 [60]. Elevated cyclin E-Cdk2 activity resulting from persistent activation of the Ras–MAPK pathway is also responsible for increased level of phosphorylated H1 in oncogene-transformed mouse fibroblasts [61].

Studies with native and reconstituted chromatin show that phosphorylated H1 destabilize chromatin structure [62,63]. Phosphorylation of H1 increases the protein's mobility in the nucleus and weakens its interaction with chromatin [64,65]. In studies with avian fibroblasts transfected with H5 (an H1 variant), H5 was shown to inhibit proliferation in normal fibroblasts but not in transformed cells, in which H5 was phosphorylated. Aubert *et al.* proposed that phosphorylated H5 lacked the ability to condense chromatin [66]. The greater affinity of the dephosphorylated relative to phosphorylated H1 for chromatin may interfere with the performance of chromatin remodeling agents. For example, dephosphorylated, but not phosphorylated, H1 inhibited chromatin remodeling by SWI/SNF [67]. Thus, these and other studies provide support for the idea that an increase in the phosphorylation of H1 leads to destabilization of the chromatin [24].

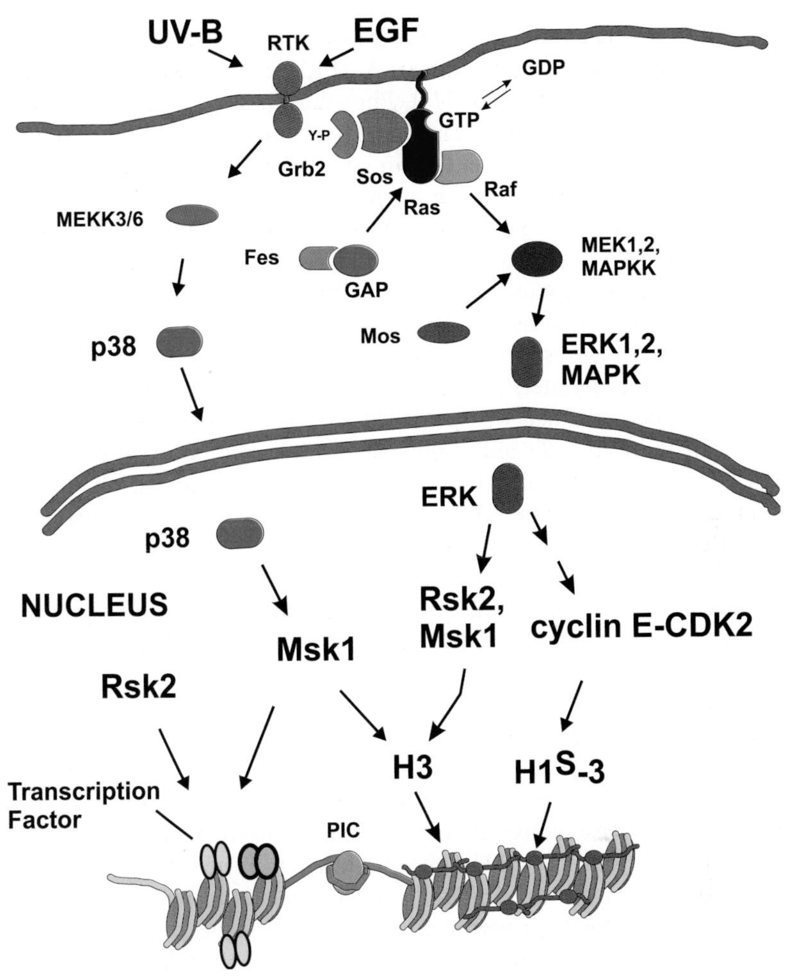

Fig. 6. Role of signal transduction pathways in phosphorylating H3 at Ser-10 and Ser-28. The Ras–MAPK (mitogen activated protein kinase) pathway is activated by EGF (epidermal growth factor) and TPA (12-*O*-tetradecanoylphorbol-13-acetate). UV-B activates both the Ras–MAPK pathway and the p38 kinase pathway (for more information about the signal transduction pathways see http://kinase.oci.utoronto.ca/signallingmap.html).

2.3. H3 phosphorylation and transcriptional regulation

Phosphorylation of H3 is not limited to mitosis and also occurs in G1 phase of the cell cycle. Activation of the Ras–Raf–MEK–ERK signal transduction pathway and/or of the p38 stress kinase pathway when cells are treated with epidermal growth factor (EGF), 12-*O*-tetradecanoylphorbol-13-acetate (TPA), anesomycin, okadiac acid, and stresses such as UV irradiation induces the rapid phosphorylation of H3 at Ser-10 and/or Ser-28 [68–72] (Figs. 5 and 6). Inhibition of the MEK1,2 activity with PD98059 prevents the activation of ERK and TPA-induced H3 phosphorylation in $10T\frac{1}{2}$ and Ciras-3 (*ras*-transformed) cells [73]. PD98059 does not inhibit the activation of protein kinase A (PKA); thus, PKA is not the TPA-stimulated H3 kinase in $10T\frac{1}{2}$ cells [74]. Phosphorylation of H3 is involved in the establishment of transcriptional competence of immediate-early response genes. H3 phosphorylation is concurrent with the transcriptional activation of the immediate early response genes, e.g., c-*fos* and c-*jun* [70]. TPA-induced phosphorylated H3 is located in numerous small foci scattered throughout the interphase nuclei of TPA treated mouse $10T\frac{1}{2}$ fibroblasts; the foci are found outside condensed chromatin regions [70]. Highly acetylated H3 is also observed in similarly positioned numerous small foci, which agrees with the observation that H3 phosphorylation is restricted to a small fraction of H3 histones that are dynamically highly acetylated [75,76]. Using the chromatin immunoprecipitation (ChIP) assay, direct evidence was provided that the newly phosphorylated H3 is associated with induced c-*fos* and c-*myc* genes [70]. The observation of numerous foci of newly phosphorylated H3 in TPA treated cells suggests that many other induced genes such as those described by Brown and colleagues are associated with phosphorylated H3 [77].

Stimulation of ovarian granulosa cells with follicle-stimulating hormone results in the phosphorylation of H3 at Ser-10 [74]. However, neither the Ras–Raf–MEK–ERK signal transduction nor the p38 stress kinase pathways are involved in this response.

2.4. H3 kinases and phosphatase

The steady state level of H3 phosphorylation is dependent upon a balance of phosphatase and kinase activities in the cell. Congruent with mitosis, protein phosphatase 1 appears to be the H3 phosphatase [70]. The kinases phosphorylating H3 in G2/M (aurora B) are different from those phosphorylating H3 in G1 (Rsk2, MSK1, JIL-1, Snf1, protein kinase A (PKA)) [41,42,69,74,78–80]. Allis and colleagues have presented evidence that the activity of Rsk2, a member of the $pp90^{rsk}$ kinases, is required for the mitogen stimulated phosphorylation of H3 [80]. Coffin-Lowry patients have a mutation in the Rsk2 gene. Fibroblasts from these patients do not exhibit EGF- or TPA-stimulated phosphorylation of H3 and, interestingly, growth factor-induced expression of the immediate early c-*fos* gene is severely impaired. However, studies by others failed to demonstrate that Rsk2 was the H3 kinase. In an *in vitro* assay with purified Rsk2 the enzyme efficiently phosphorylated H2B, which has two Rsk consensus sequences (RXXS), but failed to phosphorylate H3, which has an RXS sequence at Ser-10 and Ser-28 [73]. Mahadevan and colleagues presented evidence that MSK1 is the H3 kinase [69]. Both Rsk2 (MAPKAP kinase-1β) and MSK1 are members of a subfamily of MAPK-activated protein kinases, which have two distinct protein kinase domains. MSK1 is activated by Ras–MAPK and p38 stress kinase pathways [81]; both pathways when stimulated result in the phosphorylation of H3 (Fig. 6). MSK1, but neither ERKs nor Rsk2, is inhibited by H89, a protein kinase inhibitor. H89 inhibits TPA- and EGF-stimulated H3 phosphorylation and expression of immediate early genes, including c-*fos*, c-*jun*, c-*myc*, and urokinase plasminogen activator [69,73]. Further, MSK1 is the kinase phosphorylating Ser-10 and Ser-28 in ultraviolet B treated mouse epidermal JB6 cells [82] (Fig. 6).

JIL-1, localized to euchromatic interband regions of polytene chromosomes, is the *Drosophila* interphase H3 kinase [79,83]. Deletion of JIL-1 resulted in reduced viability, global changes (condensation of open regions) in chromatin structure, and a severe reduction in phosphorylated Ser-10 H3 to about 5% of wild-type levels. JIL-1 has two kinase domains, with domain I showing the greatest homology (63%) among the Ser/Thr kinase catalytic domains with domain I of MSK1. JIL-1 domain I has 47% homology with *Drosophila* RSK kinase domain I. JIL-1 interacts with the MSL protein complex, which contains MOF, a H4 Lys-16 specific histone acetyltransferase [84]. The MSL complex, which is required for dosage compensation of X-linked genes in males, associates with X-chromosome bound roX RNA, which directs the MSL complex to the X-chromosome [85]. The recruitment of JIL-1 and MSL complex would result in the acetylation of H4 and phosphorylation of H3, which jointly may remodel chromatin favoring enhanced expression of X-linked genes [79]. However, JIL-1 is not confined to the X-chromosome and other targeting mechanisms are likely involved in recruiting this kinase to euchromatic interband regions.

Follicle-stimulating hormone (FSH) induces the Ser-10 phosphorylation of H3 in ovarian granulosa cells by activation of protein kinase A. Based on the response to various protein kinase inhibitors, FSH-stimulated phosphorylation did not involve Rsk2 or MSK1 [74].

2.5. Histone H3 phosphorylation and acetylation

Dynamically acetylated H3 is preferentially phosphorylated [76]. As with histone methylation, labelling studies suggested that acetylation was not a prerequisite for phosphorylation. ChIP assays with antibodies recognizing phosphoacetylated H3 (phosphoSer-10/acetylLys-14 or phosphoSer-10/acetylLys-9) demonstrated directly that H3 associated with the induced immediate early response genes (c-*fos* and c-*jun*) is phosphorylated and acetylated [86,87]. It is important to note that in some instances acetylation of lysines may prevent an antibody from detecting the phosphorylated site. Mahadevan and colleagues observed that their anti-phosphoSer-10 H3 antibody would not detect the phosphoacetylated H3. In contrast the antibody generated by Allis and colleagues recognizes the phosphorylated Ser-10 H3 independent of the molecule's acetylation status. Thus, occlusion of epitopes to antibody recognition may be problematic but can be used to advantage in mapping multiply modified histone distribution along a gene [88]. Phosphorylation may also block antibody recognition of neighboring acetylated residues. A popular antibody in chromatin immunoprecipitation studies is the commercially available antibody to diacetylated (K9, K14); this antibody will not detect these acetylated sites when Ser-10 is phosphorylated [88].

Ser-10 phosphorylation of H3 precedes acetylation at Lys-14 and at Lys-9 [87–89] (Fig. 7). In studies with EGF-stimulated mouse fibroblasts, preventing the

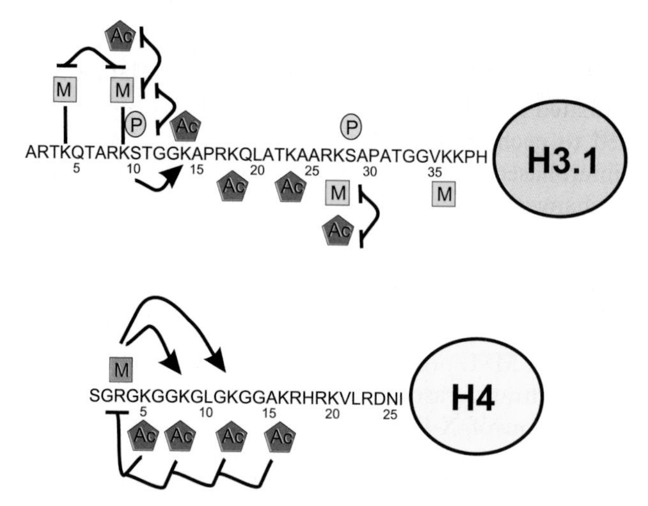

Fig. 7. Interplay between different modifications on histones H3 and H4. The modifications include methylation (M), acetylation (Ac), and phosphorylation (P). Positive and negative affects are shown.

activation of ERKs by inhibiting MEK with PD 98059 resulted in a decreased association of acetylated H3 associated with the c-*fos* gene promoter. This observation supported the timing of these two modifications, although the phosphorylation status of H3 bound to the c-*fos* promoter was not analyzed by the ChIP assay. However, preventing the activation of ERKs may prevent the phosphorylation of transcription factors required to recruit HAT and/or in the phosphorylation and activation of HATs [90,91]. ChIP analyses of the c-*jun* promoter and coding regions demonstrated that acetylation and phosphorylation of H3 were independent events [88]. Further these authors reported that the c-*jun* gene is associated with dynamically acetylated histones before the gene is induced. Further studies will be needed to decide whether H3 phosphorylation precedes acetylation.

An investigation of the substrate preference of HATs for unmodified or modified H3 demonstrated that H3 HATs (Gcn5, SAGA, ADA, CBP, p300, PCAF) preferred to acetylate the phosphorylated H3 [87,89]. Computer modeling revealed that Arg-164 of Gcn5 was positioned to interact with the oxygen molecules of phosphoSer-10 of H3. Mutation of this residue negated the enzymes preference for phosphoH3 without altering the enzyme's HAT activity. Interestingly, yeast mutant Gcn5 (R164A) lowered the activities of a subset of Gcn5-dependent promoters (e.g., *HO* and *His3*). Importantly, the activities of these same promoters were reduced when H3 was mutated (Ser10Ala). These results provide direct evidence that H3 phosphorylation is linked to gene activation.

2.6. *Histone H2AX phosphorylation and DNA damage*

There are several variants of H2A, including H2A.1, H2A.2, H2A.Z, and H2A.X. Each of these H2A variants may be phosphorylated. Mice lacking H2AX are radiation sensitive, growth retarded, immune deficient, and have frequent

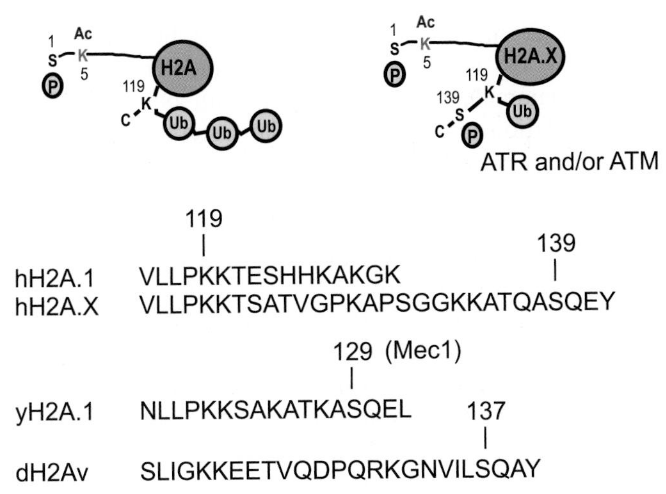

Fig. 8. Phosphorylation of human (h) H2A.X, *Drosophila* (d) H2Av, and yeast (y) H2A.1.

chromosome abnormalities [92,93]. Male mice not expressing H2AX are infertile, with $H2AX^{-/-}$ spermatocyte being arrested in pachytene stage of meiosis I. H2A.X, which makes up 2 to 25% of the H2A population in mammalian cells, is rapidly phosphorylated at Ser-139 (γ-H2AX) in response to DNA double strand breaks (DSBs) by ionizing radiation, DSB-inducing agents (neocarzinostatin, bleomycin, etoposide), DSBs in apoptotic cells, and DSBs at physiological sites of recombination in lymphocytes and germ cells [94–99] (Fig. 8). γ-H2AX appears to decondense chromatin, recruits factors involved in DNA repair and is critical in the assembly of specific DNA repair complexes in damaged DNA [100]. Repair of double strand breaks resulting from ionizing radiation involves both homologous recombination and non-homologous end-joining pathways. Brca1, Brca2, and Rad51 are DNA repair proteins involved in homologous recombination, while Rad50, Mre11, and Nbs1 (Nijmegen breakage syndrome protein) are involved in homologous recombination and non-homologous end-joining. Within minutes following the appearance of DSBs γ-H2AX forms nuclear foci, which increase for 10 to 30 min and then decline. Mre11, Rad50, Nbs1, 53bp1, and Brca1 are recruited to irradiation-induced foci. Nbs1 binds directly to γ-H2AX. The fork-head associated (FHA) and BRCA1 C-terminal (BRCT) domains located in the N-terminal domain of Nbs1 were required for interaction with γ-H2AX [101]. In lymphocytes, γ-H2AX is associated with V(D)J recombination, while in B cells, γ-H2AX has a role in class switch recombination [102,103]. Activation-induced cytidine deaminase (AID), a putative RNA-editing enzyme, is required for Nbs1 and γ-H2AX to form foci at the immunoglobulin heavy-chain constant region. Ultraviolet-induced replication in Xeroderma pigmentosum variant (XPV) accumulate Mre11/Rad50/Nbs1 and γ-H2AX in large nuclear foci at sites of stalled replication forks [104]. Inhibition of DNA replication in breast cancer cells by hydroxyurea will also induce γ-H2AX foci formation [105].

In the presence or absence of double-strand DNA breaks, fluorescent recovery after photobleaching of H2AX tagged with the green fluorescent protein demonstrated that mammalian H2AX is chromatin bound and immobile [106]. Thus, the phosphorylation and dephosphorylation events driven by kinases and phosphatases in response to DNA damage act upon chromatin bound H2AX. Mammalian H2AX has a unique C-terminus tail with a consensus site for phosphatidylinositol-3-OH kinase related kinase (PI-3KK) (Fig. 8). Treatment of cells with the PI-3KK inhibitor wortmannin prevents phosphorylation of H2AX and formation of foci with repair proteins. Current evidence is conflicting whether the PI-3 kinases, ataxia-telangietasia-Rad3-related (ATR) or ataxia-telangiectasia mutated kinase ATM, is the kinase phosphorylating H2AX [105,107]. The H2AX phosphatase appears to be protein phosphatase 1α [106].

Yeast H2A has the PI-3 kinase motif found in mammalian H2AX. This motif is important for yeast to repair DSBs and in non-homologous end joining. The yeast enzyme catalyzing phosphorylation of yeast H2AX at Ser-129 (homologous to serine 139 of mammalian H2AX) is Mec1 [100] (Fig. 8).

Drosophila melanogaster H2Av is a member of the H2A.Z family. However, *Drosophila* H2Av is phosphorylated like mammalian H2A.X in response to double-strand breaks resulting from γ-irradiation. The phosphorylation occurs at Ser-137 located in the C-terminus of H2Av [108] (Fig. 8).

2.7. *N-phosphorylation*

N-phosphorylation has been reported for H1 and H4, modifying H1 at lysine residues (N^6-phosphoryl lysine) and H4 at a histidine residue (1- or 3-phosphoryl histidine) [109]. The modification is acid-labile and alkali-stable, and is destroyed by acid-extraction procedures used in isolating histones.

Histone kinases responsible for N-phosphorylation have been isolated from regenerating rat liver [109] and Walker-256 carcinosarcoma cells [110]. One kinase with a pH optimum of 9.5 phosphorylated His-18 and His-75 of H4, while the other with a pH optimum of 6.5 phosphorylated lysine of H1. The enzyme from regenerating rat liver phosphorylated H4 at 1-phosphoryl histidine, while the carcinosarcoma enzyme phosphorylated H4 His at the position 3 [111]. Both kinases were cAMP independent [110]. Matthews and colleagues purified a 32-kDa histidine H4 kinase from yeast, *Saccharomyces cerevisiae* [112,113]. The enzyme phosphorylated His-75 (1-phosphoryl histidine) in H4. His-18 of H4 and other histidines in other core histones were not phosphorylated by this kinase [112]. Protein phosphatases 1, 2A, and 2C could dephosphorylate His-75 of H4 [114]. Applying a gel kinase approach to detect mammalian H4 histidine kinases, Besant and Attwood detected four activities in the 34–41 kDa range with extracts from porcine thymus [115].

In regenerating rat liver only pre-existing H4 was phosphorylated at the peak of DNA synthesis [116,117]. The modification has a half-life of two hours and may be involved in replication of DNA [109,116,117]. The histidine kinases in yeast and *Physarum polycephalum* nuclear extracts are unable to phosphorylate H4 in nucleosomes. His-75 resides in the histone fold part of the H4 molecule, which in the nucleosome may not be accessible to enzymes [19]. It is possible that chromatin disruptive processes (e.g., replication) present H4 to histidine kinases [117].

3. Histone methylation

Histone methylation was first reported in 1964 [118]. Core histones H2B, H3, and H4 are modified by methylation (Fig. 1). H3 and H4 are modified at lysines and arginines located primarily in the N-terminal tail. Histone methylation does change the charge of the protein at physiological pH. However, methylation does increase the hydrophobicity of the lysine residue and reduces its ability to form hydrogen bonds [119]. The sites of methylation (Lys-20 in H4 and Lys-27 in H3) are positioned at the boundary between the very basic N-terminal tail domain and the more hydrophobic sequence of the remainder of the molecule. Lys-20 of histone H4 is also in the basic region that binds to nucleosomal DNA [120]. Methylation at these sites may alter nucleosome structure [121,122].

With the exception of plants and budding yeast, H4 is methylated at Lys-20 [123–126] (Fig. 1). Lys-20 of mammalian H4 is 70–100% methylated at this site [127–130]. H4 Lys-20 may be mono-, di-, and tri-methylated, with the level of trimethylated Lys-20 increasing with age in rat liver [131]. Methylation of Lys-79 has been reported for *Physarum polycephalum* H4 [132]. H4 is also methylated at Arg-3 [133]. In human 293T cells, H4 Arg-3 is mono (N^G) methylated. In immunoblots experiments with an antibody to H4 dimethylated (N^G, N^G) Arg-3, H4 from human, yeast and chicken was methylated [133]. H3 may be methylated at lysines 4, 9, 27, 36, and 79, but the site utilization varies [134,135]. Mammalian H3 is typically methylated at lysines 9 and 27, being modified to 35 and 70–100%, respectively [134,136]. Cycad, *Chlamydomonas*, and *Tetrahymena* H3 are methylated at Lys-4 to 20, 81, and 50%, respectively. Cycad and *Chlamydomonas* H3 are also methylated at lysines 9, 27, and 36 but to varying extents (e.g., Lys-9, 100 vs. 16%; Lys-27, 50 vs. 25%). *Tetrahymena* H3 is methylated at Lys-27 (40%) but not at Lys-9 or 36. Chick H3 is methylated at lysines 9, 27, and 36 to 20, 100, and 20%, respectively. Alfalfa histone H3 variants H3.1 and H3.2 are methylated at multiple sites (lysines 4, 9, 14, 18, and 27 for H3.1; lysine residues 4, 9, 14, 18, 23, and 27 for H3.2) [137]. Microsequencing revealed the methylation of Lys-4 in *Tetrahymena* macronuclear H3, HeLa H3, and yeast H3. HeLa H3 was also methylated at lysines 9 and 27, with Lys-4 = Lys-9 > Lys-27. All H3 subtypes (H3.1, H3.2, H3.3) were methylated [138]. *Saccharomyces cerevisiae* (budding yeast) H3 is not methylated at Lys-9 or at Lys-27 [139,140]. *S. cerevisiae*, *Drosophila*, and HeLa H3 is methylated at Lys-79, which is located in the histone fold and at the surface of the nucleosome. The position of Lys-79 in the H3 loop domain joining two α-helices is near the H3/H4 tetramer and H2A/H2B dimer interface and is also near the ubiquitination site of H2B [141]. Methylation of H3 Lys-79 varies with HeLa cell cycle, being high in G1, declining in S phase, lowest in G2 and increasing during M phase [141]. Heat shock of *Drosophila melanogaster* Kc cells induces methylation of H2B at N-terminal proline residue [119,142,143]. Heat shock of these cells also results in methylation of H3 at arginine residues.

In different organs of the rat [128], Ehrlich ascites tumor cells [144], trout testis [127], calf thymus [145], and carp testis [146], H4 is modified mainly as the N^6-dimethyllysine, while H3 is modified as N^6-monomethyllysine, N^6-dimethyllysine and N^6-trimethyllysine with the N^6-dimethyllysine predominating. Pea seedling H4 is not methylated and H3 exits as N^6-mono- and N^6-dimethyllysine with N^6-trimethyllysine not being detectable [147,148].

The temporal sequence of H3 and H4 methylation after synthesis has been examined in Ehrlich ascites tumor cells [144] and trout testis [149]. Methylation lagged histone synthesis, and the histone was methylated after being bound to DNA. H4 methylation follows the stepwise acetylations and deacetylations [149]. It was suggested that methylation was involved in final arrangement of H3 and H4 on newly replicated DNA [144] and might be involved in histone interactions with other proteins such as histone kinases [149].

Histone methylation is a relatively stable modification with a slow turnover rate. However, there is evidence of methyl group turnover for HeLa H3 [150]. It remains to be shown if this histone demethylase activity is present in transformed, but not normal cells.

3.1. *Histone methylation, gene regulation, and heterochromatin*

The association of dynamically acetylated histones with transcribed chromatin suggests that methylated H3 and in some cases, methylated H4 are bound to transcriptionally active DNA [138,150,151]. With the advent of antibodies to dimethylated lysines 4 and 9 of H3 and the ChIP assay, these correlations were tested directly. Detailed analyses of the β-globin locus in erythroid cells at different stages of development and in embryonic chick brain revealed that transcribed decondensed chromatin regions were associated with acetylated H3 and highly enriched in dimethylated Lys-4 H3 [152,153]. The tight correlation between H3 acetylation and H3 methylation at Lys-4 was found for most but not all regions of the β-globin locus. The repressed condensed chromatin regions were poor in acetylated histones and associated with dimethylated Lys-9 H3 [153]. A 16-kb region upstream of the β-globin domain is condensed and associated with dimethylated Lys-9 H3, but not acetylated or dimethylated Lys-4 H3. An insulator element dividing these two chromatin domains sets a boundary that is thought to arrest the spread of methylated Lys-9 H3 from entering the β-globin domain [153] (Fig. 9). The separate placement of H3 methyl Lys-4 and H3 methyl Lys-9 support the histone code hypothesis that different histone modifications have distinct biological read outs [154].

In yeast transcriptionally active and competent chromatin regions are associated with H3 dimethylated Lys-4. However, when the gene is actively being expressed, H3 Lys-4 is trimethylated. The di- and tri-methylated Lys-4 H3 are located along the coding region of the gene, suggesting that this modification has a role in transcriptional elongation [155].

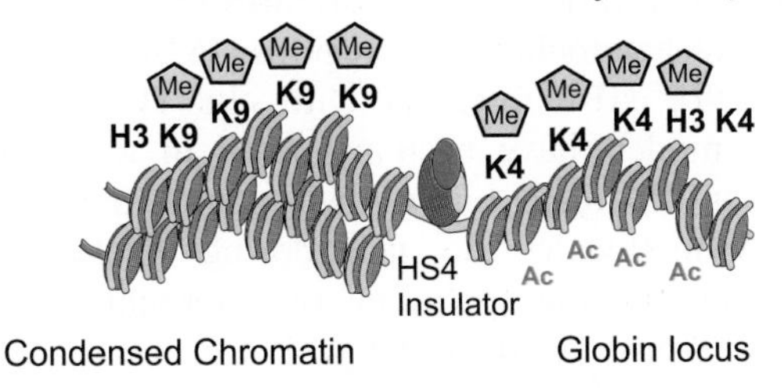

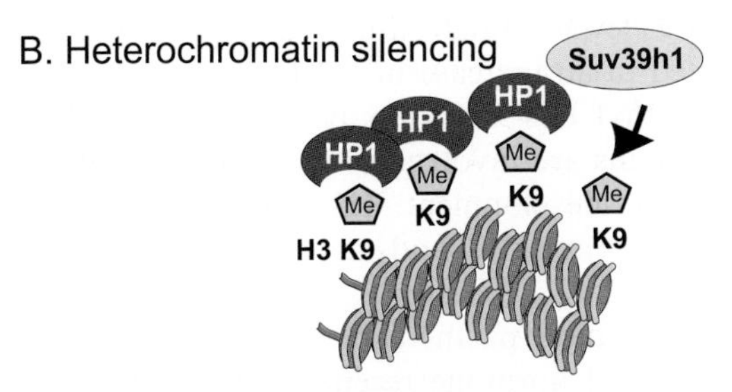

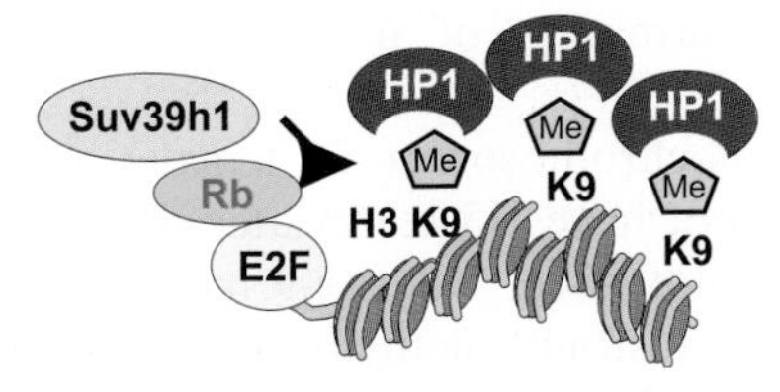

Fig. 9. Roles of histone methylation in transcriptionally active and repressed chromatin. Panel A shows a model for an insulator blocking the spread of condensed heterochromatin into the β-globin gene cluster. Histone H3 in the decondensed β-globin gene cluster is methylated at Lys-9 and acetylated. Histone H4 is also acetylated in this domain. The histones in the condensed chromatin are poorly acetylated and H3 is methylated at Lys-9. Panel B shows the heterochromatic self-maintenance by Suv39h1/HP1 complex. H3 is methylated at Lys-9 by Suv39h1. The H3 methylated at Lys-9 then recruits HP1. Panel C shows the role of the Suv39h1–HP1 complex is the transcriptional silencing of E2F regulated genes. E2F recruits Rb which in turn binds to Suv39h1. H3 methylated at Lys-9 by Suv39h1 recruits HP1.

Methylated Lys-9 H3 is localized primarily to heterochromatin regions, the histones of which are hypoacetylated [156]. ChIP assays with anti-methylated Lys-9 H3 antibodies demonstrated directly that silenced chromatin regions in *Saccharomyces pombe* (fission yeast) are associated with H3 methylated at Lys-9 [157]. Note that *Saccharomcyes cerevisiae* H3 lacks methyl Lys-9 and the H3 Lys-9 methyltransferase (Suv39h1 homolog) [139]. While H3 methyl Lys-9 and Swi6 are localized to heterochromatic regions, adjacent euchromatic regions are associated with H3 methyl Lys-4 in fission yeast. Two inverted repeats establish boundaries between the two chromatin domains [158]. Deletion of the boundary element allowed the H3 methyl Lys-9 and Swi6 to spread up to 8 kb into the adjacent domain. H3 methyl Lys-9 avidly binds to the chromodomain of heterochromatin protein 1 (HP1) and yeast Swi6 (homolog of *Drosophila* HP1), recruiting the protein to heterochromatic regions [157,159–161]. HP1 will not bind H3 methyl Lys-4 [159]. HP1 interacts with the H3 Lys-9 methyltransferase SUVAR39H1. Thus, models have been proposed in which HP1 recruited by a nucleosome bearing an H3 methyl Lys-9 will enable the HP1 bound H3 Lys-9 methyltransferase to methylate the neighboring nucleosome; hence a self-propagating mechanism for the spreading of heterochromatin is in play [159]. The boundary element stops the spread (Fig. 9). Deletion of the boundary element also resulted in the decrease of H3 methyl Lys-4, indicating a relationship between methyl Lys-4 and Lys-9 levels. Interestingly deletion of *clr4*, the H3 Lys-9 methyltransferase, resulted in an increase in H3 Lys-4 methylation [158].

In *Tetrahymena* H3 methyl Lys-9 recruits chromodomain proteins, Pdd1p and Pdd3p. Both of these proteins are involved in programmed DNA elimination that occurs during macronuclear development. H3 methylation at Lys-9 occurs only after the formation of the anlagen and at specific sites (internal eliminated sequences) [162].

SUV39H1 has several binding partners including human polycomb homolog HPC2 and unliganded thyroid hormone receptor [163,164]. It has been proposed that the interaction between SUV39H1 and HPC2 and the methylation of H3 at Lys-9 has a role in the formation of polycomb group-mediated repressive chromatin states.

The mammalian inactive X-chromosomes in females is associated with H3 methylated at Lys-9 but does not have H3 methylated at Lys-4 [165,166]. Deletion of murine Suv39h genes does not affect H3 Lys-9 methylation at the inactive X-chromosome. Thus, different histone methyltransferases are involved in the methylation of H3 Lys-9 in constitutive and facultative heterochromatin [166]. Further, HP1 is not associated with the inactive X-chromosome. Methylation of H3 Lys-9 is an early event in X inactivation in mammals [167,168]. The H3 Lys-9 methylation occurs approximately at the same time as H3 Lys-9 hypoacetylation and Lys-4 hypomethylation, and happens before transcriptional inactivation of the X-linked genes. Upstream of the Xist gene, which codes for a strictly nuclear RNA involved in X inactivation, is a constitutive hotspot of H3 methylated at Lys-9. This early event of H3 Lys-9 methylation occurs simultaneously with or immediately following the association of Xist RNA with the X-chromosome. It has been proposed that the hotspot upstream of the Xist gene serves as a nucleation site for the spreading of Xist RNA and methylated Lys-9 H3 [167].

The inactive heterochromatic X-chromosome of *C. elegans* in males and hermaphrodites is devoid of H3 methyl Lys-4. Further the male single X-chromosome has H3 methyl Lys-9, which is not observed in the X-chromosomes of the hermaphrodite [169]. Male mouse XY bodies were also found to have high levels of H3 methyl Lys-9.

In addition to heterochromatic silencing, SUV39H1 H3 methyltransferase and HP1 are involved in repression of euchromatic genes. The transcription factor E2F has a pivotal role in regulating the expression of S-phase-specific genes. Repression of these genes is through the retinoblastoma (Rb) protein which binds to E2F.

Rb recruits histone methyltransferases and histone deacetylases to repress gene expression [170–172]. Rb bound to E2F recruits SUV39H1 to the S-phase-specific gene promoter (e.g., cyclin E), which in turn recruits HP1 [170,171] (Fig. 9). Disruption of SUV39 in mouse embryonic fibroblasts increased the expression of cyclin E. Phosphorylation of Rb abolishes its association with histone deacetylase and histone H3 Lys-9 methyltransferase. The promoters of inducible inflammatory genes, *IkBα* and *IL-8*, are not associated with H3 methylated at Lys-9, but surprisingly low levels of H3 methyl Lys-9 are associated with the promoters of inducible inflammatory genes, *ELC*, *MDC*, and *IL-12p40*; genes that are activated more slowly than the first group. The H3 Lys-9 methylation at these genes was dynamic, with demethylation occurring when RNA polymerase II was recruited and Lys-9 methylation reinstated after RNA polymerse II was released [173]. It remains to be shown that there is a histone demethylase that removes the methylated Lys-9 from H3 allowing acetylation of this site and methylation at Lys-4 to occur. Elp3, a histone acetyltransferase, may harbor histone demethylase activity [174]; thus, dual function enzymes may remove methyl groups from Lys-9 while acetylating H3 lysines. Alternatively, replacement of H3 with the DNA-replication independent synthesized H3.3 variant into transcribed genes is one mechanism to get rid of the H3 with methyl Lys-9 [175–179].

Methylated Lys-20 H4 is absent or very low in transcriptionally competent chromatin regions and is localized to transcriptionally silenced condensed chromatin regions distinct from those associated with H3 methyl Lys-9 [124,180]. The level of methylated Lys-20 H4 varies throughout the cell cycle, decreasing during S phase and reaching a peak at G1 and at mitosis. In contrast, H4 acetylation at Lys-16 peaked at mid-S phase and declined to lower levels at mitosis and G1. These observations are in agreement with studies that methylation at Lys-20 inhibits acetylation at Lys-16 and vice versa [181]. However, these observations are inconsistent with the results of others [180], in that other investigators found H4 Lys-20 methylation in HeLa cells to be greatest at S phase and that H4 acetylation and Lys-20 methylation were independent. Further, studies on the effect of methylation of H4 at Lys-20 on CBP-mediated acetylation at Lys-16 are in disagreement [124,180].

3.2. Histone methyltransferases

Histone–lysine methyltransferases are chromatin-bound enzymes that catalyses the addition of methyl groups onto lysine or arginine residues of chromatin-bound H3 and H4 [151]. The methyl group is transferred enzymatically to the histone with S-adenosyl methionine as the methyl donor. Histone methylases have been isolated from HeLa S-3 cells [182], chick embryo nuclei [183], and rat brain chromatin [184]. The histone methyltransferases methylated H3 and H4 in nucleosomes [184]. Histone–lysine methyltransferase is a chromatin-bound enzyme [129,151]. Initial characterization of the *Tetrahymena* macronuclear H3 methyltransferase suggests that the enzyme has a molecular mass of 400 kDa. The enzyme preferred free histones rather than nucleosomes as substrate [138]. More recent studies have now identified the enzymes catalyzing the methylation of specific sites on different histones.

3.2.1. H3 Lys-9 methyltransferases
Drosophila Su(var)3-9 and homologs of fly protein, including human SUV39H1 and mouse Suv39h1 and Suv39h2, fission yeast Clr4 are H3 Lys-9 methyltransferases [154,185,186] (Fig. 10). Mouse Suv39h2 is expressed primarily in the testis [185]. *Drosophila* Su(var)3-9 was identified as a suppressor of position effect variegation, pointing to the enzymes role in heterochromatin formation. The SU(VAR)3-9 protein family has two domains found in chromatin regulators, the chromo and SET domains [186]. The HøøNHSC (where ø represents hydrophobic residues) within the SET domain is essential in the catalytic activity. Interestingly, mutation of Cys to Arg resulted in a hyperactive enzyme. The crystal structures of several SET domains have been reported [187–190]. However, not all SET domain proteins have histone methyltransferase activity [186]. It appears that in addition to the SET domain adjacent cysteine-rich regions are involved [157,186]. *Neurospora crassa* H3

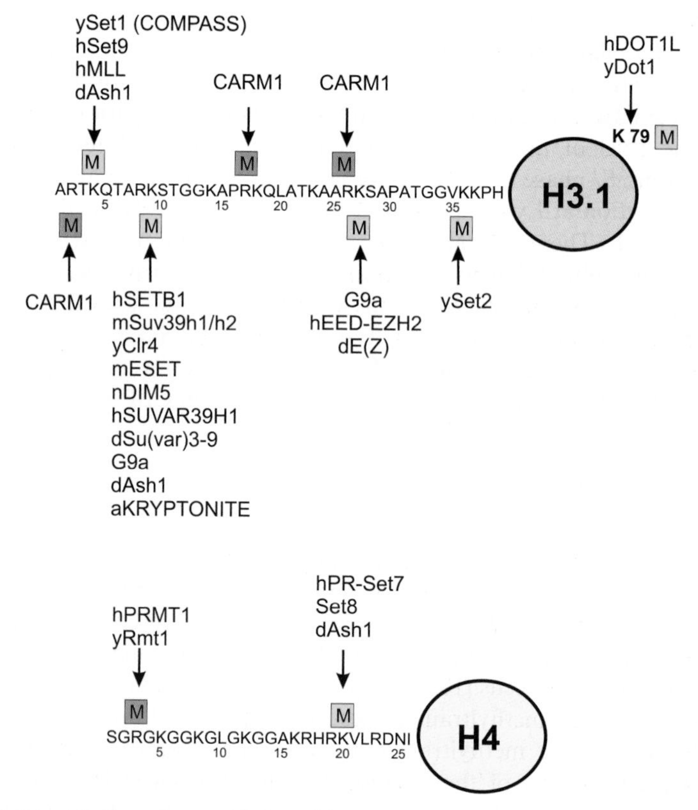

Fig. 10. Histone H3 and H4 methyltransferases. Human (h), mouse (m), yeast (y), *Drosophila* (d), *Neurospora* (n), *Arabidopsis thaliana* (a).

Lys-9 methyltransferase, DIM5, is similar in sequence to Clr4 and Su(var)3-9 and has a SET domain flanked by cysteine rich elements [191]. Murine ESET (ERG-associated protein with SET domain) has these domains and methylates free H3 [192]. This murine methyltransferase, which has high sequence similarity to human SETDB1 (SET domain, bifurcated 1), interacts with the transcription factor ERG. SETB1, a specific H3 Lys-9 methyltransferase, interacts with KAP-1 copressor, which binds to KRAB domain zinc-finger proteins [193]. SETB1 methylates Lys-9 when Lys-4 is methylated but enzymatic activity is inhibited when Ser-10 is phosphorylated or when Lys-14 is acetylated.

Both the SET and chromo-domains are required for Clr4 histone methyltransferase activity [157]. Mutations in Clr4 disrupt heterochromatin formation at centromeres and silencing at mating type loci and pericentromeric heterochromatin is reduced [194]. Similarly genomic instability, chromosome missegregation, male meiotic defects, and increased tumorigenesis (B cell lymphomas) result when murine *Suv39h1* and *Suv39h2* genes are mutated [195]. Thus, Suv39h H3 Lys-9 methyltransferases act as tumor suppressor genes. H3 methylated at Lys-9 was lost in the Suv39h deleted mice at pericentric heterochromatic regions, but not in other chromosomal regions. Further, H3 phosphorylated Ser-10 and H4 acetylated Lys-14 distribution was perturbed [195]. Knockout studies demonstrate that the murine Suv39h genes are needed for normal viability and pre- and post-natal development [195].

Mammalian G9a is a SET-domain histone methyltransferase that methylated H3 at Lys-9 and Lys-27 *in vitro* and at Lys-9 *in vivo* [196]. The consensus sequence for G9a appears to be TKXXARKS. G9a is different from Suv39h1 is several ways. G9a nuclear localization is distinct from that of Suv39h1, which locate to heterochromatic foci [197]. *Suv39h1/2* double mutant mice lose H3 Lys-9 methylation at pericentromeric heterochromatic regions but broad methylation of chromatin remains. It is the latter that is lost in *G9a*-deficient cells [196]. G9a, molecular mass about 100 kDa, methylates free H3 and nucleosomal H3 with a preference for the former; however, the presence of H1 stimulates the methylation of chromatin substrates. Suv39h1, molecular mass about 650 kDa, methylates free H3 and H3 in nucleosomes with equivalent efficiency, but when H1 is present, methylation of chromatin substrates is lessened [198].

3.2.2. H3 Lys-4 methyltransferases

Human Set9 is a 50 kDa H3 methyltransferase that methylates Lys-4 of H3. The enzyme methylated free H3 but not H3 in chromatin substrates. There is evidence that Set9 may stimulate activated transcription [198]. Set9 has the SET domain but lacks the cysteine-rich (pre-SET and post-SET) domains. Disruption of *Saccharomyces cerevisiae* and *Saccharomyces pombe* Set1 obliterates H3 methyl Lys-4 [199]. Thus this SET domain containing protein appears to be a H3 Lys-4 methyltransferase, catalyzing both di- and tri-methylation of H3 Lys-4 [155]. However, studies with recombinant Set1 failed to show histone methyltransferase activity. It has been suggested that other associated proteins may be required for the Set1 to be catalytically active [139,200]. Indeed, Set1 is associated with several other proteins including Cps60, Cps50, Cps40, Cps35, Cps30, Cps25, and Cps15, forming the multiprotein complex COMPASS (Complex Proteins Associated with Set1) [201]. In budding yeast Set1 is required for repression of RNA polymerase II transcription within the ribosomal DNA genes, silencing at telomeres and silencing at the silent mating type (HML) locus [139,200]. Thus, in these situations, H3 methyl Lys-4 has a role in gene repression rather than gene activation.

MLL (also named ALL-1, HRX, and HTRX), the human homolog of *Drosophila trithorax*, is a SET domain protein that methylates H3 at Lys-4. The enzymatic activity was enhanced with H3 acetylated at Lys-9 or Lys-14. MLL is a component of a large multiprotein complex composed of greater than 29 proteins, including TFIID, SWI/SNF remodeling complex and NuRD, a histone deacetylase complex. MLL binds to the promoter of *Hox* genes and regulates their expression [202,203].

3.2.3. H3 Lys-27 methyltransferases

HeLa cells contain an EED–EZH2 complex consisting of polycomb group proteins EZH2, SUZ12, and EED that has histone methyltransferase activity. In *Drosophila* the Extra Sex Combs (ESC)–Enhancer of Zeste [E(Z)] complex consisting of ESC, E(Z), NURF-55, and the PcG repressor, SU(Z)12 has H3 Lys-27 methyltransferase activity [204]. EZH2 and E(Z) have a SET domain which mediates the histone methyltransferase activity. The EED–EZH2 complex preferred to methylate H3 in oligonucleosomes at Lys-27 [140]. Methylation of K27 facilitated the binding between the H3 tail and polycomb, a component of the polycomb repressive complex 1, resulting in the formation of a chromatin repressive state [140].

3.2.4. H3 Lys-36 methyltransferases

Saccharomyces cerevisiae Set2, a SET domain protein, is a 175-kDa H3 methyltransferase that specifically methylated Lys-36 of nucleosomal H3 [205]. Deletion of Set2 in yeast obliterates H3 methyl Lys-36. Set2 has a role in gene repression; however, a role in expression is also possible [205]. Indeed Set2 was found to associate with phosphorylated C-terminal domain of the largest subunit of RNA polymerase II, suggesting that Set2 has a role in transcriptional elongation [206].

3.2.5. H3 Lys-79 methyltransferases

Saccharomyces cerevisiae Dot1 is the H3 Lys-79 methyltransferase [207–209]. Yeast *DOT1* (disruptor of telomeric silencing-1) has role in meiotic checkpoint control and Dot1 overexpression disrupts telomeric silencing and reduces silencing at mating type and rDNA loci. Deletion of *DOT1* results is mislocalization of silencing regulators Sir2 and Sir3 and an increase in histone acetylation levels [207,208]. Human DOT1-like (hDOT1L) prefers nucleosomal rather than free H3. Yeast DOT1 and hDOT1L lack the SET domain [141].

3.2.6. H3 Arg methyltransferases

Coactivator-associated arginine methyltransferase 1 (CARM1) is an arginine methyltransferase which will methylate H3 and to a lesser extent H2B *in vitro*. CARM1 is involved in the regulation of gene expression, including steroid hormone dependent gene expression and myogenesis [210–212] CARM1 methylates several arginines (2, 17, and 26) in H3, with Arg-17 being the major site [212,213]. ChIP assays with an antibody to H3 methyl Arg-17 demonstrated that the estrogen responsive gene pS2 is modified at this site following estrogen stimulation of breast cancer cells [212]. Further acetylation of H3 at Lys-18 and 23 precedes recruitment of CARM1 to the pS2 promoter [214].

3.2.7. H4 Lys-20 methyltransferase

PR-Set7 is a mammalian H4 methyltransferase specific for Lys-20 [124]. Mammalian SET8, distinct but highly similar to PR-Set7, is also a H4 Lys-20 methyltransferase. Both enzymes prefer nucleosomal substrates. These SET domain-containing proteins lack both the pre- and post-SET domains. The expression of PR-Set7 is cell cycle regulated. PR-Set7 levels increase throughout S phase and peak at mitosis. At metaphase and anaphase in HeLa cells PR-Set7 is bound to the mitotic chromosomes [181].

3.2.8. H4 Arg-3 methyltransferase

PRMT1, a nuclear receptor coactivator, exists as in a 330 kDa complex and is a H4 Arg-3 methyltransferase [133,215]. The enzyme appears to be a chromatin bound, and evidence from immunodepletion and knockout studies suggest that it is the principle, if not sole, H4 Arg-3 methyltransferase [133,215]. Mutation of the S-adenosyl methionine binding site in PRMT1 annihilated its nuclear receptor coactivator activity with the androgen receptor, providing evidence for the importance of the methylation event in gene expression [215]. Yeast Rmt1, which is homologous to human PRMT1, methylates Arg-3 only in free H4 [208].

3.2.9. Ash1, a multi-site histone methyltransferase

Ash1, a member of the trithorax group of epigenetic activators, methylates H3 at Lys-4 and Lys-9 and H4 at Lys-20. Ash1 is a SET domain protein that contains both pre- and post-SET domains [216].

3.3. Histone methyltransferases, HATs, HDACs, and DNA methyltransferases

Acetylated isoforms of H3 and H4 are often the targets of ongoing methylation [126,150,151,217,218]. In chicken immature erythrocytes, rapidly acetylated and deacetylated H3 and H4 are selectively methylated, while in Hela cells dynamically acetylated H3, but not H4, is methylated [150,219,220]. H4 that is slowly acetylated and deacetylated is methylated in HeLa [150]. Acetylated yeast H3 was preferentially methylated at Lys-4 [138]. These studies suggested that the processes of histone methylation and dynamic acetylation are not directly coupled, with neither modification predisposing H3 or H4 to the other [138,220].

In vitro studies with unmodified and modified N-terminal peptides of H3 demonstrated that Lys-14 acetylation did not interfere with methylation at Lys-9 by Suv39h1, while phosphorylation at Ser-10 and acetylation at Lys-9 did (Fig. 7). Further dimethylation of Lys-9 reduced enzymatic activity [186]. A Suv39h double null primary mouse embryonic fibroblasts had higher levels of Ser-10 phosphorylated H3 than wild type cells. These mutant cells had increased numbers of micro- and polynuclei. Oversized nuclei were characteristic of subpopulation of cells. The level of Lys-9 methylated H3 in wild type cells and Suv39h double null cells was similar, demonstrating that other H3 methyltransferases were involved [195]. Phosphorylation of Ser-10 by Ipl1/aurora was also studied. Acetylation at Lys-14 promoted the activity of the mitotic kinase, while dimethylation, but not acetylation at Lys-9, reduced activity of the kinase [186].

The unmodified and Lys-9 methylated H3 tail binds specifically to the HDAC complex NuRD. Methylation at Lys-4 prevents this interaction. A H3 tail methylated at Lys-9 does not affect activity of mammalian Set9. Methylation at Lys-4 does not alter activity of G9a, but does prevent the tail being a substrate for Suv39h1 [198].

PRMT1 mediated methylation of H4 Arg-3 stimulates p300 acetylation at H4 Lys-8 and Lys-12. Conversely acetylation at any of the four sites in H4, with acetylation at Lys-5 being the most extreme, prevents methylation of Arg-3 by PRMT1 [215]. These results are consistent with methylation occurring preferentially on hypoacetylated HeLa H4 [150] (Fig. 7).

Histone methyltransferases may associate or act cooperatively with either HATs or HDACs. In fission yeast the H3 Lys-14 deacetylase Clr3 interacts functionally with H3 Lys-9 methyltransferase Clr4. Clr4 methylates Lys-9 of H3, a process facilitated by Rik1, resulting in the recruitment of Swi6 and heterochromatin assembly [157,221]. In *Drosophila* SU(VAR)3-9 H3 Lys-9 methyltransferase is in complex with HDAC1 [222]. Thus, HDAC1 would deacetylate acetylated Lys-9 allowing methylation by SU(VAR)3-9 at this site to occur. CBP, a potent HAT, is associated with a histone methyltransferase that methylated H3 at Lys-9 and to a lesser extent Lys-4. H3 methylation at Lys-9 did not alter the HAT activity of CBP, and vice versa acetylation of H3 (predominantly Lys-14) did not affect the associated histone methyltransferase activity [223].

In addition to linkages between histone methyltransferases and either HATs or HDACs, histone methylation may also influence DNA methylation. In *Neurospora* H3 methylation process is linked to cytosine methylation [191]. Mutation of *dim-5* or replacement of H3 Lys-9 with Leu or Arg reduced the level of methylated DNA. KRYPTONITE, a SET domain containing protein, was identified as a H3 Lys-9 methyltransferase, which was identified in a mutant screen for suppressors of gene silencing at the *Arabidopsis thaliana* SUPERMAN (SUP) locus. Methylation of H3 Lys-9 recruits the *Arabidopsis* homolog of HP1, which in turn binds to CHROMOMETHYLASE 3 (CMT3), which methylates DNA at CpNpG. *Kryptonite* mutants reduced the level of methylated CpNpG but methylated CpG [224,225]. A decrease in DNA methylation, however, does not directly lead to a loss in H3 methyl Lys-9 [177]. Further strengthening the linkage between histone methylation and DNA methylation is the observation that the protein binding to methylated CpG dinucleotides, MeCP2, is associated with histone methyltransferase activity that is directed to Lys-9 of H3 [226].

4. Histone ubiquitination

Histone ubiquitination was first reported in 1973 [227]. H2A, H2B, H3 and their variant forms are ubiquitinated [3,125,228]. Budding yeast lacks ubiquitinated H2A (uH2A), but has uH2B [229,230]. Ubiquitinated H3 has been detected only in the elongating spermatids of rat testes [228,231]. Histone ubiquitination is a reversible modification. The carboxyl end of ubiquitin, a highly conserved 76 amino acid protein, is attached to the ε-amino group of lysine (Lys-119 in H2A; Lys-120 in H2B) [232,233] (Fig. 1). In multicellular eukaryotes, H2A is typically ubiquitinated to a greater extent than is H2B (approximately 10% of H2A versus about 1–2% of H2B). These histones are also polyubiquitinated, with H2A having the greater amounts of polyubiquitinated isoforms [234,235]. The major arrangement of ubiquitin in polyubiquitinated H2A is a chain of ubiquitin molecules joined to each other by isopeptide bonds to a ubiquitin molecule that is attached to the ε-amino group of Lys-119 of H2A [234].

Multiubiquitination of intracellular proteins is a prerequisite for the selective degradation of intracellular proteins by the ubiquitin-dependent proteolytic pathway [236,237]. Monoubiquitinated histones are not degraded by the proteasome [238,239]. Proteasome, a multisubunit complex that catalyzes both ubiquitin-dependent and ubiquitin-independent protein degradation, is found in the cytoplasm and nucleus [240–242]. Thus, it is possible that multiubiquitination of the histones may tag these proteins for degradation.

The ubiquitin ligase system consists of an ubiquitin-activating enzyme, E1, and several ubiquitin-conjugating enzymes, E2s [242,243]. E1 catalyzes the first step in the conjugation reaction. This enzyme is present in the nucleus [244,245]. However, its localization in the cell changes throughout the cell cycle. E1 is found in the nucleus in G1 and G2, but not S phase of the cell cycle [246]. In M-phase of the cell cycle this enzyme is bound to chromosomes (HeLa cells) [244]. Several E2 enzymes have been shown to catalyze the addition of ubiquitin onto histones [247–249]. In budding yeast Rad6/Ubc2 catalyzes the addition of ubiquitin onto Lys-123 of H2B (equivalent of Lys-120 in mammalian H2B) [230]. Defects in Rad6 or the ubiquitination site of yeast H2B causes defects in mitosis and meiosis [230]. *In vitro*, *Drosophila* coactivator TAFII250 ubiquitinates the linker histone H1 [250].

Ubiquitinated H2A and H2B disappear at metaphase and reappear in anaphase [251–253]. Histone ubiquitination may also be involved in cell cycle progression through S phase [254]. In dividing and non-dividing cells, the ubiquitin moiety of the ubiquitinated histones is in rapid equilibrium with a pool of free ubiquitin [238,239]. The turnover of the ubiquitinated histones is presumably catalyzed by ubiquitin-C-terminal hydrolases. Uni- and multi-cellular eukaryotes contain isopeptidase, a ubiquitin-C-terminal hydrolase which cleave uH2A into H2A and ubiquitin [255]. This enzyme is localized mostly in the cytoplasm. Another ubiquitin-C-terminal hydrolase that will process uH2A has been isolated from rat liver nucleoli [256,257].

Thermal and chemical cellular stresses result in the rapid disappearance of ubiquitinated histones [258,259]. The levels of ubiquitinated histones drops precipitously after human tumor cells are treated with proteasome inhibitors [260,261]. Deubiquitination of uH2A also occurs during apoptosis, with its lose coinciding with nuclear pyknosis and chromatin condensation [262,263].

Ubiquitinated H2A levels are elevated in cancer cells. In SV-40-transformed human fibroblasts and keratinocytes, uH2A levels are higher than the normal counterparts [264]. Trout hepatocarcinomas have greater levels of uH2A and polyubiquitinated H2A than normal trout liver [265]. It is possible that the polyubiquitinated H2A in trout hepatocarcinomas is targeted for proteolysis, implying that H2A's turnover is more rapid in the tumor cells.

Histone ubiquitination may have a role in spermatogenesis. Levels of uH2A are elevated in pachytene spermatocytes, but not round spermatids in mouse testis. In the pachytene spermatocytes uH2A was concentrated in the inactive sex body, a structure containing the heterochromatic X and Y chromosomes [231]. The presence of uH2A in the late stages of spermatogenesis has been observed in trout and chicken, suggesting that histone ubiquitination may have a role in histone to protamine replacement [266,267].

Ubiquitinated H2B and to a lesser extent ubiquitinated H2A may be associated with transcriptionally active DNA [235,268–270]. *Tetrahymena* transcriptionally active macronuclei are associated with uH2A and uH2B, while transcriptionally inert micronuclei have low levels of uH2A. Ubiquitination of H2B coincided with the transformation of an inert germinal nucleus into that of a transcriptionally active somatic nucleus, suggesting that uH2B has a role in maintaining the transcriptionally active chromatin state [271]. Ubiquitination of H2B is dependent upon ongoing transcription [268,269]. The COOH terminal sequence of H2B, but not H2A, is buried in the nucleosome [19,122,270,272]. Thus, the process of transcription may alter nucleosome structure [273–275] or transiently displace the histone octamer from DNA [276,277], allowing the COOH terminus of H2B to become accessible to the enzymes catalyzing the addition of ubiquitin. Another mechanism, which does not require ongoing transcription, is by exchange of newly synthesized uH2B and uH2A with histones that are in transcriptionally active nucleosomes [175,278]. The introduction of uH2B into the nucleosome may result in an alteration in nucleosome or higher order chromatin structure. However, analyses of nucleosomes reconstituted with ubiquitinated histones and studies on the effect of uH2A on magnesium-dependent folding of the chromatin fiber have not observed any effect of ubiquitinated histones on nucleosome structure or chromatin folding [270,279,280].

Chromatin fractionation approaches including ChIP assays have provided evidence for and against uH2A and uH2B being associated with transcriptionally active chromatin [270,281–285]. Most evidence supports uH2A being associated with transcriptionally active and inactive chromatin. Although current evidence suggests that mammalian uH2B may be associated with transcribed chromatin, the association cannot be tested directly until an antibody specific to uH2B is available. In budding yeast, the association is quite different with ubiquitination of H2B by Rad6 being involved in transcriptional silencing at various loci (telomere, *HM*, rDNA) [286]. Further, Rad6 and ubiquitination of H2B are required for repression of yeast *ARG1* transcription [287].

5. Histone ubiquitination and histone methylation—trans-histone regulatory pathway

In budding yeast, Rad-6 ubiquitination of Lys-123 of H2B is required for Set1 methylation of Lys-4 of H3 and Dot1 methylation of Lys-79 of H3 [286,288] (Fig. 11). Ubiquitination of H2B does not affect methylation at Lys-36 of H3. Methylation of Lys-4 and/or Lys-79 of H3 is not required for ubiquitination of H2B. Both H3 methylation events in yeast are involved in transcriptional silencing of genes located near telomeres and within the rDNA. Preventing the ubiquitination of H2B results in defects in silencing [287]. It has been proposed that ubiquitination of H2B perturbs nucleosome structure allowing Dot1 access to H3 Lys-79 [289]. Similarly, H2B ubiquitination may disrupt nucleosome and/or higher order chromatin structure exposing H3 Lys-4 to Set1 [286]. About 5% or lower of yeast H2B is ubiquitinated H2B, while 90% of H3 is methylated at Lys-79 and about 35% of H3 is methylated at Lys-4 [286,289]. The difference in levels of ubiquitinated H2B and methylated H3 reflect the dynamics of histone ubiquitination and the stability of histone methylation [289].

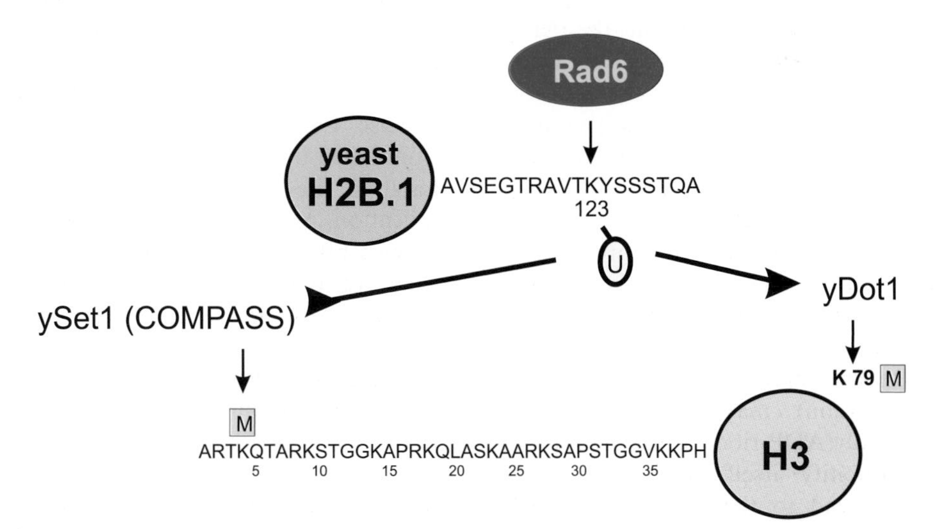

Fig. 11. Trans-histone regulatory pathway. Modification of yeast histone H3 at Lys-4 and Lys-79 is dependent upon ubiquitination of histone H2B at Lys-123.

6. *Histone ADP-ribosylation*

Histone ADP-ribosylation was first reported in 1968 [290]. Poly(ADP-ribosylation) has been implicated in several nuclear processes, including DNA replication, repair and recombination [291–294]. Histone H1 and the four core histones are modified by adenosine diphospho (ADP) ribosylation which involves the transfer of the ADP-ribose moiety of NAD^+ to the histone acceptor (Figs. 1 and 2). H1 is the principle poly(ADP-ribosylated) histone, while core histones are ADP-ribosylated to a minor extent [295–297]. H1 is modified at Glu residues 2, 14 (or 15), and 116 (or 117) and at Lys located at the C-terminus [25,298,299]. Poly(ADP-ribosylated) H1 is associated with dynamically acetylated core histones [295]. There is conflicting results as to whether poly(ADP-ribosylation) of H1 promotes chromatin decondensation [300–304].

One or more ADP-ribose groups may be transferred, resulting in core histones with mono(ADP-ribose) up to highly branched poly(ADP-ribose) residues. The acetylated isoforms of the core histones are preferentially ADP-ribosylated [292,293,305–308]. In the absence of H1, H2B is the main acceptor of polyADP ribosylated histone acceptor protein [304]. ADP-ribosylation of the core histones alters nucleosome structure as assessed by epitope availability to antibodies [304]. About 15 ADP-ribosylated isoforms were observed for histones H3.1, H3.3, H2B.1, and H2B.2 in dimethylsulfate-treated mouse myeloma cells [305]. Less than 5% of the core histones are modified by ADP-ribosylation. Further, poly ADP-ribose is rapidly turning over, having a half-life of 30 s to 10 min [291,292, 306]. For H2B, the carboxyl group of Glu-2 is the site of ADP ribosylation [309, 310]. The sites of ADP-ribosylation for the other core histones are not known. However, H2A has a carboxylate ester ADP-ribose-protein bond, while H3 and H4 appear to have arginine-linked ADP-ribose residues [307]. A note of caution, however, in realizing that the ADP-ribose polymer binds non-covalently and strongly (not removed by strong acids) to histones [311].

The enzyme catalyzing the addition of ADP-ribose units onto the histones and itself is poly(ADP-ribose) polymerase or synthetase. Poly(ADP-ribose) polymerase is a nuclear, DNA-dependent enzyme that is stimulated by DNA breaks [302]. This property of the enzyme would target its action to sites that have DNA strand breaks (regions of the genome involved in replication, repair, recombination). The enzyme is associated with chromatin areas and perichromatin regions in interphase Chinese hamster ovary cells [312]. Degradation of the ADP-ribose polymer is catalyzed by the nuclear enzyme poly(ADP-ribose) glycohydrolase and ADP-ribosyl protein lyase.

Realini and Althaus [313] have put forth the hypothesis that poly(ADP-ribosylation) may have a function in histone shuttling. They propose that poly(ADP-ribose) polymerase directed to sites of DNA strand breaks would auto-modify itself generating multiple ADP-ribose polymers. The polymers would lead to the dissociation of the histones from DNA onto the polymers. The DNA would now be free for processing (e.g., by enzymes involved in excision repair). The action of poly(ADP-ribose) glycohydrolase would degrade the

ADP-ribose polymers, leading to the release of the histones which would rebind the DNA.

Relationships between histone methylation and DNA methylation and histone acetyation and DNA methylation have been reported [191,314,315]. A similar relationship may exist between poly(ADP ribosylated) H1 and DNA methylation. Inhibition of poly(ADP-ribose) polymerase with 3-aminobenzamide increases the susceptibility of L929 mouse fibroblast nuclei to be methylated by endogenous DNA methyltransferases [316,317]. Further, there is evidence that poly(ADP ribosylation) protects CpG islands located at the 5′ end of housekeeping genes from methylation [318]. Future studies will likely reveal an interesting dynamic relationship between histone methylation, histone acetylation, and histone poly(ADP-ribosylation).

References

1. Ord, M.G. and Stocken, L.A. (1966) Biochem. J. 98, 888–897.
2. Kleinsmith, L.J., Allfrey, V.G., and Mirsky, A.E. (1966) Science 154, 780–781.
3. Van Holde, K.E. (1988) Chromatin. Springer-Verlag, New York.
4. Gurley, L.R., Walters, R.A., and Tobey, R.A. (1975) J. Biol. Chem. 250, 3936–3944.
5. Marks, D.B., Paik, W.K., and Borun, T.W. (1973) J. Biol. Chem. 248, 5660–5667.
6. Hohmann, P., Tobey, R.A., and Gurley, L.R. (1976) J. Biol. Chem. 251, 3685–3692.
7. Louie, A.J., Candido, E.P., and Dixon, G.H. (1974) Cold Spring Harb. Symp. Quant. Biol. 38, 803–819.
8. Sung, M.T. and Dixon, G.H. (1970) Proc. Natl. Acad. Sci. USA 67, 1616–1623.
9. Pantazis, P. and Bonner, W.T. (1981) J. Biol. Chem. 256, 4669–4675.
10. Takeuchi, F., Hashimoto, E., and Yamamura, H. (1992) J. Biochem. Tokyo 111, 788–792.
11. Fusauchi, Y. and Iwai, K. (1984) J. Biochem. Tokyo 95, 147–154.
12. Allis, C.D. and Gorovsky, M.A. (1981) Biochemistry 20, 3828–3833.
13. Allis, C.D., Glover, C.V.C., Bowen, J.K., and Gorovsky, M.A. (1980) Cell 20, 609–617.
14. Hashimoto, E., Takeda, M., Nishizuka, Y., Hamana, K., and Iwai, K. (1976) J. Biol. Chem. 251, 6287–6293.
15. Glass, D.B. and Krebs, E.G. (1979) J. Biol. Chem. 254, 9728–9738.
16. Glass, D.B. and Krebs, E.G. (1982) J. Biol. Chem. 257, 1196–1200.
17. Ajiro, K. (2000) J. Biol. Chem. 275, 439–443.
18. Romhanyi, T., Seprodi, J., Antoni, F., Nikolics, K., Meszaros, G., and Farago, A. (1982) Biochim. Biophys. Acta 701, 57–62.
19. Luger, K., Mader, A.W., Richmond, R.K., Sargent, D.F., and Richmond, T.J. (1997) Nature 389, 251–260.
20. Hill, C.S., Packman, L.C., and Thomas, J.O. (1990) EMBO J. 9, 805–813.
21. Hohmann, P. (1983) Mol. Cell Biochem. 57, 81–92.
22. Talasz, H., Helliger, W., Puschendorf, B., and Lindner, H. (1996) Biochemistry 35, 1761–1767.
23. Ohsumi, K., Katagiri, C., and Kishimoto, T. (1993) Science 262, 2033–2035.
24. Roth, S.Y. and Allis, C.D. (1992) Trends Biochem. Sci. 17, 93–98.
25. Parseghian, M.H., Henschen, A.H., Krieglstein, K.G., and Hamkalo, B.A. (1994) Protein Sci. 3, 575–587.
26. Krause, K., Wasner, M., Reinhard, W., Haugwitz, U., Dohna, C.L., Mossner, J., and Engeland, K. (2000) Nucleic Acids Res. 28, 4410–4418.
27. Sweet, M.T. and Allis, C.D. (1993) Chromosoma 102, 637–647.

28. Sweet, M.T., Carlson, G., Cook, R.G., Nelson, D., and Allis, C.D. (1997) J. Biol. Chem. 272, 916–923.
29. Paulson, J.R., Patzlaff, J.S., and Vallis, A.J. (1996) J. Cell Sci. 109, 1437–1447.
30. Gurley, L.R., D'Anna, J.A., Barham, S.S., Deaven, L.L., and Tobey, R.A. (1978) Eur. J. Biochem. 84, 1–15.
31. Wei, Y., Mizzen, C.A., Cook, R.G., Gorovsky, M.A., and Allis, C.D. (1998) Proc. Natl. Acad. Sci. 95, 7480–7484.
32. Wei, Y., Yu, L., Bowen, J., Gorovsky, M.A., and Allis, C.D. (1999) Cell 97, 99–109.
33. Hendzel, M.J., Wei, Y., Mancini, M.A., Van Hooser, A., Ranalli, T., Brinkely, B.R., Bazett-Jones, D.P., and Allis, C.D. (1997) Chromosoma 106, 348–360.
34. Sauve, D.M., Anderson, H.J., Ray, J.M., James, W.M., and Roberge, M. (1999) J. Cell Biol. 145, 225–235.
35. Goto, H., Tomono, Y., Ajiro, K., Kosako, H., Fujita, M., Sakurai, M., Okawa, K., Iwamatsu, A., Okigaki, T., Takahashi, T., and Inagaki, M. (1999) J. Biol. Chem. 274, 25543–25549.
36. Smith, M.M. (2002) Curr. Opin. Cell Biol. 14, 279–285.
37. Zeitlin, S.G., Shelby, R.D., and Sullivan, K.F. (2001) J. Cell Biol. 155, 1147–1157.
38. Zeitlin, S.G., Barber, C.M., Allis, C.D., and Sullivan, K. (2001) J. Cell Sci. 114, 653–661.
39. Goto, H., Yasui, Y., Nigg, E.A., and Inagaki, M. (2002) Genes Cells 7, 11–17.
40. Sugiyama, K., Sugiura, K., Hara, T., Sugimoto, K., Shima, H., Honda, K., Furukawa, K., Yamashita, S., and Urano, T. (2002) Oncogene 21, 3103–3111.
41. Hsu, J.Y., Sun, Z.W., Li, X., Reuben, M., Tatchell, K., Bishop, D.K., Grushcow, J.M., Brame, C.J., Caldwell, J.A., Hunt, D.F., Lin, R., Smith, M.M., and Allis, C.D. (2000) Cell 102, 279–291.
42. Murnion, M.E., Adams, R.R., Callister, D.M., Allis, C.D., Earnshaw, W.C., and Swedlow, J.R. (2001) J. Biol. Chem. 276, 26656–26665.
43. Adams, R.R., Eckley, D.M., Vagnarelli, P., Wheatley, S.P., Gerloff, D.L., Mackay, A.M., Svingen, P.A., Kaufmann, S.H., and Earnshaw, W.C. (2001) Chromosoma 110, 65–74.
44. Adams, R.R., Maiato, H., Earnshaw, W.C., and Carmena, M. (2001) J. Cell Biol. 153, 865–880.
45. Crosio, C., Fimia, G.M., Loury, R., Kimura, M., Okano, Y., Zhou, H., Sen, S., Allis, C.D., and Sassone-Corsi, P. (2002) Mol. Cell Biol. 22, 874–885.
46. Ota, T., Suto, S., Katayama, H., Han, Z.B., Suzuki, F., Maeda, M., Tanino, M., Terada, Y., and Tatsuka, M. (2002) Cancer Res. 62, 5168–5177.
47. MacCallum, D.E., Losada, A., Kobayashi, R., and Hirano, T. (2002) Mol. Biol. Cell 13, 25–39.
48. De Souza, C.P., Osmani, A.H., Wu, L.P., Spotts, J.L., and Osmani, S.A. (2000) Cell 102, 293–302.
49. Hashimoto, Y., Akita, H., Hibino, M., Kohri, K., and Nakanishi, M. (2002) Biochem. Biophys. Res. Commun. 293, 753–758.
50. Lee, H.L. and Archer, T.K. (1998) EMBO J. 17, 1454–1466.
51. Bhattacharjee, R.N., Banks, G.C., Trotter, K.W., Lee, H.L., and Archer, T.K. (2001) Mol. Cell Biol. 21, 5417–5425.
52. Parseghian, M.H. and Hamkalo, B.A. (2001) Biochem. Cell Biol. 79, 289–304.
53. Banks, G.C., Deterding, L.J., Tomer, K.B., and Archer, T.K. (2001) J. Biol. Chem.
54. Chadee, D.N., Allis, C.D., Wright, J.A., and Davie, J.R. (1997) J. Biol. Chem. 272, 8113–8116.
55. Mizzen, C.A., Dou, Y., Liu, Y., Cook, R.G., Gorovsky, M.A., and Allis, C.D. (1999) J. Biol. Chem. 274, 14533–14536.
56. Dou, Y., Mizzen, C.A., Abrams, M., Allis, C.D., and Gorovsky, M.A. (1999) Mol. Cell 4, 641–647.
57. Dou, Y. and Gorovsky, M.A. (2000) Mol. Cell 6, 225–231.
58. Dou, Y. and Gorovsky, M.A. (2002) Proc. Natl. Acad. Sci. USA 99, 6142–6146.
59. Chadee, D.N., Taylor, W.R., Hurta, R.A.R., Allis, C.D., Wright, J.A., and Davie, J.R. (1995) J. Biol. Chem. 270, 20098–20105.
60. Herrera, R.E., Chen, F., and Weinberg, R.A. (1996) Proc. Natl. Acad. Sci. USA 93, 11510–11515.
61. Chadee, D.N., Peltier, C.P., and Davie, J.R. (2002) Oncogene 21, 8397–8403.
62. Kaplan, L.J., Bauer, R., Morrison, E., Langan, T.A., and Fasman, G.D. (1984) J. Biol. Chem. 259, 8777–8785.

63. Hill, C.S., Rimmer, J.M., Green, B.N., Finch, J.T., and Thomas, J.O. (1991) EMBO J. 10, 1939–1948.
64. Lever, M.A., Th'ng, J.P., Sun, X., and Hendzel, M.J. (2000) Nature 408, 873–876.
65. Dou, Y., Bowen, J., Liu, Y., and Gorovsky, M.A. (2002) J. Cell Biol. 158, 1161–1170.
66. Aubert, D., Garcia, M., Benchaibi, M., Poncet, D., Chebloune, Y., Verdier, G., Nigon, V., Samarut, J., and Mura, C.V. (1991) J. Cell Biol. 113, 497–506.
67. Horn, P.J., Carruthers, L.M., Logie, C., Hill, D.A., Solomon, M.J., Wade, P.A., Imbalzano, A.N., Hansen, J.C., and Peterson, C.L. (2002) Nat. Struct. Biol. 9, 263–267.
68. Mahadevan, L.C., Willis, A.C., and Barratt, M.J. (1991) Cell 65, 775–783.
69. Thomson, S., Clayton, A.L., Hazzalin, C.A., Rose, S., Barratt, M.J., and Mahadevan, L.C. (1999) EMBO J. 18, 4779–4793.
70. Chadee, D.N., Hendzel, M.J., Tylipski, C.P., Allis, C.D., Bazett-Jones, D.P., Wright, J.A., and Davie, J.R. (1999) J. Biol. Chem. 274, 24914–24920.
71. Zhong, S.P., Ma, W.Y., and Dong, Z. (2000) J. Biol. Chem. 275, 20980–20984.
72. Zhong, S., Zhang, Y., Jansen, C., Goto, H., Inagaki, M., and Dong, Z. (2001) J. Biol. Chem. 276, 12932–12937.
73. Strelkov, I.S. and Davie, J.R. (2002) Cancer Res. 62, 75–78.
74. Salvador, L.M., Park, Y., Cottom, J., Maizels, E.T., Jones, J.C., Schillace, R.V., Carr, D.W., Cheung, P., Allis, C.D., Jameson, J.L., and Hunzicker-Dunn, M. (2001) J. Biol. Chem. 276, 40146–40155.
75. Hendzel, M.J., Kruhlak, M.J., and Bazett-Jones, D.P. (1998) Mol. Biol. Cell 9, 2491–2507.
76. Barratt, M.J., Hazzalin, C.A., Cano, E., and Mahadevan, L.C. (1994) Proc. Natl. Acad. Sci. USA 91, 4781–4785.
77. Iyer, V.R., Eisen, M.B., Ross, D.T., Schuler, G., Moore, T., Lee, J.C.F., Trent, J.M., Staudt, L.M., Hudson, J., Jr., Boguski, M.S., Lashkari, D., Shalon, D., Botstein, D., and Brown, P.O. (1999) Science 283, 83–87.
78. Lo, W.S., Duggan, L., Tolga, N.C., Emre, Belotserkovskya, R., Lane, W.S., Shiekhattar, R., and Berger, S.L. (2001) Science 293, 1142–1146.
79. Wang, Y., Zhang, W., Jin, Y., Johansen, J., and Johansen, K.M. (2001) Cell 105, 433–443.
80. Sassone-Corsi, P., Mizzen, C.A., Cheung, P., Crosio, C., Monaco, L., Jacquot, S., Hanauer, A., and Allis, C.D. (1999) Science 285, 886–891.
81. Deak, M., Clifton, A.D., Lucocq, L.M., and Alessi, D.R. (1998) EMBO J. 17, 4426–4441.
82. Zhong, S., Jansen, C., She, Q.B., Goto, H., Inagaki, M., Bode, A.M., Ma, W.Y., and Dong, Z. (2001) J. Biol. Chem. 276, 33213–33219.
83. Jin, Y., Wang, Y., Walker, D.L., Dong, H., Conley, C., Johansen, J., and Johansen, K.M. (1999) Mol. Cell 4, 129–135.
84. Jin, Y., Wang, Y., Johansen, J., and Johansen, K.M. (2000) J. Cell Biol. 149, 1005–1010.
85. Meller, V.H. and Rattner, B.P. (2002) EMBO J. 21, 1084–1091.
86. Clayton, A.L., Rose, S., Barratt, M.J., and Mahadevan, L.C. (2000) EMBO J. 19, 3714–3726.
87. Cheung, P., Tanner, K.G., Cheung, W.L., Sassone-Corsi, P., Denu, J.M., and Allis, C.D. (2000) Mol. Cell 5, 905–915.
88. Thomson, S., Clayton, A.L., and Mahadevan, L.C. (2001) Mol. Cell 8, 1231–1241.
89. Lo, W.S., Trievel, R.C., Rojas, J.R., Duggan, L., Hsu, J.Y., Allis, C.D., Marmorstein, R., and Berger, S.L. (2000) Mol. Cell 5, 917–926.
90. Ait-Si-Ali, S., Carlisi, D., Ramirez, S., Upegui-Gonzalez, L.C., Duquet, A., Robin, P., Rudkin, B., Harel-Bellan, A., and Trouche, D. (1999) Biochem. Biophys. Res. Commun. 262, 157–162.
91. Liu, Y.Z., Thomas, N.S., and Latchman, D.S. (1999) Neuroreport 10, 1239–1243.
92. Celeste, A., Petersen, S., Romanienko, P.J., Fernandez-Capetillo, O., Chen, H.T., Sedelnikova, O.A., Reina-San-Martin, B., Coppola, V., Meffre, E., Difilippantonio, M.J., Redon, C., Pilch, D.R., Olaru, A., Eckhaus, M., Camerini-Otero, R.D., Tessarollo, L., Livak, F., Manova, K., Bonner, W.M., Nussenzweig, M.C., and Nussenzweig, A. (2002) Science 296, 922–927.

93. Bassing, C.H., Chua, K.F., Sekiguchi, J., Suh, H., Whitlow, S.R., Fleming, J.C., Monroe, B.C., Ciccone, D.N., Yan, C., Vlasakova, K., Livingston, D.M., Ferguson, D.O., Scully, R., and Alt, F.W. (2002) Proc. Natl. Acad. Sci. USA 99, 8173–8178.
94. Paull, T.T., Rogakou, E.P., Yamazaki, V., Kirchgessner, C.U., Gellert, M., and Bonner, W.M. (2000) Curr. Biol. 10, 886–895.
95. Rogakou, E.P., Nieves-Neira, W., Boon, C., Pommier, Y., and Bonner, W.M. (2000) J. Biol. Chem. 275, 9390–9395.
96. Rogakou, E.P., Boon, C., Redon, C., and Bonner, W.M. (1999) J. Cell Biol. 146, 905–916.
97. Rogakou, E.P., Pilch, D.R., Orr, A.H., Ivanova, V.S., and Bonner, W.M. (1998) J. Biol. Chem. 273, 5858–5868.
98. Mahadevaiah, S.K., Turner, J.M., Baudat, F., Rogakou, E.P., de Boer, P., Blanco-Rodriguez, J., Jasin, M., Keeney, S., Bonner, W.M., and Burgoyne, P.S. (2001) Nat. Genet. 27, 271–276.
99. Talasz, H., Helliger, W., Sarg, B., Debbage, P.L., Puschendorf, B., and Lindner, H. (2002) Cell Death Differ. 9, 27–39.
100. Downs, J.A., Lowndes, N.F., and Jackson, S.P. (2000) Nature 408, 1001–1004.
101. Kobayashi, J., Tauchi, H., Sakamoto, S., Nakamura, A., Morishima, K., Matsuura, S., Kobayashi, T., Tamai, K., Tanimoto, K., and Komatsu, K. (2002) Curr. Biol. 12, 1846–1851.
102. Chen, H.T., Bhandoola, A., Difilippantonio, M.J., Zhu, J., Brown, M.J., Tai, X., Rogakou, E.P., Brotz, T.M., Bonner, W.M., Ried, T., and Nussenzweig, A. (2000) Science 290, 1962–1965.
103. Petersen, S., Casellas, R., Reina-San-Martin, B., Chen, H.T., Difilippantonio, M.J., Wilson, P.C., Hanitsch, L., Celeste, A., Muramatsu, M., Pilch, D.R., Redon, C., Ried, T., Bonner, W.M., Honjo, T., Nussenzweig, M.C., and Nussenzweig, A. (2001) Nature 414, 660–665.
104. Limoli, C.L., Giedzinski, E., Bonner, W.M., and Cleaver, J.E. (2002) Proc. Natl. Acad. Sci. USA 99, 233–238.
105. Ward, I.M. and Chen, J. (2001) J. Biol. Chem. 276, 47759–47762.
106. Siino, J.S., Nazarov, I.B., Svetlova, M.P., Solovjeva, L.V., Adamson, R.H., Zalenskaya, I.A., Yau, P.M., Bradbury, E.M., and Tomilin, N.V. (2002) Biochem. Biophys. Res. Commun. 297, 1318–1323.
107. Burma, S., Chen, B.P., Murphy, M., Kurimasa, A., and Chen, D.J. (2001) J. Biol. Chem. 276, 42462–42467.
108. Madigan, J.P., Chotkowski, H.L., and Glaser, R.L. (2002) Nucleic Acids Res. 30, 3698–3705.
109. Chen, C.C., Smith, D.L., Bruegger, B.B., Halpern, R.M., and Smith, R.A. (1974) Biochemistry 13, 3785–3789.
110. Smith, D.L., Chen, C.C., Bruegger, B.B., Holtz, S.L., Halpern, R.M., and Smith, R.A. (1974) Biochemistry 13, 3780–3785.
111. Fujitaki, J.M., Fung, G., Oh, E.Y., and Smith, R.A. (1981) Biochemistry 20, 3658–3664.
112. Huang, J.M., Wei, Y.F., Kim, Y.H., Osterberg, L., and Matthews, H.R. (1991) J. Biol. Chem. 266, 9023–9031.
113. Wei, Y.F. and Matthews, H.R. (1990) Anal. Biochem. 190, 188–192.
114. Kim, Y., Huang, J., Cohen, P., and Matthews, H.R. (1993) J. Biol. Chem. 268, 18513–18518.
115. Besant, P.G. and Attwood, P.V. (2000) Int. J. Biochem. Cell Biol. 32, 243–253.
116. Chen, C.C., Bruegger, B.B., Kern, C.W., Lin, Y.C., Halpern, R.M., and Smith, R.A. (1977) Biochemistry 16, 4852–4855.
117. Tan, E., Besant, P.G., and Attwood, P.V. (2002) Biochemistry 41, 3843–3851.
118. Murray, K. (1964) Biochemistry 3, 10–15.
119. Desrosiers, R. and Tanguay, R.M. (1988) J. Biol. Chem. 263, 4686–4692.
120. Ebralidse, K.K., Grachev, S.A., and Mirzabekov, A.D. (1988) Nature 331, 365–367.
121. Bohm, L., Briand, G., Sautiere, P., and Crane-Robinson, C. (1981) Eur. J. Biochem. 119, 67–74.
122. Rosenberg, N.L., Smith, R.M., and Rill, R.L. (1986) J. Biol. Chem. 261, 12375–12383.
123. Waterborg, J.H., Robertson, A.J., Tatar, D.L., Borza, C.M., and Davie, J.R. (1995) Plant Physiol. 109, 393–407.

124. Nishioka, K., Rice, J.C., Sarma, K., Erdjument-Bromage, H., Werner, J., Wang, Y., Chuikov, S., Valenzuela, P., Tempst, P., Steward, R., Lis, J.T., Allis, C.D., and Reinberg, D. (2002) Mol. Cell 9, 1201–1213.
125. Wu, R.S., Panusz, H.T., Hatch, C.L., and Bonner, W.M. (1986) CRC Crit. Rev. Biochem. 20, 201–263.
126. Waterborg, J.H. (1993) J. Biol. Chem. 268, 4918–4921.
127. Honda, B.M., Dixon, G.H., and Candido, E.P. (1975) J. Biol. Chem. 250, 8681–8685.
128. Duerre, J.A. and Chakrabarty, S. (1975) J. Biol. Chem. 250, 8457–8461.
129. Duerre, J.A., Wallwork, J.C., Quick, D.P., and Ford, K.M. (1977) J. Biol. Chem. 252, 5981–5985.
130. DeLange, R.J., Fambrough, D.M., Smith, E.L., and Bonner, J. (1969) J. Biol. Chem. 244, 319–334.
131. Sarg, B., Koutzamani, E., Helliger, W., Rundquist, I., and Lindner, H.H. (2002) J. Biol. Chem. 277, 39195–39201.
132. Waterborg, J.H., Fried, S.R., and Matthews, H.R. (1983) Eur. J. Biochem. 136, 245–252.
133. Strahl, B.D., Briggs, S.D., Brame, C.J., Caldwell, J.A., Koh, S.S., Ma, H., Cook, R.G., Shabanowitz, J., Hunt, D.F., Stallcup, M.R., and Allis, C.D. (2001) Curr. Biol. 11, 996–1000.
134. Brandt, W.F., Strickland, W.N., Morgan, M., and Von Holt, C. (1974) FEBS Lett. 40, 167–172.
135. Zhang, K., Tang, H., Huang, L., Blankenship, J.W., Jones, P.R., Xiang, F., Yau, P.M., and Burlingame, A.L. (2002) Anal. Biochem. 306, 259–269.
136. Ohe, Y. and Iwai, K. (1981) J. Biochem.(Tokyo) 90, 1205–1211.
137. Waterborg, J.H. (1990) J. Biol. Chem. 265, 17157–17161.
138. Strahl, B.D., Ohba, R., Cook, R.G., and Allis, C.D. (1999) Proc. Natl. Acad. Sci. USA 96, 14967–14972.
139. Briggs, S.D., Bryk, M., Strahl, B.D., Cheung, W.L., Davie, J.K., Dent, S.Y., Winston, F., and Allis, C.D. (2001) Genes Dev. 15, 3286–3295.
140. Cao, R., Wang, L., Wang, H., Xia, L., Erdjument-Bromage, H., Tempst, P., Jones, R.S., and Zhang, Y. (2002) Science 298, 1039–1043.
141. Feng, Q., Wang, H., Ng, H.H., Erdjument-Bromage, H., Tempst, P., Struhl, K., and Zhang, Y. (2002) Curr. Biol. 12, 1052–1058.
142. Camato, R. and Tanguay, R.M. (1982) EMBO J. 1, 1529–1532.
143. Arrigo, A.P. (1983) Nucleic Acids Res. 11, 1389–1404.
144. Thomas, G., Lange, H.W., and Hempel, K. (1975) Eur. J. Biochem. 51, 609–615.
145. DeLange, R.J., Hooper, J.A., and Smith, E.L. (1973) J. Biol. Chem. 248, 3261–3274.
146. Hooper, J.A., Smith, E.L., Sommer, K.R., and Chalkley, R. (1973) J. Biol. Chem. 248, 3275–3279.
147. DeLange, R.J., Fambrough, D.M., Smith, E.L., and Bonner, J. (1969) J. Biol. Chem. 244, 5669–5679.
148. Patthy, L. and Smith, E.L. (1973) J. Biol. Chem. 248, 6834–6840.
149. Honda, B.M., Candido, P.M., and Dixon, G.H. (1975) J. Biol. Chem. 250, 8686–8689.
150. Annunziato, A.T., Eason, M.B., and Perry, C.A. (1995) Biochemistry 34, 2916–2924.
151. Hendzel, M.J. and Davie, J.R. (1989) J. Biol. Chem. 264, 19208–19214.
152. Hebbes, T.R., Clayton, A.L., Thorne, A.W., and Crane-Robinson, C. (1994) EMBO J. 13, 1823–1830.
153. Litt, M.D., Simpson, M., Gaszner, M., Allis, C.D., and Felsenfeld, G. (2001) Science 293, 2453–2455.
154. Jenuwein, T. and Allis, C.D. (2001) Science 293, 1074–1080.
155. Santos-Rosa, H., Schneider, R., Bannister, A.J., Sherriff, J., Bernstein, B.E., Emre, N.C., Schreiber, S.L., Mellor, J., and Kouzarides, T. (2002) Nature 419, 407–411.
156. Dillon, N. and Festenstein, R. (2002) Trends Genet. 18, 252–258.
157. Nakayama, J., Rice, J.C., Strahl, B.D., Allis, C.D., and Grewal, S.I. (2001) Science 292, 110–113.
158. Noma, K., Allis, C.D., and Grewal, S.I. (2001) Science 293, 1150–1155.
159. Bannister, A.J., Zegerman, P., Partridge, J.F., Miska, E.A., Thomas, J.O., Allshire, R.C., and Kouzarides, T. (2001) Nature 410, 120–124.
160. Lachner, M., O'Carroll, D., Rea, S., Mechtler, K., and Jenuwein, T. (2001) Nature 410, 116–120.

161. Jacobs, S.A., Taverna, S.D., Zhang, Y., Briggs, S.D., Li, J., Eissenberg, J.C., Allis, C.D., and Khorasanizadeh, S. (2001) EMBO J. 20, 5232–5241.
162. Taverna, S.D., Coyne, R.S., and Allis, C.D. (2002) Cell 110, 701–711.
163. Sewalt, R.G., Lachner, M., Vargas, M., Hamer, K.M., den Blaauwen, J.L., Hendrix, T., Melcher, M., Schweizer, D., Jenuwein, T., and Otte, A.P. (2002) Mol. Cell Biol. 22, 5539–5553.
164. Li, J., Lin, Q., Yoon, H.G., Huang, Z.Q., Strahl, B.D., Allis, C.D., and Wong, J. (2002) Mol. Cell Biol. 22, 5688–5697.
165. Boggs, B.A., Cheung, P., Heard, E., Spector, D.L., Chinault, A.C., and Allis, C.D. (2002) Nat. Genet. 30, 73–76.
166. Peters, A.H., Mermoud, J.E., O'Carroll, D., Pagani, M., Schweizer, D., Brockdorff, N., and Jenuwein, T. (2002) Nat. Genet. 30, 77–80.
167. Heard, E., Rougeulle, C., Arnaud, D., Avner, P., Allis, C.D., and Spector, D.L. (2001) Cell 107, 727–738.
168. Mermoud, J.E., Popova, B., Peters, A.H., Jenuwein, T., and Brockdorff, N. (2002) Curr. Biol. 12, 247–251.
169. Reuben, M. and Lin, R. (2002) Dev. Biol. 245, 71–82.
170. Nielsen, S.J., Schneider, R., Bauer, U.M., Bannister, A.J., Morrison, A., O'Carroll, D., Firestein, R., Cleary, M., Jenuwein, T., Herrera, R.E., and Kouzarides, T. (2001) Nature 412, 561–565.
171. Vandel, L., Nicolas, E., Vaute, O., Ferreira, R., Ait-Si-Ali, S., and Trouche, D. (2001) Mol. Cell Biol. 21, 6484–6494.
172. Luo, R.X., Postigo, A.A., and Dean, D.C. (1998) Cell 92, 463–473.
173. Saccani, S. and Natoli, G. (2002) Genes Dev. 16, 2219–2224.
174. Chinenov, Y. (2002) Trends Biochem.Sci. 27, 115–117.
175. Hendzel, M.J. and Davie, J.R. (1990) Biochem. J. 271, 67–73.
176. Ahmad, K. and Henikoff, S. (2002) Mol. Cell 9, 1191–1200.
177. Johnson, L.M., Cao, X., and Jacobsen, S.E. (2002) Curr. Biol. 12, 1360–1367.
178. Goll, M.G. and Bestor, T.H. (2002) Genes Dev. 16, 1739–1742.
179. Bannister, A.J., Schneider, R., and Kouzarides, T. (2002) Cell 109, 801–806.
180. Fang, J., Feng, Q., Ketel, C.S., Wang, H., Cao, R., Xia, L., Erdjument-Bromage, H., Tempst, P., Simon, J.A., and Zhang, Y. (2002) Curr. Biol. 12, 1086–1099.
181. Rice, J.C., Nishioka, K., Sarma, K., Steward, R., Reinberg, D., and Allis, C.D. (2002) Genes Dev. 16, 2225–2230.
182. Lee, H.W., Paik, W.K., and Borun, T.W. (1973) J. Biol. Chem. 248, 4194–4199.
183. Greenaway, P.J. and Levine, D. (1974) Biochim. Biophys. Acta 350, 374–382.
184. Wallwork, J.C., Quick, D.P., and Duerre, J.A. (1977) J. Biol. Chem. 252, 5977–5980.
185. O'Carroll, D., Scherthan, H., Peters, A.H., Opravil, S., Haynes, A.R., Laible, G., Rea, S., Schmid, M., Lebersorger, A., Jerratsch, M., Sattler, L., Mattei, M.G., Denny, P., Brown, S.D., Schweizer, D., and Jenuwein, T. (2000) Mol. Cell Biol. 20, 9423–9433.
186. Rea, S., Eisenhaber, F., O'Carroll, D., Strahl, B.D., Sun, Z.W., Schmid, M., Opravil, S., Mechtler, K., Ponting, C.P., Allis, C.D., and Jenuwein, T. (2000) Nature 406, 593–599.
187. Min, J., Zhang, X., Cheng, X., Grewal, S.I., and Xu, R.M. (2002) Nat. Struct. Biol. 9, 828–832.
188. Zhang, X., Tamaru, H., Khan, S.I., Horton, J.R., Keefe, L.J., Selker, E.U., and Cheng, X. (2002) Cell 111, 117–127.
189. Wilson, J.R., Jing, C., Walker, P.A., Martin, S.R., Howell, S.A., Blackburn, G.M., Gamblin, S.J., and Xiao, B. (2002) Cell 111, 105–115.
190. Trievel, R.C., Beach, B.M., Dirk, L.M., Houtz, R.L., and Hurley, J.H. (2002) Cell 111, 91–103.
191. Tamaru, H. and Selker, E.U. (2001) Nature 414, 277–283.
192. Yang, L., Xia, L., Wu, D.Y., Wang, H., Chansky, H.A., Schubach, W.H., Hickstein, D.D., and Zhang, Y. (2002) Oncogene 21, 148–152.
193. Schultz, D.C., Ayyanathan, K., Negorev, D., Maul, G.G., and Rauscher, F.J., III (2002) Genes Dev. 16, 919–932.
194. Grewal, S.I. and Elgin, S.C. (2002) Curr. Opin. Genet. Dev. 12, 178–187.

195. Peters, A.H., O'Carroll, D., Scherthan, H., Mechtler, K., Sauer, S., Schofer, C., Weipoltshammer, K., Pagani, M., Lachner, M., Kohlmaier, A., Opravil, S., Doyle, M., Sibilia, M., and Jenuwein, T. (2001) Cell 107, 323–337.
196. Tachibana, M., Sugimoto, K., Nozaki, M., Ueda, J., Ohta, T., Ohki, M., Fukuda, M., Takeda, N., Niida, H., Kato, H., and Shinkai, Y. (2002) Genes Dev. 16, 1779–1791.
197. Tachibana, M., Sugimoto, K., Fukushima, T., and Shinkai, Y. (2001) J. Biol. Chem. 276, 25309–25317.
198. Nishioka, K., Chuikov, S., Sarma, K., Erdjument-Bromage, H., Allis, C.D., Tempst, P., and Reinberg, D. (2002) Genes Dev. 16, 479–489.
199. Noma, K.I. and Grewal, S.I. (2002) Proc. Natl. Acad. Sci. USA.
200. Bryk, M., Briggs, S.D., Strahl, B.D., Curcio, M.J., Allis, C.D., and Winston, F. (2002) Curr. Biol. 12, 165–170.
201. Krogan, N.J., Dover, J., Khorrami, S., Greenblatt, J.F., Schneider, J., Johnston, M., and Shilatifard, A. (2002) J. Biol. Chem. 277, 10753–10755.
202. Milne, T.A., Briggs, S.D., Brock, H.W., Martin, M.E., Gibbs, D., Allis, C.D., and Hess, J.L. (2002) Mol. Cell 10, 1107–1117.
203. Nakamura, T., Mori, T., Tada, S., Krajewski, W., Rozovskaia, T., Wassell, R., Dubois, G., Mazo, A., Croce, C.M., and Canaani, E. (2002) Mol. Cell 10, 1119–1128.
204. Muller, J., Hart, C.M., Francis, N.J., Vargas, M.L., Sengupta, A., Wild, B., Miller, E.L., O'Connor, M.B., Kingston, R.E., and Simon, J.A. (2002) Cell 111, 197–208.
205. Strahl, B.D., Grant, P.A., Briggs, S.D., Sun, Z.W., Bone, J.R., Caldwell, J.A., Mollah, S., Cook, R.G., Shabanowitz, J., Hunt, D.F., and Allis, C.D. (2002) Mol. Cell Biol. 22, 1298–1306.
206. Li, J., Moazed, D., and Gygi, S.P. (2002) J. Biol. Chem.
207. Ng, H.H., Feng, Q., Wang, H., Erdjument-Bromage, H., Tempst, P., Zhang, Y., and Struhl, K. (2002) Genes Dev. 16, 1518–1527.
208. Lacoste, N., Utley, R.T., Hunter, J.M., Poirier, G.G., and Cote, J. (2002) J. Biol. Chem. 277, 30421–30424.
209. van Leeuwen, F., Gafken, P.R., and Gottschling, D.E. (2002) Cell 109, 745–756.
210. Koh, S.S., Li, H., Lee, Y.H., Widelitz, R.B., Chuong, C.M., and Stallcup, M.R. (2002) J. Biol. Chem. 277, 26031–26035.
211. Ma, H., Baumann, C.T., Li, H., Strahl, B.D., Rice, R., Jelinek, M.A., Aswad, D.W., Allis, C.D., Hager, G.L., and Stallcup, M.R. (2001) Curr. Biol. 11, 1981–1985.
212. Bauer, U.M., Daujat, S., Nielsen, S.J., Nightingale, K., and Kouzarides, T. (2002) EMBO Rep. 3, 39–44.
213. Schurter, B.T., Koh, S.S., Chen, D., Bunick, G.J., Harp, J.M., Hanson, B.L., Henschen-Edman, A., Mackay, D.R., Stallcup, M.R., and Aswad, D.W. (2001) Biochemistry 40, 5747–5756.
214. Daujat, S., Bauer, U.M., Shah, V., Turner, B., Berger, S., and Kouzarides, T. (2002) Curr. Biol. 12, 2090–2097.
215. Wang, H., Huang, Z.Q., Xia, L., Feng, Q., Erdjument-Bromage, H., Strahl, B.D., Briggs, S.D., Allis, C.D., Wong, J., Tempst, P., and Zhang, Y. (2001) Science 293, 853–857.
216. Beisel, C., Imhof, A., Greene, J., Kremmer, E., and Sauer, F. (2002) Nature 419, 857–862.
217. Waterborg, J.H. (1993) J. Biol. Chem. 268, 4912–4917.
218. Reneker, J. and Brotherton, T.W. (1991) Biochemistry 30, 8402–8407.
219. Hendzel, M.J. and Davie, J.R. (1991) Biochem. J. 273, 753–758.
220. Hendzel, M.J. and Davie, J.R. (1992) Biochem. Biophys. Res. Commun. 185, 414–419.
221. Hall, I.M., Shankaranarayana, G.D., Noma, K., Ayoub, N., Cohen, A., and Grewal, S.I. (2002) Science 297, 2232–2237.
222. Czermin, B., Schotta, G., Hulsmann, B.B., Brehm, A., Becker, P.B., Reuter, G., and Imhof, A. (2001) EMBO Rep. 2, 915–919.
223. Vandel, L. and Trouche, D. (2001) EMBO Rep. 2, 21–26.
224. Jackson, J.P., Lindroth, A.M., Cao, X., and Jacobsen, S.E. (2002) Nature 416, 556–560.
225. Paro, R. (2000) Nature 406, 579–580.

226. Fuks, F., Hurd, P.J., Wolf, D., Nan, X., Bird, A.P., and Kouzarides, T. (2003) J. Biol. Chem.
227. Orrick, L.R., Olson, M.O., and Busch, H. (1973) Proc. Natl. Acad. Sci. USA 70, 1316–1320.
228. Chen, H.Y., Sun, J.-M., Zhang, Y., Davie, J.R., and Meistrich, M.L. (1998) J. Biol. Chem. 273, 13165–13169.
229. Swerdlow, P.S., Schuster, T., and Finley, D. (1990) Mol. Cell Biol. 10, 4905–4911.
230. Robzyk, K., Recht, J., and Osley, M.A. (2000) Science 287, 501–504.
231. Baarends, W.M., Hoogerbrugge, J.W., Roest, H.P., Ooms, M., Vreeburg, J., Hoeijmakers, J.H., and Grootegoed, J.A. (1999) Dev. Biol. 207, 322–333.
232. Goldknopf, I.L. and Busch, H. (1977) Proc. Natl. Acad. Sci. USA 74, 864–868.
233. Thorne, A.W., Sautiere, P., Briand, G., and Crane Robinson, C. (1987) EMBO J. 6, 1005–1010.
234. Nickel, B.E. and Davie, J.R. (1989) Biochemistry 28, 964–968.
235. Nickel, B.E., Allis, C.D., and Davie, J.R. (1989) Biochemistry 28, 958–963.
236. Chau, V., Tobias, J.W., Bachmair, A., Marriott, D., Ecker, D.J., Gonda, D.K., and Varshavsky, A. (1989) Science 243, 1576–1583.
237. Van Nocker, S. and Vierstra, R.D. (1993) J. Biol. Chem. 268, 24766–24773.
238. Wu, R.S., Kohn, K.W., and Bonner, W.M. (1981) J. Biol. Chem. 256, 5916–5920.
239. Seale, R.L. (1981) Nucleic Acids Res. 9, 3151–3158.
240. Seufert, W. and Jentsch, S. (1992) EMBO J. 11, 3077–3080.
241. Martinez, C.K. and Monaco, J.J. (1993) Mol. Immunol. 30, 1177–1183.
242. Glickman, M.H. and Ciechanover, A. (2002) Physiol Rev. 82, 373–428.
243. Hershko, A. (1988) J. Biol. Chem. 263, 15237–15240.
244. Cook, J.C. and Chock, P.B. (1991) Biochem. Biophys. Res. Commun. 174, 564–571.
245. Schwartz, A.L., Trausch, J.S., Ciechanover, A., Slot, J.W., and Geuze, H. (1992) Proc. Natl. Acad. Sci. USA 89, 5542–5546.
246. Grenfell, S.J., Trausch-Azar, J.S., Handley-Gearhart, P.M., Ciechanover, A., and Schwartz, A.L. (1994) Biochem. J. 300, 701–708.
247. Haas, A.L., Bright, P.M., and Jackson, V.E. (1988) J. Biol. Chem. 263, 13268–13275.
248. Jentsch, S., Seufert, W., Sommer, T., and Reins, H.A. (1990) Trends. Biochem. Sci. 15, 195–198.
249. Haas, A.L., Reback, P.B., and Chau, V. (1991) J. Biol. Chem. 266, 5104–5112.
250. Pham, A.D. and Sauer, F. (2000) Science 289, 2357–2360.
251. Matsui, S.-I., Seon, B.K., and Sandberg, A.V. (1979) Proc. Natl. Acad. Sci. USA 76, 6386–6390.
252. Mueller, R.D., Yasuda, H., Hatch, C.L., Bonner, W.M., and Bradbury, E.M. (1985) J. Biol. Chem. 260, 5147–5153.
253. Raboy, B., Parag, H.A., and Kulka, R.G. (1986) EMBO J. 5, 863–869.
254. Mori, M., Eki, T., Takahashi-Kudo, M., Hanaoka, F., Ui, M., and Enomoto, T. (1993) J. Biol. Chem. 268, 16803–16809.
255. Matsui, S.-I., Sandberg, A.A., Negoro, S., Seon, B.K., and Goldstein, G. (1982) Proc. Natl. Acad. Sci. USA 79, 1535–1539.
256. Andersen, M.W., Goldknopf, I.L., and Busch, H. (1981) FEBS Lett. 132, 210–214.
257. Andersen, M.W., Ballal, N.R., Goldknopf, I.L., and Busch, H. (1981) Biochemistry 20, 1100–1104.
258. Bond, U., Agell, N., Haas, A.L., Redman, K., and Schlesinger, M.J. (1988) J. Biol. Chem. 263, 2384–2388.
259. Parag, H.A., Raboy, B., and Kulka, R.G. (1987) EMBO J. 6, 55–61.
260. Mimnaugh, E.G., Chen, H.Y., Davie, J.R., Celis, J.E., and Neckers, L. (1997) Biochemistry 36, 14418–14429.
261. Yunmbam, M.K., Li, Q.Q., Mimnaugh, E.G., Kayastha, G.L., Yu, J.J., Jones, L.N., Neckers, L., and Reed, E. (2001) Int. J. Oncol. 19, 741–748.
262. Mimnaugh, E.G., Kayastha, G., McGovern, N.B., Hwang, S.G., Marcu, M.G., Trepel, J., Cai, S.Y., Marchesi, V.T., and Neckers, L. (2001) Cell Death. Differ. 8, 1182–1196.
263. Marushige, Y. and Marushige, K. (1995) Anticancer Res. 15, 267–272.
264. Vassilev, A.P., Rasmussen, H.H., Christensen, E.I., Nielsen, S., and Celis, J.E. (1995) J. Cell Sci. 108, 1205–1215.

265. Davie, J.R., Delcuve, G.P., Nickel, B.E., Moyer, R., and Bailey, G. (1987) Cancer Res. 47, 5407–5410.
266. Nickel, B.E., Roth, S.Y., Cook, R.G., Allis, C.D., and Davie, J.R. (1987) Biochemistry 26, 4417–4421.
267. Agell, N., Chiva, M., and Mezquita, C. (1983) FEBS Lett. 155, 209–212.
268. Davie, J.R. and Murphy, L.C. (1990) Biochemistry 29, 4752–4757.
269. Davie, J.R. and Murphy, L.C. (1994) Biochem. Biophys. Res. Commun. 203, 344–350.
270. Jason, L.J., Moore, S.C., Lewis, J.D., Lindsey, G., and Ausio, J. (2002) BioEssays 24, 166–174.
271. Davie, J.R., Lin, R., and Allis, C.D. (1991) Biochem. Cell Biol. 69, 66–71.
272. Hatch, C.L., Bonner, W.M., and Moudrianakis, E.N. (1983) Science 221, 468–470.
273. Van Holde, K.E., Lohr, D.E., and Robert, C. (1992) J. Biol. Chem. 267, 2837–2840.
274. Morse, R.H. (1992) Trends. Biochem. Sci. 17, 23–26.
275. Kireeva, M.L., Walter, W., Tchernajenko, V., Bondarenko, V., Kashlev, M., and Studitsky, V.M. (2002) Mol. Cell 9, 541–552.
276. Clark, D.J. and Felsenfeld, G. (1992) Cell 71, 11–22.
277. Studitsky, V.M., Clark, D.J., and Felsenfeld, G. (1994) Cell 76, 371–382.
278. Jackson, V. (1990) Biochemistry 29, 719–731.
279. Davies, N. and Lindsey, G.G. (1994) Biochim. Biophys. Acta 1218, 187–193.
280. Jason, L.J., Moore, S.C., Ausio, J., and Lindsey, G. (2001) J. Biol. Chem. 276, 14597–14601.
281. Barsoum, J. and Varshavsky, A. (1985) J. Biol. Chem. 260, 7688–7697.
282. Levinger, L. and Varshavsky, A. (1982) Cell 28, 375–385.
283. Huang, S.-Y., Barnard, M.B., Xu, M., Matsui, S.-I., Rose, S.M., and Garrard, W.T. (1986) Proc. Natl. Acad. Sci. USA 83, 3738–3742.
284. Dawson, B.A., Herman, T., Haas, A.L., and Lough, J. (1991) J. Cell Biochem. 46, 166–173.
285. Jasinskiene, N., Jasinskas, A., and Langmore, J.P. (1995) Dev. Genet. 16, 278–290.
286. Sun, Z.W. and Allis, C.D. (2002) Nature 418, 104–108.
287. Turner, S.D., Ricci, A.R., Petropoulos, H., Genereaux, J., Skerjanc, I.S., and Brandl, C.J. (2002) Mol. Cell Biol. 22, 4011–4019.
288. Briggs, S.D., Xiao, T., Sun, Z.W., Caldwell, J.A., Shabanowitz, J., Hunt, D.F., Allis, C.D., and Strahl, B.D. (2002) Nature 418, 498.
289. Ng, H.H., Xu, R.M., Zhang, Y., and Struhl, K. (2002) J. Biol. Chem. 277, 34655–34657.
290. Nishizuka, Y., Ueda, K., Honjo, T., and Hayaishi, O. (1968) J. Biol. Chem. 243, 3765–3767.
291. Boulikas, T. (1989) Proc. Natl. Acad. Sci. USA 86, 3499–3503.
292. Boulikas, T. (1990) J. Biol. Chem. 265, 14638–14647.
293. Boulikas, T. (1993) Toxicol. Lett. 67, 129–150.
294. Boulikas, T., Bastin, B., Boulikas, P., and Dupuis, G. (1990) Exp. Cell Res. 187, 77–84.
295. Wong, M. and Smulson, M. (1984) Biochemistry 23, 3726–3730.
296. Smith, J.A. and Stocken, L.A. (1973) Biochem. Biophys. Res. Commun. 54, 297–300.
297. Ueda, K., Omachi, A., Kawaichi, M., and Hayaishi, O. (1975) Proc. Natl. Acad. Sci. USA 72, 205–209.
298. Riquelme, P.T., Burzio, L.O., and Koide, S.S. (1979) J. Biol. Chem. 254, 3018–3028.
299. Ogata, N., Ueda, K., Kagamiyama, H., and Hayaishi, O. (1980) J. Biol. Chem. 255, 7616–7620.
300. Wong, M., Malik, N., and Smulson, M. (1982) Eur. J. Biochem. 128, 209–213.
301. Poirier, G.G., de Murcia, G., Jongstra-Bilen, J., Niedergang, C., and Mandel, P. (1982) Proc. Natl. Acad. Sci. USA 79, 3423–3427.
302. de Murcia, G., Huletsky, A., and Poirier, G.G. (1988) Biochem. Cell Biol. 66, 626–635.
303. de Murcia, G., Huletsky, A., Lamarre, D., Gaudreau, A., Pouyet, J., Daune, M., and Poirier, G.G. (1986) J. Biol. Chem. 261, 7011–7017.
304. Huletsky, A., de Murcia, G., Muller, S., Hengartner, M., Menard, L., Lamarre, D., and Poirier, G.G. (1989) J. Biol. Chem. 264, 8878–8886.
305. Boulikas, T. (1988) EMBO J. 7, 57–67.
306. Boulikas, T. (1990) J. Biol. Chem. 265, 14627–14631.
307. Golderer, G., Loidl, P., and Grobner, P. (1991) Eur. J. Cell Biol. 55, 183–185.

308. Golderer, G. and Grobner, P. (1991) Biochem. J. 277, 607–610.
309. Burzio, L.O., Riquelme, P.T., and Koide, S.S. (1979) J. Biol. Chem. 254, 3029–3037.
310. Ogata, N., Ueda, K., and Hayaishi, O. (1980) J. Biol. Chem. 255, 7610–7615.
311. Panzeter, P.L., Realini, C.A., and Althaus, F.R. (1992) Biochemistry 31, 1379–1385.
312. Fakan, S., Leduc, Y., Lamarre, D., Brunet, G., and Poirier, G.G. (1988) Exp. Cell Res. 179, 517–526.
313. Realini, C.A. and Althaus, F.R. (1992) J. Biol. Chem. 267, 18858–18865.
314. Cervoni, N. and Szyf, M. (2001) J. Biol. Chem. 276, 40778–40787.
315. Cervoni, N., Detich, N., Seo, S.B., Chakravarti, D., and Szyf, M. (2002) J. Biol. Chem. 277, 25026–25031.
316. Zardo, G., D'Erme, M., Reale, A., Strom, R., Perilli, M., and Caiafa, P. (1997) Biochemistry 36, 7937–7943.
317. Zardo, G., Marenzi, S., and Caiafa, P. (1998) Biol. Chem. 379, 647–654.
318. Zardo, G. and Caiafa, P. (1998) J. Biol. Chem. 273, 16517–16520.

J. Zlatanova and S.H. Leuba (Eds.) *Chromatin Structure and Dynamics: State-of-the-Art*

DOI: 10.1016/S0167-7306(03)39010-6

CHAPTER 10

The role of histone variability in chromatin stability and folding

Juan Ausió[1] and D. Wade Abbott

Department of Biochemistry and Microbiology, University of Victoria, Victoria, P.O. Box 3055, Petch Building, 220 Victoria, British Columbia, Canada V8W 3P6

[1]*Tel.: +250-721-8863; Fax: +250-721-8855; E-mail: jausio@uvic.ca*

1. Introduction

In the eukaryotic cell, DNA exists as a nucleoprotein complex, which is known as chromatin. The major protein components of this assembly are histones, which can be grouped into two major categories: "core histones" and "linker histones". Core histones (histones H2A, H2B, H3, and H4) are arranged into a heterotypic globular protein octamer $[(H3–H4)_2 \cdot 2(H2A–H2B)]$ [1], which serves as a structural core around which 146–180 bp of DNA are wrapped in approximately two left handed superhelical turns. In the chromatin fiber, the nucleosome [2] subunits resulting from such complexes are connected by variable stretches of linker DNA. "Linker histones" bind to these DNA connecting regions and together with the core histone "tails" [3] play a critical role in the folding of the chromatin fiber.

From the early days of chromatin research it was initially assumed that histones had a mere passive structural role [4], which would participate in the packing of DNA through "tight" protein–DNA interactions. The discovery of the nucleosome (see Ref. [5] for a review) did not do much to dispel such misconception, and it was not until much later that the concept of chromatin as a dynamic modulator of gene activity started to emerge [6–8]. Indeed, chromatin provides the substrate upon which some of most important biological functions of the cell take place, such as DNA replication, transcription, recombination, and repair. These processes involve quick, dynamic changes of DNA conformation and stability, which are most likely mediated by changes in chromatin folding and nucleosome stability.

In recent years it has become increasingly apparent that the compositional heterogeneity of the chromatin fiber through histone variants, histone post-translational modifications and chemical modifications of DNA (methylation) [9] play an important role in all these processes. The different sources of compositional heterogeneity are described in Sections 2–4 of this review.

The combinatorial effects of such chemical variability of chromatin, can be used as an informational tool or to directly modulate the physical and thermodynamic constraints of this nucleoprotein assembly. In the first instance a chemical "signature" can be used as a targeting mechanism to allow the recognition of regulatory regions by *trans*-acting factors or by ATP-dependent (i.e., SWI/SNF) or independent chromatin remodeling complexes (see Refs. [10,11] for a review). Alternatively or simultaneously, chemical modifications can have structural implications for both the stability of the nucleosome and the folding of the chromatin fiber in ways that are described in Sections 5 and 6 of this review.

2. Brief introduction to histone variants

Core histone variants are present in two major classes, homomorphous and heteromorphous [12]. Homomorphous variants are subtypes that differ only by a few residues (Fig. 1). Their chemical similarity results in the co-migration of isotypes during conventional SDS-polyacrylamide gel electrophoresis (PAGE) [13], and requires acetic acid-urea gels in the presence of the non-ionic detergent Triton X-100 to resolve distinct banding patterns [14,15]. Although the similar primary structures of these proteins do not likely confer notable alterations to nucleosome structure or stability; however, some evidence suggests that these variants are developmentally regulated [16].

Heteromorphous variants differ from homomorphous histones in the sense that they display significant alterations in amino acid sequence composition and length that allows them to be resolved by SDS-PAGE. Also in contrast to the homomorphous variants, their genes are usually replication independent, non-clustered, frequently contain introns and their mRNAs are often polyadenylated. In recent years, there has been rekindled interest in heteromorphous histone variants (referred to below as simply histone variants) as their structural and functional implications for the modulation of chromatin architecture are beginning to be unraveled [17]. The next section will highlight what it is currently known about the functional roles of several H2A and centromeric core histone variants, as well as those of H1 heterogeneity, with special emphasis on their structural implications for chromatin.

2.1. Histone H2AX

H2AX (see Fig. 2A) is a unique H2A isoform in the sense that it can be phosphorylated at its C-terminal end at the highly conserved Ser129 in yeast [18] and Ser139 in mammals [19] (the phosphorylated isotype is referred to as γ-H2AX). This reversible event has been linked to the repair of DNA double strand breakage following DNA injury and physiologically regulated cleavage events [17,20]. The generation of a phosphoserine specific γ-H2AX antibody [21] has enabled the *in situ* examination of many cellular events, which involve this modification. To date these events include double strand break (DSB) repair [19,22], meiotic synapsis [23], apoptosis [24,25] and the immunologically relevant class switching [26] and V(D)J recombination [27].

Phosphorylation of H2AX appears to be the product of three regulated signaling pathways involving the kinases DNA-PK, ATM (Ataxia Telengiectasia Mutated), and ATR (AT-Rad3 related). These enzymes display both redundancy and

HOMOMORPHOUS CORE HISTONES

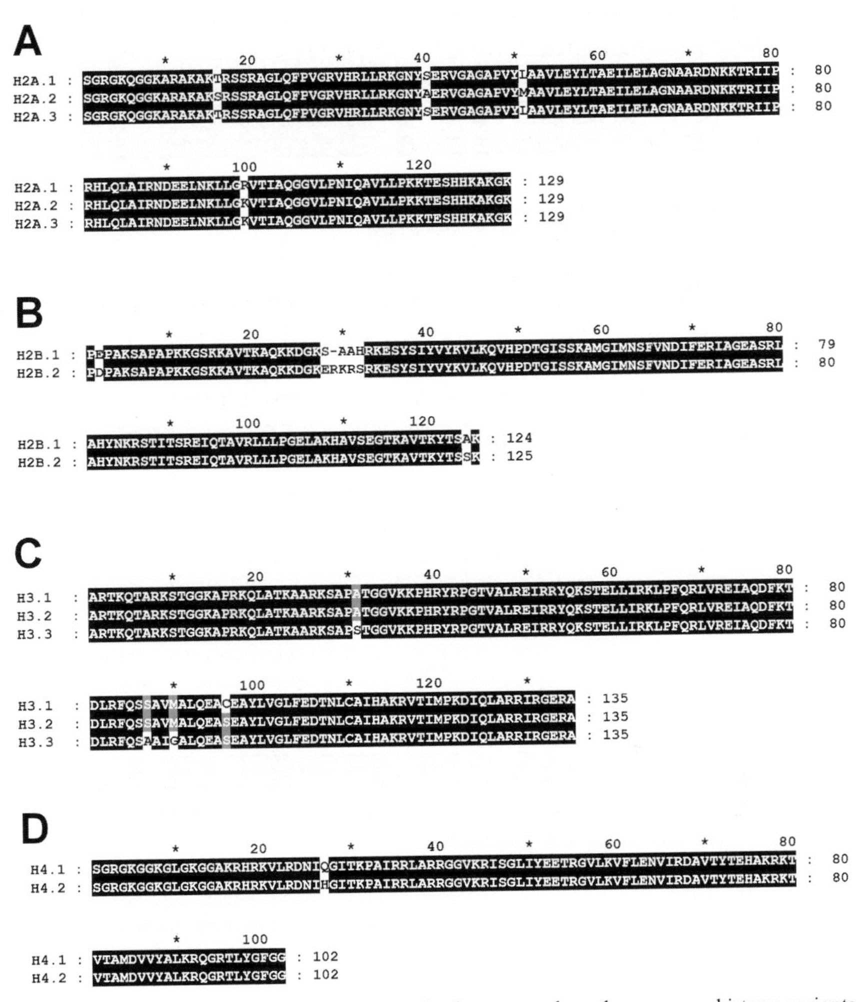

Fig. 1. Amino acid sequence of several representative homomorphous human core histone variants [12]: **A**. Histone H2A; **B**. Histone H2B; **C**. Histone H3; and **D**. histone H4 variants. The amino acid residues are shaded with intensity proportional to the extent of identity shared among the compared sequences.

specificity in response to definitive stimuli [17]. For example, ATR is exclusively induced following DSB formation at sites of replication arrest [28].

Although the effects of γ-H2AX foci are still poorly understood, two theories have been proposed that may not be mutually exclusive [17,18]. The first model suggests that the post-translational modification may impart structural transitions to DSB domains, and the second implicates the histone variant as a signaling intermediate in the repair pathway.

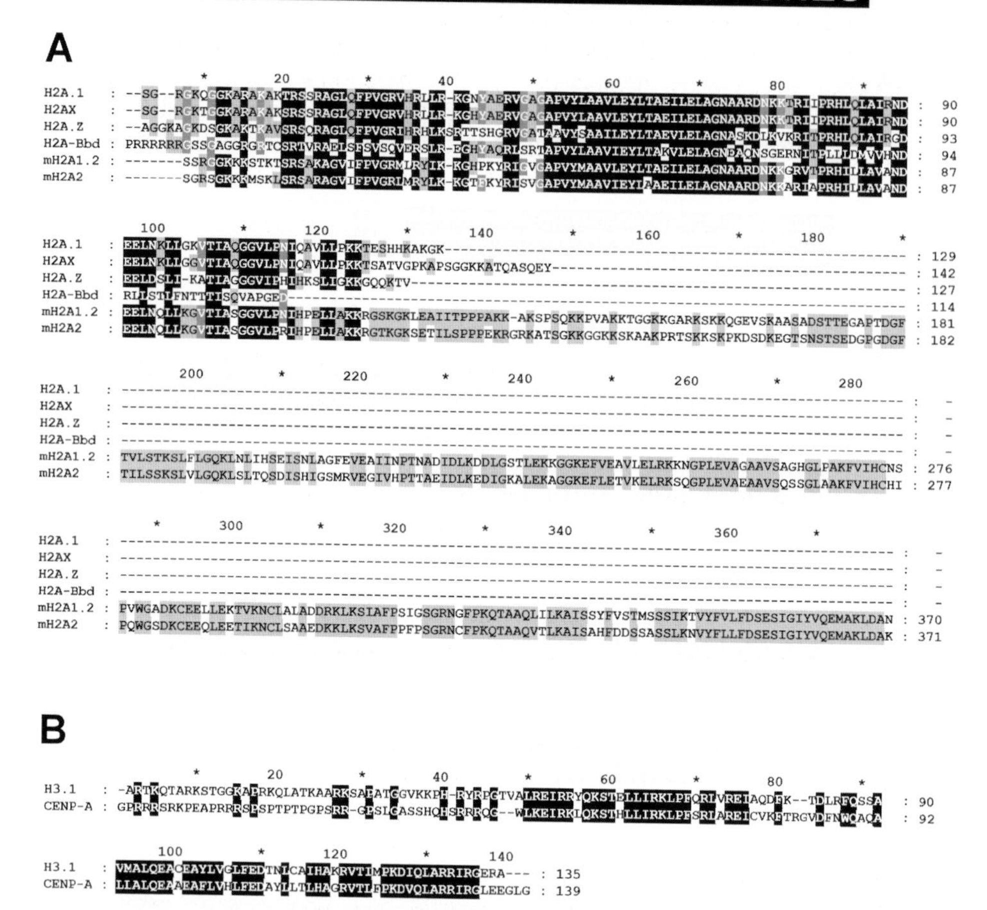

Fig. 2. Amino acid sequence of several representative heteromorphous human core histone variants [12]: **A**. Histone H2A variants in comparison to histone H2A.1; **B**. CENP-A in comparison to H3.1. The amino acid residues are shaded with intensity proportional to the extent of identity shared among the compared sequences.

The addition of a phosphate group, with its two negative charges, at the entry and exit sites of the DNA in the nucleosome most likely imparts a noticeable electrostatic repulsion between the octamer and the nucleosomal DNA [29]. A disruption to the nucleoprotein interface is a prerequisite to accessibility of the damaged substrate, as the presence of nucleosomes has been shown to repress DNA repair [30,31]. The occurrence of such structural effects has been demonstrated *in vivo*. In yeast, a general correlation was observed between the ectopic expression of a Ser129Glu mutant and nuclease hypersensitivity [18]. This mutated form of H2AX serves to chemically mimic the charge state of γ-H2AX. Therefore, the resulting accessibility of the nuclease to the chromatin substrate suggests that phosphorylating H2AX may also disrupt histone–DNA architecture and lead to chromatin decondensation and instability [17,18].

The second role of γ-H2AX foci may be informational in nature. Using immunochemistry, H2AX phosphorylation has been shown to precede the localization of repair proteins [22] and meiotic synapsis factors [23,32]. In this scenario, the modified histone may propagate signals during repair cascades [21] and be instrumental in recruiting DSB repair complexes to sites of DNA breakage [33].

2.2. *Histone H2A.Z*

H2A.Z (see Fig. 2A) has received considerable attention in the recent literature. The interest in this protein stems from the fact it is the only H2A variant that is essential for development [34,35] and viability in *Drosophila* [36,37]. Interestingly, physiologic roles attributed to H2A.Z include both transcriptional activation [38,39] and silencing [40]. With the recent characterization of the crystal structure of H2A.Z containing nucleosomes [41] important progress has been made in the understanding of the possible implications of this variant for chromatin structure. H2A.Z amino acids that diverge from the major H2A sequence map to prominent portions of the nucleosome. Significantly, a Gln104 to Gly106 transition appears to destabilize the interface between the H2A.Z–H2B dimer and $(H3–H4)_2$ tetramer, an effect that has been substantiated by physical studies [42]. The lability of this particle may confer specialized properties to chromatin that is poised for gene expression [17,38,43], and undergoing constitutive rRNA synthesis [40,44].

H2A.Z also has structural implications for the higher order structure of chromatin. By performing cassette-swapping experiments, the essential novelty of the variant has been localized to the C-terminal tail [37,39]. Significantly, this portion of H2A is disposed at the surface of the nucleosome [41] and may affect the binding of linker histones and chromatin remodeling complexes [41,45]. Furthermore, the presence of a dihistidine metal ion coordination pocket and an extended acidic patch may play a role in facilitating contacts between such *trans*-acting protein factors and H2A.Z containing nucleosomes [41].

2.3. *MacroH2A*

MacroH2A1 and 2 (mH2A) display N-terminal homology (65%) to the complete sequence of H2A, plus a large non-histone C-terminal fusion, which comprises two-thirds of its molecular mass (see Fig. 2A). Although the role of the novel COOH terminus remains to be elucidated, the conserved histone portion has been described to interact with H2B [46]; and recombinant forms of the full-length protein have been successfully reconstituted into mononucleosomes [47], and shown to repress *in vitro* transcription [48]. Immunodetection studies have identified that mH2A is enriched in the inactivated X-chromosome of mammalian females [49], and the testes of adult mammalian males [50], which suggests that the variant may participate in the formation of highly specialized chromatin domains involved in transcriptional silencing. A recent study has described the existence of a homologous section of amino acids between the non-histone region of mH2A and segments of *Sinbus* and *Rubella* viral proteins [51]. In *Sinbus*, this domain has been shown to interact with RNA [52]. Accordingly, the fusion of a similar RNA targeting polypeptide to the COOH-terminus of H2A may bridge nucleosome interaction with regulatory RNA transcripts [51].

Transcriptional silencing in females of vertebrates is a mechanism utilized by the cell to conserve energy and restrict gene product dosages to equivalent levels as the male. Repressing extensive portions of the X-chromosome, which exists as the Barr body during interphase, requires a complement of regulatory events involving both the post-translational modification of histones and mobilization of *trans*-acting factors. It has been observed that delayed replication, methylation of CpG islands, hypoacetylation of core histone H3–H4 tetramers, mH2A localization, the hybridization of *Xist* (X inactivating specific transcript) RNA, and methylation of H3 Lys9 are involved in the inactivation process [53,54]. However, it has yet to be confirmed if the silencing machinery operates in coordinated or exclusive pathways, and if the systems vary between cell types. One intriguing possibility is that mH2A may interact with *Xist* RNA through its RNA coupling domain that has been defined in viral proteins [51]. Csankovszki *et al.* [55] observed that the generation of mH2A chromatin assemblies at the Xi is subject to *Xist* activity; however, inactivation will persist in the absence of proper mH2A targeting. In mammalian spermatogenesis, a similar process may be involved in the inactivation of the X-chromosome during meiosis [56]. In this regard, the deposition of mH2A was recently characterized by immunolabeling of the XY compartment [57], which suggests again a possible link between the heterochromatinization of the XY body by *Xist* scaffolding to macroH2A [58].

2.4. *H2A-Bbd*

H2A-Bbd (Barr body deficient) is the most recently identified H2A variant. This subtype displays the largest degree of primary structure divergence from H2A (48%), with the greatest regions of similarity mapping to the histone fold domains [59] (see Fig. 2A). Significantly, the isoform consists of a C-terminal tail truncation, which eliminates the ubiquitination site, and an arginine rich amino terminus that lacks an acetylation substrate [29,59] (see Fig. 2A). The fact that these sites are not available for post-translational modifications suggests that the variant may confer intrinsic regulatory properties to novel nucleosomal assemblies. For example, the integrity of the nucleosome consisting of this histone variant may be compromised by the C-terminal truncation of H2A-Bbd [17], as an earlier study defined the prominent role of the H2A C-terminal tail in nucleosome stability [60].

H2A-Bbd is found only in the active areas of the nucleus and displays an exclusive deposition pattern with mH2A [61]. Furthermore, fluorescence immunochemistry has shown that this H2A variant co-localized with acetylated H4 [59]. Although the structural properties of H2A-Bbd containing nucleosomes remain to be characterized, preliminary observations suggest that the particle may be specialized for activating transcription.

2.5. Centromeric variants

The centromeres of chromatin define a distinct region involved in the assembly of the kinetochore and microtubule machinery responsible for the polarization of chromosomes during cell division [62,63]. A specialized protein family, responsible for autoimmune diseases such as CREST (Calcinosis, Raynaud phenomenon, Esophageal dysmotility, Sclerodactyly, Telangiectasiae) syndrome, is linked to the centromeric structure, nucleation and maturation of the kinetochore plate and dynamics of microtubule motorization [64]. Collectively, these proteins are referred to as the CENP (Centromere Protein) family. Further nomenclature is based upon the centromere-specific nature of the polypeptide. CENP-A is a histone variant that provides the causal link between the formation of centromeres and the organization of regional DNA into nucleosomes [65,66]. CENP-B is a modulating protein with an affinity for the α-satellite DNA CENP-B box regulatory element [67]. This protein factor is believed to orchestrate nucleosomal phasing by positional contacts with linker DNA [68,69]. CENP-C localizes to the inner kinetochore plate [70], facilitates chromosome segregation during metaphase [71], and interacts with the CENP-A/B complex to define a specialized centromeric chromatin particle [69,72]. CENP-E is active in regulating microtubule depolymerization [73] and localizes only at active kinetochores [74]. CENP-F is involved in assembly of the kinetochore and dissociates from the centromeric assembly during the maturation of the complex [75]. CENP-G is a scaffolding protein required for stabilization of centromeres through interactions with satellite repeats [76,77]. CENP-H colocalizes with CENP-A and CENP-C and may play a role in the hierarchical organization of centromeric nucleosomes [78,79]. Although all the CENPs help to define specialized chromatin structures, only the role of CENP-A will be discussed further based upon its contributions to core variant nucleosome assemblies.

CENP-A is an H3 variant displaying 62% sequence homology to its major H3 counterpart [80] (see Fig. 2B). As with most histone variants, the greatest regions of sequence similarity map to the histone fold domains, which are critical for the stability of the nucleosome core particle (NCP) [45,81,82]. This protein is an essential modifier of chromatin complexes at the centromere, as gene knockout leads to chromosome fragmentation and death in mice [83]. The identification of an essential N-terminal domain (END) partially explains the indispensability of the protein, as this region has been connected to nuclear localization, centromere targeting and interactions with kinetochore machinery [84]. However, the histone fold domains of CENP-A are also uniquely suited to facilitate histone–histone and histone–DNA interactions within the highly repetitive DNA environment of the centromere [85].

The recent reconstitution of CENP-A containing nucleosomes confirms the ability of these histones to form nucleosome assemblies in specialized chromatin domains [66]; although the stability and structure of these nucleosome particles remain to be defined. At first glance, the dynamic nature of centromeres during chromosome motility suggests that CENP-A may contribute to enhance the stability of the resulting nucleoprotein complexes to ensure the integrity of centromeric chromatin and limit DNA breakage events. Surprisingly, *in vitro* experiments suggest however that DNA is more loosely bound at the terminal ends of the mononucleosome [66]. This may be explained by the fact that CENP-A interacts with other CENP proteins *in vivo* [69,72], which most likely contribute to the stability of centromeric nucleosomes.

2.6. Histone H1 micro- and macroheterogeneity

H1 histones, also referred to as linker histones, are structurally and functionally distinct from the core histones. They represent a highly heterogeneous family of developmentally regulated proteins [86]. At the structural level they have a tripartite organization in which a central globular domain is flanked by extended N- and C-terminal tails. The somatic H1 family displays a significant degree of sequence microheterogeneity [86,87], which maps mainly to the N- and C-terminal tails (see Fig. 3). These signature domains are rich in basic amino acids, display little secondary and tertiary structure in solution, but can acquire α-helical conformation upon interaction with DNA [88–91]. Possibly, the nature of these tails is responsible for the differential expression and developmental roles of the different isoforms [92]. The central region of these histones consists of a highly evolutionarily conserved motif [93] whose structure has recently been determined by X-ray crystallography and NMR spectroscopy [94–96]. The crystallographic analysis of histone H5 (a highly specialized linker histone which is found in the nucleated

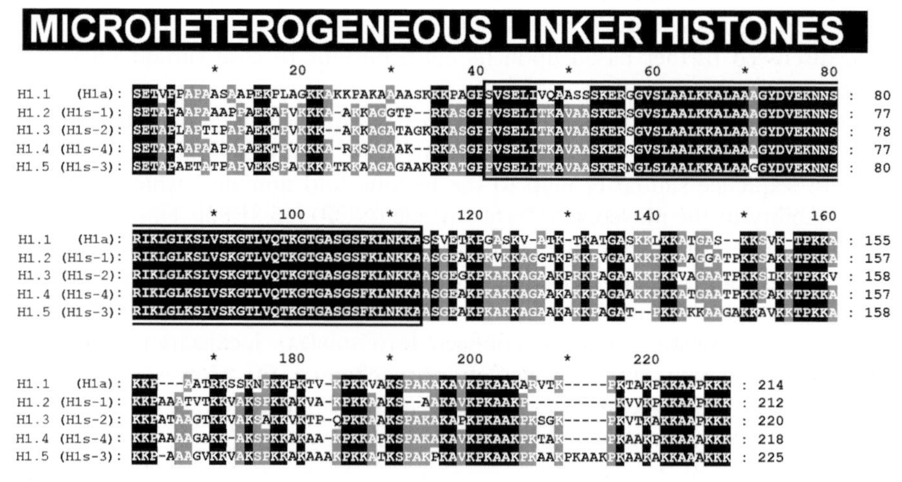

Fig. 3. Amino acid sequence of several somatic human histone H1 proteins to illustrate the microheterogeneity of linker histones. The sequences for human histone H1 variants (H1.1–H1.4) were obtained from Ref. [373] and H1-5 was from Ref. [412]. The nomenclature followed for the designation of these histone variants was Doenecke (e.g., see Ref. [412]). The nomenclature of Parseghian *et al.* [373] is shown in parentheses. The regions corresponding to the trypsin-resistant (winged helix motif [96]) which is characteristic of the protein members of the histone H1 family are indicated by a boxed inset.

erythrocytes from birds [97], see Fig. 4), revealed that this globular core consists of three α-helixes and three antiparallel β-sheets, folded into a structure termed the "winged helix" [98]. This domain has been identified in organisms as diverse as bacteria, protists, and higher eukaryotes [93].

The structural and functional implications of H1 microheterogeneity are still puzzling and in many instances the different histone H1 isoforms appear to be redundant or dispensable for the survival of the organism [99]. Various mutagenic studies in *Tetrahymena* [100,101], *Ascobolus* [102], *Aspergillus* [103], and *Saccharomyces* [104,105] have documented that the expression of distinct variant forms of somatic linker histones are not essential for survival. Indeed, H1 molecules appear to be highly promiscuous as different isoforms can be upregulated to compensate for altered dosages during deletion and transgene replacement experiments [106,107]. Similar effects have also been observed in mouse H1° [108], H1t [109,110]; and chicken H1 [111].

In addition to somatic microheterogeneity, which is characteristic of linker histones, the histone H1 family also contains a group of highly specialized tissue-specific macroheterogeneous variants (see Fig. 4). Examples of such H1 molecules include H1t, a testis specific linker histone found in a variety of vertebrate species [112–114]; the sperm-specific histone H1 from invertebrates [115]; H1fo (previously H1oo) (Fig. 4A) which is exclusively present during early embryonic maturation [116]; H5 (Fig. 4), which is restricted to the terminally differentiated nuclei of birds, reptiles, amphibians, and fish [97,114] and histone H1° that accumulates in terminally differentiated cells [117]. Also included in this group of macroheterogeneous variants are the arginine/lysine-rich protamine like (PL-I) proteins which are present in the sperm of many invertebrate and vertebrate organisms [118,119], such as for instance the EM1/6 protein from the razor clam *Ensis minor* [120] (Fig. 4B).

3. Brief introduction to post-translational modifications

As described above, histones are much more than passive structural players within chromatin. Dynamic post-translational modifications of these proteins confer specialized chemical proprieties to chromatin of both informational and structural nature with important functional implications. The highly conserved sites for acetylation, methylation, phosphorylation, ADP-ribosylation, and ubiquitination events on histone tails appear to orchestrate functional activities that range from transcriptional activation and repression to DNA repair and recombination.

There is a bounty of recent information indicating that these modifications can operate in a combinatorial fashion to provide a "histone code" that generates informational markers involved in the regulation of the assembly of *trans*-acting factors and chromatin remodeling complexes [121–124] (see Fig. 5). Contrastingly, the direct structural effects of these post-translational modifications on chromatin folding and stability, that may be important by contributing to a functional response, remain largely unknown by comparison. In addition, while the histone

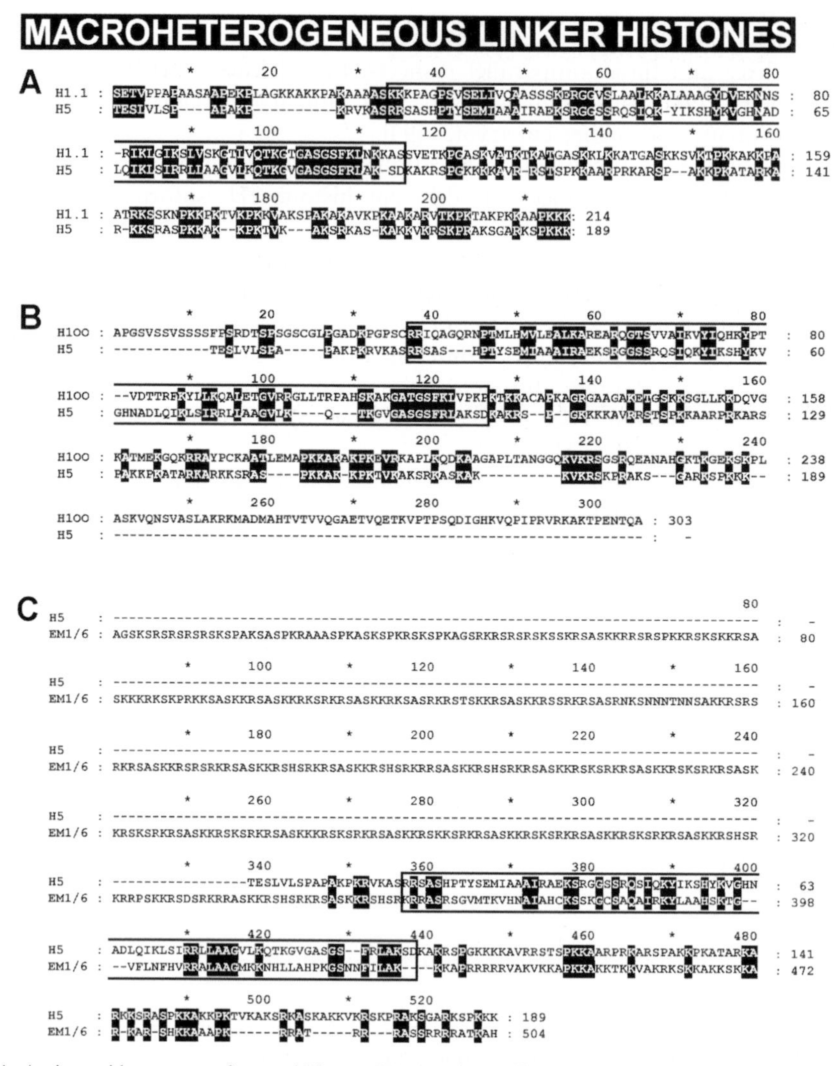

Fig. 4. Amino acid sequence of several histone H1 proteins to illustrate the macroheterogeneity of linker histones. Amino acid sequence of two highly specialized development-specific members of the histone H1 family. **A**. Oocyte specific mammalian histone H1fo (previously H1oo) [116]. **B**. PL-I (EM-1/6) protein from the sperm of the razor clam *Ensis minor* [120]. These two sequences are shown in comparison to the highly specialized histone H5 from chicken erythrocytes. The regions corresponding to the trypsin-resistant (winged helix motif [96]) which is characteristic of the protein members of the histone H1 family are indicated by a box and have been aligned to show the sequence similarity.

code hypothesis can account for the localized effects of DNA transitions, extensive histone post-translational modifications also occur across kilobase stretches of DNA sequence [125–127]. Examples of this broad reaching process include both methylation and acetylation. Although global methylation has been unequivocally correlated to heterochromatin assembly, the complete functional implications of

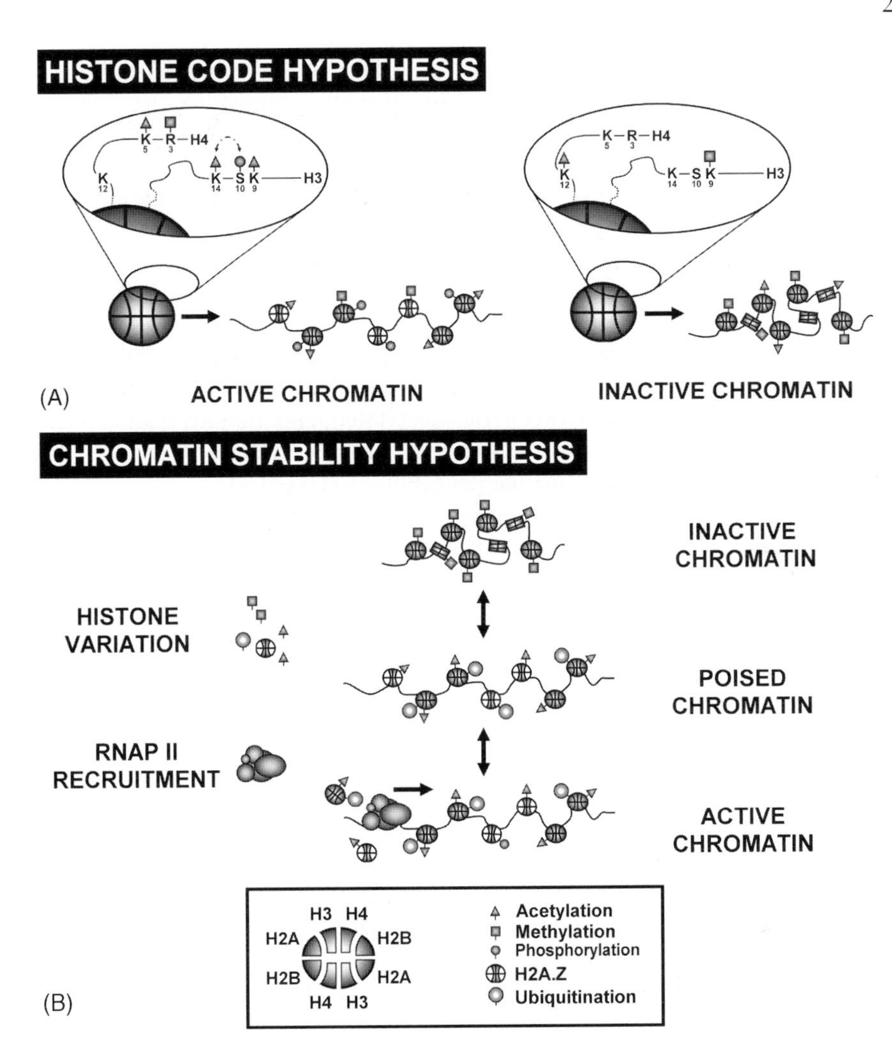

Fig. 5. Schematic representation to illustrate the coding and physical mechanisms utilized by histone variability to affect the structural and functional potential of chromatin. **A**. In the coding hypothesis [121,123,165] different combinations of histone post-translational modifications (and possibly histone variants) operate to create a "histone code" that is recognized by different regulatory *trans*-acting factors that can either repress (fold) or activate (unfold) the chromatin fiber. Two examples of the specific patterning of histone post-translation modifications during the epigenetic regulation of chromatin are shown. Within active chromatin, H3 is diacetylated at Lys9 and 14, and phosphorylated at Ser10 with a synergism observable between Ser10 and Lys14 (represented by the dashed arrow); and H4 can also be acetylated at Lys5 and methylated at Arg3. Contrastingly, during chromatin inactivation, H3 is methylated at Lys9, and H4 acetylated at Lys12 [165]. **B**. In the chromatin stability hypothesis, the synergistic or independent structural (folding or unfolding) effect on chromatin structure is directly exerted by the histone variability itself [29]. In this model, covalent modifications direct the remodeling of chromatin into either an open conformation during activation, or a condensed state during transcriptional repression. For example, H2A.Z deposition is enriched at genes that are poised for expression. It is important to note that these two models are not mutually exclusive and it is likely possible that in several instances they operate in a concerted effort.

global acetylation have yet to be defined. Regardless of their function the structural effects of these far reaching modifications provide support to the "chromatin stability" hypothesis (see Fig. 5). In this theory the chemical and structural variability of histones exert a direct effect on nucleosome stability and chromatin folding. The contributions of distinct post-translational modifications to the stability hypothesis will be further explained below.

3.1. Histone acetylation

Histone acetylation is a reversible amidation reaction involving defined ε-amino groups of lysine residues (see Fig. 6) at the N-terminal tails of core histones. The highly dynamic equilibrium between the acetylated and non-acetylated states of lysine is maintained by two enzymatic groups, referred to as histone acetyltransferases (HATs) and histone deacetylases (HDACs).

The correlation between histone acetylation and eukaryotic transcription were recognized many years ago [128,129]. However, it has not been until very recently, with the discovery that both HATs [130–133] and HDACs [134–138] are an integral part of the basal transcriptional machinery, that the molecular link for this correlation was established. This discovery has rekindled interest in this post-translational histone modification with implications ranging from basic chromatin research to applied medical investigations. Indeed, histone acetylation has been linked to cancer [139–144] and certain types of HDAC inhibitors are already being used to treat certain forms of cancer [145].

Beyond the modulation of eukaryotic gene expression, histone acetylation has also been functionally linked to histone deposition during DNA replication (see Ref. [146] for a review) and in the displacement/replacement of histones by protamines during spermiogenesis in those vertebrate (see Ref. [147] for a review) and invertebrate organisms [148] whose sperm chromatin consists of protamines [118].

Despite all these well-established functional implications, the structural involvement of this chemical modification in the chromatin changes involved in these processes has remained largely elusive [29].

From a structural perspective the effects of acetylation can be classified into "local", which affect a few nucleosomes, and "global" effects that affect chromatin domains spanning over several kilobases of DNA. Local effects include, but are not restricted to, the histone acetylation that occurs at regulatory regions [149] such as gene promoters [150,151]. In many instances these modifications act synergistically with other histone modifications [121] (see Fig. 5). During transcription, how genes are selectively acetylated has yet to be defined, however, two possibly overlapping models (general promoter targeting and specific promoter targeting) [152] have been proposed to explain localized HAT specificity. According to the first model, histone acetylation is targeted to promoters non-specifically. In the second model, acetylation is targeted to defined promoters by *trans*-acting factors that recognize and bind to specific sequences.

The global effects of acetylation have long been recognized [153] and studied in the chromatin field [126,127,154–156]. However, whether such effects are the

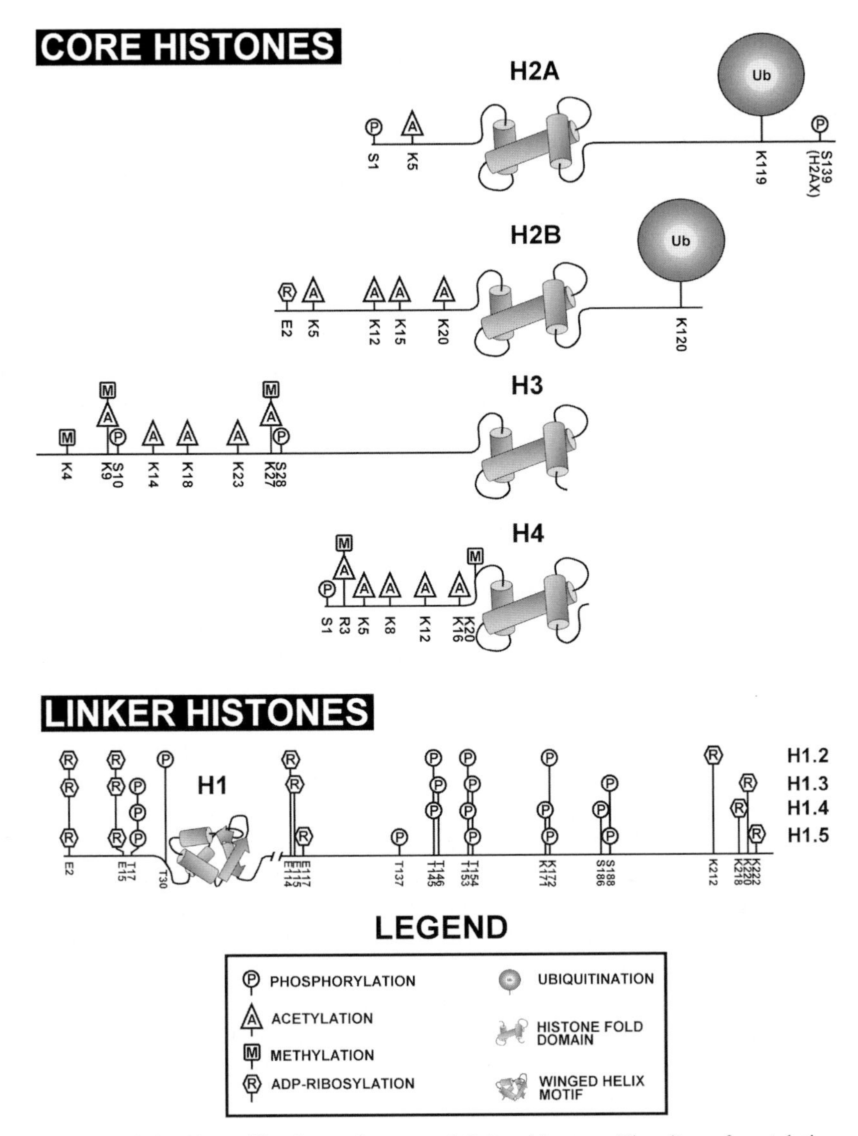

Fig. 6. Post-translational modifications of core and linker histones. The sites of acetylation, phosphorylation, poly-ADP ribosylation, methylation, and ubiquitination are indicated by numbers that correspond to the amino acid position from the N-termini of the molecules. The nomenclature of histone H1 variants is as in Fig. 3. The length of C- and N-terminal tails is in relative scale between core histones to illustrate primary structural differences between these proteins.

exclusive result of untargeted acetylation [152] or whether both specific and non-specific acetylation can simultaneously occur in a system dependent manner (organism or genes affected) [157], still requires further analysis. In this regard, the term "long-range effect" used to refer to acetylation of long stretches of chromatin (encompassing one or more genes) [154–156], in contrast to "global effect" which involves the majority of an organism's genome [127] may be useful in the distinction.

Regardless of the "local" or "global" nature of histone acetylation, how these two effects affect chromatin structure still remains an open question. As it will be discussed in Section 6.2, proving the initial hypothesis that histone acetylation "weakens" the histone–DNA interactions in a way that facilitates chromatin unfolding has not been that simple.

3.2. Histone phosphorylation

The potential substrates for histone phosphorylation include N-terminal serine and threonine hydroxyl groups of H2A, H3, and H4; the N- and C-terminal tails of H1; and the unique C-terminal of H2AX [19,29] (see Fig. 6). Similar to acetylation, phosphorylation appears to be a dynamic modification that transduces on/off signals to nuclear modulators. Enzymes implicated in regulating this pathway include the cyclin-dependent kinases and mitogen activated protein kinases, and the antagonistic phosphatase 1 [158,159].

The functional significance of histone phosphorylation appears to be multifaceted ranging from transcriptional activation to chromosome condensation. Indeed, phosphorylation of H3 Ser10 has been linked to the induction of heat shock genes in *Drosophila* [160], and mitotic chromosomal condensation events [161]. The apparent paradox of these opposing functional effects may be explained in part by the combinatorial effects of other regulatory events [122,162]. For example, H3 is phosphoacetylated during the activation of *c-fos* and *c-jun* expression [163]. Alternatively, H3 Ser10 phosphorylation may also serve as a signal for the incorporation of the centromeric H3 variant CENP-A [164,165]. Such "coded" messages may be at the heart of the epigenetic regulation of DNA, and also impart synergistic effects to localized chromatin structures [29].

Other examples of histone phosphorylation involve diphosphorylation of H3 Ser10 and Ser28 [166,167], and hyperphosphorylation of H1 [161,168] during mitosis. Global and systematic H3 phosphorylation begins during late G2 phase in transcriptionally silent domains, and spreads through the genome, peaking in late prophase [169]. Accordingly, dephosphorylation is correlated with chromosome decondensation, beginning in anaphase and completing by telophase [169]. Although these phosphorylation patterns described above are observed in wild-type systems, both linker histone knockouts [100] and H3 Ser10 mutants [170] are able to undergo mitosis. A possible explanation for this intriguing result is that other histone tails such as H2B can alternatively operate as phosphorylation substrates during chromosome assembly [170,171]. In this instance, histone phosphorylation may also have a structural role that transcends the histone code.

Phosphorylation of specialized linker histones (such as histone H5 or sperm-specific H1 histones) has also been shown to have a major role in the chromatin folding processes leading to the highly condensed chromatin structure which is present in the nuclei of terminally differentiated cells such as bird erythrocytes [172] and histone-containing sperm nuclei [115,118]. In this later instance, phosphorylation also appears to participate in the events involved in chromatin decondensation processes undergone by the male pronucleus immediately after fertilization [115,173].

Although much effort has gone into determining the structural implications of H1/H3 phosphorylation for the chromatin fiber, this issue has yet to be resolved. Mutagenic experiments in *Tetrahymena* suggest that H1 phosphorylation may possibly generate a charge patch, which may increase the dissociation constants of modified linker histones [174,175]. Furthermore, mitotic H3 phosphorylation takes place at the N-terminal end of this molecule, which has been shown to have a major role in chromatin folding [176,177]. These results seem to indicate that histone H1 and histone H3 N-terminal phosphorylation may be involved in processes leading to chromatin unfolding. This involvement of histone phosphorylation in unraveling chromatin architecture fits well with the notion that the double negative charge of the phosphate groups would be expected to induce electrostatic repulsion of those regions of the histones close to DNA contacts within nucleosomes. However, integrating these structural effects with chromosome condensation during mitosis described above [178,179] appears counter intuitive. A current model to explain this apparent dichotomy proposes that H1 and H3 mitotic phosphorylation unfolds chromatin, which allows SMC (Structural Maintenance of Chromosome) molecules access to their previously occluded binding sites [180,181]. Dimerizing SMC complexes then facilitate the packaging of fibers into higher orders of chromosomal structure. This line of reasoning appears to reconcile the antagonistic structural transitions of mitotic chromosome condensation and permissive gene activation following histone phosphorylation.

The implicated structural roles for histone phosphorylation in chromatin unfolding resulting from charge repulsion between the histone phosphorylation site and adjacent DNA is also supported by studies with phosphorylated H2AX. As indicated earlier (see Section 2.1), this modification takes place at the C-terminal end of the molecule, a region which is close to the entry and exit sites of the DNA to the nucleosome [45,182]. Charge mimicry of phosphorylated H2AX by substituting glutamic acid for Ser129 generated genomic instability and nuclease hypersensitivity [18] both results being consistent with chromatin unfolding. However, structural and biophysical studies *in vitro* have yet to substantiate the molecular mechanisms involved in the histone phosphorylation-mediated decondensation processes [183].

3.3. Histone methylation

Histone methylation is a chemical modification that primarily affects arginine and lysine residues of the N-terminal tails of histones H3 and H4. Arginines are enzymatically modified by the addition of single or dimethyl groups in a symmetrical or asymmetrical fashion, as compared to lysine residues that are mono-, di-, or trimethylated at the ε-amino group [184]. Historically, this reaction has proven very difficult to study because of the initial lack of electrophoretic resolving techniques and immunological reagents [29,121]. Therefore, efforts to determine the structural and informational nature of histone methylation have relied on other biochemical techniques, such as radioactive labeling and mass spectrometry [29]. As opposed to the dynamic nature of acetylation and phosphorylation, the methylation of histones is an irreversible modification. In support of this claim, enzymes responsible for the demethylation of histones have yet to be identified. Thus, it appears that histone methylation is an epigenetic marker involved in decisive regulatory events such as cell differentiation and heterochromatin assembly.

Fundamental to the organization of chromatin within the nucleus and maintenance of cell differentiation is the formation of the terminally silent heterochromatin domains. These regions of the genome were originally identified as DNA that remained condensed outside mitosis, and more recently have been associated with satellite DNA sequences and transcriptionally repressed chromatin that may or may not be defined by cytological techniques [185]. There is increasing evidence for a specific role for H3 Lys9 methylation in the assembly of heterochromatin [123], and X-chromosome inactivation [53,186]. Methylation at this site is preferentially bound through the chromodomain (chromatin organizer modifier) of heterochromatin protein 1 (HP1) [187], a protein responsible for the autocatalytic propagation of heterochromatin facilitated through self-dimerization. However, this reaction is not independent of other regulatory events as heterochromatinization appears to be a concerted process involving other post-translational modifications, such as histone deacetylation and H3 Ser10 phosphorylation [123,188].

The human SUV39H1 and mouse Suv39h1 genes encode heterochromatin proteins that are homologous to the *Drosophila* Su(var)3-9 family (Suppressor of variegation) of histone methyl transferases (HMTs). These enzymes methylate histones by virtue of their catalytic SET (Su(var)3-9, Enhancer of the Zeste and Trithorax) and neighboring pre-SET and post-SET cysteine rich domains [165]. Interestingly, a recent study has implicated the tumor suppressor protein Rb (Retinoblastoma) in regulating Su(var)39H1 activity, providing a causal link between DNA surveillance, cell cycle control and histone methylation [189,190].

Gene activation has been linked to arginine methylation by HMT CARM1 (coactivator-associated arginine methyltransferase-1), based upon its interaction with the p160 family of transcription factors, and that mutation of its S-adenosylmethionine binding capabilities restrict both its coactivator and HMT activity [191]. A second arginine methyl transferase, PRMT1 (Predominant cellular aRginine N-Methyltransferase of Type 1), has been shown to facilitate the p300 acetylation of H4 by methylating H4 at Arg3 [192]. Likewise, synergistic activity has also been observed between p300 and CARM1 in response to estrogen-induced RNA synthesis [193].

In a supplementary pathway, links between histone H3 Lys4 methylation and the upregulation of RNA synthesis have also been made. This discrete modification colocalizes with acetylated histone residues and is enriched in the transcriptionally active macronucleus of *Tetrahymena* [194]. Histone methylation at H3 Lys4 has been recently attributed to the novel HMT SET9, which contains the conserved SET catalytic domain, and noticeably lacks the juxtaposed pre- and post-SET domains [195] and SET7 [196]. Two functional roles in gene activation have been associated with SET9 and SET7 mediated methylation of H3 [195]. Firstly, it precludes H3 Lys9 methylation, which prevents the binding of HP1 and the formation of heterochromatin. Secondly, it disrupts the binding of the NuRD (Nucleosome Remodeling and histone Deacetylation) histone deacetylase complex [197], which may allow for subsequent histone acetylation. Thus, it is very likely that, as with other post-translational modifications, the structural effect on chromatin of histone methylation may involve the concerted action of several such modifications [29].

3.4. Histone ubiquitination

Ubiquitination involves the conjugation of the globular signaling protein ubiquitin to substrates involved in extensive physiologic processes. One such event is the tagging of mature or denatured proteins for degradation and recycling by the 26S proteosome [198,199]. Terminally tagged proteins are labeled by polyubiquitin chains, which are formed by repetitive adduction reactions catalyzed by the ubiquitin family of conjugating enzymes (E1–E3) [200,201]. In the final step of the reaction, ubiquitin is transferred from E3 to its protein substrate by the formation of an isopeptide linkage between Gly76 of ubiquitin and Lys ε-amino groups of target proteins. In the case of histone ubiquitination, the modification appears to be primarily dependent on a subset of E2 isozymes, including Rad6p/Ubc2p and Cdc34p/Ubc3p in yeast [202,203], which can successfully ubiquinate histones *in vitro* in the absence of E3 [202,204].

In vivo ubiquitination is primarily restricted to histones H2A (uH2A) and H2B (uH2B) at Lys119 and Lys120, respectively (see Fig. 6); however, H3 (uH3) and H1 (uH1) have also recently been shown to be modified at undefined sites [203,205,206]. In addition, H2A and H2B also display different patterns of ubiquitination as H2A has been found to be polyubiquitinated, and H2B only monoubiquitinated [207].

Within the nucleosome, addition of ubiquitin to H2A occurs near the entry and exit sites of DNA and the binding site of H1 [203]. Therefore, this post-translational modification is expected to have implications for both the stability of the particle and higher order structure of chromatin [45,203]. The C-terminal end of H2B and its ubiquitination site on the other hand is located at the opposite side of the nucleosome [45]. Incorporation of an ubiquitin adduct into the nucleosome at this site may have significant implications for the trajectory of the DNA and the integrity of the particle. In this regard there have been multiple biochemical results substantiating a role of H2B ubiquitination in transcriptional activation [207–210].

Ubiquitinated histones have been suggested to destabilize the interface between the H2A–H2B dimers and the H3–H4 tetramer [211], and to be depleted from highly condensed mitotic chromosomes and enriched in H1 deficient chromatin [212]. In *Drosophila*, the inducible *hsp70* and *copia* genes are ubiquitinated, which represents a marked increase over the repressed satellite sequences from the same fraction [213]. Likewise, regulatory regions in sea urchin and mouse are enriched with uH2A at the histone H3 [214] and dihydrofolate reductase gene [215], respectively. Preferential localization for uH2A and especially uH2B was also observed in the macronucleus of *Tetrahymena*, as compared with the transcriptionally silent micronucleus [209].

Intriguingly, uH2A and uH2B have also been shown to have a non-specific [216,217] and repressive effect [218] on transcription. The functional dichotomy of histone ubiquitination suggests that the structural or informational contributions of ubiquitin to the C-terminal tails of H2A and H2B may impinge upon the interactions of other modulating signals. Two recently proposed models [203] suggest that ubiquitinated histones are either a recruitment signal for remodeling complexes, or part of a synergistic mechanism to facilitate nucleosome disruption. These proposals were put forward based upon the lack of conformational changes observed in the characterization of reconstituted ubiquitinated mononucleosomes and nucleosome arrays [29,219–222] (see Section 6.3). Nevertheless, the structural basis for the correlation between histone ubiquitination and transcriptional activation/repression still remains to be elucidated.

3.5. Histone polyADP-ribosylation

The ADP-ribosylation of histones is an unusual chemical modification in the sense that it involves cascading reactions, which result in the accumulation of a massive ADP-ribosyl (ADPr) homopolymer. *In vitro*, ADP ribosylated proteins have been observed to contain in the excess of 200 ADP ribosyl subunits arranged in a linear or branched array [223,224]. In distinct conjugation reactions, the adduct is covalently transferred from β-NAD$^+$ substrates to specific glutamic acid residues located in the N-terminus of H2B and both the N- and C-termini of linker histones (see Fig. 6) [29]. This is a reversible reaction that is controlled by the coordinated interplay between poly(ADP-ribose) polymerase (PARP), which also undergoes auto(ADP)-ribosylation, and the antagonizing enzyme poly(ADP-ribose) glycohydrolase (PARG). In addition to covalent modifications, the full complement of core histones and H1 can also interact non-covalently with branched polymers of ADPr with varying affinities [225]; however, the structural and functional implications of such interactions remain to be defined.

In the case of H1 variants, linker histones selectively bind ADPr homopolymers over competitor DNA [223]. Furthermore, H1t displays a high degree of affinity for the ADPr subunits even in the presence of salt [223]. Interestingly, this testis specific variant interacts with DNA the least tightly, and has been implicated in fiber decondensation [25,226,227]. This result suggests that potential interactions between H1 molecules and ADPr are specific and not just the bi-product of electrostatic attractions. In this regard, specificity for the ADPr subunits may facilitate removal of H1 from chromatosomal DNA, and initiate an unraveling of the 30 nm fiber required for DNA activation or repair. Unfortunately, the relationship does not appear to be that simple. Previous studies showed that the reversible ADP-ribosylation of chromatin fibers facilitated decondensation and recondensation transitions without histone H1 displacement [228,229].

From a different perspective, circumstantial evidence suggests that ADPr may have a functional role in the activation of transcription. PARP copurifies with $TF_{II}C$ [230] and upregulates AP-2 (Activator Protein 2) controlled transcription. However, these results need to be interpreted cautiously, as a molecular mechanism for ADP-ribosylation of targeted histones has yet to be identified.

ADP-ribosylation has also been implicated as a proteolytic antagonist during embryonic development [231]. Following fertilization in sea urchin, sperm-specific histones are degraded by the sperm-histone-selective (SpH) protease and subsequently replaced by cleavage stage histone variants. During this process, the maternal replacement histones are protected from proteolysis by ADP-ribosylation.

The ADP-ribosylation of histones may also have significant effects for the repair of damaged DNA. Nucleotide excision repair, a system responsible for the removal of bulky adducts and helical distortions from DNA, has been implicated in a poly(ADP)-ribose mediated "histone-shuttling" mechanism that controls the unfolding of damaged and refolding of repaired DNA substrates [232,233]. In this mechanism, histones may be stripped from nucleosomal assemblies by ADP-ribosylation, which facilitates the targeting of repair proteins and generates a permissive environment for DNA recovery. In this regard increased ADP-ribosylation of H1 proteins in damaged heptoma [234] and mammary tumor [235] cells has been documented. Not surprisingly, the self-modification of PARP has proven to be an important step in DNA repair. Indeed, activation of the enzyme by auto(ADP)-ribosylation is a preliminary step in many repair responses, and parallels the mobilization of DNA-PK (DNA-dependent protein kinase), ATM, and p53 [236]. Beyond repair, it appears evident that this post-translational modification may be responsible for further nuclear functions *in vivo* [29]. Defining the potentiating effects of ADP-ribosylation in the modulation of chromatin structure may be critical to determining these roles.

4. Brief introduction to DNA methylation

The methylation of DNA at CpG islands has also turned out to be an important regulator for cell development, the differentiated proteome and the regulation of cell survival [237,238]. Indeed the implications of this chemical modification have been linked to DNA accessibility, chromatin fluidity and cell transformation [239,240]. DNA methylation is required for genomic stability and believed to act as an inert epigenetic marker in germinal cells and preimplantation embryos [238]. Presumably, DNA methylation is required for the heritable transmission of chromatin structure, which prevents the expression of terminally silenced genes in differentiated tissues, and provides a host-defense mechanism against parasitic transposable elements [241].

How the selective modification of DNA is regulated is still poorly understood, but two models have been put forward. The first hypothesis implies that promiscuous DNA methyl transferases (DNMTases) globally transfer methyl groups to CpG islands, and it is only the steric hindrance of complexed proteins at regulatory sites, such as SP1, that inhibit the reaction in euchromatic domains [242,243]. The second theory suggests that DNMTases, are specifically targeted to *cis*-acting elements through interactions with transcriptional repressors, such as Rb and HDACs [244–246]. However, evidence supporting *de novo* methylation in this regard has yet to be provided [241].

DNA methylation levels and gene expression appear to be inversely related, as silenced genes and heterochromatic domains display a pattern of hypermethylation [241], and the posied regulatory elements of active genes are hypomethylated [247]. During methylation events, consensus 5′-CpG-3′ (Cytosine-phosphate-Guanine) islands and their antisense 3′-GpC-5′ complements are both substrates for DNMTs. The symmetrical modification of these sites generates a DNA structure that transforms the three-dimensional character of the major groove [238,248], and increases the hydrophobicity of the nucleic acid. Possibly, these changes may work to repress transcription by decreasing the affinity of transcription factors for their cognate regulatory elements [249], thereby impeding the assembly of initiation complexes.

Although chemically modifying DNA have distinctive implications for chromatin transitions and fiber structure in the presence of H1 [250], *in vivo* these effects appear to work in concert with chromosomal proteins. 5′-Methylcytosines are specifically bound by members of the MBD (methyl-CpG-binding-domain) family, such as MeCP2 (Methyl-Cytosine binding Protein 2) and MBD1. These proteins have been shown to interact with HDACs and provide a casual link between DNA methylation, histone deacetylation and transcriptional repression [251–253].

Recently, a connection between histone methylation and DNA methylation has been made in *Neuospora crassa* [254,255]. This discovery outlines a redundant relationship between two distinct mechanisms for the selective repression of DNA and formation of heterochromatin, as both of these effects have been shown to facilitate the formation of restrictive chromatin architectures [256]. Importantly, this significant observation has implications for proviral repression [257], and physiologic development and disease prevention [258].

5. *Chromatin folding and dynamics*

A significant portion of the histone variability described in the previous sections affects the electrostatic charge or hydrophobic character of these proteins particularly at the N- and C-terminal regions of the molecule. While some of these modifications appear to be used as a coding mechanism [121,123,165] (see histone code hypothesis, Fig. 5), the changes in the polarity also most likely play an additionally important role in the modulation of the histone–DNA interactions. Changes in charge or hydrophobicity resulting from such variation affect the stability of the nucleosome complex and the folding ability of the chromatin fiber. It is speculated that these processes are important for the metabolic activity of the chromatin fiber (see chromatin stability hypothesis, Fig. 5). In this regard, it is particularly important to distinguish between the "short-range" or "local" effects, which involve only one or a few nucleosomes and the "long-range", or "global" effects such as those that involve modifications that span several kilobases [126,127]. In the following sections we are going to discuss the factors involved in nucleosome stability as pertaining to the short-range effects. This may also have important implications for the yet poorly understood chromatin alterations produced by remodeling complexes [10,259–262]. We are also going to briefly examine the mechanisms involved in chromatin folding as pertaining to the long-range effect, which are important for transcriptional elongation and for DNA replication and repair.

5.1. Nucleosome stability

As it had already been anticipated from early biophysical characterization of the nucleosome using techniques such as melting profiles [263] and analytical ultracentrifugation [264] only a few of the overall available charges present in the DNA and histone components of the nucleosome participate in the maintenance of the nucleoprotein complex. Recent crystallographic characterization has provided direct evidence in this regard [45,265]. This probably reflects the highly dynamic nature of the complex needed to accommodate the manifold functional demands of the genetic activity it supports [266].

Isolated nucleosomes exhibit a highly dynamic behavior in solution as it was initially revealed by sedimentation velocity analysis [267,268]. Under ionic strength conditions corresponding to the physiological range, nucleosomes sediment as a double boundary [267,268]. The slow moving boundary corresponds to free DNA whereas the fast moving boundary consists of intact nucleosome core particles. In what was shown to be a reversible equilibrium, the free DNA accumulates with increasing temperature and ionic strength and decreases with the concentration of the sample (Fig. 7A–D) [264,269–272] (see Fig. 7E) [264]. A fraction of the undissociated nucleosomes associates with the octamers corresponding to the freely sedimenting DNA [264,267]. Indeed it has been shown that at 0.5–0.6 M NaCl nucleosome core particles can bind more than an additional histone octamer [273]. Divalent ions can also affect nucleosome stability [274] and hence caution should be taken in carefully choosing the concentration and components of the buffers used when studying the interactions of nucleosomes with DNA-binding transacting factors and remodeling complexes, such as for instance SWI/SNF [275,276].

Of important note, native nucleosomes are intrinsically heterogeneous both in terms of their DNA sequence and histone compositions and both of which can contribute to the stability of the complex.

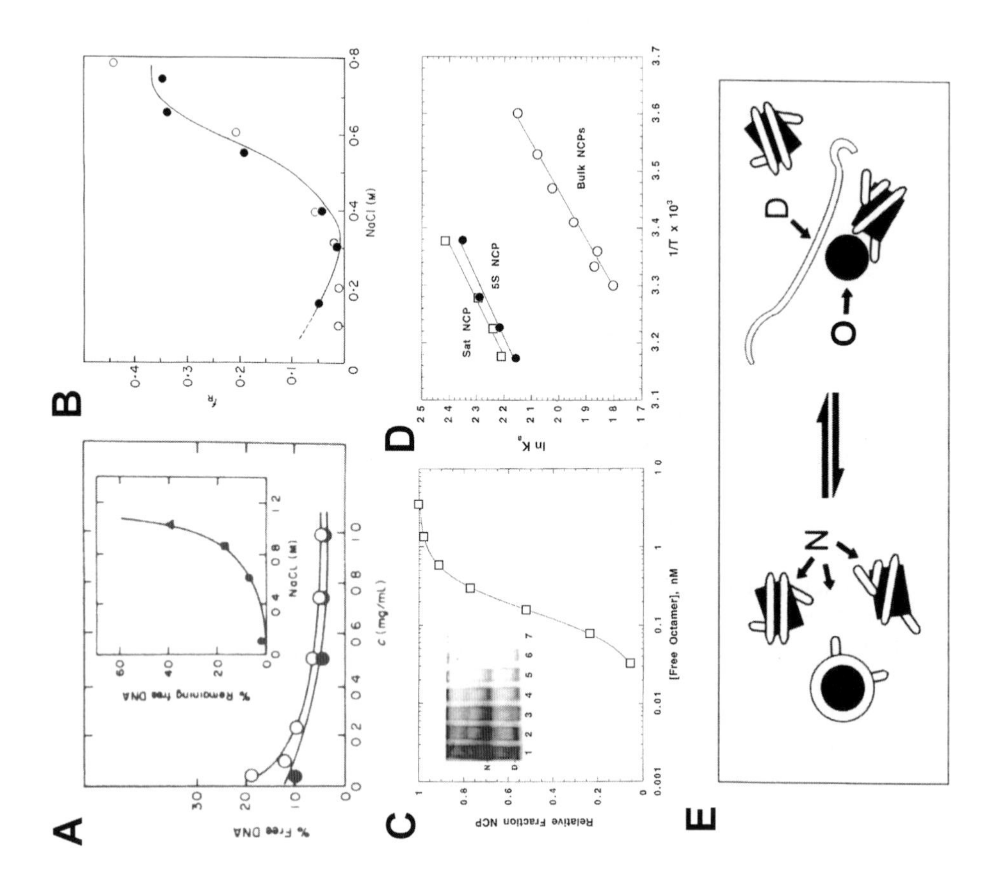
A
% Free DNA
c (mg/mL)
% Remaining free DNA
NaCl (M)
B
NaCl (M)
C
Relative Fraction NCP
[Free Octamer], nM
D
Sat NCP
5S NCP
Bulk NCPs
1/T x 10^3
E
N
D
O

Fig. 7. Nucleosome core particles in solution can reversibly dissociate into their constitutive core histone and DNA components. **A**. Percentage of free dissociated DNA as a function of the nucleosome concentration in 0.6 M NaCl at 10 °C (●) and at 20 °C (○) as determined by sedimentation velocity analysis [264]. The inset shows the percentage of free DNA (in the limit of high nucleosome core particle concentration) as a function of the NaCl concentration [264]. **B**. Fraction of DNA (f_R) becoming freed of the nucleosome constraints induced by its specific interaction with histones as a function of the salt concentration. (○) Wholly trypsinized and (●) native nucleosome core particles (146 bp) [305]. **C**. Determination of the binding affinity of the histone octamer at 37 °C for α-satellite DNA as determined by EMSA of a serial dilution of the nucleosome core particles (see inset) and phosphorimager analysis [277]. **D**. van't Hoff plots of $\ln K_a$ versus 1/temperature (in Kelvin degrees) for reconstituted α-satellite (Sat NCP) (□) and 5S DNA (5S NCP) (●) nucleosome core particles in comparison to bulk nucleosome core particles (Bulk NCPs) (○) [264] (146 bp) (see Table 1 for more details). From data such as that shown in panels **A** and **C** it is possible to calculate the apparent dissociation constants involved in this process. Under conditions of reversible equilibrium these values can be used to measure the free energy of the process $\Delta G^\circ = -RT \ln K_a$, where K_a is the equilibrium constant and T is the absolute temperature. The temperature dependence of the DNA dissociation can be represented as van't Hoff plots (panel **C**). In these plots, the slope corresponds to $-\Delta H^\circ/R$ ($R = 1.987$) so that the enthalpy (ΔH°) and the entropy (ΔS°) can be determined $-\Delta G^\circ = -H^\circ + T\Delta S^\circ$. **E**. Schematic representation of the reversible dissociation equilibrium corresponding to the results shown in **A**–**D**. Note that in this model the dissociated DNA is bound to a nucleosome core particle [264,267]. D = DNA; N = nucleosome core particle; O = histone octamer. [Parts A, B and C–D were respectively reproduced from Ausió, J. *et al.* (1984) J. Mol. Biol. 176, 77–104; Gottesfeld, J.M. and Luger, K. (2001) Biochemistry 40, 10927–10933, with permission from Academic Press Inc., and The American Chemical Society.]

Table 1

Enthalpic and entropic components of nucleosome core particle (NCP) free energies [277]

NCP	$-\Delta G^{\circ}$ at 23°C (Kcal/mol)[a]	$-\Delta H^{\circ}$ (Kcal/mol)[a]	ΔS° at 23°C (cal/mol)[a]
Reconstituted from α-Satellite DNA[b] (146 bp) + recombinant histones[c]	14.25	20.39	20.7
Reconstituted from 5S rRNA gene DNA[d] (146 bp) + recombinant histones	13.83	19.29	18.5
Native NCPs (146 bp)	12.08	23.64	39.1

[a]Calculated as described in the legend to Fig. 5.
[b]Palindromic human satellite DNA [279].
[c]Recombinant *Xenopus laevis* histones [279].
[d]146 bp DNA fragment derived from *Lytechinus variegatus* 5S rRNA gene [279].
[Reproduced from Gottesfeld, J.M. and Luger, K. (2001) Biochemistry 40, 10927–10933, with permission from Academic Press Inc., and The American Chemical Society.]

5.1.1. The role of DNA

Nucleosome stability has been recently been revisited analyzed in a series of elegant experiments by Gottesfeld and Luger [277] using EMSA (Electrophoretic Mobility Shift Assay) analysis of reconstituted nucleosomes consisting of different sequence defined DNA templates (Fig. 7C–D). Using nucleosomes reconstituted onto α-human satellite DNA [278] and the *Lytechinus variegatus* 5S r-RNA gene [279] (two strongly nucleosome positioning sequences) they were able to determine the apparent dissociation constant (K_d) of these particles. The values determined for these two DNA-sequence defined nucleosomes at 50 mM NaCl, 0.03 and 0.06 nM, were lower than that of 1.2 nM previously determined for bulk nucleosomes [264] on the basis of the experimental work described above. Furthermore, the particular DNA sequences chosen for this study were found to be approximately 2 Kcal more stable as determined from the slopes of the van't Hoff plots shown in Fig. 7D [277]. Interestingly, it was found that while the enthalpies of the bulk and sequence defined nucleosomes were similar; the bulk nucleosomes exhibited much higher entropy suggesting that DNA-sequence contributed to nucleosome stability. This suggests that nucleosome stability is enhanced by the entropy decrease, which probably results from the intrinsic curvature (bending) of the sequence-defined DNA templates used in that work.

All this leads to the question of how DNA affects nucleosomal stability at the molecular level. DNA bendability has been repeatedly put forward as a candidate to play an important role [280–282]. This property is related to the persistence length of DNA [283]. A study on the characterization of nucleosomes reconstituted onto methylated DNA [poly(dG-m^{5}dC) · poly (dG-m^{5}dC)] [284] provides support to this hypothesis. Poly(dG-m^{5}dC) · poly (dG-m^{5}dC) DNA can be induced to change from its B to its Z conformation in the presence of millimolar amounts of divalent ions [285] such as $MgCl_2$. The persistence length of poly(dG-m^{5}dC) · poly (dG-m^{5}dC) in the Z form in high salt was found to be 208 nm (ca. 612 bp) [286], a value much lower than that of the same polymer in the B form (93.8 nm, 276 bp) at low salt in the B form or with that of 33 nm (ca. 100 bp) for the linear *Col*E1 DNA [287]. It was found that upon induction of the Z conformation, the nucleosome particles reconstituted onto this DNA template became completely destabilized [284], a fact that was initially ascribed to the changes in persistence length.

The term persistence length, which was originally used in polymer chemistry to define the average sum of the projection of all bonds $j \geq i$ on bond i in an indefinitely long chain [288]. This term while appropriate to account for the bendability properties of long polymers may not necessarily be the best parameter to characterize sequence defined nucleosomal DNA fragments (146–200 bp). In this regard, Anselmi *et al.* [289–291] using a statistical mechanical approach have recently proposed that nucleosome stability is controlled by the intrinsic DNA curvature of these molecules and by their flexibility. According to these authors, DNA curvature helps to decrease the free energy of DNA distortion resulting from the wrapping around the histone octamer. It also increases the energy cost that the free DNA spends to release part of the hydration spine displaced by the interaction with histones. Similarly, DNA flexibility contributes to the decrease in the distortion energy and also by increasing the entropy difference that exists between the free DNA in solution and the DNA when constrained in the nucleosome complex. In agreement with earlier experimental evidence [292] and with the results by Gottesfeld and Luger [277] discussed above they conclude that the entropy contribution is the most important [291]. The theoretical model proposed by Anselmi *et al.* [289–291] for the prediction of sequence-dependent thermodynamic stability of nucleosomes can satisfactorily predict the experimental free-energy difference determined from competitive nucleosome reconstitution experiments carried out in other laboratories [274,281,293–299].

While the statistical mechanical approach [289–291] appears to be able to account for the experimental data, the question still remains on how does the DNA mechanistically associate and dissociate from the nucleosomal protein–DNA complex. The determination of the crystallographic structure of the nucleosome [45] has allowed the identification of the different kinds of interactions involved which suggests that DNA dissociation would occur stepwise in a semi-cooperative way beginning at the entry and exit sites of the DNA in the nucleosome core particle [45,279]. Direct evidence for what could be at the basis of a reversible stepwise uncoiling of DNA from the nucleosome has been recently provided by single-molecule optical trapping [300]. Unfortunately however, this information does not provide much insight into the dynamic aspects involved in solution. One of the most popular current models is that of "site-exposure" which is essentially based on experimental data involving the interaction of restriction enzymes with nucleosomal DNA (see Refs. [283,301] for recent reviews and references therein) and which appears to be consistent with a stepwise release of DNA. An alternative elastic model for this dissociation has also been proposed by Manning and coworkers [302–304].

As recognized by Widom [301], the equilibrium stability of the nucleosome creates an apparent paradox because the disassembly and reassembly processes take place under ionic conditions for which some of the nucleosome components are not in free exchange [301]. One of the possibilities to account for this paradox is that the histone subunits may have a biased diffusion back to their originating nucleosomes. This would be the result of the large negative charge of DNA in conjunction with the large positive charge of histones, which mainly involves their N-terminal tails [301]. However, the reaction does not appear to be this simple as removal of the histone tail regions does not significantly alter nucleosome stability [279,305] (see also Fig. 7B). Clearly, the detailed molecular mechanisms involved in the dynamic association of the histone octamer with the nucleosomal DNA and the role of these dynamics within the more physiologically relevant context of the chromatin fiber undoubtedly requires further investigation [306].

5.1.2. The role of histones

From a structural perspective, histones can be considered to consist of a tri-partite organization in which a central "histone fold" domain [81] is flanked by N- and C-terminal regions "tails" with a low level of folding. The histone fold is responsible for the "histone bundle" and "handshake" interactions that hold the histone octamer together [45,265,307,308] while the tails do not appear to have any major role in the stability of the octamer subunits [309].

At the nucleosome level the histone N-terminal tails and part of the C-terminal tails can be selectively removed using trypsin [305] or by the use of reconstituted nucleosomes consisting of recombinant histones lacking these domains [279]. While this removal was observed to affect the melting of the nucleosomal DNA under very low ionic strength conditions [305] it did not significantly affect the stability of chimeric nucleosomes (see Fig. 7B) [277,279,305,310]. It can be concluded from these studies that the effect of histone N-terminal tails of core histones on nucleosome stability are relatively small. It is important to make the distinction here between N- and C-terminal tails because neither the treatment of nucleosomes with trypsin nor the recombinant truncated forms of the histones used in these particles had completely removed C-terminal tails [5,279]. In fact, and in contrast to the lack of effect of the N-terminal regions, removal of the last 15 C-terminal tail amino acids of histone H2A by an endogenous protease [311] has been shown to substantially lower the affinity of the histone H2A–H2B dimer for the H3–H4 tetramer [60].

From all the above it would be expected that only histone variability (variants or post-translational modification) affecting the histone fold or C-terminal domains would have a major effect on nucleosome stability in agreement with the experimental results that will be described in the following sections.

A good example of the effects of histone variability on nucleosome stability is provided by the yeast histones, which are very divergent from their vertebrate counterparts [312]. Physical [313], biochemical [314], and crystallographic [315] analyses have produced evidence for a lower nucleosome stability resulting from the association of these histones with DNA. The lower stability conferred by yeast histones to the nucleosome when compared to vertebrate histones is in line with the more dynamic metabolic demand of this organism as most of its genome exists as euchromatin [316].

5.2. Chromatin fiber folding

The location of linker histones in the nucleosome ([317–320], see Ref. [321] for a review) and their ability to determine the exiting and entering orientation of the DNA from the complex results in a three dimensional zig-zag-like chromatin organization which is abolished by histone H1 depletion [322–324]. Under ionic conditions close to physiological values (100–150 mM NaCl) the polynucloesomal array folds into a higher order superstructure to form a fiber of a 300–400 Å diameter average. This folding is very important for it defines the architectural organization of the chromatin substrate upon which important genetic processes such as transcription or DNA repair take place. In the past, attempts to ascertain the internal organization of the fiber have resulted in a plethora of models such as the super bead [325,326], solenoid [327], the helical ribbon [328], twisted-ribbon [329] layered zig-zag [330] and crossed-linker [331,332], just to mention a few (reviewed in Refs. [323,333]). However, none of these models have been able to satisfactorily account for all the experimental data available, such as the differential accessibility of linker histones and linker DNA to proteases [176], antibodies [334,335] or nucleases [336,337] or even for all the physical parameters determined by X-ray diffraction of chromatin fibers [338–340] and other experimental techniques. The reason(s) for this involves the potential for enormous variability of the core and linker histones (see Section 2.6) as well as the heterogeneity in length of the linker DNA, which results in irregular folding [341–343] that contrasts with the highly regular chromatin fibers still depicted in most textbooks.

It has been shown that as the ionic strength of the medium increases from low non-physiological values to physiological concentrations adjacent nucleosomes in the three dimensional zig-zag-like organization can rotate to a different extent depending on the linker DNA length thus resulting in an irregular fiber [323,341] (see Fig. 8). Whether this rotating linker DNA remains straight [344–346] or it bends [347] continuously wrapping about the histone octamer [348] still remains controversial. In this regard, nucleosome removal or shifting resulting from the action of chromatin remodeling complexes [10,261] should be viewed as an alteration of linker length, which may result in a local unfolding of chromatin (see Ref. [349] for a review) as required by the different demands of genetic activity.

The ionic strength dependence of the folding of the chromatin fiber has been extensively characterized using analytical ultracentrifugation [350–352], X-ray diffraction [353], neutron scattering [354], electron microscopy [322,327] and more recently by cryoelectron microscopy [345] and atomic force microscopy (AFM) [344]. In recent years folding of the chromatin fiber has been characterized using nucleosome arrays reconstituted onto the synthetic 208-12 DNA template. This DNA template was constructed by Robert Simpson [355] and consists of 12 identical copies of a 208 bp nucleosome positioning fragment obtained from the 5S rRNA gene of the sea urchin *Lytechiniuis variegatus*. This highly homogeneous chromatin system has gained enormous popularity in recent years, both for physicochemical characterization (see Refs. [266] and [349] for reviews, and see also

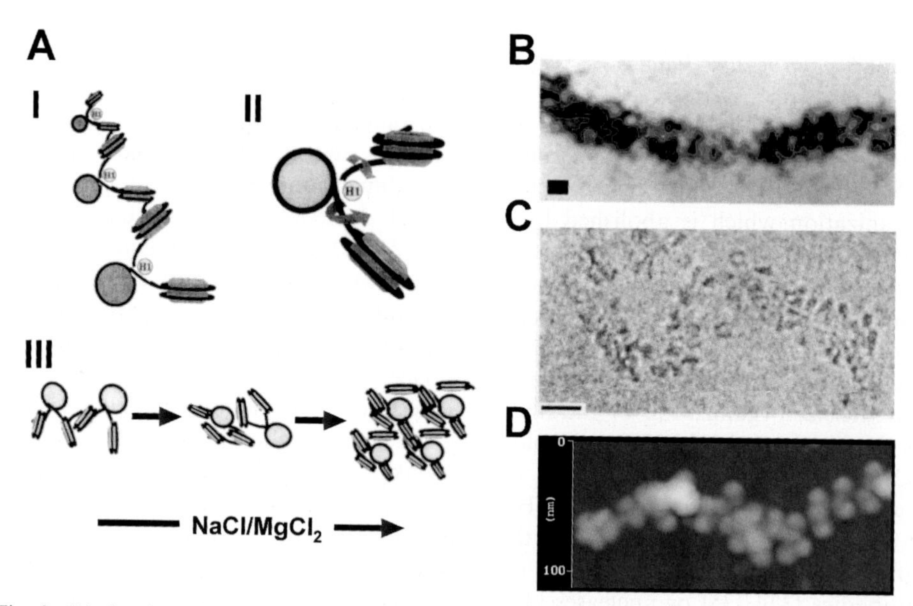

Fig. 8. (**A**) A schematic representation of the folding of the chromatin fiber caused by continuous unwinding of the internucleosomal linker domains. (**I**) Histone H1 (linker histones) bind to a region close to the entry and exit sites of DNA in the nucleosome resulting in a zig-zag-like organization of the chromatin fiber. (**II**) Folding of the chromatin fiber is mediated by the rotation between adjacent nucleosomes [413]. (**III**) Given the heterogeneity in length of the linkers this results in irregular chromatin fibers of a 30–40 nm diameter such as those seen in panels (**B**) to (**D**). The folding transition can be analyzed *in vitro* by increasing the concentration of NaCl in the range of 0–100 mM or by increasing the $MgCl_2$ concentration in the range 0–0.8 mM [352]. (**B**) Electron micrograph of a sea cucumber (*Holothuria tubulosa*) sperm chromatin fiber embedded in araldite-epon [330]. The black rectangle indicates the profile of a nucleosome on the same scale. (**C**) Electron cryomicroscopy visualization of a chromatin fiber from COS-7 cells [345]. The bar corresponds to 300 Å. (**D**) Tapping-mode scanning force microscopy (SFM) image of a glutaraldehyde-fixed chicken erythrocyte chromatin fiber (obtained as in Ref. [344]) (the image was kindly provided to us by Dr. Sanford Leuba). Notice the irregular folding of the fibers, which appears to be independent of the chromatin biological source and experimental approach used in the visualization. [Parts A, B and C were respectively reproduced from Krajewski, W.A. and Ausió, J. (1996) Biochem J. 316, 395–400 [414]; Subirana, J.A. *et al.* (1985) Chromosoma 91, 377–390; Bednar, J. *et al.* (1998) Proc. Natl. Acad. Sci. USA 95, 14173–14178, with permission from Portland Press/Biochemical Society, Springer Verlag, and The National Academy of Sciences.]

Ref. [298] for a recent application using a similar system) and functional analysis [356,357] of the chromatin fiber. The experimental data obtained with these nucleosome arrays, both in the presence of mono- and divalent cations, agrees very well with earlier data obtained with more heterogeneous native counterparts [350,351,358,359]. The ionic strength-dependent folding of the chromatin fiber, when analyzed using Manning's polyelectrolyte theory [360,361], reveals that the nature of the fiber folding mechanism is mainly electrostatic [362–364].

Of the three major structural components that can affect chromatin folding: linker DNA length, histone tails and linker histones, the last two are critical for the dynamic aspects of the process and hence it would be expected that histone variability may play an important role. In this respect the use of the sequence-defined 208-12 becomes particularly useful for this analysis as it largely circumvents the problem of linker length heterogeneity.

Experimental results regarding the role of the histone tails indicate that these histone domains play a critical role in chromatin folding [358,365]. Removal as well as the modification (acetylation) of the lysine amino acids within these regions produces an imbalance of the electrostatic interactions, which results in a hierarchically impaired folding ability (H3/H4–H2A/H2B > H3/H4 > H2A/H2B) of the chromatin fiber [358,366–369]. Therefore, sources of histone tail variability (histone variants and post-translational modifications other than lysine acetylation) are also likely to alter the extent of folding of chromatin.

The involvement of linker histones in the stabilization of chromatin folding is in itself highly dynamic. Early structural studies indicated that linker histones can easily exchange between chromatin fibers [370]. These results have been corroborated by recent experiments using fluorescence recovery after photobleaching (FRAP), which have shown that a rapid exchange of linker histones can occur in the chromatin of living cells [371] with a residence time of several minutes [372]. Unfortunately, information regarding the role of linker histone variability on chromatin folding and dynamics is still very scarce. However, both functional observations [373] and the differences in the charge distribution of the histone H1 isoforms present in a given cell or organism [374] (see Fig. 3), suggest that they may also contribute to this process. In this regard, the role of the post-translational modifications of these histones such as phosphorylation (see Fig. 6) appears very limited [183] and needs to be reevaluated.

Finally, it is possible that factors affecting the individual stability of the nucleosome (see Section 5.1.2) may also indirectly affect the folding of the chromatin fiber. The characterization of chromatin complexes consisting of core histone variants (see Section 6.1) may be very useful in this regard.

Also, local changes in the structural and chemical variation of DNA may have important effects on the overall extent of chromatin folding. For instance, transitions from the B to the Z form of DNA will result in nucleosome dissolution (as discussed earlier) and this could affect the folding of the fiber. As well, chemical modifications of the bases such as methylation have been shown to increase the folding of the chromatin fiber when linker histones are present [250] although the mechanism involved in this later case remains to be elucidated.

6. Histone variation and chromatin stability. A few selected examples

In the sections that follow, we are going to describe a few representative examples of how histone variability (histone variants and their post-translational modifications) can affect nucleosome stability and folding. This data supports the notion that while some of this variability may be exclusively used to provide an informational code [121,123,165] it can also have important implications for structural aspects involved in the highly dynamic nature of the chromatin fiber.

6.1. Histone H2A.Z

The biological relevance of this histone variant has already been discussed in Section 2.2. As it was pointed out there, the recent crystallographic structure [41] of a nucleosome containing recombinant H2A.Z (rH2A.Z) revealed structural features that pointed towards a nucleosome destabilizing role for this histone variant (see Fig. 9A–B). Two recent papers [42,375] have pursued the preliminary findings established by this paper and proceeded with the structural characterization of chromatin complexes in solution.

In the first approach it was found that nucleosome core particles reconstituted with rH2A.Z and native histones H2B, H3, and H4 (a non-recombinant complement) exhibited a marked lack of stability when compared to similarly reconstituted nucleosome core particles consisting of recombinant rH2A.1. This is demonstrated by the marked drop in the salt dependent sedimentation coefficient exhibited by the rH2A.Z nucleosome core particles when compared to their rH2A.1 counterparts (see Fig. 9C). The (8.3 S) value of the sedimentation coefficient of these particles at 0.6 M NaCl suggested the loss of a histone rH2A.Z-H2B dimer [42]. It was also found that sample manipulation, such as for instance concentration using Centricon concentration devices, led to a significant increase in the electrophoretic mobility of the rH2A.Z nucleosome core particles (Fig. 9D) which was proven to be the result of the loss of rH2A.Z-H2B dimers (Fig. 9E) as predicted from the sedimentation data. These results clearly support the crystallographic suggestions [41] and provide direct evidence that rH2A.Z nucleosome destabilization can be attributed to more labile binding of the histone rH2A.Z-H2B dimer (see Fig. 9C–D).

The effects of histone rH2A.Z in the folding of nucleosome arrays have thus far provided what appear to be contradictory results [42,375]. Abbott *et al.* [42] studied the NaCl dependent folding of reconstituted 208-12 nucleosome arrays consisting of a histone octamer mixture of recombinant rH2A.Z and native histone counterpart (as above) (see Fig. 10A). Using analytical ultracentrifuge analysis, they found that the saturated rH2A.Z 208-12 complexes had a significantly impaired folding ability (10% lower) at NaCl concentrations approaching physiological values (see Fig. 10B) when compared to 208-12 complexes similarly reconstituted but using recombinant H2A.1 instead of rH2A.Z. Importantly the rH2A.1 208-12 oligonuclosome complexes exhibited a folding that was undistinguishable from that of 208-12 complexes reconstituted with native histone octamers (see Fig. 10B). It was also found that the lower extent of folding of the rH2A.Z 208-12 complexes did not preclude their association into higher folding complexes which is characteristic of the native histone recontituted 208-12 complexes when analyzed in 100–150 mM NaCl [266,358] or in the presence of divalent ions [349].

In contrast, the results of Fan *et al.* [375] using 208-12 nucleosome arrays that had been reconstituted with a full set of recombinant histones showed that rH2A.Z enhanced the intramolecular folding of these complexes while simultaneously inhibiting the formation of higher folding structures that result from intermolecular

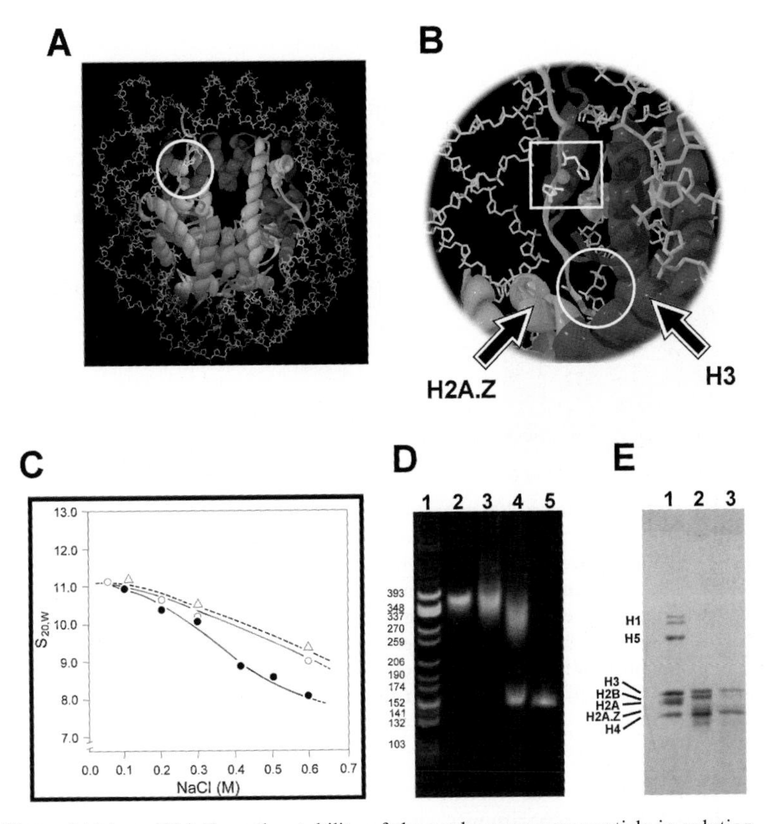

Fig. 9. Effects of histone H2A.Z on the stability of the nucleosome core particle in solution. **A**. Three-dimensional image of the nucleosome core particle containing the histone variant H2A.Z generated using the coordinates deposited by Suto *et al.* [41]. Core histones H2A, H2B, H3, and H4 are differentially shaded for clarity. The circular inset displayed in the top left-hand corner is blown up in **B**, which highlights a portion of the essential C-terminal domain of H2A.Z. In this figure, H4 is removed to allow for easier visualization of the two regions of interest. The boxed inset depicts the His112–His114 binding pocket with a coordinated manganese ion. The positioning of this unique structure at the surface of the particle has implications for the binding of *trans*-acting factors, such as H1 and remodeling proteins, and therefore, the higher order structure of chromatin. The circular inset illustrates the novel Gly106 residue of H2A.Z (substituted for Gln104 in H2A.1), which interacts with the neighboring histone H3. This glycine residue has been implicated in destabilizing the interaction between H2A.Z–H2B dimers and the H3–H4 tetramer [41,42]. **C**. Ionic strength dependent variation of the sedimentation coefficient ($s_{20,w}$) of H2A.Z-containing (●) and H2A.1-containing (○) reconstituted nucleosome core particles in comparison to chicken erythrocyte nucleosome core particles (△) [42]. **D**. Native (4%) polyacrylamide gel electrophoresis of native chicken erythrocyte nucleosomes (lane 2), recombinant H2A.1-containing reconstituted nucleosome core particles (lane 3) and recombinant H2A.Z-containing reconstituted nucleosome core particles (lane 4) after a five fold concentration using a Centricon YM-10 filtration device (Amicon-Millipore) [42]. Lanes 1 and 4 respectively show a *Hha* I digested pBr 322 used as a molecular standard and the 146 bp chicken erythrocyte random sequence DNA used in the reconstitution of the particles shown in lanes 3 and 4. **E**. SDS-polyacrylamide gel electrophoresis depicting the destabilizing properties of the H2A.Z–H2B dimer. Lane 1 shows a standard histone marker, purified from chicken erythrocytes. Lane 2 displays the histone octamers used to reconstitute mononucleosomes in (**D**), consisting of two copies of H2A.Z, H2B, H3, and H4. Lane 3 exhibits the same octamers observed in lane 2 after concentration through a centrifugal filtration device (Centricon YM-50, Millipore Corp., Bedford, MA). This protein gel suggests that the apparent decrease in particle mass revealed in (**D**) lane 4 and implied by solution studies (**C**), can be attributed to the preferential dissociation of H2A.Z–H2B dimers during particle analysis.

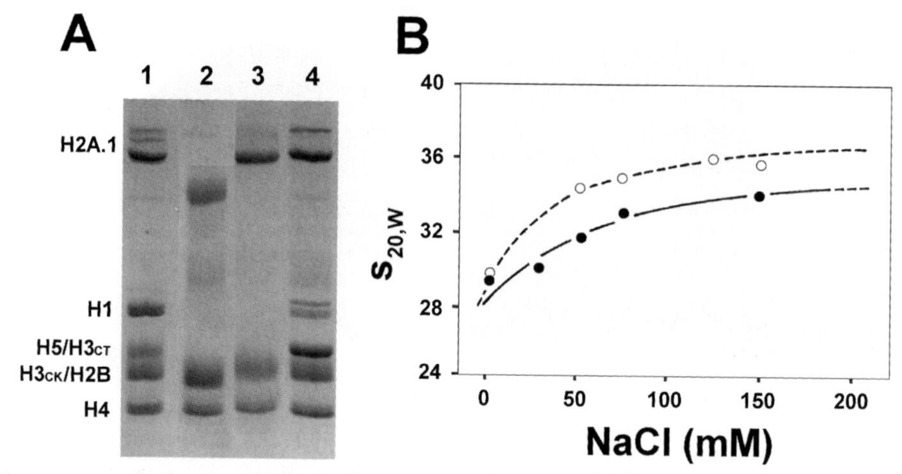

Fig. 10. **A**. Acetic acid-urea-triton-X-100 polyacrylamide gel electrophoresis [15] of the histones used to reconstitute 208-12 nucleosome arrays consisting of recombinant H2A.Z (lane 2) or recombinant H2A.1 (lane 3). Lanes 1 and 4 respectively are chicken erythrocyte and calf thymus histones used as markers [42]. **B**. Ionic strength (NaCl concentration) dependence of the average sedimentation coefficient ($s_{20,w}$) of reconstituted 208-12 nucleosome arrays containing either H2A.1 (○) or H2A.Z (●) [42]. The dotted line represents the behavior of a 208-12 complex reconstituted with chicken erythrocyte histones [406]. [Reproduced from Abbott D.W. *et al.* (2001) J. Biol. Chem. 276, 41945–41949, with permission from The American Society for Biochemistry and Molecular Biology.]

association of the arrays. However, while the sedimentation velocity and magnesium chloride solubility experiments shown in this paper appear to support these conclusions the micrococcal nuclease digestion of their reconstituted complexes (Fig. 1B from Ref. [375]) does not. Indeed it appears that the rH2A.Z 208-12 complexes are digested much faster than the rH2A counterpart, a result that would support a less folded conformation of the H2A.Z complexes as indicated by the sedimentation velocity analysis described in Ref. [42].

6.2. Histone acetylation

Histone acetylation is without a doubt one of the most thoroughly characterized post-translational modifications of histones where both the functional (see Section 3.1) and structural implications for chromatin have been explored. In the sections that follow we are going to summarize the major structural effects of this post-translational modification as they pertain to the nucleosome and the chromatin fiber.

6.2.1. The structure of the acetylated nucleosome core particle

The early characterization of acetylated nucleosomes suggested that this histone modification imparted some important structural changes at the level of the nucleosome core particle. It was found that the flanking DNA regions at the entry

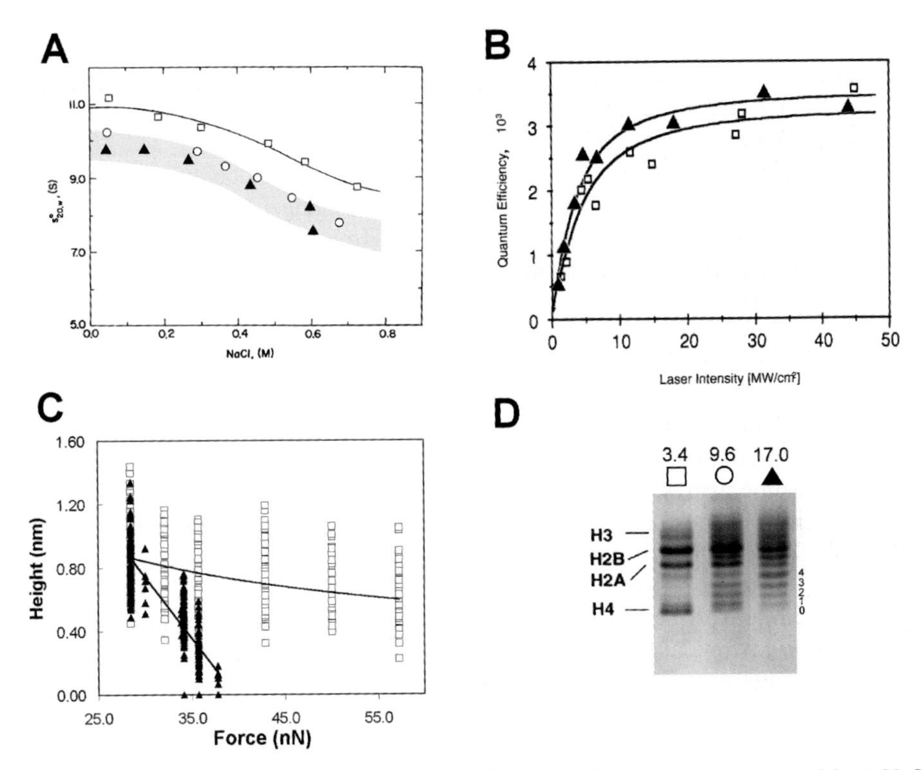

Fig. 11. Effects of histone acetylation on the folding and stability of the nucleosome core particle. **A**. NaCl dependence of the sedimentation coefficient ($s_{20,w}$) of the nucleosome core particles with different extent of acetylation [(□,○,▲), see **D**] [379]. **B**. Dose–response curves obtained with hypo- (□) and hyperacetylated (▲) nucleosomes in 100 mM NaCl after being subjected to UV-laser induced histone–DNA crosslinking [381]. **C**. Height (nanometers, nm) versus force (nanoNewtons, nN) for hypo- (□) and hyperacetylated (▲) nucleosome core particles determined using AFM in the contact mode at different force values [387]. **D**. Acetic acid (5%) urea (2.5 M) polyacrylamide gel electrophoresis analysis [415,416] of the histone components from the nucleosome core particles used in the experiments shown in A–C. The small numbers on the right indicate the number of acetyl residues present in histone H4. The numbers on top indicate the average number of moles of acetyl per mole of histone octamer [379]. [Parts A, B and C were respectively reproduced from Ausió, J. and van Holde, K.E. Biochemistry, 316 (1986), 395–400; Mutskov *et al.*, Mol. Cell. Biol. 18 (1998), 6293–6304; Dunker A.K. *et al.*, J. Mol. Graph. Model. 19 (2001), 26–59, with permission from American Chemical Society, American Society for Microbiology, and Elsevier Science Inc.]

and exit sites of the nucleosome were digested more readily by both DNase I and staphylococcal nuclease [376] and the nucleosome was described to adopt a relaxed "open" structure [377,378]. Upon closer inspection [379] hydrodynamic characterization revealed that the unfolding of the nucleosome core particle was not as extensive as suggested [377]. Hyperacetylated nucleosome core particles exhibited a 10% decrease in sedimentation coefficient (see Fig. 11A). This change in the frictional parameters (increased asymmetry) of the particle coincided with an increase of similar magnitude in the molar ellipticity in the DNA bands and with a 20% increase in the area corresponding to the first melting transition of the nucleosome [379], which in both instances are indicative of DNA being freed from histone binding constraints. Using neutron scattering it was subsequently confirmed that the structural change was indeed less significant than originally anticipated [380].

The question still remains however as to where the 10% change in asymmetry of the nucleosome core particle arises from. This change was originally ascribed to the partial release of the histone tails [379]. However, using UV-laser induced histone–DNA crosslinking [381] it has been recently shown (see Fig. 11B) that below and under physiological salt concentrations (ca. 100–150 mM NaCl), the acetylated tails of the histones are persistently bound to the nucleosomal DNA. This result although unexpected is not completely surprising as histone acetylation affects only lysines residues within these regions and thus it does not alter the arginines which most likely are not released from their DNA interaction until much higher ionic strengths [382]. As it will be discussed in the following section, work with nucleosome arrays has indicated that the increase in the frictional properties and ellipticity are most likely the result of a partial release of the flanking DNA ends of the core particle, in agreement with the above mentioned enhanced nuclease accessibility of these regions [376].

Given the relatively small conformational changes undergone by the nucleosome particle upon histone acetylation, and considering that the majority of the histone tail firmly adheres to the nucleosomal DNA, it is thus not surprising that acetylation was found to have no apparent role in facilitating the accessibility of transcription factors to nucleosomal organized DNA [383,384].

We have analyzed so far the effects of histone acetylation on the structure of the nucleosome core particle, but how does this histone modification affect the stability of the complex? As shown in Fig. 11A, nucleosome core particles with different extents of acetylation (Fig. 11D) exhibit a salt dependence of the sedimentation coefficient that is identical to that displayed by control non-acetylated particles. However, results from electron microscopy visualization of highly acetylated nucleosome complexes consistently revealed the presence of a larger particle asymmetry [385,386] than that determined by the other biophysical approaches. This was taken as an indication that hyperacetylation causes nucleosome core particles to become more sensitive to the forces encountered during sample preparation [385,386] and hinted at the possibility that hyperacetylated nucleosome core particles could be easily destabilized by external forces. To further test this possibility AFM has recently been performed on these nucleoprotein complexes [387]. It is possible with this technique to control the amount of force that is applied to the particles during the viewing process. Furthermore, in contrast to electron microscopy the samples can be studied in a hydrated state at atmospheric pressure. As shown in Fig. 11C histone hyperacetylation has a dramatic effect on the stability of the nucleosome core particles that start falling apart at much lower forces than those required to disrupt the native non-acetylated counterparts when analyzed by this technique [387]. These results while confirming previous observations [385,386] clearly show that histone hyperacetylation poises the particle for destabilization by external forces by providing it with an open structure that facilitates transcription [388,389]. While the molecular aspects involved in these dynamics are not completely clear this finding has important physiological implications. A good candidate for such destabilizing forces *in vivo* could be the torsional stress resulting from DNA supercoiling arising from the movement of RNA polymerase during transcription elongation [390–392].

6.2.2. Is the acetylated chromatin fiber unfolded?

From the beginning, structural work carried out with acetylated chromatin fibers revealed that in contrast to what occurs with the acetylated nucleosome a very small effect is observed within higher order structures [393,394]. These small contributions of histone acetylation to folding [395] contrast with enhanced solubility of the chromatin fiber [153,395]. Such increase in solubility has been shown to depend directly on core histone acetylation and is not related to a differential binding affinity of linker histones nor to an alteration of the relative amounts of the histone H1 variants [395]. All this suggests that while acetylation may have little effect at the intramolecular level it may play an important role in preventing intermolecular interactions and in maintaining the unaggregated state of euchromatin.

It has now been well documented that although the 300 Å conformation of the chromatin fiber is lost in the absence of linker histones (see Fig. 8B–D and Fig. 13B(I)), under physiological ionic strength concentrations the fiber still retains a partly folded organization [100,358,396] (see Fig. 13(II)). In contrast, core histone acetylation in the absence of linker histones produces a highly extended fiber conformation resulting from the unwrapping of the DNA regions flanking the nucleosome [369] (see Fig. 12 and Fig. 13(III)). DNA unwrapping of these regions is fully consistent with the finding that histone acetylation reduces the nucleosome core particle linking number change, as reported by Norton *et al.* [397] using oligonucleosome complexes reconstituted onto circular DNA templates. The physiological relevance of these observations stems from the well established relation existing between histone acetylation and transcription ([129], see Ref. [398] for a review) as well as with the evidence for linker histone depletion ([399], see Refs. [400,401] for a review) in transcriptionally active regions. In addition, the partial DNA release from the flanking regions of the nucleosome can account for the asymmetry of the nucleosome core particle described in the previous section. Although the detailed molecular determinants of this DNA release are not known they could arise from the protraction of the histone tails resulting from the increased helical content of these regions upon acetylation [402].

The major structural effects of histone acetylation that affect both the nucleosome core particle and the chromatin fiber are schematically summarized in Fig. 13.

6.3. Histone H2A ubiquitination

The bulky nature of the ubiquitin adduct [403] and the site in histone H2A (K-119) where ubiquitination takes place (see Fig. 5) has fueled speculation that this

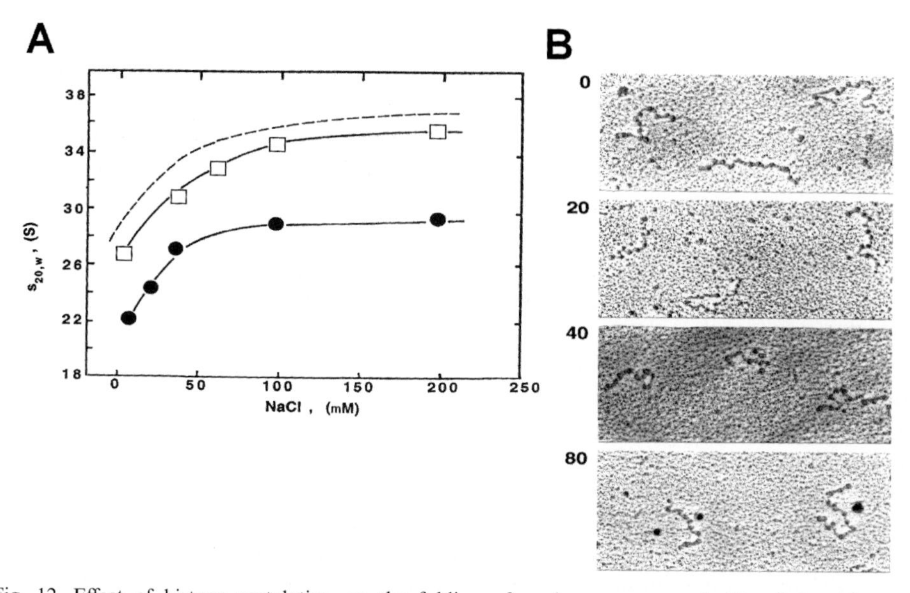

Fig. 12. Effect of histone acetylation on the folding of nucleosome arrays lacking linker histones. **A**. Effect of the ionic strength (mM NaCl concentration) on the average sedimentation coefficient ($s_{20,w}$) of (208-12) oligonucleosome arrays reconstituted with HeLa cell native histone octamers [solid line, □], chicken erythrocyte histone octamers (broken line) [369], or hyperacetylated. HeLa cell histones 208-12 oligonucleosome complexes reconstituted with hyperacetylated HeLa cell histones (●) [369]. **B**. Effect of the ionic strength on hyperacetylated 208-12 nucleosome arrays as visualized by electron microscopy. The numbers to the left indicate the milimolar NaCl concentration [369]. [Reproduced from Garcia-Ramirez M. *et al.* (1995) J. Biol. Chem. 270, 17923–17928, with permission from The American Society for Biochemistry and Molecular Biology.]

post-translational modification must have important structural consequences for both the nucleosome and chromatin fiber [203]. Indeed the C-terminal tail of histone H2A occupies a region of the nucleosome close to the binding site of linker histones [45,182,320] and therefore it could be expected to significantly affect the binding of these histones and hence affect the higher order architecture.

In the early work on this topic, Kleinschmidt and Martinson [219] used reconstituted nucleosome core particles consisting of two uH2A molecules to find that these particles had structural features that were almost indistinguishable from those of particles reconstituted with native histones or from native nucleosome core particles. A more exhaustive and detailed study in the same topic carried out a few years later [220] came in corroboration of these results and showed that the same observations could be extended to nucleosomes reconstituted with uH2B. Furthermore, the NaCl dependent stability of the uH2A reconstituted nucleosome core particles is identical to that exhibited by the native counterpart [222].

In an attempt to study the role of this histone post-translational modification on the folding of the chromatin fiber, Jason *et al.* [221] performed a series of experiments using nucleosome arrays reconstituted onto the 208-12 DNA template.

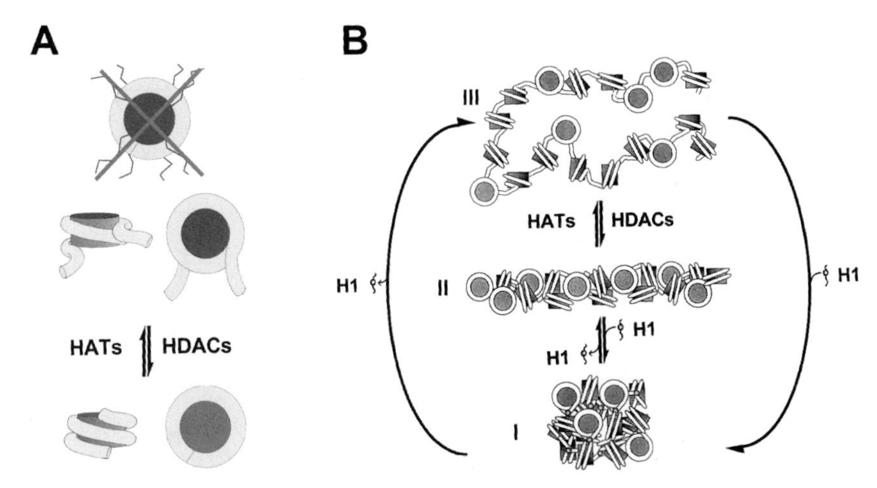

Fig. 13. **A**. Schematic representation of the effects exerted by histone acetylation on the nucleosome core particle. A view from top (right) and from the side (left) are represented. As shown in the model, dynamic histone acetylation resulting from the concerted action of histone acetyltransferases (HATs) and histone deacetylases (HDACs) *in vivo* can result in a partial unfolding of the nucleosome core particle, such an unfolding results from the release of the nucleosomal DNA at the entry and exit sites of the particle but NOT from the release of the histone tails. This phenomenon is often portrayed misleadingly in the literature. **B**. Schematic representation of the effects of histone acetylation on the folding dynamics of the chromatin fiber. (**I**) In the presence of linker histones (Histone H1) the chromatin fiber retains a folded conformation, which is almost identical regardless of the acetylation extent of the core histones. (**II**) Removal or displacement of the linker histones results in a partial unfolding of the chromatin fiber in which, under physiological ionic conditions, nucleosomes are still clustered together. (**III**) The action of HATs either preceding or following linker histone displacement results in a highly unfolded chromatin fiber resulting from the unwinding of the DNA regions flanking the nucleosome particle.

Analyses of the reconstituted complexes by quantitative agarose gel electrophoresis [404,405] and analytical ultracentrifugation [266,406] in the presence of $MgCl_2$ showed that the arrays were able to fold in a way that is almost indistinguishable from complexes reconstituted with major histones (see Fig. 14A–B). Despite this, it was found that histone H2A ubiquitination affects the $MgCl_2$ solubility of the reconstituted complexes (see Fig. 14C) suggesting that this modification may play a role in enhancing the intermolecular associations between chromatin fibers [221].

Although the effect of uH2A on chromatin folding in the presence of linker histones has not yet been characterized, preliminary results ([222] and Jason, L.J.M., unpublished results) suggest that the binding of linker histones to uH2A nucleosomes is not adversely affected by the presence of ubiquitinated H2A as it could have been anticipated. This observation is in agreement with earlier crosslinking studies carried out *in vivo* [407] and *in vitro* [408] which showed that in both instances uH2A could be crosslinked to histone H1 arguing for a close proximity between these two histones.

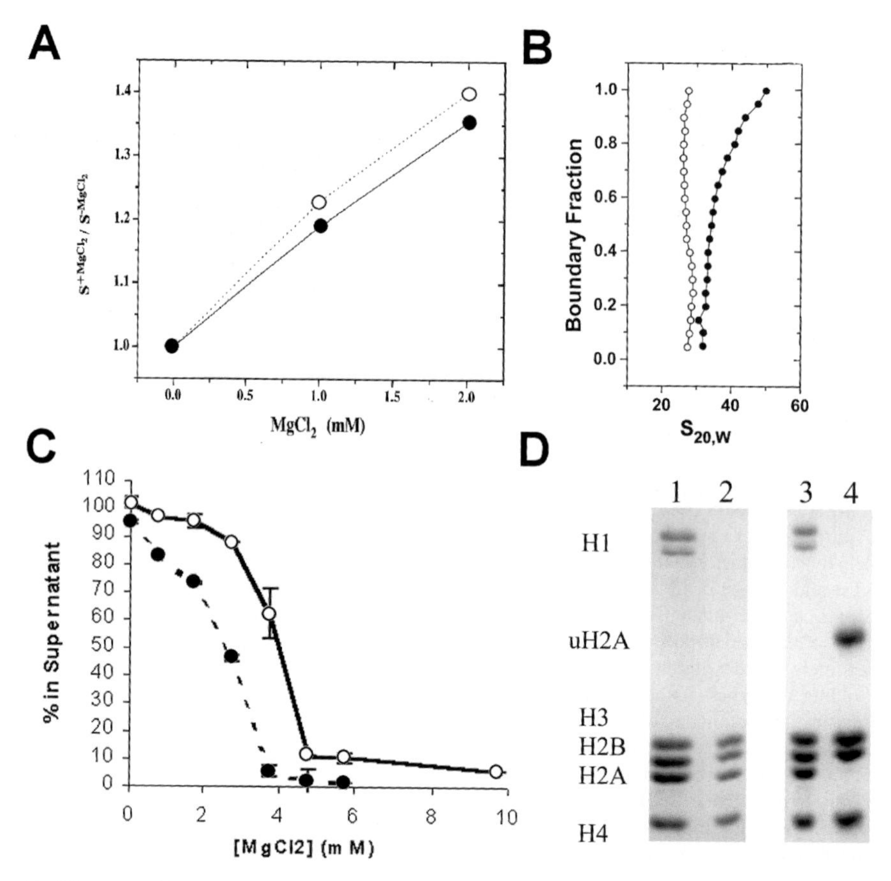

Fig. 14. Effects of histone H2A ubiquitination on the folding and solubility of chromatin [221]. **A**. Magnesium chloride dependence of the sedimentation coefficient ($s_{20,w}$) of 208-12 nucleosome arrays. Sedimentation coefficients at a given magnesium chloride concentration (S^{+MgCl2}) are plotted relative to the sedimentation coefficient in the starting buffer [10 mM Tris–HCl (pH 7.5)] in the absence of magnesium (S^{-MgCl2}). Arrays reconstituted with native histones are designated by (○); the average of two independent experiments using arrays reconstituted with uH2A is designated by (●). **B**. Integral distribution of the sedimentation coefficient of uH2A containing nucleosome arrays in the absence (○) or in the presence (●) of 2 mM $MgCl_2$. **C**. $MgCl_2$ solubility of 208-12 nucleosome arrays reconstituted with native (○) and with uH2A (●) histones. **D**. SDS polyacrylamide gel electrophoresis [13] of the histones used to reconstitute 208-12 nucleosome arrays consisting of native H2A (lane 2) or uH2A (lane 4). Lanes 1 and 3 are calf thymus histones used as markers. [Reproduced from Jason, L.J.M. *et al.*, J. Biol. Chem. 276 (2001), 14597–14601, with permission from The American Society for Biochemistry and Molecular Biology.]

The still limited structural information on this intriguing post-translational modification seems to suggest that at least in the case of uH2A its most important role may be merely informational. Indeed, it has been suggested that histone ubiquitination could label specific chromatin domains [409–411] and as such could be a component of the histone code [121,123,165]. The structural results obtained so far with uH2A nucleosomes and nucleosome arrays clearly support this notion.

7. *Concluding remarks*

In this chapter we have presented a brief structural overview of the modifications of the primary structure and at the post-translational level that contribute to histone variability. We have also discussed the different contributions of histones and DNA to the stability and dynamics of the chromatin nucleoprotein complex. In Section 6 we have integrated both topics using several representative examples from experimental results available from recent and past literature. The results described in this section provide a good example of the dual informational and structural role of histone variability.

The exploration of the informational "coding hypothesis" has grown almost exponentially over the past few years and the results obtained have proven to be very rewarding in terms of connecting histone structure to function. The availability of a powerful DNA template (208-12) [355] in combination with the use of recombinant histones [279] allows the reconstitution of highly homogeneous chromatin structures that behave like native complexes of equivalent nucleosome loading [358]. Such complexes are amenable to analysis by a wide variety of biophysical techniques providing an invaluable tool in ascertaining the role played by histone variability on chromatin structure (see Refs. [356,357] for recent examples). As these structural results keep coming forward, the understanding of the detailed aspects involved in the modulation of chromatin function should prove to be equally rewarding.

References

1. Eickbush, T.H. and Moudrianakis, E.N. (1978) Biochemistry 17, 4955–4964.
2. Kornberg, R.D. (1974) Science 184, 868–871.
3. Lilley, D.M., Howarth, O.W., Clark, V.M., Pardon, J.F., and Richards, B.M. (1976) FEBS Lett. 62, 7–10.
4. Wilkins, M.M.F. (1959) In: Stoops, R. (ed.) Nucleoproteins. Interscience Publishers Inc., New York, pp. 45–76.
5. van Holde, K.E. (1989) In: Rich, A. (ed.) Chromatin. Springer Verlag, New York, pp. 1–497.
6. Grunstein, M. (1990) Annu. Rev. Cell Biol. 6, 643–678.
7. Grunstein, M. (1990) Trends Genet. 6, 395–400.
8. Wolffe, A.P. (1992) FASEB J. 6, 3354–3361.
9. Peterson, C.L. (2001) Biochem. Cell Biol. 79, 219–225.
10. Vignali, M., Hassan, A.H., Neely, K.E., and Workman, J.L. (2000) Mol. Cell. Biol. 20, 1899–1910.
11. Marmorstein, R. and Berger, S.L. (2001) Gene 272, 1–9.
12. West, M.H.P. and Bonner, W.M. (1980) Biochemistry 19, 3238–3245.
13. Laemmli, U.K. (1970) Nature 227, 680–685.
14. Zweidler, A. (1978) Methods Cell Biol. 17, 223–233.
15. Bonner, W.M., West, M.H., and Stedman, J.D. (1980) Eur. J. Biochem. 109, 17–23.
16. Newrock, K.M., Cohen, L.H., Hendricks, M.B., Donnelly, R.J., and Weinber, E.S. (1978) Cell 14, 327–336.
17. Ausió, J. and Abbott, D.W. (2002) Biochemistry 41, 5945–5949.
18. Downs, J.A., Lowndes, N.F., and Jackson, S.P. (2000) Nature 408, 1001–1004.

19. Rogakou, E.P., Pilch, D.R., Orr, A.H., Ivanova, V.S., and Bonner, W.M. (1998) J. Biol. Chem. 273, 5858–5868.
20. Redon, C., Pilch, D., Rogakou, E., Sedelnikova, O., Newrock, K., and Bonner, W. (2002) Curr. Opin. Genet. Dev. 12, 162–169.
21. Rogakou, E.P., Boon, C., Redon, C., and Bonner, W.M. (1999) J. Cell. Biol. 146, 905–916.
22. Paull, T.T., Rogakou, E.P., Yamazaki, V., Kirchgessner, C.U., Gellert, M., and Bonner, W.M. (2000) Curr. Biol. 10, 886–895.
23. Mahadevaiah, S.K., Turner, J.M.A., Baudat, R., Rogakou, E.P., de Boer, P., Blanco-Rodriguez, J., Jasin, M., Keeney, S., Bonner, W.M., and Burgoyne, P.S. (2001) Nat. Genet. 27, 271–276.
24. Rogakou, E.P., Nieves-Neira, W., Boon, C., Pommier, Y., and Bonner, W.M. (2000) J. Biol. Chem. 275, 9390–9395.
25. Talasz, H., Helliger, W., Sarg, B., Debbage, P.L., Puschendorf, B., and Lindner, H. (2002) Cell Death Differ. 9, 27–39.
26. Peterson, S., Casellas, R., Reina-San-Martin, B., Chen, H.T., Difilippantonio, M.J., Wilson, P.C., Hanitsch, L., Celeste, A., Muramatsu, M., Pilch, D.R., Redon, C., Ried, T., Bonner, W.M., Honjo, T., Nussenzweig, M.C., and Nussenzweig, A. (2001) Nature 414, 660–665.
27. Chen, H.T., Bhandoola, A., Difilippantonio, M.J., Zhu, J., Brown, M.J., Tai, X., Rogakou, E.P., Brotz, T.M., Bonner, W.M., Ried, T., and Nussenzweig, A. (2000) Science 290, 1962–1965.
28. Ward, I.M. and Chen, J. (2001) J. Biol. Chem. 276, 47759–47762.
29. Ausió, J., Abbott, D.W., Wang, X., and Moore, S.C. (2001) Biochem. Cell Biol. 79, 693–708.
30. Green, C.M. and Almouzni, G. (2002) EMBO Rep. 3, 28–33.
31. Thoma, F. (1999) EMBO J. 18, 6585–6598.
32. Hunter, N., Valentin Borner, G., Lichten, M., and Kleckner, N. (2001) Nat. Genet. 27, 236–238.
33. Celeste, A., Petersen, S., Romanienko, P.J., Fernandez-Capetillo, O., Chen, H.T., Sedelnikova, O.A., Reina-an-Martin, B., Coppola, V., Meffre, E., Difilippantonio, M.J., Redon, C., Pilch, D.R., Olaru, A., Eckhaus, M., Camerini-Otero, R.D., Tessarollo, L., Livak, F., Manova, K., Bonner, W.M., Nussenzweig, M.C., and Nussenzweig, A. (2002) Science 296, 922–927.
34. Jackson, J.D. and Gorovsky, M.A. (2000) Nucleic Acids Res. 27, 3811–3816.
35. Faast, R., Thonglairoam, V., Schulz, T.C., Beall, J., Wells, J.R.E., Taylor, H., Matthaei, K., Rathjen, P.D., Tremethick, D.J., and Lyons, I. (2001) Curr. Biol. 11, 1183–1187.
36. van Daal, A. and Elgin, S.C.R. (1992) Mol. Biol. Cell 3, 593–602.
37. Clarkson, M.J., Wells, J.R.E., Gibson, F., Saint, R., and Tremethick, D.J. (1999) Nature 399, 694–697.
38. Santisteban, M.S., Kalashnikova, T., and Smith, M.M. (2000) Cell 103, 411–422.
39. Adam, M., Robert, F., Larochelle, M., and Gaudreau, L. (2001) Mol. Cell. Biol. 21, 6270–6279.
40. Dhillion, N. and Kamakaka, R.T. (2000) Mol. Cell 6, 769–780.
41. Suto, R.K., Clarkson, M.J., Tremethick, D.J., and Luger, K. (2000) Nat. Struct. Biol. 7, 1121–1124.
42. Abbott, D.W., Ivanova, V.S., Wang, X., Bonner, W.M., and Ausió, J. (2001) J. Biol. Chem. 276, 41945–41949.
43. Leach, T.J., Mazzeo, M., Chotkowski, H.L., Madigan, J.P., Wotring, M.G., and Glaser, R.L. (2000) J. Biol. Chem. 275, 23267–23272.
44. Allis, C.D., Ziegler, Y.S., Gorovsky, M.A., and Olmsted, J.B. (1982) Cell 31, 131–136.
45. Luger, K., Mader, A.W., Richmond, R.K., Sargent, D.F., and Richmond, T.J. (1997) Nature 389, 251–260.
46. Lee, Y., Hong, M., Kim, J.W., Hong, Y.M., Choe, Y.-K., Chang, S.Y., Lee, K.S., and Choe, I.S. (1998) Biochim. Biophys. Acta 1399, 73–77.
47. Changolkar, L.N. and Pehrson, J.R. (2002) Biochemistry 41, 179–184.
48. Perche, P.Y., Vourc'h, C., Konecny, L., Souchier, C., Robert-Nicoud, M., Dimitrov, S., and Khochbin, S. (2000) Curr. Biol. 10, 1531–1534.
49. Costanzi, C. and Pehrson, J.P (1998) Nature 393, 599–601.
50. Rasmussen, T., Huang, T., Mastrangelo, M.-A., Loring, J., Panning, B., and Jaenisch, R. (1999) Nucleic Acids Res. 27, 3685–3689.

51. Pehrson, J.P. and Fuji, R.N. (1998) Nucleic Acids Res. 26, 2837–2842.
52. LaStarza, M.W., Lemm, J.A., and Rice, C.M. (1994) J. Virol. 68, 5781–5791.
53. Boumil, R.M. and Lee, J.T. (2001) Hum. Mol. Genet. 10, 2225–2232.
54. Mermoud, J.E., Popova, B., Peters, A.H., Jenuwein, T., and Brockdorff, N. (2002) Curr. Biol. 12, 247–251.
55. Csankovszki, G., Panning, B., Bates, B., Pehrson, J., and Jaenisch, R. (1999) Nat. Genet. 22, 323–324.
56. Richler, C., Ast, G., Goitein, R., Wahrman, J., Sperling, R., and Sperling, J. (1994) Mol. Biol. Cell. 5, 1341–1352.
57. Richler, C., Dhara, S.K., and Wahrman, J. (2000) Cytogent. Cell Genet. 89, 118–120.
58. Hoyer-Fender, S., Costanzi, C., and Pehrson, J.P. (2000) Exp. Cell Res. 258, 254–260.
59. Chadwick, B.P. and Willard, H.F. (2001) J. Cell Biol. 152, 375–384.
60. Eickbush, T.H., Godfrey, J.E., Elia, M.C., and Moudrianakis, E.N. (1988) J. Biol. Chem. 263, 18972–18978.
61. Chadwick, B.P. and Willard, H.F. (2001) Hum. Mol. Genet. 10, 1101–1113.
62. Saffery, R., Irvine, D.V., Griffiths, B., Kalitsis, P, Wordeman, L., and Choo, K.H.A. (2000) Hum. Mol. Genet. 9, 175–185.
63. Pidoux, A. and Allshire, R.C. (2000) Curr. Opin. Cell Biol. 12, 308–319.
64. Sullivan, B.A., Schwartz, S., and Willard, H.F. (1996) Environ. Mol. Mutat. 28, 128–191.
65. Palmer, D.K., O'Day, K., Wener, M.H., Andrews, B.S., and Margolis, R.L. (1987) J. Cell Biol. 104, 805–815.
66. Yoda, K., Ando, S., Morishita, S., Houmura, K., Hasimoto, K., Takeyas, K., and Okazaki, T. (2000) Proc. Natl. Acad. Sci. USA 97, 7266–7271.
67. Masumoto, H., Masukata, H., Muro, Y., Nozaki, N., and Okazaki, T. (1989) J. Cell Biol. 109, 1963–1973.
68. Zhang, X.-Y., Fittler, F., and Horz, W. (1983) Nucleic Acids Res. 11, 4275–4306.
69. Ando, S., Yang, J., Nozaki, N., Okazaki, T., and Yoda, K. (2002) Mol. and Cell. Biol. 22, 2229–2241.
70. Saitoh, H., Tomkiel, J., Cooke, C.A., Ratrie, H., Maurer, M., Rothfield, N., and Earnshaw, W.C. (1992) Cell 70, 115–125.
71. Tomkiel, H.E., Cook, C.A., Saitoh, H., Bernat, R.L., and Earnshaw, W.C. (1994) J. Cell Biol. 125, 531–545.
72. Meluh, P.B. and Strunnikov, A.V. (2002) Eur. J. Biochem. 269, 2300–2314.
73. Lombillo, V.A., Nislow, C., Yen, T.J., Gelfand, V.I., and McIntosh, J.R. (1995) J. Cell Biol. 128, 107–115.
74. Sullivan, B.A. and Schwartz, S. (1995) Hum. Mol. Genet. 4, 2189–2197.
75. Rattner, J.B., Rao, A., Fritzler, M.J., Valencia, D.W., and Yen, T.J. (1993) Cell Motil. Cytoskel. 26, 214–226.
76. He, D., Zeng, C., Woods, K., Zhong, L., Turner, D., Busch, R.K., Brinkley, B.R., and Busch, H. (1998) Chromosoma 107, 189–197.
77. Warburton, P.E. (2001) Trends Genet. 17, 243–247.
78. Sugata, N., Li, S., Earnshaw, W.C., Yen, T.J., Yoda, K., Maumoto, H., Munekata, E., Warburton, P.E., and Todokoro, K. (2000) Hum. Mol. Genet. 9, 2916–2919.
79. Fukagawa, T., Mikami, Y., Nishihashi, A., Regnier, V., Haraguchi, T., Hiraoka, Y., Sugata, N., Todokoro, K., Brown, W., and Ikemura, T. (2001) EMBO J. 20, 4603–4617.
80. Sullivan, B.A., Hechenberger, M., and Masri, K. (1994) J. Cell Biol. 127, 581–592.
81. Arents, G. and Moudrianakis, E.N. (1995) Proc. Natl. Acad. Sci. USA 92, 11170–11174.
82. Glowczewski, L., Yang, P., Kalashnikova, T., Santisteban, S.M., and Smith, M. (2000) Mol. Cell. Biol. 20, 5700–5711.
83. Howman, E.V., Fowler, K.J., Newson, A.J., Redward, S., MacDonald, A.C., Kalitsis, P., and Choo, K.H.A. (2000) Proc. Natl. Acad. Sci. USA 97, 1148–1153.
84. Chen, Y., Baker, R.E., Keith, K.C., Harris, K., Stoler, S., and Fitzgerald-Hayes, M. (2000b) Mol. Cell. Biol. 20, 7037–7048.

85. Keith, K.C., Baker, R.E., Chen, Y., Harris, K., Stoler, S., and Fitzgerald-Hayes, M. (1999) Mol. Cell. Biol. 19, 6130–6139.
86. Cole, R.D. (1984) Anal. Biochem. 136, 24–30.
87. Cole, R.D. (1987) Pept. Protein Rev. 30, 440–449.
88. Hill, C.S., Martin, S.R., and Thomas, J.O. (1989) EMBO J. 8, 2591–2599.
89. Clark, D.J., Hill, C.S., Martin, S.R., and Thomas, J.O. (1988) EMBO J. 7, 69–75.
90. Vila, R., Ponte, I., Collado, M., Arrondo, J.L., Jimenez, M.A., Rico, M., and Suau, P. (2001) J. Biol. Chem. 276, 46429–46435.
91. Vila, R., Ponte, I., Collado, M., Arrondo, J.L., and Suau, P. (2001) J. Biol. Chem. 276, 30898–30903.
92. Crane-Robinson, C. (1999) Bioessays 21, 367–371.
93. Kasinsky, H.E., Lewis, J.D., Dacks, J.B., and Ausió, J. (2001) FASEB J. 15, 34–42.
94. Zarbock, J., Clore, G.M., and Gronenborn, A.M. (1986) Proc. Natl. Acad. Sci. USA 83, 7628–7632.
95. Cerf, C., Lippens, G., Ramakrishnan, V., Muyldermans, S., Segers, A., Wyns, L., Wodak, S.J., and Hallenga, K. (1994) Biochemistry 33, 11079–11086.
96. Ramakrishnan, V., Finch, J.T., Graziano, V., Lee, P.L., and Sweet, R.M. (1993) Nature 362, 219–223.
97. Neelin, J.M., Callahan, P.X., Lamb, D.C., and Murray, K. (1964) Can. J. Biochem. 42, 1743–1752.
98. Clark, K.L., Halay, E.D., Lai, E., and Burley, S.K. (1993) Nature 364, 412–420.
99. Ausió, J. (1999) Bioessays 22, 873–877.
100. Shen, X., Yu, L., Weir, J.W., and Gorovsky, M.A. (1995) Cell 82, 47–56.
101. Shen, X. and Gorovsky, M.A. (1996) Cell 86, 475–483.
102. Barra, J.L., Rhounim, L., Rossignol, J.L., and Faugeron, G. (2000) Mol. Cell. Biol. 20, 61–69.
103. Ramon, A., Muro-Pastor, M.I., Scazzocchio, C., and Gonzalez, R. (2000) Mol. Microbiol. 35, 223–233.
104. Escher, D. and Schaffner, W. (1997) Mol. Gen. Genet. 256, 456–461.
105. Patterton, H.G., Landel, C.C., Landsman, D., Peterson, C.L., and Simpson, R.T. (1998) J. Biol. Chem. 273, 7268–7276.
106. Sun, J.M., Ali, Z., Lurz, R., and Ruiz-Carrillo, A. (1990) EMBO J. 9, 1651–1658.
107. Rabini, S., Franke, K., Saftig, P., Bode, C., Doenecke, D., and Drabent, B. (2000) Exp. Cell Res. 255, 114–124.
108. Sirotkin, A.M., Edelmann, W., Cheng, G., Klein-Szanto, A., Kucherlapati, R., and Skoultchi, A.I. (1995) Proc. Natl. Acad. Sci. USA 92, 6434–6438.
109. Lin, Q., Sirotkin, A., and Skoultchi, A.I. (2000) Mol. Cell. Biol. 20, 2122–2128.
110. Drabent, B., Saftig, P., Bode, C., and Doenecke, D. (2000) Histochem. Cell. Biol. 113, 433–442.
111. Takami, Y., Nishi, R., and Nakayama, T. (2000) Biochem. Biophys. Res. Commun. 268, 501–508.
112. Seyedin, S.M., Cole, R.D., and Kistler, W.S. (1981) Exp. Cell Res. 136, 399–405.
113. Doenecke, D., Drabent, B., Bode, C., Bramlage, B., Franke, K., Gavenis, K., Kosciessa, U., and Witt, O. (1997) Adv. Exp. Med. Biol. 424, 37–48.
114. Kochbin, S. (2001) Gene 271, 1–12.
115. Poccia, D.L. and Green, G.R. (1992) Trends Biochem. Sci. 17, 223–237.
116. Tanaka, M., Hennebold, J.D., Macfarlane, J., and Adashi, E.Y. (2001) Development 128, 655–664.
117. Zlatanova, J. and Doenecke, D. (1994) FASEB J. 8, 1260–1268.
118. Ausió, J. (1999) J. Biol. Chem. 274, 3115–3118.
119. Watson, C.E. and Davies, P.L. (1998) J. Biol. Chem. 273, 6157–6162.
120. Bandiera, A., Patel, U.A., Manfioletti, G., Rustighi, A., Giancotti, V., and Crane-Robinson, C. (1995) Eur. J. Biochem. 233, 744–749.
121. Strahl, B.D. and Allis, C.D. (2000) Nature 403, 41–45.
122. Turner, B.M. (2000) Bioessays 22, 836–845.
123. Rice, J.C. and Allis, C.D. (2001) Nature 414, 258–261.
124. Rice, J.C. and Allis, C.D. (2001) Curr. Opin. Cell Biol. 13, 263–273.
125. Thorne, A.W., Kmiciek, D., Mitchelson, K., Sautiere, P., and Crane-Robinson, C. (1990) Eur. J. Biochem. 193, 701–713.

126. Hebbes, T.R., Clayton, A.L., Thorne, A.W., and Crane-Robinson, C. (1994) EMBO J. 13, 1823–1830.
127. Vogelauer, M., Wu, J., Suka, N., and Grunstein, M. (2000) Nature 408, 495–498.
128. Phillips, D.M.P. (1963) Biochem. J. 87, 258–263.
129. Allfrey, V.G., Faulkner, R.M., and Mirsky, A.E. (1964) Proc. Natl. Acad. Sci. USA 51, 786–794.
130. Brownell, J.E. and Allis, C.D. (1996) Curr. Opin. Genet. Dev. 6, 176–184.
131. Brownell, J.E., Zhou, J., Ranalli, T., Kobayashi, R., Edmondson, D.G., Roth, S.Y., and Allis, C.D. (1996) Cell 84, 843–851.
132. Spencer, V.A. and Davie, J.R. (1999) Gene 240, 1–12.
133. Brown, C.E., Lechner, T., Howe, L., and Workman, J.L. (2000) Trends Biochem. Sci. 25, 15–19.
134. Taunton, J., Hassig, C.A., and Schreiber, S.L. (1996) Science 272, 408–411.
135. Johnson, C.A. and Turner, B.M. (1999) Semin. Cell Dev. Biol. 10, 179–188.
136. Ng, H.H. and Bird, A. (2000) Trends Biochem. Sci. 25, 121–126.
137. Cress, W.D. and Seto, E. (2000) J. Cell. Physiol. 184, 1–16.
138. Khochbin, S., Verdel, A., Lemercier, C., and Seigneurin-Berny, D. (2000) Curr. Opin. Genet. Dev. 11, 162–166.
139. Archer, S.Y. and Hodin, R.A. (1999) Curr. Opin. Genet. Dev. 9, 171–174.
140. Davie, J.R., Samuel, S.K., Spencer, V.A., Holth, L.T., Chadee, D.N., Peltier, C.P., Sun, J.M., Chen, H.Y., and Wright, J.A. (1999) Biochem. Cell. Biol. 77, 265–275.
141. Lindner, H., Sarg, B., Grunicke, H., and Helliger, W. (1999) J. Cancer Res. Clin. Oncol. 125, 182–186.
142. Davie, J.R. and Spencer, V.A. (2000) Prog. Nucleic Acids Res. Mol. Biol. 65, 299–340.
143. Jacobson, S. and Pillus, L. (1999) Curr. Opin. Genet. Dev. 9, 175–184.
144. Gray, S.G. and Teh, B.T. (2001) Curr. Mol. Med. 1, 401–429.
145. Conley, B.A., Egorin, M.J., Tait, N., Rosen, D.M., Sausville, E.A., Dover, G., Fram, R.J., and Van Echo, D.A. (1998) Clin. Cancer Res. 4, 629–634.
146. Annuziato, A.T. and Hansen, J.C. (2000) Gene Expr. 9, 37–61.
147. Oliva, R. and Dixon, G.H. (1991) Prog. Nucleic Acids Res. Mol. Biol. 40, 25–94.
148. Wouters-Tyrou, D., Martin-Ponthieu, A., Sautiere, P., and Biserte, G. (1981) FEBS Lett. 128, 195–200.
149. Emerson, B. (2002) Cell 109, 267–270.
150. Parekh, B.S. and Maniatis, T. (1999) Mol. Cell 3, 125–129.
151. Hassan, A.H., Neely, K.E., and Workman, J.L. (2001) Cell 104, 817–827.
152. Struhl, K. (1998) Genes Dev. 12, 599–606.
153. Perry, M. and Chalkley, R. (1982) J. Biol. Chem. 257, 7336–7347.
154. Clayton, A.L., Hebbes, T.R., Thorne, A.W., and Crane-Robinson, C. (1993) FEBS Lett. 336, 23–26.
155. Smith, E.R., Allis, C.D., and Lucchesi, J.C. (2001) J. Biol. Chem. 276, 31483–31586.
156. Litt, M.D., Simpson, M., Recillas-Targa, F., Prioleau, M.-N., and Felsenfeld, G. (2001) EMBO J. 20, 2224–2235.
157. Myers, F.A., Evans, D.R., Clayton, A.L., Thorne, A.W., and Crane-Robinson, C. (2001) J. Biol. Chem. 276, 20197–20205.
158. Davie, J.R. and Chadee, D.N. (1998) J. Cell. Biochem. 30, 203–213.
159. Spencer, V.A. and Davie, J.R. (1999) Gene 240, 1–12.
160. Nowak, S.J. and Corces, V.G. (2000) Genes Dev. 14, 3003–3013.
161. Bradbury, E.M. (1992) Bioessays 14, 9–16.
162. Berger, S.L. (2000) Nature 408, 412–414.
163. Clayton, A.L., Rose, S., Barratt, M.J., and Mahadevan, L.C. (2000) EMBO J. 19, 3714–3726.
164. Zeitlin, S.G., Barber, C.M., Allis, C.D., Sullivan, K.F., and Sullivan, K. (2001) J. Cell Sci. 114, 653–661.
165. Jenuwein, T. and Allis, C.D. (2001) Science 293, 1074–1080.
166. Van Hooser, A., Goodrich, D.W., Allis, C.D., Brinkley, B.R., and Mancini, M.A. (1998) J. Cell Sci. 111, 3497–3506.

167. de la Barre, A.E., Gerson, V., Gout, S., Creaven, M., Allis, C.D., and Dimitrov, S. (2000) EMBO J. 19, 379–391.
168. Bradbury, E.M., Inglis, R.J., Matthews, H.R., and Sarner, N. (1973) Eur. J. Biochem. 33, 131–139.
169. Hans, F. and Dimitrov, S. (2001) Oncogene 20, 3021–3027.
170. Hsu, J.Y., Sun, Z.W., Li, X., Reuben, M., Tatchell, K., Bishop, D.K., Grushcow, J.M., Brame, C.J., Caldwell, J.A., Hunt, D.F., Lin, R., Smith, M.M., and Allis, C.D. (2000) Cell 102, 279–291.
171. Cheung, P., Allis, C.D., and Sassone-Corsi, P. (2000) Cell 103, 263–271.
172. Wagner, T.E., Hartford, J.B., Serra, M., Vandegrift, V., and Sung, M.T. (1977) Biochemistry 16, 286–290.
173. Green, G.R. and Poccia, D.L. (1985) Dev. Biol. 108, 235–245.
174. Dou, Y. and Gorovsky, M.A. (2000) Mol. Cell 6, 225–231.
175. Dou, Y., Mizzen, C.A., Abrams, M., Allis, C.D., and Gorovsky, M.A. (1999) Mol. Cell 4, 641–647.
176. Marion, C., Roux, B., Pallotta, L., and Coulet, P.R. (1983) Biochem. Biophys. Res. Commun. 114, 1169–1175.
177. Leuba, S.H., Bustamante, C., van Holde, K., and Zlanatova, J. (1998) Biophys. J. 74, 2830–2839.
178. Th'ng, J.P., Guo, X.W., Swank, R.A., Crissman, H.A., and Bradbury, E.M. (1994) J. Biol. Chem. 269, 9568–9573.
179. Swank, R.A., Th'ng, J.P., Guo, X.W., Valdez, J., Bradbury, E.M., and Gurley, L.R. (1997) Biochemistry 36, 13761–13768.
180. Roth, S.Y. and Allis, C.D. (1992) Trends Biochem. Sci. 17, 93–98.
181. Ball, A.R., Jr. and Yokomori, K. (2001) Chromosome Res. 9, 85–96.
182. Usachenko, S.I., Bavykin, S.G., Gavin, I.M., and Bradbury, E.M. (1994) Proc. Natl. Acad. Sci. USA 91, 6845–6849.
183. Kaplan, L.J., Bauer, R., Morrison, E., Langan, T.A., and Fasman, G.D. (1984) J. Biol. Chem. 259, 8777–8785.
184. Zhang, Y. and Reinberg, D. (2001) Genes Dev. 15, 2343–2360.
185. Hennig, W. (1999) Chromosoma 108, 1–9.
186. Heard, E., Rougeulle, C., Arnaud, D., Avner, P., Allis, C.D., and Spector, D.L. (2001) Cell 107, 727–738.
187. Lachner, M., O'Carroll, D., Rea, S., Mechtler, K., and Jenuwein, T. (2001) Nature 410, 116–120.
188. Rea, S., Eisenhaber, F., O'Carrol, D., Strahl, B.D., Sun, Z.-W., Schmid, M., Opravil, S., Mechtier, K., Ponting, C.P., Allis, C.D., and Jenuwein, T. (2000) Nature 406, 593–599.
189. Ferreira, R., Naguibneva, I., Pritchard, L.L., Ait-Si-Ali, S., and Harel-Bellan, A. (2001) Oncogene 20, 3128–3133.
190. Nielsen, S.J., Schneider, R., Bauer, U.-M., Bannister, A.J., Morrison, A., O'Carroll, Firestein, R., Cleary, M., Jenuwein, T., Herrera, R.E., and Kouzarides, T. (2001) Nature 412, 561–565.
191. Chen, D., Ma, H., Hong, H, Koh, S.S., Huang, S.M., and Schurter, B.T. (1999) Science 284, 2174–2177.
192. Wang, H., Huang, Z.Q., Xia, L., Feng, Q., Erdjument-Bromage, H., Strahl, B.D., Briggs, S.D., Allis, C.D., Wong, J., Tempst, P., and Zhang, Y. (2001b) Science 293, 853–857.
193. Koh, S.S., Chen, D., Lee, Y.H., and Stallcup, M.R. (2001) J. Biol. Chem. 276, 1089–1098.
194. Strahl, B.D., Ohba, R., Cook, R.G., and Allis, C.D. (1999) Proc. Natl. Acad. Sci. USA 96, 14967–14972.
195. Nishioka, K., Chuikov, S., Sarma, K., Erdjument-Bromage, H., Allis, C.D., Tempst, P., and Reinberg, D. (2002) Genes Dev. 16, 479–489.
196. Wang, H., Cao, R., Xia, L., Erdjument-Bromage, H., Borchers, C., Tempst, P., and Zhang, Y. (2001) Mol. Cell 8, 1207–1217.
197. Zegerman, P., Canas, B., Pappin, D., and Kouzarides, T. (2002) J. Biol. Chem. 277, 11621–11624.
198. Finley, D. and Chau, V. (1991) Ann. Rev. Cell Biol. 7, 25–69.
199. Jennissen, H.P. (1995) Eur. J. Biochem. 231, 1–30.
200. Pickart, C.M. and Rose, I.A. (1985) J. Biol. Chem. 260, 1573–1581.
201. Hershko, A. and Ciechanover, A. (1998) Ann. Rev. Biochem. 67, 425–479.
202. Hass, A., Reback, P.B., and Chau, V. (1991) J. Biol. Chem. 266, 5104–5112.

203. Jason, L.J.M., Moore, S.C., Lewis, J.D., Lindsey, G., and Ausió, J. (2002) Bioessays 24, 166–174.
204. Goebl, M.G., Yochem, J., Jentsch, S., McGrath, J.P., Varshavsky, A., and Byers, B. (1988) Science 241, 1331–1335.
205. Chen, H.Y., Sun, J.-M., Zhang, Y., Daview, J.R., and Meistrich, M.L. (1998) J. Biol. Chem. 273, 13165–13169.
206. Pham, A.-D. and Sauer, F. (2000) Science 289, 2357–2360.
207. Nickel, B.E. and Davie, J.R. (1989) Biochemistry 28, 953–963.
208. Ridsdale, J.A. and Davie, J.R. (1987) Nucleic Acids Res. 15, 1081–1096.
209. Davie, J.R., Lin, R., and Allis, C.D. (1991) Biochem. Cell Biol. 69, 66–71.
210. Davie, J.R. and Murphy, L.C. (1994) Biochem. Biophys. Res. Commun. 203, 344–350.
211. Li, W., Nagaraja, S., Delcuve, G.P., Hendzel, M.J., and Davie, J.R. (1993) Biochem. J. 296, 737–744.
212. Davie, J.R. and Nickel, B.E. (1987) Biochim. Biophys. Acta 909, 183–189.
213. Levinger, L. and Varshavsky, A. (1982) Cell 28, 375–385.
214. Jasinskiene, N., Jasinskas, A., and Langmore, J.P. (1995) Dev. Genet. 16, 278–290.
215. Barsoum, J. and Varshavsky, A. (1985) J. Biol. Chem. 260, 7688–7697.
216. Parlow, M., Haas, A.L., and Lough, J. (1990) J. Biol. Chem. 265, 7507–7512.
217. Dawson, B.A., Herman, T., Haas, A.L., and Lough, J. (1991) J. Cell. Biochem. 46, 166–173.
218. Ballal, N.R., Kang, Y.J., Olson, M.O., and Busch, H. (1975) J. Biol. Chem. 250, 5921–5925.
219. Kleinschmidt, A.M. and Martinson, H.G. (1981) Nucleic Acids Res. 9, 2423–2431.
220. Davies, N. and Lindsey, G.G. (1994) Biochim. Biophys. Acta 1218, 187–193.
221. Jason, L.J.M., Moore, S.C., Lindsey, G., and Ausió, J. (2001) J. Biol. Chem. 276, 14597–14601.
222. Moore, S.C. Jason, L. and Ausió, J. Biochem. Cell Biol. 80, 311–319.
223. Malanga, C., Atorino, L., Tramontano, F., Farina, B., and Quesada, P. (1998) Biochim. Biophys. Acta 1399, 154–160.
224. D'Armours, D., Desnoyers, S., D'Silva, I., and Poirier, G.G. (1999) Biochem. J. 342, 249–268.
225. Realini, C.A. and Althaus, F.R. (1992) J. Biol. Chem. 264, 18858–18865.
226. De Lucia, F., Faraone-Mennella, M.R., D'Erme, M., Quesada, P., Caiafa, P., and Farina, B. (1994) Biochem. Biophys. Res. Commun. 198, 32–39.
227. Khadake, J.R. and Rao, M.R. (1995) Biochemistry 34, 15792–15801.
228. Poireir, G.G., de Murcia, G., Jongstra-Bilen, J., Niedergang, C., and Mandel, P. (1982) Proc. Natl. Acad. Sci. USA 79, 3423–3427.
229. De Murcia, G., Huletsky, A., Lamarre, D., Gaudreau, A., Pouyett, J., Daune, M., and Porier, G.G. (1986) J. Biol. Chem. 261, 7011–7017.
230. Slattery, E., Dignam, D.J., Matsui, T., and Roeder, R.G. (1983) J. Biol. Chem. 258, 5955–5959.
231. Morin, B., Diaz, F., Montecino, M., Fothergill-Gilmore, L, Puchi, M., and Imschenetzky, M. (1999) Biochem. J. 343, 95–98.
232. Althaus, F.R. (1992) J. Cell Sci. 102, 663–670.
233. Althaus, F.R., Hofferer, L., Kleczkowska, H.E., Malanga, M., Naegeli, H., Panzeter, P.L., and Realini, C.A. (1994) Mol. Cell Biochem. 138, 53–59.
234. Kreimeyer, A., Wielckens, K., Adamietz, P., and Hilz, H. (1984) J. Biol. Chem. 259, 890–896.
235. Tanuma, S., Yagi, T., and Johnson, G.S. (1985) Arch. Biochem. Biophys. 237, 38–42.
236. Herceg, Z. and Wang, Z.-Q. (2001) Mutat. Res. 477, 97–110.
237. Bird, A.P. and Wolffe, A.P. (1999) Cell 99, 451–454.
238. Nakao, M. (2001) Gene 278, 25–31.
239. Leonhardt, H. and Cardoso, M.C. (2000) J. Cell. Biochem. 35, S78–S83.
240. Rountree, M.R., Bachman, K.E., Herman, J.G., and Baylin, S.B. (2001) Oncogene 20, 3156–3165.
241. Jones, P.A. and Takai, D. (2001) Science 293, 1068–1070.
242. Macleod, D., Charlton, J., Mullins, J., and Bird, A.P. (1994) Genes Dev. 8, 2292.
243. Brandeis, M., Frank, D., Keshet, I., Siegfried, Z., Mendelsohn, M., Nemes, A., Temper, V., Razin, A., and Cedar, H. (1994) Nature 371, 435–438.
244. Fuks, F., Burgers, W.A., Brehm, A., Hughes-Davies, L., and Kouzarides, T. (2000) Nat. Genet. 24, 88–91.

245. Robertson, K.D., Ait-Si-Ali, S., Yokochi, T., Wade, P.A., Jones, P.L., and Wolffe, A.P. (2000) Nat. Genet. 25, 338–342.
246. Rountree, M.R., Bachman, K.E., and Baylin, S.B. (2000) Nat. Genet. 25, 269–277.
247. Ng, H.H. and Bird, A. (1999) Curr. Opin. Genet. Dev. 9, 158–163.
248. Ohki, I., Shimotake, N., Fujita, N., Jee, J., Ikegami, T., Nakao, M., and Shirakawa, M. (2001) Cell 105, 487–497.
249. Becker, P.B., Ruppert, S., and Schutz, G. (1987) Cell 51, 435–443.
250. Karymov, M.A., Tomschik, M., Leuba, S.H., Caiafa, P., and Zlatanova, J. (2001) FASEB J. 15, 2631–2641.
251. Jones, P.L., Veenstra, G.J., Wade, P.A., Vermaak, D., Kass, S.U., Landsberger, N., Strouboulis, J., and Wolffe, A.P. (1998) Nat. Genet. 19, 187–191.
252. Nan, X., Ng, H.H., Johnson, C.A., Laherty, C.D., Turner, B.M., Eisenman, R.N., and Bird, A. (1998) Nature 393, 386–389.
253. Gregory, R.I., Randall, T.E., Johnson, C.A., Khosla, S., Hatada, I., O'Neill, L.P., Turner, B.M., and Feil, R. (2001) Mol. Cell. Biol. 21, 5426–5436.
254. Tamaru, H. and Selker, E.U. (2001) Nature 414, 277–283.
255. Jackson, J.P., Lindroth, A.M., Cao, X., and Jacobsen, S.E. (2002) Nature 416, 556–560.
256. Wade, P.A. (2001) Bioessays 23, 1131–1137.
257. Lorincz, M.C., Schubeler, D., and Groudine, M. (2001) Mol. Cell. Biol. 21, 7913–7922.
258. El-Osta, A. and Wolffe, A.P. (2000) Gene Expr. 9, 63–75.
259. Pazin, M.J. and Kadonaga, J.T. (1997) Cell 88, 737–740.
260. Aalfs, J.D. and Kingston, R.E. (2000) TIBS 25, 548–555.
261. Peterson, C.L. (2000) FEBS Lett. 476, 68–72.
262. Narlikar, G.J., Fan, H.-Y., and Kingston, R.E. (2002) Cell 108, 475–487.
263. McGhee, J.D. and Felsenfeld, G. (1980) Nucleic Acids Res. 8, 2751–2769.
264. Ausió, J., Seger, D., and Eisenberg, H. (1984) J. Mol. Biol. 176, 77–104.
265. Harp, J.M., Hanson, B.L., Timm, D.E., and Bunick, G.J. (2000) Acta Crystallogr. D Biol. Crystallogr. 12, 1513–1534.
266. Ausió, J. (2000) Biophys. Chem. 86, 141–153.
267. Stein, A. (1979) J. Mol. Biol. 130, 103–134.
268. Eisenberg, H. and Felsenfeld, G. (1981) J. Mol. Biol. 150, 537–555.
269. Stacks, P.C. and Schumaker, V.N. (1979) Nucleic Acids Res. 7, 2457–2467.
270. Cotton, R.W. and Hamkalo, B. (1981) Nucleic Acids Res. 9, 445–457.
271. Vassilev, L., Russev, G., and Tsanev, R. (1981) Int. J. Biochem. 13, 1247–1255.
272. Yager, T.D. and van Holde, K.E. (1984) J. Biol. Chem. 259, 4212–4222.
273. Voordow, G. and Eisenberg, H. (1978) Nature 273, 446–448.
274. Godde, J.S. and Wolffe, A.P. (1996) J. Biol. Chem. 271, 15222–15229.
275. Sengupta, S.M., VanKanegan, M., Persinger, J., Logie, C., Cairns, B.R., Peterson, C.L., and Bartholomew, B. (2001) J. Biol. Chem. 276, 12636–12644.
276. Hill, D.A. and Imbalzano, A.N. (2000) Biochemistry 39, 11649–11656.
277. Gottesfeld, J.M. and Luger, K. (2001) Biochemistry 40, 10927–10933.
278. Yang, T.P., Hansen, S.K., Oishi, K.K., Ryder, O.A., and Hamkalo, B.A. (1982) Proc. Natl. Acad. Sci. USA 79, 6593–6597.
279. Luger, K., Rechsteiner, T.J., Flaus, A.J., Waye, M.M., and Richmond, T.J. (1997) J. Mol. Biol. 272, 301–311.
280. Thastrom, A., Lowary, P.T., Widlund, H.R., Cao, H., Kubista, M., and Widom, J. (1999) J. Mol. Biol. 288, 213–229.
281. Shrader, T.E. and Crothers, D.M. (1989) Proc. Natl. Acad. Sci. USA 86, 7418–7422.
282. Travers, A. and Drew, H. (1997) Biopolymers 44, 423–433.
283. Widom, J. (2001) Quart. Rev. Biophys. 34, 269–324.
284. Ausió, J., Zhou, G., and van Holde, K.E. (1987) Biochemistry 26, 5595–5599.
285. Behe, M. and Felsenfeld, G. (1981) Proc. Natl. Acad. Sci. USA 78, 1619–1623.
286. Thomas, T.J. and Bloomfield, V.A. (1983) Nucleic Acids Res. 11, 1919–1930.

287. Ausió, J., Borochov, N., Zvi Kam Reich, M., Seger, D., and Eisenberg, H. (1983) In: Hélène, C. (ed.) Structure, Dynamics, Interactions, and Evolution of the Biological Macromolecules. Reidel, Dordrecht, The Netherlands, pp. 89–100.
288. Flory, P.J. (1969) Statistical Mechanics of Chain Molecules. Interscience Publishers (John Wiley and Sons), New York, p. 111.
289. Anselmi, C., De Santis, P., Paparcone, R., Savino, M., and Scipioni, A. (2002) Biophys. Chem. 95, 23–47.
290. Anselmi, C., Bocchinfuso, G., De Santis, P., Savino, M., and Scipioni, A. (2002) Biophys. J. 79, 601–613.
291. Anselmi, C., Bocchinfuso, G., De Santis, P., Savino, M., and Scipioni, A. (1999) J. Mol. Biol. 286, 1293–1301.
292. Bina, M., Sturtevant, J.M., and Stein, A. (1980) Proc. Natl. Acad. Csi. USA 77, 4044–4047.
293. Godde, J.S., Kass, S.U., Hirst, M.C., and Wolffe, A.P. (1996) J. Biol. Chem. 271, 24325–24328.
294. Widlund, H.R., Cao, H., Simonsson, S, Margusson, E., Simonsson, T., Nielsen, P.E., Kahn, J.D., Crothers, D.M., and Kubista, M. (1997) J. Mol. Biol. 267, 807–817.
295. Cacchione, S.M., Cerone, A., and Savino, M. (1997) FEBS Lett. 400, 37–41.
296. Lowary, P.T. and Widom, J. (1998) J. Mol. Biol. 276, 19–42.
297. Rossetti, L., Cacchione, S., Fuà, M., and Savino, M. (1998) Biochemistry 37, 6727–6737.
298. Dal Cornò, M., De Santis, P., Sampaolese, B., and Savino, M. (1998) FEBS Lett. 431, 66–70.
299. Cao, H., Widlund, H.R., Simonsson, T., and Kubista, M. (1998) J. Mol. Biol. 281, 253–260.
300. Brower-Toland, B.D., Smith, C.L., Yeh, R.C., Lis, J.T., Peterson, C.L., and Wang, M.D. (2002) Proc. Natl. Acad. Sci. USA 99, 1960–1965.
301. Widom, J. (1999) Methods Mol. Biol. 119, 61–77.
302. Manning, G.S. (1985) Cell Biophys. 7, 177–184.
303. Marky, N.L. and Manning, G.S. (1991) Biopolymers 31, 1543–1557.
304. Marky, N.L. and Manning, G.S. (1995) J. Mol. Biol. 254, 50–61.
305. Ausió, J., Dong, F., and van Holde, K.E. (1989) J. Mol. Biol. 206, 451–463.
306. Hayes, J.J. and Hansen, J.C. (2002) Proc. Natl. Acad. Sci. USA 99, 1752–1754.
307. Burlingame, R.W., Love, W.E., Wang, B.C., Hamlin, R., Nguyen, H.X., and Moudrianakis, E.N. (1985) Science 228, 546–553.
308. Arents, G., Burlingame, R.W., Wang, B.C., Love, W.E., and Moudrianakis, E.N. (1991) Proc. Natl. Acad. Sci. USA 88, 10148–10152.
309. Karantza, V., Freire, E., and Moudrianakis, E.N. (2001) Biochemistry 40, 13114–13123.
310. Widlund, H.R., Vitolo, J.M., Thiriet, C., and Hayes, J.J. (2000) Biochemistry. 39, 3835–3841.
311. Eickbush, T.H., Watson, D.K., and Moudrianakis, E.N. (1976) Cell 9, 785–792.
312. Baxevanis, A.D. and Landsman, D. (1998) Nucleic Acids Res. 26, 372–375.
313. Lee, K.P., Baxter, H.J., Guillemette, J.G., Lawford, H.G., and Lewis, P.N. (1982) Can. J. Biochem. 60, 379–388.
314. Piñeiro, M., Puerta, C., and Palacián (1991) Biochemistry 30, 5805–5810.
315. White, C.L., Suto, R.K., and Luger, K. (2001) EMBO J. 20, 5207–5218.
316. Waterborg, J. (2000) J. Biol. Chem. 275, 13007–13011.
317. Simpson, R.T. (1978) Biochemistry 17, 5524–5531.
318. Allan, J., Hartman, P.G., Crane-Robinson, C., and Aviles, F.X. (1980) Nature 288, 675–679.
319. Ramakrishnan, V. (1997) Crit. Rev. Eukaryot. Gene Expr. 7, 215–230.
320. Zhou, Y.B., Gerchman, S.E., Ramakrishnan, V., Travers, A., and Muyldermans, S. (1998) Nature 395, 402–405.
321. Travers, A. (1999) Trends Biochem. Sci. 24, 4–7.
322. Thoma, F., Koller, T., and Klug, A. (1979) J. Cell Biol. 83, 403–427.
323. van Holde, K.E. and Zlanatova, J. (1995) J. Biol. Chem. 270, 8373–8376.
324. Rydberg, B., Holley, W.R., Mian, S.I., and Chatterjee, A. (1998) J. Mol. Biol. 284, 71–84.
325. Renz, M., Nehls, P., and Hozier, J. (1977) Proc. Natl. Acad. Sci. USA 74, 1879–1883.
326. Zentgraf, H. and Franke, W.W. (1984) J. Cell Biol. 99, 272–286.
327. Finch, J.T. and Klug, A. (1976) Proc. Natl. Acad. Sci. USA 73, 1897–1901.

328. Worcel, A., Strogatz, S., and Riley, D. (1981) Proc. Natl. Acad. Sci. USA 78, 1461–1465.
329. Woodcock, C.L., Frado, L.L., and Rattner, J.B. (1984) J. Cell Biol. 99, 42–52.
330. Subirana, J.A., Muñoz-Guerra, S., Aymami, J., Radermacher, M., and Frank, J. (1985) Chromosoma 91, 377–390.
331. Williams, S.P., Athey, B.D., Muglia, L.J., Schappe, R.S., Gough, A.H., and Langmore, J.P. (1986) Biophys. J. 49, 233–248.
332. Widom, J. (1989) Ann. Rev. Biophys. Chem. 18, 365–395.
333. Staynov, D.Z. (1981) Int. J. Biol. Macromol. 5, 310–311.
334. Dimitrov, S.I., Russanova, V.R., and Pashev, I.G. (1987) EMBO J. 6, 2387–2392.
335. Thibodeau, A. and Ruiz-Carrillo, A. (1988) J. Biol. Chem. 263, 16236–16241.
336. Burgoyne, L.A. and Skinner, J.D. (1981) Biochem. Biophys. Res. Commun. 99, 893–899.
337. Staynov, D.Z. and Proykova, Y.G. (1998) J. Mol. Biol. 279, 59–71.
338. Pardon, J.F. and Wilkins, M.H. (1972) J. Mol. Biol. 68, 115–124.
339. Azorín, F., Martínez, A.B., and Subirana, J.A. (1980) Int. J. Biol. Macromol. 2, 81–86.
340. Widom, J. and Klug, A. (1985) Cell 43, 207–213.
341. Subirana, J.A., Muñoz-Guerra, S., Aymamí, Radermacher, M., and Frank, J. (1983) J. Biomol. Struct. Dyn. 1, 705–714.
342. Subirana, J.A. (1992) FEBS Lett. 302, 105–107.
343. Woodcock, C.L., McEwen, B.F., and Frank, J. (1991) J. Cell Sci. 99, 107–114.
344. Leuba, S.H., Yang, C., Robert, C., Samori, B., van Holde, K., Zlatanova, J., and Bustamante, C. (1994) Proc. Natl. Acad. Sci. USA 91, 11621–11625.
345. Bednar, J., Horowitz, R.A., Grigoryev, S.A., Carruthers, L., Hansen, J.C., Koster, A.J., and Woodcock, C.L. (1998) Proc. Natl. Acad. Sci. USA 95, 14173–14178.
346. Pehrson, J.R. (1989) Proc. Natl. Acad. Sci. USA 86, 9149–9153.
347. Butler, P.J. and Thomas, J.O. (1998) J. Mol. Biol. 281, 401–407.
348. Green, G.R., Ferlita, R.R., Walkenhorst, W.F., and Poccia, D.L. (2001) Biochem. Cell Biol. 349–363.
349. Fletcher, T.M. and Hansen, J.C. (1996) Crit. Rev. Eukaryot. Gene Expr. 6, 149–188.
350. Thomas, J.O. and Butler, P.J. (1980) J. Mol. Biol. 144, 89–93.
351. Butler, P.J. and Thomas, J.O. (1980) J. Mol. Biol. 140, 505–529.
352. Ausió, J., Borochov, N., Seger, D., and Eisenberg, H. (1984) J. Mol. Biol. 177, 373–398.
353. Bordas, J., Perez-Grau, L., Koch, M.H., Vega, M.C., and Nave, C. (1986) Eur. Biophys. J. 13, 157–173.
354. Suau, P., Kneale, G.G., Braddock, G.W., Baldwin, J.P., and Bradbury, E.M. (1977) Nucleic Acids Res. 4, 3769–3786.
355. Simpson, R.T., Thoma, F., and Brubaker (1985) Cell 42, 799–808.
356. Horn, P.J., Carruthers, L.M., Logie, C., Hill, D.A., Solomon, M.J., Wade, P.A., Imbalzano, A.N., Hansen, J.C., and Peterson, C.L. (2002) Nat. Struct. Biol. 9, 263–267.
357. Horn, P.J., Crowley, K.A., Carruthers, L.M., Hansen, J.C., and Peterson, C.L. (2002) Nat. Struct. Biol. 9, 167–171.
358. Garcia-Ramirez, M., Dong, F., and Ausió, J. (1992) J. Biol. Chem. 267, 19587–19595.
359. Howe, L., Iskandar, M., and Ausió, J. (1998) J. Biol. Chem. 273, 11625–11629.
360. Manning, G.S. (1977) Biophys. Chem. 7, 95–102.
361. Manning, G.S. (1978) Quart. Rev. Biophys. 11, 179–246.
362. Clark, D.J. and Kimura, T. (1990) J. Mol. Biol. 211, 883–896.
363. Schiessel, H. (2002) Europhys. Lett. 58, 140–146.
364. Ehrlich, L., Munkel, C., Chirico, G., and Langowski, J. (1997) Comput. Appl. Biosci. 13, 271–279.
365. Allan, J., Harborne, N., Rau, D.C., and Gould, H. (1982) J. Cell Biol. 93, 285–297.
366. Moore, S.C. and Ausió, J. (1997) Biochem. Biophys. Res. Commun. 230, 136–139.
367. Tse, C. and Hansen, J.C. (1997) Biochemistry 36, 11381–11388.
368. Hansen, J.C., Tse, C., and Wolffe, A.P. (1998) Biochemistry 37, 17637–17641.
369. Garcia-Ramirez, M., Rocchini, C., and Ausió, J. (1995) J. Biol. Chem. 270, 17923–17928.
370. Caron, F. and Thomas, J.O. (1981) J. Mol. Biol. 146, 513–537.

371. Lever, M.A., Th'ng, J.P., Sun, X., and Hendzel, M.J. (2000) Nature 408, 873–876.
372. Misteli, T., Gunjan, A., Hock, R., Bustin, M., and Brown, D.T. (2000) Nature 408, 877–881.
373. Parseghian, M.H., Henschen, A.H., Krielstein, K.G., and Hamkalo, B.A. (1994) Protein Sci. 3, 575–587.
374. Parseghian, M.H., Newcomb, R.L., Winokur, S.T., and Hamkalo, B.A. (1994) Chromosome Res. 8, 405–424.
375. Fan, J.Y., Gordon, F., Luger, K., Hansen, J.C., and Tremethick, D.J. (2002) Nat. Struct. Biol. 9, 172–176.
376. Simpson, R.T. (1978) Cell 13, 691–699.
377. Bode, J. (1984) Arch. Biochem. Biophys. 228, 364–372.
378. Bode, J., Gomez-Lira, M.M., and Schroter, H. (1983) Eur. J. Biochem. 130, 437–445.
379. Ausió, J. and van Holde, K.E. (1986) Biochemistry 25, 1421–1428.
380. Imai, B.S., Yau, P., Baldwin, J.P., Ibel, K., May, R.P., and Bradbury, E.M. (1986) J. Biol. Chem. 261, 8784–8792.
381. Mutskov, V., Gerber, D., Angelov, D., Ausió, J., Workman, J., and Dimitrov, S. (1998) Mol. Cell. Biol. 18, 6293–6304.
382. Ichimura, S., Mita, K., and Zama, M. (1982) Biochemistry 21, 5329–5334.
383. Howe, L. and Ausió, J. (1998) Mol. Cell. Biol. 18, 1156–1162.
384. Panetta, G., Buttinelli, M., Flaus, A., Richmond, T.J., and Rhodes, D. (1998) J. Mol. Biol. 282, 683–697.
385. Oliva, R., Bazett-Jones, D., Mezquita, C., and Dixon, G.R. (1987) J. Biol. Chem. 262, 17016–17025.
386. Oliva, R., Bazett-Jones, D.P., Locklear, L., and Dixon, G.H. (1990) Nucleic Acids Res. 18, 2739–2747.
387. Dunker, A.K., Lawson, J.D., Brown, C.J., Williams, R.M., Romero, P., Oh, J.S., Oldfield, C.J., Campen, A.M., Ratliff, C.M., Hipps, K.W., Ausió, J., Nissen, M.S., Reeves, R., Kang, C., Kissinger, C.R., Bailey, R.W., Griswold, M.D., Chiu, W., Garner, E.C., and Obradovic, Z.J. (2001) Mol. Graph. Model. 19, 26–59.
388. Walia, H., Chen, H.Y., Sun, J.M., Holth, L.T., and Davie, J.R. (1998) J. Biol. Chem. 273, 14516–14522.
389. Morales, V. and Richard-Foy, H. (2000) Mol. Cell. Biol. 20, 7230–7237.
390. Liu, L.F. and Wang, J.C. (1987) Proc. Natl. Acad. Sci. USA 84, 7024–7027.
391. Harada, Y., Ohara, O., Takatsuki, A., Itoh, H., Shimamoto, N., and Kinosita, K., Jr. (2001) Nature 409, 113–115.
392. Sathyanarayana, U.G., Freeman, L.A., Lee, M.S., and Garrard, W.T. (1999) J. Biol. Chem. 274, 16431–16436.
393. McGhee, J.D., Nickol, J.M., Felsenfeld, G., and Rau, D.C. (1983) Nucleic Acids Res. 11, 4065–4075.
394. Dimitrov, S., Makarov, V., Apostolova, T., and Pashev, I. (1986) FEBS Lett. 197, 217–220.
395. Wang, X., He, C., Moore, S.C., and Ausió, J. (2001) J. Biol. Chem. 276, 12764–12768.
396. Hansen, J.C. and Ausió, J. (1992) Trends Biochem. Sci. 17, 187–191.
397. Norton, V.G., Imai, B.S., Yau, P., and Bradbury, E.M. (1989) Cell 57, 449–457.
398. Turner, B.M. (1998) Cell. Mol. Life Sci. 54, 21–31.
399. Kamakaka, R.T. and Thomas, J.O. (1990) EMBO J. 9, 3997–4006.
400. Zlatanova, J., Caiafa, P., and van Holde, K.E. (2000) FASEB J. 14, 1697–1704.
401. Zlatanova, J. and van Holde, K.E. (1992) J. Cell Sci. 103, 889–895.
402. Wang, X., Moore, S.C., Laszckzak, M., and Ausió, J. (2000) J. Biol. Chem. 275, 35013–35020.
403. Vijay-Kumar, S, Bugg, C.E., and Cook, W.J. (1987) J. Mol. Biol. 194, 531–544.
404. Fletcher, T.M., Serwer, P., and Hansen, J.C. (1994) Biochemistry 33, 10859–10863.
405. Hansen, J.C., Kreider, J.I., Demeler, B., and fletcher, T.M. (1997) Methods 12, 62–72.
406. Ausió, J. and Moore, S.C. (1998) Methods 14, 333–342.
407. Bonner, W.M. and Stedman, J.D. (1979) Proc. Natl. Acad. Sci. USA 76, 2190–2194.
408. Nelson, P.P., Albright, S.C., Wiseman, J.M., and Garrard, W.T. (1979) J. Biol. Chem. 254, 11751–11760.

409. Mueller, R.D., Yasuda, H., Hatch, C.L., Bonner, W.M., and Bradbury, E.M. (1985) J. Biol. Chem. 260, 5147–5153.
410. Goldknopf, I.L., Wilson, G., Ballal, N.R., and Busch, H. (1980) J. Biol. Chem. 255, 10555–10558.
411. Mimnaugh, E.G., Kayastha, G., McGovern, N.B., Hwang, S.G., Marcu, M.G., Trepel, J., Cai, S.Y., Marchesi, V.T., and Neckers, L. (2001) Cell Death Differ. 8, 1182–1196.
412. Albig, W., Meergans, T., and Doenecke, D. (1997) Gene 184, 141–148.
413. Woodcock, C.L. and Horowitz, R.A. (1995) Trends Cell Biol. 5, 272–277.
414. Krajewski, W.A. and Ausió, J. (1996) Biochem. J. 316, 395–400.
415. Panyim, S. and Chalkley, R. (1969) Arch. Biochem. Biophys. 130, 337–346.
416. Hurley, C.K. (1977) Anal. Biochem. 80, 624–626.

J. Zlatanova and S.H. Leuba (Eds.) *Chromatin Structure and Dynamics: State-of-the-Art*

DOI: 10.1016/S0167-7306(03)39011-8

CHAPTER 11

Nucleosome modifications and their interactions; searching for a histone code

Bryan M. Turner

Chromatin and Gene Expression Group, Institute of Biomedical Research, University of Birmingham Medical School, Birmingham B15 2TT, UK. Tel.: +44-(0)121-414-6824; Fax: +44-(0)121-414-6815; E-mail: b.m.turner@bham.ac.uk

Since the nucleosome was first described, it has generally been assumed that its primary function is to package DNA into the interphase nucleus and mitotic chromosomes [1]. Although the nucleosome itself provides only a small fraction of the several thousand-fold length reduction required for compaction into chromosomes, it is presumably an essential first step that enables higher-order structures to assemble. The nucleosome's packaging job became more interesting with the realization that compaction of DNA into chromatin is a crucial element in eukaryotic gene regulation [2]. Appropriately located nucleosomes can hinder assembly of transcription initiation complexes and polymerase progression, and over recent years, whole families of chromatin remodeling enzymes have been identified whose primary purpose is to reorganize histone–DNA interactions so as to facilitate (or repress) transcription [3–5]. These enzymes disrupt the nucleosome in an ATP-dependent manner that often results in movement of the histone core relative to the DNA, thereby exposing recognition sequences for DNA-binding activators or repressors. However, even these insights still leave the nucleosome as a passive inhibitor of transcription, something to be moved aside so transcription can proceed.

A true appreciation of the subtle and complex ways in which the nucleosome can influence gene expression, has come only recently, largely through studies of the post-translational modifications to which all histones are subject and of the enzymes that add and remove these modifications. It has been known for many years that the histone N-terminal tails are exposed on the surface of the nucleosome and that selected amino acid residues are subject to a variety of enzyme-catalyzed, post-translational modifications. These include acetylation of lysines, phosphorylation of serines, and methylation of lysines and arginines ([6,7], see also chapters by Davie, and Ausio and Abbott, this volume). The locations of the histone N-terminal tails in the nucleosome and the residues that can be modified are shown in Fig. 1.

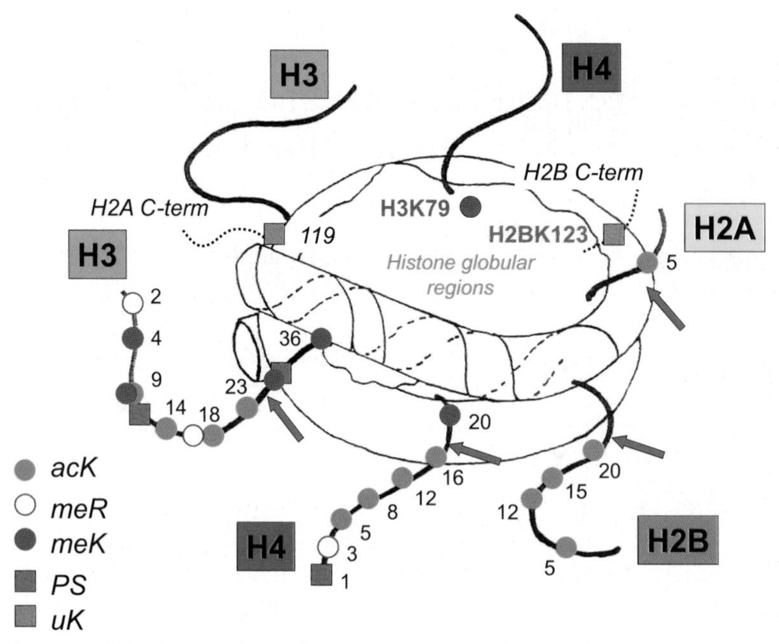

Fig. 1. Histone modifications on the nucleosome core particle. The nucleosome core particle showing 6 of the 8 core histone N-terminal tail domains and 2 C-terminal tails. Sites of post-translational modification are indicated by coloured symbols that are defined in the key (lower left); acK = acetyl lysine, meR = methyl arginine, meK = methyl lysine, PS = phosphoryl serine, and uK = ubiquitinated lysine. Residue numbers are shown for each modification. Note that H3 lysine 9 can be *either* acetylated or methylated. The C-terminal tail domains of one H2A molecule and one H2B molecule are shown (dashed lines) with sites of ubiquitination at H2A lysine 119 (most common in mammals) and H2B lysine 123 (most common in yeast). Modifications are shown on only one of the two copies of histones H3 and H4 and only one tail is shown for H2A and H2B. Sites marked by green arrows are susceptible to cutting by trypsin in intact nucleosomes. Note that the cartoon is a compendium of data from various organisms, some of which may lack particular modifications e.g., there is no H3meK9 in *S. cerevisiae*. (From Ref. [7].)

1. The importance of residue-specific modifications

The association between a histone tail modification and a particular functional state of chromatin, came with the demonstration that transcriptionally active chromatin fractions were enriched in acetylated histones, firstly by biochemical co-fractionationation ([8,9] and references therein) and then by Chromatin ImmunoPrecipitation, ChIP [10]. Subsequently, regions of transcriptionally silent constitutive and facultative heterochromatin, were shown, by immunofluorescence microscopy, to be under-acetylated [11,12]. This supported the idea that acetylation of the histone tails, with the associated loss of positive charge and reduction in DNA-binding constant, somehow caused chromatin to become more "open" (or less "condensed") and thereby more conducive to transcription. While this is likely to be an important contributory factor, it has now become clear that the

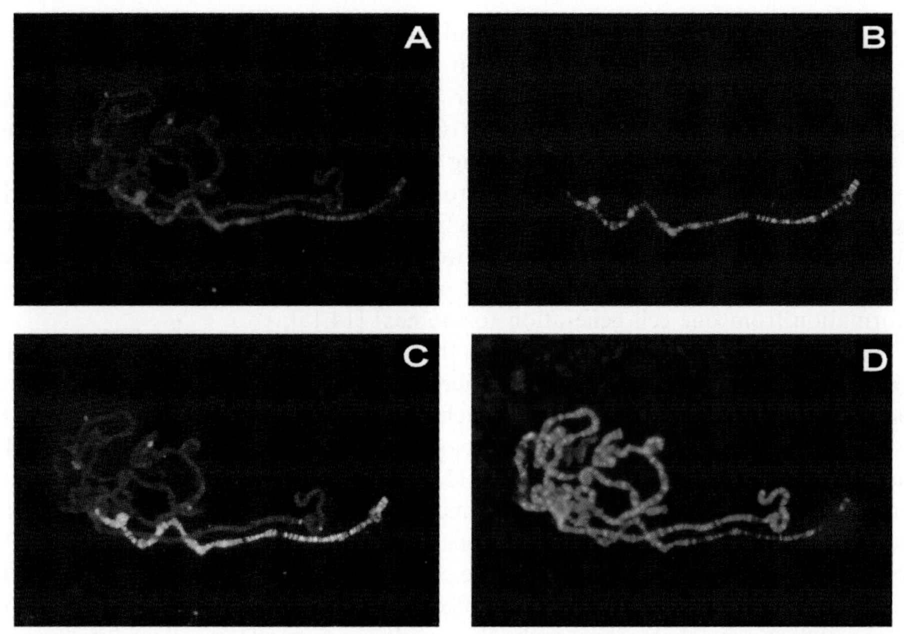

Fig. 2. Polytene chromosomes from a male third instar *Drosophila* larvae immunostained to show co-localization of H4acK16 and the dosage-compensation protein MLE on the X chromosome: (A) Staining with antibodies to H4acK16. The male X chromosome stains intensely, with only occasional patches of staining on the autosomes. In chromosome spreads from female larvae, the X chromosome and the autosomes all stain weakly. (B) Staining with antibodies to MSL1 (Male Specific Lethal 1), a component of the *Drosophila* dosage compensation complex that specifically associates with the X chromosome only in male cells. (C) Double exposure to show the co-localization of H4acK16 and MSL1. Superimposition of red (H4acK16) and green (MSL1) images gives a yellow signal. (D) Chromosomal DNA stained with Hoechst 33258. (Data from Ref. [83]).

story is much more complex, and that the functional effects of tail modifications depend not just on charge-mediated structural changes, but on the specific amino acids that are modified.

The first indication that modification of specific tail residues were linked to chromatin functional states, came from immunostaining of *Drosophila* polytene chromosomes with antibodies specific for H4 acetylated at defined lysines [13]. As shown in Fig. 2A, H4 acetylated at lysine 16 (H4acK16) was found almost exclusively on the transcriptional hyperactive male X chromosome (Fig. 2). (Genes on the *Drosophila* male X are transcribed twice as fast as their female counterparts so as to equalize levels of X-linked gene products between XY males and XX females.) In addition, H4 lysine 12 was found to remain acetylated in centric heterochromatin, while lysines 5, 8, and 16 were all under-acetylated [13]. These observations led to the suggestion that the histone N-terminal tails constitute nucleosome surface markers that can be recognized by non-histone proteins in a modification-dependent manner to alter the functional state of chromatin [13].

Related work raised the possibility that tail modifications may be involved not only in the ongoing up- and downregulation of transcription, but also in carrying information regarding the *potential* for transcription (in response to specific factors). Importantly, the specific H4acK16 mark was found to be retained on the male X through mitosis [13a], indicating that it is not simply a by-product of ongoing (hyper)transcription (minimal through mitosis), but a mechanism by which the potential for (hyper)transcription is carried through mitosis into the next interphase. In light of these findings, it was proposed that the nucleosome has a role not only in DNA packaging, but also in the transmission of epigenetic information from one cell generation to the next [14,15].

The amount of epigenetic information that can be carried in the histone tails is enormous. For example, there are 50 different acetylated isoforms of the four core histones (H2B, H3, and H4 have 16 each and H2A has 2). These isoforms can be modified further by methylation of selected lysines and arginines (H3 and H4) and phosphorylation of serine (H2A, H2B, H3, and H4). Further, methylation can involve attachment of 1, 2, or 3 methyl groups [16] and there are other, chemically more complex, modifications, such as ubiquitination and ADP-ribosylation ([6], see chapters by Davie, and Ausio and Abbott, this volume). The total number of possible histone isoforms, carrying different combinations of tail modifications, that can mark the nucleosome surface, runs into many thousands. This has given rise to the idea that the tail modifications constitute a histone code [17,18] or epigenetic code [19] which is set and maintained by tail modifying and de-modifying enzymes and read by non-histone proteins.

2. *Dynamics of histone modification*

The modifications to the N-terminal histone tails in chromatin are put in place and maintained by the action of families of modifying and de-modifying enzymes. For example, most histone acetates are undergoing a continuous cycle of acetylation and deacetylation and turn over with half lives ranging from just a few minutes to several hours ([20–22,22a] and references therein). Thus, the level of acetylation of a particular lysine in a particular histone, is a steady-state maintained by the ongoing activities of histone acetyltransferases, HATs, and deacetylases, HDACs [23,24]. While this makes continuous demands on the cell's energy resources, it also offers an advantage to those genes whose expression may need to be up or down regulated relatively quickly, such as those that respond to hormonal cues or growth factor stimulation. Removal or inhibition of either HATs or HDACs will precipitate a rapid change in acetylation, and hence transcription or transcriptional potential. In contrast, where levels of gene expression must be stably maintained over long periods, the requirement is to conserve patterns of histone modification. In these situations patterns can be stabilized by removing (or inhibiting) the HATs and HDACs targeted to that genomic region. In the case of the imprinted gene *U2af1-rs1* in mice, differential acetylation levels on the maternal and paternal alleles are maintained dynamically early in development, but become static (operationally defined by resistance to HDAC inhibitors) as development proceeds [25].

In contrast, lysine methylation seems to be an exceptionally stable modification. Early studies showed that turnover of histone methyl groups was even slower than turnover of the histones themselves (e.g., [26,27]). No conclusive evidence has yet been found for histone demethylating enzymes, and they may not exist [28]. It may be that removal of methylated histones mostly occurs passively, through post-replication chromatin assembly and replacement of old, methylated histones with new, unmethylated ones. However, the possibility remains that local methylation patterns may be more dynamic and may involve novel mechanisms for removal of methylated tails [28].

3. *Modifications interact in cis*

Selected lysines and arginines can be methylated in the N-terminal tail domains of H3 and H4 (Fig. 1). The enzymes responsible, the histone methyl transferases (HMT), are either lysine or arginine specific and several have now been identified and characterized [16,29,30]. In higher eukaryotes, two H3 lysines, K4 and K9, are commonly methylated. They seem to have complementary functions, with methyl K4 being enriched in transcriptionally active regions and methyl K9 in silent regions, particularly heterochromatin [31,32]. *In vitro* binding assays and immunolocalization, have provided strong evidence that heterochromatin formation is mediated by the preferential binding of the heterochromatin protein HP1 to chromatin in which H3 tails are methylated at K9 [33–36]. Methylation of H3 K9 in mammals is carried out by the enzyme SUV39H [30] and in fission yeast by the homologous enzyme Clr4 [37]. An enzyme isolated from HeLa cells, designated Set9 and homologous to the budding yeast enzyme Set1, catalyzes the methylation of H3 specifically at lysine 4 [38].

Using *in vitro* peptide binding assays, methylation of H3 K4 has been shown to have two potentially important functional effects. First, the preferential association of the NuRD chromatin remodelling and deacetylase complex with the H3 tail is inhibited by K4 (but not K9) methylation [38,39]. Second, the *in vitro* methylation of H3 K9 by purified SUV39H complexes is strongly inhibited if the H3 peptide substrate is methylated at K4. Thus, methylation of H3 K4 by Set9 has the ability to block both chromatin remodelling/deacetylation and methylation of H3 K9 by SUV39H, thereby preventing the placement of a silencing mark to be read by HP1. The data complements earlier work showing that phosphorylation of H3 S10 and, not surprisingly, acetylation of H3 K9, both prevented K9 methylation by SUV39H *in vitro* [30]. These interactions are summarized in Fig. 3. In the same assays, acetylation of H3 K14 caused only a modest reduction in K9 methylation. However, fission yeast mutants deficient in the H3 K14-specific histone deacetylase (HDAC) Clr3, also showed reduced H3 K9 methylation [40], leaving open the possibility of a functional interaction between these two modifications. It is interesting to note that there is evidence of physical interaction between the

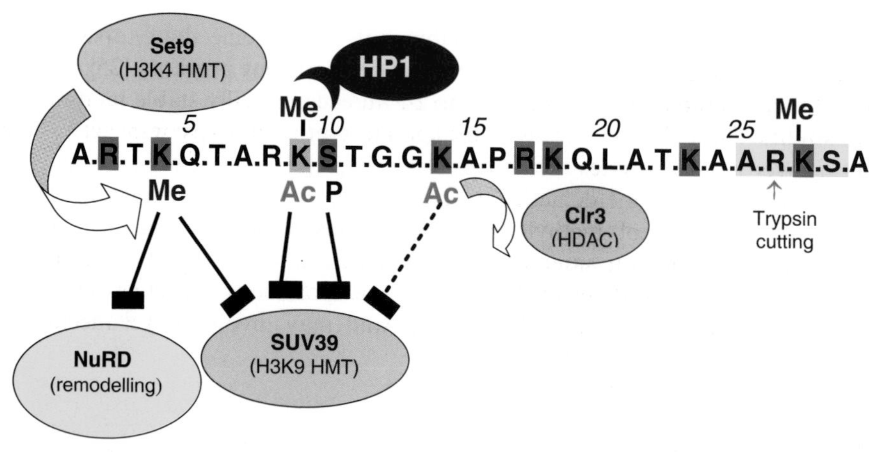

Fig. 3. Modifications of the histone H3 N-terminal tail interact through their effects on chromatin modifying enzymes. The amino acid sequence of the H3 N-terminal tail domain is shown in single letter code. Residues on the N-terminal side of the trypsin cutting site (arrowed) are accessible on the nucleosome surface. Only those residues highlighted in yellow have been located in nucleosome core particle crystals [96]. Lysines (K) and arginines (R) that can be methylated are blocked in blue, lysines that can be acetylated in red and the serine (S) that can be phosphorylated in green. Lysine 9 can be either acetylated or methylated and is blocked in violet. The enzyme Set9 methylates H3 lysine 4 specifically. Methylation of this residue prevents association of the H3 tail with the NuRD chromatin remodelling and deacetylase complex and the SUV39H histone methyl transferase. Methylation of H3 K9 by SUV39H is also prevented or inhibited by acetylation of K9, phosphorylation of S10 and, possibly, acetylation of K14. Methylation of K9 is enhanced by removal of K9 and K14 acetates by deacetylases (HDAC) such as the yeast enzyme Clr3. Methylation of H3 K9 facilitates binding of the heterochromatin protein HP1. For references, see text. (From Ref. [7].)

Drosophila HMT SU(VAR)3-9 and a histone deacetylase [41] and also between the HAT CBP and a methyl transferase activity [42].

Further evidence for a possible interaction between methylation and acetylation comes from studies on the enzyme SET8 that methylates H4 specifically at lysine 20 [43–45]. The native enzyme is highly specific for *nucleosomal* H4 K20. Some results suggest that the functional effects of H4meK20 may be modulated by acetylation of H4 K16, or *vice versa*. However, there are discrepancies in the results presented by different groups working on this modification, possibly attributable to differences in the detailed specificities of the different antisera used. Until these are resolved, the way in which H4meK20 and H4acK16 interact remains unclear.

4. Modifications interact in trans

Experiments in yeast to investigate how H3 K4 methylation itself is regulated, have revealed a surprising relationship. In budding yeast, methylation of H3 K4 requires the *SET1* gene, and strains carrying mutations of *SET1* not only lack H3 K4 methylation but show a slow growth phenotype with loss of silencing of rDNA genes [46]. A similar phenotype is seen in cells in which H3 K4 (but not K9) is substituted with arginine or alanine. Several genes in yeast, in addition to *SET1*, are known to be required for efficient telomeric and rDNA silencing and mutations in just one of these, *RAD6*, not only disrupt silencing, but also completely abolish H3 K4 methylation detected by western blotting [47,48].

The Rad6 protein is a ubiquitin conjugating enzyme involved in many cellular processes, including gene silencing, DNA repair and sporulation. Rad6 attaches the 76 amino acid peptide ubiquitin to selected protein lysines via an isopeptide bond, thereby targeting these proteins for degradation by the proteasome system [49]. Certain histone lysines are also targets for ubiquitination and Rad6 in yeast is the major activity responsible for attachment of ubiquitin to lysine 123 in the C-terminal tail of H2B (Fig. 1). This activity seems to provide the link to H3 methylation because substitution of H2B lysine 123 with an alanine, thereby preventing ubiquitination, also abolished methylation of H3 lysine 4 [47].

The H2B K123 → A substitution mutant also showed defective silencing at telomeres. This effect may be related to a second remarkable property of H2B K123 substitution and *rad6* deletion mutants, namely the complete absence of methylation at H3 lysine 79 [50]. H3 K79 is located in the globular domain of the histone, far away from the N-terminal tail, but is exposed (i.e., potentially accessible *in vivo*) on the upper and lower faces of the core particle (Fig. 1). The modification is common, with around 90% of H3 molecules being methylated at this site in wild type cells [51]. H3 K79 methylation is carried out exclusively by the enzyme Dot1, an unusual HMT that lacks a SET domain [51–53]. *Dot1* mutants are defective in telomeric silencing. Thus, Rad6-catalyzed ubiquitination of H2B K123 seems to be essential for methylation of both H3 K4 and H3 K79 and, consequently, for appropriate patterns of gene silencing. A third H3 methylation site in *S. cerevisiae*, lysine 36, shows no change in methylation level in either *rad6* or H2B K123 substitution mutants, so there is no generic effect of H2B ubiquitination on histone methylation [47].

H3 K4 methylation and H2B ubiquitination, occur on the same nucleosome, but there is a large overall difference in frequency between the two modifications, with about 35% of H3 molecules being methylated at K4 and less than 5% of H2B molecules ubiquitinated at K123 [47]. It has been suggested that ubiquitination of H2B on a single nucleosome acts as a "wedge" to open up a stretch of chromatin, making other nucleosomes accessible to the Set1 methylating enzyme. It is also possible that the different frequencies of methylation and ubiquitination reflect their different turnover rates. Methylation is usually a stable modification and could be retained long after the ubiquitin, that allowed its attachment in the first place, has been lost.

A second puzzle is highlighted by mass spectrometric analysis of H3 K79 methylation levels [51]. About 90% of H3 molecules in wild type yeast are methylated at K79, with most (about 50%) being trimethylated, a finding that argues against a role for this modification itself as a silencing signal at telomeres, or anywhere else. This high frequency of methylation is consistent with earlier work on H3 and H4 methylation levels in higher eukaryotes [26,54,55]. So what role can there be for such a common and widely distributed histone modification? One possibility is that silenced regions are marked by *low* levels of H3 K79 methylation. It has been proposed that H3 K79 methylation blocks the binding to active chromatin of "sticky" proteins, such as the silencing proteins Sir2 and Sir3 [51]. Absence of H3 K79 methylation in *dot1* mutant cells will allow promiscuous binding of such proteins across the genome, with consequent depletion of their levels at silent regions. Overexpression of Dot1, leading to methylation (almost exclusively trimethylation) of 98% of H3 molecules at K79, was found to diminish silencing, presumably by forcing H3 K79 methylation in potentially silent regions that are sheltered from methylation in wild type cells. The hypothesis is consistent with ChIP data and the properties of *dot1* and H3 K79 substitution mutants, but there is as yet, no biochemical data to show how H3 K79 methylation might influence, directly or indirectly, binding of Sir proteins to the nucleosome.

The *trans* interactions discussed so far all relate to the mechanisms by which patterns of histone modification are set. Recent results show that they can also determine how histone modifications are read [56]. The *Drosophila* epigenetic activator ASH1 (*Absent, Small, or Homeotic discs 1*), is a histone methyltransferase with a rather broad substrate specificity, methylating H3 at lysines 4 and 9 and H4 at lysine 20. Genetic and biochemical evidence shows that this combination of methylated lysines both facilitates binding of the activator BRAHMA (an ATPase component of a SWI/SNF-like remodeling complex) and inhibits binding of epigenetic repressors such as polycomb [57] and HP1. Thus, in the presence of H3meK4 and H4meK20, H3meK9 is not a determinant of silencing. Instead, it is part of a combination of marks that constitutes a binding platform for a BRAHMA-containing remodeling complex. BRAHMA is a member of the Trithorax (Trx) group of proteins, whose function is to maintain the transcriptionally active state of selected genes through development [58].

5. *Interplay between histone modifications and DNA methylation*

The histone tail modifications are likely to act in concert with a rather more widely known mediator of chromatin structure and gene expression, namely methylation of cytosine residues in CpG dimers [59,60]. Long-term silencing, as found in imprinted genes or the female inactive X chromosome, is generally associated with relatively high levels of CpG methylation [60]. An attractive feature of CpG methylation is that the symmetry of the site (i.e., 5′CpG/GpC5′) allows for its propagation after DNA replication by CpG-specific DNA methyltransferases, Dnmt [61]. This model is supported by the finding that some mammalian Dnmt preferentially methylate *hemimethylated* CpG/GpC sites [62]. The mechanism(s) by which CpG methylation leads to gene silencing remain to be defined, but there is evidence that in at least some situations, histone modifications are involved. For example, the methyl DNA binding protein MeCP2 can bind histone deacetylases, thereby targeting them to methylated DNA [63]. The resultant local histone deacetylation leads to suppression of transcription. Studies of the mouse imprinted gene *U2af1-rs1* support this general mechanism, but show that the deacetylation induced by methyl DNA binding proteins can be selective for specific histones [64]. Thus, on the maternal *U2af1-rs1* allele, which is silent and heavily methylated, H3, at all lysines tested, is underacetylated compared to H3 on the active and unmethylated paternal allele. In contrast, H4 acetylation at lysines 8, 12, and 16 is the same on both alleles. Only H4 lysine 5 shows reduced acetylation on the silent, maternal allele [64]. In transgenic mice engineered so that both maternal and paternal *U2af1-rs1* alleles were equally methylated, H3 acetylation on *U2af1-rs1* fell to very low levels (i.e., both alleles were underacetylated), while H4 remained acetylated at all lysines, including lysine 5. Thus, while CpG methylation seems sufficient to drive H3 deacetylation on *U2af1-rs1*, H4 deacetylation must require other factors.

Conversely, experiments in the filamentous fungus *Neurospora crassa*, have shown that DNA methylation can be dependent upon methylation of histone H3 lysine 9 [65]. In *Neurospora* and related fungi, DNA methylation is a rare event and its primary role seems to be as part of a defense mechanism for the marking, and subsequent destruction, of repeated sequences found in mobile genetic elements [66,67]. Various mutations have been found to interfere with DNA methylation in *Neurospora.* One of these, *dim-5*, is a SET-domain protein with histone methyltransferase activity specific for H3 K9 [66]. Biochemical and genetic evidence suggests that H3meK9 provides a mark that targets the DNA methylation machinery to defined regions of the *Neurospora* genome and that allows regional DNA methylation to be retained through DNA replication. The molecular mechanisms by which this link between histone and DNA methylation operates remain unknown, nor do we yet know whether a similar link is found in higher eukaryotes. It is worth remembering that, in contrast to mammals, most cytosine methylation in *Neurospora* is at non-CpG sites [66]. Although these sites lack the intrinsic heritability of methylCpG, patterns of DNA methylation in *Neurospora* are clearly heritable through at least several cell generations [66,67]. Perhaps H3meK9 is important for mainenance of DNA methylation only at non-CpG sites.

6. Short-term changes; transcription initiation

Changes in histone modifications, particularly acetylation, have long been associated with short-term aspects of transcriptional regulation, particularly initiation. Recent data strongly suggest that transcriptional initiation involves the progressive setting of precisely-defined patterns of histone modification. For example, activation of the endogenous *IFN-β* gene in response to viral infection is accompanied by the progressive acetylation of H4 K8 then H3 K9 and K14 [68]. *In vitro* experiments with immobilized mono-nucleosomes carrying an *IFN-β* promoter fragment, showed that H4acK8 was necessary for recruitment of the SWI/SNF complex and that H3acK9 and H3acK14 were necessary for recruitment of TFIID. Acetylation and transcription complex assembly, required the HAT GCN5/PCAF, but occurred as normal in extracts immuno-depleted of another HAT, CBP. The results lead to a model in which transcription initiation involves sequential modifications at defined sites on the histone tails, each of which facilitates recruitment/binding of the next element in the assembly pathway [68]. A similar model can accommodate the changes that occur on activation of the *pS2* promoter by estrogen. Here, CBP is recruited within 15 min of estrogen stimulation, leading to acetylation of H3 K18. This is followed by acetylation of H3 K23, recruitment of the arginine HMT CARM1, methylation of H3 R17 and transcription complex assembly [69].

7. Long-term effects

One of the most consistent and widespread associations between histone modification and chromatin structure, is the general under-acetylation of all four core histones in constitutive and facultative heterochromatin [11,12,70]. This underacetylation is not reversible by HDAC inhibitors such as butyrate or TSA, suggesting that the underacetylated state is static and that HATs at least are absent from heterochromatin. The facultative heterochromatin of the inactive X chromosome in female mammals (Xi), is also deficient in H3 di-methylated at lysine 4 [71,72], but appears to retain H3 di-methylated at lysine 9 [72]. (Studies on H3 K9 methylation have been complicated by the recent finding the some antisera to H3meK9 cross-react with H3meK27 (the two sites share the ARKS peptide motif (see below, BMT unpublished). Until this is resolved, it is possible that chromatin regions defined as rich in H3meK9, i.e., centric and facultative heterochromatin, contain H3meK27 as well, or even H3meK27 alone).

Thus, Xi exemplifies both extremes of the histone code, namely the general deacetylation of all four core histones and the specific retention of H3 dimethylated at lysine 9/27. It also illustrates the great stability of epigenetic signals carried by the histone tails. The general underacetylation of Xi, the undermethylation of H3 lysine 4 and the selective retention of H3 dimethylated at lysine 9/27, are all maintained in human x hamster somatic cell hybrids in which Xi is the only human chromosome and therefore exists in an almost completely foreign (i.e., hamster) cellular environment [73].

It may be that the general deacetylation of all four histones provides a molecular or structural context in which the specific methylation signal can be most effectively read. Perhaps deacetylated tails encourage binding of a protein that recognizes H3 dimethylated at lysine 9/27, just as deacetylation of the H4 tail encourages binding of the Sir2/Sir3 proteins that assemble heterochromatin-like structures in yeast [74]. This hypothetical protein remains to be identified. It seems not to be HP-1, which has not been detected on Xi [75]. The fact that HP-1 binds to constitutive centric heterochromatin, but not to Xi, even though both are underacetylated and apparently retain H3 dimethylated at K9 (but see above), suggests that there is an additional determinant of HP-1 binding.

Other examples of long-term, stable silencing have come from studies of the Polycomb group (PcG) proteins. These were first identified in *Drosophila* on the basis of their ability to associate with genes and gene clusters silenced during development and thereby to maintain the silent state through subsequent developmental stages [57]. E(Z) (*Enhancer of Zeste*), a component of PcG silencing complexes isolated from *Drosophila* embryos, has histone methyltransferase activity [76,77]. Mutations in the E(Z) SET domain remove this activity and also disrupt silencing of HOX genes *in vivo* [77]. With nucleosome substrates (which it prefers), the E(Z) HMT methylates H3 K27 exclusively [77], but can methylate both H3 K27 and H3 K9 when presented with peptide substrates [76]. A complex containing the human homologue of E(Z), EZH2, is completely specific for nucleosomal H3 K27 [78]. H3 methylated by the E(Z) complex binds specifically to Polycomb protein [76], arguing for a direct relationship between H3 methylation by E(Z) and assembly of the PcG silencing complex. Further, by chromatin immunoprecipitation (ChIP), H3meK27 co-localizes with E(Z) at binding sites for PcG proteins (Polycomb Response Elements) within the *Utrabithorax* (*Ubx*) locus and correlates with repression of transcription [78]. Other data, from immunofluorescence labeling of polytene chromosomes, suggest that PcG silencing complexes co-localize with H3tri-methylK9 [76]. This apparent discrepancy may reflect subtle differences in the specificities of the antisera used. Nevertheless, whatever lysines may be involved, the results show that silencing complexes containing E(Z) and PcG proteins, have the potential to place a self-sustaining epigenetic mark.

8. A molecular memory mechanism

It has been known for some time that certain bromodomains, sequence elements found in many chromatin associated proteins and most HATs [79], bind preferentially to acetylated peptides in *in vitro* binding assays, leading to speculation that acetylated histone tails could form targets for the binding of bromodomain-containing proteins *in vivo* [80,81]. Recent experiments provide direct evidence for this.

Nucleosome arrays attached to magnetic beads have been used to study the binding of the SWI/SNF chromatin remodelling complex and the SAGA and NuA4 HAT complexes to chromatin *in vitro* [82] The SWI/SNF complex is attracted to chromatin by the activator Gal4-VP16 (the assembled arrays contain a block of four Gal4 binding sites). Prior acetylation of the array with SAGA or NuA4 complexes does not enhance SWI/SNF binding, but does allow retention of the complex when the activator Gal4-VP16 is removed by treatment with competing oligonucleotides. This retention requires the bromodomain of the Swi2/Snf2 subunit of the SWI/SNF complex. Further, the SAGA complex itself is anchored to acetylated arrays following removal of VP16, but only if the bromodomain of the Gcn5 subunit is intact. In contrast, the NuA4 complex, which lacks a bromodomain, is not retained following removal of Gal4-VP16. To determine the *in vivo* relevance of Swi2/Snf2 bromodomain, ChIP was used to study the binding of SWI/SNF to the promoter of the inducible *SUC2* gene. The wild type Swi2/Snf2 showed enhanced association with the *SUC2* promoter under inducing conditions, whereas mutant Swi12/Snf2 lacking a bromodomain, did not. The model that emerges from these experiments is that the SAGA–HAT complex binds to acetylated nucleosomes, through its Gcn5 bromodomain, and provides a self-perpetuating, epigenetic mark that attracts and tethers other chromatin remodelers (i.e., SWI/SNF) to a small chromatin domain.

9. What should we expect of a histone code?

It is now a decade since evidence was first presented that the modification (acetylation) of particular residues on the histone tails was functionally significant [13,14]. Since then, it has been established beyond all reasonable doubt that specific modifications and combinations thereof, mediate protein–protein interactions crucial for the long-term and short-term regulation of transcriptional activity. Some of these interactions involve the modifying enzymes themselves, and recent data shows how particular histone tail modifications can interact, both to put in place defined *patterns* of modification and to control their recognition. Examples are presented in Table 1. But do these findings support the existence of a histone code?

In addressing this, it is useful to make two important distinctions. First, are we dealing with short-term or long-term transcriptional effects (the two have rather different regulatory requirements) and second, are the possible functions we are considering concerned with the process of transcription itself, or with its heritability? Earlier, we considered the rapid transcriptional upregulation of two inducible genes, *IFN-β* [68] and *pS2* [69]. In both cases, upregulation involves a

Table 1
Examples of residue-specific histone modifications associated with defined functional chromatin states

Modification	Organism	Modific'n set by	Modific'n read by	Chromatin state	Refs.
H4acK16	*Drosophila*	MOF	ISWI binding decreased	Decondensation of male X, transcription up-regulated	13,83–86
H3meK9	*Drosophila*, mammals	Su(var)3-9, SUV39H	HP1 binding	Heterochromatin formation	33–36,40
H3meK27	*Drosophila*, mammals	E(Z), EZH2	PC binding	Maintenance of silent state	76–78
H3meK4, K9 + H4meK20	*Drosophila*	ASH1	BRAHMA binding	Maintenance of transcriptionally active state	56

cascade of events, each dependent on the one preceding it and each involving specific histone tail modifications. While it is clear that specific modifications play important roles in this sequence of events, the final pattern of modifications may represent no more than the end result of the regulatory cascade, with no significance in itself, for upregulation of transcription. Nor is the final pattern of modifications likely to have a heritable function. High levels of induced transcriptional activity are usually, necessarily, transient, with no requirement that they be passed on from one cell generation to the next. It is noteworthy that induction of the *pS2* and *IFN-β* genes involves quite different sets of histone modifications. This is perhaps not surprising given that different HATs are involved, and further studies may yet reveal other modifications shared by the two systems, but they provide no evidence that a universal pattern, or sequence, of histone modifications is involved in induced transcriptional upregulation.

A histone code could be particularly valuable where there is a requirement for long-term maintenance of a transcriptional state and where it could be involved in both the transcriptional process itself and in its inheritance. What could be the nature of this long-term code? Despite the examples listed in Table 1, the idea that a *single* modification of a particular tail residue can, *unfailingly*, specify a particular functional effect becomes, at least as a general principle, increasingly unlikely. Specific modifications certainly have the *potential* to exert such effects, but whether they do or not is largely dependent on the context in which they are presented. For example, H3 methylated at K9 has the potential to initiate chromatin condensation and silencing, in part through its ability to bind proteins such as HP1 and possibly PcG proteins (see above). However, in the context of a nucleosome also methylated at H3 K4 and H4 K20, it provides part of a binding platform for the chromatin remodeling component BRAHMA and a mark for the long-term maintenance of a transcriptionally *active* state [56]. The silencing potential of H3meK9 may also be overridden by more transient modification of adjacent sites, such as phosphorylation of the adjacent residue (S10) or acetylation of K14 [30].

Perhaps the most enduring example of a single modification linked to a functional effect remains the hyperacetylation of H4 K16 that distinguishes the male X chromosome in *Drosophila* [13]. As shown in Fig. 2, H4acK16 co-localizes with the protein–RNA complex containing various components of the dosage compensation system [83], including MOF (Males absent On the First), a histone acetyltransferase specific for H4 K16 [84–86]. In polytene chromosomes from the salivary glands of third instar larvae, the male X is noticeably less condensed than the autosomes or the two female Xs, consistent with its two-fold higher rate of transcription. A genetic approach has provided evidence that the selective accumulation of H4 acetylated at lysine 16 on the *Drosophila* male X chromosome counteracts the *condensing* effect of the chromatin remodeling ATPase ISWI, thus allowing the chromosome to expand [87]. Related work shows that the site on chromatin recognized by ISWI consists of the H4 tail bound to DNA and that acetylation of H4 at lysine 16 reduces the ability of ISWI to interact productively with its substrate [87,88]. Together, these results provide a plausible mechanism by which H4 acetylated at lysine 16 can lead to decondensation and transcriptional upregulation of the *Drosophila* male X chromosome. However, whether H4acK16 alone is sufficient to trigger this sequence of events remains unclear. The protein kinase JIL-1 also localizes preferentially to the male X chromosome, leading to increased phosphorylation of H3 S10 [89–91]. It is also not clear what role, if any, is played by histone modifications in the maintenance of the dosage compensated state from one cell generation to the next. Hyperacetylation of H4 K16 on the male X persists through mitosis [13a], when transcription is repressed, but whether it plays a key role in S-phase, directing assembly of the dosage compensation complex on newly replicated DNA, is uncertain.

In considering long-term aspects of gene silencing, histone modifications can be seen as part of a sequence of events, possibly involving structural and catalytic proteins and RNAs, [92,93] whose end result is a functionally stable chromatin state. The different events may be mechanistically linked, but do not have to be. Thus, silencing of X-linked genes in female ES cells as they differentiate in culture involves a well-defined sequence of events, including coating with *Xist* RNA (days 1–2), global deacetylation of core histones (days 3–5) and cytosine methylation at selected CpGs (days 14–21) [94]. While these changes act synergistically to provide the extreme stability required of this silent state [95], their interdependence remains uncertain. In human x hamster somatic cell hybrids, *Xist* RNA does not closely co-localise with the human Xi, as it does in normal human cells, and yet the chromosome remains transcriptionaly silent, underacetylated and undermethylated at H3 K4 ([73] and BMT unpublished). While *Xist* RNA may be absolutely necessary for the early stages of X inactivation, it may become redundant once other (later) epigenetic changes have been put in place [95]. The *heritable* component of such a stable transcriptional state could usefully be defined as an epigenetic code. This heritable code would involve both DNA methylation, which is stable and has a defined maintenance mechanism [59] along with histone modifications such as methylation and acetylation, both of which have the potential to self-perpetuate through binding of HATs and HMTs to specifically-modified tail domains.

References

1. Kornberg, R.D. and Lorch, Y. (1999) Twenty-five years of the nucleosome, fundamental particle of the eukaryote chromosome. Cell 98, 285–294.
2. Grunstein, M. (1992) Histones as regulators of genes. Sci. Am. 267, 68–74B.
3. Langst, G. and Becker, P.B. (2001) Nucleosome mobilization and positioning by ISWI-containing chromatin remodelling factors. J. Cell. Sci. 114, 2561–2568.
4. Davie, J.R. and Moniwa, M. (2000) Control of chromatin remodeling. Crit. Rev. Eukaryot. Gene Expr. 10, 303–325.
5. Peterson, C.L. (2002) Chromatin remodeling: nucleosomes bulging at the seams. Curr. Biol. 12, R245–R247.
6. Hansen, J.C., Tse, C., and Wolffe, A.P. (1998) Structure and function of the core histone N-termini. More than meets the eye. Biochem. 37, 17637–17641.

7. Turner, B.M. (2002) Cellular memory and the histone code. Cell 111, 285–291.
8. Boffa, L.C., Walker, J., Chen, T.A., Sterner, R., Mariani, M.R., and Allfrey, V.G. (1990) Factors effecting nucleosome structure in transcriptionally active chromatin. Histone acetylation, nascent RNA and inhibitors of RNA synthesis. Eur. J. Biochem. 194, 811–823.
9. Spencer, V.A. and Davie, J.R. (2001) Dynamically acetylated histone associated with transcriptionally active and competent genes in the avian adult beta-globin gene domain. J. Biol. Chem. 276, 34810–34815.
10. Hebbes, T., Thorne, A.W., and Crane-Robinson, C. (1988) A direct link between core histone acetylation and transcriptionally active chromatin. EMBO J. 7, 1395–1403.
11. Jeppesen, P., Mitchell, A., Turner, B.M., and Perry, P. (1991) Antibodies to defined histone epitopes reveal variations in chromatin conformation and underacetylation of centric heterochromatin in human metaphase chromosomes. *Chromosoma* 100, 322–332.
12. Jeppesen, P. and Turner, B.M. (1993) The inactive X chromosome in female mammals is distinguished by lack of histone H4 acetylation, a marker for gene expression. Cell 74, 281–289.
13. Turner, B.M., Birley, A.J., and Lavender, J.S. (1992) Histone H4 isoforms acetylated at specific lysine residues define individual chromosomes and chromatin domains in *Drosophila* polytene nuclei. Cell 69, 375–384.
13a. Lavender, J.S., Birley, A.J., Palmer, M.J., Kuroda, M.I., and Turner, B.M. (1994) Histone H4 acetylated at lysine 16 and other components of the *Drosophila* dosage compensation pathway colocalize on the male X chromosome through mitosis. Chrom. Res. 2, 398–404
14. Turner, B.M. (1993) Decoding the nucleosome. Cell 75, 5–8.
15. Tordera, V., Sendra, R., and Pérez-Ortin, J.E. (1993) The role of histones and their modifications in the informative content of chromatin. Experientia 49, 780–788.
16. Zhang, Y. and Reinberg, D. (2001) Transcription regulation by histone methylation:interplay between different covalent modifications of the core histone tails. Genes Dev. 15, 2343–2360.
17. Strahl, B.D. and Allis, C.D. (2000) The language of covalent histone modifications. Nature 403, 41–45.
18. Jenuwein, T. and Allis, C.D. (2001) Translating the histone code. Science 293, 1074–1080.
19. Turner, B.M. (2000) Histone acetylation and an epigenetic code. BioEssays 22, 836–845.
20. Covault, J. and Chalkley, R. (1980) The identification of distinct populations of acetylated histones. J. Biol. Chem. 255, 9110–9116.
21. Waterborg, J.H. (2001) Dynamics of histone acetylation in Saccharomyces cerevisiae. Biochemistry 40, 2599–2605.
22. Waterborg, J.H. (2002) Dynamics of histone acetylation in vivo. A function for acetylation turnover? Biochem. Cell Biol. 40, 363–378.
22a. Sun, J.-M., Chen, H.Y., and Davie, J.R. (2001) Effect of oestradiol on histone acetylation dynamics in human breast cancer cells. J. Biol. Chem. 276, 49435–49442
23. Ng, H.-H. and Bird, A. (2000) Histone deacetylases: silencers for hire. TIBS 25, 121–126.
24. Grant, P.A. and Berger, S.L. (1999) Histone acetyltransferase complexes. Sem. Cell Dev. Biol. 10, 169–178.
25. Gregory, R.I. *et al.* (2002) Inhibition of histone deacetylases alters allelic chromatin conformation at the imprinted *U2af1-rs1* locus in mouse embryonic stem cells. J. Biol. Chem. 277, 11728–11734.
26. Thomas, G., Lange, H.W., and Hempel, K. (1975) Kinetics of histone methylation in vivo and its relation to the cell cycle in Ehrlich ascites tumor cells. Eur. J. Biochem. 51, 609–615.
27. Waterborg, J.H. (1990) Dynamic methylation of alfalfa histone H3. J. Biol. Chem. 268, 4918–4921.
28. Bannister, A., Schneider, R., and Kouzarides, T. (2002) Histone methylation: dynamic or static? Cell 109, 801–806.
29. Kouzarides, T. (2002) Histone methylation in transcriptional control. Curr. Opin. Gen. Dev. 12, 198–209.
30. Rea, S. *et al.* (2000) Regulation of chromatin structure by site-specific histone H3 methyltransferases. Nature 406, 593–599.

31. Litt, M.D., Simpson, M., Gaszner, M., Allis, C.D., and Felsenfeld, G. (2001) Correlation between histone lysine methylation and developmental changes at the chicken beta-globin locus. Science 293, 2453–2455.
32. Noma, K., Allis, C.D., and Grewal, S.I.S. (2001) Transitions in distinct histone H3 methylation patterns at the heterochromatin domain boundaries. Science 293, 1150–1155.
33. Bannister, A.J. *et al.* (2001) Selective recognition of methylated lysine 9 on histone H3 by the HP1 chromodomain. Nature 410, 120–124.
34. Lachner, M., O'Carroll, D., Rea, S., Mechtler, K., and Jenuwein, T. (2001) Methylation of histone H3 lysine 9 creates a binding site for HP1 proteins. Nature 410, 116–120.
35. Nielsen, P.R. *et al.* (2002) Structure of the HP1 chromodomain bound to histone H3 methylated at lysine 9. Nature 416, 103–107.
36. Jacobs, S.A. and Khorasanizadeh, S. (2002) Structure of HP1 Chromodomain Bound to a Lysine 9-Methylated Histone H3 Tail. Science 295, 2080–2083.
37. Horita, D.A., Ivanova, A.V., Altieri, A.S., Klar, A.J., and Byrd, R.A. (2001) Solution structure, domain features and structural implications of mutants of the chromo domain from the fission yeast histone methyltransferase Clr4. J. Mol. Biol. 307, 861–870.
38. Nishioka, K., Chuikov, S., Sarma, K., Erdjument-Bromage, H., Allis, C.D., Tempst, P., and Reinberg, D. (2002) Set9, a novel histone H3 methyltransferase that facilitates transcription by precluding histone tail modifications required for heterochromatin formation. Genes Dev. 16, 479–489.
39. Zegerman, P., Canas, B., Pappin, D., and Kouzarides, T. (2002) Histone H3 lysine 4 methylation disrupts binding of nucleosome remodelling and deacetylase (NuRD) repressor complex. J. Biol. Chem. 277, 11621–11624.
40. Nakayama, J., Rice, J.C., Strahl, B.D., Allis, C.D., and Grewal, S.I.S. (2001) Role of histone H3 lysine 9 methylation in epigenetic control of heterochromatin assembly. Science 292, 110–113.
41. Czermin, B. *et al.* (2001) Physical and functional association of SU(VAR)3-9 and HDAC1 in Drosophila. EMBO Rep. 2, 915–919.
42. Vandel, L. and Trouche, D. (2000) Physical association between the histone acetyl transferase CBP and a histone methyl transferase. EMBO Rep. 2, 21–26.
43. Fang, J., Feng, Q., Ketel, C.S., Wang, H., Cao, R., Xia, L., Erdjument-Bromage, H., Tempst, P., Simon, J.A., and Zhang, Y. (2002) Purification and functional characterization of SET8, a nucleosomal histone H4-lysine 20-specific methyltransferase. Curr. Biol. 12, 1086–1099.
44. Nishioka, K., Rice, J.D., Sarma, K., Erdjument-Broamge, H., Werner, J., Wang, Y., Chuikov, S., Valenzuela, P., Tempst, P., Steward, R., Lis, J.T., Allis, C.D., and Reinberg, D. (2002) PR-Set7 is a nucleosome-specific methyltransferase that modifies lysine 20 of histone H4 and is associated with silent chromatin. Mol. Cell 9, 1201–1213.
45. Rice, J.D., Nishioka, K., Sarma, K., Steward, R., Reinberg, D., and Allis, C.D. (2002) Mitotic-specific methylation of histone H4 Lys 20 follows increased PR-Set7 expression and its localization to mitotic chromosomes. Genes Dev. 16, 2225–2230.
46. Bryk, M., Briggs, S.D., Strahl, B.D., Curcio, M.J., Allis, C.D., and Winston, F. (2002) Evidence that Set1, a factor required for methylation of histone H3, regulates rDNA silencing in *S. cerevisiae* by a Sir2-independent mechanism. Curr. Biol. 12, 165–170.
47. Sun, Z-W. and Allis, C.D. (2002) Ubiquitination of histone H2B regulates H3 methylation and gene silencing in yeast. Nature 418, 104–108.
48. Dover, J., Schneider, J., Boateng, M.A., Wood, A., Dean, K., Johnston, M., and Shilatifard, A. (2002) Methylation of histone H3 by COMPASS requires ubiquitination of histone H2B by Rad6. J. Biol. Chem. 277, 28368–28371.
49. Jason, L.J.M., Moore, S.C., Lewis, J.D., Lindsey, G., and Ausio, J. (2002) Bioessays 24, 166–174.
50. Briggs, S.D., Xiao, T., Sun, Z.W., Caldwell, J.A., Shabanowitz, J., Hunt, D.F., Allis, C.D., and Strahl, B.D. (2002) Gene silencing: trans-histone gene regulatory pathway in chromatin. Nature 418, 498.
51. Van Leeuwen, F., Gafken, P.R., and Gottschling, D.E. (2002) Dot1p modulates silencing in yeast by methylation of the nucleosome core. Cell 109, 745–756.

52. Feng, Q., Wang, H., Ng, H.H., Erdjument-Bromage, H., Tempst, P., Struhl, K., and Zhang, Y. (2002) Methylation of H3-lysine 79 is mediated by a new family of HMTases without a SET domain. Curr. Biol. 12, 1052–1058.
53. Ng, H.H., Feng, Q., Wang, H., Erdjument-Bromage, H., Tempst, P., Zhang, Y., and Struhl, K. (2002) Lysine methylation within the globular domain of histone H3 by Dot1 is important for telomeric silencing and Sir protein association. Genes Dev. 16, 1518–1527.
54. Duerre, J.A. and Chakrabarty, S. (1975) Methylated basic amino acid composition of histones from the various organs from the rat. J. Biol. Chem. 250, 8457–8461.
55. Strahl, B.D., Ohba, R., Cook, R.G., and Allis, C.D. (1999) Methylation of histone H3 at lysine 4 is highly conserved and correlates with transcriptinoally active nuclei in Tetrahymena. Proc. Natl. Acad. Sci. USA 96, 14967–14972.
56. Beisel, C., Imhof, A., Greene, J., Kremmer, E., and Sauer, F. (2002) Histone methylation by the *Drosophila* epigenetic regulator Ash1. Nature 419, 857–862.
57. Pirrotta, V. (1997) PcG complexes and chromatin silencing. Curr. Opin. Genet. Dev. 7, 249–258.
58. Poux, S., Horard, B., Sigrist, C.J.A., and Pirrotta, V. (2002) The Drosophila Trithorax protein is a coactivator required to prevent re-establishment of Polycomb silencing. Development 129, 2483–2493.
59. Bird, A. (2002) DNA methylation patterns and epigenetic memory. Genes Dev. 16, 6–21.
60. Bird, A. and Wolffe, A.P. (1999) Methylation-induced repression—Belts, braces and chromatin. Cell 99, 451–454.
61. Bestor, T.H. (2000) The DNA methyltransferases of mammals. Hum. Mol. Genet. 9, 2395–2402.
62. Pradhan, S., Bacolla, A., Wells, R.D., and Roberts, R.J. (1999) Recombinant human DNA (cytosine-5) methyltransferase I. Expression, purification and comparison of de novo and maintenance methylation. J. Biol. Chem. 274, 33002–33010.
63. Nan, X., Ng, H.-H., Johnson, C.A., Laherty, C.D., Turner, B.M., Eisenman, R.N., and Bird, A. (1998) Transcriptional repression by the methyl-CpG binding protein MeCP2 involves a histone deacetylase complex. Nature 393, 386–389.
64. Gregory, R.I., Randall, T.E., Johnson, C.A., Khosla, S., Hatada, I., O'Neill, L.P., Turner, B.M., and Feil, R. (2001) DNA methylation is linked to deacetylation of histone H3, but not H4, on the imprinted genes *Snrpn* and *U2af1-rs1*. Mol. Cell Biol. 21, 5426–5436.
65. Tamaru, H. and Selker, E.U. (2001) A histone H3 methyltransferase controls DNA methylation in *Neurospora crassa*. Nature 414, 277–283.
66. Selker, E.U., Freitag, M., Kothe, G.O., Margolin, B.S., Rountree, M.R., Allis, C.D., and Tamaru, H. (2002) Induction and maintenance of non-symmetrical DNA methylation in *Neurospora*. Proc. Natl. Acad. Sci. USA 99, 16485–16490.
67. Selker, E.U. (2002) Repeat-induced gene silencing in fungi. Adv. Genet. 46, 439–450.
68. Agalioti, T., Chen, G., and Thanos, D. (2002) Deciphering the transcriptional histone acetylation code for a human gene. Cell 111, 381–392.
69. Daujat, S., Bauer, U-M., Shah, V., Turner, B.M., Berger, S., and Kouzarides, T. (2002) Cross-talk between CARM1 methylation and CBP acetylation on histone H3. Curr. Biol., in press.
70. Maison, C. *et al.* (2002) Higher-order structure in pericentric heterochromatin involves a distinct pattern of histone modification and an RNA component. Nat. Genet. 30, 329–334.
71. Boggs, B.A., Cheung, P., Heard, E., Spector, D.L., Chinault, C., and Allis, C.D. (2002) Differentially methylated forms of histone H3 show unique association patterns with inactive human X chromosomes. Nat. Genet. 30, 73–76.
72. Heard, E. *et al.* (2001) Methylation of histone H3 at lys-9 is an early mark on the X chromosome during X inactivation. Cell 107, 727–738.
73. Spotswood, H.T. and Turner, B.M. (2002) An increasingly complex code. J. Clin. Invest. 110, 577–582.
74. Hecht, A., Laroche, T., Strahl-Bosinger, S., Gasser, S.M., and Grunstein, M. (1995) Histone H3 and H4 N-termini interact with the Silent Information Regulators Sir3 and Sir4: a molecular model for the formation of heterochromatin in yeast. Cell 80, 583–592.

75. Cowell, I.G. *et al.* (2002) Heterochromatin, HP1 and methylation at lysine 9 of histone H3 in animals. Chromosoma 111, 22–36.
76. Czermin, B., Melfi, R., McCabe, D., Seitz, V., Imhof, A., and Pirrotta, V. (2002) *Drosophila* Enhancer of Zeste/ESC complexes have a histone H3 methyltransferase activity that marks chromosomal Polycomb sites. Cell 111, 185–196.
77. Müller, J., Hart, C.M., Francis, N.J., Vargas, M.L., Sengupta, A., Wild, B., Miller, E.L., O'Connor, M.B., Kingston, R.E., and Simon, J.A. (2002) Histone methyltransferase activity of a *Drosophila* Polycomb Group repressor complex. Cell 111, 197–208.
78. Cao, R., Wang, L., Wang, H., Xia, L., Erdjument-Bromages, H., Tempst, P., Jones, R.S., and Zhang, Y. (2002) Role of histone H3 lysine 27 methylation in Polycomb-group silencing. Science 298, 1039–1043.
79. Zeng, L. and Zhou, M.M. (2002) Bromodomain: an acetyl-lysine binding domain. FEBS Lett. 513, 124–128.
80. Dhalluin, C., Carlson, J.E., Zheng, L., He, C., Aggarwal, A.K., and Zhou, M.-M. (1999) Structure and ligand of a histone acetyltransferase bromodomain. Nature 399, 491–496.
81. Jacobson, R.H., Ladurner, A.G., King, D.S., and Tjian, R. (2000) Structure and function of a human TAFII250 double bromodomain module. Science 288, 1422–1425.
82. Hassan, A.H., Prochasson, P., Neely, K.E., Galasinski, S.C., Chandy, M., Carrozza, M.J., and Workman, J.L. (2002) Function and selectivity of bromodomains in anchoring chromatin-modifying complexes to promoter nucleosomes. Cell 111, 369–379.
83. Bone, J.R., Lavender, J.S., Richman, R., Palmer, M.J., Turner, B.M., and Kuroda, M.I. (1994) Acetylated histone H4 on the male X chromosome is associated with dosage compensation in *Drosophila.* Genes Dev. 8, 96–104.
84. Smith, E.R., Pannuti, A., Gu, W., Steurnagel, A., Cook, R.G., Allis, C.D., and Lucchesi, J.C. (1999) The *Drosophila* MSL complex acetylates histone H4 at lysine 16, a chromatin modification linked to dosage comensation. Mol. Cell. Biol. 20, 312–318.
85. Akhtar, A and Becker, P.B. (2000) Activation of transcription through histone H4 acetylation by MOF, an acetyltransferase essential for dosage compensation. Mol. Cell 5, 367–375.
86. Akhtar, A. and Becker, P.B. (2001) The histone H4 acetyltransferase MOF uses a C2HC zinc finger for substrate recognition. EMBO Rep. 2, 113–118.
87. Corona, D.F., Clapier, C.R., Becker, P.B., and Tamkun, J.W. (2001) Modulation of ISWI function by site-specific histone acetylation. EMBO Rep. 3, 242–247.
88. Clapier, C.R., Nightingale, K.P., and Becker, P.B. (2002) A critical epitope for substrate recognition by the nucleosome remodeling ATPase ISWI. Nuc. Acids Res. 30, 649–655.
89. Jin, Y., Wang, Y., Walker, D.L., Dong, H., Conley, C., Johansen, J., and Johansen, K.M. (1999) JIL-1: a novel chromosomal tandem kinase implicated in transcriptional regulation in Drosophila. Mol. Cell 4, 129–135.
90. Jin, Y., Wang, Y., and Johansen, K.M. (2000) JIL-1: a chromosomal kinase implication in regulation of chromatin structure associates with the male specific lethal (MSL) dosage compensation complex. J. Cell Biol. 149, 1005–1010.
91. Wang, Y., Zhang, W., Jin, Y., Johansen, J., and Johansen, K.M. (2001) The JIL-1 tandem kinase mediates histone phosphorylation and is required for maintenance of chromatin structure in Drosophila. Cell 105, 433–443.
92. Volpe, T.A., Kidner, C., Hall, I.M., Teng, G., Grewal, S.I.S., and Martienssen, R.A. (2002) Regulation of heterochromatic silencing and histone H3 lysine-9 methylation by RNAi. Sciences 297, 1833–1837.
93. Richards, E.J. and Elgin, S.C.R. (2002) Epigenetic codes for heterochromatin formation and silencing: rounding up the usual suspects. Cell 108, 489–500.
94. Keohane, A.M., Belyaev, N.D., Lavender, J.S., O'Neill, L.P., and Turner, B.M. (1996) X-inactivation and histone H4 acetylation in embryonic stem cells. Dev. Biol. 180, 618–630.
95. Csankovszki, G., Nagy, A., and Jaenisch, R. (2002) Synergism of Xist RNA, DNA methylation, and histone hypoacetylation in maintaining X chromosome inactivation. J. Cell Biol. 153, 773–783.
96. Luger, K., Mader, A.W., Richmond, R.K., Sargent, D.F., and Richmond, T.J. (1997) Crystal structure of the nucleosome core particle at 2.8Å resolution. Nature 389, 251–260.

J. Zlatanova and S.H. Leuba (Eds.) *Chromatin Structure and Dynamics: State-of-the-Art*

DOI: 10.1016/S0167-7306(03)39012-X

CHAPTER 12

DNA methylation and chromatin structure

Jordanka Zlatanova[1], Irina Stancheva[2], and Paola Caiafa[3]

[1]*Department of Chemical and Biological Sciences and Engineering, Polytechnic University, Brooklyn, NY 11201, USA. Tel.: 718-260-3176; Fax: 786-524-5899; E-mail: jzlatano@duke.poly.edu and jzlatanova@hotmail.com*

[2]*Department of Biomedical Sciences, University of Edinburgh, Edinburgh EH8 9XD, UK. E-mail: istanche@staffmail.ed.ac.uk*

[3]*Department of Cellular Biotechnology and Hematology, University of Rome 'La Sapienza', 00161 Rome, Italy. E-mail: caiafa@bce.uniroma1.it*

1. *Introduction*

There is increasing evidence that epigenetic modifications play a crucial role in determining the dynamic structure and function of chromatin and that epigenetic changes at the chromatin level control the ordered flow of decisions in cells. Post-synthetic modifications of both DNA and chromatin proteins are involved in achieving a regulated access to the underlying DNA in order to perform DNA replication, to initiate appropriate transcriptional programs, and to carry out repair of DNA damage. These epigenetic modifications are often correlated: in fact, links have been found between histone acetylation and histone methylation, histone acetylation and DNA methylation, histone methylation and DNA methylation, and poly(ADP-ribosyl)ation and DNA methylation. Most probably, all these modifications work in concert to reach the final aim, chromatin structural changes that silence or activate gene transcription.

DNA methylation is a specific post-synthetic, enzymatic modification of DNA that, in eukaryotic cells, appears to play an important role in the epigenetic modulation of gene expression. It is the major DNA modification that is achieved through the transfer of methyl groups from S-adenosyl methionine to cytosine (C), converting it into 5-methylcytosine (5mC), a new base in the DNA [1]. Although in vertebrates 5mC has occasionally been found in CpC, CpA, or CpT dinucleotides, the best substrate for the addition of methyl group is cytosine located in the CpG dinucleotide [2]. Recently, methylation of the internal cytosine within the sequence CC(A/T)GG has also been recognized in endogenous mammalian or retroviral genes (reviewed in Ref. [3]). In plants DNA methylation has been found at both CpG dinucleotides and CpNpG trinucleotides (reviewed in Ref. [4]).

DNA methylation patterns are heritably propagated in somatic lineage cells, but in some circumstances methylation of DNA is a dynamic process, especially during early embryogenesis. Remodeling of DNA methylation levels and patterns have been observed during development of many vertebrate species [5–9], the most dramatic example being the rapid demethylation of the mouse sperm DNA few hours after fertilization, followed by a second wave of demethylation in the pre-implantation mouse embryo (Fig. 1). DNA methylation levels are restored by *de novo* methylation after implantation, a highly selective process leading to the establishment of specific non-random patterns of 5mC (described further). Later, during gastrulation, some genes become demethylated in a lineage-dependent manner, to accommodate the need for differential gene expression in specialized tissues [5,10]. Global demethylation/remethylation, however, is not an obligatory requirement for vertebrate development; no changes in the levels of 5mC have been observed during zebrafish embryogenesis [11], and a more limited drop in methylation occurs in the frog *Xenopus laevis* [12,10]. A genome-wide demethylation is also seen in primordial germ cells during the proliferative stages of spermatogenesis and oogenesis [13,14]. Finally, the methylation of specific genes may change in connection with their expression status within a given cell [15].

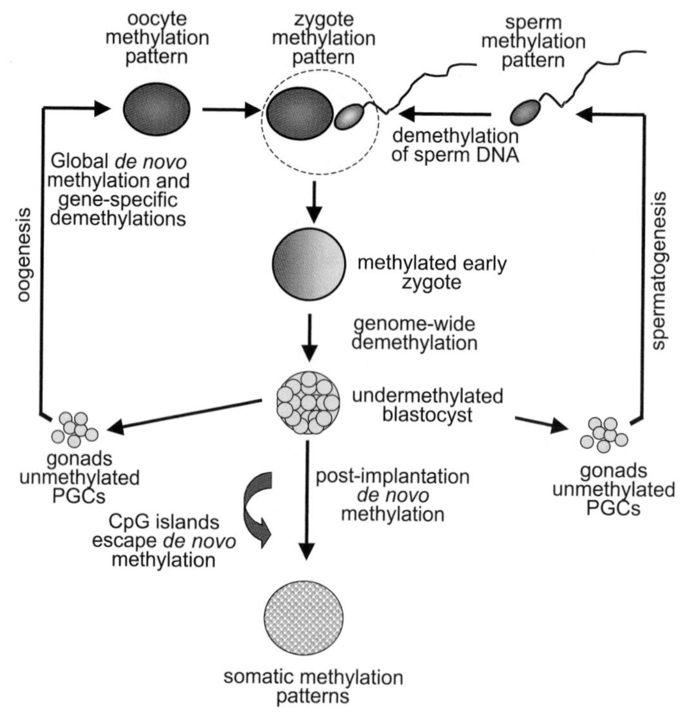

Fig. 1. Dynamics of DNA methylation levels during mouse development. The methylation patterns of the oocyte and the rapidly demethylated after fertilization sperm create the combined methylation patterns in the early mouse zygote. During the first two to three cleavage divisions, the 5mC levels decrease further and stay low through the blastula stage. Post-implantation, the mouse embryo genome is methylated *de novo*; the CpG islands remain mostly unmethylated. The primordial germ cells remain unmethylated. During gametogenesis specific parental (maternal or paternal) patterns of DNA methylation are established at imprinted loci (for further details see Refs. [13, 14]) (re-drawn from Ref. [4]).

2. DNA methylation: the biology

2.1. The distribution of methylated CpGs in the genome is not random

Seventy-eighty percent of all CpG dinucleotides are symmetrically methylated in vertebrate somatic cells. The distribution of methylated CpGs, however, follows a non-random pattern (Fig. 2): most methylated CpGs are scattered throughout the genome, and it is unclear whether they contribute to transcriptional silencing (see below). The majority of the unmethylated CpGs are concentrated within particular DNA regions, from 0.5 to 2 kbp in size, termed "CpG islands" [16] (Table 1). CpG islands are usually found in the promoter regions of housekeeping genes that are constitutively expressed, and may overlap the structural portions of these genes to variable extents [16,17]. Evidence exists that transcription of genes containing CpG islands is inhibited when these regions are methylated [18]. When such anomalous DNA methylation occurs *in vivo*, in promoters of tumor-suppressor genes, it silences these genes and leads to tumorigenesis. There are numerous examples of such genes inactivated by CpG island methylation: *Rb* in retinoblastomas, *VHL* in renal carcinoma, *16INK4a* in melanomas and other tumors, *hMHH1* in colorectal carcinoma, *BRCA1* in breast cancer, to mention a few well studied cases (for reviews on DNA methylation and cancer, see Refs. [19–21]). It is interesting to note that some CpG islands are located within tissue-specific and imprinted genes; methylation of these CpG islands does not block transcription [22,23].

The fact that CpG dinucleotides exist in an unmethylated state in CpG islands is amazing, given that the frequency of CpGs in such islands is six to ten times higher than in bulk DNA [17] (Table 1). A great deal of investigation has been and is being

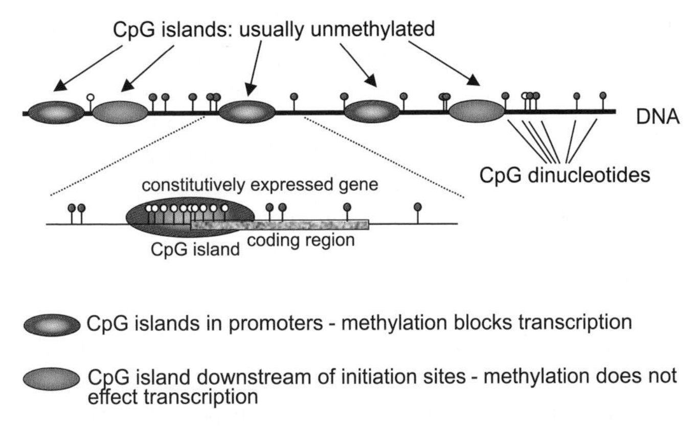

Fig. 2. Distribution of methylated CpGs in genomic DNA and their effect on transcription. 70–80% of all CpG dinucleotides are methylated in vertebrate genomes. mCpGs (filled lollipops) are randomly distributed throughout the genome but are excluded from regions with unusually high CpG density—CpG islands [16,17]. Most of the CpG islands are associated with gene promoters and maintained unmethylated (white lollipops) in all types of somatic cells. Aberrant methylation of CpG islands occurs in cancer cells and leads to silencing of tumor-suppressor and other essential genes [19–21].

Table 1
Distribution of CpG and mCpG in eukaryotic DNA

	Total genomic DNA (99%)	CpG islands (1%)
G+C content	40%	>60%
Level of CpG (observed/randomly expected)	0.2	>0.6
Methylation level	High (60–70% of all CpGs)	Unmethylated
Number of CpG island in the human genome	29,000[a]	29,000[a]
DNA length	—	0.2–2 kbp
Localization	—	Promoter regions of constitutively expressed gene; intragenic[b]

[a]Based on the sequence of the human genome
[b]CpG islands can remain methylation-free even when the associated gene is silent [23].

performed in order to clarify the mechanism by which CpG islands are protected from methylation (for a review, see Ref. [23]). That the CpG islands are not by themselves unmethylatable has been demonstrated by *in vitro* experiments [24]. How do the CpG islands remain untouched by the action of DNA methyltransferases despite their localization in the promoter regions of housekeeping genes, which, being located in decondensed chromatin, are permanently accessible to transcriptional factors and enzymes? Several mechanisms, which fall into two different groups, have been proposed to explain how CpG islands maintain or lose their unmethylated state. One group of mechanisms invokes control of specific genes, whereas the other concentrates on a more general mechanism for aberrant genome methylation.

As in the case of all gene-specific control mechanisms, the specificity resides in the recognition and binding of specific proteins to defined nucleotide sequences in control gene regions. In some cases, the protein recognizes a specific DNA target in a gene and recruits DNA methyltransferases onto these specific DNA regions. An example of this scenario is PML–RAR, an oncogenic transcription factor found in acute promyelocytic leukemias, that binds the RARβ2 promoter containing a CpG island; the binding of this protein silences gene transcription through recruiting DNA methyltransferases and inducing DNA hypermethylation [25] (Fig. 3a). In other cases, sequence-specific protein factors such as Sp1 may have a protective role, successfully competing with methyltransferases for sites within the CpG islands, thus precluding methylation. Indeed, most CpG islands contain Sp1 sites and mutation of Sp1 sites in transgenic mice leads to methylation of the transgene CpG island [26,27] (Fig. 3b). Although this mechanism involves sequence-specific binding of a transcription factor, the widespread appearance of Sp1 sites in CpG islands may put this process in the category of global protection from methylation.

Transcription factors other than Sp1 and/or chromatin remodeling proteins that support the maintenance of transcriptionally favorable chromatin conformation

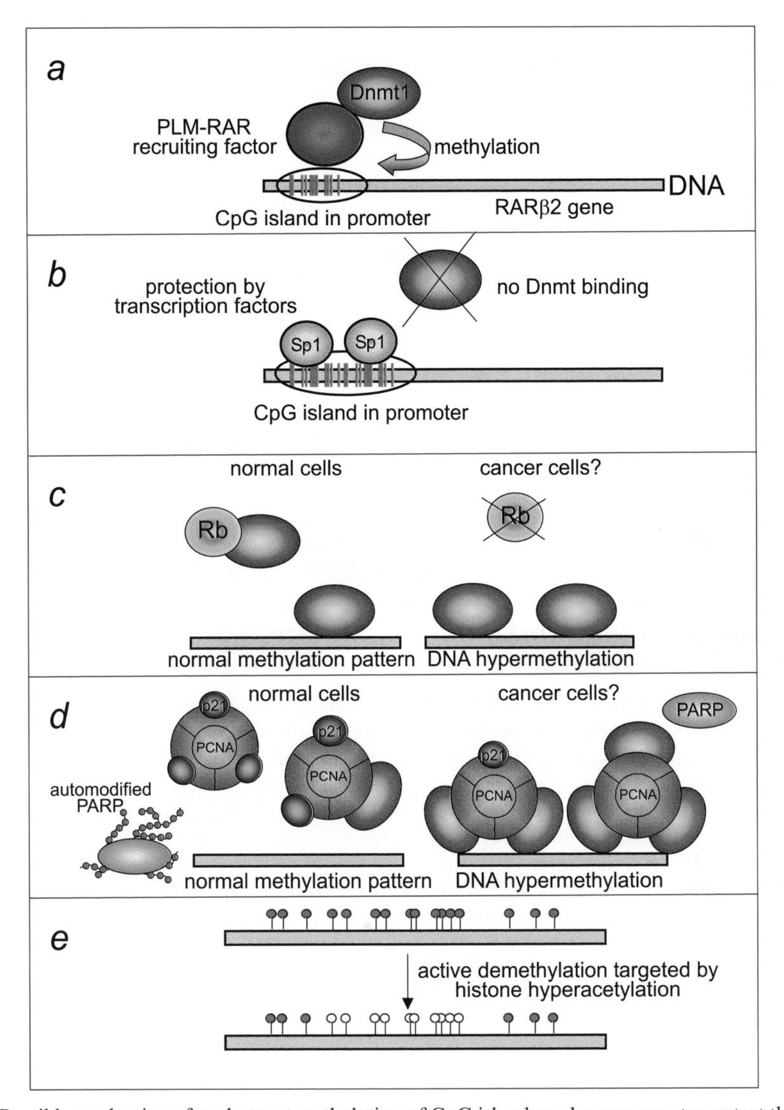

Fig. 3. Possible mechanisms for aberrant methylation of CpG islands and some ways to protect them from DNA methyltransferases. (*a*) Aberrant methylation of a promoter region of a specific gene in cancer. The oncogenic transcription factor PML-RAR binds to the promoter CpG island and silences transcription through recruitment of DNA methyltransferase and methylating the island [25]. (*b*) Protection of a CpG island from methylation by binding of the transcription factor Sp1 to its recognition sites within the island [28,29]. (*c*) Participation of the retinoblastoma protein pRb in protecting CpG islands from methylation in normal cells: pRb inactivates Dnmt1 by dissociating the binary DNA–Dnmt1 complex. In this scenario, DNA hypermethylation in cancer cells is due to changes in pRb levels [36]. (*d*) In this model, CpG island methylation is regulated by the intracellular levels of Dnmt1 and p21 that compete for the same binding site on PCNA [38]. One possible mechanism of regulating the level of Dnmt1 could involve changes in the automodification state of PARP [poly(ADP-ribose) polymerase] [41] (see Section 5.3). (*e*) Targeted active demethylation as an explanation for the unmethylated state of CpG islands [40,43–45] (see Section 5.1).

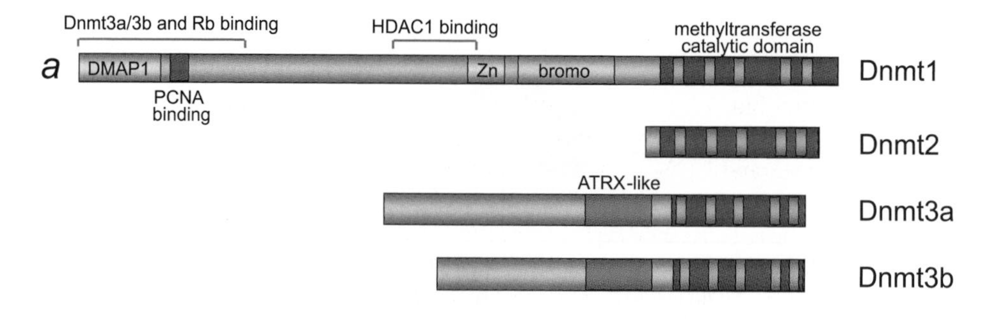

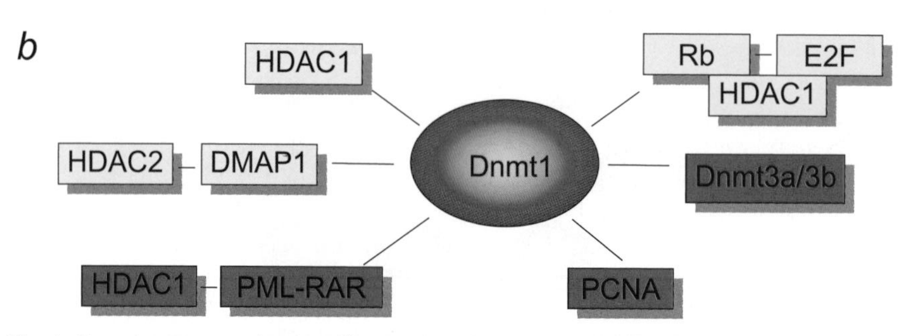

Fig. 4. Domain structure of mammalian DNA methyltransferases. (*a*) The domain structure of the known DNA methyltransferases, depicting the conserved catalytic domain (dark box) and other identified domains. Conserved aminoacid motifs in the catalytic domain are shown in lighter shade of gray. (*b*) Schematic representation of the reported protein–protein interactions of Dnmt1 with a number of regulatory proteins: interactions that modulate Dnmt1 methyltransferase activity (darker rectangles) or mediate methylation-independent transcriptional repression mechanisms (lighter rectangles). When Dnmt1 represses transcription through its enzymatic activity, it has been described to interact with some proteins: PCNA [37] and an oncogenic transcription factor PML–RAR [25]. Note that in the case of the PML–RAR transcription factor, histone deacetylase 1 (HDAC1) is also bound to the complex. When Dnmt1 represses transcription via methylation-independent pathways, it binds to HDACs either directly [34] or indirectly through other proteins: the corepressor DMAP1 [33], the retinoblastoma protein, and a gene-specific transcription factor [31].

may have a similar protective effect [28]. Indeed, *in vivo* DNA footprinting studies have demonstrated that most of the CpG islands in mouse and human cells are bound by protein factors even in cell types where the associated gene is not expressed [29]. It may also be speculated that certain histone modifications (e.g., H3 lysine 4 methylation found at actively transcribed or posed for transcription loci) may support chromatin conformation or binding of protein factors that repel *de novo* DNA methylation [30].

It is important to note that although protection from methylation by excluding methyltransferase binding is a viable scenario, the presence of methyltransferases on DNA is not always accompanied by DNA methylation; in some cases, the enzyme acts as co-repressor through some other mechanism (see Fig. 4b) [31–35].

Two interesting models have recently been put forward to explain the non-specific aberrant methylation of housekeeping genes. The first one attributes an important role to the retinoblastoma protein pRb, since its tight association with the major maintenance DNA methyltransferase Dnmt1 (Table 2) inactivates the enzyme by dissociating the binary DNA–Dnmt1 complex. pRb–Dnmt1 complex formation is regulated during the cell-cycle so that some Dnmt1 is free to perform maintenance DNA methylation in S phase [36] (Fig. 3c, left panel). Therefore, the aberrant pattern of DNA methylation which characterizes tumor cells could be due to alterations in the metabolic pathway of pRb or to pRb mutations that would abolish its binding to Dnmt1; such situations are generally encountered in tumor cells.

The second model is based on research showing that Dnmt1 and p21 compete for the same binding site on PCNA [37]. (PCNA is the protein that loads Δ- and ε-DNA polymerases onto the replication foci; it also recruits Dnmt1 to these sites, thus coordinating maintenance methylation with DNA replication. p21 is a tumor-suppressor protein that controls cell proliferation by inhibiting cyclin-dependent kinase and/or by directly inhibiting the activity of PCNA and, hence, DNA replication.) Baylin [38] suggested a model in which a change in the intracellular level of p21 or Dnmt1 could favor the binding of one over the other to PCNA (Fig. 3d). Hyperexpression of Dnmt1 is not a general condition in tumor cells [39], but the expression of the enzyme is cell-cycle dependent [40]. Thus, for example, an anomalous expression of Dnmt1 in early S phase may explain the anomalous methylation of onco-suppressor genes, since it is in this phase that regions of DNA enriched in unmethylated CpG dinucleotides (active genes) are replicated [17]. However, Dnmt1 does not display methylation activity towards unmethylated DNA *in vivo*, as it became apparent lately, and the above two scenarios would be possible only assuming that there is a cooperation between the maintenance and *de novo* DNA methyltransferases in the establishment of aberrant patterns of 5mC (see Section 2.2).

What is unclear in this scenario is the molecular event that causes deregulation of Dnmt1 expression in tumor cells. Recent data suggest that inhibition of poly(ADP-ribose) polymerases *in vivo* may lead to deregulation of Dnmt1 expression [41]. Several experimental strategies have shown that competitive inhibition of poly(ADP-ribose) polymerases, through treatment of mammalian cells with 3-aminobenzamide, leads to *in vivo* DNA hypermethylation (a more detailed account of these findings will be given in a later section; also reviewed in detail in Ref. [42]). The observed hypermethylation was due to increased mRNA and protein levels of Dnmt1; this resulted in a greater PCNA–Dnmt1 complex formation, which, in turn, led to DNA hypermethylation (Fig. 3d). The connection between poly(ADP-ribosyl)ation and DNA methylation may be still another mechanism to explain the aberrant methylation in cancer.

Finally, targeted active demethylation has been put forward as an explanation for the unmethylated state of CpG islands (Fig. 3e). In this model, features of active chromatin structure (e.g., histone hyperacetylation) cause active demethylation of associated sequences [43,44]. Based on a broad analysis of existing data, Szyf

Table 2

Mammalian DNA methyltransferases

Enzyme	Function	Main expression pattern	Phenotype of homozygous deletion mutants
Dnmt1	Maintenance (functions after replication on hemimethylated DNA to restore existing methylation patterns; *in vitro* can methylate non-methylated DNA); associates with histone deacetylase	Adult/Embryo	DNMT1$^{-/-}$ die *in utero*; DNA in these embryos is hypomethylated, imprinted genes are biallelically expressed; ES cells from DNMT1$^{-/-}$ mice are viable and capable of *de novo* methylation
Dnmt3a[a]	*De novo* DNA methyltransferase	Adult/Embryo	DNMT3a$^{-/-}$ mice die at 4 weeks; ES cells from DNMT3a$^{-/-}$ mice are viable and capable of *de novo* methylation; DNMT3a$^{-/-}$ embryos and ES cells exhibit demethylation of centromeric satellite repeats
Dnmt3b[a,b]	*De novo* DNA methyltransferase	Adult/Embryo	DNMT3b$^{-/-}$ mice die *in utero*; ES cells from DNMT3b$^{-/-}$ mice are viable and capable of *de novo* methylation; DNMT3b$^{-/-}$ embryos and ES cells do not exhibit demethylation of centromeric satellite repeats

[a]DNMT3a$^{-/-}$ and DNMT3b$^{-/-}$ double ES mutants exhibit no *de novo* methylase activity; double-mutant mice fail to develop beyond gastrulation. This observation, taken together with the ability of single mutants to support *de novo* methylation, show redundancy between the two genes.

[b]Heterozygous mutations affecting the catalytic domain of Dnmt3b have been identified in ICF syndrome in humans.

For the original work, see Refs. [50–53].

suggests that an incorrect pattern of methylation is established *only* if the epigenetic information embodied in the chromatin structure is lost or changed [44,45]. This new notion will be discussed more in-depth below (see Section 5.1).

It is clear that we still do not fully understand how the CpG moieties of the CpG islands resist the action of DNA methyltransferases or become vulnerable to it, thus maintaining or losing their characteristic pattern of methylation. More research is definitely needed to answer this intriguing and important question.

2.2. The enzymatic machinery involved in governing the DNA methylation status

The specific DNA methylation pattern results from a combination of maintenance and *de novo* methylation and demethylation processes. The maintenance methylase Dnmt1 (Table 2) (Fig. 4) was first identified in mammalian cells, through homology with bacterial methyltransferases. The enzyme recognizes and modifies hemi-methylated sites generated during DNA replication, thus preserving the tissue-specific methylation pattern [46,47]. Proteins performing similar function have been described in other vertebrates, plants, fungi, and even in *Drosophila*, which was for a long time believed to completely lack DNA methylation [4,47–49]. In higher eukaryotes, the enzymatic methylation process takes place within a minute or two after replication [50], when Dnmt1 is in its active form bound to PCNA [37]. Numerous other proteins that directly interact with Dnmt1 have been identified (illustrated in Fig. 4b). Some of these proteins modulate the enzymatic activity of Dnmt1, and thus change the methylation status of DNA. Other Dnmt1-binding proteins (e.g., histone deacetylases) function to directly affect chromatin structure; in these cases Dnmt1 acts as transcriptional co-repressor, without involving DNA methylation.

The *de novo* methylation process explains how the pattern of methylation can be changed by introducing new methyl groups onto DNA at sites in which neither strand was previously methylated. In fact, this process plays an important role during early stages of embryonic development, when the bimodal pattern of methylation is defined (see above, and Fig. 1). The CpG islands at the 5′ end of housekeeping genes somehow escape the massive *de novo* methylation process that affects most other genomic sequences during post-implantation; the CpG islands remain constitutively unmethylated, ensuring the constitutive transcription of the linked genes [26].

Until recently, only one mammalian DNA methyltransferase (Dnmt1) was known and this enzyme was believed to perform both the maintenance and the *de novo* modifications. Now the scenario has changed [51], since Dnmt1 does not seem to be involved in the *de novo* methylation *in vivo*; other methyltransferases (Dnmt3a and Dnmt3b) found in embryonic stem cells and cancer cells are thought to carry out the *de novo* methylation process [52]. That Dnmt3a and 3b are indeed the *de novo* methylases has been shown by their preference to use non-methylated DNA as substrate rather than hemimethylated one [52,53]. All three Dnmts play very important roles *in vivo* (Table 2). Homozygous deletions of *Dnmt1* or *Dnmt3b* in mouse germ lines are embryonic lethal, whereas *Dnm3a*$^{-/-}$ mice die within four weeks after birth [54,55]. Two other proteins, Dnmt2 and Dnmt3L, share high degree of homology with the members of the Dnmt family but do not possess methyltransferase enzymatic activity [56–58]. While Dnmt3L is involved in the establishment of maternal methylated imprints (see Section 4), the function of Dnmt2 is still unclear.

It should be noted that other proteins containing regions of homology with the methyltransferase domain of eukaryotic DNA methyltransferases may recognize and specifically bind unmethylated CpGs; methylation of such sites actually precludes protein binding. A recently described example is the proto-oncoprotein MLL, whose binding *in vivo* is suggested to discriminate against methylation of the appropriate recognition sequence [59]. The biological role of such binding remains to be established.

The faithful propagation of DNA methylation patterns during cell division does not exclusively rely on the function of maintenance DNA methyltransferase Dnmt1. Methylation of CpG-rich satellite DNA and repetitive sequences such as Alu and L1 elements requires cooperative action of Dnmt1 and Dnmt3a/3b [60,61] as well as auxiliary chromatin remodeling factors of Snf2 family ATPases [62,63]. Indeed, Dnmt1 associates with Dnmt3a and 3b *in vivo*, in human cells, through its N-terminal region [64]. A simultaneous knockout of Dnmt1 and Dnmt3b results in almost complete loss of DNA methylation in human HTC 116 cells, while the absence of either Dnmt1 or Dnmt3b alone has very little effect [65,66]. The discovery that DNA methyltransferases require chromatin-remodeling factors in order to access DNA came as a great surprise. A genetic screen for mutations that cause defects in DNA methylation in *Arabidopsis thaliana* apart from DNA methyltransferase enzymes yielded a protein called DDM1 (decrease in DNA methylation). DDM1 turned out to be a DNA helicase related to the Snf2 family of chromatin remodeling activities [62]. Subsequently, the mouse Lsh and human PASG/Hells homologs of DDM1 were identified [63,67]. Loss-of-function mutations in these proteins lead up to 30% decrease of genomic DNA methylation, mainly at repetitive sequences in mammals and transposons and introduced transgenes in plants [68,69]. Thus, one-way DNA methylation patterns can be modulated and controlled is through the accessibility of chromatin structure.

Three different mechanisms have been described to explain the demethylation process. The first one involves a proteic 5mC glycosylase activity that is able to remove the 5mC base and to substitute it with cytosine [70]. The second one removes the entire mCpG dinucleotide and substitutes it with unmethylated CpG dinucleotide [71,72]. In either case, the removal of 5mC is followed by the synthesis of DNA repair patches. The third mechanism reportedly transforms 5mC into C by catalyzing replacement of the methyl group with a hydrogen atom derived from water and releasing the methyl group in the form of methanol [73,74]. The enzyme actively demethylating DNA in this reaction has been identified as MBD2, one member of the family of methyl-DNA binding proteins (see below). It should be noted that this demethylation mechanism is still an object of dispute, as the data could not be reproduced in other laboratories [75,76].

Independently of what the exact enzymatic activities involved in methylation/demethylation are, it is important to realize that methylation may be very specific with respect to the affected sequences, and is thus probably regulated in a sequence-specific manner. Recent detailed analysis of the changes in the methylation pattern of exogenously modified proviral gene sequences introduced into the genome of cultured cell into predetermined genome locations by an ingenious targeting procedure [77] showed that a methylated gene sequence remained methylated through many cell division cycles, with the notable exception of CpGs in the enhancer region that became consistently demethylated, and several CpGs in the promoter and the reporter gene that were sporadically demethylated [78]. Interestingly, regardless of the methylation status of the introduced provirus, no methylation was detected in the region upstream of the introduced integration cassette, suggesting that spreading of methylation did not occur at this integration site. Similar region-specific demethylation has been observed in integrated copies of the β-globin gene, with *in vivo* removal of methyl groups only from the enhancer region of the introduced gene sequence [79]. The enhancer became demethylated and hyperacetylated, and DNase I hypersensitive sites formed; nevertheless, the modified enhancer was unable to activate the methylated and hypoacetylated reporter gene.

2.3. Methyl-CpG binding proteins

An important breakthrough in the DNA methylation field came from the identification of different methyl-DNA binding proteins [80,51]. These are listed in Table 3, together with their most important properties. MeCP2 (for Methyl-CpG binding Protein 2) was the very first protein reported to bind with high affinity to a single symmetrically methylated CpG dinucleotide *in vitro* through a specific sequence motif—methylated DNA binding domain (MBD) [81,82]. Subsequently four other proteins were found by EST database search that contained the conserved MBD motif, and these are presently known as MBD1, MBD2, MBD3, and MBD4 [80]. The MBD proteins are highly conserved in mice, humans, and most vertebrate species (*Xenopus laevis*, *Danio rerio*), with some homologs in insects, including *Drosophila* [83–85]. Individual proteins of the MBD family show very little homology to each other outside the MBD domain, with the exception of MBD2 and MBD3, and their affinity to methylated DNA substrates *in vitro* is similar to that of MeCP2—most MBD proteins bind to a single symmetrically methylated CpG on various double-stranded oligonucleotides without any sequence specificity [80]. Interestingly, mammalian MBD3, but not the *Xenopus* homolog, despite having over 70% amino acid similarity to the MBD2 protein has lost its ability to bind methylated DNA [80,76].

Three of the MBD family members, namely MeCP2, MBD1, and MBD2, have been shown to act as transcriptional repressors *in vitro* and in cell culture assays *in vivo* [86–89,75,76]. Initially it was proposed that methyl-CpG binding proteins inhibit transcription directly, by binding to methylated DNA and blocking the access of transcription factors to their sites at gene promoters. It was soon

Table 3

Methyl-CpG binding proteins: properties and function

Protein	Features of protein structure	Effect on transcription Other properties	*In vivo* expression and localization
MeCP2[a]	Contains conserved MBD motif and a transcription repression domain	Represses transcription from a methylated promoter *in vitro* and *in vivo*; participates in several co-repressor complexes: Sin3A/HDAC1-2, NCoR/Ski, Rest/CoRest	Expressed in somatic tissues and in embryonic stem (ES) cells; co-localizes with heavily methylated satellite DNA in mouse cells
MBD1[b]	Contains MBD domain and 1-3 CxxCxxC motifs; has several splice variants	Represses transcription from a methylated promoter *in vitro* and *in vivo*	Expressed in somatic tissues but not in ES cells
MBD2a	Contain MBD domain and (glycine–arginine)$_{11}$ domain	Transcriptional repressor, a component of the Mi2/NuRD deacetylase complex; MBD2b was reported to have demethylase activity with direct removal of the methyl group	Co-localizes with heavily methylated satellite DNA in mouse cells; expressed in somatic tissues but not ES cells
MBD2b	a truncated version of MBD2a (translation stars at a second methionine codon in MBD2a) that lacks the (GR)$_{11}$ domain		
MBD3	Contains MBD domain and a C-terminal stretch of 12 glutamic acid residues; mammalian protein does not bind methylated DNA *in vivo* or *in vitro*	A component of the Mi2/NuRD and SMRT/HDAC5-7 deacetylase complexes	Expressed in somatic tissues and in ES cells
MBD4	Contains MBD domain and a "repair" domain (T–G mismatch glycosylase)	Thymine glycosylase that binds to the product of deamination at methylated CpG sites	Co-localizes with heavily methylated satellite DNA in mouse cells; expressed in somatic tissues and in ES cells
Kaiso	Contains POZ domain and three Zn-fingers; binds methylated DNA in a sequence specific context via Zn-fingers	Inhibits transcription *in vitro* and *in vivo*; partner of p120 cathenin when localized to the cytoplasm	Cytoplasmic and nuclear; expressed in somatic tissues and in ES cells

[a]Mutation of X chromosome MeCP2 gene results in neurological defects in human females (Rett syndrome) and mice.

[b]The solution structure of the methyl–CpG binding domain of human MBD1 complexed with a methylated oligonucleotide has been recently presented [164]. The structure indicates how MBD may access nucleosomal DNA without encountering steric hindrance from the histone octamer.

For the original work, see Refs. [79–94,182].

recognized that DNA methylation was often accompanied by deacetylation of core histone N-terminal tails that leads to the establishment of chromatin structure non-permissive for transcription. This activity was localized to amino acid sequences outside the MBD motifs of MeCP2 and MBD2, to the transcriptional repression domain (TRD). Both MeCP2 and MBD2 proteins were found to associate with large multi-subunit complexes that recruit class I histone deacetylase activities (HDAC1/2) [84,86,75,88]. MeCP2 apparently can be a member of several different complexes, which most often include Sin3A co-repressor protein [86,90], but it also can repress transcription via direct interaction with TFIID, a component of the basic transcription machinery [91]. Research showing the capability of MeCP2 to form exclusive complexes with HDAC1 and Dnmt1 and to bind hemimethylated DNA [92] suggests an additional role for MeCP2–Dnmt1 complex in driving the maintenance methylation at replicative foci.

MBD2 together with MBD3 (previously known as MeCP1) are part of the Mi2/NuRD co-repressor complex [76,89]. So far, no molecular partner proteins have been identified for MBD1; this protein has multiple splice variants and can repress transcription through HDAC-dependent, as well as HDAC-independent mechanisms [87,88].

MBD4 is the only family member with a different function; in addition to the MBD motif it has a glycosylase domain and is mainly involved in DNA repair removing the thymidine from the TdG, resulting from the spontaneous deamination of 5mC (see Section 2.4) [93].

A new fifth methyl-binding protein, Kaiso, has recently been discovered due to its methylation-dependent binding to a region within the Multiple Tumor Suppressor gene [94]. Similar to MeCP2, MBD2, and MBD1, Kaiso is a methylation-dependent transcriptional repressor; when phosphorylated, Kaiso is released from an interaction with the cytoplasmic domain of the transmembrane adhesion molecule p120 catenin and translocates to the nucleus, where it is able to bind methylated DNA in a sequence specific context via a zinc-finger motif. Thus, this methyl-binding protein may link signal transduction events at the cell surface to gene silencing in the nucleus and, as such, is an excellent candidate for a metastasis-related transcriptional repressor. This new addition to the list of methyl-CpG binding proteins hints at the existence of other, not yet identified, proteins that might be able to recognize methylated DNA and perform specific function. Moreover, all presently known methyl-CpG binding proteins, except MBD3, are dispensable for mammalian development [95–98], suggesting that *in vivo* their function may be to some extent redundant. The fact that usually more than one methyl-CpG protein is present in every cell type speaks about the importance of this function. In mouse cells, MBD proteins usually co-localize with the heavily methylated satellite DNA [80], but very little is known about the specificity of individual methyl-CpG binding proteins towards particular sequences or endogenous gene promoters *in vivo*. Recent data, obtained by chromatin immunoprecipitation (ChIP) showed that methylated promoters of tumor suppressor genes and methylated marks in some imprinted genes are enriched in MBD2 and MeCP2 [30,99,100].

2.4. DNA methylation and human disease

It is widely recognized that cancer cells are characterized by significant aberrations in DNA methylation (reviewed in Refs. [19–21]). The overall decrease in DNA methylation level is accompanied by region-specific hypermethylation of the CpG islands of certain genes that become silenced by the modification. The list of affected genes includes a whole battery of tumor-suppressor genes (see above), and genes that are involved in cell cycle control, cell signalling, cell survival, and maintenance of genome stability. In addition, 5mC is predisposed to spontaneous deamination, with the formation of T. This chemical change is not easily recognized and repaired by the cell repair machinery (T is a naturally occurring DNA base, whereas U, the deamination product of C, does not occur in DNA and is efficiently repaired by the highly abundant and efficient uracil DNA glycosylase). Thus, 5mC can be considered as a mutational "hot-spot", i.e., 5mC is an endogenous mutagenic factor that leads to elevated C to T transversion rates [101,102]; this mutagenic effect may contribute to the induction and progression of the malignant state. Another possible involvement of DNA methylation in carcinogenesis has been suggested by recent *in vivo* experiments showing that the presence of 5mC may facilitate the binding of some carcinogens to DNA [103]; furthermore, the pre-existence of 5mC on DNA may lead to the introduction of additional methyl groups [104]. Finally, deregulation of MBD2 seems to be involved in tumorigenesis [105].

Recently, mutations in methyl-DNA binding proteins or in the enzymatic activities involved in DNA methylation/demethylation were found in non-cancerous human disease (for a review, see Ref. [106]). Mutations in the methyltransferase gene *Dnmt3b* have been identified in patients with immunodeficiency, centromere instability, and facial anomalies syndrome (ICF), and mutations in MeCP2 have been found to cause a very common (1:10,000) neuro-developmental disorder in human females known as Rett syndrome. These important discoveries clearly point to a role played by methyl-DNA binding or modifying proteins in the correct interpretation of gene expression programs and in gene-specific control.

3. DNA methylation and transcriptional regulation: the phenomenology

DNA methylation negatively affects gene expression and is thus important in the control of development, cellular proliferation, differentiation, malignant transformation, and gene imprinting in mammals and plants.

The connection between DNA methylation and gene expression was first noticed when undifferentiated cells in culture were treated with the methylation inhibitor 5-azacytidine: inhibition of methylation led to the appearance of stably inherited differentiated cell types and activation of previously silenced genes [107,108]. Ectopic DNA methylation inhibits gene expression [109], whereas drug-inhibition of methylation induces the expression of certain genes [110]. Transfection experiments revealed that the extent of inhibition of promoter activity depends on the promoter strength and the density of methylated CpGs in the promoter [111]. In addition, recent studies have demonstrated the repressive effect of DNA methylation *in vivo* [112]. Finally, it has been shown that inhibition of Dnmt1 by antisense RNA in *Xenopus* leads to premature initiation of zygotic transcription by all three RNA polymerases (Pol I, II, and III), two to three cell cycles earlier than it normally occurs—after the 12th cell division at the mid-blastula transition [12]; subsequently, it was reported that knockout deletion of *dnmt1* affects the transcription of at least 10% of all genes expressed in mouse embryonic fibroblasts [113]. It is now clear that DNA methylation can inhibit transcription of chromatin templates by additional ~50-fold (chromatin structure itself reduces the basal level of transcription of naked DNA by ~50-fold) (Fig. 5a).

Two possible molecular mechanisms have been proposed to explain the transcriptional inhibition caused by DNA methylation (Fig. 5b and c). The presence of methyl groups in the major groove of DNA alters its appearance to transcription factors, and they may not recognize or bind to sites containing methylated CpGs, thus precluding transcription initiation [114,115]. The other mechanism involves the interaction with methylated CpGs of methyl-DNA binding proteins, which, in turn, recruit other factors that may affect chromatin structure in a way to make it non-permissive to transcription (see Section 2.3).

A large number of studies have addressed in detail the molecular mechanisms of gene silencing mediated by DNA methylation [116–122]. As a most recent example, Curradi and colleagues [116] injected differentially methylated templates into *Xenopus* oocytes and, after allowing chromatin structure to assemble on the DNA, studied their methylation status and transcriptional activity. They demonstrated that even a short region of methylated DNA not covering any regulatory sequences and containing only few modified CpGs at a physiological density could severely affect the expression of the associated gene. Insertion of stretches of unrelated DNA, between 100 and 700 bp in length, between the methylated sequence and the promoter, or downstream from the coding sequence, demonstrated that the silencing mechanism could spread from the methylation site in both directions; this spreading was, however, a function of the number of methylated sites and did not occur when the methylated region contained only three or four methylated CpGs. The contribution of histone deacetylation and other histone modifications to the transcriptional inhibition by methyl-CpG binding protein is the most likely mechanism of spreading of repressive chromatin structure (see below). Finally, a competition between transcriptional activators and methyl-DNA binding proteins for the establishment of permissive or repressive chromatin conformation has been inferred.

In a more natural situation, DNA methylation and MBD proteins are thought to be involved in the long-term maintenance of the inactive state of the gene established by other factors, rather than actively enforcing it. The prevailing evidence indicates that DNA methylation in mammalian cells usually affects genes that have already been silenced: for example, retroviral transcription is shut down in ES cells two days post-infection but *de novo* methylation of pro-viral sequences

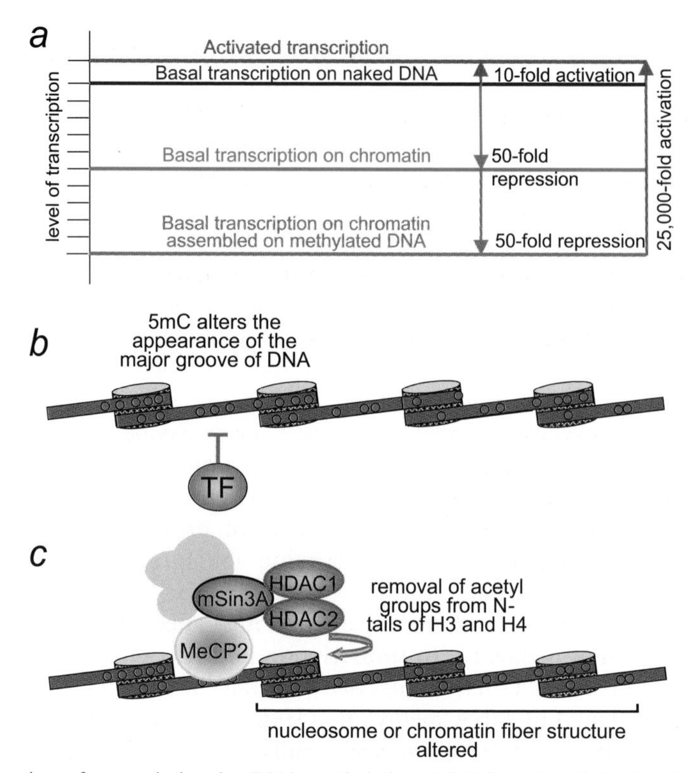

Fig. 5. Repression of transcription by DNA methylation. (*a*) Schematic of the basal and regulated transcription in chromatin (redrawn from Ref. [51]). The basal level of transcription on naked DNA can be activated still further by transcription factors, or can be repressed by chromatin formation. Methylation of chromatin DNA represses transcription still further. The combination of all these factors allows a wide range of modulation of transcriptional activity. (*b*) DNA methylation can directly interfere with transcription factor binding: cytosine methylation alters the width of the DNA major groove, and may thus inhibit binding of transcription factors to sequences that contain CpGs. In a recently reported example, methylation of a single CpG dinucleotide within the upstream control element of the rDNA promoter abrogates rDNA transcription through inhibition of UBF binding to nucleosomal rDNA [149]. (*c*) DNA methylation can inhibit transcription indirectly via recruitment of methyl-CpG binding protein(s) [51]. MeCP2 possesses a transcriptional repressor domain that interacts with the transcriptional corepressor mSin3A, a multiprotein complex that includes histone deacetylases [86,84]. The removal of acetyl groups from the N-tails of histones H3 and H4 affects chromatin structure in a way that makes it refractive to transcription. The changes in chromatin structure caused by histone acetylation are not well understood (see Ausio and Abbott, this book, p. 241).

occurs much later, in ~15 days [123]. The idea that DNA methylation is not required for the initiation of retroviral silencing is also substantiated by the observation that silencing occurs with the same efficiency in Dnmt3a/3b deficient cells [124]. Clearly DNA methylation is in this case a secondary event. In the same line of argumentation, methylation of genes that are already silent was observed during X chromosome inactivation in mammals [125]. An attractive model has recently been put forward suggesting that *de novo* DNA methylation is guided by histone modifications at repressed chromosomal loci (see Section 5.2).

4. *DNA methylation, insulators, and boundaries of chromatin domains*

Studies over the past decade have uncovered the existence of DNA regions with specific properties, known as insulators, which mark the boundaries of chromatin domains so that the individual genes within these domains can follow their specific expression program without being influenced by neighboring genes (reviewed in Ref. [126]). Insulators are operationally defined as having the ability to protect against position effects and/or to block distal enhancer activity [126]. There are two properties of insulator action, which are directly linked to DNA methylation. We will describe them in some detail in view of their novelty and importance.

The first property concerns the ability of insulator elements to contribute to the proper regulation of two adjacent genes with different expression programs during differentiation. In a well-studied example in chicken [127], the folate receptor gene and the adjacent genes in the β-globin locus are differentially regulated during erythroid differentiation. Both genes are erythroid-specific but the folate gene is expressed in erythroid precursor cells in the developing embryo, whereas the β-globin genes are expressed much later in the differentiation program. At a time when the folate receptor gene is active, the 16-kbp region between the β-globin locus and the folate receptor gene is organized in a micrococcal nuclease-resistant chromatin structure, and the DNA is fully methylated (Fig. 6a). The spreading of condensed chromatin in both directions is impeded by boundary elements, the upstream one of which is the well-recognized insulator of the β-globin gene locus (characterized as DNase hypersensitive site 4 which marks the positions of binding sites for the ubiquitous DNA binding-protein CTCF [128]); the DNA in the boundary elements is unmethylated. In the process of differentiation, only few sites lose methylation along the condensed chromatin domain, partial *Hpa*II digestion being observed at every single site (Fig. 6a). This partial demethylation is probably not accompanied by decondensation of the region. The unmethylated boundaries are maintained as such. What are the mechanisms that allow this selective methylation/demethylation pattern on adjacent regions to be established and maintained? Evidently, more research on this interesting system will reveal the missing pieces of the methylation puzzle.

The second property concerns the participation of insulators in the control of the differential expression of imprinted genes. Imprinted genes are expressed according to their origin—paternal or maternal—and DNA methylation has long been thought to serve as an epigenetic mark to discern the two alleles (for a review on imprinting, see Ref. [129]). Three independent laboratories [130–132] have used a well-studied imprinted system, that of the adjacent insulin-like growth factor 2 gene (*Igf2*) and the *H19* gene in the mouse, to show that a differentially methylated region (DMR) between the two genes serves as an insulator. *Igf2* and *H19* share an enhancer, but *H19* is expressed only from the maternal allele, whereas *Igf2* only from the paternal one (Fig. 6b). The epigenetic mark required for the imprinting of these genes has been recognized as a region upstream of *H19* (the so-called ICR, imprinted control region) that is methylated only in the paternal allele. This region

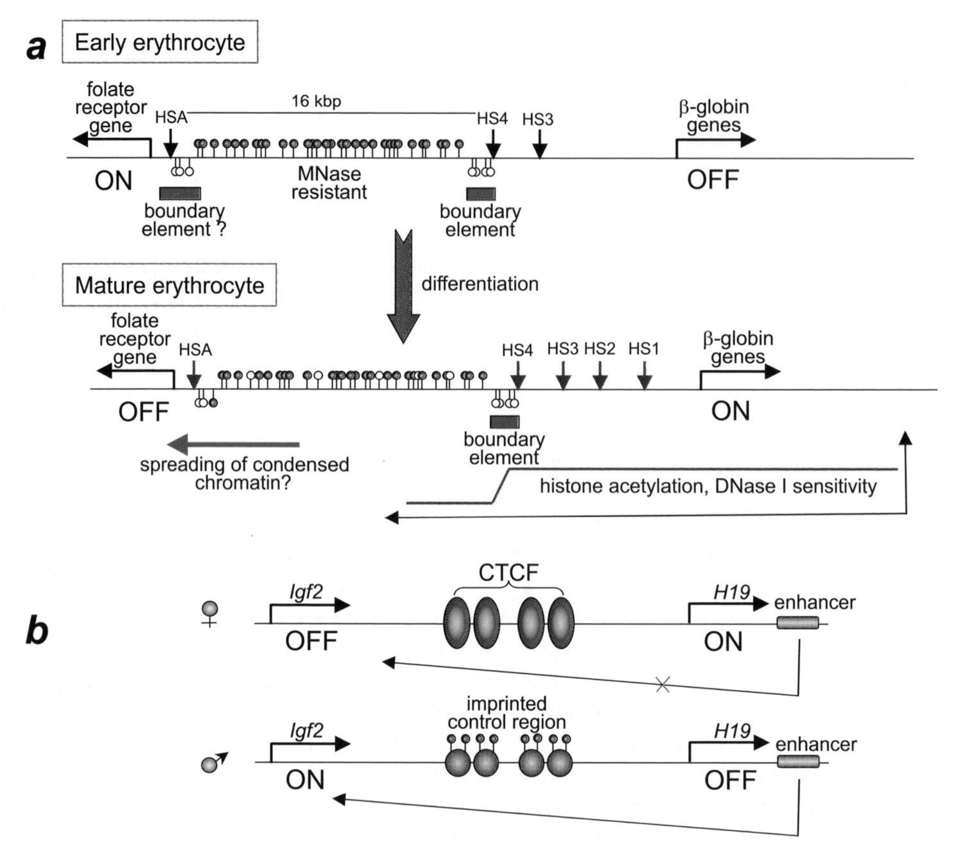

Fig. 6. DNA methylation and boundaries of chromatin domains. (*a*) Schematic of the changes in the DNA methylation pattern and chromatin organization of the locus containing the folate receptor gene and the β-globin gene cluster during erythroid differentiation in chicken (redrawn from Ref. [128]). At the early stages of differentiation, the *Hpa*II sites in the region between the two genes are fully methylated (filled lollipops), whereas the sites near the two DNase hypersensitive sites (the insulator HS4 and HSA) are unmethylated (white lollipops). Differentiation is accompanied by a random demethylation of a few methylated sites but the 3′ boundary function is preserved. Despite this partial DNA demethylation, the condensed chromatin structure of the region established earlier in development is most probably preserved. (*b*) Allele-specific regulation of the imprinted *Igf2*/*H19* locus in the mouse. On the maternal allele, the enhancer action on *Igf2* gene is blocked by binding of CTCF to the unmethylated CpG-containing binding sites within the insulator. On the paternal allele, the CTCF sites are methylated, CTCF does not bind, and the insulator is inactivated. This leaves the enhancer in the position to activate *Igf2* expression from the paternal allele. The inactivation of *H19* expression on the paternal allele involves a separate mechanism. (Adapted from Ref. [129].)

contains an insulator that blocks the enhancer action on *Igf2*, and the insulator is only active when bound to CTCF. Methylation of CpGs within the CTCF-binding sites precludes factor binding *in vitro*, and deletion of the sites results in loss of the enhancer-blocking activity *in vivo*. Thus, the activity of the insulator is restricted to the maternal allele only, and the discrimination of the two alleles is accomplished by the methylation of the insulator. This work demonstrates for the first time that the boundary function of some insulators can be modified in a potentially regulatable fashion, and that DNA methylation is involved in this regulation.

How are the differential patterns of methylation in the imprinted alleles established and maintained? Some evidence exists that differential methylation at the paternal and maternal imprinted alleles are established independently. Knockout of Dnmt3L, a protein that shares homology with *de novo* DNA methyltransferase enzymes but does not possess methyltransferase activity, results in a complete loss of the maternally imprinted DNA methylation marks [57,58]. The precise mechanisms by which the paternal methylation imprints are established are yet to be determined.

5. Chromatin and DNA methylation

Histones, as well as many other proteins in the cell, are subject to post-synthetic modifications such as acetylation, methylation, phosphorylation, ADP-ribosylation, and ubiquitylation. Histone modifications may affect chromatin conformation and therefore have either positive or negative effect on transcription. It has been known for years that DNA methylation pattern and gene activity are highly correlated to chromatin structure, the transcriptionally active chromatin domains being hypomethylated, and the inactive regions being hypermethylated [133,117]. Research in the past two decades has demonstrated unequivocally that DNA methylation represses transcription, to a large extent through the establishment of specific chromatin environment that is refractive to transcription (for reviews, see Refs. [5,117]). That methylated genes are in an unusual chromatin conformation insensitive to nuclease digestion has been experimentally shown as early as 1986 [118] (see also [119–121]). Microinjection of HSV thymidine kinase gene—unmethylated or methylated—into living cells in culture has shown that methylation affects the transcriptional activity of the methylated construct only after a significant lag of time, needed to assemble the introduced gene into chromatin [122]. Microinjection of the same gene into *Xenopus* oocyte nuclei has led to similar conclusions [121]. Finally, ingenious experiments in the laboratory of Roger Adams using patch-methylated constructs for transfection revealed that the repressive chromatin structure spreads from the foci of initial methylation [120].

5.1. The histone acetylation link

The past few years witnessed some important developments that established a molecular link between CpG methylation and the status of histone acetylation. It was shown that the methyl-binding protein MeCp2 contains a transcriptional repressor domain that interacts directly with the multiprotein corepressor complex mSin3A [84,86]. Antibodies against MeCP2 coprecipitated mSin3A, HDAC1 and HDAC2, and the deacetylase inhibitor trichostatin A (TSA) relieved the repression.

These results were further elaborated in two experimental systems. In the first system, *in vitro* methylated or unmethylated constructs were injected into *Xenopus* oocytes [121,134]. Following assembly of chromatin on the constructs, the methylated gene was silenced whereas the unmethylated one remained transcriptionally active, in agreement with earlier results [122] (see above). TSA treatment restored the transcriptional activity of the methylated construct and caused reappearance of DNase I hypersensitive sites. The *Xenopus* MeCP2, similar to its mammalian homolog, co-fractionated biochemically with both mSin3A and HDACs, and anti-MeCP2 antibodies coimmunoprecipitated mSin3A and the HDACs.

The second experimental system made use of cells in culture stably transfected with *in vitro* methylated or unmethylated gene constructs [135]. PCR analysis for the presence of the gene in hyperacetylated nucleosomal fraction revealed that the gene was present in hyperacetylated structures when active, i.e., in cells containing the unmethylated constructs. When cells transfected with the methylated construct were treated with TSA, the acetylation level at the gene was elevated, and the region became DNase I sensitive; importantly, TSA treatment was accompanied by increase in the transcriptional activity of the gene.

Several recent reports introduce an important element of clarification to the acetylation issue. TSA-mediated activation of silenced genes seems to be dependent on the density of methylated CpGs: heavily methylated gene promoters fail to be depressed by TSA treatment [78,100,136]. In a more detailed study from Landsberger's group [116], histone deacetylation was found to be the main cause of gene repression only when methylation does not reach a level sufficient to establish stable, spreadable heterochromatic structure. On the contrary, when a large number of methylated CpGs exert a long-distance inhibitory effect on transcription, histone acetylation cannot relieve the repression. One possible explanation for this effect is that deacetylation of histone tails that accompanies DNA methylation allows additional histone modifications (see Section 5.2) to take place, which may reinforce heterochromatin stability.

Finally, the deacetylation of histones in methylated gene regions may be selective with respect to the histone affected. Chromatin immunoprecipitation analysis of the histone acetylation status of proviral sequences targeted to defined genome loci in MEL cells showed that the *in vitro* methylated proviral sequences that were transcriptionally silent *in vivo*, were hypoacetylated only with respect to histone H3; in contrast, the acetylation status of histone H4 was comparable on the methylated and unmethylated provirus [78]. Moreover, the H3 acetylation status correlated closely with the location of methylated CpGs along the proviral sequence.

As mentioned earlier, MeCP2 is not the only protein that associates with HDAC activities. Both MBD2 and MBD1 repress transcription in histone deacetylation-dependent manner. Interestingly, MBD2 was found associated not only with class I histone deacetylases (HDAC1 and 2) from the Mi2/NuRD corepressor complex but also with class II histone deacetylases in the silencing mediator of retinoic and thyroid receptor (SMRT)/HDAC 5,7 complex [137].

a Inhibition of histone acetylation prevents active DNA demethylation and transcription

b Histone acetylation allows demethylase binding, DNA demethylation, and transcription

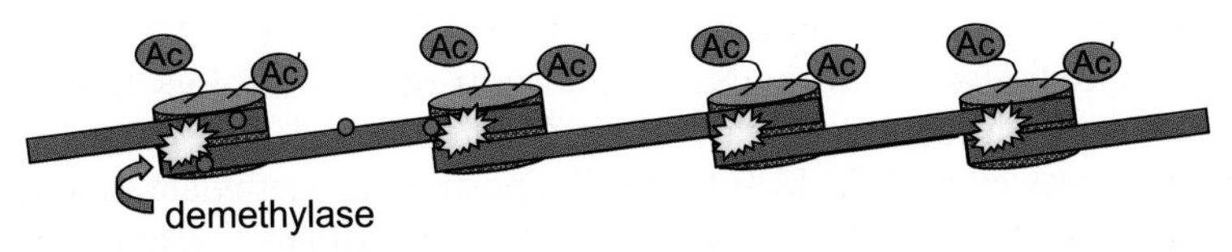

Fig. 7. The model proposed by Szyf to explain the unmethylated status of active genes (for references and further details see Section 5.1). (*a*) Inhibition of acetylation of the tails of core histones prevents active demethylation. (*b*) When histone tails are acetylated, as in transcriptionally active chromatin regions, the demethylase binding to the regions is enhanced, and the DNA is actively demethylated.

Collectively, the results of these studies suggest a dominant role for DNA methylation, which dictates the acetylation status of the histones, and thereby, chromatin structure (Fig. 5c). The cells jealously preserve and inherit DNA methylation patterns, and respectively chromatin conformation, through mitosis.

A totally different view on the connection between DNA methylation and histone acetylation status (read chromatin structure) was recently proposed by Szyf and co-workers [40,43–45] (Fig. 7). The group logically argues that the above model explains well the compartmentalization of the genome into active (hypomethylated) and inactive (hypermethylated), and its inheritance during somatic cell division, but it fails to address the important issue of how genes are demethylated during their transcriptional activation in response to changing physiological conditions. Szyf and co-workers propose a model where transcriptionally active chromatin structure leads to active demethylation of the associated sequences, explaining both the unmethylated status of active genes and its maintenance through cell division (Fig. 7). Using a transient transfection approach, Cervoni and Szyf [43] show that the methylation status is not fixed in somatic cells but is dynamically modulated, in concert with the activity of the gene; moreover, active demethylation is directed by histone hyperacetylation. Thus, it was reported that ectopically methylated genes when residing downstream from a strong promoter did lose some of their methyl groups, whereas the same gene when linked to a weak promoter kept its highly methylated state. In other words, identical genes can be differentially demethylated depending on their state of expression. Furthermore, TSA treatment increased the demethylation of the ectopically methylated reporter gene, and chromatin immunoprecipitation assay using anti acetyl-histone H3 antibody directly demonstrated elevated histone acetylation in the activated, demethylated gene. The authors speculate that the binding of demethylase activity to the region is enhanced by the acetylation of the histone tails. In support of this model, Cervoni *et al.* [44] have further demonstrated that the oncoprotein Set/TAF-1β, a known inhibitor of histone acetyltransferase, inhibited active demethylation of DNA. It may be relevant to mention that TSA treatment of the fungus *Neurospora* led to selective derepression of certain (but not all) genes, and that these genes underwent selective demethylation [138].

5.2. The histone methylation link

Methylation of specific lysines in the N-terminal tails of core histones has recently attracted considerable attention as part of the molecular mechanism involved in gene regulation (reviewed in Refs. [139,140]). The position of methylated lysine(s) in histone N-terminal tails is recognized by interacting proteins; protein binding results in a change in the accessibility of packaged DNA to transcriptional activator proteins. Thus, methylation of lysine 9 (K9) or 27 of histone H3 tail marks transcriptionally inactive loci, while methylation of lysine 4 (K4) of the same histone is associated with actively transcribed genes. Methylated H3K9 is believed to provide a binding site for chromodomain-containing heterochromatin protein HP1, which is involved in chromatin condensation [141,142]. Acetylation of H3 K9, which occurs in actively transcribed chromatin, is incompatible with methylation of the same amino acid and, therefore, requires the action of histone deacetylases which are often brought about by transcriptional repressors including the methyl-CpG binding proteins. Indeed, in mouse cells MeCP2 was found in addition to HDACs to recruit histone methylase activity with a preference for H3K9 to methylated DNA [143]. Additionally, a certain class of histone methyltransferases in human cells were found to contain a potential MBD domain [144]. These data suggest an attractive model where MBD-containing proteins or histone methyltransferases introduce methyl groups onto the histone tails via binding to methylated DNA, and thus generate a second layer of silencing through the sequential binding of HP1 proteins (Fig. 8a). Is, then, the flow of epigenetic information always directed from methylation of DNA to the modification of core histones?

In fact, the cross-talk between histone methylation and DNA methylation has been experimentally demonstrated for the first time in *Neurospora* [145] (see commentary by Bird [146]) and proved the second scenario. Using the power of *Neurospora* genetics, Tamaru and Selker [145] have identified a gene which, when mutated, abolishes methylation of all tested DNA sequences. This gene, *dim-5*, is different from *dim-2*, the previously identified gene that encodes the only DNA methyltransferase in *Neurospora* responsible for all known cytosine methylation in this fungus. The *dim-5* gene encodes a histone H3 methyltransferase: the deduced polypeptide sequence contains a SET domain with sequence similarity to some known histone methyltransferases; moreover, the recombinant DIM-5 protein exhibits histone methyltransferase activity *in vitro*. Additional *in vivo* experiments

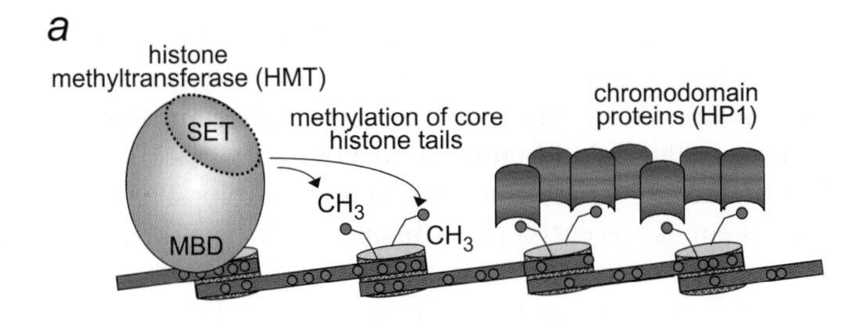

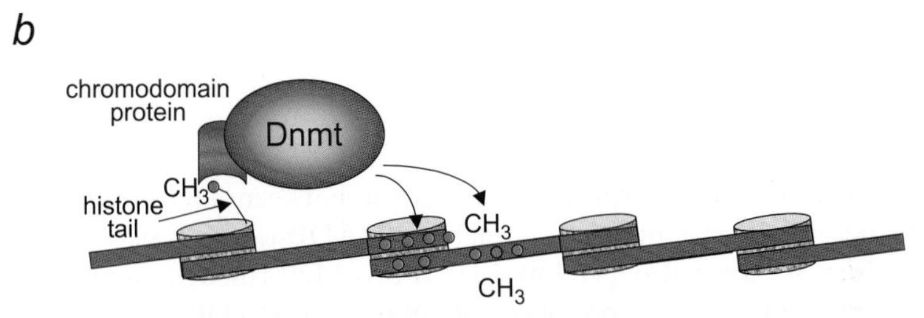

Fig. 8. Proposed models that link histone methylation to DNA methylation (for details see Section 5.2). Methylated cytosines attract histone methyltransferases that contain a methyl-binding domain or a methyl-CpG binding protein (MeCP2) that recruits histone methylase activities; these introduce methyl groups into the histone tails. The binding of chromodomain HP1 proteins to H3 tails methylated at lysine 9 generates a secondary layer of repressive chromatin structure. (*b*) In a reverse scenario, methylated histone tails attract chromodomain-binding proteins, which in turn recruit Dnmts to methylate adjacent DNA sequences.

prove the case still further. Thus, a mutation in a gene for histone methyltransferase phenocopies a mutation in a gene encoding a cytosine methyltransferase.

Further evidence for the link between DNA methylation and histone methylation came from genetic experiments in *Arabidopsis thaliana* and more recently in mammalian cells [30,147–149]. In *Arabidopsis*, the loss-of-function mutations in a Snf2-like protein, DDM1 (decrease in DNA methylation), not only abolished DNA methylation at transposons and silenced transgenes but also led to replacement of H3K9 methylation with H3K4 methylation [147]. Subsequently, Jackson and colleagues [148] have shown that the KRYPTONITE gene in *Arabidopsis* that encodes a H3K9 methylase is responsible not only for the enzymatic histone methylation but also for the maintenance of DNA methylation at CpNpG trinucleotides. Furthermore, the *Arabidopsis* homolog of HP1 directly interacts with CHROMOMETHYLASE3, a methyltransferase that methylates cytosine at CpNpG [148]. These results strongly suggest that in plant genomes DNA methylation is guided by histone H3 methylation. Such a model (Fig. 8b) not only places histone methylation upstream of DNA methylation but is also in agreement with the notion that DNA methylation is not the primary mechanism for gene silencing (see Section 3).

Santoro and Grummt, using as experimental model the nucleolar remodeling complex NoRC and rDNA transcription in mammalian cells, have shown that changes in histone tail modifications may be responsible for changes in DNA methylation patterns [149]. In this model, the methylated H3-lysine 9 recruits DNA methyltransferase allowing DNA methylation of neighboring DNA sequences; afterwards MBD proteins might bind methylated DNA, promoting recruitment of remodeling chromatin factors capable of inducing a stably silenced chromatin. Whatever the initial event is, recent studies in mammals have demonstrated that differentially methylated sequences of Snrp, Igf2r and U2af1-rs1 imprinted genes are accompanied by differentially methylated histone H3 [30]. Thus methylated DNA in various species was always found associated with methylated H3K9. Evidently, the field is wide-open for new research and future studies will help to establish the chain of events that leads to stable and heritable gene silencing.

5.3. The poly(ADP-ribosyl)ation link

Another important and unexpected connection between DNA methylation and another post-synthetic modification, namely poly(ADP-ribosyl)ation of proteins has also been recently unraveled [41,150–153]. Poly(ADP-ribosyl)ation is a dynamic process guaranteed by the presence of many poly(ADP-ribose) polymerases (PARP1, PARP2, VPARP, Tankyrase1 and 2) and one glycohydrolase (PARG), an enzyme which hydrolyzes and detaches ADP-ribose units from the polymers—whose half-life is calculated to be one minute--when their nuclear level exceeds 15 mM (reviewed in Refs. [154–156]). The enzyme poly(ADP-ribose) polymerase catalyzes the transfer of ADP-ribose units from NAD to a variety of nuclear proteins, including histones and itself. The best known enzyme is PARP1, which is responsible for about 90% of all the ADP-ribose polymers found in cells. When PARP1 automodifies itself, it builds long and branched ADP ribose polymers of ~100–200 units attached to numerous automodification sites in its central domain. The molecule becomes enzymatically active following its binding to DNA breaks in damaged cell and increases its catalytic activity by about 500 times following nicks in both single and double strand DNA. Heteromodification occurs when ADPR polymers are introduced onto different chromatin proteins. Automodified PARP1 or ADP-ribose polymers are by themselves able to interact *in vitro* with chromatin proteins also through non-covalent binding, which is so strong that the association between protein and ADPR polymers cannot be reversed either by drastic chemical conditions or by elevated concentration of single or double strand DNA [157]. Thus, important chromatin proteins undergo covalent and uncovalent poly(ADP-ribosyl)ation and sometimes the same protein—e.g. H1 histone [157,158]—is substrate for both modifications.

Enzymatically active PARP molecules have also been found involved in transcription in undamaged cells. Recent research on *Drosophila* has demonstrated that active puff loci are rich in PARP1 and ADPR polymers, this being the necessary condition to induce chromatin decondensation and gene expression [159]. Mechanism of PARP1-mediated puffing and chromatin remodeling reinforces the idea previously suggested by Althaus to explain the role played by PARP1 in DNA repair [160].

The assertion that poly(ADP-ribosyl)ation is involved in the control of the DNA methylation pattern is based on results from numerous experimental strategies: (i) *In vivo* inhibition of PARP activity, through treatment of cells with 3-aminobenzamide (3-ABA) leads to a change in the methylation pattern of CpG islands, as evidenced by the disappearance of the "tiny HpaII fragments" [150]. (In normal cells, the unmethylated CpG islands produce such fragments upon digestion of genomic DNA with the methylation-sensitive restriction enzyme HpaII, [16].) (ii) 3-ABA treatment leads to aberrant methylation of the promoter regions of the *HTF9* gene [150]. (iii) Transfection of a plasmid whose sequence is CpG-island-like into cells lacking active ADP-ribosylation leads to its methylation [151]. (iv) Microscopic examination of control and 3-ABA-treated cells following anti-5mC-antibody staining shows an increase in the staining intensity and the number of stained loci in the treated cells [152]. (v) An elevated methylation level of DNA in the 3-ABA-treated cells was found also through the use of *in vitro* methyl-accepting ability assay [153].

Therefore, there is no doubt that inhibition of PARP leads to DNA hypermethylation *in vivo*. But what is the molecular mechanism behind this phenomenon? To get an insight into this mechanism, PARPs were inhibited at different phases of the cell cycle [41]. Both the mRNA and protein levels of Dnmt1 were precociously increased in G1/S phase. These elevated Dnmt1 levels, affecting the normal equilibrium between p21 and Dnmt1 in this phase, led to higher than normal levels of the PCNA–Dnmt1 complex, and thus, to higher DNA methylation of the sequences replicated in early S phase (CpG islands are early replicating) (see Fig. 3d).

Yet another hypothesis considers the somatic variant H1e of histone H1, in its poly(ADP-ribosyl)ated isoform, as a nuclear *trans*-acting factor involved in maintaining the methylation pattern on DNA [161].

Thus, the poly(ADP-ribosyl)ation process is involved in determining and/or maintaining the methylation patterns on DNA. Considering the importance for cells to preserve these patterns, further research will be performed to find additional proof to convalidate one or another mechanism or to identify other possibilities.

5.4. DNA methylation and chromatin structure: beyond the post-synthetic modifications of histones and other proteins

The notion that DNA methylation is an epigenetic mechanism that acts in conjunction with post-synthetic modifications of histones and other proteins is nowadays widely accepted and is the focus of major research effort worldwide. It must be noted though that this is probably not the only way through which methylation affects chromatin structure, and that DNA methylation may directly modulate some basic structural features of nucleosomes or the chromatin fibers they form.

As already mentioned on several occasions, chromatin containing higher levels of methylated Cps is more condensed than that containing less number or lower density of modified CpGs. On the other hand, the chromatin structure of the unmethylated CpG islands is nuclease-sensitive, is depleted of histone H1, and contains highly acetylated histones H3 and H4 [162].

The accepted involvement of CpG methylation in transcription regulation has guided various lines of research into the direct effects methylation may have on chromatin structure. As early as 1982, Felsenfeld and co-workers looked for effects of DNA methylation on the affinity of DNA for histone octamers: they reported a slight increase in this parameter [163]. Nightingale and Wolffe [164] using mononucleosome reconstitution on a specific sequence, the 5S rDNA from *X. borealis*, failed to see such an effect. More recently, Travers (personal communication) has seen clear-cut differences in the affinity of the histone octamer for a number of methylated versus unmethylated sequences. The fine structure of the core particle was, however, unaffected by the methylation status of the DNA [165].

A number of laboratories have studied the effect of methylation on nucleosome placement and/or positioning. Methylation of several sequences did not seem to affect this parameter: sequences from the tetracycline resistance gene [165], Alu repeats from human DNA [166], 5S rDNA from *X. borealis* [164]. On the other hand, nucleosomes were excluded from a specific region in the β-globin gene, when methylated [167], indicating the importance of the DNA sequence. Earlier data by Keshet *et al.* [118] showed disappearance of nucleosome-free regions (i.e., formation of nucleosomes) in methylated promoters *in vivo*. This effect, however, may be only indirectly connected to the methylation status of the promoters.

Since linker histones are involved in compacting the chromatin fiber, it was of interest to see whether the methylation-mediated chromatin compaction could be explained by a higher affinity of linker histones for methylated DNA. The literature on this point is highly controversial: some reports demonstrated strong preference of linker histones for methylated DNA [168,169] whereas others found that linker histone binding was indifferent to the methylation status of the DNA [164,170,171].

As far as the distribution of 5mC in chromatin is concerned, early data showed that nucleosomes that contain histone H1 are enriched in methylated DNA [172]. The case for co-localization/co-operation of 5mC and linker histones is further strengthened by the observation that the unmethylated CpG islands are characterized by absence of histone H1 [162]. Furthermore, the inhibitory effect of linker histones on *in vitro* transcription depends on the density of methyl groups in the promoter regions [168,173]. The possible collaboration between linker histones and DNA methylation in creating transcriptionally non-permissive chromatin structure is seen in the ability of strong activators, such as GAL4-VP16, to counteract the inhibitory effect of H1 on transcription: whereas GAL4-VP16 can alleviate the H1-mediated repression on unmethylated chromatin templates [174], it is not in the position to do so on methylated chromatin templates [121].

The possible structural involvement of linker histones in the methylation-mediated chromatin condensation has been recently addressed by using the Atomic Force Microscope [175]. This extensive study combined *in vivo* and *in vitro* approaches to demonstrate that DNA methylation could cause chromatin compaction only in co-operation with linker histone binding. Finally, it may be relevant to note that MeCP2 is capable of displacing histone H1 in order to gain access to its preferred binding site on a methylated template [176].

It should be noted that the link between linker histone binding and DNA methylation may not be functioning in lower eukaryotes. The fungus *Ascobolus immerses* possesses a "normal" histone H1; mutant strains lacking the protein display normal methylation-associated gene silencing [177]. Interestingly, the H1-depleted mutants exhibit global DNA hypermethylation (the methylation level rises from 8.4% in wild-type strains to ~13% in the knockouts), which may be consistent with the hypothesis that H1 protects DNA from methylation (see below). It is clear that much more research is needed to understand the intricate interplay between linker histones, methyl-DNA binding proteins and DNA methylation in the establishment of compacted chromatin structure.

Another intriguing and ill-understood feature concerns the distribution of methylated CpGs in chromatin: the linker regions are hypomethylated with respect to the "core" particle. This decreased level of linker DNA methylation is not due to an intrinsic CpG deficiency in linker DNA but may be ascribed to an inhibitory effect of linker histone on the enzymatic DNA methylation [178]. *In vitro* experiments have demonstrated that among histones, only H1 inhibits DNA methylation, and within the H1 histone family only the H1e variant is inhibitory. H1e is the only linker histone subtype that binds to CpG island-like regions; moreover, H1e protects genomic DNA from full methylation only when poly(ADP-ribosyl)ated and ADP-ribose polymers are by themselves inhibitory on *in vitro* DNA methylation [153]. The latter observation may be of particular importance, since it extends further the connection between poly(ADP-ribosyl)ation and DNA methylation described above. Multiple observations bear on this connection: (i) poly(ADP-ribosyl)ation occurs in transcribing regions [179–181]; (ii) all histone H1 subtypes are able to bind long and branched ADP-ribose polymers in a non-covalent way [161]; (iii) H1e in its poly(ADP-ribosyl)ated form is present in decondensed chromatin [161]; (iv) active poly(ADP-ribosyl)ation process is involved in inducing chromatin decondensation in undamaged cells [159].

All these experimental observations can be unified in a model in which automodified PARP1 is involved in protecting CpG islands against anomalous methylation in chromatin. In this model, the modified H1e plays a role in marking CpG islands (for a more detailed description of the entire set of data and the model see Ref. [42]).

6. *Concluding remarks*

The more we learn about the biology and biochemistry of DNA methylation, the more fascinating and challenging the field becomes. The last decade has witnessed a significant progress in our knowledge: we learned that there are many enzymes involved in methylation/demethylation of DNA; we appreciated the discovery of proteins that bind specifically to methylated DNA sequences to recruit numerous other proteins with a variety of functions; we realized that the effect of DNA methylation is exerted through direct or indirect changes in chromatin structure. The connection between epigenetic factors and human disease has been unraveled. And yet, we have just seen the tip of the iceberg. The new understanding has given rise to numerous new questions that need to be addressed experimentally. The complexity of the phenomena requires the concerted action of many laboratories worldwide, and the application of new methodologies. It is certain that the future will be full of challenge, but also of rewards, and that the basic knowledge that we acquire will lead to new discoveries and new tools to diagnose and treat human disease.

Acknowledgements

The work by J.Z. is supported by Polytechnic University, Brooklyn, USA; the work by I.S. is funded by Cancer Research UK, and the work by P.C. is supported by Ministero Italiano dell'Universita (60%, Progetti di Ateneo, COFIN, FIRB), by Ministero della Sanita, by Istituto Pasteur Fondazione Cenci Bolognetti and by AIRC (Associazione Italiana Ricerca sul Cancro).

References

1. Bestor, T.H. and Ingram, V.M. (1983) Proc. Natl. Acad. Sci. 80, 5559–5563.
2. Gruenbaum, Y., Szyf, M., Cedar, H., and Razin, A. (1983) Proc. Natl. Acad. Sci. USA 80, 4919–4921.
3. Lorincz, M.C. and Groudine, M. (2001) Proc. Natl. Acad. Sci. USA 98, 10034–10036.
4. Finnegan, E.J. and Kovac, K.A. (2000) Plant Mol. Biol. 43, 189–201.
5. Razin, A. and Shemer, R. (1999) Results Probl. Cell Differ. 25, 189–204.
6. Mayer, W., Niveleau, A., Walter, J., Fundele, R., and Haaf, T. (2000) Nature 403, 501–502.
7. Dean, W., Santos, F., Stojkovic, M., Zakhartchenko, V., Walter, J., Wolf, E., and Reik, W. (2001) Proc. Natl. Acad. Sci. USA 98, 13734–13738.
8. Bourc'his, D., Le Bourhis, D., Patin, D., Niveleau, A., Comizzoli, P., Renard, J.P., and Viegas-Pequignot, E. (2001) Curr. Biol. 11, 1542–1546.
9. Wilmut, I., Beaujean, N., de Sousa, P.A., Dinnyes, A., King, T.J., Paterson, L.A., Wells, D.N., and Young, L.E. (2002) Nature 419, 583–586.
10. Stancheva, I., El-Maarri, O., Walter, J., Niveleau, A., and Meehan, R.R. (2002) Dev. Biol. 243, 155–165.
11. Macleod, D., Clark, V.H., and Bird, A. (1999) Nat Genet 23, 139–140.
12. Stancheva, I. and Meehan, R.R. (2000) Genes Dev. 14, 313–327.
13. Tada, M., Tada, T., Lefebvre, L., Barton, S.C., and Surani, A.M. (1997) EMBO J. 16, 6510–6520.
14. Reik, W., Dean, W., and Walter, J. (2001) Science 293, 1089–1093.
15. Cervoni, N. and Szyf, M. (2001) J. Biol. Chem. 276, 40778–40787.
16. Bird, A.P. (1986) Nature 321, 209–213.
17. Antequera, F. and Bird, A. (1999) Curr. Biol. 9, R661–R667.
18. Keshet, I., Yisraeli, J., and Cedar, H. (1985) Proc. Natl. Acad. Sci. USA 82, 2560–2564.
19. Robertson, K.D. and Jones, P.A. (2000) Carcinogenesis 21, 461–467.
20. Baylin, S.B. and Herman, J.G. (2000) Trends Genet. 16, 168–174.
21. Costello, J.F. and Plass, C. (2001) J. Med. Genet. 38, 285–303.
22. Jones, P.A. (1999) Trends Genet. 15, 34–37.

23. Bird, A. (2002) Genes Dev. 16, 6–21.
24. Bestor, T.H., Gundersen, G., Kolsto, A.B., and Prydz, H. (1992) Genet. Anal. Tech. Appl. 9, 48–53.
25. Di Croce, L., Raker, V.A., Corsaro, M., Fazi, F., Fanelli, M., Faretta, M., Fuks, F., Lo Coco, F., Kouzarides, T., Nervi, C., Minucci, S., and Pelicci, P.G. (2002) Science 295, 1079–1082.
26. Brandeis, M., Frank, D., Keshet, I., Siegfried, Z., Mendelsohn, M., Nemes, A., Temper, V., Razin, A., and Cedar, H. (1994) Nature 371, 435–438.
27. Mummaneri, P., Yates, P., Simpson, J., Rose, J., and Turker, M.S. (1998) Nucleic Acids Res. 26, 5163–5169.
28. Voo, K.S., Carlone, D.L., Jacobsen, B., Flodin, A., and Skalnik, D.G. (2000) Mol. Cell. Biol. 20, 2108–2121.
29. Cuadrado, M., Sacristan, M., and Antequera, F. (2001) EMBO Rep. 2, 586–592.
30. Fournier, C., Goto, Y., Ballestar, E., Delaval, K., Hever, A.M., Esteller, M., and Feil, R. (2002) EMBO J. 21, 6560–6570.
31. Robertson, K.D., Ait-Si-Ali, S., Yokochi, T., Wade, P.A., Jones, P.L., and Wolffe, A.P. (2000) Nat. Genet. 25, 338–342.
32. Rountree, M.R., Bachman, K.E., and Baylin, S.B. (2000) Nat. Genet. 25, 269–277.
33. Fuks, F., Burgers, W.A., Brehm, A., Hughes-Davis, L., and Kouzarides, T. (2000) Nat. Genet. 24, 88–91.
34. Fuks, F., Burgers, W.A., Godin, N., Kasai, M., and Kouzarides, T. (2001) EMBO J. 20, 2536–2544.
35. Bachman, K.E., Rountree, M.R., and Baylin, S.B. (2001) J. Biol. Chem. 276, 32282–32287.
36. Pradhan, S. and Kim, G.-D. (2002) EMBO J. 21, 1–10.
37. Chuang, L.S.-H., Ian, H.-I., Koh, T.-W., Ng, H.-H., Xu, G., and Li, B.F.L. (1997) Science 277, 1996–2000.
38. Baylin, S.B. (1997) Science 277, 1948–1949.
39. Warnecke, P.M. and Bestor, T.H. (2000) Curr. Opin. Oncol. 12, 68–73.
40. Szyf, M. (2001) Front. Biosci. 6, 599–609.
41. Zardo, G., Reale, A., Passananti, C., Pradhan, S., Buontempo, S., De Matteis, G., Adams, R.L.P., and Caiafa, P. (2002) FASEB J. 16, 1319–1321.
42. Reale, A., Zardo, G., Malanga, M., Zlatanova, J. and Caiafa, P. (2004) In: Szyf, M. (ed.) DNA Methylation and Cancer. Eurekah.com and Kluwer Academic Press/Plenum Publishers, Georgetown, in press.
43. Cervoni, N. and Szyf, M. (2001) J. Biol. Chem. 276, 40778–40787.
44. Cervoni, N., Detich, N., Seo, S.-B., Chakravarti, D., and Szyf, M. (2002) J. Biol. Chem. 277, 25026–25031.
45. Szyf, M. and Detich, N. (2001) Prog. Nucleic Acids Res. Mol. Biol. 69, 47–79.
46. Razin, A. and Riggs, A.D. (1980) Science 210, 604–610.
47. Bestor, T., Laudano, A., Mattaliano, R., and Ingram, V. (1988) J. Mol. Biol. 203, 971–983.
48. Tajima, S., Tsuda, H., Wakabayashi, N., Asano, A., Mizuno, S., and Nishimori, K. (1996) J. Biochem. 117, 1050–1057.
49. Kimura, H., Ishihara, G., and Tajima, S. (1996) J. Biochem. 120, 1182–1189.
50. Araujo, F.D., Knox, J.D., Ramchandani, S., Pelletier, R., Bigey, P., Price, G., Szyf, M., and Zannis-Hadjopoulos, M. (1999) J. Biol. Chem. 274, 9335–93341.
51. Bird, A.P. and Wolffe, A.P. (1999) Cell 99, 451–454.
52. Okano, M., Xie, S., and Li, E. (1998) Nat. Genet. 19, 219–220.
53. Yokochi, T. and Robertson, K.D. (2002) J. Biol. Chem. 277, 11735–11745.
54. Li, E., Bestor, T.H., and Jaenisch, R. (1992) Cell 69, 915–926.
55. Okano, M., Bell, D.W., Haber, D.A., and Li, E. (1999) Cell 99, 247–257.
56. Yoder, J.A. and Bestor, T.H. (1998) Hum. Mol. Genet. 7, 279–284.
57. Bourc'his, D., Xu, G.L., Lin, C.S., Bollman, B., and Bestor, T.H. (2001) Science 294, 2536–2539.
58. Hata, K., Okano, M., Lei, H., and Li, E. (2002) Development 129, 1983–1993.
59. Birke, M., Schreiner, S., Garcia-Cuellar, M.-P., Mahr, K., Titgemeyer, F., and Slany, R.K. (2002) Nucleic Acids Res. 30, 958–965.

60. Liang, G., Chan, M.F., Tomigahara, Y., Tsai, Y.C., Gonzales, F.A., Li, E., Laird, P.W., and Jones, P.A. (2002) Mol. Cell. Biol. 22, 480–491.
61. Fatemi, M., Hermann, A., Gowher, H., and Jeltsch, A. (2002) Eur. J. Biochem. 269, 4981–4984.
62. Jeddeloh, J.A., Stokes, T.L., and Richards, E.J. (1999) Nat. Genet. 22, 94–97.
63. Geiman, T.M., Tessarollo, L., Anver, M.R., Kopp, J.B., Ward, J.M., and Muegge, K. (2001) Biochim. Biophys. Acta 1526, 211–220.
64. Kim, G.D., Ni, J., Kelesoglu, N., Roberts, R.J., and Pradhan, S. (2002) EMBO J. 21, 4183–4195.
65. Rhee, I., Jair, K.W., Yen, R.W., Lengauer, C., Herman, J.G., Kinzler, K.W., Vogelstein, B., Baylin, S.B., and Schuebel, K.E. (2000) Nature 404, 1003–1007.
66. Rhee, I., Bachman, K.E., Park, B.H., Jair, K.W., Yen, R.W., Schuebel, K.E., Cui, H., Feinberg, A.P., Lengauer, C., Kinzler, K.W., Baylin, S.B., and Vogelstein, B. (2002) Nature 416, 552–556.
67. Lee, D.W., Zhang, K., Ning, Z.Q., Raabe, E.H., Tintner, S., Wieland, R., Wilkins, B.J., Kim, J.M., Blough, R.I., and Arceci, R.J. (2000) Cancer Res. 60, 3612–3622.
68. Dennis, K., Fan, T., Geiman, T., Yan, Q., and Muegge, K. (2001) Genes Dev. 15, 2940–2944.
69. Singer, T., Yordan, C., and Martienssen, R.A. (2001) Genes Dev. 15, 591–602.
70. Jost, J.P., Siegmann, M., Sun, L., and Leung, R. (1995) J. Biol. Chem. 270, 9734–9739.
71. Weiss, A., Keshet, I., Razin, A., and Cedar, H. (1996) Cell 86, 709–718.
72. Swisher, J.F., Rand, E., Cedar, H., and Pyle, M.A. (1998) Nucleic Acids Res. 26, 5573–5580.
73. Bhattacharya, S.K., Ramchandani, S., Cervoni, N., and Szyf, M. (1999) Nature 397, 579–583.
74. Cervoni, N., Bhattacharya, S., and Szyf, M. (1999) J. Biol. Chem. 274, 8363–8366.
75. Ng, H.-H., Zhang, Y., Hendrich, B., Johnson, C.A., Turner, B.M., Erdjument-Bromage, H.E., Tempst, P., Reinberg, D., and Bird, A. (1999) Nat. Genet. 23, 58–61.
76. Wade, P.A., Gegonne, A., Jones, P.L., Ballestar, E., Aubry, F., and Wolffe, A.P. (1999) Nat. Genet. 23, 62–66.
77. Schubeler, D., Lorincz, M.C., and Groudine, M. (2001) Sci. STKE (83):PL1.
78. Lorincz, M.C., Schubeler, D., and Groudine, M. (2001) Mol. Cell. Biol. 21, 7913–7922.
79. Schubeler, D., Lorincz, M.C., Cimbora, D.M., Telling, A., Feng, Y.-Q., Bouhassira, E.E., and Groudine, M. (2000) Mol. Cell. Biol. 20, 9103–9112.
80. Hendrich, B. and Bird, A. (1998) Mol. Cell. Biol. 18, 6538–6547.
81. Meehan, R.R., Lewis, J.D., and Bird, A.P. (1992) Nucleic Acids Res. 20, 5085–5092.
82. Nan, X., Meehan, R.R., and Bird, A. (1993) Nucleic Acids Res. 21, 4886–4892.
83. Hendrich, B., Abbott, C., McQueen, H., Chambers, D., Cross, S., and Bird, A. (1999) Mamm. Genome 10, 906–912.
84. Jones, P.L., Veenstra, G.J., Wade, P.A., Vermaak, D., Kass, S.U., Landsberger, N., Strouboulis, J., and Wolffe, A.P. (1998) Nat Genet. 19, 187–191.
85. Tweede, S., Ng, H.-H., Barlow, A.L., Turner, B.M., Hendrich, B., and Bird, A. (1999) Nat. Genet. 23, 389–390.
86. Nan, X., Ng, H.-H., Johnson, C.A., Laherty, C.D., Turner, B.M., Eisenman, R.N., and Bird, A. (1998) Nature 393, 386–389.
87. Ng, H.-H., Jeppesen, P., and Bird, A. (2000) Mol. Cell. Biol. 20, 1394–1406.
88. Fujita, N., Shimotake, N., Ohki, I., Chiba, T., Saya, H., Shirakawa, M., and Nakao, M. (2000) 20, 5107–5118.
89. Zhang, Y., Ng, H.-H., Erdjument-Bromage, H., Tempst, P., Bird, A., and Reinberg, D. (1999) Genes Dev. 13, 1924–1935.
90. Kokura, K., Kaul, S.C., Wadhwa, R., Nomura, T., Khan, M.M., Shinagawa, T., Yasukawa, T., Colmanares, C., and Ishii, S. (2001) J. Biol. Chem. 276.
91. Kaludov, N.K. and Wolffe, A.P. (2000) Nucleic Acids Res. 28, 1921–1928.
92. Kimura, H. and Shiota, K. (2003) J. Biol. Chem. 278, 4806–4812.
93. Hendrich, B., Hardeland, U., Ng, H.H., Jiricny, J., and Bird, A. (1999) Nature 401, 301–304.
94. Prokhortchouk, A., Hendrich, B., Jorgensen, H., Ruzov, A., Wilm, M., Georgiev, G., Bird, A., and Prokhortchouk, E. (2001) Genes Dev. 15, 1613–1618.
95. Guy, J., Hendrich, B., Holmes, M., Martin, J.E., and Bird, A. (2001) Nat. Genet. 27, 322–326.

96. Chen, R.Z., Akbarian, S., Tudor, M., and Jaenisch, R. (2001) Nat. Genet. 27, 327–331.
97. Hendrich, B., Guy, J., Ramsahoye, B., Wilson, V.A., and Bird, A. (2001) Genes Dev. 15, 710–723.
98. Millar, C.B., Guy, J., Sansom, O.J., Selfridge, J., MacDougall, E., Hendrich, B., Keightley, P.D., Bishop, S.M., Clarke, A.R., and Bird, A. (2002) Science 297, 403–405.
99. Nguyen, C.T., Gonzales, F.A., and Jones, P.A. (2001) Nucleic Acids Res. 29, 4598–4606.
100. Magdinier, F. and Wolffe, A.P. (2001) Proc. Natl. Acad. Sci. USA 98, 4990–4995.
101. Rideout, W.M.I., Coetzee, G.A., Olumni, A.F., and Jones, P.A. (1990) Science 249, 1288–1290.
102. Denissenko, M.F., Pao, A., Tang, M., and Pfeifer, G.P. (1996) Science 274, 430–432.
103. Chen, J.X., Zheng, Y., West, M., and Tang, M.S. (1998) Cancer Res. 58, 2070–2075.
104. Christman, J.K., Sheikhnejad, G., Marasco, C.J., and Sufrin, J.R. (1995) Proc. Natl. Acad. Sci. USA 92, 7347–7351.
105. Slack, A., Bovenzi, V., Bigey, P., Ivanov, M.A., Ramchandani, S., Bhattacharya, S., TenOever, B., Lamrihi, B., Scherman, D., and Szyf, M. (2002) J. Gene Med. 4, 381–389.
106. Hendrich, B. (2000) Curr. Biol. 10, R60–R63.
107. Taylor, S.M. and Jones, P.A. (1979) Cell 17, 771–779.
108. Jones, P.A., Taylor, S.M., Mohandas, T., and Shapiro, L.J. (1982) Proc. Natl. Acad. Sci. USA 79, 1215–1219.
109. Stein, R., Razin, A., and Cedar, H. (1982) Proc. Natl. Acad. Sci. USA 79, 3418–3422.
110. Jones, P.A. (1985) Cell 40, 485–486.
111. Boyes, J. and Bird, A. (1992) EMBO J. 11, 327–333.
112. Siegfried, Z., Eden, S., Mendelsohn, M., Feng, X., Tsuberi, B.Z., and Cedar, H. (1999) Nat. Genet. 22, 203–206.
113. Jackson-Grusby, L., Beard, C., Possemato, R., Tudor, M., Fambrough, D., Csankovszki, G., Dausman, J., Lee, P., Wilson, C., Lander, E., and Jaenisch, R. (2001) Nat. Genet. 27, 31–39.
114. Watt, F. and Molloy, P.L. (1988) Genes Dev. 2, 1136–1143.
115. Eden, S. and Cedar, H. (1994) Curr. Opin. Genet. Dev. 4, 255–259.
116. Curradi, M., Izzo, A., Badaracco, G., and Landsberger, N. (2002) Mol. Cell. Biol. 22, 3157–3173.
117. Kass, S.U., Pruss, D., and Wolffe, A.P. (1997) Trends Genet. 12, 444–449.
118. Keshet, I., Lieman-Hurwitz, J., and Cedar, H. (1986) Cell 44, 535–543.
119. Graessmann, A. and Graessmann, M. (1998) Gene 74, 135–137.
120. Kass, S.U., Goddard, J.P., and Adams, R.L.P. (1993) Mol. Cell. Biol. 13, 7372–7379.
121. Kass, S.U., Landsberger, N., and Wolffe, A.P. (1997) Curr. Biol. 7, 157–165.
122. Buschhausen, G., Wittig, B., Graessmann, M., and Graessmann, A. (1987) Proc. Natl. Acad. Sci. USA 84, 1177–1181.
123. Gautsch, J.W. and Willson, M.C. (1983) Nature 301, 32–37.
124. Pannel, D., Osborne, C.S., Yao, S., Sukonnik, T., Pasceri, P., Kariskakis, A., Okano, M., Li, E., Lipshitz, H.D., and Ellis, J. (2000) EMBO J. 19, 5884–5894.
125. Lock, L.F., Takagi, N., and Martin, G.R. (1987) Cell 48, 39–46.
126. Bell, A.C., West, A.G., and Felsenfeld, G. (2001) Science 291, 447–450.
127. Pioleau, M.-N., Nony, P., Simpson, M., and Felsenfeld, G. (1999) EMBO J. 18, 4035–4048.
128. Bell, A.C., West, A.G., and Felsenfeld, G. (1999) Cell 98, 387–396.
129. Tilghman, S.M. (1999) Cell 96, 185–193.
130. Szabo, P., Tang, S.H., Rentsendorj, A., Pfeifer, G.P., and Mann, J.R. (2000) Curr. Biol. 10, 607–610.
131. Bell, A.C. and Felsenfeld, G. (2000) Nature 405, 482–485.
132. Hark, A.T., Schoenherr, C.J., Katz, D.J., Ingram, R.S., Levorse, J.M., and Tilghman, S.M. (2000) Nature 405, 486–487.
133. Razin, A. and Cedar, H. (1977) Proc. Natl. Acad. Sci. USA 74, 2725–2728.
134. Jones, P.L., Veenstra, G.J.C., Wade, P.A., Vermaak, D., Kass, S.U., Landsberger, N., Strouboulis, J., and Wolffe, A.P. (1998) Nat. Genet 19, 187–191.
135. Eden, S., Hashimshony, T., Keshet, I., Cedar, H., and Thorne, A.W. (1998) Nature 394, 842.
136. Cameron, E.E., Bachman, K.E., Myohanen, S., Herman, J.G., and Baylin, S.B. (1999) Nat. Genet. 21, 103–107.

137. Downes, M., Ordentlich, P., Kao, H.Y., Alvarez, J.G., and Evans, R.M. (2000) Proc. Natl. Acad. Sci. USA 97, 10330–10335.
138. Selker, E.U. (1998) Proc. Natl. Acad. Sci. USA 95, 9430–9435.
139. Kouzarides, T. (2002) Curr. Opin. Genet. Dev. 12, 198–209.
140. Jenuwein, T. and Allis, C.D. (2001) Science 293, 1074–1080.
141. Lachner, M., O'Carroll, D., Rea, S., Mechtler, K., and Jenuwein, T. (2001) Nature 410, 116–120.
142. Bannister, A.J., Zegerman, P., Partridge, J.F., Miska, E.A., Thomas, J.O., Allshire, R.C., and Kouzarides, T. (2001) Nature 410, 120–124.
143. Fuks, F., Hurd, P.J., Wolf, D., Nan, X., Bird, A.P., and Kouzarides, T. (2002) J. Biol. Chem. 278, 4035–4040.
144. Mabuchi, H., Fujii, H., Calin, G., Alder, H., Negrini, M., Rassenti, L., Kipps, T.J., Bullrich, F., and Croce, C.M. (2001) Cancer Res. 61, 2870–2877.
145. Tamaru, H. and Selker, E.U. (2001) Nature 414, 277–283.
146. Bird, A. (2001) Science 294, 2113–2155.
147. Gendrel, A.-V., Lippman, Z., Yordan, C., Colot, V., and Martienssen, R. (2002) Science 297, 1871–1873.
148. Jackson, J.P., Lindroth, A.M., Cao, X., and Jacobsen, S.E. (2002) Nature 416, 556–560.
149. Santoro, R., Li, J., and Grummt, I. (2002) Nature Genet. 32, 393–396.
150. Zardo, G. and Caiafa, P. (1998) J. Biol. Chem. 273, 16517–16520.
151. Zardo, G., Marenzi, S., Perilli, M., and Caiafa, P. (1999) FASEB J. 13, 1518–1522.
152. de Capoa, A., Febbo, F.R., Giovannelli, F., Niveleau, A., Zardo, G., Marenzi, S., and Caiafa, P. (1999) FASEB J. 13, 89–93.
153. Zardo, G., D'Erme, M., Reale, A., Strom, R., Perilli, M., and Caiafa, P. (1997) Biochemistry 36, 7937–7943.
154. D'Amours, D., Desnoyers, S., D'Silva, I., and Poirier, G.G. (1999) Biochem. J. 342, 249–268.
155. Burkle, A. (2001) BioEssays 23, 795–806.
156. Davidovic, L., Vodenicharov, M., Affar, E.B., and Poirier, G.G. (2001) Exp. Cell Res. 268, 7–13.
157. Panzeter, P.L., Realini, C.A., and Althaus, F.R. (1992) Biochemistry 31, 1379–1385.
158. Panzeter, P.L., Zweifel, B., Malanga, M., Waser, S.H., Richard, M., and Althaus, F.R. (1993) J. Biol. Chem. 268, 17662–17664.
159. Tulin, A. and Spradling, A. (2003) Science 299, 560–562.
160. Althaus, F.R. (1992) J. Cell Sci. 102, 663–670.
161. Zardo, G., Marenzi, S., and Caiafa, P. (1998) Biol. Chem. 353, 647–654.
162. Tazi, J. and Bird, A. (1990) Cell 60, 909–920.
163. Felsenfeld, G., Nickol, J., Behe, M., McGhee, J., and Jackson, D. (1982) Cold Spring Harb. Symp. Quant. Biol. 47, 577–584.
164. Nightingale, K. and Wolffe, A.P. (1995) J. Biol. Chem. 270, 4197–4200.
165. Drew, H.R. and McCall, M.J. (1987) J. Mol. Biol. 197, 485–511.
166. Englander, E.W., Wolffe, A.P., and Howard, B.H. (1993) J. Biol. Chem. 268, 19565–19573.
167. Davey, C., Pennings, S., and Allan, J. (1997) J. Mol. Biol. 267, 276–288.
168. Levine, A., Yeivin, A., Ben-Asher, E., Aloni, Y., and Razin, A. (1993) J. Biol. Chem. 268, 21754–21759.
169. McArthur, M. and Thomas, J.O. (1996) EMBO J. 15, 1705–1714.
170. Higurashi, M. and Cole, R.D. (1991) J. Biol. Chem. 266, 8619–8625.
171. Campoy, F.J., Meehan, R.R., McKay, S., Nixon, J., and Bird, A. (1995) J. Biol. Chem. 270, 26473–26481.
172. Ball, D.J., Gross, D.S., and Garrard, W.T. (1983) Proc. Natl. Acad. Sci. USA 80, 5490–5494.
173. Johnson, C.A., Goddard, J.P., and Adams, R.L.P. (1995) J. Biochem. 305, 791–798.
174. Laybourn, P.J. and Kadonaga, J.T. (1991) Science 254, 238–245.
175. Karymov, M.A., Tomschik, M., Leuba, S.H., Caiafa, P., and Zlatanova, J. (2001) FASEB J. 15, 2631–2641.
176. Meehan, R.R., Lewis, J.D., McKay, S., Kleiner, E.L., and Bird, A.P. (1989) Cell 58, 499–507.
177. Barra, J.L., Rhounim, L., Rossignol, J.-L., and Faugeron, G. (2000) Mol. Cell. Biol. 20, 61–69.

178. Strom, R., Santoro, R., D'Erme, M., Mastrantonio, S., Reale, A., Marenzi, S., Zardo, G., and Caiafa, P. (1995) Gene 157, 253–256.
179. Fakan, S., Leduc, Y., Lamarre, D., Brunet, G., and Poirier, G.G. (1988) Exp. Cell Res. 179, 517–526.
180. Meisterernst, M., Stelzer, G., and Roeder, R.G. (1997) Proc. Natl. Acad. Sci. USA 94, 2261–2265.
181. Oei, S.L., Griesenbeck, J., Schweiger, M., and Ziegler, M. (1998) J. Biol. Chem. 273, 31644–31647.
182. Lunyak, V.V., Burgess, R., Prefontaine, G.G., Nelson, C., Sze, S.H., Chenoweth, J., Schwartz, P., Pevzner, P.A., Glass, C., Mandel, G., and Rosenfeld, M.G. (2002) Science 298, 1747–1752.

J. Zlatanova and S.H. Leuba (Eds.) *Chromatin Structure and Dynamics: State-of-the-Art*

DOI: 10.1016/S0167-7306(03)39013-1

CHAPTER 13

Chromatin structure and function: lessons from imaging techniques

David P. Bazett-Jones and Christopher H. Eskiw

Programme in Cell Biology, The Hospital for Sick Children, 555 University Avenue, Toronto, ON M5G 1X8, Canada. Tel.: 416 813-2181; Fax: 416 813-2235; E-mail: dbjones@sickkids.ca

1. Introduction

Structural information is central to an understanding of the molecular mechanisms involved in biochemical processes. Elucidation of chromatin and chromosome structure have relied on biochemical approaches, physico-chemical techniques such as X-ray and neutron scattering and crystallography, and direct imaging techniques such as fluorescence microscopy, electron microscopy, and atomic force microscopy (AFM). Whereas the central role that direct imaging techniques have played in the chromatin field has a long history, the important role for imaging was dramatically realized in the discovery of the chromatin subunit. Biochemical data that implied a repeating structure could be readily interpreted by the images of "nu bodies", observed by electron microscopy [1]. Shortly thereafter, electron microscopy was used to demonstrate the ability of the 10-nm beads-on-a-string fiber to fold reversibly into a higher-order 30-nm structure in a salt-dependent manner [2].

The 10-nm fiber and the 30-nm structure represent the organization of DNA at the higher levels of spatial resolution. Microscopy, however, has made significant contributions at other levels of chromatin organization *in situ*. The positioning of largely inactive, condensed heterochromatin domains at the nuclear envelope and on the periphery of the nucleolus could only be revealed by microscopic approaches. Similarly, information about the physical localization of transcriptional activity on the outer edges of condensed chromatin was obtained by microscopical techniques [3,4]. Most recently, the ability to create fusions between fluorescent proteins and specific nuclear proteins of interest, permits visualizing dynamic events involving chromatin organization in nuclei of living cells.

2. Microscopy: a complementary approach

The many advantages that microscopical imaging offers have placed microscopical methods in the center of chromatin research, complementing biochemical and physico-chemical techniques. Whereas microscopy techniques generally cannot rival X-ray crystallographic information, such as that attained in the 2.8 Å map of the fundamental nucleosomal subunit [5], imaging techniques are not limited by the same constraints as crystallography or magnetic resonance spectroscopy, for example. Microscopy of biological macromolecules does not require crystallization, with the result that molecules or complexes that have not been or cannot be crystallized, are amenable to structural studies with sufficient resolution to provide meaningful insights into function and molecular mechanisms. A second advantage is that small amounts of sample are required, and the sample can be studied at physiologically relevant concentrations. Another advantage that microscopy offers is the ability to characterize inhomogenous specimens. Whereas biochemical techniques generally measure the average behavior of the members of a population, imaging techniques provide the opportunity to characterize the structure and frequency of representation of members of an ensemble. Finally, although electron microscopy can only provide static snapshots of biochemical events, fluorescence microscopy of structures *in situ* and even in live cells has become routine, offering the ability to visualize dynamic events at the sub-organelle level. The spatial resolution achieved cannot be rivalled by fractionation techniques. In addition, the temporal resolution after a stimulus or at a defined point in the cell cycle can be very precise. For example, differences in acetylation levels of chromatin in two daughter cells in late mitosis/early GI can be observed, a result of one of the cells moving into interphase ahead of the other. Biochemical analysis of cells with such temporal resolution is not achievable. Indeed, imaging techniques provide a means to treat an individual cell as a test tube.

2.1. Microscopical approaches

Advances in technology and improvements in imaging techniques have provided many approaches for imaging chromatin under *in vitro* conditions and *in situ* in fixed or living cells. Some of these techniques are described below.

2.1.1. Transmission electron microscopy (TEM)

The TEM has played an important role in chromatin research, providing images that supported other indications of a repeating chromatin subunit. Conventional approaches for molecular imaging have relied on staining and shadowing of chromatin in order to provide sufficient image contrast in the bright field mode. Sample preparation has required dehydration of the specimen by air drying, critical-point drying, or freeze drying. The samples, therefore, are visualized in a completely dehydrated state. Recently, however, electron cyro-microscopy (ECM) has gained acceptance. The primary advantage of ECM is that the sample can be visualized in a completely hydrated state, albeit in a layer of vitreous ice. Though the ice layer greatly reduces image contrast, chromatin subunits and fiber can be visualized without the requirements of heavy atom staining. Such contrast agents limit spatial resolution, and can generate misconceptions about the molecular structure as a result of unpredictable and differential binding of the stains (e.g., uranyl acetate) to particular biochemical components, or differential binding to particular biochemical entities in distinct chemical environments within macromolecular complexes. A second advantage of ECM that is particularly relevant to complexes with fiber morphology, is that the complexes retain their three-dimensional orientations without being absorbed onto a two-dimensional support film. Although the three-dimensional information is lost in a single image projection, stereo images or a tilt tomographic series can be applied to gain the three-dimensional information.

A second advance in transmission electron microscopy that is relevant to studies of DNA–protein complexes is an analytical technique based on electron energy loss spectroscopy, known as energy filtered transmission electron microscopy (EFTEM) or electron spectroscopic imaging (ESI). The principle of the technique is that some incident electrons, which pass through the sample, will lose energy due to excitations and ionizations of the specimen's atoms. The energy losses of the incident electrons provide a finger-print of the elemental chemistry of the specimen. With the introduction of an electron spectrometer into the microscope, a device which is capable of separating electrons according to energy loss while simultaneously correcting for image aberrations caused by the spectrometer, an "energy filtered" image can be obtained. Because such images are element-enhanced, they provide spatial maps of the distribution of particular chemical elements. When applied to the imaging of nucleoprotein complexes, for example, net phosphorus images reveal the location of the nucleic acid in the complex, because of the relatively very high concentration of phosphorus in the nucleic acid compared to the associated protein (reviewed in Ref. [6]).

Structural and functional relationships between chromatin domains and subnuclear structure has also advanced over the years by exploiting the advantages of high resolution provided by the transmission electron microscope with gold-labelled immunological reagents for locating particular biochemical components. Fine examples of this approach are too numerous; an example, however, is the application in mapping domains within the nucleolus [7]. In another study, immuno-cytochemistry at the electron microscope level has been combined with energy filtered transmission electron microscopy, thereby simultaneously benefiting from the advantages of each technique [8].

2.1.2. Scanning force microscopy

Scanning force microscopy (SFM) is one of several scanning probe imaging techniques, developed over the past 20 years [9]. SFM can provide three-dimensional information of chromatin fiber under potentially less damaging conditions than those required for conventional TEM. Although the sample has to be adsorbed onto a two-dimensional substrate, it does not have to be dried in a vacuum. Instead, chromatin fiber can be retained and imaged at approximately 50% relative humidity. The greatest success in imaging chromatin has been achieved with the tapping mode of operation of the microscope. In this mode, a stiff cantilever oscillates at its resonance frequency near the surface of the specimen. Forces between the tip and the specimen will act to dampen the oscillation.

The force required to resist the dampening can be detected by a feedback mechanism and related to topographical or structural information as the tip is scanned laterally over the specimen. The major advantage of this mode of operation is that the specimen experiences a minimum of lateral distortion by the scanning tip.

2.1.3. Fluorescence microscopy

Electron cryo-microscopy and scanning force microscopy are emerging as important tools in structural biology, where they can be applied to studies of purified components imaged *in vitro*. The organization of DNA as chromatin *in situ* has relied on both conventional transmission electron microscopy and epi-fluorescence microscopy. Instrumentation and mathematical algorithms have led to improvements in both resolution and signal-to-noise, by the introduction of confocal microscopy and deconvolution techniques to remove out-of-focus haze in optical sections. Fluorescence microscopy approaches have relied extensively on the development of fluorescently-labeled immunological reagents for determining the location and distribution of particular proteins within the nucleus. In addition, hybridization techniques, with labelled nucleotides that can be detected directly or indirectly with fluorescent tags, have been developed to detect the presence and location of particular genetic loci or the sites of synthesis of nascent RNA transcripts in fixed cells and tissues. Whereas these techniques are based on permeabilized, fixed, and sometimes harshly treated cells, the advent of fluorescent tags that can be used in live cells has greatly increased the range of questions that can be addressed related to chromatin organization, and the relationships of chromatin structure and composition to subnuclear domains. In particular, fusion proteins with Green Fluorescent Protein (GFP) provide a means to determine the localization of a protein of interest in the cell, or to follow a translocation event after stimulation, or to determine a protein's location at a defined time point in the cell cycle. More recently, fusion proteins with GFP have been used to determine the dynamic behavior of chromatin components; fluorescence recovery after photobleaching (FRAP) or fluorescence loss in photobleaching (FLIP) are becoming routine techniques for studying nuclear events. The technique involves bleaching a well-defined region of the nucleus with an intense laser. Subsequently, the recovery of the fluorescence signal in the bleached region by the diffusion or movement of unbleached molecules into the bleached region is followed over time. The rate and extent of recovery of fluorescence provide a measure of the average mobility of the molecule of interest, and a measure of the fraction of molecules in the nucleus that are mobile.

Some of the insights that these techniques have brought to the field of chromatin research are described in this chapter. In some cases, microscopy has served simply to confirm conclusions obtained from less direct, but powerful biochemical approaches. In other cases, however, studies with imaging techniques are described where the information could only be obtained by microscopy, because of the spatial or temporal resolution that the microscope can provide.

3. Imaging of modified and functionally engaged nucleosomes

The high degree of homogeneity of nucleosome core particles that can be assembled *in vitro* provided the basis for a high resolution structure determination by X-ray crystallography [5]. Important features of the structure include the non-uniform wrapping of DNA on a histone ramp, and the position of the core histone tails. Whereas this information has great value in our understanding of the role of the nucleosome in DNA compaction, the structural basis for nucleosome remodelling, critical in nuclear events such as transcription, will likely remain intractable by crystallographic methods. Highly modified or remodelled core particles are too heterogeneous to form crystals. Although imaging techniques often do not provide structural detail below 1 nm, they can reveal useful information at ~ 2 nm resolution, sufficient to provide useful information on the organization of DNA by chromosomal proteins.

3.1. Imaging acetylated nucleosomes

Post-translational modifications of core histone tails have been correlated with functionally important chromatin states. Acetylation of specific lysine residues in histones H3 and H4 have been known to be associated with transcriptionally active or competent chromatin for several years [10]. A direct role for acetylation of core histones in promoter function followed from the discovery that some transcriptional activators possess histone acetyltransferase activity (reviewed in Ref. [11]). A result of histone acetylation may be a reduced contact between DNA and the histones. Although biochemical and physico-chemical techniques have not revealed major structural rearrangements within the core particle, electron spectroscopic imaging has revealed a unique feature of nucleosomes containing hyperacetylated core histones. Whereas the structure of such nucleosomes appear indistinguishable from bulk chromatin nucleosomes core particles when imaged under low salt conditions, hyperacetylated nucleosomes imaged in the presence of salt have a propensity to unfold. Under physiological salt conditions, the hyperacetylated nucleosomes appear normal, but clearly are less stable and can become more "open" under conditions of low salt (Fig. 1).

3.2. Imaging transcriptionally active nucleosomes

A similar structural feature, the propensity of the nearly two turns of DNA to open up like a stretched spring, is also a characteristic of nucleosomes isolated by Hg affinity chromatography [12] and hyperacetylated nucleosomes isolated from cultured cells treated with butyrate [13]. Chromatography with this resin has been shown to be effective in enriching for nucleosomes associated with transcriptionally active chromatin [14]. Binding to the Hg column is thought to result from the accessibility of sulfhydryls on histone H3, located near the core particle's dyad axis, which become accessible through this propensity to unfold. Energy filtered electron microscope images reveal an increased length : width ratio, consistent with

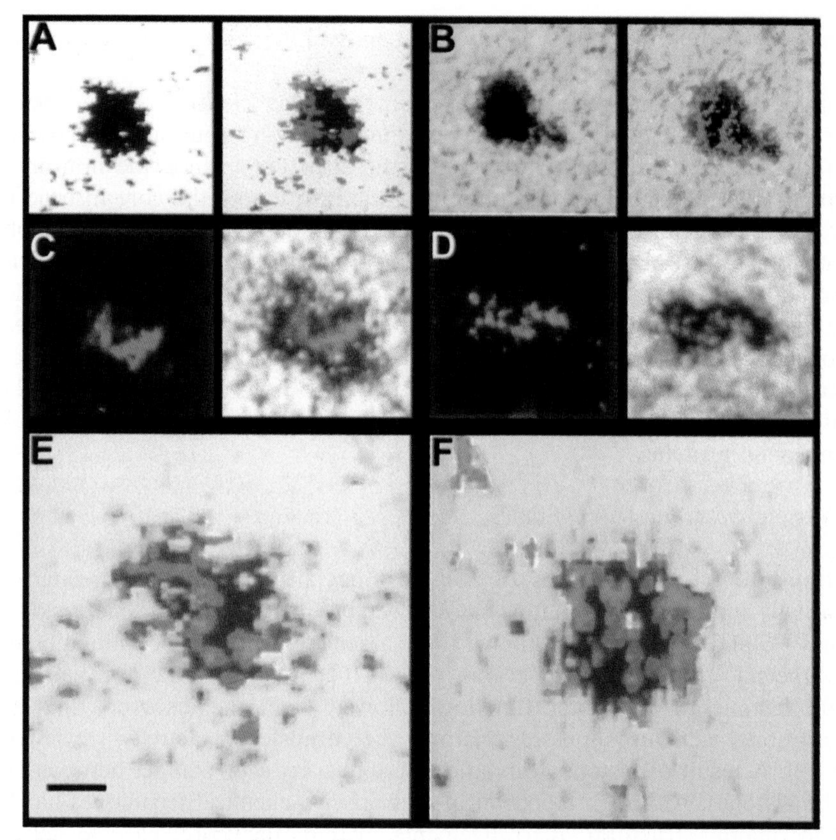

Fig. 1. Electron spectroscopic images of nucleosomes. Grey levels represent molecular mass, and red represents the phosphorus distribution. (A) *En face* view of a nucleosome reconstituted *in vitro* with HeLa core histones onto 215 bp *Xenopus* rRNA somatic gene DNA. The phosphorus map (red overlay) indicates that DNA is restricted to the periphery of the particle, consistent with being wound around a protein-based core. (B) Edge-on view of a nucleosome core particle. The phosphorus distribution (red overlay) is consistent with two gyres of the DNA supercoil, wound around a protein core. (C,D) Hyperacetylated nucleosomes (3 acetyl residues per histone H4) prepared from HeLa cells (from Ref. [13]). (E,F) Nucleosomes isolated by Hg-affinity chromatography (from Ref. [12]). The hyperacetylated nucleosomes and those isolated by Hg-affinity chromatography have a high length : width ratio, and the phosphorus distribution indicates that the DNA is unwound like a stretched spring. Scale bar 10 nm in A–D and 7 nm in E, F.

an opening at the dyad. In addition, phosphorus mapping reveals that the DNA is unwound like a stretched spring (Fig. 1). ESI also permits the estimation of stoichiometric relationships; some nucleosomes isolated by Hg affinity chromatography display a lower protein : DNA ratio on one side of the mirror symmetry axis. This is consistent with a transient displacement of histone from one side of the nucleosome to the other, which may be required for polymerase passage. The low-mass side of the particles has a higher DNA : protein ratio, consistent with a loss of protein from one side of the particle. This may be the result of a loss of an H2A–H2B dimer.

3.3. Imaging remodeled nucleosomes

Imaging reveals a very different reorganization of the core particle when remodelled under *in vitro* conditions by the SWI/SNF nucleosome remodeling complex, purified from yeast [15]. Energy filtered images revealed that SWI/SNF creates loops in the DNA, independent of ATP, and indicating that the complex has multiple DNA binding sites on its surface. This brings otherwise distant sites into close proximity. ATP-dependent SWI/SNF action then appears to disrupt nucleosomes that are constrained in these loop domains. Stoichiometric analysis of nucleosomes outside of these loop domains have a normal histone content, $102 + 15$ kDa, and a normal DNA content of $155 + 17$ bp. In contrast, nucleosomes in the constrained loops have a protein content of 101 ± 26 kDa, consistent with canonical nucleosomes, but only $101 + 17$ bp of DNA. The lower DNA content is consistent with a loss of nearly one supercoil turn. These imaging results are consistent with a disruption by the remodelling complex that results in a loss of DNA from nucleosomes without a loss of histone protein, implying a mechanism by which SWI/SNF unwraps part of the nucleosomal DNA.

The destabilization of histone : DNA contacts by chromatin remodeling complexes such as SWI/SNF, acetylation of core histones, or activities found in complexes such as the transcriptional elongation complex FACT [16], may be required for binding of trans-acting regulatory factors to promoter elements, or for unhindered elongation by RNA polymerases through nucleosome particles. How the nucleosome itself acts as a barrier to transcription elongation has been studied with a combination of biochemical approaches and electron cryo-microscopy [17]. In this study, ECM images at subnucleosomal resolution provide direct evidence that the histone core remains associated with the DNA during polymerase passage. In addition, "open transcriptional intermediates" can be detected in which the DNA has been partially displaced from the histones. The intermediates, with the exception of a looped structure in which the DNA behind the polymerase forms a transient interaction with the exposed surface of the octamer deduced from the biochemical experiments, can be visualized and directly interpreted. This study shows the power of combining a direct imaging approach with more routine biochemical assays. In other studies, for example, nucleosomes in front of the polymerase are released and transferred to negatively supercoiled DNA behind the polymerase [18], or where the octamer can step around the transcribing polymerase [19]. In these studies and that of the interaction of SWI/SNF binding to nucleosome templates [15], the histone octamer is shown to remain bound to the DNA. This would indicate that dynamic changes in chromatin structure can occur very rapidly and efficiently, without the loss of the histone octamer core from the DNA. However, the issue of the fate of the nucleosome associated with polymerase passage is complex and appears to depend on the experimental system and conditions. In another elegant imaging experiment, the data clearly indicate that nucleosomes are completely removed from the transcribed region of a nucleosomal template transcribed *in vitro* by T7 polymerase, independent of the template's topological state [20]. Nevertheless, an activity can be added to the reaction that is responsible for persistence of nucleosomes on the template, supporting a mechanism for nucleosome transfer that may operate *in situ*.

3.4. Imaging exchange of core and linker histones

Exchange of both core histones and linker histones can now be accomplished in live cells under a variety of conditions and through the cell cycle. Green-fluorescent protein (GFP) tagged histones can be used for the visualization of chromatin dynamics. GFP-tagged histones bind with affinity similar to those of wild-type histones and can be used to examine properties of these proteins such as binding affinity and exchange rates with free nucleoplasmic, core nucleosomal components. GFP–histone fusion proteins have been used to examine chromatin structure and dynamics at specific points within the cell cycle without disrupting normal events. GFP-tagged histones were first used to examine double minute segregation in mitotic cells [21]. The ability to examine chromatin structure and dynamics *in vivo* allows for confirmation of previous biochemical data, but has also provided new insights that may not have been attainable from *in vitro* methods.

Normally histone gene expression is tightly coupled to DNA synthesis. However, expression of GFP-tagged histones can be driven by a constitutively active promoter, thus causing the expression of GFP–histones outside of S phase. This provided a unique opportunity to examine the histone dynamics *in vivo* [22]. Using this strategy, H2A/H2B dimers in transfected HeLa cells expressing fusion GFP–histone fusions exchange during interphase outside of S phase. Heterochromatic regions of transfected HeLa cells show labeling similar to the DNA specific dye Heochst 33342, demonstrating that the GFP–H2B has been incorporated into the chromatin. FRAP experiments show that exchange of GFP–H2B can occur outside of S phase, indicating that there is a propensity for the outer H2A/H2B dimers to dissociate from the core nucleosomal octamer and to be replaced by soluble dimers of H2A/H2B.

Duplicating the GFP–H2B experiments with both GFP–H3 and GFP–H4 vectors, very little H3 and H4 exchange outside of S phase was observed. Fluorescence of GFP–H3 and GFP–H4 in HeLa cells in G1 recovered rapidly. The extremely rapid recovery rate is similar to that of a diffuse soluble protein, indicating that at this stage of the cell cycle, GFP-tagged H3 and -H4 are not incorporated into chromatin. FRAP experiments of transfected cells in S or G2 show that there is very little recovery of GFP–H3 or –H4 fluorescence. The fluorescence imaging indicates that once the H3 and H4 proteins are incorporated into the chromatin, they are essentially immobile for the remainder of the cell cycle. Unlike histones H2A and H2B, which associate as dimers in the nucleosome histone octamer, there is very little exchange of the components of the H3/H4 tetramer.

The dynamics of histone H1 *in vivo* has also been described by taking advantage of GFP fusions in combination with FRAP [23]. From previous biochemical

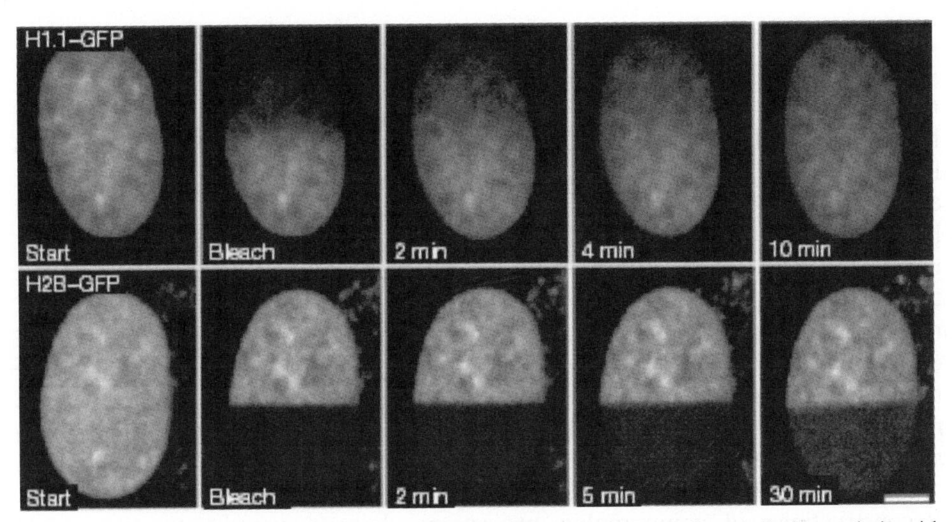

Fig. 2. Exchange of histones H1.1 and H2B from chromatin in interphase cells by analysis with fluorescence recovery after photobleaching (FRAP). Half of a nucleus of an SK-N cell expressing GFP–H1.1 was bleached (upper panel), and the recovery monitored over the times shown. Similarly, a region of a nucleus of an SK-N cell stably expressing H2B–CFP was bleached (lower panel), and the recovery monitored over the times shown. Whereas unbleached H1 molecules move into the bleached region after a few minutes, the H2B histones are much less mobile, since the bleached region shows no recovery (from Ref. [23]). Scale bar 5 μm.

studies, it has been generally accepted that H1 binds less tightly to chromatin than the other histone chromosomal proteins. When the H1.1 gene fused to GFP (H1.1–GFP) was expressed in human cells in culture, Lever *et al.* demonstrated that exchange of H1.1 occurs rapidly on both condensed and de-condensed chromatin (Fig. 2). This result implies that a more stably bound linker histone is not the basis for maintaining a condensed heterochromatin state through interphase. Treatment of these cells with drugs to inhibit general protein phosphorylation, greatly inhibits the exchange of H1 between the soluble and bound forms. Although all of the substrates for phosphorylation have not been identified in this experiment, it may be that phosphorylation of histone H1 itself is required for exchange on both heterochromatin and euchromatin. Exchange of the linker histone does not require a constant source of energy since depletion of ATP does not affect exchange rates. Misteli *et al.* also found that histone H1 exchanged continuously between both heterochromatin and euchromatm regions [24]. In contrast to the Lever *et al.* study, they detected a less mobile fraction as well. Perhaps this corresponds to a phosphorylated form of histone H1 observed in the Lever study. Interestingly, the result of inducing levels of core histone acetylation leads to higher levels of H1 exchange. Both studies support a dynamic interaction of histone H1 with chromatin. Whereas the exchange with euchromatin may have been expected, a rapid exchange with heterochromatin regions is a surprise and an important result that will have to be studied further.

4. *Perspectives on the organization of the 30-nm fiber from imaging approaches*

It is thought that the 10-nm nucleosome fiber is further folded or compacted *in situ* into a higher order 30-nm fiber. In spite of much effort, a firm consensus has not emerged on the structure of this nucleosome arrangement. Early studies involved biochemical and imaging studies of isolated fiber [2]. The main feature of the first model, the so-called solenoid model, was a 1-start helix of nucleosomes, with the linker DNA coiled between adjacent nucleosomes. Other so-called helical "zig-zag" models have emerged; these are based on the compaction of a zig-zag conformation of nucleosomes, which is observed at intermediate ionic strength [25,26]. More recent imaging studies, however, provide little support for any of these helical models. In one of these studies, 30-nm chromatin fiber were studied in the starfish sperm nuclei, prepared under conditions of low temperature embedding. Enhanced contrast was obtained with the DNA-specific stain osmium ammine-B [27]. Reconstructions of tomographic projections do not support a symmetrical arrangement of nucleosomes within the fiber. Instead, the chromatin fiber exhibit smoothly bending regions interrupted by sharp turns and bends (Fig. 3), and the nucleosomes themselves are on the periphery of the fiber, with the linkers projecting towards the interior. The model most consistent with the images is not a helical arrangement. Instead, a two nucleosome-wide ribbon is randomly bent and twisted, with little or no face-to-face contacts between adjacent nucleosomes. Striking similarity to the fiber observed *in situ* results from modelling of the fiber if just two parameters are considered, the entry–exit angle of the linker DNA and the linker length. If the entry–exit angle is maintained at 60°, and the linker length is kept constant, a symmetrical fiber with a circular cross-section is obtained [28]. If, however, the linker length is allowed to vary, the fiber structure prevails, but its 3D trajectory is no longer straight. With further variability introduced into the linker lengths, typical of that observed *in vivo*, abrupt changes in the trajectory result. Moreover, regions where the nucleosomes appear ordered are seen, interspersed with no apparent order. Local variations in fiber diameter are also observed. This irregular fiber model is not only supported by TEM [27], but also by tapping-mode SFM. When chicken erythrocyte chromatin fiber, containing linker histones H1 and H5, were prepared and imaged at 10 mM NaCl *in vitro*, an irregular three-dimensional array, with little or no regular fiber morphology, was also observed [29]. Whereas a solenoidal structure may exist in stretches of uniform, regularly spaced nucleosomes, direct imaging approaches indicate so far that such regularity in fiber structure does not extend over long lengths.

The principles whereby a chain of nucleosomes can compact to form a 30 nm chromatin fiber are still not well understood. Nevertheless, important aspects of this process are becoming clear from imaging studies, employing both ECM and SFM. When isolated chicken erythrocyte chromatin or chromatin reconstituted onto six tandem 208 bp nucleosome positioning units were examined by ECM, a linker DNA stem-like architectural motif was observed at the entry–exit sites (Fig. 4) [30]. Particles consistent with an octamer are surrounded with 1.7 turns of DNA, a linker

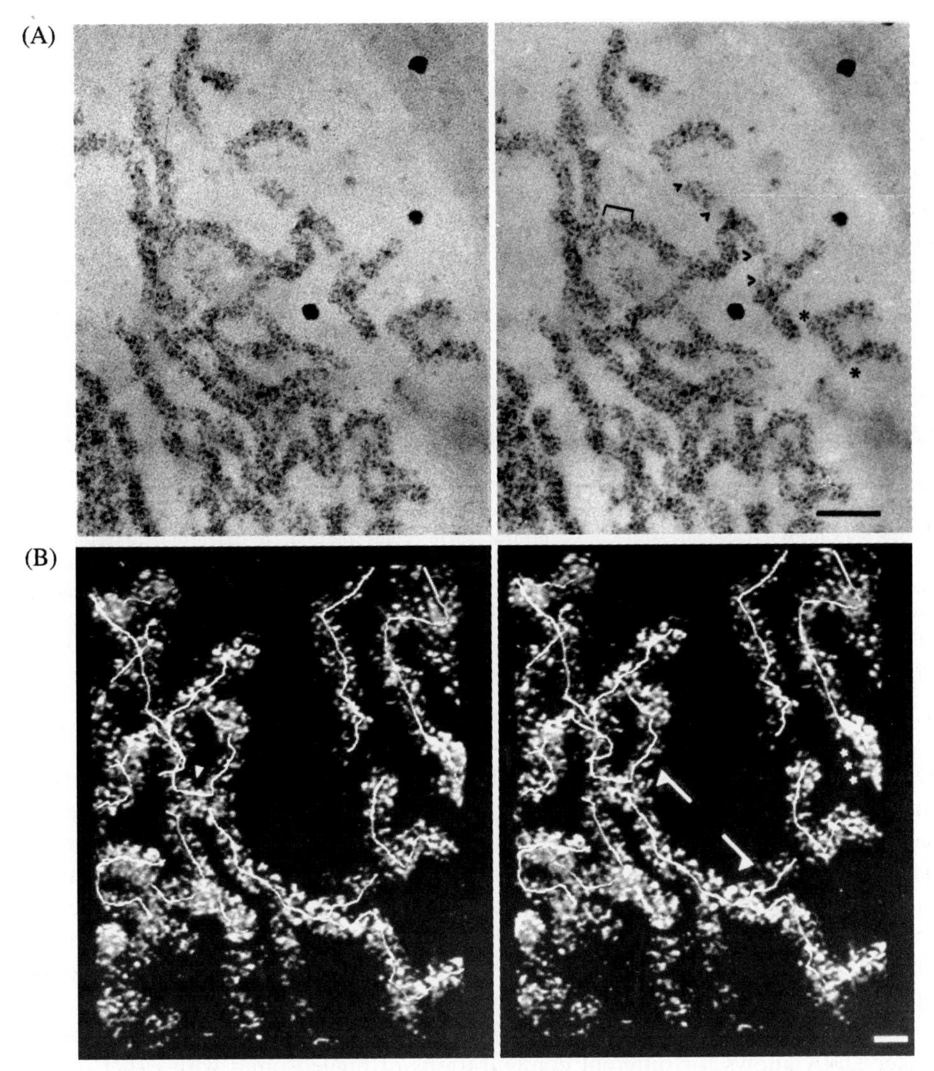

Fig. 3. (A) Stereo pair of starfish sperm chromatin fibers stained with osmium ammine-B. Bracket indicates the fine deposition of stain on nucleosomes and linker DNA. Arrowheads indicate where fibers enter and exit the plane of the section, and asterisks indicate sharp bends in the fibre axis. (B) Stereo pair of a reconstructed volume of a starfish sperm head by EM tomography. Axes of some fibers have been marked. Arrowhead indicates where individual fibers cannot be distinguished (from Ref. [27]). Scale bar 100 nm.

segment consisting of an "intersection" zone approximately 8 nm from the nucleosome center is observed, and extends over a length of 3–5 nm. Subsequently, the two linker segments diverge. These observations support a very different model from that of the chromatosome, a structure in which 166 bp or two complete turns of DNA are protected by the histone octamer core [31]. Instead of the extra linker

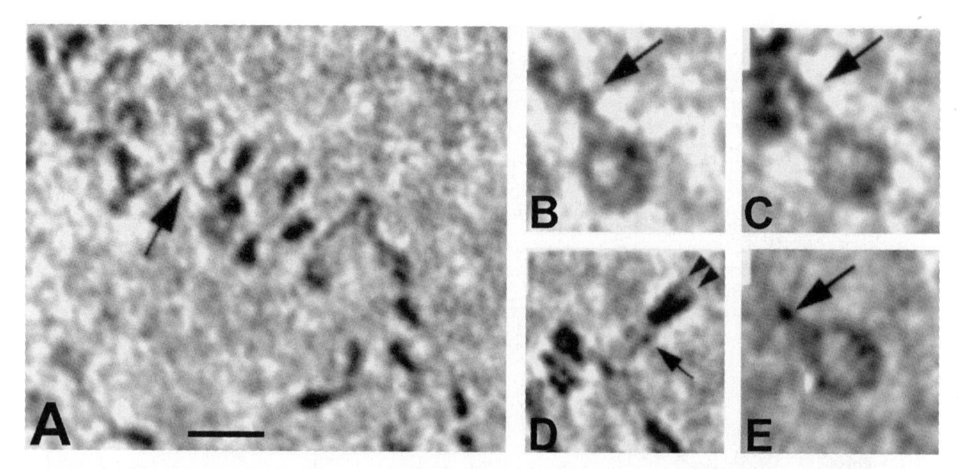

Fig. 4. Images of unfixed and unstained chromatin in a frozen and hydrated state. All samples shown contain linker histone H5. (A) Soluble chromatin prepared from chicken erythrocyte nuclei. Arrow indicates a nucleosome with a linker histone "stem" conformation. (B–E) Chromatin reconstituted onto an array of the 5S rDNA nucleosome positioning sequence. *En face* views (B–D) of nucleosomes show the linker DNA entering and exiting the nucleosome tangentially, before interacting and remaining associated for 3–5 nm before separating (arrows). An edge-on view (E) shows the two gyres of DNA (arrow heads) and the apposed linker DNA (arrow) (from Ref. [30]). Scale bar 20 nm (A) and 10 nm (B–E).

DNA wrapped on the histone core, the ECM data indicate that it is complexed with linker histories in a stem-like architectural motif. Chromatin fiber imaged under low salt conditions reveal a 3D zig-zag structure that clearly involves the stem motif. This zig-zag arrangement is still observed when the chromatin fiber compact with increased concentrations of salt. The nearest neighbor nucleosomes are much closer, leading to a more compact arrangement, and this results from a reduction in the linker DNA entry–exit angle, from approximately 85° to 45°. An interesting feature of this compaction model is that the transition from an extended nucleosomal array to a compact 30 nm fiber does not result in a topological change. The enter/exiting linker DNA allows for different configurations of nucleosomes with either crossed or non-crossed linkers, and so on an average, is topologically neutral. Supercoiling of linker DNA in helical models, such as an the solenoid model, would require large changes in linking number, such changes are not observed.

Insights into the molecular determinants that affect the entry–exit angle have been gained from SFM imaging of chromatin fiber under conditions of mild digestion of histone tails. The globular domain of linker histones has a profound effect on the entry–exit angle, and preserves the zig-zag morphology [32]. On the other hand, the linker histone tails affect the internucleosomal distances. In addition, the N-terminal tails of histone H3 stabilize the 3D structure of the fiber, since cleavage results in a flattening of the fiber (Fig. 5). Support for a redundancy in function between the tails of the linker histones and the N-terminal tails of H3 has also been obtained by SFM [33].

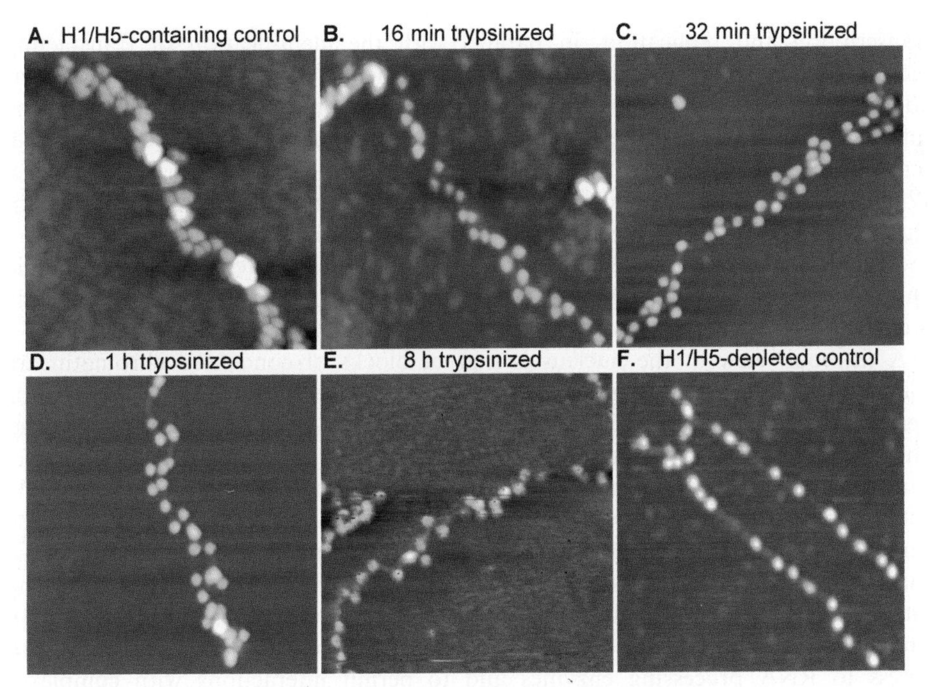

Fig. 5. SFM images of chicken erythrocyte chromatin fibers. (A) Untrypsinized, linker histone-containing control fibers, and (B) linker histone-stripped fibers. The stripping of linker histones destroys both the three-dimensional interactions of adjacent nucleosomes and the zig-zag arrangement of consecutive nucleosomes. Trypsinization of the N-terminal histone tails of the linker histones and core histone H3 result in the loss of the three-dimensional association of the consecutive nucleosomes, but does not destroy the zig-zag configuration. Imaging of fibers deposited onto mica was performed in air under conditions of ambient humidity and temperature (from Ref. [32]). Full width of each image corresponds to 500 nm.

5. *Chromatin organization in the nucleus*

Our knowledge of chromatin is largely based on studies of material isolated from intact cells, through fractionation techniques, and by isolation and purification of components and intact fragments by chromatographic techniques and immunological reagents. Further characterization of chromatin and its interactions with transcription, replication, or remodelling machinery has been derived from studies on chromatin complexes reconstituted *in vitro*. A complete understanding of how chromatin is organized in the nucleus and how it participates in nuclear events will only be attained through studies that directly probe its organization *in situ*. Imaging techniques continue to play a pivotal role in this endeavor.

Chromatin organization and its relation to function in the intact cell have become part of a broader area of investigation, the organization of the cell nucleus itself. The success of modern molecular biology, which has focussed on the characterization of cis-elements in the DNA that affect nuclear events such as transcription and replication, in parallel with the identification of trans-acting factors that interact with these regulatory elements, has precluded studies that address whether there might be a central role for sub-nuclear organization in gene function. The advent of new microscopical techniques, however, has led to a renewed interest in chromatin organization in the context of sub-nuclear domains. Whereas a model of the nucleus as a membrane-bound, disordered collection of chromatin and other molecules was adequate during the molecular biology revolution of the past few decades, it is now appropriate to recognize the dynamic nature of chromatin through the cell cycle and its relation to the sub-nuclear architecture.

A textbook view of the nucleus is that of blocks of condensed chromatin are found throughout the nucleoplasm, with higher concentrations along the nuclear periphery and on the periphery of the nucleolus. This condensed so-called "heterochromatin" is thought to be predominantly transcriptionally inactive. Euchromatin, on the other hand, constituting transcriptionally active or potentially active genes transcribed by RNApolymerase II, is thought to be segregated to the surfaces of the blocks of heterochromatin. The less condensed chromatin on these surfaces would have access to regulatory complexes that move about the nucleoplasm through interchromatin channels (reviewed in Refs. [34,35]). The channels would also provide a passage for nascent transcripts to gain access to RNA processing enzymes and to permit interactions with complexes involved in transport of pre-messenger RNA to the cytoplasm via nuclear pores. Superimposed on this rather crude level of organization is the distribution of the DNA into chromosome-based territories. An example of this was demonstrated by fluorescence *in situ* hybridization (FISH) labeling of 42 positions on chromosome 2L in nuclei *of a Drosophila* embryo [36]. Nuclear envelope interactions were observed at discrete sites over the length of the chromosome, interactions that are established after telophase. The mapping revealed that a mechanism exists to specifically target loci either to the nuclear envelope or to the interior of the nucleus, since each locus occupies a well defined position in the interphase nucleus. Positioning of loci specifically to the interior or periphery of the nucleus indicates that long distance movements of chromosome segments can take place.

Following the movement of a pericentromeric locus in yeast clearly demonstrates that chromatin is capable of significant diffusional mobility, though the motion is constrained to radius of $\sim$0.25 μm [37]. In mammalian cells, specific chromosomal loci tagged with lac operator sequences demonstrate a range of mobilities [38]. Some loci are capable of high diffusional rates, though lower than those seen in yeast, indicating more resistance to movement, perhaps through interactions with a nuclear substructure. The radius of movement in human cells is approximately two-fold greater than that observed in yeast. Not all loci, however, are mobile. Those associated with either the nuclear periphery or the nucleolus are essentially immobile, though the latter can become mobile with nucleolur disruption by the action of the drug DRB. The structure of such nuclear compartments may impose steric limitations on large scale movement.

5.1. Nuclear positioning of transcribed genes

Precise nuclear positioning through the movement of loci has been demonstrated by generating an artificially amplified chromosome segment, which can be visualized in live cells [39]. The large amplified chromosome region, containing a lac operator/repressor for tagging, produces a homogeneously staining region (HSR). The locus undergoes a pattern of condensation and decondensation during interphase, and these changes are accompanied by reproducible changes in nuclear positioning of the HSR. For example, HSRs with condensed morphologies are predominantly associated with the nuclear envelope. More "open" HSRs, which are not replicating, assayed through BrdU incorporation, are also associated with the nuclear envelope. However, open HSRs that are incorporating BrdU, show a striking correlation with a location towards the interior of the nucleus. It is possible that the movement of the locus towards the interior is the result of loss of nuclear envelope contacts followed by diffusion to the interior. Alternatively, chromosome regions may be specifically targeted to specific intranuclear locations where replication occurs [40].

A non-random organization of chromatin may apply to transcribing chromatin as well as to chromatin involved in replication. If active genes are randomly distributed throughout the nucleoplasm, then this should be reflected in a random localization of highly acetylated chromatin. At the biochemical level, it appears that highly acetylated chromatin is organized differently than bulk chromatin. Nuclear fractionation demonstrates that highly acetylated chromatin fragments are enriched in an insoluble "nuclear matrix" preparation [41], and approximately 50% of histone acetyltransferase or histone deacetylase activities are also associated with an insoluble nuclear component [42]. Support for a non-random organization of transcriptionally active chromatin is strongly supported in direct imaging studies. A spatial relationship between intranuclear structures known as interchromatin granule clusters (IGCs) and transcribed loci has been observed [43–46]. The implication that transcription may be occurring on the periphery of IGCs, though not exclusively at these sites, is further supported by imaging experiments that employed antibodies directed against highly acetylated core histones H3 and H4. Highly acetylated chromatin is localized on the periphery of IGCs and histone acetyltransferases and histone deacetylases also show an enriched concentration at these sites (Fig. 6). These enzymes are functionally engaged at these sites, since inhibition of histone deacetylases with the drug trychostatin-A (TSA) causes a dramatic enrichment of acetylated chromatin on the periphery of IGCs. Deconvolution microscopy reveals a ring-like distribution of hyperacetylated chromatin on the immediate periphery of IGCs. Highly acetylated chromatin is also notably absent from the nuclear periphery and perinucleolar domains.

5.2. Role of sub-nuclear domains in establishing nuclear activity

Specific domains that are enriched in transcription activators and co-activators would lead to an accumulation of transcribed loci to these sites, through a random

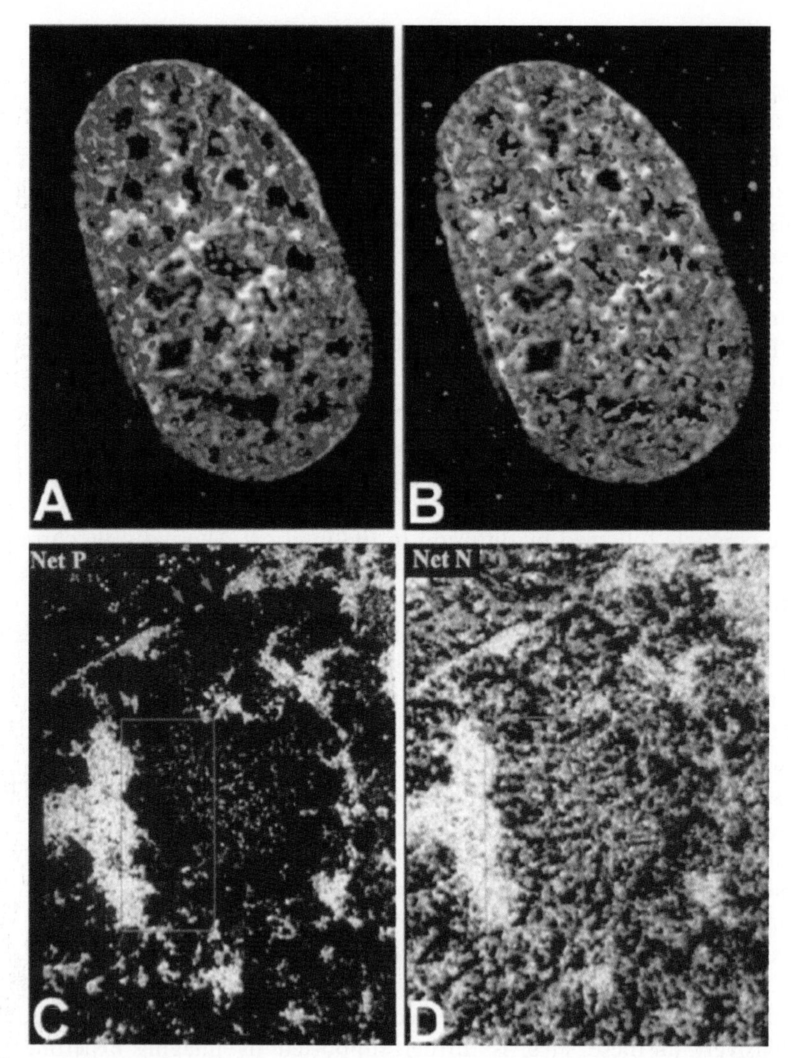

Fig. 6. Immuno-fluorescence and electron spectroscopic images of interchromatin granule clusters (IGCs). (A, B) Relative organization of the TAFII250 histone acetyltranferase and substrate chromatin in the fibroblast cell nucleus. A digital optical section of an Indian muntjac fibroblast cell nucleus co-stained with anti-TAFII250, antiacetylated histone H3, and DAPI is shown. Spatial relationship between DAPI-stained bulk chromatin (white) and highly acetylated chromatin. Images were subjected to thresholding, and the extent of overlap and independence was determined for the highly acetylated H3 distribution. Regions that are blue indicate regions where DAPI signal was detected. Regions that are red indicate regions where no DAPI signal was detected after digital deconvolution. Spatial relationship between the TAFII250 and chromatin distributions. The DAPI distribution is shown in white. Regions that contain TAFII250 but neither DAPI nor acetylated histone H3 are shown as green. Regions that contain both TAFII250 and highly acetylated histone H3 are shown as red, whereas regions that contain both TAFII250 and DAPI but no detectable acetylated histone H3 are shown as blue (from Ref. [48]). (C) Net phosphorus image of an IGC (white structures on black background). Blocks of chromatin surround the core or phosphorus-rich granules (RNP-based structures) in the interior. Small red arrows indicate regions of nuclear pores, seen as gaps in the condensed chromatin of the nuclear periphery. (D) Net nitrogen image of the same region in C. The core of the IGC contains protein-based structures in which the RNP-rich granules are embedded. Protein fibers appear to connect the granules in the interior and the chromatin on the periphery. Scale bar represents 5 μm in A and B, 350 nm in C and D.

slow diffusion of chromatin or through a directed movement of chromatin by molecular motors [47]. Indeed an accumulation of transcription regulatory molecules has been demonstrated at IGCs [48]. With antibodies directed against the $TAF_{II}250$ transcription factor and histone acetyltransferase, a large fraction of this protein, perhaps as much as 50% is localized in the interior of IGCs (Fig. 6). This is surprising since there is no DNA in the cores of these domains [49]. Electron spectroscopic imaging, however, reveals that the core of IGCs is composed of a protein-based structure that connects the cluster of RNA-rich granules in the interior with the chromatin domains on the periphery. The protein-based domain may have a propensity for accumulating factors involved in either transcription or pre-mRNA processing. PML (promyelocytic leukemia) nuclear bodies may also serve to concentrate regulatory factors that service transcriptional events on their surfaces [50]. Genes that are implicated in requiring PML nuclear bodies for expression are early viral genes [51–53], and possibly genes important for immune function, including MHC class I and II [54–56]. A parallel between PML bodies and IGCs is that the core of PML bodies can also accumulate regulatory factors. It is tempting to speculate that the concentration gradient of regulatory factors around nuclear bodies contributes to the recruitment of specific gene loci to these subnuclear domains.

A state of equilibrium of histone acetyltransferases (HATS) and deacetylases (HDACs) between a chromatin-bound domain and a protein-based domain in the nucleoplasm could provide a mechanism for regulating the overall level of chromatin acetylation in gene domains throughout interphase. Evidence that the state of histone acetylation in mitosis is regulated by compartmentalization rather than biochemical inactivation of HATS or HDACs is provided by an imaging study [57]. Although promoter-specific acetylation and deacetylation has received much attention, this study was directed at broader levels of chromatin acetylation as cells enter and exit mitosis. In contrast to the hypothesis that HATs and HDACs remain bound to mitotic chromosomes to provide an epigenetic imprint for post-mitotic reactivation of specific loci, HATS and HDACs were observed to reorganize spatially, and to be displaced from condensing chromatin throughout mitosis, beginning in prophase. Although HATS and HDACs are fully catalytically active in mitosis, when assayed *in vitro* after isolation from mitotic cells, they are unable to acetylate or deacetylate chromatin *in situ* because they have translocated to non-chromatin domain. The subsequent re-association of HATs and HDACs with chromatin in late telophase precedes a global restoration of chromatin acetylation, which itself precedes synthesis of nascent transcripts, monitored through fluorouridine incorporation.

Evidence for a mechanism to exchange regulatory factors between the chromatin template and a nucleoplasmic compartment has been provided from imaging transcriptionally active chromatin loci *in situ* in live cells. Such studies have provided much information on the dynamics of histones and regulatory factors, as well as the large-scale organization of the loci in the context of the nuclear environment.

5.3. Dynamics of exchange of regulatory factors with transcriptionally active genes

It is thought that families of transcription regulatory factors, including the steroid nuclear receptors, modulate transcription through the recruitment of basal transcription factors and co-activators/co-repressors, as well as chromatin-modifying and -remodeling activities. From binding analysis carried out *in vitro*, the prevailing view has emerged that a receptor binds to a cis-element and remains stably bound if the ligand is present. To determine whether this indeed occurs in the intact nucleus, a cell line was engineered that contains a large tandem array of the mouse mammary tumor virus (MMTV) promoter/ras reporter construct [58]. With approximately 200 copies of the viral long terminal repeat (LTR), there are approximately 1000 glucocorticoid receptor (GR) binding sites. This repeating locus is easily observed in the fluorescence microscope with the aid of an expressed GFP–GR fusion. The question addressed in this study was whether GR remains stably bound to the promoter DNA with ligand present, or does it participate in a hit-and-run mechanism. To address this, both fluorescence recovery after photo-bleaching (FRAP) and fluorescence loss in photo-bleaching (FLIP) techniques were employed to measure the dynamics of GFP–GR at the MMTV promoter.

The observation was that bleached GFP–GR is rapidly replaced with unbleached GFP–GR (FRAP) (Fig. 7), and if GFP–GR is bleached elsewhere in the nucleus, GFP–GR is replaced with non-fluorescent GR molecules (either bleached GFP–GR or endogenous unlabelled GR) on the array (FLIP). The conclusion from these observations is that GFP–GR exchanges at a high rate between the array promoters and the nucleoplasm. Observations made in live cells has led to a completely

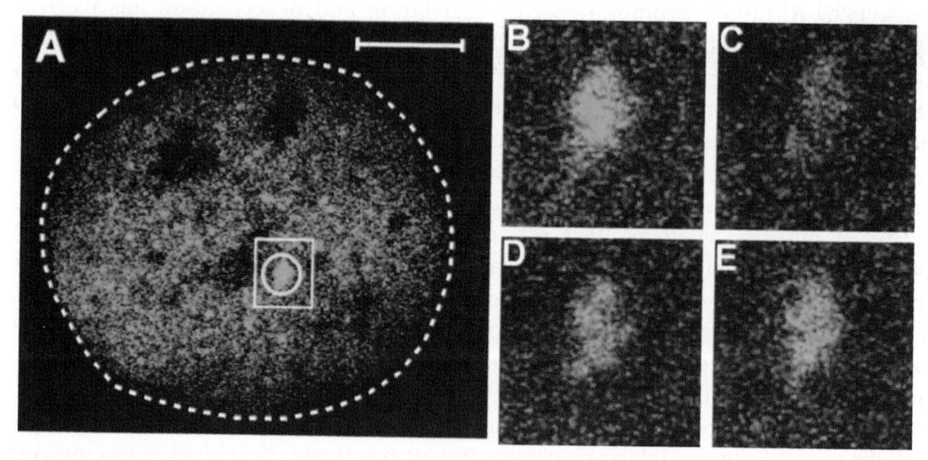

Fig. 7. GFP–GR bound to the MMTV array was analyzed by fluorescence recovery after photobleaching (FRAP). The bleached region is indicated in the image of the pre-bleached nucleus (A). The pre-bleach array is shown in (B), the post-bleach image (C), and at 4.1 s (D) and 11.6 s (E) post-bleach. This analysis, along with Fluorescence Loss in Photobleaching (FLIP) experiments, show that GFP–GR undergoes rapid exchange with the array (from Ref. [58]). Scale bar 5 μm.

new interpretation of previous, more traditional approaches. Apparently, DNase I footprints detected either *in vitro* or *in vivo* do not necessarily imply a continuous occupancy state. An imaging approach circumvents the need for cell disruption, including isolation of nuclei or chromatin extraction, which can interupt dynamic exchange processes. A rapidly exchanging model for GR action, and for other factors as well, opens up the possibility for other levels of fine tuning of promoter function. A rapidly exchanging nuclear receptor, for example, could be frequently exposed to post-translational machinery in the nucleoplasm, including a protein-based compartment, where modifications such as phosphorylation/dephosphorylation could take place. This would provide the promoter with a constantly up-graded receptor molecule with a modified activity or role at the promoter, independent of the constant presence of ligand.

In previous imaging studies, several observations indicated that a highly organized and dynamic sub-nuclear environment provides a framework for receptor function, in contrast to a model of activity based solely on freely diffusing molecules [59,60]. In a dynamic, live cell imaging study [61], the recruitment of co-activators, including a histone acetyltransferase (Creb-binding protein, CBP) was investigated. Estrogen receptor (ER) in the presence of ligand, is thought to recruit the steroid receptor co-activator 1 (SRC-1) and CBP, in order to increase transcription levels [62–64]. As a result, ER, SRC-1, and CBP are thought to exist in a large molecular complex. With the use of a lac operator array, it was shown that a CFP–LacER fusion could be targetted to the array [61]. In addition, YFP–SRC-1 and YFP–CBP could also be targetted to these arrays shortly after the addition of hormone. In contrast to the study described above [58], a FRAP experiment indicates that CFP–LacER is stably bound on the promoters of the array. YFP–SRC-1, on the other hand, initially exchange with the nucleoplasmic compartment, but later become more tightly bound. In contrast, YFP–CBP is never stably bound, and exchanges rapidly throughout the entire time course after hormone exposure. Once again, an imaging approach reveals a surprisingly high degree of molecular exchange with a component of the transcriptional assembly apparatus; in particular, a co-activator and potential chromatin modifying activity.

5.4. Imaging condensation/decondensation as a function of gene activity

Imaging techniques have frequently been employed to examine the relationship between the state of chromatin condensation and transcriptional activity. Condensed heterochromatin, for example, has been considered to be transcriptionally silent (reviewed in Ref. [65]). Likewise, decondensation of chromatin and transcription have been thought to be related because transcriptionally active genes are hypersensitive to nucleases [66,67], which implies not only a change in the 30 nm organization of chromatin, but also an alteration or disruption of the nucleosome itself. Live cell imaging permits the direct visualization of large-scale chromatin organization and re-organization as a function of transcriptional activation. Again, a tandem array of MMTV promoters were visualized by expression of a GFP–GR fusion protein [68]. As a function of time after hormone treatment, the number

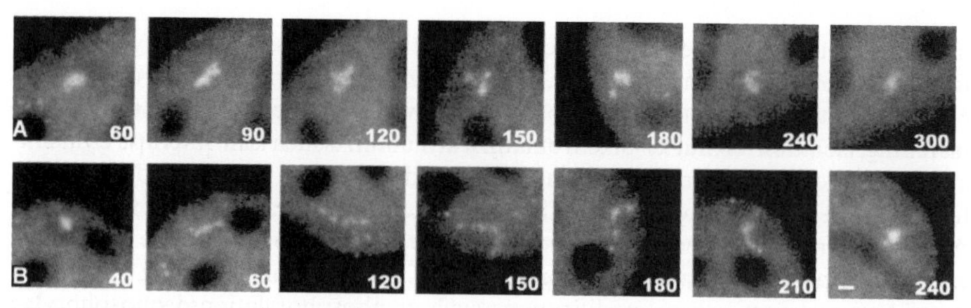

Fig. 8. Time lapse images of two MMTV arrays visualized with GFP–GR. Time in minutes after addition of 100 nM dexamethasone is indicated. Some arrays become very extended (3–10 μm, B), whereas some cells exhibit a less extended array (≤ 3 μm, A) (from Ref. [68]). Scale bar 1 μm.

of large, apparently decondensed arrays, increases in observations of hundreds of cells over the first three hours. This was confirmed in time-lapse imaging of single cells as a function of time after hormone addition, where the formation of decondensed arrays can be observed in the first few hours (Fig. 8). The largest arrays produce the highest levels of transcripts as determined by RNA FISH analysis. This correlation between the decondensation of the array and transcription could mean that transcription induces decondensation, or that decondensation facilitates transcription. To test the relationship, cells were treated with hormone and the transcription inhibitor DRB. The number of large arrays decreases with transcription inhibitor, indicating that transcription itself keeps the array in a more open or decondensed configuration. The decondensation that was observed with this approach is a transition from a pattern of thick, bright "blobs", to smaller beads connected with intervening strands, indicative of a transition to a linear unravelling into domains of variable DNA packing densities. Overall, a relatively high DNA packing density, of > 1300 was estimated, a value significantly greater than that observed in Balbiani puffs, where values < 10 have been estimated by electron microscopy [69,70]. Elucidation of the precise changes in chromatin structure that occur at these arrays will also have to make use of electron microscopy, in order to complete our structural understanding of this decondensed chromatin state.

Large-scale chromatin decondensation was also observed in an array construct to which the acidic activation domain (AAD) of the potent transcriptional activator VP 16 was targeted [71]. The array contains multiple copies of the lac operator, and a fusion of lac repressor with the VP 16 AAD was transiently expressed. Under such circumstances, the homogeneous staining region (HSR), otherwise seen as a very condensed domain, opens up into two types of decondensed configurations. One is a "ball-shaped" structure with peripheral localization of lac repressor, and the other is an open, fibrillar complex, similar to that described in other studies [61,68]. The former is probably the result of an exaggerated accumulation and self-association of nuclear proteins that are recruited to this site. The decondensation appears to occur in flanking regions between the VP16 targeting domains, propagating over distances of 100s and even 1000s of kilo bases. Such large-scale decondensation may support a chromonema fiber organization [72,73], but does not necessarily rule out loop domain organization models [74,75]. The protein-based architecture that would be required for the loop domains may be transient and susceptible to re-organization throughout interphase, and in response to transcriptional activation.

The VP16 AAD has been shown to function with the transcriptional co-activator GCN5 in yeast, and GCN5's transcriptional activity correlates with its histone acetyltransferase activity. It is thought, therefore, that VP16 functions, at least in part, by recruiting histone acetyltransferases. The chromatin opening of the lac operator HSR regions through the action of VP16 indeed correlates with the recruitment of P300, CBP, P/CAF molecules, all of which are potential histone acetyltransferases [71]. The decondensed state, however, is not dependent on transcription itself, since treatment with the transcriptional inhibitor α-amanitin, has no effect on the decondensation event itself or on maintenance of the decondensed state. This result differs drammatically from that obtained with the MMTV promoter array where transcription was required to cause a decondensed configuration [68]. The difference may be a reflection that different promoter systems may open differently, or that the unusual potentcy and concentration of the VP16 activator may create an artificial environment that does not reflect more typical situtations.

6. Summary

Direct visualization of chromatin structures, from single nucleosome particles to large-scale chromatin fiber in relatively unperturbed nuclei, has had a major impact on our understanding of how DNA is organized in eukaryotic cells. Without images of chromatin at all levels of organization, our interpretations of more indirect biochemical data would be impaired. We predict that advances in chromatin structure and function research will continue to rely on developments in imaging techniques.

Imaging of purified complexes studied *in vitro* will benefit from continued development of electron cryomicroscopy techniques. Better detectors and the removal of inelastically scattered electrons by energy filtering will improve the image contrast, thereby providing a means to examine single nucleosomes and complexes of nucleosomes with transcription and replication machinery, and with remodelling complexes. Improvements in the manufacture of tips for scanning force microscopes will also improve resolution, perhaps allowing the visualization of the DNA duplex in chromatin complexes visualized using *in vitro* approaches. The efforts in these microscopical approaches will pay off due to the advantage that they offer, of being able to visualize structures in a hydrated environment. The disadvantage of *in vitro* approaches are that the complexes that are studied after reconsitution may not represent the structures that are native in the living cell.

The potential of live cell imaging techniques, which overcome the drawbacks of *in vitro* techniques, is just beginning to be realized. A major advance that is required, however, is the generation of better fluorescent probes for tagging molecules of interest. Green fluorescent protein is relatively large in size, and thus may interfere sterically with some molecular interactions. Photo-damage through the by-products of fluorescence decay of these probes can also be toxic to living cells. Moreover, overexpression of fusion proteins can lead to accumulations that are not physiologically relevant, and may disrupt the nuclear events that are being studied. Therefore, better vectors for regulated expression of tagged molecules are also required.

We think that developments in both light and electron microscopy will make correlative imaging more routine and practical to carry out. The advantage of live cell imaging is that events that are rare or are transient can easily be detected. Searching for structures represented by rare or transient events at high resolution in the EM would be impossible without correlative approaches. The visualization of chromosome arrays under defined conditions of gene expression, for example, could be studied at much higher resolution than has so far been attempted. Thin sections imaged with electron energy loss spectroscopy would be possible without resolution-limiting heavy metal contrast agents. Moreover, with the advent of computer controlled stages in modern electron microscopes, it is more practical than in the past to do tomographic reconstructions routinely. By collecting images at various tilt angles of the specimen, three-dimensional information within the section is possible. Combining tomogaphy with ESI would permit the direct visualization and relationships of chromatin fiber, of nascent transcripts and of protein-based structures. In addition, the combination of these techniques with metal tagged antibodies or metal tags expressed as fusion proteins, will permit high resolution mapping of specific proteins in macromolecular complexes imaged *in situ*. The ultimate goal here, which we hope will be achieved in the next few years, is to identify and visualize macromolecular complexes and machines in a functionally active configuration on chromatin templates in the native nuclear environment.

References

1. Olins, A.L. and Olins, D.E. (1974) Spheroid chromatin units (v bodies). Science 25; 183(122), 330–332.
2. Thoma, F., Koller, T., and Klug, A. (1979) Involvement of histone H1 in the organization of the nucleosome and of the salt-dependent superstructures of chromatin. J. Cell. Biol. 83(2 Pt 1), 403–427.
3. Wansink, D.G., Sibon, O.C., Cremers, F.F., van Driel, R., and de Jong, L. (1996) Ultrastructural localization of active genes in nuclei of A431 cells. J. Cell. Biochem. 62(1), 10–18.
4. Fakan, S. and Bernhard, W. (1971) Localisation of rapidly and slowly labelled nuclear RNA as visualized by high resolution autoradiography. Exp. Cell. Res. 67(1), 129–141.
5. Luger, K., Rechsteiner, R.J., Flaus, A.J., Waye, M.M., and Richmond, T.J. (1997) Characterization of nucleosome core particles containing histone proteins made in bacteria. J Mol. Biol. 26; 272(3), 301–311.
6. Bazett-Jones, D.P. and Hendzel, M.J. (1999) Electron spectroscopic imaging of chromatin. Methods 17(2), 188–200.
7. Roussel, P., Andre, C., Masson, C., Geraud, G., and Hernandez-Verdun, D. (1993) Localization of the RNA polymerase I transcription factor hUBF during the cell cycle. J. Cell. Sci. 104(Pt 2), 327–337.
8. Biggiogera, M., Malatesta, M., Abolhassani-Dadras, S., Amalric, F., Rothblum, L.I., and Fakan, S. (2001) Revealing the unseen: the organizer region of the nucleolus. J. Cell. Sci. 114(Pt 17), 3199–3205.

9. Bustamante, C.D., Keller, D., and Yang, G. (1993) Scanning force microscopy of nucleic acids and nucleoprotein assemblies. Curr. Opin. Struct. Biol. 3, 363–372.
10. Vidali, G., Boffa, L.C., Bradbury, E.M., and Allfrey, V.G. (1978) Butyrate suppression of histone deacetylation leads to accumulation of multiacetylated forms of histones H3 and H4 and increased DNase I sensitivity of the associated DNA sequences. Proc. Natl. Acad. Sci. 75(5), 2239–2243.
11. Wade, P.A. and Wolffe, A.P. (1997) Histone acetyltransferases in control. Curr. Biol. 1; 7(2), R82–R84.
12. Bazett-Jones, D.P., Mendez, E., Czarnota, G.J., Ottensmeyer, F.P., and Allfrey, V.G. (1996) Visualization and analysis of unfolded nucleosomes associated with transcribing chromatin. Nucleic Acids Res. 15; 24(2), 321–329.
13. Oliva, R., Bazett-Jones, D.P., Locklear, L., and Dixon, G.H. (1990) Histone hyperacetylation can induce unfolding of the nucleosome core particle. Nucleic Acids Res. 18(9), 2739–2747.
14. Chen-Cleland, T.A., Boffa, L.C., Carpaneto, E.M., Mariani, M.R., Valentin, E., Mendez, E., and Allfrey, V.G. (1993) Recovery of transcriptionally active chromatin restriction fragments by binding to organomercurial-agarose magnetic beads. A rapid and sensitive method for monitoring changes in higher order chromatin structure during gene activation and repression. J. Biol. Chem. 5; 268(31), 23409–23416.
15. Bazett-Jones, D.P., Cote, J., Landel, C.C., Peterson, C.L., and Workman, J.L. (1999a) The SWI/SNF complex creates loop domains in DNA and polynucleosome arrays and can disrupt DNA–histone contacts within these domains. Mol. Cell. Biol. 19(2), 1470–1478.
16. Orphanides, G., LeRoy, G., Chang, C.H., Luse, D.S., and Reinberg, D. (1998) FACT, a factor that facilitates transcript elongation through nucleosomes. Cell 9; 92(1), 105–116.
17. Bednar, J., Studitsky, V.M., Grigoryev, S.A., Felsenfeld, G., and Woodcock, C.L. (1999) The nature of the nucleosomal barrier to transcription: direct observation of paused intermediates by electron cryomicroscopy. Mol. Cell. 4(3), 377–386.
18. Clark, D.J. and Feisenfeld, G. (1991) Formation of nucleosomes on positively supercoiled DNA. EMBO J. 10(2), 387–395.
19. Studitsky, V.M., Clark, D.J., and Felsenfeld, G. (1994) A histone octamer can step around a transcribing polymerase without leaving the template. Cell 76, 371–382.
20. ten Heggeler-Bordier, B., Schild-Poulter, C., Chapel, S., and Wahli, W. (1995) Fate of linear and supercoiled multinucleosomic templates during transcription. EMBO J. 14(11), 2561–2569.
21. Kanda, T., Sullivan, K.F., and Wahl, G.M. (1998) Histone–GFP fusion protein enables sensitive analysis of chromosome dynamics in living mammalian cells. Curr. Biol. 26; 8(7), 377–385.
22. Kimura, H. and Cook, P.R. (2001) Kinetics of core histones in living human cells: little exchange of H3 and H4 and some rapid exchange of H2B. J. Cell. Biol. 5; 153(7), 1341–1353.
23. Lever, M.A., Th'ng, J.P., Sun, X., and Hendzel, M.J. (2000) Rapid exchange of histone H1.1 on chromatin in living human cells. Nature 14; 408(6814), 873–876.
24. Misteli, T., Gunjan, A., Hock, R., Bustin, M., and Brown, D.T. (2000) Dynamic binding of histone H1 to chromatin in living cells. Nature 408(6814), 877–881.
25. Williams, S.P., Athey, B.D., Muglia, L.J., Schappe, R.S., Gough, A.H., and Langmore, J.P. (1986) Chromatin fiber are left-handed double helices with diameter and mass per unit length that depend on linker length. Biophys. J. 49(1), 233–248.
26. Woodcock, C.L., Frado, L.L., and Rattner, J.B. (1984) The higher-order structure of chromatin: evidence for a helical ribbon arrangement. J. Cell. Biol. 99(1 Pt 1), 42–52.
27. Horowitz, R.A., Agard, D.A., Sedat, J.W., and Woodcock, C.L. (1994) The three-dimensional architecture of chromatin *in situ*: electron tomography reveals fiber composed of a continuously variable zig-zag nucleosomal ribbon. J. Cell. Biol. 125(1), 1–10.
28. Woodcock, C.L., Grigoryev, S.A., Horowitz, R.A., and Whitaker, N. (1993) A chromatin folding model that incorporates linker variability generates fiber resembling the native structures. Proc. Natl. Acad. Sci. 1; 90(19), 9021–9025.
29. Leuba, S.H., Yang, G., Robert, C., Samori, B., van Holde, K., Zlatanova, J., and Bustamante, C. (1994) Three-dimensional structure of extended chromatin fibers as revealed by tapping mode scanning force microscopy. Proc. Natl. Acad. Sci. 22; 91(24), 11621–11625.

30. Bednar, J., Horowitz, R.A., Grigoryev, S.A., Carruthers, L.M., Hansen, J.C., Koster, A.J., and Woodcock, C.L. (1998) Nucleosomes, linker DNA, and linker histone form a unique structural motif that directs the higher-order folding and compaction of chromatin. Proc. Natl. Acad. Sci. 24; 95(24), 14173–14178.
31. McGhee, J.D., Nickol, J.M., Felsenfeld, G., and Rau, D.C. (1983) Higher order structure of chromatin: orientation of nucleosomes within the 30 nm chromatin solenoid is independent of species and spacer length. Cell 33(3), 831–841.
32. Leuba, S.H., Bustamante, C., Zlatanova, J., and van Holde, K. (1998) Contributions of linker histones and histone H3 to chromatin structure: scanning force microscopy studies on trypsinized fibers. Biophys. J. 74(6), 2823–2829.
33. Leuba, S.H., Bustamante, C., van Holde, K., and Zlatanova, J. (1998) Linker histone tails and N-tails of histone H3 are redundant: scanning force microscopy studies of reconstituted fibers. Biophys. J. 74(6), 2830–2839.
34. Kruhlak, M.J., Lever, M.A., Fischle, W., Verdin, E.., Bazett-Jones, D.P., and Hendzel, M.J. (2000) Reduced mobility of the alternate splicing factor (ASF) through the nucleoplasm and steady state speckle compartments. J. Cell. Biol. 10; 150(1), 41–51.
35. Hendzel, M.J., Boisvert, F., and Bazett-Jones, D.P. (1999) Direct visualization of a protein nuclear architecture. Mol. Biol. Cell. 10(6), 2051–2062.
36. Marshall, W.F., Dernburg, A.F., Harmon, B., Agard, D.A., and Sedat, J.W. (1996) Specific interactions of chromatin with the nuclear envelope: positional determination within the nucleus in Drosophila melanogaster. Mol. Biol. Cell. 7(5), 825–842.
37. Marshall, W.F., Straight, A., Marko, J.F., Swedlow, J., Dernburg, A., Belmont, A., Murray, A.W., Agard, D.A., and Sedat, J.W. (1997) Interphase chromosomes undergo constrained diffusional motion in living cells. Curr. Biol. 7(12), 930–939.
38. Chubb, J.R., Boyle, S., Perry, P., and Bickmore, W.A. (2002) Chromatin motion is constrained by association with nuclear compartments in human cells. Curr. Biol. 12(6), 439–445.
39. Li, G., Sudlow, G., and Belmont, A.S. (1998) Interphase cell cycle dynamics of a late-replicating, heterochromatic homogeneously staining region: precise choreography of condensation/decondensation and nuclear positioning. J. Cell. Biol. 9; 140(5), 975–989.
40. Jackson, D.A. and Cook, P.R. (1995) The structural basis of nuclear function. Int. Rev. Cytol. 162A, 125–149.
41. Hendzel, M.J., Delcuve, G.P., and Davie, J.R. (1991) Histone deacetylase is a component of the internal nuclear matrix. J. Biol. Chem. 15; 266(32), 21936–21942.
42. Davie, J.R. (1996) Histone modifications, chromatin structure and the nuclear matrix. J. Cell. Biochem. 62(2), 149–157.
43. Fakan, S. and Nobis, P. (1978) Ultrastructural localization of transcription sites and of RNA distribution during the cell cycle of synchronized CHO cells. Exp. Cell. Res. 113(2), 327–337.
44. Huang, S. and Spector, D.L. (1991) Nascent pre-mRNA transcripts are associated with nuclear regions enriched in splicing factors. Genes. Dev. 5(12A), 2288–2302.
45. Clemson, C.M., McNeil, J.A., Willard, H.F., and Lawrence, J.B. (1996) XIST RNA paints the inactive X chromosome at interphase: evidence for a novel RNA involved in nuclear/chromosome structure. J. Cell. Biol. 132(3), 259–275.
46. Smith, K.P., Moen, P.T., Wydner, K.L., Coleman, J.R., and Lawrence, J.B. (1999) Processing of endogenous pre-mRNAs in association with SC-35 domains is gene specific. J. Cell. Biol. 22; 144(4), 617–629.
47. Heun, P., Laroche, T., Shimada, K., Furrer, P., and Gasser, S.M. (2001) Chromosome dynamics in the yeast interphase nucleus. Science. 7; 294(5549), 2181–2186.
48. Hendzel, M.J., Kruhlak, M.J., and Bazett-Jones, D.P. (1998) Organization of highly acetylated chromatin around sites of heterogeneous nuclear RNA accumulation. Mol. Biol. Cell. 9(9), 2491–2507.
49. Thiry, M. (1995) Behavior of interchromatin granules during the cell cycle. Eur. J. Cell. Biol. 68(1), 14–24.
50. Boisvert, F.M., Hendzel, M.J., and Bazett-Jones, D.P. (2000) Promyelocytic leukemia (PML) nuclear bodies are protein structures that do not accumulate RNA. J. Cell. Biol. 24; 148(2), 283–292.

51. Guldner, H.H., Szostecki, C., Grotzinger, T., and Will, H. (1992) IFN enhance expression of Sp100, an autoantigen in primary biliary cirrhosis. J. Immunol. 15; 149(12), 4067–4473.
52. Chelbi-Alix, M.K., Pelicano, L., Quignon, F., Koken, M.H., Venturini, L., Stadler, M., Pavlovic, J., and de Degos, L. (1995) The H. Induction of the PML protein by interferons in normal and APL cells. Leukemia 9(12), 2027–2033.
53. Ishov, A.M., Stenberg, R.M., and Maui, G.G. (1997) Human cytomegalovirus immediate early interaction with host nuclear structures: definition of an immediate transcript environment. J. Cell. Biol. 14 138(1), 5–16.
54. Kretsovali, A., Agalioti, T., Spilianakis, C., Tzortzakaki, E., Merika, M., and Papamatheakis, J. (1998) Involvement of CREB binding protein in expression of major histocompatibility complex class II genes via interaction with the class II transactivator. Mol. Cell. Biol. 18(11), 6777–6783.
55. Zheng, P., Guo, Y., Niu, Q., Levy, D.E., Dyck, J.A., Lu, S., Sheiman, L.A., and Liu, Y. (1998) Proto-oncogene PML controls genes devoted to MHC class I antigen presentation. Nature 26; 396(6709), 373–376.
56. Shiels, C., Islam, S.A., Vatcheva, R., Sasieni, R., Sternberg, M.J., Freemont, P.S., and Sheer, D. (2001) PML bodies associate specifically with the MHC gene cluster in interphase nuclei. J. Cell. Sci. 114(Pt 20), 3705–3716.
57. Kruhlak, M.J., Hendzel, M.J., Fischle, W., Bertos, N.R, Hameed, S., Yang, X.J., Verdin, E., and Bazett-Jones, D.P. (2001) Regulation of global acetylation in mitosis through loss of histone acetyltransferases and deacetylases from chromatin. J. Biol Chem. 12; 276(41), 38307–38319.
58. McNally, J.G., Muller, W.G., Walker, D., Wolford, R., and Hager, G.L. (2000) The glucocorticoid receptor: rapid exchange with regulatory sites in living cells. Science 18; 287(5456), 1262–1265.
59. Stenoien, D., Sharp, Z.D., Smith, C.L., and Mancini, M.A. (1998) Functional subnuclear partitioning of transcription factors. J. Cell. Biochem. 1; 70(2), 213–221.
60. Stein, G.S., van Wijnen, A.J., Stein, J.L., Lian, J.B., Pockwinse, S., and McNeil, S. (1998) Interrelationships of nuclear structure and transcriptional control: functional consequences of being in the right place at the right time. J. Cell. Biochem. 1; 70(2), 200–212.
61. Stenoien, D.L., Nye, A.C., Mancini, M.G., Patel, K., Dutertre, M., O'Malley, B.W., Smith, C.L., Belmont, A.S., and Mancini, M.A. (2001) Ligand-mediated assembly and real-time cellular dynamics of estrogen receptor alpha-coactivator complexes in living cells. Mol. Cell. Biol. 21(13), 4404–4412.
62. Onate, S.A., Tsai, S.Y., Tsai, M.J., and O'Malley, B.W. (1995) Sequence and characterization of a coactivator for the steroid hormone receptor superfamily. Science 270(5240), 1354–1357.
63. Chakravarti, D., LaMorte, V.J., Nelson, M.C., Nakajima, T., Schulman, I.G., Juguilon, H., Montminy, M., and Evans, R.M. (1996) RM. Role of CBP/P300 in nuclear receptor signalling. Nature 5; 383(6595), 99–103.
64. Kamei, Y., Xu, L., Heinzel, T., Torchia, J., Kurokawa, R., Gloss, B., Lin, S.C., Heyman, R.A., Rose, D.W., Glass, C.K., and Rosenfeld, M.G. (1996) A CBP integrator complex mediates transcriptional activation and AP-1 inhibition by nuclear receptors. Cell. 3; 85(3), 403–414.
65. Hennig, W. (1999) Heterochromatin. Chromosoma 108(1), 1–9.
66. Weintraub, H. and Groudine, M. (1976) Chromosomal subunits in active genes have an altered conformation. Science 3; 193(4256), 848–856.
67. Wood, W.I. and Felsenfeld, G. (1982) Chromatin structure of the chicken beta-globin gene region. Sensitivity to DNase I, micrococcal nuclease, and DNase II. J. Biol. Chem. 10; 257(13), 7730–7736.
68. Müller, W.G., Walker, D., Hager, G.L., and McNally, J.G. (2001) Large-scale chromatin decondensation and recondensation regulated by transcription from a natural promoter. J. Cell. Biol. 9; 154(1), 33–48.
69. Lamb, M.M. and Daneholt, B. (1979) Characterization of active transcription units in Balbiani rings of Chironomus tentans. Cell. 17(4), 835–848.
70. Olins, D.E., Olins, A.L., Levy, H.A., Durfee, R.C., Margle, S.M., Tinnel, E.P., and Dover, S.D. (1983) Electron microscope tomography: transcription in three dimensions. Science 29; 220(4596), 498–500.

71. Tumbar, T., Sudlow, G., and Belmont, A.S. (1999) Large-scale chromatin unfolding and remodeling induced by VP16 acidic activation domain. J. Cell. Biol. Jun 28; 145(7), 1341–1354.
72. Belmont, A.S., Braunfeld, M.B., Sedat, J.W., and Agard, D.A. (1989) Large-scale chromatin structural domains within mitotic and interphase chromosomes *in vivo* and *in vitro*. Chromosoma 98(2), 129–143.
73. Belmont, A.S. and Bruce, K. (1994) Visualization of G1 chromosomes: a folded, twisted, supercoiled chromonema model of interphase chromatid structure. J. Cell. Biol. 127(2), 287–302.
74. Manuelidis, L. (1990) A view of interphase chromosomes. Science. 14; 250(4987), 1533–1540.
75. Boy de la Tour, E. and Laemmli, U.K. (1988) The metaphase scaffold is helically folded: sister chromatids have predominantly opposite helical handedness. Cell. 23; 55(6), 937–944.

J. Zlatanova and S.H. Leuba (Eds.) *Chromatin Structure and Dynamics: State-of-the-Art*

DOI: 10.1016/S0167-7306(03)39014-3

CHAPTER 14

Chromatin structure and dynamics: lessons from single molecule approaches

Jordanka Zlatanova[1] and Sanford H. Leuba[2]

[1]*Department of Chemical and Biological Sciences and Engineering, Polytechnic University, 6 MetroTech Center, Brooklyn, NY 11201, USA. Tel.: 718-260-3176; Fax: 786-524-5899; E-mail: jzlatano@duke.poly.edu*

[2]*Department of Cell Biology and Physiology, University of Pittsburgh School of Medicine, Hillman Cancer Center, UPCI Research Pavilion, Pittsburgh, PA 15213, USA. Tel.: 412-623-7788; Fax: 412-623-4840; E-mail: leuba@pitt.edu*

1. Introduction

The recent resurrection of interest in chromatin structure and dynamics stems from our understanding of the roles chromatin plays in regulation of processes that need access to the genetic information stored in the DNA double helix: transcription, replication, repair, and recombination. For the DNA to be accessible to the enzymatic machineries involved in all these processes, the compacted chromatin fiber has to undergo unraveling [1], followed by temporary removal of the histone octamers from the DNA. Regulation of gene expression at the level of transcription apparently also involves some kind of dynamic alterations to the structure [2–4] so as to allow binding of sequence-specific transcription factors to their recognition sequences.

The emergence of single-molecule methods [5,6] has provided a powerful set of tools to approach chromatin structure and dynamics in an unprecedented way, allowing real-time observations on the behavior of individual chromatin fibers, and assessing the variability among individual representatives of a fiber population. These approaches will undoubtedly help us understand the dynamic changes in chromatin during processes like transcription and replication and assess the contributions of post-synthetic modifications of both the chromatin proteins and the DNA to the regulation of gene expression. The majority of the single-molecule chromatin work has been done using the atomic force microscope (AFM), both for visualization and micromanipulation, but recently several other single-molecule techniques have proved useful: optical tweezers, magnetic tweezers, and fluorescence videomicroscopy. Here we will discuss the first meaningful results obtained through AFM imaging and through mechanically manipulating single chromatin fibers, comparing the different techniques in terms of their capabilities and limitations.

2. *Atomic force microscope imaging of chromatin fibers*

The AFM [7] uses a sharp tip mounted at the end of a flexible cantilever to probe a number of properties of the sample, including its topographical features and its mechanical characteristics. Interaction forces, both attractive and repulsive, between atoms on the AFM tip and atoms on the sample cause deflections of the flexible cantilever (for a detailed description of the interaction forces sensed by the AFM, see Ref. [8]). These deflections are registered by a laser beam reflected off of the back of the cantilever onto a photodiode position detector (Fig. 1a);

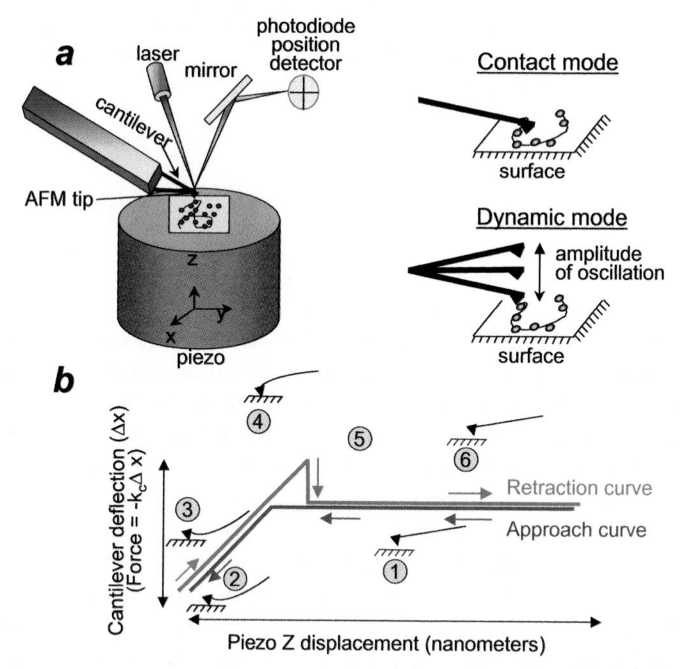

Fig. 1. The Atomic Force Microscope. (*a*) Schematic (see text). Enlargements on the right-hand side are schematics of the contact and dynamic modes of action of the cantilever. In both modes, the tip/cantilever raster-scans the biological sample on an atomically flat surface. In the contact mode, the tip is in continuous contact with the surface, and a topographic image is created from the changes in the laser signal caused by the deflections of the cantilever, which, in turn, are caused by tip/sample interactions. In the dynamic mode, the cantilever is oscillated at a certain frequency above the sample. Tip/sample interactions reduce the amplitude of the freely oscillating cantilever; this dampening of oscillation is transformed into a topographical image. The dynamic mode is preferred for imaging of soft biological samples as it diminishes the deformation inflicted to the sample by the dragging tip. (*b*) schematic of a typical force-extension curve. In the approach curve, the tip/cantilever has not yet come into contact with the surface: as a result, there is no deflection of the cantilever (position 1). As the tip engages with the surface and is pushed into it, the cantilever is deflected (position 2). The retraction curve starts at position 3, with the cantilever beginning to gradually straighten. If the tip sticks to the surface (as in position 4), then the cantilever is deflected in the opposite direction. At some point (position 5), the retracting tip 'snaps off' the surface (see vertical downward arrow and the dramatic change in the deflection depicted on the y-axis). The retraction curve ends just as the approach curve had started, with the tip not touching the surface, the cantilever straight, and zero deflection recorded on the y-axis.

the signal from the position detector is further used to produce either a topographic image (the probe is raster-scanned in the x–y direction), or the so-called force curves (the probe is moved in the z-direction only, upwards and downwards) (Fig. 1b). The AFM provides unique possibilities to image soft biological materials with a nanometer lateral resolution under conditions much less harsh than those needed for electron microscopy (EM) imaging [9]. The AFM is unique among high-resolution imaging methods in that it allows visualization of individual macromolecules and their complexes under nearly physiological conditions, i.e., in buffers. Imaging in air is done under ambient humidity conditions, usually higher than 30% relative humidity, which allows the preservation of a thin layer of liquid water above the sample. Thus, even air-imaged samples are certain to contain at least the structurally important water molecules bound to the macromolecules. The AFM can also be used to mechanically stretch macromolecules, an application to which we will return later.

2.1. *AFM assessment of chromatin organization*

AFM imaging of nucleosomes and chromatin fibers has contributed significantly to our understanding of how chromatin is structured. A list of the most significant papers published since 1992, the year of the first reported chromatin AFM images [10] is presented in Table 1. The imaged samples range from fibers spread out of hypotonically-lysed nuclei to isolated soluble chromatin fibers, to nucleosome arrays reconstituted from naked DNA templates and core histones (with or without linker histones, LH), to isolated or reconstituted mono- or di-nucleosomal particles. The most important results and conclusions from all this work will be briefly discussed in the following paragraphs.

In a series of three papers [11–13], we have studied chromatin fibers solubilized from chicken erythrocyte nuclei by micrococcal nuclease treatment. The fibers were deposited on mica or glass from low ionic strength solutions, briefly rinsed with water and fluxed with nitrogen gas. In the absence of added salt, the fibers exhibited a loose, rather irregular, three-dimensional organization of individual, well-resolved nucleosomes (Fig. 2a); beads-on-a-string fibers were observed only in LH-depleted fibers (Fig. 2b). In an attempt to elucidate the molecular determinants of such irregular fibers, mathematical models were built based on known structural parameters, and assuming straight conformation of linker DNA (Fig. 2c). (Similar modeling was first reported by Woodcock *et al.* [14], with very similar outcome.) In order to relate these mathematical models to actual AFM images, the modeled fibers were transformed into simulated AFM images by first simulating the surface deposition process, and then introducing the broadening effect of the AFM tip. Such simulations resulted in "fake" AFM images of the modeled fibers that were very similar to actual AFM images of real chromatin fibers (for details see Refs. [12,13,15]). These similarities between models and actual images gave credence to our views that chromatin fibers were organized in a much more irregular way than envisaged by the canonical

Table 1
Major results and conclusions from AFM imaging

Imaged substrate	Imaging conditions	Major results	Refs.
Chicken erythrocyte (CE)[a] chromatin	Mica Contact mode Air/propanol	First beads-on-a-string AFM images	10
Nucleosomal arrays reconstituted from histone octamers and 208-18[b] DNA	Glass Contact mode Air	Beads-on-a-string morphology; center-to-center distances of ~37 nm	36
Hypotonically spread CE nuclei	Glass Contact mode Air	Beads-on-a-string morphology; center-to-center distances of ~30 nm	84
CE chromatin fibers at different salt concentrations	Mica Tapping mode Air	Loose, three-dimensional, 30 nm structures even in the absence of salt; At 10 mM NaCl the fiber condenses slightly; At 80 mM NaCl highly compacted, irregularly segmented fibers	11
Unfixed or glutaraldehyde-fixed CE chromatin fibers; native or LH-depleted	Mica/glass Tapping mode Air	Fibers reveal three-dimensional irregular organization at low ionic strength; Beads-on-a-string fibers seen only in H1/H5-depleted fibers; Theoretical models of chromatin fibers and their simulated AFM images support the view that extended chromatin fibers are irregular three-dimensional arrays of nucleosomes connected by straight linkers	12
CE chromatin fibers, native, or LH-depleted	Mica Tapping mode Air	Comparisons between simulated AFM images of theoretical models and actual AFM images of native or LH-depleted fibers indicate that LH removal leads to a partial release of DNA from the histone core and corresponding lengthening of linker DNA: thus, LH close the nucleosome, fixing the entry-exit angle of DNA	13
Hypotonically spread CE nuclei; Detergent-treated nuclei from human B lymphocytes; native, dry, or rehydrated samples	Glass Contact mode Air or buffer	The hypotonic spreads are beads-on-a-string; The detergent spreads are supranucleosomal chains; Spatial resolution is reduced in liquid even on glutaraldehyde-fixed samples	85

rDNA minichromosomes from *Tetrahymena thermophila*	Mica Tapping mode Air at less than 10% relative humidity	Condensed 30 nm fibers near center of mica; Extended fibers at the mica periphery with partially dissociated nucleosomes; clusters of smaller particles within these nucleo somes suggested to be individual histone molecules	28
Hypotonically spread CE nuclei	Glass Contact mode Air	Beads-on-a-string morphology; Image processing (extraction of cross-sections of nucleosomes at half-maximum height) reveals ellipsoid shape of nucleosomes with an aspect ratio of 1.2–1.4 and a relatively smooth perimeter	30
Hypotonically spread CE nuclei	Glass Contact mode Air	Image processing as above; The orientation of the virtual ellipsoid cross-sections of nucleosomes was correlated with the direction of the fiber axis, with $>50\%$ of nucleosomes aligned with the axis (could be partly due to interaction with glass and/or drying)	31
Hypotonically spread CE nuclei	Mica Tapping mode Air	Beads-on-a-string coexisting with superbeads on same fibers. Claim of occasional solenoidal conformation (JZ and SHL are not convinced of this interpretation).	86
Isolated di- and trinucleosomes; Hyper-acetylated CE chromatin fibers isolated in the presences of the histone deacetylase inhibitor sodium butyrate	Glass/mica Tapping mode Air/Buffer	20 nm for center to center distances for dinucleosomes in 5 mm triethanolamine supplemented with 0 or 20 mM NaCl. Median DNA entry–exit angle for trinucleosomes containing linker histone was 86 °, whereas without linker histone it was 101°. Fibers prepared in butyrate appeared better adhere to glass coverslip, lending themselves to more stable AFM imaging	87
Progressively trypsinized CE chromatin fibers	Mica Tapping mode Air	Cleavage of LH tails is associated with fiber lengthening whereas cleavage of H3 N-tails results in fiber flattening; Zig-zag fiber morphology persists at later stages of digestion and is attributed to retention of the globular domain of LH in fiber	43
Reconstitution of CE chromatin fibers depleted of LH or of LH and the N-tails of H3 with either intact H5 or its isolated globular domain	Mica Tapping mode Air	The three-dimensional organization of nucleosomes in extended (low ionic strength) chromatin fibers requires the globular domain of LHs and either the tails of LH or the N-terminal tails of H3	44
LH-stripped mono-, di-, and oligonucleosomes from CE	Mica Tapping mode Air	Occasional visualization of the DNA wrapped around the histone octamer and of linker DNA; Occasional superbeads observed	29

(*Continued*)

Table 1
Continued

Imaged substrate	Imaging conditions	Major results	Refs.
Hypotonically spread CE nuclei	Mica Tapping mode Air	AFM can visualize nucleosome positioning; Addition of H1 reportedly compacts the dinucleosome; suggested stem-structure formation by H1	88
Subsaturated reconstituted nucleosomal arrays from 208-12 DNA fragment and HeLa control or hyperacetylated core histones	Mica Tapping mode Air	Beads-on-a-string; Frequency distributions of number of nucleo somes/template vary with average nucleosome loading: between 4 and 8 nucleosomes/template the distributions are broader than random and contain peaks/shoulders, indicating a tendency for correlated nucleosome loading along templates; Hyperacetylated reconstitutes show slightly broader distributions; In highly loaded reconstitutes, hyperacetylated arrays look less compact	37
Subsaturated reconstituted nucleosomal arrays from 172-12 DNA fragment and HeLa control or hyperacetylated core histones	Mica Tapping mode Air	The mid-range non-random features of frequency distributions (see Ref. [37], above) are similar for short-repeat length arrays (172-12) reconstituted with control histones and long repeat-length arrays (208-12) containing hyperacetylated histones; Short-repeat length and acetylation lead to similar changes in chromatin structure; It is thermodynamically more difficult for hyperacetylated histones than for control histones to assemble into subsaturated arrays	38
Chromatin fibers from control or poly(ADP-ribosyl)ated CE nuclei; *In vitro* poly(ADP-ribosyl)ated fibers	Mica Tapping mode Air	Poly(ADP-ribosyl)ation induces decondensation of chromatin structure which remains significantly decondensed even in the presence of Mg ions: Mg ions cannot substitute for linker histones to induce compaction	50

HeLa mononucleosomes; Nucleosome arrays reconstituted from modified 208-12 and core histones; The arrays remodeled with hSWI/SNF	Mica and carbon nanotube tips Tapping mode Air	Dimers form from mononucleosomes with ~60 bp more weakly bound by histones than in control mononucleosomes; Control arrays with evenly spaced nucleosomes are disorganized by SWI/SNF; compact dimers within these arrays could not be positively identified	39
Chromatin fibers isolated form cells with normal or elevated levels of m^5C; Nucleosome arrays reconstituted from either unmethylated or *in vitro* methylated 208-12 and core histones; additional reconstitution of LH	Mica Tapping mode Air	DNA methylation induced chromatin fiber compaction only in the presence of bound LH; AFM results substantiated by MNase digestion patterns and sucrose gradient centrifugation; AFM imaging can visualize alternative nucleosome positioning on adjacent 208-bp repeats (the distribution of center-to-center distances on 208-12 is bimodal)	40
Nucleosome arrays reconstituted from 208-18 and either histone octamers, H3/H4 tetramers or the histone-fold protein HMf from *Archaea*	Mica Tapping mode Air	The HMf-nucleoprotein complexes can be considered *bona fide* chromatin structures; The HMf-containing mononucleosomes are less stable than the canonical octasomes	41
Nucleosomal arrays reconstituted from a 5.4 kbp circular template and control or totally-tailless recombinant histones	Mica Tapping mode Air	Beads-on-a-string; Center-to-center distance frequency distributions indistinguishable for the control and tailless reconstitutes	42

[a]CE, chicken erythrocyte

[b]208-18, a tandemly repeated DNA sequence that has a nucleosome positioning 208-bp-sequence repeated 18 times [89].

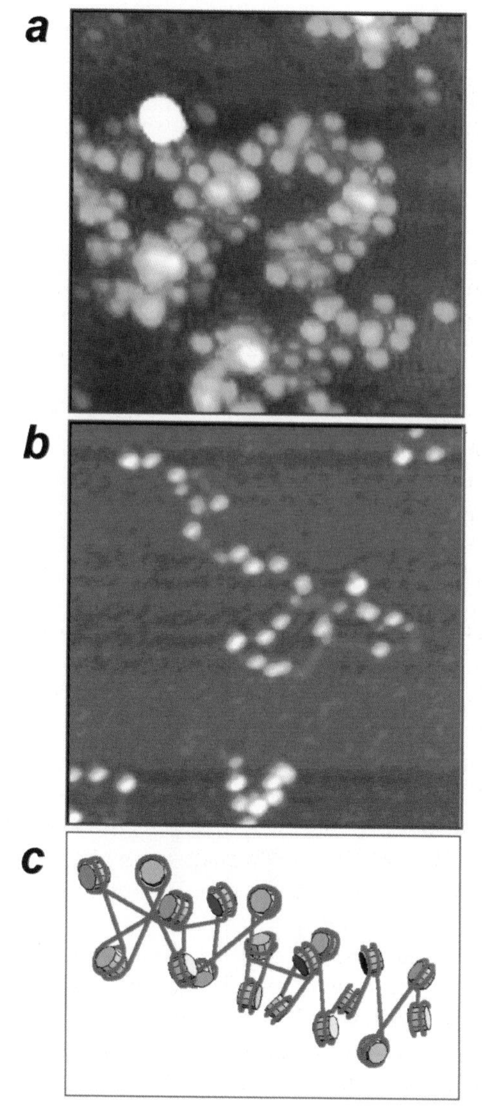

Fig. 2. AFM images of chromatin fibers. The fibers were deposited on glass or mica and imaged in air [12]. (*a*) Chicken erythrocyte chromatin fiber solubilized by micrococcal nuclease treatment of isolated crude nuclei. (*b*) Same fibers following LH depletion. Image sizes in *a* and *b* are 400 nm × 400 nm with the height depicted in color on a range from 0 to 15 nm. (*c*) Mathematical model of a chromatin fiber [12]. This three-dimensional geometric model was generated by using the dimensions of the nucleosome core particle from the crystal structure (1.75 superhelical wraps of 146 bp DNA around the histone octamer) combined with straight linker DNA stretches whose length was allowed to vary randomly within the range between 51 and 72 bp (this range reflects the biochemically determined variability in linker length in chicken erythrocyte chromatin fibers). The angle between successive linkers was maintained at 45° to reflect the contribution of linker histone maintaining this angle. Using these constraints, irregular three-dimensional chromatin fibers are generated, in agreement with earlier modeling [14] and the cryo-EM imaging experiments [19].

solenoidal model [16,17], and that the linker DNA is most probably straight, giving extended fibers a zig-zag appearance (for discussion on the straight linker issue, see Ref. [1]).

It is worthwhile noting that although these views on the organization of the chromatin fiber are still not universally accepted, very similar notions have been put forward by Woodcock's laboratory, on the basis of both mathematical modeling [14] and direct cryo-EM observations [18,19]. That similar images have been obtained by both AFM and cryo-EM is of particular importance since this negates the possibility that the irregular fiber morphology in AFM images may be an artifact caused by sample/surface interactions (in cryo-EM, the sample is freely suspended in vitrified water, so such surface artifacts are excluded [20]).

In addition, these early AFM reports helped to define the structural role of LHs in fixing the entry/exit angle of DNA at the nucleosome. Exactly how the LHs perform this function is not clear, especially in view of the recently reported stem structure formed at the entry-exit point of the nucleosome by LH binding [19,21]. In this structure, the incoming and outgoing DNA helices are brought together by the unstructured tails of LH, and diverge from each other only at a distance of ~30 bp from the surface of the nucleosome. A highly intriguing observation has been recently reported from Woodcock's laboratory [22]. Using cryo-EM to visualize chromatin fibers isolated from either proliferating cells in culture or terminally-differentiated cells, it was noticed that at 20 mM salt linker DNA stems were clearly discernable in fibers from differentiated cells, but were not seen in fibers from proliferating cells. This amazing difference between the two types of fibers was correlated with their propensity for lateral self-association: the stem-less fibers failed to form the fold-back structures typically seen in stemmed fibers (for an overview, see Ref. [23]). The formation of the nucleosome stem by LH binding has to be taken into account in all attempts to understand and model chromatin fiber structure (e.g., [24–27]).

A couple of reports suggest that it may be possible, upon further improvements of instrumentation, deposition, and imaging techniques to study subnucleosomal entities and their dynamics. Thus, Martin *et al.* [28] reported visualization of individual histone molecules within partially dissociated nucleosomes, and Zhao *et al.* [29] have occasionally observed the DNA superhelix on the outside of the nucleosome particle.

The insight from AFM images may be greatly boosted by sophisticated image analysis. Fritzsche and Henderson [30,31] have extracted cross-sections of nucleosomes at half-maximum height and have fitted them to virtual ellipsoids. These ellipsoids had relatively smooth perimeter and an aspect ratio of 1.2–1.4; moreover, the orientation of the ellipsoids was correlated with the direction of the fiber axis, with more than 50% of nucleosomes aligned with the axis. While this orientation effect may result from surface interactions, as discussed by the authors themselves, it may also represent an actual, and structurally important, feature of fiber structure. Ellipsoid-shaped nucleosomes have been reported in electron EM studies [32,33], and have been predicted in models of chromatin fibers that assume that fiber structure is governed by the elastic properties of DNA [34,35].

2.2. *AFM studies of biochemically manipulated or reconstituted chromatin fibers*

As already noted above, AFM imaging and quantitations on LH-stripped chromatin fibers revealed extended beads-on-a-string structures, in agreement with early EM results [17]. Nucleosomal arrays reconstituted from naked DNA fragments and histone octamers had similar appearance [36–42].

Important insights into the molecular determinants of the three-dimensional organization of extended (low ionic strength) chromatin fibers have been obtained by a combination of biochemical manipulations and AFM imaging. Leuba *et al.* [43] subjected isolated chicken erythrocyte chromatin fibers to progressive mild trypsinolysis and observed that cleavage of the LH tails resulted in fiber lengthening (in retrospect, this lengthening might reflect the destruction of the linker DNA stem structures described above). The somewhat later removal of a portion of the N-terminal unstructured tails of histone H3 led to fiber flattening on the mica surface. Even at later points of digestion, the fibers retained a zig-zag morphology, attributed to the continued presence of the LH globular domain on the fiber.

A series of reconstitution experiments confirmed and extended these observations [44]. In one set of experiments, chromatin fibers were stripped of LH, and then reconstituted with either intact H5 or its isolated globular domain (GH5). In another set of experiments, the reconstitution with H5 or GH5 was performed on chromatin fibers that had been stripped of LH, and whose H3 had been cleaved to remove its N-terminal tail (see schematic in Fig. 3a). The data showed that a stable, three-dimensional organization of the chromatin fiber has two absolute requirements: (i) the presence of bound LH globular domain, and (ii) the presence of either the unstructured tails of LH, or the unstructured N-tails of histone H3 (Fig. 3b). Since the tails of the other core histones were intact and could not substitute for the missing H3 tail, it was suggested that the H3 tail has a unique role in maintaining the proper structure of the extended chromatin fiber, in addition to its known role in salt-induced chromatin fiber folding (for references see Introduction in Leuba *et al.* [43]). The latter result was unexpected and significant, since it showed for the first time the structural involvement of the H3 tails in the maintenance of the extended fiber structure. A possible explanation for the unique structural role of the H3 tails may be seen in the diagram in Fig. 3c. (It must be noted that although the extended fiber structure is experimentally achieved by non-physiologically low salt concentrations, it is probably the structure of transcriptionally active chromatin [1].)

The contribution of individual core histone N-tails to transcriptional regulation has been recently addressed in a reconstitution study [42]. Varied combinations of recombinant and mutant cores histones have been assembled on circular plasmids, and the resultant nucleosomal arrays were tested for transcriptional

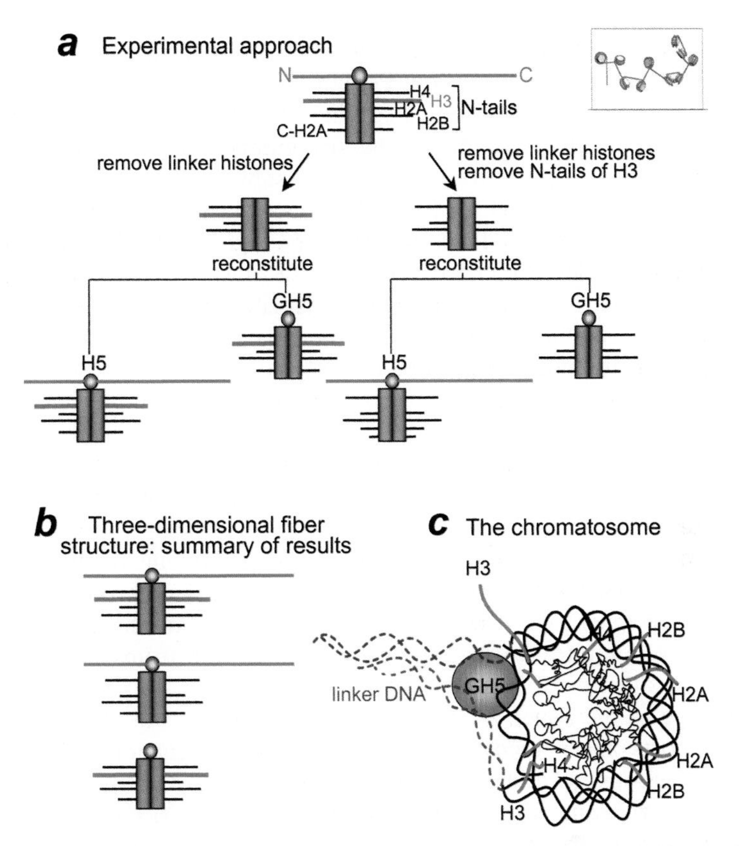

Fig. 3. Structural redundancy of the LH tails and the N-terminal tails of histone H3 in the three-dimensional organization of nucleosomes in the extended chromatin fiber. (*a*) Experimental approach (for details see Leuba *et al.* [44]). (*b*) Schematic of results: the three schematic drawings of nucleosomes show the three cases in which irregular, three-dimensionally organized chromatin fibers can be observed by AFM imaging and quantitations. The proper fiber structure requires the presence of the globular domain of LH, and either the tails of the LH, or the N-tails of histone H3. If both of these tails are missing, the fibers lose their three-dimensionality, flattening on the AFM surface. (*c*) Possible explanation of the results based on the crystal structure of the nucleosome [81]. A two dimensional trace of the crystal structure is shown in black. The portions of the N-terminal core histone tails (and C-terminal tails of H2A) that are visible in the crystal structure are depicted in color and labeled appropriately. The location of the LH globular domain is indicated by the spherical object labeled GH5. A possible path for linker DNA is indicated by dashed blue lines. The N-terminal tail of core histone H3 protrudes from the particle at a location close to the dyad axis; such a location would allow this tail to interact with the incoming and exiting linker DNA helices, neutralizing their negative charges. Such a scenario would occur in cases when the LH tails are not strongly bound to the linker DNA, e.g., when they are highly phosphorylated.

activity and histone acetylation by the transcriptional activator VP-16 and the histone acetylase-containing co-activator p300. AFM imaging was employed to visualize templates that contained intact histones and those that were assembled with completely-tailless histones: the two were indistinguishable both morphologically and by center-to-center internucleosomal distance measurements. This result indicates that the core histone tails are not needed for chromatin assembly, at least in the case of ACF-mediated reaction. Perhaps even more importantly, the tails are not needed for the proper positioning/spacing of histone octamers on DNA.

Finally, AFM imaging and measuring the number of nucleosomes per DNA fragment have been applied to the study of subsaturated nucleosomal arrays, believed to be a faithful model for functionally-active chromatin fibers that contain nucleosome-free sites at specific key regulatory regions [37,38]. At intermediate nucleosome loading densities (between 4 and 8 nucleosomes per template that contains 12 potential histone octamer-binding sites), there was a tendency for correlated nucleosome loading along the template. The molecular mechanism behind such correlations remains to be elucidated.

2.3. AFM assessment of structural effects of histone post-translational modifications

The demonstrated involvement of histone acetylation in regulation of gene expression (Refs. [45,46], Turner, this volume [47]) has prompted numerous attempts to decipher, at the molecular level, the mechanism behind the phenomenology. Despite the massive effort, we are far from understanding the structural consequences of histone acetylation (see Ausio, this volume [48], for a recent coverage of the issue).

Several reports have used AFM imaging, in conjunction with traditional biochemical methods, to address this question (it must be noted that an in-depth AFM study of this issue is still to be performed). The AFM data suggest thus far that hyperacetylated histones isolated from cells treated with histone deacetylase inhibitors produce beads-on-a string structures somewhat more extended than those obtained using histones purified from control cells [37,38], in agreement with EM imaging results obtained on circular chromatin templates [49]. In addition, acetylation seems to enhance the non-random nucleosome-loading behavior seen in subsaturated nucleosomal arrays (see above, and Table 1).

An example in-depth study combining AFM imaging and biochemical manipulation of chromatin fibers has addressed the structural consequences of histone poly(ADP-ribosyl)ation [50]. In this work, chromatin fibers isolated from control chicken erythrocyte nuclei were compared to fibers isolated from nuclei modified *in vitro*; additional experiments involved solubilized fibers that were directly modified *in vitro* by purified poly(ADP-ribose) polymerase. The data convincingly showed poly(ADP-ribosyl)ation-induced decondensation of fiber structure, confirming previous studies (e.g., Ref. [51]); moreover, the modified fibers remained significantly decondensed even in the presence of magnesium ions. Another important result indicates that magnesium ions cannot substitute for LH to induce compaction, since H1/H5 depleted fibers remained less compact than native fibers in 10 mM Mg^{2+} (the center-to-center internucleosome distances were <9 nm for native-Mg-fibers, whereas this distance was ~ 25 nm for the LH-stripped-Mg fibers). These data are especially important because of assertions that physiologically-relevant fiber compaction can be achieved by ions in the absence of LH (e.g., Refs. [52,53]).

2.4. AFM visualization of salt-induced chromatin fiber compaction

The structure of the condensed chromatin fiber is still under discussion [1,23,54], with two competing models: the original solenoid model of Finch and Klug [16], and the straight-linker model [12,14,55]. Assessing the structure *in vivo* or *in situ* has proven impossible thus far, due to technical limitations. Chromatin fibers released from nuclei into solution by nuclease treatment have been widely used as models for fiber structure: such fibers are extended at low ionic strength and condensed at ionic strengths believed to be close to those found *in vivo* (~150 mM Na^+ or 0.35 mM Mg^{2+}). The salt-induced fiber compaction has been extensively studied in the past but is still poorly understood in terms not only of the details of the structure but also in terms of the molecular mechanisms of the compaction process.

AFM images of salt-compacted chromatin fibers have unfortunately contributed little to the resolution of the controversy. Only a couple of studies have addressed this structure because of limitations to the lateral resolution achievable for soft biological samples. Salt compaction leads to a loss of individual nucleosome resolution in the fiber, leaving us with images of rather low informational content. Examples of chicken erythrocyte chromatin fibers, imaged after glutaraldehyde fixation at various concentrations of either NaCl of $MgCl_2$, can be found in Zlatanova *et al.* [11,56]. (Imaging has to be performed on fixed material, since salts interfere with imaging and have to be removed following sample deposition: fixation avoids changing of fiber conformation during the washing steps. An additional factor precluding direct imaging of unfixed samples at increased ionic strengths is the fact that the interaction of the fiber with the imaging substrate leads to dissociation of the histones from the DNA even at salt concentrations as low as 10 mM [12,36]. Both sodium and magnesium ions led to the formation, under these conditions, of highly compact, irregularly segmented structures with an apparent diameter of ~45 nm; the segmentation was recognizable already at 20 mM NaCl. These structures are highly reminiscent of the superbeads reported about two decades ago by EM [57] but later dismissed as artifacts (for further discussion and references, see Refs. [56,58,59]. It still remains to be determined whether the segmentation observed in AFM images is not caused by interactions with the mica surface.

d'Erme *et al.* [50] have reported data on native chicken erythrocyte chromatin fibers fixed in the absence or presence of 10 mM $MgCl_2$, and then imaged in the absence of ions. The magnesium fibers were much more compact by both visual inspection of images or by quantitative measurements of fiber parameters (width, center-to-center internucleosomal distance, and number of nucleosomes per unit fiber length). The irregular segmentation pattern ("superbeads") reported earlier was also discernable. As mentioned above, an important result indicated that magnesium ions cannot substitute for LH to induce compaction, since H1/H5 depleted fibers remained less compact than native fibers in 10 mM Mg^{2+}.

3. Chromatin fiber assembly under applied force

Chromatin assembly *in vivo* takes place massively during DNA replication; in addition, nucleosomes have to assemble in the wake of the transcriptional machinery since the transcribing RNA polymerase removes nucleosomes in its way (by itself or with the help of other factors) (see, for example, Jackson and references therein, this volume [60]). The naked DNA stretches have to quickly reform chromatin, so that the roles of chromatin in both compacting the DNA and regulating its functions are restored. It is essential to realize that this reformation of nucleosomes in the wake of RNA polymerase (probably in the wake of other DNA-tracking enzymes as well) is a process that takes place while the DNA molecule is still under tension as a result of the pulling exerted by the stationary RNA polymerase [61] on the transcribed DNA. Both RNA and DNA polymerases have been shown to be amongst the strongest molecular motors, developing forces of up to 30–40 pN [62,63]. If the forces measured *in vitro* are also developed *in vivo*, then the question arises whether the DNA under tension can be assembled into nucleosomes and what the force dependence of the assembly process is.

The group of Viovy in France has performed videomicroscopy experiments in which a single λ-DNA molecule (48.5 kbp, 16.4 μm contour length) was attached at one end to a glass surface, the other end being free so that it could be stretched by a flow (Fig. 4a) [64]. The assembly of chromatin on such single DNA molecules was achieved by the flowing in of *Xenopus* or *Drosophila* cell-free extracts and the shortening of the molecules as a result of the assembly was observed by real-time fluorescence microscopy (DNA was fluorescently labeled by intercalation of YOYO-1). There was a clear dependence of the rate of chromatin assembly on the extract dilution, and more importantly, on the shear rate, hence on the force applied to the DNA. Assembly could proceed, albeit at a much-reduced rate, up to forces of 12 pN.

Similar findings have been reported by using an optical tweezers set-up to follow assembly on a single λ-DNA driven by *Xenopus* egg extract [65]. In this study, a single DNA molecule was suspended between two micron-sized beads, one held by a pipette, and the other one in an optical trap (Fig. 4b). Since the presence of cell debris in the extract precluded the use of the optical trap for force measurements, the optical trap was switched off during the assembly and the instrument was used in the laminar-flow mode. Forces were estimated using either the parameters of the laminar flow (Stokes' law) or by measuring the Brownian motion of the freely suspended bead. The addition of the extract led to visible shortening of the distance between the two beads, reflecting chromatin formation. The kinetics was strongly dependent on the applied force, with complete inhibition

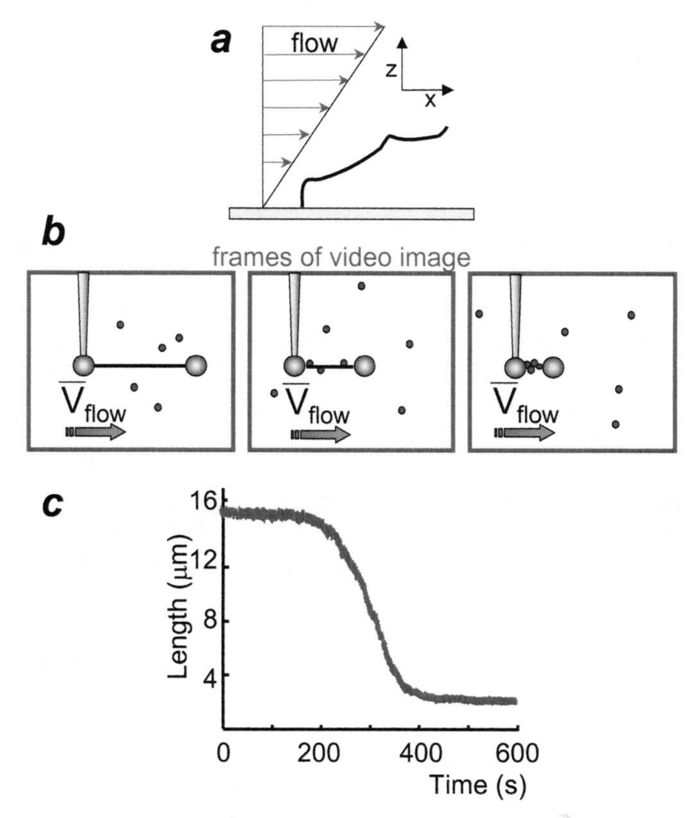

Fig. 4. Schematic of the single-molecule chromatin assembly experiments. (*a*) The fluorescence videomicroscopy experiments of Ladoux *et al.* [64] (for details see text). (*b*) The chromatin assembly experiments of Bennink *et al.* [65]: a single λ-DNA molecule suspended between two micron-sized beads is being assembled with the help of *Xenopus* egg extract containing core histones and assembly factors (for details see text). (*c*) Example assembly curve at ~1 pN (redrawn from Ref. [65]).

of assembly at forces exceeding 10 pN. An example assembly curve is presented in Fig. 4c.

A more detailed study of single chromatin fiber assembly has been carried out in our laboratories using magnetic tweezers, the kind of instrument pioneered by David Bensimon and Vincent Croquette in Paris [66,67]. In magnetic tweezers, the DNA molecule to be studied is suspended between the surface of a glass cuvette and a paramagnetic bead, which is manipulated by permanent external magnets (Fig. 5a). The force applied to the bead, and hence to the suspended DNA molecule, can be manipulated by changing the parameters of the magnetic field (e.g., the strength of the magnet, its distance to the cuvette, the size of the bead, etc.). The force is relatively conveniently estimated by measuring only two distance parameters: the extension l of the DNA tether in the z-direction and the Brownian motion of the bead in the x–y plane (Fig. 5a).

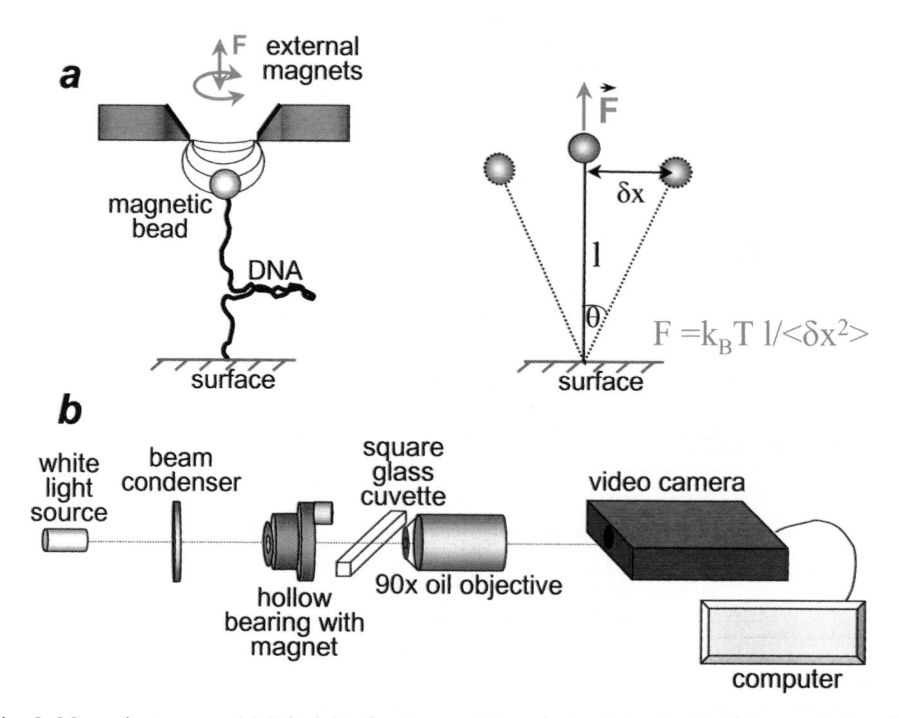

Fig. 5. Magnetic tweezers. (*a*) Principle of action and formula to determine the force applied to the tethered bead (for details see text). (*b*) Horizontal version of the instrument used in the laboratories of SHL and JZ.

We constructed a simple horizontal version of a magnetic-tweezers set-up, in which a single permanent magnet is placed at an angle with respect to the path of the light beam (Fig. 6b). This placement allows easy observation of the movement of the bead across the video camera screen; the successive bead positions can be estimated using the Pythagorean theorem, as illustrated in Fig. 6c. (The actual shortening of the DNA can also be determined from the known position of the DNA attachment point to the surface, and hence the angle between the DNA and the surface, see schematic in Fig. 6d.)

In these experiments, a λ-DNA molecule is suspended between a glass surface and the bead, and assembly is driven by the addition of a solution of purified histone octamers and Nucleosome Assembly Protein 1 (NAP-1) (Figs. 6b and 6b). An example assembly curve is presented in Fig. 7a. The overall dependence of the assembly rate on the tension applied to the DNA molecule agrees with the results of Ladoux *et al.* [64] and Bennink *et al.* [65]. In addition, the instrumental set-up allows controlled step-wise adjustment of the force during a single round of assembly by controlled step-wise movement of the external magnet away from the cuvette. In such an experiment, the assembly rate changes from slower rates at higher forces, to faster rates at lower forces (Fig. 7b). These observations

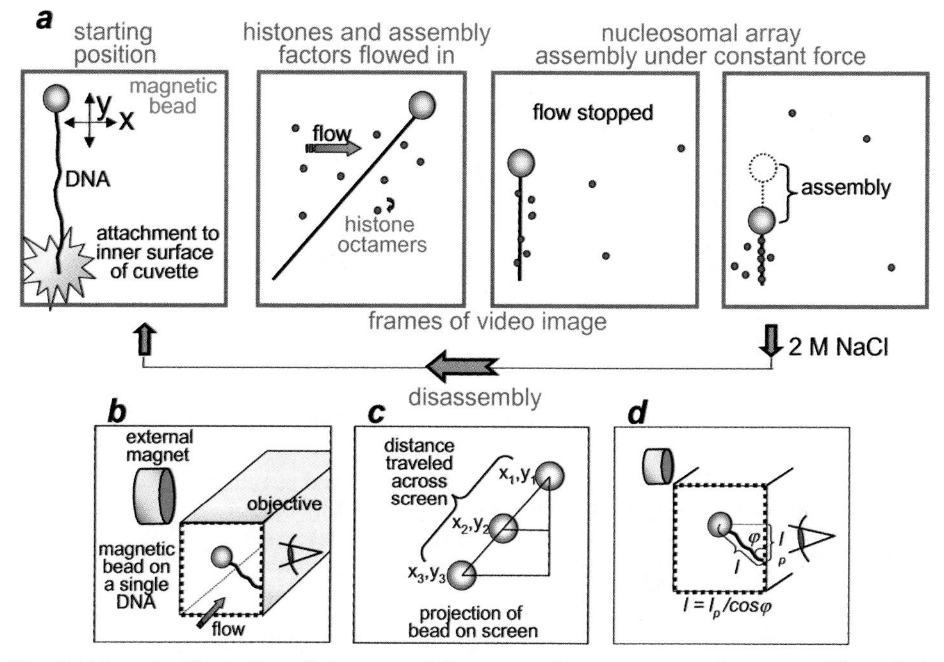

Fig. 6. Schematic illustration of the stopped-flow magnetic tweezers experiments to follow single chromatin fiber assembly. (*a*) Flow diagram of how the experiment was performed. (*b*) A blow-up of the cuvette, with the bead attached to its side: note that the DNA tether is not normal to the wall of the cuvette because of the position of the external magnet, i.e., the z direction is out of the plane of the video frame. (*c*) A schematic explaining the calculation of the distance traveled by the bead across the videoscreen. The x- and y-coordinates of the bead position on each successive video frame are used to calculate the projected traveled distance. (*d*) The actual shortening of the fiber can be calculated from the projected shortening (travel of bead across screen) and the cosine of the angle theta.

allowed the first-ever assessment of the speed with which the system responded to changing forces: the response of the assembly reaction to changes in the tension in the DNA molecule is instantaneous. This observation provides an unprecedented insight of how changes in the *in vivo* rate of transcription (and hence in the tension experienced by the DNA being threaded through the RNA polymerase) may regulate the rate of reformation of nucleosomes in the wake of the polymerase.

Finally, the data gave a clear indication of the reversibility of the assembly process. As illustrated in the magnified portion of the assembly curve in Fig. 7a (see inset), the path of the curve was not smooth but exhibited occasional upward stretches, indicating that in an overall assembly process there were occasional dissociation steps involving a few nucleosomes at a time. Thus, we believe that we see the first real-time demonstration of the dynamic equilibrium between an assembled and a disassembled state of individual nucleosomes in the fiber context; such equilibrium has been previously suggested on the basis of

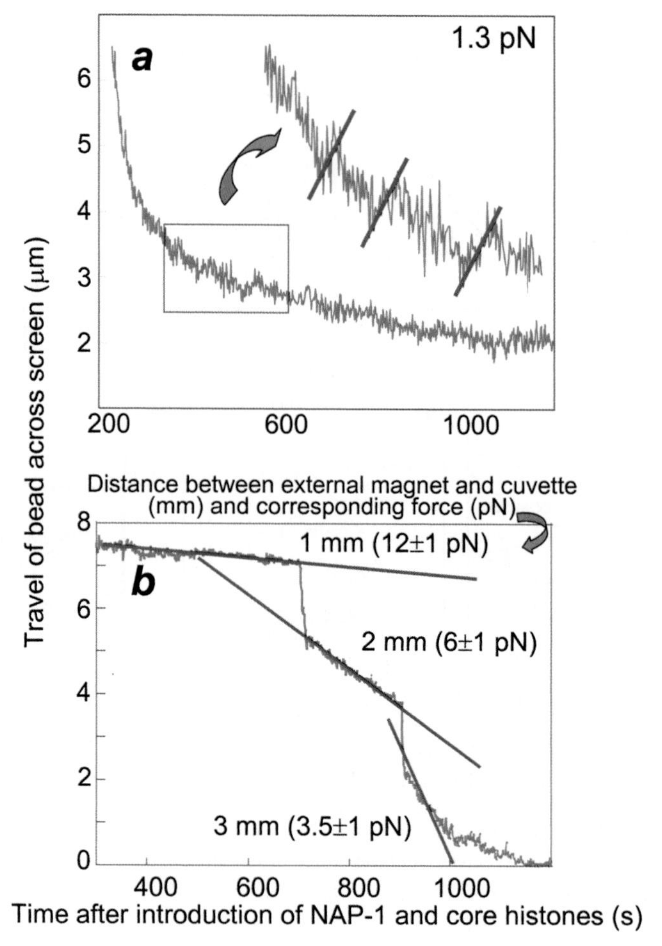

Fig. 7. (*a*) Example chromatin assembly curve recorded at 1.3 pN force applied to the magnetic bead. A magnification of the squared portion of the curve is also presented. (*b*) Assembly curve of the same DNA molecule, subjected to difference forces in the course of a single assembly round (forces were adjusted by adjusting the distances between the cuvette and the external magnet as indicated). The lines drawn indicate that the assembly rate changes as a function of force. Note the immediate response of the assembly reaction to the changing force.

a few biochemical experiments performed at the mononucleosomal level, and theory [58,68].

It is obvious that the single-molecule approaches can reveal features of the dynamics of chromatin assembly and its force dependence that could not be observed using the standard biochemical assays to study assembly: following changes in superhelicity in closed circular DNA molecules, looking at changes in protection against enzymes like micrococcal nuclease, or the accessibility of restriction enzymes.

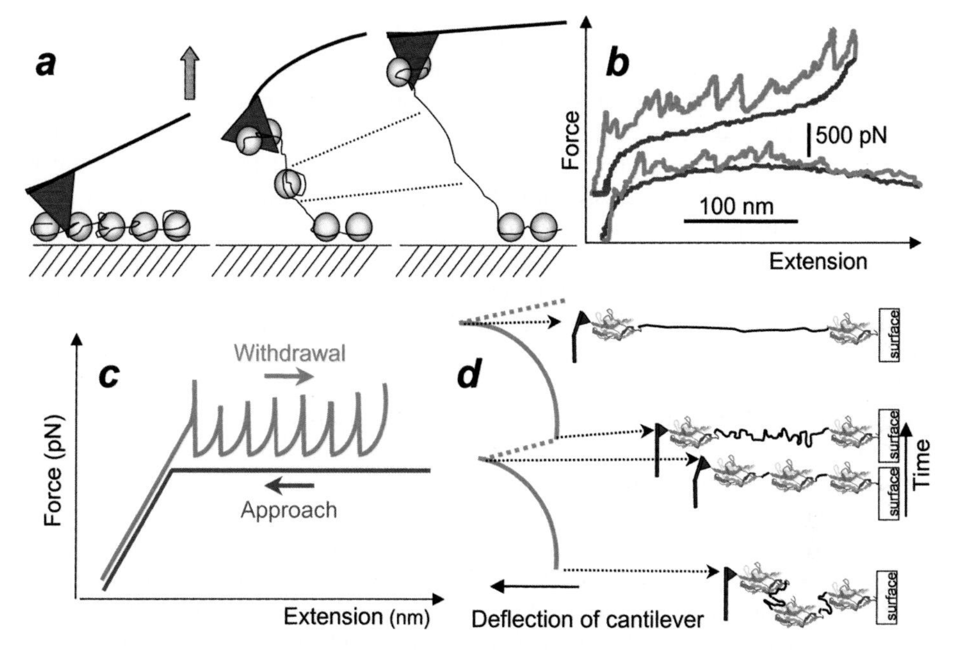

Fig. 8. (*a*) Schematic of the AFM pulling experiments and expected unraveling of an individual nucleosome as a result of pulling on the DNA. (*b*) Example force-extension curves on isolated chicken erythrocyte chromatin fibers redrawn from Ref. [69]. (*c*) Idealized schematic of a typical force-extension curve obtained on pulling single titin molecules, as in the experiments of Rief *et al.* [71]. (*d*) Explanation of the titin force curve by successive unfolding of individual protein domains (see text).

4. *Chromatin fiber disassembly under applied force*

Chromatin fiber disassembly under applied force has been successfully studied so far only with optical tweezers.

Earlier attempts to use the AFM for mechanically stretching chromatin fibers have run into a rather unexpected artifact. Long native chromatin fibers isolated from chicken erythrocytes, or fibers assembled *in vitro* from purified histones and relatively short, tandemly repeated DNA sequences were deposited on mica or glass surfaces and pulled with the AFM tip [69,70]. In such stretching experiments the scanning of the sample in the x- and y-direction used for imaging was disabled, and the cantilever-mounted tip was allowed to move only in the z-direction, i.e., upwards and downwards, away and towards the surface. When the AFM tip is pushed into the sample, it may attach to the sample by non-specific adsorption; upon retraction it stretches the sample and force-extension curves are recorded (see Fig. 1b for an explanation of a typical force curve).

The experimental approach used to mechanically stretch a chromatin fiber with the AFM is depicted schematically in Fig. 8a, and some example curves obtained with native chicken erythrocyte chromatin fibers are presented in Fig. 8b. These curves exhibited a saw-tooth pattern, similar to the patterns obtained upon stretching of multi-domain proteins like titin [71] or tenascin [72] (Fig. 8c). Each of the peaks in these patterns arises because the initial entropic stretching of the polymer chain, accompanied by a gradual build-up of tension in the molecule, is followed by enthalpic changes in the internal organization (unraveling) of a certain individually-folded domain: as a result of the unraveling of a domain the chain elongates in a jump, which in turn leads to an abrupt fall in the tension of the molecule (Fig. 8d). Successive unraveling events of individual domains in a multi-domain polypeptide chain lead to the appearance of multiple peaks in the force-extension curve, hence, the saw-tooth pattern.

The force-extension curves of both chromatin fibers extracted from cells and fibers reconstituted *in vitro* exhibited this multi-peak appearance (Fig. 8b). A closer look at the distances between individual peaks, however, made it clear that the sought-after unraveling of individual nucleosomes as a result of mechanical stretching of the DNA did not occur, despite the relatively high forces applied, in the range of 300–600 pN. Control experiments with glutaraldehyde-fixed fibers (in which unraveling is precluded because of the fixation) and analysis of AFM images of the fibers being stretched suggested that the jumps in the force curves corresponded to removal of successive intact nucleosomes from the glass surface, followed by stretching of the naked DNA between the nucleosomes attached to the tip and the surface. Earlier work on stretching pieces of naked double-helical DNA with either optical tweezers [73], optical fibers [74], or AFM [75] has demonstrated a structural transition from the B-form DNA to the so-called S (stretched)-form, in which DNA is extended to $\sim$2-fold over its original contour length. The force causing the B- to S-transition was measured to be $\sim$70 pN in both the optical tweezers and optical fibers experiments, and $\sim$120 pN in the AFM work. Despite the apparent inconsistency among results from different laboratories (for further discussion, see Ref. [8]), it is clear that the forces applied in the chromatin stretching experiments [69,70] were by far exceeding those needed for the B- to S-transition. Thus, the scenario explaining the observed saw-tooth pattern by a succession of nucleosomes popping-off from the surface followed by stretching of linker DNA seems plausible.

Importantly, the AFM experiments set a lower limit of $\sim$350 pN for the unraveling of individual nucleosomes. We will come back to the significance of these results later.

Three papers have reported results on stretching individual chromatin fibers with optical tweezers, with consistent results. Optical traps make use of the property of a laser beam focused through an objective to hold small dielectric beads suspended in the focus of the beam (Fig. 9a). Photons hitting the bead refract and change momentum; by conservation of momentum, the bead experiences force from light and is pushed into the direction of the focus. A bead held in an optical trap can be used as a handle to hold an attached macromolecule, which in turn can be subjected to pulling and/or twisting through its other end (usually by holding it by suction in a pipette) (Fig. 9b). Since the external force applied to the molecule is equal in magnitude and opposite in direction to that applied to the bead by the optical trap, the displacement of the bead from its focal position can be used to calculate the force applied to the molecule.

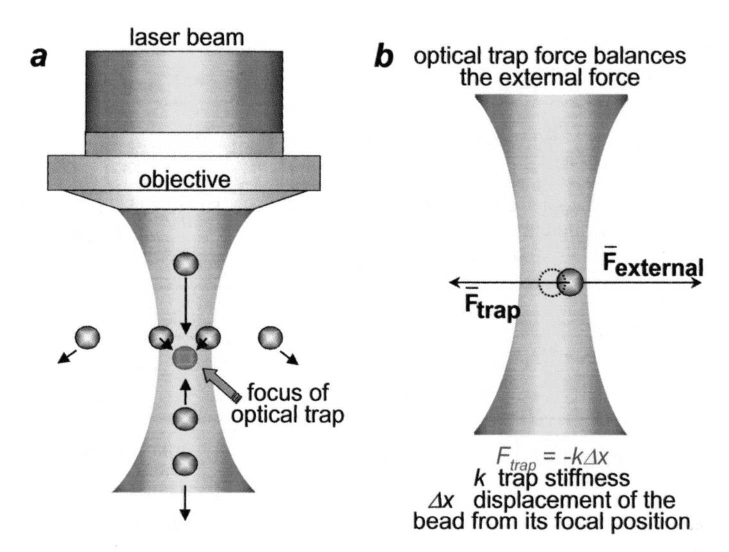

Fig. 9. (*a*) Optical trapping; (*b*) The use of an optical trap to estimate extensions and forces in single macromolecules. Adapted from Bennink [82]. For details see text.

Using optical traps, Cui and Bustamante [76] stretched isolated chicken erythrocyte fibers, and Bennink *et al.* [77] pulled on fibers directly reconstituted in the flow cell from λ-DNA and purified histones with the help of *Xenopus* extracts (see Fig. 10a for a schematic of the latter experiment). Up to 20 pN, the fibers underwent reversible stretching, but applying stretching forces above 20 pN led to irreversible alterations, interpreted in terms of removal of histone octamers from the fibers with recovery of the mechanical properties of naked DNA.

The high speed of data acquisition in the Bennink *et al.* [77] experiments allowed recording of force curves in which discrete, sudden drops in force could be observed upon fiber stretching, reflecting discrete opening events in fiber structure (Fig. 10b). These opening events were quantized at increments in fiber lengths of ~65 nm and were attributed to unwrapping of individual nucleosomal particles. The high-resolution data allowed the first measurements of the forces needed to unspool individual nucleosomes: these forces ranged between 20 and 40 pN.

Recently, Michelle Wang and colleagues [78] used a feedback-enhanced optical trap to stretch 208-17 nucleosomal arrays pre-assembled in bulk by the salt-dialysis method and then suspended between a glass surface and a polystyrene microbead (a commentary appeared in the same issue [79]). Following trapping of the bead, the array was stretched by piezo-controlled movements of the glass (Fig. 11a). The force-extension curves (Fig. 11b) exhibited 17 peaks, corresponding to successive unraveling of the 17 nucleosomes in the array. The low rate of stretching (28 nm/s, compare with 1 μm/s in the Bennink *et al.* [77] experiments) allowed observation of a step-wise release of the DNA from its interactions with the histone core,

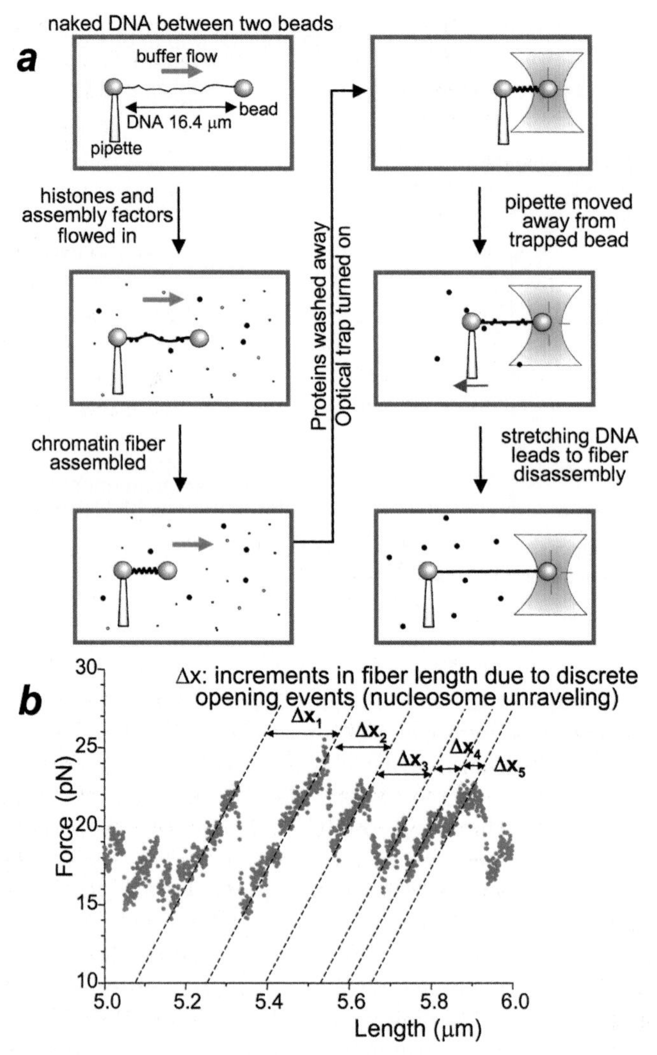

Fig. 10. (*a*) Flow-chart of the optical tweezers experiments of Bennink *et al.* [77]. (*b*) A blow-up of a portion of the force-extension curve, with discrete opening events (unraveling of individual nucleosomes) clearly visible (reprinted with permission from Bennink *et al.* [77]). The length increments in the fiber accompanying each opening event are indicated as Δx_n. The different Δx_n distances correspond to unraveling of one nucleosome at a time, or of two or three nucleosomes simultaneously. These values are quantized with a unit step of ~65 nm which represents the lengthening of the fiber upon dissociation of one single nucleosome.

each step corresponding to breaking of interactions of different chemical stability (Fig. 11c).

The authors also provide reversibility data: repeated stretching and relaxation of the some fiber showed reformation of *some* (by no means all) nucleosomes during the relaxation portion of the cycle (Fig. 12a). This nucleosome reformation was a

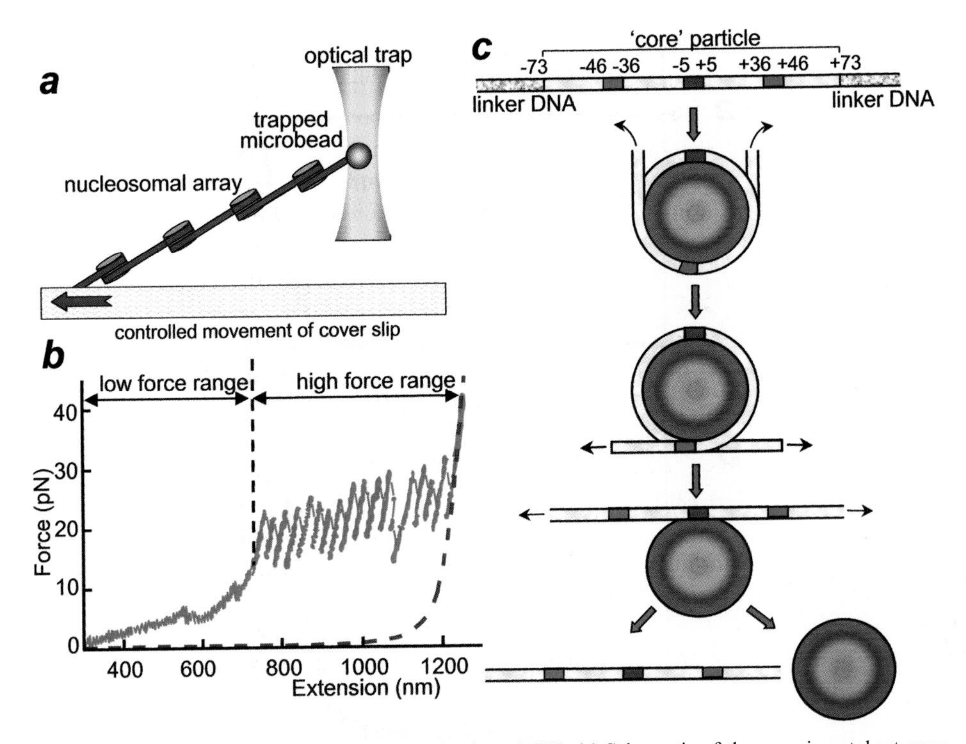

Fig. 11. The pulling experiments of Brower-Toland *et al.* [78]. (*a*) Schematic of the experimental set-up: a pre-assembled single nucleosomal array is attached to a cover slip and to a bead held in an optical trap. The force is applied by moving the cover slip, see arrow. (*b*) An example for force-extension curve: note the 17 discrete peaks corresponding to successive unraveling of the 17 nucleosomes in this nucleosomal array (redrawn from Brower-Toland *et al.* [78]). (*c*) Linear map of the sites on the nucleosomal DNA with strongest DNA–histone interactions (according to Ref. [83]), and a three-stage model for the mechanical disruption of the nucleosomes (adapted from Brower-Toland *et al.* [78]). In this model, the application of low forces leads to partial unwrapping of the DNA from the histone core (first stage); the strong DNA–histone interactions at positions +4 and −4 present a barrier to the unraveling process and higher forces are needed to break these interactions (stage two); at stage three, the final remaining contacts between the histones and the DNA around the dyad axis of the particle will be broken.

function of the force applied: if the force was below that of the overstretching DNA (B–S) transition, some nucleosomes reformed. In contrast, when the initial stretching force approached the B–S transition force (~60 pN) the reversibility was impeded. The authors interpret these results as "contrasting significantly" with those of Bennink *et al.* [77]. A more careful analysis, however, reveals that the two types of experiments are not directly comparable: the maximum force exerted in the reversibility experiments of Bennink *et al.* (where the second and all further stretchings were taken all the way to the overstretched DNA length) was such that it distorted the double helical DNA structure (Fig. 12b). This would, of course, lead to complete dissociation of the octamer from the DNA, as actually stated by Brower-Toland *et al.* [78] themselves. As far as the lack of reversibility following

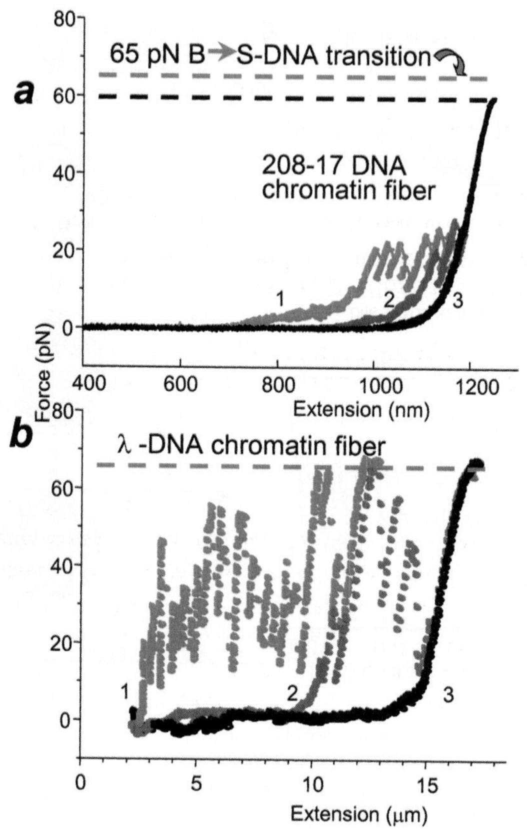

Fig. 12. The reversibility experiments of (*a*) Brower-Toland *et al.* (redrawn from Ref. [78]) and (*b*) Bennink *et al.* (adapted from Ref. [78]). In these experiments the fiber was first stretched to a certain force, allowed to relax, and then restretched. Note in *a* that when the force applied during the initial stretching did not reach that needed for the B–S overstretching transition (~65 pN), there was some reformation of nucleosomes during the relaxation: the second stretching curve did unravel a couple of nucleosomes. When, in contrast, the stretching was taken all the way to the overstretching force of ~65 pN, the next stretching curve was that of naked DNA (see curves 2 and 3 in *b*). The first stretch in *b* was only partial, leaving some nucleosomes not unraveled; the latter unraveled during the second stretch that was taken to higher force. Note that the sum of curves 1 and 2 would be equivalent to a complete stretching curve, if the overstretching force is reached. The lack of reformation of nucleosomes during the relaxation portion of the first stretch may be due to the high loading rate (see text).

the very first stretch in Bennink *et al.* [77] is concerned (Fig. 12b), it could be attributed to the high rate of pulling (~1 μm/s, as contrasted to the 28 nm/s pulling rate in Ref. [78]). Thus, the "contrasting" results of Bennink *et al.* [77] are *non*-contrasting at all: just the experiments are not directly comparable!

As mentioned earlier, the AFM data [70] set *a lower* limit of ~350 pN for the de-spooling process, while the OT data obtained in three different laboratories using three very different biological samples consistently gave a dissociation value of ~20 pN. Why are the forces measured in the AFM and in the optical tweezers so different? One of the reasons is no doubt connected to the very different loading rates dF/dt: while in the Bennink *et al.* [77] experiments the loading rates were $\sim$100–150 pN/s, the AFM data were collected at rates between 35×10^3 and 10^6 pN/s. (The loading rates in the other two papers were $\sim$1–5 pN/s [76], and typically $\sim$10 pN/s [78].) The difference in the loading rates can hardly be the only explanation since a similar loading rate difference resulted in a much smaller force difference in titin stretching experiments [80]. Another factor may be connected to the length over which the tension is applied: the lengths of the fibers in the AFM experiments of Leuba *et al.* [70] and OT experiments of Bennink *et al.* [77] differed by $\sim$20 fold (the fibers stretched by Cui and Bustamante were of undefined, heterogeneous length). Additional unidentified factors must be involved, though, since the Brower-Toland *et al.* [78] optical tweezer work was performed on relatively short fibers, comparable to those used in the Leuba *et al.* [70] AFM work, with different results.

Thus, kinetic effects should always be taken into account in determining the consequences of the forces applied to nucleosomes by molecule motors, such as polymerases. Meaningful interpretations would also depend on knowledge of the length of molecules to which tension is applied. It may be of relevance to mention here that these forces are in the same range as the stall forces developed by RNA and DNA polymerases [62,63], the enzymes that encounter nucleosomes while reading the information in the DNA. This may mean that the polymerases are by themselves capable of removing the nucleosomes out of their way.

5. *Summary*

This brief overview of the single-molecule approaches to understanding chromatin fiber structure and dynamics convincingly shows the potential these methods have for providing insights unachievable up to now. The AFM studies have given us new insights into the static chromatin structure, whereas the other single-molecule techniques (e.g., optical and magnetic tweezers) have revealed some dynamic features of the structure. We believe that the dynamic studies are where the most exciting new work is to be found, and where the most significant and original contributions of these techniques are being made. The application of these methods to chromatin studies is still in its infancy but will undoubtedly constitute the research focus of more and more laboratories in the future.

Acknowledgements

We thank Drs. M. Tomschik and M. Karymov for assistance with figures.

References

1. van Holde, K. and Zlatanova, J. (1996) Proc. Natl. Acad. Sci. USA 93, 10548–10555.
2. Wu, C. (1997) J. Biol. Chem. 272, 28171–28174.
3. Workman, J.L. and Kingston, R.E. (1998) Annu. Rev. Biochem. 67, 545–579.
4. Fry, C.J. and Peterson, C.L. (2001) Curr. Biol. 11, R185–R197.
5. Bustamante, C., Macosko, J.C., and Wuite, G.J. (2000) Nat. Rev. Mol. Cell. Biol. 1, 130–136.
6. Leuba, S.H. and Zlatanova, J. (eds.) (2001) Biology at the Single-Molecule Level, Pergamon, Amsterdam.
7. Binnig, G., Quate, C.F., and Rohrer, C. (1986) Phys. Rev. Lett. 56, 930–933.
8. Zlatanova, J., Lindsay, S.M., and Leuba, S.H. (2000) Prog. Biophys. Mol. Biol. 74, 37–61.
9. Bustamante, C., Keller, D., and Yang, G. (1993) Curr. Opin. Struct. Biol. 3, 363–372.
10. Vesenka, J., Hansma, H., Siegerist, C., Siligardi, G., Schabtach, E., and Bustamante, C. (1992) SPIE 1639, 127–137.
11. Zlatanova, J., Leuba, S.H., Yang, G., Bustamante, C., and van Holde, K. (1994) Proc. Natl. Acad. Sci. USA 91, 5277–5280.
12. Leuba, S.H., Yang, G., Robert, C., Samori, B., van Holde, K., Zlatanova, J., and Bustamante, C. (1994) Proc. Natl. Acad. Sci. USA 91, 11621–11625.
13. Yang, G., Leuba, S.H., Bustamante, C., Zlatanova, J., and van Holde, K. (1994) Nat. Struct. Biol. 1, 761–763.
14. Woodcock, C.L., Grigoryev, S.A., Horowitz, R.A., and Whitaker, N. (1993) Proc. Natl. Acad. Sci. USA 90, 9021–9025.
15. van Holde, K. and Zlatanova, J. (1995) J. Biol. Chem. 270, 8373–8376.
16. Finch, J.T. and Klug, A. (1976) Proc. Natl. Acad. Sci. USA 73, 1897–1901.
17. Thoma, F., Koller, T., and Klug, A. (1979) J. Cell Biol. 83, 403–427.
18. Bednar, J., Horowitz, R.A., Dubochet, J., and Woodcock, C.L. (1995) J. Cell Biol. 131, 1365–1376.
19. Bednar, J., Horowitz, R.A., Grigoryev, S.A., Carruthers, L.M., Hansen, J.C., Koster, A.J., and Woodcock, C.L. (1998) Proc. Natl. Acad. Sci. USA 95, 14173–14178.
20. Bednar, J. and Woodcock, C.L. (1999) Methods Enzymol. 304, 191–213.
21. Hamiche, A., Schultz, P., Ramakrishnan, V., Oudet, P., and Prunell, A. (1996) J. Mol. Biol. 257, 30–42.
22. Grigoryev, S.A., Bednar, J., and Woodcock, C.L. (1999) J. Biol. Chem. 274, 5626–5636.
23. Grigoryev, S.A. (2001) Biochem. Cell Biol. 79, 227–241.
24. Katritch, V., Bustamante, C., and Olson, W.K. (2000) J. Mol. Biol. 295, 29–40.
25. Hammermann, M., Toth, K., Rodemer, C., Waldeck, W., May, R.P., and Langowski, J. (2000) Biophys. J. 79, 584–594.
26. Schiessel, H., Gelbart, W.M., and Bruinsma, R. (2001) Biophys. J. 80, 1940–1956.
27. Schiessel, H. (2002) Europhys. Lett. 58, 140–146.
28. Martin, L.D., Vesenka, J.P., Henderson, E., and Dobbs, D.L. (1995) Biochemistry 34, 4610–4616.
29. Zhao, H., Zhang, Y., Zhang, S.B., Jiang, C., He, Q.Y., Li, M.Q., and Qian, R.L. (1999) Cell Res. 9, 255–260.
30. Fritzsche, W. and Henderson, E. (1996) Biophys. J. 71, 2222–2226.
31. Fritzsche, W. and Henderson, E. (1997) Scanning 19, 42–47.
32. Oliva, R., Bazett-Jones, D.P., Locklear, L., and Dixon, G.H. (1990) Nucleic Acids Res. 18, 2739–2747.
33. Zabal, M.M., Czarnota, G.J., Bazett-Jones, D.P., and Ottensmeyer, F.P. (1993) J. Microsc. 172, 205–214.
34. Bishop, T.C. and Hearst, J.E. (1998) J. Phys. Chem. B 102, 6433–6439.
35. Bishop, T.C. and Zhmudsky, O.O. (2002) J. Biomol. Struct. Dyn. 19, 877–887.
36. Allen, M.J., Dong, X.F., O'Neill, T.E., Yau, P., Kowalczykowski, S.C., Gatewood, J., Balhorn, R., and Bradbury, E.M. (1993) Biochemistry 32, 8390–8396.
37. Yodh, J.G., Lyubchenko, Y.L., Shlyakhtenko, L.S., Woodbury, N., and Lohr, D. (1999) Biochemistry 38, 15756–15763.

38. Bash, R.C., Yodh, J., Lyubchenko, Y., Woodbury, N., and Lohr, D. (2001) J. Biol. Chem. 276, 48362–48370.
39. Schnitzler, G.R., Cheung, C.L., Hafner, J.H., Saurin, A.J., Kingston, R.E., and Lieber, C.M. (2001) Mol. Cell Biol. 21, 8504–8511.
40. Karymov, M.A., Tomschik, M., Leuba, S.H., Caiafa, P., and Zlatanova, J. (2001) FASEB J. 15, 2631–2641.
41. Tomschik, M., Karymov, M.A., Zlatanova, J., and Leuba, S.H. (2001) Struct. Fold. Des. 9, 1201–1211.
42. An, W., Palhan, V.B., Karymov, M.A., Leuba, S.H., and Roeder, R.G. (2002) Mol. Cell 9, 811–821.
43. Leuba, S.H., Bustamante, C., Zlatanova, J., and van Holde, K. (1998) Biophys. J. 74, 2823–2829.
44. Leuba, S.H., Bustamante, C., van Holde, K., and Zlatanova, J. (1998) Biophys. J. 74, 2830–2839.
45. Allfrey, V.G., Faulkner, R., and Mirsky, A.E. (1964) Proc. Natl. Acad. Sci. USA 51, 786–794.
46. Roth, S.Y., Denu, J.M., and Allis, C.D. (2001) Annu. Rev. Biochem. 70, 81–120.
47. Turner, B. (2002) In: Zlatanova, J. and Leuba, S.H. (eds.) Chromatin Structure and Dynamics: State-of-the-Art, Elsevier, Amsterdam, pp. 291–308.
48. Ausio, J. (2002) In: Zlatanova, J. and Leuba, S.H. (eds.) Chromatin Structure and Dynamics: State-of-the-Art, Elsevier, Amsterdam, pp. 241–290.
49. Loyola, A., LeRoy, G., Wang, Y.H., and Reinberg, D. (2001) Genes Dev. 15, 2837–2851.
50. d'Erme, M., Yang, G., Sheagly, E., Palitti, F., and Bustamante, C. (2001) Biochemistry 40, 10947–10955.
51. Poirier, G.G., de Murcia, G., Jongstra-Bilen, J., Niedergang, C., and Mandel, P. (1982) Proc. Natl. Acad. Sci. USA 79, 3423–3427.
52. Hansen, J.C., Ausio, J., Stanik, V.H., and van Holde, K.E. (1989) Biochemistry 28, 9129–9136.
53. Fletcher, T.M. and Hansen, J.C. (1996) Crit. Rev. Eukaryot. Gene Expr. 6, 149–188.
54. Widom, J. (1998) Annu. Rev. Biophys. Biomol. Struct. 27, 285–327.
55. Staynov, D.Z. (1983) Int. J. Biol. Macromol. 5, 3–9.
56. Zlatanova, J., Leuba, S.H., and van Holde, K. (1998) Biophys. J. 74, 2554–2566.
57. Zentgraf, H. and Franke, W.W. (1984) J. Cell Biol. 99, 272–286.
58. van Holde, K.E. (1988) Chromatin, Springer-Verlag, New York.
59. Tsanev, R., Russev, G., Pashev, I., and Zlatanova, J. (1992) Replication and Transcription of Chromatin, CRC Press, Boca Raton.
60. Jackson, V. (2002) In: Zlatanova, J. and Leuba, S.H. (eds.) Chromatin Structure and Dynamics: State-of-the-Art, Elsevier, Amsterdam, pp. 467–491.
61. Cook, P.R. (1999) Science 284, 1790–1795.
62. Wang, M.D., Schnitzer, M.J., Yin, H., Landick, R., Gelles, J., and Block, S.M. (1998) Science 282, 902–907.
63. Wuite, G.J., Smith, S.B., Young, M., Keller, D., and Bustamante, C. (2000) Nature 404, 103–106.
64. Ladoux, B., Quivy, J.P., Doyle, P., du Roure, O., Almouzni, G., and Viovy, J.L. (2000) Proc. Natl. Acad. Sci. USA 97, 14251–14256.
65. Bennink, M.L., Pope, L.H., Leuba, S.H., de Grooth, B.G., and Greve, J. (2001) Single Mol. 2, 91–97.
66. Strick, T.R., Allemand, J.F., Bensimon, D., Bensimon, A., and Croquette, V. (1996) Science 271, 1835–1837.
67. Strick, T., Allemand, J., Croquette, V., and Bensimon, D. (2000) Prog. Biophys. Mol. Biol. 74, 115–140.
68. Widom, J. (1999) In: Becker, P.B. (ed.) Methods Mol. Biol, Humana Press, Totowa, New Jersey, pp. 61–77.
69. Leuba, S.H., Karymov, M.A., Liu, Y., Lindsay, S.M., and Zlatanova, J. (1999) Gene Ther. Mol. Biol. 4, 297–301.
70. Leuba, S.H., Zlatanova, J., Karymov, M.A., Bash, R., Liu, Y.-Z., Lohr, D., Harrington, R.E., and Lindsay, S.M. (2000) Single Mol. 1, 185–192.
71. Rief, M., Gautel, M., Oesterhelt, F., Fernandez, J.M., and Gaub, H.E. (1997) Science 276, 1109–1112.
72. Oberhauser, A.F., Marszalek, P.E., Erickson, H.P., and Fernandez, J.M. (1998) Nature 393, 181–185.

73. Smith, S.B., Cui, Y., and Bustamante, C. (1996) Science 271, 795–799.
74. Cluzel, P., Lebrun, A., Heller, C., Lavery, R., Viovy, J.L., Chatenay, D., and Caron, F. (1996) Science 271, 792–794.
75. Noy, A., Vezenov, D.V., Kayyem, J.F., Meade, T.J., and Lieber, C.M. (1997) Chem. Biol. 4, 519–527.
76. Cui, Y. and Bustamante, C. (2000) Proc. Natl. Acad. Sci. USA 97, 127–132.
77. Bennink, M.L., Leuba, S.H., Leno, G.H., Zlatanova, J., de Grooth, B.G., and Greve, J. (2001) Nat. Struct. Biol. 8, 606–610.
78. Brower-Toland, B.D., Smith, C.L., Yeh, R.C., Lis, J.T., Peterson, C.L., and Wang, M.D. (2002) Proc. Natl. Acad. Sci. USA 99, 1960–1965.
79. Hayes, J.J. and Hansen, J.C. (2002) Proc. Natl. Acad. Sci. USA 99, 1752–1754.
80. Evans, E. and Ritchie, K. (1999) Biophys. J. 76, 2439–2447.
81. Luger, K., Mader, A.W., Richmond, R.K., Sargent, D.F., and Richmond, T.J. (1997) Nature 389, 251–260.
82. Bennink, M.L. (2000) Ph.D. Thesis, University of Twente, Twente, The Netherlands.
83. Luger, K. and Richmond, T.J. (1998) Curr. Opin. Struct. Biol. 8, 33–40.
84. Fritzsche, W., Schaper, A., and Jovin, T.M. (1994) Chromosoma 103, 231–236.
85. Fritzsche, W., Schaper, A., and Jovin, T.M. (1995) Scanning 17, 148–155.
86. Qian, F.L., Liu, Z.X., Zhou, M.Y., Xie, H.Y., Jiang, C., Yan, Z.J., Li, M.Q., Zhang, Y., and Hu, J. (1997) Cell Res. 7, 143–150.
87. Bustamante, C., Zuccheri, G., Leuba, S.H., Yang, G., and Samori, B. (1997) Methods 12, 73–83.
88. Sato, M.H., Ura, K., Hohmura, K.I., Tokumasu, F., Yoshimura, S.H., Handoka, F., and Takeyasu (1999) FEBS Lett. 452, 267–271.
89. Simpson, R.T., Thoma, F., and Brubaker, J.M. (1985) Cell 42, 799–808.

J. Zlatanova and S.H. Leuba (Eds.) *Chromatin Structure and Dynamics: State-of-the-Art*

DOI: 10.1016/S0167-7306(03)39015-5

CHAPTER 15

Theory and computational modeling of the 30 nm chromatin fiber

Jörg Langowski[1] and Helmut Schiessel[2]

[1] *Division Biophysics of Macromolecules (B040), Deutsches Krebsforschungszentrum, Im Neuenheimer Feld 580, D-69120 Heidelberg, Germany. E-mail: jl@dkfz.de*
[2] *Max Planck Institute for Polymer Research, Theory group, PO Box 3148, D-55021 Mainz, Germany. E-mail: heli@mpip-mainz.mpg.de*

1. *Introduction*

For its fundamental importance in gene regulation and epigenetics, the physical chemistry of structural changes in the chromatin fiber has become a major focus of interest in recent years. It is now clear that the mechanism of chromatin remodeling, opening and closing of the structure during transcription, and many other biological processes related to the higher order structure of the genome cannot be understood without fundamental knowledge of the arrangement of nucleosomes and DNA in the chromatin fiber and its variations during different physiological states of the cell.

The "bead-chain" structure of nucleosomes spaced regularly on DNA will compact into a higher order structure under physiological conditions. The generally accepted view is that the first stage of compaction of the nucleosome chain is a fiber-like structure with a diameter of approximately 30 nm, the so-called "30-nm fiber", although alternative structures have been proposed (e.g., the "superbeads" seen in electron microscopy images by Zentgraf and Franke [1]). Since the discovery of the bead-chain structure of chromatin by Olins and Olins [2,3] and Woodcock [4], many attempts have been made to construct models for the path of the DNA inside this fiber and its possible conformations. Although many new insights have been obtained, the picture is not yet conclusive and the precise arrangement of DNA and histones inside the 30 nm fiber is still controversial [5–8]. Mainly two competing classes of models have been discussed: the *solenoid* models [9–11]; and the *zig-zag* or *crossed-linker* models [12–17]. In the solenoid model (Fig. 1a) it is assumed that the chain of nucleosomes forms a helical structure with the nucleosome axis being almost perpendicular to the solenoid axis. The DNA entry–exit side faces inward towards the axis of the solenoid. The linker DNA is required to be bent in order to connect neighboring nucleosomes in the solenoid. The other class of models assumes *straight* linkers that connect nucleosomes located on opposite sides of the fiber. The axis of the nucleosomal disk is slightly tilted with respect to the fiber direction. The resulting linker geometry shows a three-dimensional zig-zag-like pattern (Fig. 1b).

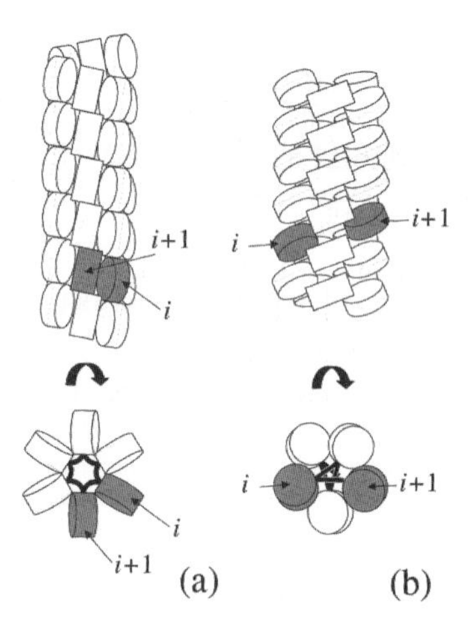

Fig. 1. The two competing classes of models for the 30-nm fiber can be distributed into (a) solenoid models and (b) crossed-linker models. For both fiber types the side and the top view are shown. Two nucleosomes that are directly connected via DNA linker are shaded in grey. In the solenoid these nucleosomes are located on the same side of the fiber requiring the linker to be bent. In the crossed-linker case they sit on opposite sides of the fiber and are connected via a straight linker.

Images obtained by cryo electron microscopy should in principle be able to distinguish between the structural features proposed by the different models mentioned above [16]. The micrographs show a zig-zag motif at lower salt concentrations and they indicate that the chromatin fiber becomes more and more compact when the ionic strength is raised towards the physiological value (i.e., about 150 mM monovalent ions).

From the physics point of view, the system that we deal with here—a semiflexible polyelectrolyte that is packaged by protein complexes regularly spaced along its contour—is of a complexity that still allows the application of analytical and numerical models. For quantitative prediction of chromatin properties from such models, certain physical parameters must be known such as the dimensions of the nucleosomes and DNA, their surface charge, interactions, and mechanical flexibility. Current structural research on chromatin, oligonucleosomes, and DNA has brought us into a position where many such elementary physical parameters are known. Thus, our understanding of the components of the chromatin fiber is now at a level where predictions of physical properties of the fiber are possible and can be experimentally tested.

Computational modeling can be a very powerful tool to understand the structure and dynamics of complex supramolecular assemblies in biological systems. We need to sharpen the definition of the term "model" somewhat, designating a procedure that allows us to *quantitatively* predict the physical properties of the system. In that sense, the simple geometrical illustrations in Fig. 1 only qualify if by some means experimentally accessible parameters can be calculated. As an example, a quantitative treatment of DNA bending in the solenoid model would only be possible if beyond the mechanical and charge properties of

DNA and nucleosomes the energetics of linker DNA–histone and nucleosome-nucleosome interaction necessary to overcome the elastic stress in the DNA due to the bend were quantitatively known. The straight linker configuration offers the advantage for modeling that the geometry and the energetics of the fiber are controlled to a large extent by the linker DNA, whose conformations and mechanical properties are very well understood, and its energetics are favorable because no energy is required to bend the DNA. It therefore comes as no surprise that most recent numerical and analytical models assume straight linkers (at elastic equilibrium, notwithstanding thermal fluctuations, v.i.), which is also supported by current experimental data.

In this chapter we attempt to give an overview of current computational modeling approaches of the 30 nm fiber, their capabilities and limitations. In order to setup a quantitative model of this supramolecular assembly, we first need to understand the physical properties of its subcomponents: nucleosomes and DNA.

2. *Physical properties of nucleosomes and DNA*

DNA, on a length scale beyond some tens of base pairs, can be described as a stiff polymer chain with an electrostatic charge. In the most basic view, four parameters are sufficient to describe this chain with sufficient precision: the diameter, the elastic constants for bending and torsion and the electrostatic potential.

The average DNA helix diameter used in modeling applications such as the ones described here includes the diameter of the atomic-scale B-DNA structure and—approximately—the thickness of the hydration shell and ion layer closest to the double helix [18]. Both for the calculation of the electrostatic potential and the hydrodynamic properties of DNA (i.e., the friction coefficient of the helix for viscous drag) a helix diameter of 2.4 nm describes the chain best [19–22]. The choice of this parameter was supported by the results of chain knotting [23] or catenation [24], as well as light scattering [25] and neutron scattering [26] experiments.

The bending elasticity of the DNA chain can be expressed as the *bending persistence length* L_p, the distance over which the average cosine of the angle between the two chain ends has decayed to 1/e. Molecules shorter than L_p behave approximately like a rigid rod, while longer chains show significant internal flexibility. The bending elasticity A—the energy required to bend a polymer segment of unit length over an angle of 1 radian—is related to the persistence length by $L_p = A/k_B T$, k_B being Boltzmann's constant and T the absolute temperature. For DNA, L_p has been determined in number of experiments (for a recent compilation, see Ref. [27]). While some uncertainties remain as regards the value at very high or low salt concentrations, the existing data agree on a consensus value of $L_p = 45$–50 nm (132–147 bp) at intermediate ionic strengths (10–100 mM NaCl and/or 0.1–10 μM Mg^{2+}).

The torsional elasticity C, defined as the energy required to twist a polymer segment of unit length through an angle of 1 radian, may be related in an analogous way to a *torsional persistence length* L_T, which is the correlation length of the torsional orientation of a vector normal to the chain axis. Again, C is related to L_T by $L_T = C$ where $k_B T$. C has been measured by various techniques, including fluorescence polarization anisotropy decay [28–30] and DNA cyclization [31–33], and the published values converge on a torsional persistence length of 65 nm (191 bp).

The stretching elasticity of DNA has been measured by single molecule experiments [34,35] and also calculated by molecular dynamics simulations [36,37]. One may estimate the stretching elasticity of DNA to be given by a stretching modulus σ of about 1500 pN, where $\sigma = F \cdot \Delta L / L_0$ (ΔL being the extension of a chain of length L_0 by the force F). We may safely assume that DNA stretching does not play a significant role in chromatin structural transitions, since much smaller forces are already causing large distortions of the 30 nm fiber (see below).

The structure of the second important component of the chromatin fiber, the histone octamer [38] respectively the nucleosome [39], has been determined by X-ray crystallography to atomic resolution. From this structure, one can approximate the overall dimensions of the nucleosome as a flat disk of 11 nm diameter and 5.5 nm thickness. The mechanical properties of the nucleosome are not known, and current models take it as non-deformable, which is probably a good approximation in the range of forces usually employed in single-molecule experiments on chromatin fibers (Sivolob *et al.* showed quite a high stability of nucleosomes even when the DNA was under considerable superhelical stress [40], however the $(H3\text{–}H4)_2$ tetramer showed a structural transition [41]). Also, the nucleosomal DNA is usually assumed as being rigidly attached to the histone core while the elasticity of the DNA at the point where the linker leaves the nucleosome is the same as that for free DNA; the unwrapping of the DNA from the histone core under tension, which is experimentally proven [42,43], is not yet included in current models, mainly because of lack of quantitative thermodynamic data on core particle-DNA binding. Very recently a theory has been developed that accounts for tension-induced nucleosome unwrapping; the theoretical predictions of typical unwrapping forces are in good agreement with the experimental observations (I. Kulic and H. Schiessel, preprint cond-mat/0302188).

The interaction between nucleosomes plays an important role for the stability of the 30 nm fiber; recent experiments on liquid crystals of mononucleosomes [44–47] and also less concentrated mononucleosome solutions [48,49] show an attractive interaction that can be parameterized by an anisotropic Lennard-Jones type potential [50]. Also, an electrostatic interaction potential has been computed using the crystallographic structure of the nucleosome [51]. The influence of these potentials on the structure of the fiber is discussed below together with the corresponding models.

3. *Computational implementation*

The computational description of a large biomolecular complex such as the chromatin fiber requires techniques that are different from the widely applied molecular dynamics methods used to simulate biopolymers at atomic resolution.

The current limits (beginning of 2003), for which one can still solve Newton's equations of motion explicitly for each atom in its force fields in a reasonable amount of computer time, are a system size of 10^5–10^6 atoms (including the water) and simulation lengths of some nanoseconds. Chromatin fragments of biologically interesting size (i.e., more than 12 nucleosomes) are much more complex, and the times at which typical processes such as nucleosome opening, DNA slipping, etc. take place, are much longer. It is evident that in such a case the biological macromolecule must be described by some approximation.

4. Energetics: coarse-graining and interaction potentials

All such approximations are based on "coarse-graining": subunits are defined that contain many atoms, but behave like rigid objects on the time and length scales considered. As an example, let us look at a chain of "naked" DNA. If molecular detail is not required, DNA may be approximated by a chain of segments that are substantially shorter than its bending or torsional persistence length; segments up to 50 bp constitute a safe choice. The segments or other subunits interact through appropriate force fields, which can in principle be derived from the interatomic potentials. With the development of all-atom molecular dynamics (MD) in recent years, detailed studies have been possible on the molecular basis of DNA flexibility and on its sequence dependence. While a comprehensive overview of that work falls outside the scope of this article, we would like to mention some recent studies that can serve as starting points to the reader looking for information in this field: normal-mode analyses of the sequence dependence of DNA flexibility [52,53] or work from our own laboratory in which DNA flexibility constants are extracted from molecular dynamics trajectories of DNA oligonucleotides [36,37]. A review of this field is given in Ref. [54]. Since the absolute reliability of force constants determined from MD simulations is still limited, however, one usually employs parameters that are determined separately by experiment.

For coarse-grained models of linear biopolymers—such as DNA or chromatin—two types of interactions play a role. The connectivity of the chain implies stretching, bending, and torsional potentials, which exist only between directly adjacent subunits and are harmonic for small deviations from equilibrium. As mentioned above, these potentials can be directly derived from the experimentally known persistence length or by directly measuring bulk elastic properties of the chain.

The second type of interactions necessary for modeling the chromatin fiber occurs between non-adjacent subunits of the chain, i.e., contact or other long-range interactions between nucleosomes and/or DNA segments. DNA–DNA interactions in a typical biological environment are mainly due to electrostatic repulsion between the negatively charged backbone phosphates. For intermediate ionic strengths (1 mM–1 M) and distances larger than about the DNA double helix diameter of 2 nm, these interactions are rather well characterized. A number of simulations simply used an effective hard-core radius, below which the DNA segments are not allowed to approach each other, and which decreases with increasing ionic strength [23,55–59]. Other work described the electrostatic interaction through a screened Coulomb potential in the Debye-Hückel approximation, using a renormalized surface charge density for the DNA in order to account for ion condensation [18]. This has the advantage that the dependence of the interaction on the ionic strength does not have to be calibrated empirically, but is contained implicitly in the form of the potential. DNA models based on such potentials have been developed and applied widely in the last two decades, and their predictive power for average solution structural properties, intramolecular interaction kinetics and other globally measurable parameters is usually very good [19–22,26,60–64]. One can assume that the approximation of a DNA chain as an elastic filament with Debye-Hückel electrostatics constitutes a safe choice for obtaining a realistic picture of the solution structure.

Less is known about the interaction of the nucleosomes between themselves or with free DNA. The nucleosome-nucleosome interaction has recently been parameterized by using the surface charge density of the known crystal structure [39] in a point-charge model [51]. While in that work only electrostatic interactions were considered and the quantitative influence of the histone tails on the interaction potential still remains obscure, simulations based on this potential allowed to predict an ionic-strength dependent structural transition of a 50-nucleosome chromatin fragment that occurred at a salt concentration compatible with known experimental data (Ref. [65], see below).

Nucleosome–nucleosome interaction potentials can be calibrated by comparison with the characteristics of liquid crystals of mononucleosomes at high concentrations. Under suitable conditions, nucleosome core particles form a hexagonal-columnar phase with a distance of 11.55 ± 1 nm between the columns and a mean distance of 7.16 ± 0.65 nm between the particles in one column [44,46]. These distances may be assumed to correspond to the positions of the minima of an attractive internucleosomal potential. The depth of the interaction potential (i.e., the binding energy per nucleosome) was estimated in the stretching experiments of Cui and Bustamante [66] to 2.6–3.4 kT. A slightly lower potential minimum of 1.25 kT is obtained by a comparison of the stability of the nucleosome liquid crystal phase with simulations [50].

The DNA–nucleosome interaction parameters are not known at present. In most of the theoretical work it is deemed negligible compared to the DNA–DNA and nucleosome–nucleosome interaction, except for a hard-core excluded volume interaction. Nevertheless, recent work on the mechanism of nucleosome repositioning [67] assumes that the DNA can dynamically detach from the nucleosome surface and reattach in different conformations, such that it is conceivable that distant DNA segments may also transiently bind to "open regions" of the DNA-binding surface of the nucleosome.

5. *Mechanics of the chromatin fiber*

The conformations of the "zig-zag" model first proposed by Woodcock *et al.* [12] and van Holde and Zlatanova [5] are determined by the linker DNA length, the nucleosome shape and two angles: the opening angle θ of the linker DNAs at the nucleosome and the twisting angle φ between successive nucleosomes on the chain. Since for a given nucleosome spacing (and therefore linker length) the geometry of the resulting chain is uniquely determined by θ and φ, this model is also called the "two-angle" model of the chromatin chain. Simply generating conformations of polynucleosome chains with varying values of θ and φ will create chains that are already very similar to typical conformations of chromatin fibers as seen in cryo-electron microscopy [12] or scanning force microscopy [15].

For the solenoid conformation, few attempts at a quantitative description exist. One notable exception is the work presented by Bishop and Hearst [68] and Bishop and Zhmudsky [69]. In their approach the chromatin fiber is described as a continuous "coiled-coil" filament of an elastic polymer, in which the DNA forms a continuous 11 nm diameter spiral that is wound into a 30 nm superhelix. The nucleosomes are not assigned to fixed positions, but viewed as bound in a delocalized manner, leading—on average—to a "fluid-like" structure for the 30 nm fiber. Under these conditions, the chromatin elasticity mostly depends on the mechanical properties of the DNA coil, and the histones may be regarded as some viscous fluid that sticks non-specifically to the DNA surface. This view is rationalized because in its biological function the energy landscape of nucleosome positioning is "fairly smooth with multiple local minima", because of the rather low specificity of nucleosome positioning. The authors calculate the elastic constants for stretching, shearing, and torsion, the linear mass density and moment of inertia for linear DNA and the DNA/histone coil at various degrees of compaction. While qualitative conclusions could be drawn about the relative elasticities of DNA and the chromatin coil, the absolute values of the elasticity parameters differ quite strongly from the known experimental data in that model. However, with more knowledge about the histone–DNA interaction, the approach by Bishop and co-workers could be valuable for comparing the energetics of zig-zag and solenoid conformations.

For the moment, however, we concentrate here on variants of the two-angle model. We will first give a systematic account on the possible fiber geometries based on straight linkers, then show how the mechanical properties of the model can be understood as being a result of the fiber geometry. It will become clear why it is indispensible to go beyond purely theoretical descriptions, which focus mainly on the geometry and energetics of the linker DNA, to numerical computer models in order to obtain the full picture. It turns out that in determining the fiber properties the interplay between linker DNA stiffness and nucleosome–nucleosome interaction is crucial.

5.1. The "two angle" model—basic notions

Since the linker DNA is assumed straight and the nucleosome non-deformable, the fiber geometry of the two-angle model is completely determined by the entry–exit angle of the linker DNA at each nucleosome and by the rotational angle

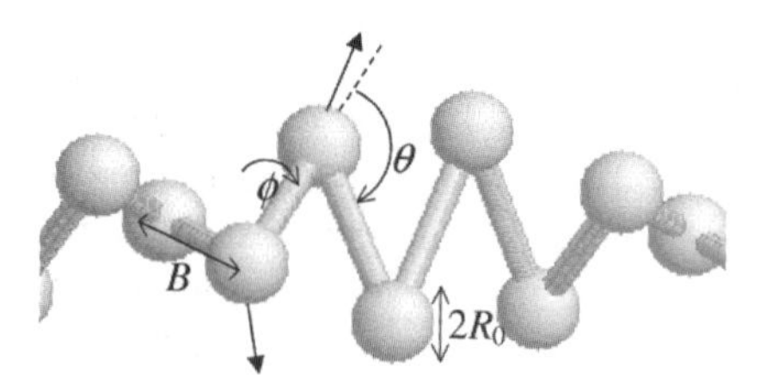

Fig. 2. The geometry of the two-angle fiber can be characterized by the deflection angle θ and the dihedral angle ϕ. Also indicated is B, characterizing the linker length, and the nucleosome diameter $2R_0$. For simplicity, the nucleosomes are represented by spheres connected via links, the linker DNA. The arrows indicate the nucleosomal superhelix axis.

between neighboring nucleosomes. Depending on the values of these angles and their variation, the structures obtained are either completely regular or more random fibers that resembled real fibers at lower ionic strength. As far as the linker geometry can be detected via cryo-electron microscopy [16] or scanning force microscopy [15,70], this model indeed describes the geometry of the 30-nm fiber adequately.

Schiessel *et al.* [17] introduced a mathematical description for the different possible folding pathways in the two-angle model. At the simplest level, it was assumed that the geometric structure of the 30-nm fiber can be obtained from the intrinsic, single-nucleosome structure where the incoming and outgoing linker chains make an angle θ with respect to each other. In the presence of the histone H1 (or H5) the in- and outcoming linker are in close contact forming a stem before they diverge [16].

Next, there is a rotational (dihedral) angle ϕ between the axis of neighboring histone octamers along the necklace (see Fig. 2). Because nucleosomes are rotationally positioned along the DNA, the angle ϕ is a periodic function of the linker length B, with the 10 bp repeat length of the helical twist of DNA as the period. There is experimental evidence that the linker length is preferentially quantized by integral multiples of this helical twist [71], i.e., there is a preferred value of ϕ.

The geometrical structure of the chain in Fig. 2 is determined entirely by θ, ϕ, and B. The model only describes the linker geometry and does not account for excluded volume effects and other forms of nucleosome–nucleosome interaction; it assumes that the core particles are point-like and that they are located at the joints of the linkers, which are straight rods.

An overview of the possible two-angle fibers is provided in Fig. 3, where θ and ϕ are varied over the range 0–180° (for a more thorough discussion of the possible structures, see Schiessel *et al.* [17].

Various examples of two-angle fibers were already displayed by Woodcock *et al.* [12] in their Fig. 2, namely fibers with $\theta = 150°$ and many different values of ϕ, corresponding to a vertical trajectory on the right-hand side of Fig. 3. Three different configurations with a fixed value of ϕ and different values of θ are displayed in Fig. 3c in another paper by these authors [16]. Schiessel *et al.* [17] were able to obtain analytical expressions for all the geometrical quantities that characterize

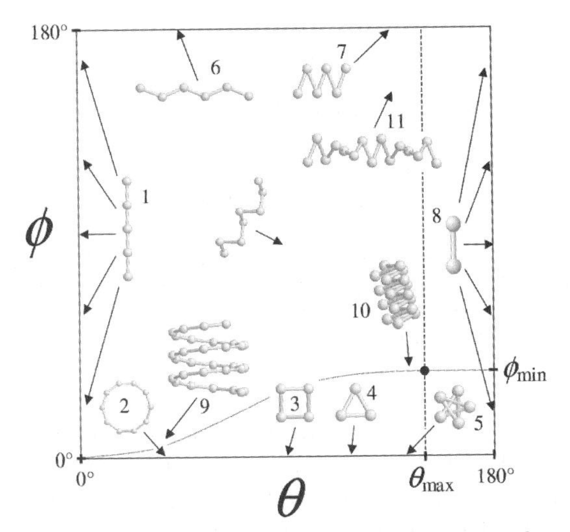

Fig. 3. The full range of the two-angle structures. Shown are some example configurations with the arrows indicating their position in the (θ, ϕ)-plane. The lines denote the boundaries to the forbidden structures where the "nucleosomes" would overlap. To the right of the dashed line nucleosomes i and $i+2$ would overlap ("short-range excluded volume"). Below the dotted line nucleosomes further apart with respect to their chemical distance would overlap ("long range excluded volume"). For instance, for the circle, structure "2", nucleosomes collide after one turn (10 nucleosomes).

the resulting fiber as a function of θ, ϕ, and B (cf. Ref. [72]). Such quantities are, for instance, the diameter of the fiber, the nucleosome line density or the nucleosomal tilt angle.

If one takes into account the excluded volume of the core particles, then certain areas in that phase diagram, Fig. 3, are forbidden—reminiscent of the familiar Ramachandran plots used in the study of protein folding. This is indicated in Fig. 3 by dashed and dotted lines. Except for these regions, however, the diagram in Fig. 3 by itself does not favor any structure over another, since the energetics are not included. Schiessel *et al.* [17] suggested that the optimum structure for the 30-nm fiber might be a balance between (*i*) *maximum compaction* and (*ii*) *maximum accessibility*, in order to fit the DNA chain into the limited nuclear volume (cf. Ref. [73] to see how severe this packing problem actually is) and to keep local accessibility, which is required for transcription.

In order to attain maximum compaction one needs structures that lead to high bulk densities. A comparison of the densities of all possible structures shows that fibers with internal linkers have highest densities [17]. As detailed there, the highest density is achieved for the largest possible value of θ and the smallest possible value of ϕ that is still in accordance with the excluded volume condition, corresponding to the black dot in Fig. 3 where the dotted curve and the dashed line cross each other. This unique set of angles is given by $\theta_{max} = 2 \arccos(R_0/B)$ and $\phi_{min} \cong (8/\pi)(R_0/B)$. It was suggested in Ref. [17] that maximum accessibility is achieved for structures that, for a given entry–exit angle $\pi-\theta$ of a highly compacted structure, lead to the maximum reduction in nucleosome line density for a given small change of the angle θ. Interestingly, that the such defined accessibility is maximized at the same unique pair of angles (θ_{max}, ϕ_{min}). The corresponding fiber shows a crossed-linker geometry as depicted in Fig. 1b (cf. also structure "10" in Fig. 3). For reasonable values of the linker length it was found [17] that $\theta_{max} = 151°$, $\phi_{min} = 36°$ together with a nucleosome line density $\lambda = 6.9$ nucleosomes per 11 nm [72] and a diameter of the order of 30 nm, values that are close to the experimental ones reported by Bednar *et al.* [16] for chicken erythrocyte chromatin fibers.

The local accessibility can be controlled *in vitro* by changing the salt concentration. Bednar *et al.* [16] report, for example, that θ decreases with decreasing ionic strength, namely $\theta \approx 145°$ at 80 mM, $\theta \approx 135°$ at 15 mM and $\theta \approx 95°$ at 5 mM [16]. In the biochemical context the change of θ can be accomplished by other mechanisms, especially by the depletion of linker histones and the acetylation of core histone tails (cf., for instance [74]), both of which occur in transcriptionally active regions of chromatin (for details compare [75]). These mechanisms lead effectively to a decrease of θ causing the linear nucleosome density to decrease $\lambda \approx 6.0$ (80 mM salt) via $\lambda \approx 3.2$ (15 mM) to $\lambda \approx 1.5$ (5 mM) [16], in good agreement with theoretical predictions [17].

Beyond pure geometry, the two-angle model is also useful to predict some of the *physical properties* of the 30-nm fiber, for instance, its response to elastic stress [17]. In an independent study on the two-angle model by Ben-Haim *et al.* [76] this question has been the major focus, and as demonstrated by Schiessel [72], the elastic properties of the two-angle model as a function of θ and ϕ are analytically solved completely by combining the results from both papers.

The elastic stress may be external or internal. External stresses are exerted on the chromatin during the cell cycle when the mitotic spindle separates chromosome pairs. The 30-nm fiber should be both highly flexible and extensible to survive these stresses. The *in vitro* experiments by Cui and Bustamante demonstrated that the 30-nm fiber is indeed very "soft" [66]. The 30-nm fiber is also exposed to internal stresses. Attractive or repulsive forces between the nucleosomes will deform the linkers connecting the nucleosomes. For instance, electrostatic interactions, either repulsive (due to the net charge of the nucleosome core particles) or attractive (bridging via the lysine-rich core histone tails [49]) could lead to considerable structural rearrangements.

The purpose of this chapter is not to present further mathematical details of the two-angle fiber. We merely mention that its mechanical properties can be described by four moduli, namely the stretching, bending, and twisting moduli as well as a twist-stretch coupling constant [76]. Thus the two-angle fiber behaves as an *extensible* worm-like chain as compared to naked DNA that is inextensible for moderate forces. The extensibility of two-angle fibers can be easily understood as a result of the DNA bending/twisting. A special case is depicted in Fig. 4: the planar zig-zag fiber. The external tension F induces an increase of the contour length from the unperturbed value L_0 (Fig. 4a) to the new value $L > L_0$ (Fig. 4b). In the zig-zag case this is accomplished via the bending of the linker DNA only. For more general geometries fiber stretching involves linker twisting as well.

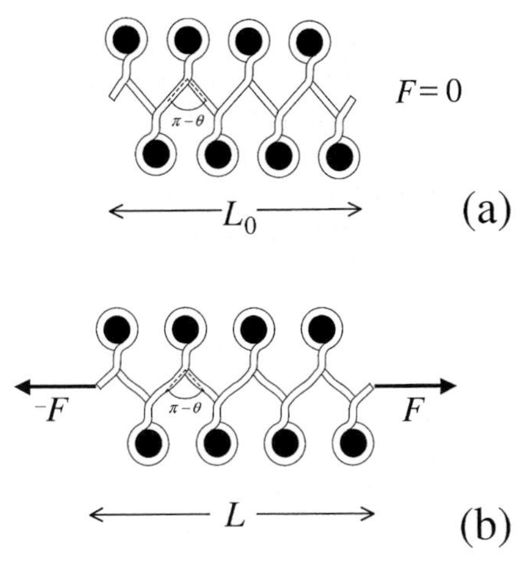

Fig. 4. Two-angle fibers can be easily deformed via the bending and twisting of their linkers. This can be most easily seen for the special case of a planar zig-zag fiber under an external tension F, which extends the fiber via the bending of its linkers from its unperturbed state with contour length L_0 (a) to a stretched state of length $L > L_0$ (b).

The linker backbone is predicted to be very soft: For instance, the stretching modulus of fibers with crossed linkers and zig-zag chains is of the order one (per nm) for an effective linker length of 20 bp as compared to the much larger value of 300 nm^{-1} for naked DNA. Of course, depending on the values of θ, ϕ, and linker length, this value varies over a wide range. Also the other mechanical parameters of the two-angle fiber indicate an extremely soft structure, as long as the elastic properties are determined by the DNA backbone alone. It is thus evident that the presence of the nucleosomes must play a crucial role in determining many of the mechanical properties of the 30-nm fiber. For instance, the excluded volume will not allow a strong fiber bending that would lead to overlapping nucleosomes, and the nucleosome–nucleosome attraction counteracts the stretching of the fiber under external tension.

To account for these effects, one has to go beyond the simple geometrical two-angle model and must include nucleosome–nucleosome interactions. On a purely theoretical level this is a hard task, and there have only been approximate estimates to the tension-induced condensation–decondensation transition of the fiber [17]. Numerical computer models are of great help to clarify the picture since it is relatively straightforward to build in linker elasticity, nucleosome/nucleosome interactions and other force fields as they become necessary to describe structural details. Furthermore, while the two-angle model is of great help in exploring those regions of conformational space that are accessible to the chromatin chain, it generates very regular structures. Local variations in nucleosome positioning and, most importantly, thermal fluctuations at room temperature will cause random deviations from this regular structure; to understand the dynamics of the chromatin chain, it is very important to include these fluctuations into the model. The same issue has been discussed in the DNA field some time ago: while elastic rod models were successful in early theoretical descriptions of DNA supercoiling (e.g., [77–79], for review see Ref. [80]), the computation of real-life structural properties of superhelical DNA is only possible with numerical models that include thermal fluctuations (for a discussion of this point, see also Ref. [81]). Here we will therefore now consider methods that create configurational ensembles which reflect the structure of the chromatin chain at a finite temperature.

6. The chromatin chain at thermodynamic equilibrium

To find the equilibrium structure of the chromatin chain, one could try and minimize the total energy of the two-angle model computed numerically from appropriate elastic tension, electrostatic attraction/repulsion and other interaction potentials. This energy would then be minimized by standard minimization procedures, leading to a conformation at elastic equilibrium. If the molecule is small enough, its structural fluctuations will be small compared to its overall size, and then this simple energy minimization can in principle find a structure close to the structure free in solution at room temperature. However, in the case of a complex polymer chain typically only one or a few minimum energy conformations are predicted. As the temperature becomes larger than absolute zero (certainly at physiological temperatures), thermal motion leads to a multitude of possible conformations with almost the same energy; as long as the energy difference between two conformations is less than kT, both will be present at equilibrium with significant probability. The chance for the minimum elastic energy conformation to occur at thermodynamic equilibrium is actually very close to zero in this case.

Thus, to reflect correctly the state of the system at a certain temperature, one needs a simulation technique that generates an ensemble of chain conformations whose statistical properties reflect those of the "real" chain at thermodynamic equilibrium. The chromatin fiber models that are discussed in the following use one or the other variation of such techniques.

6.1. Metropolis Monte-Carlo

Since minimum *energy* structures do not adequately describe the average structure of large biopolymers in an ensemble at thermodynamic equilibrium, one must minimize the *free energy*, which includes the configurational entropy of the structure. Depending on whether time dependent information is to be calculated, different approaches can be taken for such a minimization. A very popular method to sample the thermodynamic equilibrium of a system is known under the name of "importance-sampling Monte-Carlo" and has been described half a century ago by Metropolis *et al.* [82]. The "Metropolis Monte-Carlo" algorithm works as follows:

(1) Starting from a given initial conformation with energy E_1, we generate a new "trial" conformation by a random variation (e.g., in the position or

orientation of a subunit) and calculate its energy E_2. For a model chromatin chain, the total energy of each particular conformation can be very easily calculated if all the interaction energies between the constituents, DNA and nucleosomes, are known (as is the case for the interactions discussed above).

(2) If $E_2 \leq E_1$, then we accept the new conformation into the ensemble.

(3) If $E_2 > E_1$, we calculate the Boltzmann factor p for the increase in energy, $p = e^{-(E_2 - E_1)/kT}$, and then compare this factor to a random number x between 0.0 and 1.0. If $p > x$, we accept the new conformation into the ensemble, otherwise we reject it and add the old conformation once more to the ensemble.

This algorithm will generate the ensemble of conformations at thermodynamic equilibrium, if the conformational variations introduced in step 1 (the so-called "moves") are sufficient to cover all possible regions of conformational space and are locally reversible. It is easy to understand that feature of the Metropolis procedure: the transition probability into the higher energy state is given by the Boltzmann factor, the transition probability into the lower energy state is one. The relative population of the two states is then the ratio of the transition probabilities, thus the Boltzmann factor; this is the outcome expected from thermodynamics.

6.2. Brownian dynamics simulation

For calculating the *time-dependent* properties of biopolymers, the equations of motion of the molecule in a viscous medium (i.e., water) under the influence of thermal motion must be solved. This can be done numerically by the method of Brownian dynamics (BD) [83]. Allison and co-workers [61,62,84] and later others [85–88] have employed BD calculations to simulate the dynamics of linear and superhelical DNA; BD models for the chromatin chain will be discussed below.

Like in the case of Monte-Carlo, the internal energy of the chain is computed from the interaction potentials. Forces are then calculated from the derivatives of these energies with respect to the coordinates, and allowed to act during a certain time step (typically a few nanoseconds). This procedure will relax a given initial conformation into a state that fluctuates around the thermodynamic equilibrium. Since the motions at each step of the simulation come from the solution of equations of motion, including real forces, viscosities, and thermal energies, the BD approach has the advantage that the calculated trajectories describe the real-time motion of the molecule in its solvent. Thus quantities related to the motions of the polymer like translational and rotational diffusion coefficients or internal relaxation modes can be obtained from the model using only the known physical properties of the constituents, as discussed above.

7. *Monte-Carlo modeling of the chromatin fiber*

The first Monte-Carlo model of the chromatin chain was introduced by Katritch *et al.* [89] to interpret the stretching experiments on chromatin fibers by Cui and Bustamante [66]. Their model approximates the nucleosome as a spherical particle with an effective diameter between 11 and 20 nm and an optional attractive interaction. The linker DNA was modeled as a chain of rigid segments held together through bending and twisting potentials; the angle of the linker DNA at the nucleosome was varied as one of the parameters used to fit the experimental data. Different from the two-angle model described above, the authors did not fix the value for the rotational angle ϕ between each pair of neighboring nucleosomes. The energy function used in the Monte-Carlo model included a constant stretching force. The best fit to the experimental force-extension curves at low salt (5 mM) was obtained with an effective nucleosome diameter of 14 nm, a free linker length of 40 base pairs and a linker DNA opening angle of 50°. Other parameters of the model, such as the twist between nucleosomes, do not influence the computed force-extension characteristics significantly. To reproduce the response of chromatin fibers to extension at high salt conditions, it was necessary to introduce an attractive potential between closely spaced nucleosomes, which was a simple step function and radially symmetric. The equilibrium conformations of chromatin fibers generated by the model without an applied force, however, are quite irregular and do not resemble the typical 30 nm fiber; rather, even small ($<2\,kT$) internucleosomal attractions collapse the chain into a highly condensed form.

To overcome the restrictions of the Katritch model due to the approximation of the nucleosome as a sphere, Wedemann and Langowski [50] developed a Monte-Carlo model for the 30 nm fiber in which the nucleosome core particles are represented by oblate ellipsoids. The linker DNA was again described as a segmented elastic polymer, whose segments are connected by elastic bending, torsional, and stretching springs. The electrostatic interaction between the DNA segments was described by the Debye-Hückel approximation.

An important addition compared to previous models was the parameterization of the internucleosomal interaction potential in the form of an anisotropic attractive potential of the Lennard-Jones form, the so-called Gay-Berne potential [90]. Here, the depth and location of the potential minimum can be set independently for radial and axial interactions, effectively allowing the use of an ellipsoid as a good first-order approximation of the shape of the nucleosome. The potential had to be calibrated from independent experimental data, which exists, e.g., from the studies of mononucleosome liquid crystals by the Livolant group [44,46] (see above). The position of the potential minima in axial and radial direction were obtained from the periodicity of the liquid crystal in these directions, and the depth of the potential minimum was estimated from a simulation of liquid crystals using the same potential.

In order to connect the chromatin chain and to model the effect of the linker histone, DNA and chromatosomes were linked either at the surface of the chromatosome or through a rigid nucleosome stem 2 nm long. A Metropolis-Monte Carlo algorithm was then used to generate equilibrium ensembles of 100-nucleosome chains at physiological ionic strength. For a DNA linked at the nucleosome stem and a nucleosome repeat of 200 bp, the simulated fiber diameter of 32 nm and the mass density of 6.1 nucleosomes per 11 nm fiber length are in excellent agreement with the canonical value of 6 for the solenoid fiber as well as with experimental values from the literature, e.g., the neutron scattering data reported by Gerchman and Ramakrishnen [91] (other references given in Ref. [50]). The experimental value of the inclination of DNA and nucleosomes to the fiber axis [92] could also be reproduced.

One result of the simulations was that the torsion angle φ between successive nucleosomes determines the properties of the structure to a great extent (as also predicted by the two-angle model). While a variation in the internucleosome interaction potential by a factor of four changes that simulated mass density by only about 5%, this quantity is very sensitive to variations in twist angle (see Fig. 6 in Ref. [50]).

The linker DNA connects chromatosomes on opposite sides of the fiber, and the overall packing of the nucleosomes leads to a helical aspect of the structure. The persistence length of simulated fibers with 200 bp repeat and stem is 265 nm. For more random fibers where the tilt angles between two nucleosomes are chosen according to a Gaussian distribution along the fiber, the persistence length decreases to 30 nm with increasing width of the distribution, while the other observable parameters such as the mass density remain unchanged. Polynucleosomes with repeat lengths of 212 bp also form fibers with the expected experimental properties. Systems with even larger repeat length form fibers, but the mass density is significantly lower than the measured value. While a nucleosome chain without a stem (i.e., DNA and nucleosomes are connected at the core particle) and a repeat length of 192 bp gives stable fibers with linear mass densities in range with the experimental values, chains without a stem and a repeat length of 217 bp do not form fibers.

The persistence length computed from the bending fluctuations of the computed conformations shows an increase for shorter linker lengths, which confirms the tendency predicted by the simple two-angle model. However, the absolute values of the persistence length are between 60 and 260 nm, much higher than in the two-angle model, indicating that the nucleosome–nucleosome interactions are essential for controlling the mechanical properties of chromatin. Also, highly ordered chromatin structures are stiffer than more irregular ones: High values for the persistence length (200–300 nm) were obtained when the twist angle between adjacent nucleosomes was constant; when this twist was varied randomly, the persistence length decreased significantly.

7.1. Simulation of chromatin stretching

Cui and Bustamante [66], Bennink *et al.* [43], Brower-Toland *et al.* [42], and Leuba and Zlatanova (unpublished results) have reported single molecule stretching experiments on chromatin. For the force-extension curves, generally two regimes with different behavior can be distinguished: at low forces no structural transitions occur and the extension of the chain is determined solely by its elasticity, while at forces above 10–20 pN individual nucleosomes start to disintegrate. The dissociation of the NCPs that occurs at higher forces gives rise to distinct "jumps" in the force-extension curve, whose amplitude is directly related to the length of DNA liberated during the dissociation [42,43]. The energetics of DNA unbinding from the histones have not been characterized in detail, although some estimates exist [93,94]. Therefore, current computer models of the chromatin fiber do not yet include the "unrolling" of the DNA from the nucleosome core during stretching. We hope that this will become possible soon since a theoretical understanding of this process has now been achieved (I. Kulic and H. Schiessel, manuscript in preparation).

While structural transitions at the level of nucleosomes are still outside the scope of current models, applying a constant force to both ends of the fiber allows to simulate the low end of the force-extension curve. First results from our own work (Aumann, Caudron, Wedemann, and Langowski, manuscript in preparation) are shown in Fig. 5. In this particular example, a nucleosome repeat of 200 bp and a linker DNA length of 11 bp was used.

The stretching modulus of the chromatin fiber can be extracted from the simulations in Fig. 5. Through a comparison with the third-power dependency predicted by the two-angle model [17] we can estimate that for linker lengths of 11–24 bp, stretching moduli of 50–5 pN are obtained, in line with the existing single molecule stretching data. A more detailed analysis of the dependence of the stretching rigidity on the local geometry of the chromatin fiber and on the internucleosomal interaction is under work.

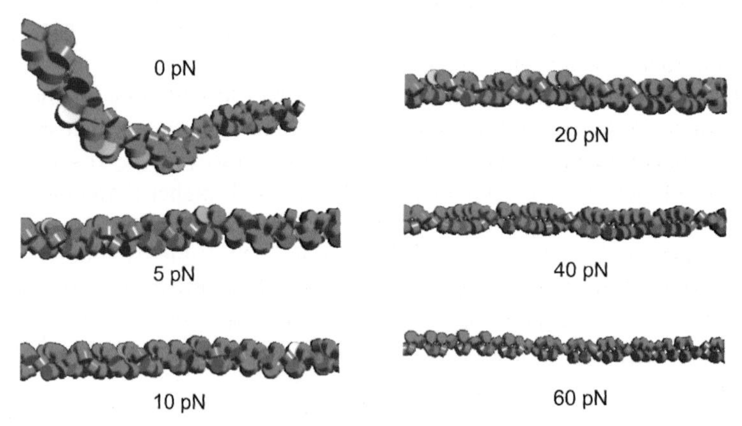

Fig. 5. Monte-Carlo simulation of the stretching of a chromatin chain. A 100 nucleosome chain with 200 bp repeat and 11 bp linker DNA was first equilibrated and then exposed to a pulling force as indicated below the drawings. For displaying the full chain, the scale of the picture was changed with increasing force. An unwinding of the chromatin fiber and sliding of the nucleosomes can be readily observed.

8. *Dynamic simulations of the chromatin fiber*

For modeling time-dependent structural changes in the chromatin chain, time scales need to be used that are by far larger than the typical nanosecond range that is used in all-atom molecular dynamics. Conformational changes, internal motions, etc. of a large biomolecule occur on a millisecond time scale or longer. In order to reach this time scale in a simulation, the molecule has to be described by a coarse-grained model (see above), and the embedding solvent also needs to be approximated by a homogeneous fluid with given viscosity, dielectric constant and ion composition. The Brownian motion of the biomolecule, which is caused by the random thermal "pushes" of the solvent molecules, is described by a random force.

In this type of modeling, called Brownian dynamics [83], an ensemble of conformations at thermodynamic equilibrium is generated in a way very analogous to the Monte-Carlo model: starting from an initial conformation, the model relaxes in steps towards the minimum free energy state. Other than in the Monte-Carlo method, however, the displacements of the subunits (beads and DNA segments) at each step are determined by the forces that act on each subunit, their viscous drag in the aqueous medium and the random force that corresponds to the thermal motion. In this way, real-time dynamics may be calculated for large systems, such as superhelical DNA [21,95], chromatin chains [65,96] or entire interphase chromosomes [97], over time scales of several hundred milliseconds.

8.1. *Brownian dynamics models of the chromatin fiber*

In an early attempt to model the dynamics of the chromatin fiber, Ehrlich and Langowski [96] assumed a chain geometry similar to the one used later by Katritch *et al.* [89]: nucleosomes were approximated as spherical beads and the linker DNA as a segmented flexible polymer with Debye-Hückel electrostatics. The interaction between nucleosomes was a steep repulsive Lennard-Jones type potential; attractive interactions were not included.

With this model, dynamics of dinucleosomes could be simulated for trajectories of 50 μs within a couple of hours of CPU time. From these trajectories, diffusion coefficients were extracted and compared with the light scattering data of Yao *et al.* [98]. The salt-dependent increase of the diffusion coefficient could be reproduced quantitatively, using a linker length of 76 bp, an effective hydrodynamic radius of the nucleosome of 5.95 nm and a linker DNA bending angle of 40°. Also, the simulated sedimentation coefficients showed good agreement with the experiments of Butler and Thomas [99]. The ionic strength dependence of the diffusion coefficient was interpreted by a change in the conformation of the linker DNA opening angle rather than changed bending of the linker DNA.

The missing internucleosome attraction, however, led to problems when longer nucleosome chains were simulated. Starting from an extended zig-zag conformation, folding of a 25-nucleosome chain occurred within 200 μs and the diameter of the resulting fiber-like structure was 45 nm in approximate agreement with values measured for chicken erythrocyte chromatin. On the other hand, the structure showed no regular helical arrangement of the nucleosomes, and the mass density of 1.3 nucleosomes/11 nm was much less than typical experimental values.

A more detailed view of the dynamics of a chromatin chain was achieved in a recent Brownian dynamics simulation by Beard and Schlick [65]. Like in previous work, the DNA is treated as a segmented elastic chain; however, the nucleosomes are modeled as flat cylinders with the DNA attached to the cylinder surface at the positions known from the crystallographic structure of the nucleosome. Moreover, the electrostatic interactions are treated in a very detailed manner: the charge distribution on the nucleosome core particle is obtained from a solution to the non-linear Poisson-Boltzmann equation in the surrounding solvent, and the total electrostatic energy is computed through the Debye-Hückel approximation over all charges on the nucleosome and the linker DNA.

Using this model, the authors performed Brownian dynamics simulations with a time step of 2 ps and maximum simulation times of 50 ns for a 48-nucleosome chain and 100 ns for di- and trinucleosomes. Like in the previous work of Ehrlich *et al.* [96] the oligonucleosome trajectories were used to compute diffusion coefficients and compare them to the experimental data by Yao *et al.* [98] and Bednar *et al.* [16]. The study does predict a salt-dependent compaction of the oligonucleosomes, which manifests itself in an increased diffusion coefficient; within the limits of salt concentration studied, the authors obtained good quantitative agreement with the experimental data. The simulations on the 48-nucleosome chain show a stable 30 nm diameter zig-zag fiber at 50 mM salt, which unfolds when the salt concentration is decreased to 10 mM. Due to the computational requirements of the dynamics simulation, only the initial phase of the unfolding could be simulated and the equilibrium configuration was obtained through a Monte-Carlo simulation using the same energetics as the BD model.

9. *The flexibility of the 30 nm fiber*

For understanding the folding of the chromatin chain into interphase and metaphase chromosomes, an important quantity to be known is its flexibility, expressed as the bending persistence length. Distance distribution analysis for genetic marker pairs in human fibroblast nuclei [100–102] suggested a wormlike chain conformation for the 30 nm fiber with a persistence length $L_p = 100$–140 nm. On the other hand, scanning force microscopy analysis of end-to-end distances of chromatin fibers on a mica surface gave $L_p = 30$–50 nm (Castro [103] as cited by Houchmanzadeh [104]), but the binding conditions of the fiber to the mica will influence the measured persistence length to a great extent (Bussiek, Mücke, and Langowski [109]). At low salt concentrations, stretching experiments of single chromatin fibers from chicken erythrocytes with laser tweezers yielded $L_p = 30$ nm [66], but no data for the persistence length at physiological salt was given there. Two *in vivo* studies have been published, where recombination frequencies in human cells [105] or formaldehyde crosslinking probabilities in yeast [106] have been used to measure relative end-to-end distances. These studies report rather small persistence lengths of 30–50 nm. Thus, the experimental values for L_p in the literature span a rather large range of 30–140 nm, where the smaller values were obtained either on chromatin fibers in low salt or on chromosomes that were constrained in the volume of a nucleus. Our simulations of the 30 nm chromatin fiber structure reported above suggest a rather stiff chain of $L_p \approx 250$ nm for very regular nucleosome spacing and short linkers, which decreases for more irregularly spaced nucleosomes or longer linkers.

The small persistence length obtained *in vivo* by recombination or cross-linking experiments, however, may correspond to a chromatin fiber stiffness several times higher than that estimated from the measured L_p alone. The persistence length is computed from distance measurements assuming an unconstrained self-crossing random walk. Since this condition is only fulfilled for the interphase chromatin fiber over rather short distances, the measured apparent L_p will depend, for a given chain flexibility, on the folding topology and the region or genomic separation for which it is calculated. The apparent L_p decreases with increasing compaction of the chromatin fiber relative to a random walk, because the assumption of a free wormlike chain breaks down. In recent model calculations of the arrangement of a 30 nm fiber with a *free* persistence length of $L_p = 150$ nm in interphase chromosomes (T. A. Knoch, Ph.D. thesis, University of Heidelberg, Knoch, and Langowski, manuscript in preparation) we find that the folding topology may reduce the *apparent* persistence length quite dramatically.

This shows that the chromatin chain flexibility can only be reliably obtained from genomic distance measurements when an unconstrained random walk behavior of the chromatin fiber is assured [100,101]. On the other hand this might explain the rather big differences in experimental values for the chromatin persistence length [100–102,107,108]. The *in vivo* studies that determined the persistence length indirectly by recombination [105] or crosslinking techniques [106] probably overestimate the chromatin fiber flexibility, because the external constraints such as folding topology and finite nuclear volume have not been taken into account. At any rate, a persistence length for the unconstrained chromatin fiber that is comparable to its diameter would give rise to structures that are so irregular that the notion of a "fiber" breaks down completely. Comparison with numerical simulations, however, might allow extraction of physical properties of the chromatin fiber from sufficiently detailed experimental data.

9.1. Conclusion

The state of our understanding of the physics of chromatin folding is such that the current knowledge about the structure and interaction of the basic components of chromatin—histones and DNA—enables us to develop the first quantitative models of the structure and dynamics of the chromatin fiber. Even so, these models are still at a very rudimentary stage: data on the interaction of the histone tails with their surroundings, on DNA binding/unbinding at the nucleosome surface, nucleosome/nucleosome interactions, the role of histone modifications and other chromatin-associated proteins are badly needed. However, biophysical techniques together with computational modeling and the ever-expanding body of such quantitative data hold a promising outlook for a much more detailed picture of the chromatin fiber in the years to come.

References

1. Zentgraf, H. and Franke, W.W. (1984) Differences of supranucleosomal organization in different kinds of chromatin: cell type-specific globular subunits containing different numbers of nucleosomes. J. Cell Biol. 99, 272–286.
2. Olins, A.L. and Olins, D.E. (1973) Spheroid chromatin units (v bodies). J. Cell Biol. 59, 252a.
3. Olins, A.L. and Olins, D.E. (1974) Spheroid chromatin units (v bodies). Science 183, 330–332.
4. Woodcock, C.L.F. (1973) Ultrastructure of inactive chromatin. J. Cell Biol. 59, 368a.
5. van Holde, K. and Zlatanova, J. (1995) Chromatin higher order structure: Chasing a mirage? J. Biol. Chem. 270, 8373–8376.
6. van Holde, K. and Zlatanova, J. (1996) What determines the folding of the chromatin fiber. Proc. Natl. Acad. Sci. USA 93, 10548–10555.
7. van Holde, K.E. (1989) Chromatin. Springer, Heidelberg.
8. Widom, J. (1989) Toward a unified model of chromatin folding. Annu. Rev. Biophys. Biophys. Chem. 18, 365–395.
9. Finch, J.T. and Klug, A. (1976) Solenoidal model for superstructure in chromatin. Proc. Natl. Acad. Sci. USA 73, 1897–1901.
10. Thoma, F., Koller, T., and Klug, A. (1979) Involvement of histone H1 in the organization of the nucleosome and of the salt-dependent superstructures of chromatin. J. Cell Biol. 83, 403–427.
11. Widom, J. and Klug, A. (1985) Structure of the 300A chromatin filament: X-ray diffraction from oriented samples. Cell 43, 207–213.
12. Woodcock, C.L., Grigoryev, S.A., Horowitz, R.A., and Whitaker, N. (1993) A chromatin folding model that incorporates linker variability generates fibers resembling the native structures. Proc. Natl. Acad. Sci. USA 90, 9021–9025.
13. Horowitz, R.A., Agard, D.A., Sedat, J.W., and Woodcock, C.L. (1994) The three-dimensional architecture of chromatin in situ: electron tomography reveals fibers composed of a continuously variable zig-zag nucleosomal ribbon. J. Cell Biol. 125, 1–10.
14. Woodcock, C.L. and Dimitrov, S. (2001) Higher-order structure of chromatin and chromosomes. Curr. Opin. Genet. Dev. 11, 130–135.
15. Leuba, S.H., Yang, G., Robert, C., Samori, B., van Holde, K., Zlatanova, J., and Bustamante, C. (1994) Three-dimensional structure of extended chromatin fibers as revealed by tapping-mode scanning force microscopy. Proc. Natl. Acad. Sci. USA 91, 11621–11625.
16. Bednar, J., Horowitz, R.A., Grigoryev, S.A., Carruthers, L.M., Hansen, J.C., Koster, A.J., and Woodcock, C.L. (1998) Nucleosomes, linker DNA, and linker histone form a unique structural motif that directs the higher-order folding and compaction of chromatin. Proc. Natl. Acad. Sci. USA 95, 14173–14178.
17. Schiessel, H., Gelbart, W.M., and Bruinsma, R. (2001) DNA folding: Structural and mechanical properties of the two-angle model for chromatin. Biophys. J. 80, 1940–1956.
18. Stigter, D. (1977) Interactions of highly charged colloidal cylinders with applications to double stranded DNA. Biopolymers 16, 1435–1448.
19. Klenin, K.V., Frank-Kamenetskii, M.D., and Langowski, J. (1995) Modulation of intramolecular interactions in superhelical DNA by curved sequences: a Monte Carlo simulation study. Biophys. J. 68, 81–88.

20. Delrow, J.J., Gebe, J.A., and Schurr, J.M. (1997) Comparison of hard-cylinder and screened Coulomb interactions in the modeling of supercoiled DNAs. Biopolymers 42, 455–470.
21. Klenin, K., Merlitz, H., and Langowski, J. (1998) A Brownian dynamics program for the simulation of linear and circular DNA and other wormlike chain polyelectrolytes. Biophys. J. 74, 780–788.
22. Merlitz, H., Rippe, K., Klenin, K.V., and Langowski, J. (1998) Looping dynamics of linear DNA molecules and the effect of DNA curvature: a study by Brownian dynamics simulation. Biophys. J. 74, 773–779.
23. Rybenkov, V.V., Cozzarelli, N.R., and Vologodskii, A.V. (1993) Probability of DNA knotting and the effective diameter of the DNA double helix. Proc. Natl. Acad. Sci. USA 90, 5307–5311.
24. Rybenkov, V.V., Vologodskii, A.V., and Cozzarelli, N.R. (1997) The effect of ionic conditions on DNA helical repeat, effective diameter and free energy of supercoiling. Nucleic Acids Res. 25, 1412–1418.
25. Hammermann, M., Steinmaier, C., Merlitz, H., Kapp, U., Waldeck, W., Chirico, G., and Langowski, J. (1997) Salt effects on the structure and internal dynamics of superhelical DNAs studied by light scattering and Brownian dynamics. Biophys. J. 73, 2674–2687.
26. Hammermann, M., Brun, N., Klenin, K.V., May, R., Toth, K., and Langowski, J. (1998) Salt-dependent DNA superhelix diameter studied by small angle neutron scattering measurements and Monte Carlo simulations. Biophys. J. 75, 3057–3063.
27. Lu, Y., Weers, B., and Stellwagen, N.C. (2001) DNA persistence length revisited. Biopolymers 61, 261–275.
28. Barkley, M.D. and Zimm, B.H. (1979) Theory of twisting and bending of chain macromolecules: analysis of the fluorescence depolarization of DNA. J. Chem. Phys. 70, 2991–3007.
29. Fujimoto, B.S. and Schurr, J.M. (1990) Dependence of the torsional rigidity of DNA on base composition. Nature 344, 175–177.
30. Schurr, J.M., Fujimoto, B.S., Wu, P., and Song, L. (1992), Fluorescence studies of nucleic acids: dynamics, rigidities and structures. In: Lakowicz, J.R. (ed.) Topics in Fluorescence Spectroscopy, Vol. 3. Plenum Press, New York, pp. 137–229.
31. Shore, D. and Baldwin, R.L. (1983) Energetics of DNA twisting. I. Relation between twist and cyclization probability. J. Mol. Biol. 179, 957–981.
32. Horowitz, D.S. and Wang, J.C. (1984) Torsonal Rigidity of DNA and Length Dependence of the Free Energy of DNA Supercoiling. J. Mol. Biol. 173, 75–91.
33. Taylor, W.H. and Hagerman, P.J. (1990) Application of the method of phage T4 DNA ligase-catalyzed ring-closure to the study of DNA structure I.NaCl-dependence of DNA flexibility and helical repeat. J. Mol. Biol. 212, 363–376.
34. Cluzel, P., Lebrun, A., Heller, C., Lavery, R., Viovy, J.-L., Chatenay, D., and Caron, F. (1996) DNA: An extensible molecule. Science 271, 792–794.
35. Smith, S.B., Cui, Y., and Bustamante, C. (1996) Overstretching B-DNA: the elastic response of individual double-stranded and single-stranded DNA molecules. Science 271, 795–799.
36. Lankas, F., Sponer, J., Hobza, P., and Langowski, J. (2000) Sequence-dependent elastic properties of DNA. J. Mol. Biol. 299, 695–709.
37. Lankas, F., Cheatham, I.T., Spackova, N., Hobza, P., Langowski, J., and Sponer, J. (2002) Critical effect of the N2 amino group on structure, dynamics, and elasticity of DNA polypurine tracts. Biophys. J. 82, 2592–2609.
38. Arents, G., Burlingame, R.W., Wang, B.-C., Love, W.E., and Moudrianakis, E.N. (1991) The nucleosomal core histone octamer at 3.1 Å resolution: A triparatite protein assembly and a left-handed superhelix. Proc. Natl. Acad. Sci. USA 88, 10148–10152.
39. Luger, K., Mäder, A.W., Richmond, R.K., Sargent, D.F., and Richmond, T.J. (1997) Crystal structure of the nucleosome core particle at 2.8 Å resolution. Nature 389, 251–260.
40. Sivolob, A., DeLucia, F., Révet, B., and Prunell, A. (1999) Nucleosome dynamics. II. High flexibility of nucleosome entering and exiting DNAs to positive crossing. An ethidium bromide fluorescence study of mononucleosomes on DNA minicircles. J. Mol. Biol. 285, 1081–1099.

41. Sivolob, A. and Prunell, A. (2000) Nucleosome dynamics V. Ethidium bromide versus histone tails in modulating ethidium bromide-driven tetrasome chiral transition. A fluorescence study of tetrasomes on DNA minicircles. J. Mol. Biol. 295, 41–53.
42. Brower-Toland, B.D., Smith, C.L., Yeh, R.C., Lis, J.T., Peterson, C.L., and Wang, M.D. (2002) Mechanical disruption of individual nucleosomes reveals a reversible multistage release of DNA. Proc. Natl. Acad. Sci. USA. 99, 1960–1965.
43. Bennink, M.L., Leuba, S.H., Leno, G.H., Zlatanova, J., de Grooth, B.G., and Greve, J. (2001) Unfolding individual nucleosomes by stretching single chromatin fibers with optical tweezers. Nat. Struct. Biol. 8, 606–610.
44. Leforestier, A. and Livolant, F. (1997) Liquid crystalline ordering of nucleosome core particles under macromolecular crowding conditions: evidence for a discotic columnar hexagonal phase. Biophys. J. 73, 1771–1776.
45. Livolant, F. and Leforestier, A. (2000) Chiral discotic columnar germs of nucleosome core particles. Biophys. J. 78, 2716–2729.
46. Leforestier, A., Dubochet, J., and Livolant, F. (2001) Bilayers of nucleosome core particles. Biophys. J. 81, 2414–2421.
47. Mangenot, S., Leforestier, A., Durand, D., and Livolant, F. (2003) X-ray diffraction characterization of the dense phases formed by nucleosome core particles. Biophys. J. 84, 2570–2584.
48. Mangenot, S., Leforestier, A., Vachette, P., Durand, D., and Livolant, F. (2002) Salt-induced conformation and interaction changes of nucleosome core particles. Biophys. J. 82, 345–356.
49. Mangenot, S., Raspaud, E., Tribet, C., Belloni, L., and Livolant, F. (2002) Interactions between isolated nucleosome core particles: A tail-bridging effect? Eur. Phys. J. E 7, 221–231.
50. Wedemann, G. and Langowski, J. (2002) Computer simulation of the 30-nanometer chromatin fiber. Biophys. J. 82, 2847–2859.
51. Beard, D.A. and Schlick, T. (2001) Modeling salt-mediated electrostatics of macromolecules: the discrete surface charge optimization algorithm and its application to the nucleosome. Biopolymers 58, 106–115.
52. Matsumoto, A. and Go, N. (1999) Dynamic properties of double-stranded DNA by normal mode analysis. J. Chem. Phys. 110, 11070–11075.
53. Matsumoto, A. and Olson, W.K. (2002) Sequence-dependent motions of DNA: a normal mode analysis at the base-pair level. Biophys. J. 83, 22–41.
54. Cheatham, T.E., 3rd and Young, M.A. (2000) Molecular dynamics simulation of nucleic acids: successes, limitations, and promise. Biopolymers 56, 232–256.
55. Klenin, K.V., Vologodskii, A.V., Anshelevich, V.V., Dykhne, A.M., and Frank-Kamenetskii, M.D. (1988) Effect of excluded volume on topological properties of circular DNA. J. Biomol. Struct. Dyn. 5, 1173–1185.
56. Vologodskii, A.V., Levene, S.D., Klenin, K.V., Frank-Kamenetskii, M., and Cozzarelli, N.R. (1992) Conformational and thermodynamic properties of supercoiled DNA. J. Mol. Biol. 227, 1224–1243.
57. Langowski, J., Kapp, U., Klenin, K., and Vologodskii, A. (1994) Solution structure and dynamics of DNA topoisomers. Dynamic light scattering studies and Monte-Carlo simulations. Biopolymers 34, 639–646.
58. Rybenkov, V.V., Vologodskii, A.V., and Cozzarelli, N.R. (1997) The effect of ionic conditions on the conformations of supercoiled DNA .1. Sedimentation analysis. J. Mol. Biol. 267, 299–311.
59. Huang, J., Schlick, T., and Vologodskii, A. (2001) Dynamics of site juxtaposition in supercoiled DNA. Proc. Natl. Acad. Sci. USA 98, 968–973.
60. Fujimoto, B.S. and Schurr, J.M. (2002) Monte Carlo simulations of supercoiled DNAs confined to a plane. Biophys. J. 82, 944–962.
61. Allison, S.A. and McCammon, J.A. (1984) Transport properties of rigid and flexible macromolecules by Brownian dynamics simulation. Biopolymers 23, 167–187.
62. Allison, S.A. (1986) Brownian dynamics simulation of wormlike chains. Fluorescence depolarization and depolarized light scattering. Macromolecules 19, 118–124.

63. Allison, S.A., Sorlie, S.S., and Pecora, R. (1990) Brownian dynamics simulations of wormlike chains: Dynamic light scattering from a 2311 base pair DNA fragment. Macromolecules 23, 1110–1118.
64. Gebe, J.A., Allison, S.A., Clendenning, J.B., and Schurr, J.M. (1995) Monte-Carlo simulations of supercoiling free-energies for unknotted and trefoil knotted DNAs. Biophys. J. 68, 619–633.
65. Beard, D.A. and Schlick, T. (2001) Computational modeling predicts the structure and dynamics of chromatin fiber. Structure 9, 105–114.
66. Cui, Y. and Bustamante, C. (2000) Pulling a single chromatin fiber reveals the forces that maintain its higher-order structure. Proc. Natl. Acad. Sci. USA 97, 127–132.
67. Schiessel, H., Widom, J., Bruinsma, R.F., and Gelbart, W.M. (2001) Polymer reptation and nucleosome repositioning. Phys. Rev. Lett. 86, 4414–4417.
68. Bishop, T.C. and Hearst, J.E. (1998) Potential function describing the folding of the 30 nm fiber. J. Phys. Chem. B 102, 6433–6439.
69. Bishop, T.C. and Zhmudsky, O.O. (2002) Mechanical model of the nucleosome and chromatin. J. Biomol. Struct. Dyn. 19, 877–887.
70. Zlatanova, J., Leuba, S.H., and van Holde, K. (1998) Chromatin fiber structure: morphology, molecular determinants, structural transitions. Biophys. J. 74, 2554–2566.
71. Widom, J. (1992) A relationship between the helical twist of DNA and the ordered positioning of nucleosomes in all eukaryotic cells. Proc. Natl. Acad. Sci. USA 89, 1095–1099.
72. Schiessel, H. (2003) The physics of chromatin. J. Phys: Condens. Matter, in press.
73. Daban, J.R. (2000) Physical constraints in the condensation of eukaryotic chromosomes. Local concentration of DNA versus linear packing ratio in higher order chromatin structures. Biochemistry 39, 3861–3866.
74. vanHolde, K. and Zlatanova, J. (1996) What determines the folding of the chromatin fiber? Proc. Natl. Acad. Sci. USA 93, 10548–10555.
75. Schiessel, H. (2002) How short-ranged electrostatics controls the chromatin structure on much larger scales. Euro. Lett. 58, 140–146.
76. Ben-Haim, E., Lesne, A., and Victor, J.M. (2001) Chromatin: a tunable spring at work inside chromosomes. Phys. Rev. E Stat. Nonlin. Soft Matter. Phys. 64, 051921.
77. Fuller, F.B. (1971) The writhing number of a space curve. Proc. Natl. Acad. Sci. USA 68, 815–819.
78. Benham, C.J. (1978) The statistics of superhelicity. J. Mol. Biol. 123, 361–370.
79. Benham, C.J. (1977) Elastic model of supercoiling. Proc. Natl. Acad. Sci. USA 74, 2397–2401.
80. Olson, W.K. and Zhurkin, V.B. (2000) Modeling DNA deformations. Curr. Opin. Struct. Biol. 10, 286–297.
81. Langowski, J., Olson, W.K., Pedersen, S.C., Tobias, I., Westcott, T.P., and Yang, Y. (1996) DNA supercoiling, localized bending and thermal fluctuations. Trends Biochem. Sci. 21, 50.
82. Metropolis, N., Rosenbluth, A.W., Rosenbluth, M.N., Teller, A.H., and Teller, E. (1953) Equation of state calculations by fast computing machines. J. Chem. Phys. 21, 1087–1092.
83. Ermak, D.L. and McCammon, J.A. (1978) Brownian dynamics with hydrodynamic interactions. J. Chem. Phys. 69, 1352–1359.
84. Allison, S.A., Austin, R., and Hogan, M. (1989) Bending and twisting dynamics of short linear DNAs-analysis of the triplet anisotropy decay of a 209-base pair fragment by Brownian simulation. J. Chem. Phys. 90, 3843–3854.
85. Chirico, G. and Langowski, J. (1992) Calculating hydrodynamic properties of DNA through a second-order Brownian dynamics algorithm. Macromolecules 25, 769–775.
86. Chirico, G. and Langowski, J. (1994) Kinetics of DNA supercoiling studied by Brownian dynamics simulation. Biopolymers 34, 415–433.
87. Chirico, G. and Langowski, J. (1996) Brownian dynamics simulations of supercoiled DNA with bent sequences. Biophys. J. 71, 955–971.
88. Garcia de la Torre, J., Navarro, S., and Lopez Martinez, M.C. (1994) Hydrodynamic properties of a double-helical model for DNA. Biophys. J. 66, 1573–1579.
89. Katritch, V., Bustamante, C., and Olson, V.K. (2000) Pulling chromatin fibers: Computer simulations of direct physical micromanipulations. J. Mol. Biol. 295, 29–40.

90. Gay, J.G. and Berne, B.J. (1981) Modification of the overlap potential to mimic a linear site–site potential. J. Chem. Phys. 74, 3316–3319.
91. Gerchman, S.E. and Ramakrishnan, V. (1987) Chromatin higher-order structure studied by neutron scattering and scanning transmission electron microscopy. Proc. Natl. Acad. Sci. USA 84, 7802–7806.
92. Dimitrov, S.I., Makarov, V.L., and Pashev, I.G. (1990) The chromatin fiber: structure and conformational transitions as revealed by optical anisotropy studies. J. Biomol. Struct. Dyn. 8, 23–35.
93. Widlund, H.R., Vitolo, J.M., Thiriet, C., and Hayes, J.J. (2000) DNA sequence-dependent contributions of core histone tails to nucleosome stability: Differential effects of acetylation and proteolytic tail removal. Biochemistry 39, 3835–3841.
94. Thastrom, A., Lowary, P.T., Widlund, H.R., Cao, H., Kubista, M., and Widom, J. (1999) Sequence motifs and free energies of selected natural and non-natural nucleosome positioning DNA sequences. J. Mol. Biol. 288, 213–229.
95. Wedemann, G., Münkel, C., Schoppe, G., and Langowski, J. (1998) Kinetics of structural changes in superhelical DNA. Phys. Rev. E 58, 3537–3546.
96. Ehrlich, L., Munkel, C., Chirico, G., and Langowski, J. (1997) A Brownian dynamics model for the chromatin fiber. Comput. Appl. Biosci. 13, 271–279.
97. Münkel, C. and Langowski, J. (1998) Chromosome structure described by a polymer model. Phys. Rev. E 57, 5888–5896.
98. Yao, J., Lowary, P.T., and Widom, J. (1990) Direct detection of linker DNA bending in defined-length oligomers of chromatin. Proc. Natl. Acad. Sci. USA 87, 7603–7607.
99. Butler, P.J. and Thomas, J.O. (1998) Dinucleosomes show compaction by ionic strength, consistent with bending of linker DNA. J. Mol. Biol. 281, 401–407.
100. van den Engh, G., Sachs, R., and Trask, B.J. (1992) Estimating genomic distance from DNA sequence location in cell nuclei by a random walk model. Science 257, 1410–1412.
101. Trask, B.J., Allen, S., Massa, H., Fertitta, A., Sachs, R., van den Engh, G., and Wu, M. (1993) Studies of metaphase and interphase chromosomes using fluorescence in situ hybridization. Cold Spring Harb. Symp. Quant. Biol. 58, 767–775.
102. Yokota, H., van den Engh, G., Hearst, J., Sachs, R.K., and Trask, B.J. (1995) Evidence for the organization of chromatin in megabase pair-sized loops arranged along a random walk path in the human G0/G1 interphase nucleus. J. Cell Biol. 130, 1239–1249.
103. Castro, C. (1994). Measurement of the elasticity of single chromatin fibers: the effect of histone H1. PhD Thesis, University of Oregon, Eugene.
104. Houchmandzadeh, B., Marko, J.F., Chatenay, D., and Libchaber, A. (1997) Elasticity and structure of eukaryote chromosomes studied by micromanipulation and micropipette aspiration. J. Cell. Biol. 139, 1–12.
105. Ringrose, L., Chabanis, S., Angrand, P.O., Woodroofe, C., and Stewart, A.F. (1999) Quantitative comparison of DNA looping *in vitro* and *in vivo*: chromatin increases effective DNA flexibility at short distances. EMBO J. 18, 6630–6641.
106. Dekker, J., Rippe, K., Dekker, M., and Kleckner, N. (2002) Capturing chromosome conformation. Science 295, 1306–1311.
107. Ostashevsky, J.Y. and Lange, C.S. (1994) The 30 nm chromatin fiber as a flexible polymer. J. Biomol. Struct. Dyn. 11, 813–820.
108. Sachs, R.K., van den Engh, G., Trask, B., Yokota, H., and Hearst, J.E. (1995) A random-walk/giant-loop model for interphase chromosomes. Proc. Natl. Acad. Sci. USA 92, 2710–2714.
109. Bussiek, M., Mücke, N., and Langowski, J. (2003) Polylysine-coated mica can be used to observe systematic changes in the supercoiled DNA conformation by scanning force microscopy in solution. Nucleic Acids Res. 31, 137.

J. Zlatanova and S.H. Leuba (Eds.) *Chromatin Structure and Dynamics: State-of-the-Art*

DOI: 10.1016/S0167-7306(03)39016-7

CHAPTER 16

Nucleosome remodeling

Andrew A. Travers[1] and Tom Owen-Hughes[2]

[1]*MRC Laboratory of Molecular Biology, Hills Road, Cambridge CB2 2QH, UK.*
Tel.: +44-1223-402419; Fax: +44-1223-412142; E-mail: aat@mrc-lmb.cam.ac.uk
[2]*Division of Gene Regulation and Expression, Wellcome Trust Biocentre,*
University of Dundee, Dundee DD1 5EH, UK.
Tel.: +44-1382-345796; Fax: +44-1382-348072; E-mail: t.a.owenhughes@dundee.ac.uk

1. Introduction

The many-fold compaction of the eukaryotic genome in a nucleus implies that the essential requirements for packaging and accessibility of the DNA be efficiently reconciled. The fundamental unit of packaging is the nucleosome core particle in which DNA is tightly wrapped in 1.65 left-handed superhelical turns around the histone octamer [1]. Transcription, replication, recombination, and DNA repair all require access to DNA, but at the same time compaction, at least to a limited extent must be maintained. These requirements can be met by the destabilization of nucleosome structure. Such destabilization could enhance the movement of nucleosomes along DNA or simply increase access to histone-bound sequences without changing the translational position of the octamer. By the same token there will be other situations, for example, in the inactive X chromosome (Barr body), where stability rather than accessibility, is the prime requirement. Biological processes that alter nucleosome stability and/or position are generically termed nucleosome remodeling, a topic which has been extensively reviewed elsewhere [2–24].

2. Nucleosome mobility

The ability of nucleosomes to slide along a DNA molecule has long been apparent. This passive and apparently spontaneous sliding of the histone octamer is favored by both higher temperatures and higher ionic strength [25–32] but is reduced by 2 mM $MgCl_2$ [33]. This motion relocates nucleosomes to the most energetically favorable positions which can be at the end of a DNA fragment [29]. It appears that all eukaryotes possess enzymes capable of increasing nucleosome mobility in an ATP-dependent fashion. The first biochemical demonstration of ATP-dependent chromatin remodeling resulted from an analysis of GAGA transcription factor mediated disruption of an *in vitro* reconstituted nucleosome array [34]. Subsequently multiprotein complexes containing the Swi2p/Snf2p (SWI = Switch; SNF = Sucrose Non-Fermenting) DNA-stimulated ATPase, which had already been characterized genetically as a transcriptional regulator [35–37] were shown to alter chromatin structure in an ATP-dependent manner [38–46]. Many ATP-dependent chromatin remodeling activities have subsequently been shown to alter nucleosome positioning [47–53].

All ATP-dependent chromatin remodeling assemblies contain a defining catalytic subunit with high homology to the yeast Swi2p/Snf2p protein. These subunits share domain consisting of a series of conserved motifs that are also found within an extended family of "helicase-related" proteins [54]. Amongst these helicase-related proteins Swi2p/Snf2p defines a distinct grouping of polypeptides many of which have been identified in different classes of remodeling complexes (Fig. 1; Table 1). These can be further classified as being most similar to Swi2p/Snf2p [35–37,40], ISWI (Imitation Switch) [55–59], Ino80 [60,61], or Mi-2/NRD/NURD (named from the identity of the ATPase subunit with the major dermatomyositis-specific Mi-2 autoantigen [62] and the Nucleosome Remodeling and histone Deacetylation activities of the complex [63–67] and its close relative CHD1 (CHD = Chromodomain Helicase DNA binding) [68]. Other complexes with remodeling activity have also been isolated but their relationship to these classes is unclear [69–72]. In addition, there are some proteins that are closely related to the ATP-dependent remodeling enzymes but have not been shown to possess chromatin remodeling activity [73–76] (Table 1).

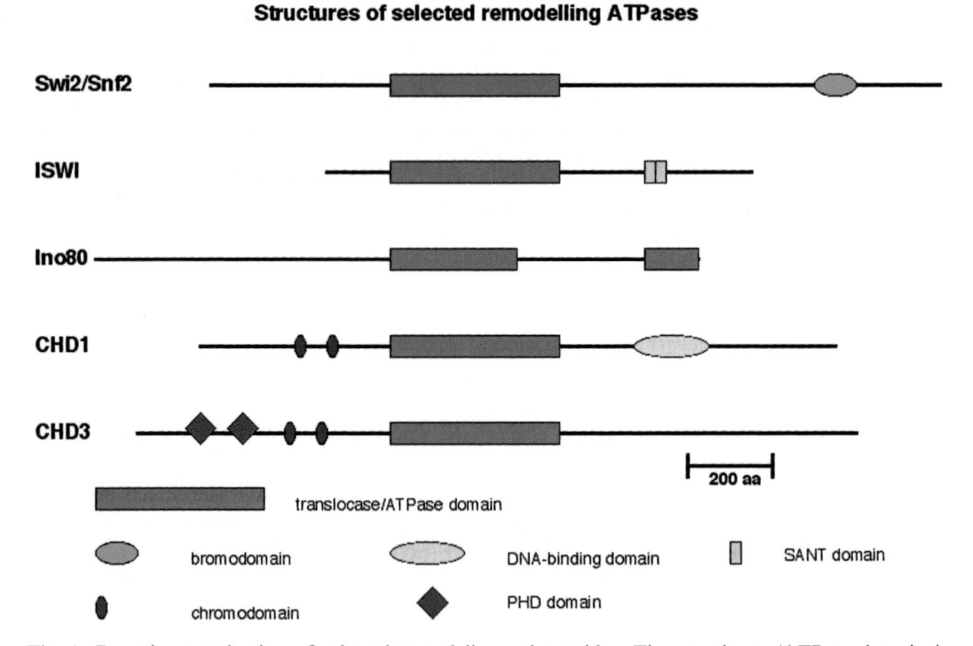

Fig. 1. Domain organization of selected remodeling polypeptides. The translocase/ATPase domain in Ino80 is split into two regions [60].

Table 1

Subunit composition and functions of chromatin remodeling complexes

The Swi2p/Snf2p sub-family			
Complex	Catalytic subunit/ species	Other identified subunits	Biological functions
SWI/SNF	Swi2p/Snf2p *S. cerevisiae*	Swi1p, Snf5p, Snf6p, Snf11p, Swi3p, Swp29p/Taf$_I$4p/TFG-3p, Swp82p, Swp59p/Arp9p, Swp61p/Arp7, Swp73p/Snf12p	Non-essential, but affects either the activation or repression of around 6% of yeast genes [224,225]. Genetic studies show that the requirement for SWI/SNF can be suppressed by mutations to chromatin components [264] and suggest functional redundancy with H2A.Z [215]. Can be recruited by through interactions with transcription factor activation domains [265]. Can function in conjunction with HAT activities to activate transcription [239,265–268].
RSC	Sth1p/Nsp1p *S. cerevisiae*	Sfh1p, Rsc8p/Swh8p, Rsc6p, Rsc11p/ Arp7p, Rsc12p/Arp9p, Rsc1p, Rsc2p, Rsc3p, Rsc5p, Rsc7p, Rsc9p, Rsc10p, Rsc13-15p	STH1 and several other genes encoding RSC components are essential for viability in yeast [35,77]. RSC is present in at least 10-fold higher concentration than that of SWI/ SNF in the cell [77]. RSC may function in the activation and repression of pol II and pol III transcription [142,252], it is also implicated in kinetochore function [269].
SWI/SNF	Not known [270] *Arabidopsis*	AtSWI3B, AtSNF5/BSH	BSH interacts with FCA, a regulator of flowering time in *Arabidopsis* [271].
SWI/SNF	PSA-4 *C. elegans*	PSA-1	Required for asymmetric cell division during *C. elegans* development [272].
Brahma	BRM *Drosophila*	MOR, Bap111, Bap60, Bap55, Bap47, Bap45, OSA, p170, p26	The brm gene was isolated as a dominant suppressor of Pc mutations suggesting that it plays a role in the activation of homeotic genes [37]. It may play a general role in pol II transcription [273].
BRG1	BRG1 Zebrafish	Not known	Retinal cell differentiation [274].
BRG1/BRM	BRG1/BRM Mouse	Likely to be similar to human complexes	SNF5/INI1 component is required for embryonic development and tumor suppression [275,276]. Loss of BRG1 is required pre-implantation, although other cell types can survive in the absence of BRG1. BRG1 heterozygotes are pre-disposed to tumors [277]. BRG1 is involved in the differentiation of CD4 and CD8 T-cell lineages [136].

(*Continued*)

Table 1
Continued

Complex	Catalytic subunit/ species	Other identified subunits	Biological functions
PBAF (SWI/SNF complex B)	hBRG1 Human	BAF180, BAF170, BAF155, BAF60a, BAF57, BAF53, β-actin, BAF47, INI1/hSnf5 Some complexes contain BRCA1 and ENL [279,280] Note that the subunit composition varies [44,82]. More similar to the yeast RSC complex.	BRG1 is involved in the activation of transcription many genes including the GR and estrogen-responsive promoters [246,281], the β-interferon enhanceosome [282], genes involved in muscle differentiation [283], TCF-β-catenin signaling [250] and interferon-α-inducible genes [284]. Signaling pathways can regulate the association of BRG1 with the chromatin matrix [148]. However BRG1 also participates in the repression of transcription, e.g., that mediated by Rb [285] and c-fos [286]. It may also regulate the elongation of transcription [260]. Enl is a translocation partner for MLL that is a common target for chromosomal translocations in human acute leukemia. Truncating mutations of hSNF5/INI1 are associated with aggressive pediatric cancer [287].
BAF (SWI/SNF complex A)	hBRM or BRG1 Human	BAF270/p270, No BAF180, otherwise as for PBAF More similar to yeast SWI/SNF complex [82].	In many cases it has not been determined whether functional effects result from either the PBAF or BAF complexes. However, the situation is complicated as there is heterogeneity in the composition of these complexes. In some cases it appears that both hBRM and BRG1 containing complexes play similar roles in gene regulation [283], but in others there is evidence for distinct functions [249].
mSin3A-BRG1/ BRM	hBRM or hBRG1 Human	mSin3a, HDAC1, HDAC2, RbAp48	Some BRG1 and hBRM complexes have been observed to contain components of the mSin3A deacetylase complex [84].
E-RC1	hBRG1 Human	Baf170, Baf155, Baf47, Baf57	E-RC1 acts as a co-activator for tissue specific transcriptional regulation of the chromatin assembled erythroid β-globin gene by EKLF *in vitro* [288].

ISWI sub-family			
Complex	Catalytic subunit/ species	Other subunits	Biological functions
ISW1a	Isw1p *S. cerevisiae*	Ioc3p	Partial redundancy with CHD1, ISW2 and ISW1b [94,96]. Isw1 functions in the repression of transcription following recruitment to promoters by Cbf1 [289], it is not known which ISW1 complexes are involved.
ISW1b	Isw1p *S. cerevisiae*	Ioc2p-Ioc4p	Partial redundancy with CHD1, ISW2 and ISW1a [94,96].
ISW2	Isw2p *S. cerevisiae*	Itc1p (p140)	Partial redundancy with CHD1 and ISW1 [94]. Repression of early meiotic genes in parallel with the Sin3-Rpd3 histone deacetylase complex [235,236].
NURF	ISWI *Drosophila*	Nurf-55, Nurf-301, Nurf-38-inorganic pyrophosphatase	Although the bulk of Drosophila ISWI does not co-localize with RNA polymerase [233], the NURF complex functions in the activation of transcription [234].
ACF	ISWI *Drosophila*	Acf1 [59]	ISWI is essential for Drosophila development. Most ISWI protein does not co-localize with RNA polymerase [233], but the function of the Drosophila ACF complex in vivo is not clear. However, the equivalent human complex appears to function in DNA replication (see below).
dCHRAC	ISWI *Drosophila*	Acf1, CHRAC-14, CHRAC-16, [91]. Not Topoisomerase II [92]	As for ACF, but it has been proposed that this complex might play a specialized role in early embryos [91].
xISWI	XISWI *Xenopus*	xAcf-1	Not known. XISWI is found to be present in four complexes one of which contains xAcf1 [95]. May play a role in nuclear reprogramming [290].
RSF	HSnf2h Human	P325	Not known [291].
NoRC	Snf2h Mouse	TIP5	Interacts with RNA polymerase I transcription termination factor I (TTF-1), localized to nucleoli [292] and implicated in repression of ribosomal gene transcription [293].
hACF/RSF	hSnf2h Human	hACF1 [294]	Mouse homolog of hACF1 localizes to heterochromatin [295], suggesting a role for this complex in repression. Found to facilitate replication through heterochromatin [237].

(*Continued*)

Table 1
Continued

Complex	Catalytic subunit/ species	Other subunits	Biological functions
hCHRAC	hSnf2h Human	HuCHRAC-15, HuCHRAC-17, hACF1 [90]	It seems possible that the small subunits could be present in other Snf2h containing complexes but not detected due to their low Mw. The function of the small subunits is not known.
WCRF	hSnf2h Human	WCRF180	WCRF180 is closely related to WSTF a gene deleted in Williams syndrome [296].
WICH	hSNF2h (xISWI) mouse (Xenopus)	Williams Syndrome Transcription Factor (WSTF)	Localized to heterochromatic replication foci [297]. Some ISWI remains associated with chromosomes during mitosis [298].
SNF2h/cohesin/ NuRD complex	hSnf2h hRAD21 Human	SMC1, SMC3, SA1/SA2, MTA1, MTA2, HDAC1, HDAC2, MBD2, MBD3	Loading of cohesion complex components onto chromatin, chromosome cohesion [299].

The CHD sub-family

Complex	Catalytic subunit/ species	Other subunits	Biological functions
CHD1	Chd1p *S. cerevisiae*	Not known	In yeast involved in the regulation of 2-4% of yeast genes. Approximately 1000 molecules/cell. Chd1p is partially redundant with SWI/SNF complex, ISW1p and ISW2p [94,98]. Chd1p may function during transcript elongation [255].
HRP1	Hrp1 *S. pombe*	Not known	There is evidence suggesting Hrp1 plays a role in chromosome separation [300] and transcription elongation [100].
CHD1	CHD1 mouse/Human	Not known	Excluded from heterochromatin [99].
PICKLE	PICKLE *Arabidopsis*	Not known	Repression of gene expression programs in tissues or at times when they are not necessary [301–303].
NURD	CHD-3, CHD-4 *C. elegans*	RBA-1, LIN-53, HAD-1	Antagonizes Ras-induced vulval development [304]. Involved in regulation of germline-soma distinctions [305].

Complex	Catalytic subunit/species	Other subunits	Biological functions
Mi-2	Mi-2/CHD4 *Drosophila*	dRPD3, likely to contain other subunits, native MW 1Mda [51].	Essential for development [306]. Interacts with dHunchback protein. Functions in both Hunchback and Polycomb mediated repression in vivo [63] Interacts with the transcriptional repressor tramtrack69 [106].
Mi-2	Mi-2/CHD4 *Xenopus*	RPD3 (HDAC1/2), RbAp 48 MBD3, MTA1 like, p66	May be associated with methylation induced gene silencing. [102]
Mi-2	CHD4 but also CHD3 Mouse	HDAC1/2, RbAP48, Ikaros1, Ikaros2, Ikaros7, Aiolos	Involved in T cell development. Ikaros proteins target NuRD to heterochromatic regions. [307]
NuRD/NURD/ Mi-2	Mi-2β/CHD4 Mi-2α/CHD3 in some complexes Human	HDAC 1 and 2 RbAp46 and 48 MBD3, MTA1 MTA2, MTA3. (subunit composition varies)	Initially characterized as an autoantigen in a human connective tissue disease [62]. May be associated with methylation induced gene silencing and heterochromatin [64,65,103]. Interacts with a component of DNA repair machinery [308]. Functions in estrogen dependent repression of Snail. Aberrant *Snail* expression results in loss of expression of the cell adhesion molecule E-cadherin, an event associated with invasive growth of breast cancers [309].

Snf2 family members that do not fit into the Snf2, ISWI or CHD subfamilies

Complex	Catalytic subunit/species	Other subunits	Biological functions
INO80.com	Ino80p *S. cerevisiae*	Arp8p, p100p, p90p, Arp5p, Arp4p, Rvb1p, Rvb2p, Act1p, p32p, p28p, p26p 1–1.5 Mda	INO80 deletions cause reductions in the levels of INO1, PHO5, and Ty1 gene expression and have DNA repair phenotypes [60,61]. May facilitate TBP recruitment [107].
Mot1	Mot1 *S. cerevisiae*	TBP	Stimulates ATP-dependent displacement of TBP from DNA in vitro [310]. May assist recruitment of TBP in vivo [311].
Taf172	Taf172 Human	TBP	Similar to Mot1 [312]
MOM1	MOM1 *Arabidopsis*	Not known	Required for silencing of methylated genes [313].
DDM1	DDM1 *Arabidopsis*	Not known	Required for maintenance of DNA methylation [314] and histone H3 methylation [315].
Lsh	Lsh Mouse	Not known	Required for maintenance of DNA methylation [316].
Yfr038wp	*S. cerevisiae*	Not known	Hypothetical ORF.

(*Continued*)

Table 1
Continued

Complex	Catalytic subunit/species	Other subunits	Biological functions
PASG	PASG Human	Not known	A human Snf2p homolog altered in leukemia [317].
ATRX	ATRX Human	Not known	Mutations in the gene that code hATRX cause severe syndromal mental retardation associated with α-thalassemia. ATRX is localized to heterochromatin including rDNA repeats where it plays a role in conferring patterns of DNA methylation [318].
ARIP4	ARIP4 Human	Not known	Interacts with androgen receptor [69].
Rad26	Rad26p *S. cerevisiae*	Def1p, Associates with RNA pol II	Transcription coupled nucleotide excision repair [319,320].
CSB/ERCC6	CSB Human	Associates with a subset of Pol II complexes	Mutation causes Cockayne syndrome, a defect in transcription coupled nucleotide excision repair [321,322]. Human Rad26p homolog
Rad16	Rad16p *S. cerevisiae*	Rad7p, Abf1p	Global genome nucleotide excision repair [76,323,324].
Rad5	Rad5p *S. cerevisiae*	Not Known	Rad6p dependent post replicative repair [325]
Rad54	Rad54p *S. cerevisiae*	Possibly none	Recombination repair [326] alters DNA topology in an ATP-dependent reaction [327] assists strand pairing [328], also alters chromatin structure [329–331].
RapA	RapA *E. coli*	RNA polymerase core or RNA polymerase holoenzyme	Recycling of RNA polymerase [332].

More than one type of both the SWI/SNF and the ISWI polypeptides are found in most eukaryotic cells. Thus, the yeast representatives of the SWI/SNF complex are the SWI/SNF complex itself containing the Swi2p/Snf2p ATPase and the much more abundant RSC complex for which Sth1p is the ATPase subunit [35,77]. Both these complexes are very large ($\sim$2 MDa) and contain 10–15 polypeptides [77–79]. Homologous assemblies are also found in vertebrates [80,81]: BAF (otherwise known as hSwi/Snf-B) and PBAF (otherwise hSwi/Snf-A) being homologous to the yeast RSC and SWI/SNF complexes, respectively [82]. In addition, both the mammalian SWI/SNF and yeast RSC complexes can exist in functionally distinct forms [44,83,84].

The ISWI ATPase was first identified in *Drosophila* [55] and in contrast to the yeast Swi2p/Snf2p ATPase contains two SANT domains [85]. In *Drosophila* it is a component of three complexes, NURF (Nucleosome Remodeling Factor) [56], ACF (ATP-utilizing and Chromatin assembly and remodeling Factor) [59,86] and CHRAC (Chromatin Accessibility Complex) [58]. NURF contains three polypeptides in addition to ISWI [56]; a 301 kDa polypeptide (NURF 301), a 38 kDa pyrophosphatase and a 55 kDa WD40 protein that is also a component of the chromatin assembly factor CAF-1 [87–89]. The two other *Drosophila* ISWI-containing complexes are closely related. ACF contains one additional polypeptide Acf1, which is also found in CHRAC [59,90]. CHRAC contains in addition a further two small histone fold proteins [91,92]. Although topoisomerase II was also found to copurify with CHRAC [59] it is now doubtful whether this enzyme is a bona fide component of the complex [92]. All the ISWI containing complexes catalyse nucleosome sliding but while ACF and CHRAC confer regularity on nucleosome spacing NURF can be disruptive [58,59,93]. Other organisms, for example yeast and *Xenopus*, contain more than one ISWI-containing complex that may contain different variants of the ISWI polypeptide [94–96].

The third class of remodeling complexes is defined by the CHD ATPases [97]. Typically these contain a pair of chromodomains in addition to the "helicase-related" domain (Fig. 1). Chd1p from budding yeast purifies as an apparent dimer of a single polypeptide [98]. *Drosophila* CHD1 is found in active decondensed chromatin [99], while a related protein, Hrp1p, in fission yeast is believed to be involved in remodeling associated with transcription termination [100]. The other members of the CHD family, CHD3 (Mi-2α), and CHD4 (Mi-2β) contain two PHD fingers in addition to the chromodomains [101]. Mi-2 is a component of the NuRD remodeling complex. In addition to the ATPase subunit this complex contains the histone deacetylases HDAC1/2 together with the histone H4 interacting proteins RbAp46/48 [52,102]. Other identified proteins copurifying with the NURD complex include MTA-2, a metastasis-associated antigen essential for efficient deacetylase activity [55,103] MBD3 [102,103], a protein similar to the methyl-cystosine binding protein MBD2, which itself associates with NURD to form the MeCP1 complex [104,105] and the *Drosophila* DNA binding repressor Tramtrack69 [106]. It remains an open question whether these proteins are all present in a single assembly or whether the complex as isolated is heterogeneous.

Yet another type of ATP-dependent remodeling complex is represented by the INO80 complex [60,61]. This assembly contains the Ino80 protein in which the helicase-related motif comprises a large proportion of the coding region [60]. Ino80p also lacks the SANT domains found in ISWI ATPases and contains instead two other short motifs (Fig. 1). This complex contains 12–13 polypeptides including two proteins (Tih1p and Tih2p) related to the RuvB helicase as well as actin and three other actin-related proteins. The mammalian orthologues of the RuvB-related proteins physically associate with the TATA-binding protein (TBP) while in yeast certain mutations that affect TBP binding to the TATA box cause synthetic growth defects in a *tih1* background suggesting that Tih1p might facilitate the recruitment of TBP at a subset of promoters [107]. Mutants in *INO80* show hypersensitivity to DNA damage in addition to defects in transcription [60,61], suggesting that chromatin remodeling dependent on the INO80 complex may be involved in both transcription and DNA damage repair [60,61].

In addition to the complexes containing an ATP-dependent helicase-related motif other assemblies can mediate nucleosome remodeling. In an *in vitro* model system the SP6 RNA polymerase and eukaryotic RNA polymerase III facilitate translocation in *cis* of a single nucleosome [108–110], while RNA polymerase II promotes the loss of a single H2A/H2B dimer without translocation [111]. Another class of complex that facilitates transcript elongation through nucleosomes is the FACT complex [112]. This complex can contain either two or three (in the equivalent yeast complex SPN) subunits [113–115] and interacts directly with nucleosomes [114]. Although there is no evidence that this type of complex moves nucleosomes structural parallels with the ATP-dependent remodeling complexes suggest that it may be involved in the activation or "priming" of remodeling (see below for further discussion) [20].

3. Interactions of remodeling complexes

Although, the abundance of different chromatin remodeling complexes varies greatly, it is likely that the action of most of these complexes is targeted to specific regions within genomes. There are two major classes of mechanism by which this could occur. The first involves the interaction of complexes with a bound sequence-specific DNA-binding protein or the incorporation of such a protein into the complex itself. The second involves interactions with molecular features of the chromatin that are restricted to a specific region of a genome. These potentially include specific patterns of post-translational histone modifications, chromatin containing histone variants that interact with remodeling activities and the association of non-core histone proteins capable of recruiting remodeling activities. There is good evidence that these different mechanisms for the recruitment of remodeling activities are not mutually exclusive (see below). In addition to regulation at the stage of recruitment, it is also possible that an activation step may precede remodeling prior to or subsequent to recruitment. For example, during mitosis, SWI/SNF complexes are phosphorylated and inactivated [116,117]. Active remodeling can normally be effected by the ATPase subunits of the complex in isolation [49,51,52,118–121]. However, this process is generally much less efficient than remodeling mediated by the corresponding complexes (but see refs. [49,120]). This loss of function could reflect, amongst other possibilities, an inefficiency in the activation process, a loss of nucleosome-complex interactions that might impair the topological coupling between remodeling and DNA translocation (for further discussion see below) or alterations to the structure of the catalytic subunit in the absence of additional subunits.

3.1. Common motifs and subunits in remodeling complexes

Both the ATPase and other subunits of the remodeling complexes contain some protein motifs that are likely involved in protein–protein interactions and others that potentially bind DNA. However, in many cases the true targets of the protein–protein interaction motifs remain to be rigorously identified even though some these motifs could potentially interact with components of the nucleosome. Once such motif is the bromodomain, which binds with high specificity to acetylated lysine residues [122–124]. One and two copies respectively of this domain are found towards the C-terminus of the yeast Swi2p/Snf2p and Sth1p polypeptides. In the ISWI containing complexes both the Acf1 subunit of the ACF complex and the p301 subunit of the NURF complex contain a single bromodomain [85,86]. By contrast the RSC complex contains two subunits, Rsc1p and Rsc2p, with two bromodomains [81] and the vertebrate homologues of RSC contain subunits with multiple bromodomains, five and six in the chicken and human homologues, respectively [125,126]. In both the polybromo polypeptides, Rsc1p and Rsc2p another motif, also present in Sir3p, the bromo-adjacent domain (BAH) is found [79]. Cells with a mutation in either of RSC1 or RSC2 genes exhibit slow mitotic growth while loss of function of both is lethal indicating that the two proteins share an essential common function.

Another protein motif which is represented in a number of remodeling complexes is the SANT domain [84]. This motif is present in a class of proteins known or assumed to interact with nucleosomes and was proposed on the basis of its structural homology with c-myb to act as a DNA-binding domain. Proteins containing this domain include components of remodeling complexes (yeast Swi3p, the ISWI polypeptides, yeast Rsc8p and human BAF150/170 (*Drosophila* Moira)), histone acetyltransferase complexes (yeast and human Ada2p), histone deacetylase complexes (mouse N-CoR, Co-REST, and Mta1-L2), histone methyltransferase complexes (enhancer of zeste) as well as yeast TFIIIB (SANT is an acronym derived from "Swi3p, Ada2p, N-CoR and TFIIIB") [84,127–130]. The SANT domains of both Swi3p and Rsc8p are essential for SWI/SNF and RSC function respectively *in vivo* although no effect of deletion of the SANT domain was observed *in vitro* [131]. The SANT domain is also essential for ADA2 function *in vivo*. Interestingly mutation of Ada2p residues homologous to those known to be required for DNA binding by c-myb did not affect *in vivo* function. One interpretation of these results is that the SANT domain has a general role in functional interaction with the histone N-terminal tails [131].

Several subunits of chromatin remodeling complexes contain motifs that actually or potentially interact with DNA. Among the more commonly occurring of these motifs are HMG domains [132] and HMGA motifs (aka AT hooks) or closely related sequences [133]. Examples of complexes containing subunits with HMG domains include the mammalian BAF [134] and the *Drosophila* BRM (*brahma*) complexes [135] containing respectively the HMG-domain proteins BAF57 and BAP111. None of these proteins appears to be essential for remodeling, but loss of BAP111 results in a significant reduction of function [135]. However, mutations in the BAF57 subunit impair the function of the BAF complex *in vivo* in both the silencing of the CD4 locus and the activation of the CD8 locus [136]. Nevertheless the BAF57 HMG domain is dispensable for tethering BAF complexes to the CD4 silencer or other chromatin loci suggesting that BAF-dependent chromatin remodeling *in vivo* requires HMG-induced DNA bending [136]. HMGA domains are found in both Rsc1p and Rsc2p from yeast, *Drosophila* ISWI, SWI2p/SNF2p, mammalian BRM/SNF2α, the N-terminal domain of p301 of the NURF complex and the N-terminal region of *Drosophila* Mi-2 [86,137–139]. In the last example these basic sequences are juxtaposed with highly acidic regions. However, a major component of the DNA-binding activity of dMi-2 has been ascribed to the two chromodomains [139]. These belong to the Myst subfamily of chromodomains, one example of which is found in the *Drosophila* Mof histone acetylase and binds RNA [140]. A DNA-binding function has also been ascribed to the C-terminal region of mouse CHD1 but no recognizable motif has been ascribed to this function [98]. Yet another DNA-binding motif, the non-sequence specific ARID domain (<u>A</u>T-rich <u>I</u>nteracting <u>D</u>omain) [141], found in a number of transcription factors, occurs in several subunits of complexes. These include Rsc9p in the yeast RSC complex, the Swi1/Adr6 subunit in the yeast SWI/SNF complex, p270 in a human SWI/SNF complex and Osa (aka Eyelid) in the *Drosophila* BRM complex [142–146]. Although the ARID domain in Swi1p is reportedly dispensable for biological function [141] both it and Osa are necessary for both the repression and activation of different target genes [143,146–148].

In addition to the common occurrence of different protein motifs actin and actin-related proteins (Arps) are found in an number of remodeling complexes. The yeast INO80 complex contains actin itself, together with three actin-related proteins Arp4p, Arp5p, and Arp8p [61]. β-Actin is a functional component of the mammalian BAF complex and the SWI/SNF-like p400 complex which both also contain the Arp protein BAF53 [69,149]. In yeast Arp7p and Arp9p are shared by the yeast SWI/SNF and RSC complexes [150]. Interestingly actin and Arps are also subunits of the yeast NuA4 and Tip60 histone acetyltransferase complexes, which contain in addition Arp4p and ArpBAF53, respectively [151,152]. The functional role of actin and the Arp components of remodeling complexes is not known. They appear to be essential for the activity of the INO80 complex and could be involved in the regulation of remodeling function by phosphatidylinositol polyphosphates (see below).

3.2. *Interactions between remodeling complexes and nucleosomes*

Most chromatin remodeling complexes are involved in extensive interactions with nucleosomes although the nature of these interactions is in general ill defined. However, the ATPase activities of the different complexes clearly respond to different structural features of nucleosomes. Whereas the ATPase activities of the ISWI and Mi-2 polypeptides are stimulated to a greater extent by nucleosomes compared to free DNA, Snf2 homologues are stimulated by both [51,118,120–122, 154]. In this respect the ligand preferences of the individual translocases qualitatively reflect those of the NURF and SWI/SNF complexes respectively and therefore these properties must be determined at least in part by the translocase components themselves. Both the ISWI-containing complexes NURF and CHRAC, as well as the isolated ISWI polypeptide itself, require the N-terminal tail of histone H4 in particular the sequence 16-KRHR-19, but not the N-terminal tails of the other histones for ATP-dependent nucleosome sliding [155,156]. This contrasts with the dMi-2 ATPase which does not require any of the histone N-terminal tails for activity [51,52]. Differences in the selectivity of complete complexes containing the same ATPase could result from different recognition components being provided by other polypeptides in the NURF complex. In particular, the component of NURF with homology to the mammalian protein RbAp48 is known to bind helix 1 of histone H4 [157]. In addition three regions of another subunit of NURF, p301, interact with nucleosomes, a region at the N-terminus containing HMGA motifs, a central region and a C-terminal region containing two PHD fingers and a bromodomain [87]. Deletion of the HMGA motifs impairs remodeling function.

4. *Mechanism of remodeling*

4.1. *The mechanics of remodeling*

The process of nucleosome remodeling, with the exception of that mediated by RNA polymerase II [111], often leads to the displacement of the histone octamer from one segment of the DNA to another. For the ISWI and dMi-2 containing complexes transfer only in *cis*, but not in *trans* has been observed [48].

By contrast, under certain conditions the SWI/SNF and RSC complexes can mediate transfer in *trans* to a molecule of competing DNA [47,118,158,159]. Many experiments also suggest that *in vitro* SWI/SNF complexes can generate a stable "remodeled" state of the nucleosome that is distinct from the canonical state [160–166]. Remodeling of nucleosome arrays in the presence of topoisomerase I leads to a loss of constrained negative superhelicity [50,167] suggesting that the bound DNA is at least partially unwrapped. This conclusion is consistent with both the increased DNase I [168] and restriction enzyme accessibility to internal DNA sites [166] and also the loss of close contacts between nucleosomal DNA and the Swi2p/Snf2p and Rsc4p subunits after remodeling [169]. Analysis of the stable remodeled state has revealed two structures, the association of two remodeled nucleosomes to form a "dinucleosome" particle [7,163–165] and a structure containing a single octamer with the DNA looped between the exit and entry points on the octamer surface [166]. These structures are not mutually exclusive since the site of rebinding of the initially displaced DNA will depend in part on the concentration of octamers; high concentrations would be expected to favor dimer formation and low concentration monomer formation. A key issue that needs to be addressed is whether either of these structures is of relevance during the remodeling of nucleosomal arrays when nucleosomes will not be located close to DNA ends. In this respect it may be significant that atomic force and electron microscopy of initially regularly spaced nucleosome arrays after exposure to the SWI/SNF complex reveal disorganization accompanied by a significant number of closely abutting nucleosome pairs [165,170]. These dimers contain 60–80 bp of DNA that is more weakly bound than in normal nucleosomes. The relative ability to remodel arrays or mononucleosomes is specific to the type of Swi2/Snf2 ATPase [171]. Thus, while all ATP-dependent remodeling complexes appear to be capable of altering nucleosome positioning, the SWi/SNF related subset may be capable of generating additional forms of altered nucleosome.

In contrast to the RSC and SWI/SNF complexes, ISWI complexes appear to slide nucleosomes along DNA without significant disruption to nucleosome structure [48,49,172,173] and again, unlike SWI/SNF, they do not increase the accessibility of internal sites to restriction endonucleases [172]. These observations argue for a sliding mechanism and suggest that remodeling by ISWI does not involve the persistent unwrapping of a substantial portion of nucleosomal DNA. Instead the observations favor a structural dislocation in the DNA, either a change in twist or in writhe, or even components of both, that transits the surface of the octamer [173] and could be occluded by the remodeling complex itself.

Several models for nucleosome remodeling by ATP-dependent complexes have been proposed. Among the most-commonly considered are the rotation of the histone octamer within the wrapped DNA, propagation of DNA twist around the octamer ("twist diffusion"), and the generation at one edge of the nucleosome of a small DNA loop which is then propagated around the octamer (Fig. 4) [13]. An analysis in which the DNA was treated as an elastic rod concluded that of the twist diffusion and bulge propagation mechanisms either or both could occur [174]. The bulge propagation model is congruent with a model of DNA site accessibility [175,176] in which it was proposed that DNA could transiently unwrap from the ends of the particle and then rebind at the either at the same location reforming a canonical nucleosome or at a different location forming a short bulge.

An excellent example of bulge propagation is provided by the passive repositioning of nucleosomes by the passage of SP6 RNA polymerase and of RNA polymerase III [109,110]. In this situation the polymerase essentially acts as a DNA translocase. In a closed topological domain any check to the free rotation of the polymerase around the DNA during transcription would be expected to generate positive supercoils in advance of the transcribing enzyme and negative supercoils behind [177]. Provided this condition was met, at least partially, in the experimental situation [110], this positive superhelicity could in principle facilitate the unwrapping of the negatively wrapped nucleosomal DNA while the trailing negative superhelicity, if available, would preferentially facilitate nucleosome assembly. Thus, the passage of the polymerase in a closed domain (essentially a DNA bulge) where rotation of the polymerase was constrained would create an energetically favorable situation for the directional translocation of a nucleosome. By contrast, eukaryotic polymerase II appears not to translationally displace nucleosomes in its path but instead interacts directly with the core particle [178] and displaces a single H2A/H2B dimer [111] at moderate salt concentrations which assists the progression of the polymerase around the octamer. This displacement of the H2A/H2B dimer can also be mediated by histone chaperones such as the FACT complex [112,178a], a process which requires the acidic tail of the Spt16 subunit [178a].

There may be similarities between the redistribution of nucleosomes driven by polymerases and remodeling activities. In a model *in vitro* system with a single nucleosome assembled on a 183 bp DNA fragment the SWI/SNF complex slides the octamer up to 50 bp beyond the end of the DNA [167]. The exposed DNA end then appears to reassociate with the newly exposed H2A/H2B dimer forming a DNA loop. Intriguingly the SWI/SNF mediated nucleosome redistribution stalls at positions that are very comparable to those at which yeast RNA polymerase III stalls when transcribing around a nucleosome [110] and correspond to sites in the central gyre which are less susceptible to unwrapping [179]. These positions correspond to the least mobile, and hence probably most tightly bound, regions of wrapped DNA in the crystal of the nucleosome core particle [180].

The similarities between nucleosome redistribution by polymerases and remodeling activities suggest that the propagation of a bulge around the nucleosome could be involved in remodeling driven by SWI/SNF related complexes. Additional support for this stems from the observation that the redistribution of nucleosomes driven by ATP-dependent remodeling activities can negotiate DNA junctions introduced within nucleosome core particles [53]. However, it remains possible the introduction of these structures into nucleosomes causes redistribution to occur via an alternative pathway. Previously, the alterations to nuclease accessibility and DNA topology that result from SWI/SNF remodeling were also taken as evidence that could support the existence of stable intermediates in which DNA is looped out from the surface of the nucleosome. However, the remarkable ability of SWI/SNF-related activities to move nucleosomes up to 50 bp beyond DNA ends would be expected to alter DNA accessibility throughout nucleosome core particles. Thus, the possibility remains that SWI/SNF moves nucleosomes via an incremental mechanism to positions that alter accessibility throughout nucleosome core particles.

The major alternative mechanism for nucleosome redistribution to the bulge propagation mechanism is twist diffusion. Supporting this, it is clear that the nucleosome core particle is capable of accommodating alterations to DNA twist of ± 1–2 bp per double helical turn without significant changes to the histone–DNA contacts [1,35,180]. Of possible greater significance is the ability of small DNA ligands to alter the rotational positioning of nucleosome-bound DNA [181,182]. Minimal disruption to nucleosome integrity could be achieved by twist diffusion and this is consistent with what is observed during ISWI driven redistribution [48]. The observation that nicks introduced within nucleosomes do not prevent nucleosome movement [53,173] has been used to argue against twist diffusion. However, in order for twist to be dissipated by a nick on the surface of a nucleosome one DNA strand would have to rotate around the other which would involve the disruption of histone–DNA contacts. This means it is possible that nicks will not dissipate twist on the surface of a nucleosome as efficiently as the do in free DNA.

4.2. DNA translocation and chromatin remodeling

A conserved feature of all the ATP-dependent remodeling complexes is a subunit containing a DNA helicase-related motif [35] (Fig. 2). This motif also occurs in a large superfamily of nucleic acid manipulating enzymes including DNA and RNA helicases, reverse gyrases as well as enzymes such as RecG involved in the processing of stalled replication forks [183]. In certain helicases composed of a hexameric ring of identical subunits each containing one copy of the motif its primary function is to act as an ATP-dependent motor driving the processive translocation of the enzyme along a single DNA backbone, either as a single strand form or as incorporated into a duplex. Within the motif seven short conserved sequences have been identified [54], of which conserved regions I and II correspond to the A and B boxes of the NTP-binding motif [184]. With the possible exception of conserved region IV all these sequences are conserved in the translocase motif present in the major remodeling polypeptides. However, the remodeling polypeptides lack the residues in region III responsible for the processive displacement of single-stranded DNA that have been identified in the PcrA helicase [185] and although they possess an essential DNA- and/or

	I	Ia	II	III	V	VI
Sc Swi2/Snf2	DEMGLGKT	PLSTLI	DEGH	RLILTGTPLQN	RAGGLGLNL	RAHRIGQ
Sc Sth1	DEMGLGKT	PLSTII	DEGH	RLILTGTPLQN	RAGGLGLNL	RAHRIGQ
Sc Ino80	DEMGLGKT	PASTLH	DEAQ	RLLLTGTPLQN	RAGGLGINL	RAHRLGQ
Dm ISWI	DEMGLGKT	PKSTLQ	DEAH	RLLITGTPLQN	RAGGLGINL	RAHRIGQ
Sc CHD1	DEMGLGKT	PLSTMP	DEAH	RMLITGTPLQN	RAGGLGINL	RAHRIGQ
Dm Mi-2	DEMGLGKT	PLSTLQ	DEAH	RLLLTGTPLQN	RAGGLGINL	RAHRIGQ
Sc Mot1	DDMGLGKT	PSLTGH	DEGH	RLILTGTPIGN	KVGGLCLNL	RAHRIGQ
Sc Rad16	DEMGLGKT	PSSLVN	DEGH	RVILSGTPIQN	KAGGVALNL	RVHRIGQ
Sc Rad54	DDMGLGKT	IIVCPG	DEGH	RVILSGTPIQN	KAGGCGINL	RVWRDGQ
Af reverse gyrase	APTGVGKT	PTVTLV	DDVD	LMVSTATAKPR	LVRGLDLPE	YIQGSGR
Bs PcrA	AGAGSGKT	FTNKAA	DEAQ	GDADQSIYRWR	HAAKGLE	VGITRAEE

Fig. 2. Homology of chromatin remodeling ATPases to DNA translocases. Motifs I–III, V, and VI in the Sf2 "helicase" signature in the Swi/Snf superfamily (including yeast Mot1p, Rad16p and Rad54p) are aligned and compared with the motifs in reverse gyrase Sf2-like signature and the PcrA helicase Sf1 signature. The conserved elements of the latter two are taken from Refs. [192] and [333], respectively. Sc, *Saccharomyces cerevisiae*; Dm, *Drosophila melanogaster*; Af, *Archaeoglobus fulgidus*; Bs, *Bacillus stearotemophilus*.

chromatin-dependent ATPase activity they also lack helicase function [41]. This suggests that in remodeling complexes these motors could drive translocation along a DNA duplex, implying that, if so, the complex and the DNA must rotate relative to each other in a manner analogous to the progression of RNA polymerase along a DNA template [177].

Two lines of evidence are consistent with the view that the single "helicase-related" motif is involved in DNA translocation. First, both the RSC complex and its isolated ATPase subunit Sth1p displace the third strand of a short triple helix formed on naked DNA in an ATP-dependent manner [153]. Similarly the ISWI protein is able to displace triplex-forming oligonucleotides efficiently when they are introduced at sites close to a positioned nucleosome but less efficiently when the triple helix is formed 30–60 bp outside the nucleosome [186]. In the former experiments indirect disruption of the triple-stranded region by, for example, transmission of under- or over-twisting was considered unlikely. Second, a strong prediction of a model invoking DNA translocation by the remodeling ATPases is that the ATPase activity should be dependent on the length of the DNA translocated up to the processivity limit. For RSC and Sth1p [153] as well as for ISWI [186] this has been shown to be the case. For the former activity an estimate of $\sim$80 bp for the processivity limit was determined while that for ISWI, in the presence of a nucleosome, was $\sim$40 bp. Compared with many other DNA translocating enzymes, for example RecBCD [187] and the Ø29 DNA packaging translocase [188], these distances are short. If the remodeling ATPases act as translocases how much ATP hydrolysis is required for translation? When acting as a chromatin assembly factor the ACF complex has been estimated to hydrolyse 2–4 molecules of ATP per bp assembled into chromatin [189]. This value was determined after correcting for hydrolysis observed in the presence of free DNA. If this value is indeed a measure of the energy expended during translocation it corresponds well to values of 1–3 molecules of ATP hydrolysed per bp translocated for helicases [190] and of 1 molecule of ATP hydrolysed/2 bp translocated for the Ø29 DNA translocase [191].

What is the direction of the translocation in the context of a nucleosome? In principle a motor bound to DNA close to the border of a nucleosome could by translocating towards the particle pull the DNA away from the histone octamer (Fig. 3). Alternatively the motor could translocate away from the nucleosome and so push more DNA towards the octamer. In the latter case DNA between the end of the nucleosome in its initial position and the end of the fragment becomes incorporated into the nucleosome and thus the DNA must be translocated towards the octamer (Fig. 4). A conformational change would then be required for the DNA translocated into the domain delimited by the remodeling complex and the nucleosome to be incorporated into the nucleosome. Two studies have attempted to address the issue of directionality with respect to the action of the ISWI-containing remodelling enzymes and nucleosomes. In one case the asymmetric binding of ISWI to one side of a nucleosome together with unidirectional movement was used to assign directionality [173]. However, it appears that ISWI can also interact with the other side of the nucleosome studied. In the other case, triplex

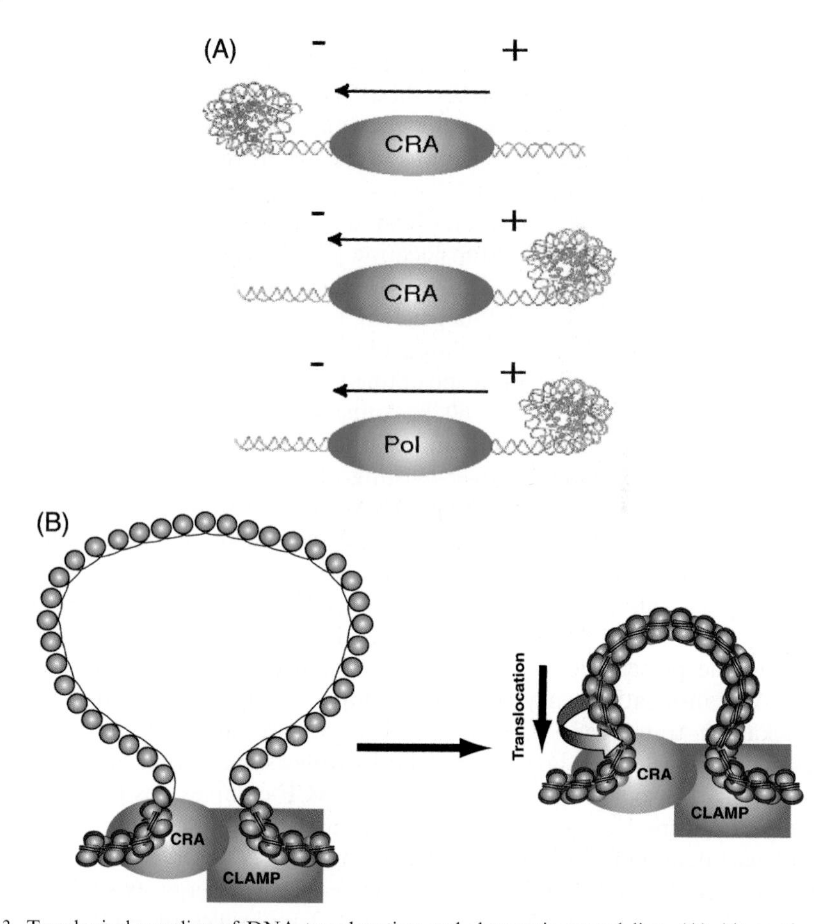

Fig. 3. Topological coupling of DNA translocation and chromatin remodeling. (A) Alternative models for remodeling of a single nucleosome driven by the translocating complexes are compared with passive remodeling driven by the SP6 RNA polymerase or RNA polymerase III [109]. Note that no superhelicity would be constrained unless rotation of the translocase and DNA ends is impeded or prevented. It is also assumed that translocation occurs in steps of less than 5bp. CRA, chromatin remodelling assembly. The arrows indicate the direction of translocation of the DNA. (B) Model for remodeling of a nucleosome array within a topologically defined domain. Adapted with permission from Ref. [119].

displacement was found to occur in a fashion consistent with translocation of the complex oriented away from the nucleosome [186]. In this case, it is difficult to rule out the possibility that transient movement of the nucleosome rather than the remodeler caused displacement. However, the specificity for the 3′–5′ strand observed is more likely to be a property of the translocase than of nucleosome movement. Both studies support the idea that translocation is oriented away from the nucleosome. Another example of such a coupling of ATP-dependent DNA translocation and a conformational change resulting in a change of DNA structure is found in the reverse gyrase from hyperthermophilic Archaea and Eubacteria where ATP hydrolysis by a "translocase" motif drives the

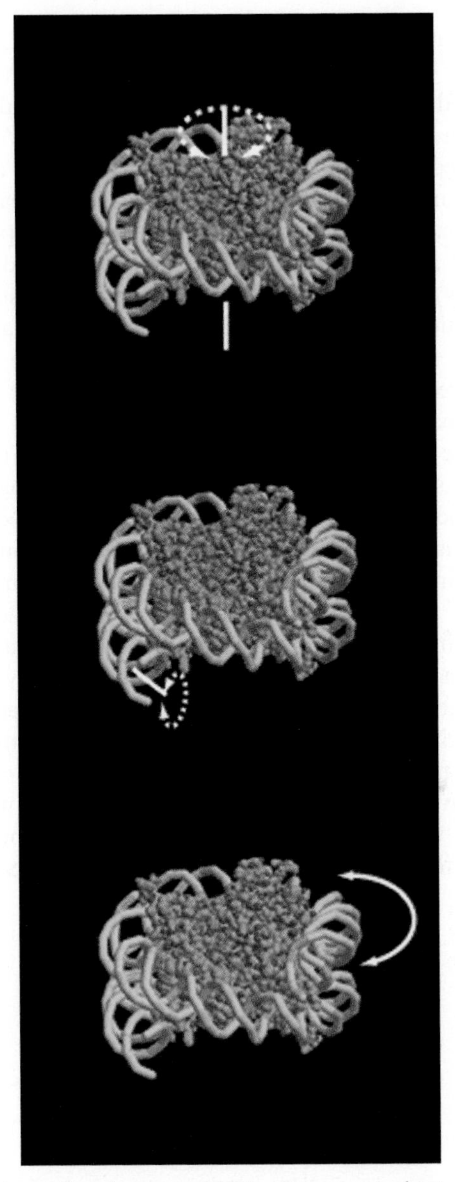

Fig. 4. Proposed mechanisms of nucleosome mobility. From top to bottom the figure show octamer rotation, twist diffusion and bulge propagation.

introduction of positive supercoils into DNA [192,193]. In the case of the remodeling enzymes, translocation would result in both the forcing of DNA into nucleosomes which could generate bulges, and the generation of twist which could be accommodated as either a change in the twist or writhe of DNA on the surface of the nucleosome.

4.3. *The DNA topology of remodeling*

It is well established that remodeling ATPases can induce ATP-dependent topological changes in DNA. The SWI/SNF complex, the *Xenopus* Mi-2 complex as well as the isolated ISWI and BRG polypeptides can all generate negative superhelical torsion on linear chromatin and, in some cases on DNA [119]. This assay depends on the establishment of constrained topological domains on the linear DNA but whether such domains were created within or between bound complexes was not established. The generation of superhelical torsion is consistent with the notion that the remodeling complexes can act as ATP-dependent DNA translocases. It does not however in itself determine the directionality of movement of DNA relative to the nucleosome nor does it establish whether in the context of chromatin remodeling the torsional stress acts directly on the wrapped DNA or the histone octamer, or indeed both (Fig. 3). Not only can the remodeling ATPases generate superhelicity but topological constraints can also restrict the remodeling activity of at least the SWI/SNF complex [194]. In particular the circularisation of a trinucleosomal array inhibits SWI/SNF induced remodeling and this inhibition can be reduced by topoisomerase which relieves torsional strain.

If the torsional strain generated by DNA translocation is utilized for mobilizing nucleosomes [195] then the direction of translocation will play an important role in determining the direction in which DNA is passed over the nucleosome surface and the orientation in which torsion is altered. Provided that rotation of the remodeling complex is restricted, incremental translocation of the motor towards the nucleosome will apply positive superhelical torsion to the particle whereas translocation away the nucleosome will apply negative superhelical torsion [119] (Fig. 3). The problem then is how to translate the motion of the translocase into translational

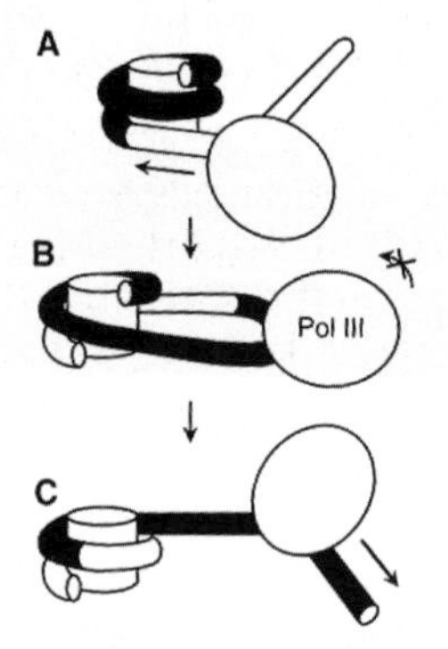

Fig. 5. Proposed mechanism of passive *cis*-displacement of a histone octamer by a transcribing SP6 RNA polymerase or RNA polymerase III. (Adapted with permission from Ref. [109].) A. The transcribing polymerase III approaches a nucleosome. B. The nucleosomal DNA is partially unwrapped, a loop containing the polymerase is formed, and the DNA behind the polymerase binds to the vacated histones. C. The octamer is reformed in a new position behind the polymerase.

movement of the octamer. For further translocation to occur the translocase would need to reset, a transition which would require, as in the PcrA helicase, a conformational change [196]. This change could be coupled to octamer sliding thus bringing the octamer closer to the remodeling complex or polypeptide. The negative torque could facilitate this process, either directly, or by inducing a local DNA distortion as either a change in twist or a small loop.

For any topologically driven mechanism preventing the translocase from rotating around DNA is required. This rotation might be constrained by direct interactions between the remodeling complex and the histone octamer. The recently identified cavities within the yeast RSC and SWI/SNF complexes might provide this [197,198]. A mechanism of this type is also consistent with an observed decrease in the linking number deficit of ~0.6 associated with the persistently remodeled state [43,50,163,167]. In contrast to ISWI, the presence of specifically located nicks in octamer-bound DNA reduced the efficiency of remodeling by both the RSC complex and the Sth1 polypeptide by 2–3 fold leading to the conclusion that twist diffusion might also contribute to remodeling in this case [153]. Strictly, if torsional stress is applied to the particle as a whole the nature of any rotational response in the particle will depend on the relative energy barriers for the processes involved and also on the contacts that define a topological domain in the broadest sense. For example, a nick in octamer-bound DNA might dissipate torque that could be utilised for rotating the whole octamer or an H2A/H2B dimer relative to the H3/H4 tetramer. Alternatively a nick just outside the octamer might lower the energy barrier for an activation process involving rotation by facilitating translocation of the motor [173].

4.4. Nucleosome "priming"

Does chromatin remodeling require an activation step? This was first mooted to explain the facilitation of ISWI induced nucleosome sliding by a specific nick [173]. A possible key observation is that at least some of the principal classes of complex either contain an HMGB protein as an associated subunit or require such a protein for maximal activity *in vitro* or *in vivo* (reviewed in Refs. [20,198,199]; see also [200,201]. Recently it has been shown that one such protein, HMGB1, potentiates remodeling by the ACF complex by promoting the binding of the complex [200]. Similarly a *Drosophila* counterpart of HMGB1, HMG-D, can increase the accessibility of the wrapped nucleosomal DNA both in the central region and to a lesser extent at one edge of the particle while restricting accessibility over the remainder of the particle [201a]. This structural change requires the acidic tail of HMG-D and suggests that this protein and others of its class, which bind at one edge of the nucleosome [201], could "prime" the nucleosome for remodeling by, for example, binding to the N-terminal tail of one of the two copies of histone H3.

Since HMGB proteins appear to impair intrinsic nucleosome mobility while at the same time enhancing enzymatically driven translocation [202] this observation is not wholly consistent with a simple unwrapping from one end of the bound DNA postulated to be a key component of the mechanism of intrinsic translocation [176]. Instead one possibility is that the HMGB protein induces a structural change in the histone octamer. Such a change could involve a rearrangement of inter-histone contacts within the octamer requiring the breakage of several histone–DNA contacts and reformation of a new inter-histone surface. This reorganization would then be reflected in a change in the path of the DNA around the octamer.

Such a mechanism implies that the HMGB protein should contact both the DNA and the histone octamer. Both HMG-D [203] and the FACT complex [113] have been shown to contact H2A, although in the latter case it is not known whether Spt16 or SSRP1, or even both, make this contact. Binding to both components, provided the linking domain was not rotationally flexible could establish a closed topological domain. HMGB proteins constrain DNA unwinding within the binding site and so, in principle could induce a region of torque with the closed domain, i.e., the situation should be similar to that of a translocase except that a small torque is generated by constraint instead of by translocation.

4.5. An active role for core histones in remodeling?

Most models for nucleosome remodeling assume that the structure of the nucleosome core particle, as typified by recent crystal structures, is essentially invariant [15,176]. An important corollary to this assumption is that the energy penalty for simultaneously disrupting many of the electrostatic bonds between the DNA and histones would be prohibitive. However, if remodeling proceeded via an obligatory "activated" (and, by implication, higher energy state) conformation of the core particle the energetics for disruption could be more favorable permitting mechanical models that might otherwise be excluded. A pertinent question is therefore whether changes in histone–histone contacts occur during remodeling. It could be argued that the loss of a single H2A/H2B dimer during transcription by RNA polymerase II represents an unusual situation. However the observation implies that remodeling mediated by RNA polymerase II destabilises the contacts between one H2A/H2B dimer and the H3/H4 tetramer. *In vitro* a nucleosome represents a significant barrier to RNA polymerase II elongation. If this barrier represents the requirement to destabilise inter-histone contacts FACT (or in yeast by the related SPN complex) [112,114] would facilitate elongation by lowering the activation energy for disruption.

Another line of evidence is the occurrence of several mutations within the central α-helical region of the histone octamer can confer a *sin* phenotype [204]. Thus mutations in histones H4 of tyrosines at or close to the interface of histone H4 with the H2A/H2B dimer (Y72G, Y88G, and Y98H) suppress inositol auxotrophy due to a *snf2* disruption but do not suppress defects in *SUC2* transcription in a similar disruption, nor do they change the overall accessibility of yeast chromatin to micrococcal nuclease [205]. Similarly, mutation of Ser73 to cysteine and of Val43, both located at or near the H2A/H2B:H3/H4 interface, confers a *sin* phenotype [206,207]. A weak *sin* phenotype is also observed with mutations in the structured α-helical domain of histone H2B in the residues involved in H2B association with histone H2A (Y40G) or with histone H4 (Y86G) [208]. These observations are consistent with the view that perturbation of the dimer–tetramer contacts in the octamer can, in certain situations, overcome a requirement for SWI/SNF function suggesting that at least a transient change in histone–histone contacts could be involved in remodeling. Whether this suppression involves changes in the intrinsic mobility of the bound octamer or simply weakens the histone–DNA interactions remains to be established.

Other *sin* mutations such as H4 R45H and R45C, H4 S47C, H3 T118I, and H3R116H are clustered on the surface of a small region of the H3–H4 tetramer in the vicinity of the nucleosome dyad [207,209]. The crystal structures of some of these core particles indicate that a relatively small number of histone DNA contacts are lost [209a]. However, *in vivo* these mutations result in increased sensitivity to micrococcal nuclease and to Dam methyltransferase [206,209] and impair the ability of nucleosomes to constrain supercoils [209]. *In vitro* nucleosomes assembled with octamers containing either R116H or T118I are more sensitive to micrococcal nuclease and DNase I cleavage and do not constrain DNA in a precise rotational position [210]. In addition these nucleosomes have a higher inherent thermal mobility [210a]. These disruptions of histone–DNA contacts occur in a similar position to the alterations induced by HMGB proteins and thus may promote mobility by affecting the most stable region of wrapped DNA [180]. It is also possible that increased nucleosome mobility might affect the ability of nucleosomal arrays to adopt more condensed structures. This may explain why some *sin* mutations also affect the higher order folding of nucleosomal arrays [211].

Another implication of a mechanism in which the interface between an H2A/H2B dimer and the H3/H4 tetramer is destabilised during remodeling is that nucleosomes containing histone variants, particularly H2A variants, which modulate the H2A/H3 interface might be more or less susceptible to remodeling. In nucleosomes with a complement of normal histones this interface, and that of H2B and H4, are stabilised by a network of bridging water molecules with only a few direct hydrogen bonds between the subunits [180]. This type of interface, which is analogous to that in activated phosphofructokinase [212], could confer rotational fluidity. The H2A/H3 interface can be altered by substitution of H2A variants. Thus the *in vivo* function of one H2A variant, H2AvD, the *Drosophila* homologue of H2A.Z, requires unique residues in the C-terminus, a region is buried within the histone core [213] altering both the contacts at the H2A/H3 interface and extending an acidic patch at the center of the top surface of the particle [214]. Genetic evidence also supports a link between chromatin remodeling and replacement of the canonical H2A with H2A.Z. The phenotype of null mutants of *HTZ1*, the gene encoding this protein in *Saccharomyces cerevisiae*, is enhanced by loss of *SNF2* function, an effect which contrasts with the *sin* phenotype exhibited by strains lacking other histones [215]. This suggests that substitution of H2A by H2A.Z in yeast partially relieves the requirement for the SWI/SNF complex for remodeling.

Another unresolved question is the extent to which passive nucleosome sliding and enzymatically induced sliding proceed via similar mechanisms. In the case of ISWI and NURF-facilitated nucleosome sliding both the ability of a nucleosome to migrate and its final position are determined by positioning signals and hence by the affinity of the octamer for a particular sequence [217,218]. These characteristics are shared with passive sliding [29]. An obvious possibility is that the fundamental mechanisms are similar but NURF lowers the activation energy both for the propagation and the initiation of sliding and imparts directionality on an otherwise random walk. Removal of the N-terminal tail of histone H2B promotes intrinsic nucleosome sliding [155]. The role of the H2B N-terminal tail in this context has not been established. One possibility is that the tail contacts nucleosomal DNA and presents a barrier to either the initiation or propagation of nucleosome sliding.

5. *Biological functions of chromatin remodeling*

Nucleosomes are the ubiquitous packaging units of the eukaryotic genome and consequently nucleosome remodeling is a pervasive phenomenon that affects virtually all aspects of enzyme-mediated DNA transactions. The concept of chromatin fluidity [7,19] envisages short-range ATP-independent nucleosome movements that allow the maintenance of packing while permitting access. Short range alterations to nucleosome positioning of this type can have profound affects on the way in which a gene responds to activating signals [219]. Typically nucleosome movements are characterised in nuclei by determining the accessibility to cleavage reagents such as micrococcal nuclease or DNase I. These reagents are powerful tools for analysing a stable nucleosome structure but where the structure of a nucleosome is altered by, for example, the loss of an H2A/H2B dimer or remodeling by a SWI/SNF type complex, accessibility to the nucleases may be increased, even though in principle, the octamer remains intact at its previous position. Another complicating phenomenon is that the translational positioning of nucleosomes in an array may be slightly different in different nuclei. For example, at many yeast promoters analysis reveals a population of nucleosome arrays related by a common rotational signal [220]. Since the nucleosomes within such arrays are often separated by only 10–20 bp any simple translational movements are likely to involve the array as a whole, as is observed during activation of the *ADH2* promoter [221]. By contrast SWI/SNF dependent remodeling at the *SUC2* [222] and *FLO1* [206] can affect chromatin structure over much more extensive regions—in the case of *FLO1* up to 7 kb [206]. There is also evidence that remodeling activities participate in alteration of chromatin structure at the level of chromatin domains in mammalian cells [223].

5.1. General functions of remodeling complexes

The functional contexts of chromatin remodeling are diverse and, with the possible exception of the NuRD complex, it no longer appears likely that a specific biological role such as transcription activation or repression can be ascribed to many of the complexes. Instead it appears more plausible that the remodeling by a complex in a particular chromatin context may be dictated more by a particular mechanism of remodeling (disruptive or non-disruptive) and/or by the ability to recruit other complexes involved in a process. Although the genes encoding the SWI/SNF complex were initially characterized as pleiotropic transcriptional activators genome-wide analysis in yeast revealed that in rich medium more genes required SWI/SNF for repression than for activation while in poor medium the converse effect is apparent [224,225]. This result is consistent with the notion that the complex is required for induced, but not constitutive, transcription of genes in yeast. In addition, the SWI/SNF complex has been implicated in DNA replication and the elongation of transcription in yeast [226,227]. Both the SWI/SNF complex and yeast Isw2p also facilitate DNA repair *in vitro* [228,229]. Similarly although yeast RSC was initially characterized as a transcriptional activator, at certain loci, e.g., *CHA1*, it is required for repression [230]. Both this complex and the INO80 complex have a role in DNA repair *in vivo* but it is unclear whether these affects are direct or indirect [61,231]. Finally, *domino* which encodes a *Drosophila* homologue of the Ino80p family was isolated in a genetic screen for mutations that cause embryonic haematopoietic defects [232].

A comparable picture emerges from studies of ISWI function. In *Drosophila* visualization of polytene chromosomes shows that the bulk of ISWI does not colocalize with RNA polymerase II while analysis of *iswi* mutant animals indicates that *Drosophila* ISWI promotes chromatin condensation *in vivo* [233]. However, the phenotypes of these same mutants suggested that ISWI is required for homeotic gene activation. Since ISWI is present in at least three remodeling complexes these apparently different functions cannot be ascribed to individual complexes. Evidence that an ISWI containing complex may be directly involved in transcription activation has been obtained from studies on mutants of the p301 subunit of NURF. These mutants are impaired in transcription activation *in vivo*, in particular in the expression of both homeotic and at least two heat shock genes [234]. They recapitulate the phenotype of *Enhancer of bithorax*, a member of the *trxG* group previously characterized as a positive regulator of the Bithorax complex and mapping to the same genetic interval as *nurf301*. By contrast to NURF the main function of the yeast ISW2 complex appears to be the repression of target genes, for example the early meiotic genes that are shut off in mitotically growing cells [235,236]. In human cells there is evidence that ISWI containing complexes play a role in assisting DNA replication through heterochromatic regions of the genome [237].

To date only repressive functions have been ascribed to the NuRD complex. It has been suggested that in *Drosophila* its interaction with the Hunchback and

Tramtrack69 transcriptional repressors may be an initial step in a process leading to the establishment of more permanent epigenetic repression [63].

Support for the notion that the requirement for a particular remodeling activity may depend in some cases on the mechanism of remodeling or on the nature of the substrate is provided by the observation that in the late G2/M phase of the cell cycle catalytic action of the yeast SWI/SNF complex is required for the recruitment of the histone acetyltransferase Gcn5p to the promoters of the genes it activates [238,239]. More crucially the expression of *GAL1*, which is normally SWI/SNF independent during interphase becomes SWI/SNF dependent when its expression is artificially shifted towards mitosis [239]. On the basis of these results it was proposed that tight repression of a promoter renders it dependent on SWI/SNF for activation. However it is also notable that in different systems the temporal order in which histone acetylation and ATP-dependent chromatin remodeling are required differs. For example at the β-interferon promoter histone acetylation of histone H4 at lysine 8 is required in concert with the assembly of an enhanceosome for recruitment of SWI/SNF [240]. This is consistent with *in vitro* observations that suggest histone acetylation can enhance the retention of SWI/SNF complexes [241]. Yet at the yeast RNR3 promoter SWI/SNF is recruited together with components of RNA polymerase and other general transcription factors with histone acetylation playing little role in the recruitment process [242]. These observations suggest there is not a universally conserved means by which SWI/SNF complexes are recruited but a range of different pathways that are presumably tailored to meet the requirements for regulation at specific genes. Although the role of histone modifications in the recruitment of SWI/SNF related complexes is most advanced, it also seems likely that covalent histone modifications will play important roles in the recruitment of other classes of ATP-dependent remodeling activity. This is likely to be especially important for complexes that have not been found to interact with sequence specific activator or repressor complexes.

5.2. Targeting of remodeling complexes

The notion that remodeling complexes could be targeted to specific genes was initially controversial. However there are a plethora of reports that targeting is a biologically relevant phenomenon. The purified yeast SWI/SNF complex interacts with the activation domains of Gcn4p, VP16, and Hap4p DNA binding proteins as well as the Hir corepressors which are themselves recruited by a transcription factor binding to the histone H2A and H2B genes [243–245]. Similarly the BRG1 complex interacts with the glucocorticoid receptor, NF1/CSF, erythroid transcription factors and the co-activator β-catenin [246–250].

The subunits of the remodeling complexes contain distinct sites for these interactions. Thus NURF301 contains four motifs that define sites for interaction with nuclear hormone receptors but also interacts with GAGA factor, heat shock factor and VP16. The binding of GAGA factor is consistent with the observations that GAGA promotes the NURF mediated remodeling of the

Drosophila hsp70 promoter and that p301 mutants are defective in the activation of heat shock genes [234]. In a similar manner there is genetic and biochemical evidence that the yeast ISW2 complex is recruited to its target promoters by the DNA-binding protein Ume6p [235].

The co-repressor KAP-1 functionally links the DNA-binding Krüppel-associated box zinc finger proteins to the NURD complex by recruiting the Mi-2α subunit. [251]. This interaction requires a tandem PHD/bromodomain motif in which the individual domains appear to act together as a functional unit. The nature of any possible acetylated lysine targets of the bromodomain remains unclear but it is not excluded that this domain could bind to an acetylated lysine in Mi-2α rather than to an acetylated histone tail.

Some chromatin remodeling complexes are highly abundant. For example the RSC complex in yeast is present at an estimated 1000–2000 copies per cell and has the potential to act at an equivalent number of loci [76]. A combination of chromatin immunoprecipitation by an antibody to the Rsc9p subunit with microarray analysis revealed that under normal growth conditions this component, and presumably the complex itself was widely distributed but was also selectively located at genes responsible for particular physiological pathways [252]. When the cells were subjected to stress, such as hydrogen peroxide treatment or rapamycin, a genome wide change in Rsc9p occupancy was observed [141,252]. Similarly, staining of *Drosophila* salivary gland polytene chromosomes with antibodies to ISWI [233] and dMi-2 [106] revealed that in both cases the motor polypeptides were present at a large number of loci. However, whereas dMi-2 colocalized to a large degree, although not completely, with the transcriptional repressor Tramtrack69 [106] there was much less colocalisation of ISWI with the GAGA transcription factor [233].

5.3. Nucleosome remodeling during transcription

Transcription of a gene requires both the regulatory region become accessible to transcription factors and the passage of the polymerase through the nucleosomes in the transcribed region be facilitated. There is now evidence that these two phases may result from distinct remodeling processes. The initial recruitment of histone modifying enzymes and remodeling activities to the promoter region is likely mediated by a transcription factor and may ultimately result in the complete unfolding of nucleosomes, for example at the *PHO5* promoter [253,254]. However remodeling associated with the act of transcription itself in many cases requires the active involvement of at RNA polymerase II. Although RNA polymerase II can displace a single H2A/H2B dimer from a nucleosome *in vitro* it has not been established whether this activity is relevant *in vivo*. However a number of proteins directly involved in nucleosome remodeling and/or histone modification are part of or can associate with complexes bound the phosphorylated C-terminal domain of the largest subunit of polymerase II. In yeast Chd1p is localized to the coding and 3′ untranslated regions of genes [100] and in *Drosophila* is associated with interbands and puffed regions [99]. The yeast protein has recently been shown to interact directly with both Rtf1p, a member of the Paf1p complex that associates with RNA polymerase II during transcription elongation and also with the elongation factors Spt4p–Spt5p and Spt16p–Pob3p [255]. This latter elongation factor is an essential component of the SPN/FACT complex [256]. However, deletion of *CHD1* in yeast has only small effects on transcription [98] and it seems likely that Chd1p may operate redundantly with Isw1p and Isw2p [100]. The function of these complexes may be facilitated by histone modification activities associated with RNA polymerase II. In yeast these include a histone acetylase, Elp3, in the elongator complex [257], the histone H3 K4 methyltransferase Set1p [258] and the histone H3 K36 methyltransferase Set2p [259]. The mouse SWI/SNF complex has also been shown to facilitate the production of full-length *hsp70* transcripts when transcription is induced by a partially disabled heat shock transcription factor, an effect which correlates with the presence of the BRG1 ATPase in the transcribed region [260].

5.4. Regulation of remodeling complexes

A hitherto neglected aspect of remodeling complexes is the regulation of their activity either by covalent modification or by small molecule effectors. Recently it was observed that mutations in the yeast *ARG82/IPK2 gene*, encoding an inositol trisphosphate kinase activity which converts inositol trisphosphate (IP_3) to inositol hexaphosphate (IP_6) as well as phosphorylating phosphatidyl-IP_3, resulted in impaired remodeling of *PHO5* promoter chromatin [261,262]. This effect is a consequence of a failure to recruit the INO80 and SWI/SNF complexes to phosphate-responsive promoters, implying a link *in vivo* between inositol polyphosphate metabolism and chromatin remodeling. *In vitro* at physiological concentrations IP_6 directly inhibits nucleosome mobilization by the *Drosophila* NURF complex and the yeast ISWI2 and INO80 complexes. However IP_6 has little effect on the yeast SWI/SNF complex whereas, in contrast, both IP_4(1,4,5,6) [but not its isomer IP_4(1,3,4,5)] and IP_5 stimulate nucleosome mobilization by this complex. Both recombinant NURF and ISWI proteins bind IP6 but the molecular consequences of these interactions have not been established. Both INO80 and SWI/SNF complexes regulate the *INO1* gene positively and ISWI2 regulates *INO1* negatively suggesting that antagonistic chromatin-remodeling activities could be regulated by the balance between the cellular levels of IP_4, IP_5, and IP_6. However, whereas *in vitro* the effective concentrations of IP_4, IP_5, and IP_6 are similar, *in vivo* the concentration of IP_6 is normally substantially greater (at least 10-fold) than that of either IP_5 or IP_4. The recent observation that conserved Plant Homeodomain (PHD) fingers of several proteins involved in chromatin metabolism bind to phosphoinositides raises the possibility that this may be a more widespread means of regulation [262a]. Another possible, but unproven, connection between inositol metabolism and chromatin remodeling is the presence of actin and nuclear Arps in certain complexes. The mammalian BAF complex binds phosphosphatidylinositol-bis(4,5)phosphate [262b] and this binding is believed to enhance the association of the complex with the actin filaments [149].

6. Endnote

All current evidence favours the view that the core of the ATP-dependent chromatin remodeling complexes is a DNA translocase activity, functionally analogous to similar engines in the RecG helicase, reverse gyrase and the Mcm translocase. Translocation provides potential means by which both DNA bulges and alterations to DNA twisting could be propagated over the surface of nucleosomes. However, it remains to be determined what the relative importance of these different types of DNA distortion play in nucleosome mobilization and how this varies for different classes of remodeling activity. Whatever the mechanism, it is clear that these complexes function in a diverse range of processes that involve the movement of nucleosomes either to increase or to decrease the accessibility of specific regions of chromatin.

Whilst this work was in preparation, it has been reported that ATP-dependent chromatin remodeling activities can cause histone dimers to be removed from nucleosomes [263]. This lends substance to the suggestion in Section 4.5 that histone dimer tetramer interactions might be disrupted during the course of remodeling reactions. It has also been shown that the yeast Swr1 complex can direct the incorporation of the histone variant H2AZ [262a,262b,263]. This new role for ATP-dependent remodeling enzymes raises the prospect of additional as yet unidentified roles in manipulating the histone composition of chromatin.

Acknowledgements

We thank Dr. Andrew Flaus for help with Fig. 3A and Janice Walker for assistance with the references. Paul Badenhorst, Andrew Flaus, and Chris Stockdale for useful suggestions. TOH is a Wellcome Trust Senior Fellow.

References

1. Luger, K., Mader, A.W., Richmond, R.K., Sargent, D.F., and Richmond, T.J. (1997) Crystal structure of the nucleosome core particle at 2.8 Å resolution. Nature 389, 251–260.
2. Workman, J.L. and Kingston, R.E. (1998) Alteration of nucleosome structure as a mechanism of transcriptional regulation. Annu. Rev. Biochem. 67, 545–579.
3. Mizuguchi, G. and Wu, C. (1999) Nucleosome remodeling factor NURF and in vitro transcription of chromatin. Methods Mol. Biol. 119, 333–342.
4. Kornberg, R.D. and Lorch, Y. (1999) Chromatin-modifying and -remodeling complexes. Curr. Opin. Genet. Dev. 9, 148–151.
5. Guschin, D. and Wolffe, A.P. (1999) SWItched-on mobility. Curr. Biol. 9, R742–R746.
6. Muchardt, C. and Yaniv, M. (1999) ATP-dependent chromatin remodelling: SWI/SNF and Co. are on the job. J. Mol. Biol. 293, 187–198.
7. Kingston, R.E. and Narlikar, G.J. (1999) ATP-dependent remodeling and acetylation as regulators of chromatin fluidity. Genes Dev. 13, 2339–2352.
8. Whitehouse, I., Flaus, A., Havas, K., and Owen-Hughes, T. (2000) Mechanisms for ATP-dependent chromatin remodeling. Biochem. Soc. Trans. 28, 376–379.

9. Felsenfeld, G., Clark, D., and Studitsky, V. (2000) Transcription through nucleosomes. Biophys. Chem. 86, 231–237.
10. Jones, K.A. and Kadonaga, J.T. (2000) Exploring the transcription-chromatin interface. Genes Dev. 14, 1992–1996.
11. Aalfs, J.D. and Kingston, R.E. (2000) What does 'chromatin remodeling' mean? Trends Biochem. Sci. 25, 548–555.
12. Längst, G. and Becker, P.B. (2001) Nucleosome mobilization and positioning by ISWI-containing chromatin-remodeling factors. J. Cell Sci. 114, 2561–2568.
13. Flaus, A. and Owen-Hughes, T. (2001) Mechanisms for ATP-dependent chromatin remodeling. Curr. Opin. Genet. Dev. 11, 148–154.
14. Goodwin, G.H. and Nicolas, R.H. (2001) The BAH domain, polybromo and the RSC chromatin remodeling complex. Gene 268, 1–7.
15. Becker, P.B. and Hörz, W. (2002) ATP-dependent nucleosome remodeling. Annu. Rev. Biochem. 71, 247–273.
16. Becker, P.B. (2002) Nucleosome sliding: facts and fiction. EMBO J. 21, 4749–4753.
17. Narlikar, G.J., Fan, H.Y., and Kingston, R.E. (2002) Cooperation between complexes that regulate chromatin structure and transcription. Cell 108, 475–487.
18. Olave, I.A., Reck-Peterson, S.L., and Crabtree, G.R. (2002) Nuclear actin and actin-related proteins in chromatin remodeling. Annu. Rev. Biochem. 71, 755–781.
19. Caserta, M., Verdone, L., and Di Mauro, E. (2002) Aspects of nucleosomal positional flexibility and fluidity. Chembiochem. 3, 1172–1182.
20. Travers, A.A. (2003) Priming the nucleosome: a role for HMGB proteins? EMBO Rep. 4, 131–136.
21. Katsani, K.R., Mahmoudi, T., and Verrijzer, C.P. (2003) Selective gene regulation by SWI/SNF related chromatin remodeling factors. Curr. Top. Microbiol. Immunol. 274, 113–142.
22. Wang, W. (2003) The SWI/SNF family of ATP-dependent chromatin remodelers: similar mechanisms for diverse functions. Curr. Top. Microbiol. Immunol. 274, 143–169.
23. Formosa, T. (2003) Changing the DNA landscape: putting a SPN on chromatin. Curr. Top. Microbiol. Immunol. 274, 171–201.
24. Feng, Q. and Zhang, Y. (2003) The NuRD complex: linking histone modification to nucleosome remodeling. Curr. Top. Microbiol. Immunol. 274, 269–290.
25. Beard, P. (1978) Mobility of histones on the chromosome of simian virus 40. Cell 15, 955–967.
26. Spadafora, C., Oudet, P., and Chambon, P. (1979) Rearrangement of chromatin structure induced by increasing ionic strength and temperature. Eur. J. Biochem. 100, 225–235.
27. Glotov, B.O., Rudin, A.V., and Severin, E.S. (1982) Conditions for sliding of nucleosomes along DNA: SV40 minichromosomes. Biochim. Biophys. Acta 696, 275–284.
28. Yager, T.D. and van Holde, K.E. (1984) Dynamics and equilibria of nucleosomes at elevated ionic strength. J. Biol. Chem. 259, 4212–4222.
29. Pennings, S., Meersseman, G., and Bradbury, E.M. Mobility of positioned nucleosomes on 5 S rDNA, J. Mol. Biol. 220,101–110.
30. Meerseman, G., Pennings, S., and Bradbury, E.M. (1992) Mobile nucleosomes—a general behavior. EMBO J. 11, 2951–2959.
31. Ura, K., Kurumizaka, H., Dimitrov, S., Almouzni, G., and Wolffe, A.P. (1997) Histone acetylation: influence on transcription, nucleosome mobility and positioning and linker-histone dependent transcriptional repression. EMBO J. 16, 2096–2107.
32. Flaus, A. and Richmond, T.J. (1998) Positioning and stability of nucleosomes on MMTV 3' LTR sequences. J. Mol. Biol. 275, 427–441.
33. Pennings, S., Meersseman, G., and Bradbury, E.M. (1994) Linker histones H1 and H5 prevent the mobility of positioned nucleosomes. Proc. Natl. Acad. Sci. USA 91, 10275–10279.
34. Tsukiyama, T., Becker, P.B., and Wu, C. (1994) ATP-dependent nucleosome disruption at a heat-shock promoter mediated by binding of GAGA transcription factor. Nature 367, 525–532.
35. Laurent, B.C., Yang, X., and Carlson, M. (1992) An essential Saccharomyces cerevisiae gene homologous to SNF2 encodes a helicase-related protein in a new family. Mol. Cell Biol. 12, 1893–1902.

36. Peterson, C.L. and Herskowitz, I. (1992) Characterization of the yeast SWI1, SWI2, and SWI3 genes, which encode a global activator of transcription. Cell 68, 573–583.
37. Tamkun, J.W., Deuring, R., Scott, M.P., Kissinger, M., Pattatucci, A.M., Kaufman, T.C., and Kennison, J.A. (1992) *brahma*: a regulator of Drosophila homeotic genes structurally related to the yeast transcriptional activator SNF2/SWI2. Cell 68, 561–572.
38. Hirschhorn, J.N., Brown, S.A., Clark, C.D., and Winston, F. (1992) Evidence that SNF2/SWI2 and SNF5 activate transcription in yeast by altering chromatin structure. Genes Dev. 6, 2288–2298.
39. Yoshinaga, S.K., Peterson, C.L., Herskowitz, I., and Yamamoto, K.R. (1992) Roles of SWI1, SWI2, and SWI3 proteins for transcriptional enhancement by steroid receptors. Science 258, 1598–1604.
40. Laurent, B.C., Treich, I., and Carlson, M. (1993) The yeast SNF2/SWI2 protein has DNA-stimulated ATPase activity required for transcriptional activation. Genes Dev. 7, 583–591.
41. Côté, J., Quinn, J., Workman, J.L., and Peterson, C.L. (1994) Stimulation of GAL4 derivative binding to nucleosomal DNA by the yeast SWI/SNF complex. Science 265, 53–60.
42. Imbalzano, A.N., Kwon, H., Green, M.R., and Kingston, R.E. (1994) Facilitated binding of TATA-binding protein to nucleosomal DNA. Nature 370, 481–485.
43. Kwon, H., Imbalzano, A.N., Khavari, P.A., Kingston, R.E., and Green, M.R. (1994) Nucleosome disruption and enhancement of activator binding by a human SWI/SNF complex. Nature 370, 481–485.
44. Wang, W., Côté, J., Xue, Y., Zhou, S., Khavari, P.A., Biggar, S.R., Muchardt, C., Kalpana, G.V., Goff, S.P., Yaniv, M., Workman, J.L., and Crabtree, G.R. (1996) Purification and biochemical heterogeneity of the mammalian SWI-SNF complex. EMBO J. 15, 5370–5382.
45. Wang, W., Xue, Y., Zhou, S., Kuo, A., Cairns, B.R., and Crabtree, G.R. (1996) Diversity and specialization of mammalian SWI/SNF complexes. Genes Dev. 10, 2117–2130.
46. Papoulas, O., Beek, S.J., Moseley, S.L., McCallum, C.M., Sarte, M., Sheam, A., and Tamkun, J.W. (1998) The *Drosophila* trithorax group proteins BRM, ASH1 and ASH2 are subunits of distinct protein complexes. Development 125, 3955–3966.
47. Whitehouse, I., Flaus, A., Cairns, B.R., White, M.F., Workman, J.L., and Owen-Hughes, T. (1999) Nucleosome mobilization catalysed by the yeast SWI/SNF complex. Nature 400, 784–787.
48. Längst, G., Bonte, E.J., Corona, D.F., and Becker, P.B. (1999) Nucleosome movement by CHRAC and ISWI without disruption or trans-displacement of the histone octamer. Cell 97, 843–852.
49. Hamiche, A., Sandaltzopoulos, R., Gdula, D.A., and Wu, C. (1999) ATP-dependent histone octamer sliding mediated by the chromatin remodeling complex NURF. Cell 97, 833–842.
50. Jaskelioff, M., Gavin, I.M., Peterson, C.L., and Logie, C. (2000) SWI-SNF-mediated nucleosome remodeling: role of histone octamer mobility in the persistence of the remodeled state. Mol. Cell. Biol. 20, 3058–3068.
51. Brehm, A., Längst, G., Kehle, J., Clapier, C.R., Imhof, A., Eberharter, A., Müller, J., and Becker, P.B. (2000) dMi-2 and ISWI chromatin remodeling factors have distinct nucleosome binding and mobilization properties. EMBO J. 19, 4332–4341.
52. Guschin, D., Wade, P.A., Kikyo, N., and Wolffe, A.P. (2000) ATP-Dependent histone octamer mobilization and histone deacetylation mediated by the Mi-2 chromatin remodeling complex. Biochemistry 39, 5238–5245.
53. Aoyagi, S. and Hayes, J.J. (2002) hSWI/SNF-catalysed nucleosome sliding does not occur solely via a twist-diffusion mechanism. Mol. Cell. Biol. 22, 7484–7490.
54. Gorbalenya, A.E. and Koonin, E.V. (1993) Helicases: amino acid sequence comparisons and structure–function relationships. Curr. Opin. Struct. Biol. 3, 419–429.
55. Elfring, L.K., Deuring, R., McCallum, C.M., Peterson, C.L., and Tamkun, J.W. (1994) Identification and characterization of Drosophila relatives of the yeast transcriptional activator SNF2/SWI2. Mol. Cell. Biol. 14, 2225–2234.
56. Tsukiyama, T. and Wu, C. (1995) Purification and properties of an ATP-dependent nucleosome remodeling factor. Cell 83, 1011–1020.
57. Tsukiyama, T., Daniel, C., Tamkun, J., and Wu, C. (1995) ISWI, a member of the SWI2/SNF2 ATPase family, encodes the 140 kDa subunit of the nucleosome remodeling factor. Cell 83, 1021–1086.

58. Varga-Weisz, P.D., Wilm, M., Bonte, E., Dumas, K., Mann, M., and Becker, P.B. (1997) Chromatin-remodeling factor CHRAC contains ATPases ISWI and topoisomerase II. Nature 388, 598–602.
59. Ito, T., Bulger, M., Pazin, M.J., Kobayashi, R., and Kadonaga, J.T. (1997) ACF, an ISWI-containing and ATP-utilizing chromatin assembly and remodeling factor. Cell 90, 145–155.
60. Ebbert, R., Birkmann, A., and Schuller, H.J. (1999) The product of the SNF2/SWI2 paralogue INO80 of Saccharomyces cerevisiae required for efficient expression of various yeast structural genes is part of a high-molecular-weight protein complex. Mol. Microbiol. 32, 741–751.
61. Shen, X., Mizuguchi, G., Hamiche, A., and Wu, C. (2000) A chromatin remodeling complex involved in transcription and DNA processing. Nature 406, 541–544.
62. Seelig, H.P., Moosbrugger, I., Ehrfeld, H., Fink, T., Renz, M., and Genth, E. (1995) The major dermatomyositis-specific Mi-2 autoantigen is a presumed helicase involved in transcriptional activation. Arthritis Rheum. 38, 1389–1399.
63. Kehle, J., Beuchle, D., Treuheit, S., Christen, B., Kennison, J.A., Bienz, M., and Müller, J. (1998) dMi-2, a hunchback-interacting protein that functions in polycomb repression. Science 282, 1897–1900.
64. Tong, J.K., Hassig, C.A., Schnitzler, G.R., Kingston, R.E., and Schreiber, S.L. (1998) Chromatin deacetylation by a chromatin-remodeling complex. Nature 395, 917–921.
65. Xue, Y., Wong, J., Moreno, G.T., Young, M.K., Côté, J., and Wang, W. (1998) NURD, a novel complex with both ATP-dependent chromatin-remodeling and histone deacetylase activities. Mol. Cell 2, 851–861.
66. Wade, P.A., Jones, P.L., Vermaak, D., and Wolffe, A.P. (1998) A multiple subunit Mi-2 histone deacetylase from *Xenopus laevis* with an associated Snf2 superfamily ATPase. Curr. Biol. 8, 843–846.
67. Zhang, Y., LeRoy, G., Seelig, H.P., Lane, W.S., and Reinberg, D. (1998) The dermatomyositis-specific autoantigen Mi-2 is a component of a complex containing histone deacetylase and nucleosome remodeling activities. Cell 95, 279–289.
68. Delmas, V., Stokes, D.G., and Perry, R.P. (1994) A mammalian DNA-binding protein that contains a chromodomain and an SNF2/SWI2-like helicase domain. Proc. Natl. Acad. Sci. USA 90, 2414–2418.
69. Rouleau, N., Domans'kyi, A., Reeben, M., Moilanen, A.M., Havas, K., Kang, Z., Owen-Hughes, T., Palvimo, J.J., and Janne, O.A. (2002) Novel ATPase of SNF2-like protein family interacts with androgen receptor and modulates androgen-dependent transcription. Mol. Biol. Cell 13, 2106–2119.
70. Fuchs, M., Gerber, J., Drapkin, R., Sif, S., Ikura, T., Ogryzko, V., Lane, W.S., Nakatani, Y., and Livingston, D.M. (2001) The p400 complex is an essential E1A transformation target. Cell 106, 297–307.
71. Olave, I., Wang, W., Xue, Y., Kuo, A., and Crabtree, G.R. (2002) Identification of a polymorphic, neuron-specific chromatin remodeling complex. Genes Dev. 16, 2509–2517.
72. Brzeski, J. and Jerzmanowski, A. (2003) Deficient in DNA methylation 1 (DDM1) defines a novel family of chromatin-remodeling factors. J. Biol. Chem. 278, 823–828.
73. Gibbons, R.J., Picketts, D.J., Villard, L., and Higgs, D.R. (1995) Mutations in a putative global transcriptional regulator cause X-linked mental retardation with alpha-thalassemia (ATR-X syndrome). Cell 80, 837–845.
74. Davis, J.L., Kunisawa, R., and Thorner, J. (1992) A presumptive helicase (MOT1 gene product) affects gene expression and is required for viability in the yeast Saccharomyces cerevisiae. Mol. Cell. Biol. 12, 1879–1892.
75. Auble, D.T., Hansen, K.E., Mueller, C.G., Lane, W.S., Thorner, J., and Hahn, S. (1994) Mot1, a global repressor of RNA polymerase II transcription, inhibits TBP binding to DNA by an ATP-dependent mechanism. Genes Dev. 8, 1920–1934.
76. Schild, D., Glassner, B.J., Mortimer, R.K., Carlson, M., and Laurent, B.C. (1992) Identification of RAD16, a yeast excision repair gene homologous to the recombinational repair gene RAD54 and to the SNF2 gene involved in transcriptional activation. Yeast 8, 385–395.

77. Cairns, B.R., Lorch, Y., Li, Y., Zhang, M., Lacomis, L., Erdjument-Bromage, H., Tempst, P., Du, J., Laurent, B., and Kornberg, R.D. (1996) RSC, an essential, abundant chromatin-remodeling complex. Cell 87, 1249–1260.
78. Peterson, C.L., Dingwall, A., and Scott, M.P. (1994) Five SWI/SNF subunits are components of a large multisubunit complex required for transcriptional enhancement. Proc. Natl. Acad. Sci. USA 91, 2905–2908.
79. Cairns, B.R., Kim, Y.J., Sayre, M.H., Laurent, B.C., and Kornberg, R.D. (1994) A multisubunit complex containing the SWI1/ADR6, SWI2/SNF2, SWI3, SNF5, and SNF6 gene products isolated from yeast. Proc. Natl. Acad. Sci. USA 91, 1950–1954.
80. Khavari, P.A., Peterson, C.L., Tamkun, J.W., Mendel, D.B., and Crabtree, G.R. (1993) BRG1 contains a conserved domain of the SWI2/SNF2 family necessary for normal mitotic growth and transcription. Nature 366, 170–174.
81. Muchardt, C. and Yaniv, M. (1993) A human homologue of Saccharomyces cerevisiae SNF2/SWI2 and Drosophila brm genes potentiates transcriptional activation by the glucocorticoid receptor. EMBO J. 12, 4279–4290.
82. Xue, Y., Canman, J.C., Lee, C.S., Nie, Z., Yang, D., Moreno, G.T., Young, M.K., Salmon, E.D., and Wang, W. (2000) The human SWI/SNF-B chromatin-remodeling complex is related to yeast Rsc and localizes at kinetochores of mitotic chromosomes. Proc. Natl. Acad. Sci. USA 97, 13015–13020.
83. Cairns, B.R., Schlichter, A., Erdjument-Bromage, H., Tempst, P., Kornberg, R.D., and Winston, F. (1999) Two functionally distinct forms of the RSC nucleosome-remodeling complex, containing essential AT hook, BAH, and bromodomains. Mol. Cell 4, 715–723.
84. Sif, S., Saurin, A.J., Imbalzano, A.N., and Kingston, R.E. (2001) Purification and characterization of mSin3A-containing Brg1 and hBrm chromatin remodelling complexes. Genes Dev. 15, 603–618.
85. Aasland, R., Stewart, A.F., and Gibson, T. (1996) The SANT domain: a putative DNA-binding domain in the SWI-SNF and ADA complexes, the transcriptional co-repressor N-CoR and TFIIIB. Trends Biochem. Sci. 21, 87–88.
86. Ito, T., Levenstein, M.E., Fyodorov, D.V., Kutach, A.K., Kobayashi, R., and Kadonaga, J.T. (1999) ACF consists of two subunits, Acf1 and ISWI, that function cooperatively in the ATP-dependent catalysis of chromatin assembly. Genes Dev. 13, 1529–1539.
87. Xiao, H., Sandaltzopoulos, R., Wang, H.M., Hamiche, A., Ranallo, R., Lee, K.M., Fu, D., and Wu, C. (2001) Dual functions of largest NURF subunit NURF301 in nucleosome sliding and transcription factor interactions. Mol. Cell. 8, 531–543.
88. Gdula, D.A., Sandaltzopoulos, R., Tsukiyama, T., Ossipow, V., and Wu, C. (1998) Inorganic pyrophosphatase is a component of the Drosophila nucleosome remodeling factor complex. Genes Dev. 12, 3206–3216.
89. Martinez-Balbas, M.A., Tsukiyama, T., Gdula, D., and Wu, C. (1998) *Drosophila* NURF-55, a WD repeat protein involved in histone metabolism. Proc. Natl. Acad. Sci. USA 95, 132–137.
90. Poot, R.A., Dellaire, G., Hulsmann, B.B., Grimaldi, M.A., Corona, D.F., Becker, P.B., Bickmore, W.A., and Varga-Weisz, P.D. (2000) HuCHRAC, a human ISWI chromatin remodeling complex contains hACF1 and two novel histone-fold proteins. EMBO J. 19, 3377–3387.
91. Corona, D.F., Eberharter, A., Budde, A., Deuring, R., Ferrari, S., Varga-Weisz, P., Wilm, M., Tamkun, J., and Becker, P.B. (2000) Two histone fold proteins, CHRAC-14 and CHRAC-16, are developmentally regulated subunits of chromatin accessibility complex (CHRAC). EMBO J. 19, 3049–3059.
92. Eberharter, A., Ferrari, S., Längst, G., Straub, T., Imhof, A., Varga-Weisz, P., Wilm, M., and Becker, P.B. (2001) Acf1, the largest subunit of CHRAC, regulates ISWI-induced nucleosome remodeling. EMBO J. 20, 3781–3788.
93. Mizuguchi, G., Tsukiyama, T., Wisniewski, J., and Wu, C. (1997) Role of nucleosome remodeling factor NURF in transcriptional activation of chromatin. Mol. Cell 1, 141–150.
94. Tsukiyama, T., Palmer, J., Landel, C.C., Shiloach, J., and Wu, C. (1999) Characterisation of the imitation switch family of ATP-dependent chromatin remodeling factors in Saccharomyces cerevisiae. Genes Dev. 13, 686–697.

95. Guschin, D., Geiman, T.M., Kikyo, N., Tremethick, D.J., Wolffe, A.P., and Wade, P.A. (2000) Multiple ISWI ATPase complexes from Xenopus laevis. Functional conservation of an ACF/CHRAC homolog. J. Biol. Chem. 275, 35248–35255.
96. Vary, J.C., Jr., Gangaraju, V.K., Qin, J., Landel, C.C., Kooperberg, C., Bartholomew, B., and Tsukiyama, T. (2003) Yeast Isw1p forms two separable complexes in vivo. Mol. Cell. Biol. 23, 80–91.
97. Woodage, T., Basrai, M.A., Baxevanis, A.D., Hieter, P., and Collins, F.S. (1997) Characterization of the CHD family of proteins. Proc. Natl. Acad. Sci. USA 94, 1472–1477.
98. Tran, H.G., Steger, D.J., Iyer, V.R., and Johnson, A.D. (2000) The chromo domain protein Chd1p from budding yeast is an ATP-dependent chromatin-modifying factor. EMBO J. 19, 2323–2331.
99. Stokes, D.G. and Perry, R.P. (1995) DNA-binding and chromatin localization properties of CHD1. Mol. Cell. Biol. 15, 2745–2753.
100. Alen, C., Kent, N.A., Jones, H.S., O'Sullivan, J., Aranda, A., and Proudfoot, N.J. (2002) A role for chromatin remodeling in transcriptional termination by RNA polymerase II. Mol. Cell 10, 1441–1452.
101. Knoepfler, P.S. and Eisenman, R.N. (1999) Sin meets NuRD and other tails of repression. Cell 99, 447–450.
102. Wade, P.A., Gegonne, A., Jones, P.L., Ballestar, E., Aubry, F., and Wolffe, A.P. (1999) Mi-2 complex couples DNA methylation to chromatin remodeling and histone deacetylation. Nat. Genet. 23, 62–66.
103. Zhang, Y., Ng, H.-H., Erdjument-Bromage, H., Tempst, P., Bird, A., and Reinberg, D. (1999) Analysis of the NuRD subunits reveals a histone deacetylase core complex and a connection with DNA methylation. Genes Dev. 13, 1924–1935.
104. Ng, H.-H., Zhang, Y., Hendrich, B., Johnson, C.A., Turner, B.M., Erdjument-Bromage, H., Tempst, P., Reinberg, D., and Bird, A. (1999) MBD2 is a transcriptional repressor belonging to the MeCP1 histone deacetylase complex. Nat. Genet. 23, 58–61.
105. Feng, Q. and Zhang, Y. (2001) The MeCP1 complex represses transcription through preferential binding, remodeling, and deacetylating methylated nucleosomes. Genes Dev. 15, 827–832.
106. Murawsky, C.M., Brehm, A., Badenhorst, P., Lowe, N., Becker, P.B., and Travers, A.A. (2001) Tramtrack69 interacts with the dMi-2 subunit of the Drosophila NuRD chromatin remodeling complex. EMBO Rep. 2, 1089–1094.
107. Ohdate, H., Lim, C.R., Kokubo, T., Matsubara, K.I., Kimata, Y., and Kohno, K. (2003) Impairment of the DNA-binding activity of the TATA-binding protein (TBP) renders the transcriptional function of Rvb2p/Tih2p, the yeast RuvB-like protein, essential for cell growth. J. Biol. Chem. 278, 14647–14656.
108. O'Donohue, M.F., Duband-Goulet, I., Hamiche, A., and Prunell, A. (1994) Octamer displacement and redistribution in transcription of single nucleosomes. Nucleic Acids Res. 22, 937–945.
109. Studitsky, V.M., Clark, D.J., and Felsenfeld, G. (1994) A histone octamer can step around a transcribing polymerase without leaving the template. Cell 76, 371–382.
110. Studitsky, V.M., Kassavetis, G.A., Geiduschek, E.P., and Felsenfeld, G. (1997) Mechanism of transcription through the nucleosome by eukaryotic RNA polymerase. Science 278, 1960–1963.
111. Kireeva, M.L., Walter, W., Tchernajenko, V., Bondarenko, V., Kashlev, M., and Studitsky, V.M. (2002) Nucleosome remodeling induced by RNA polymerase II: loss of the H2A/H2B dimer during transcription. Mol. Cell 9, 541–552.
112. Orphanides, G., LeRoy, G., Chang, C.H., Luse, D.S., and Reinberg, D. (1998) FACT, a factor that facilitates transcript elongation through nucleosomes. Cell 92, 105–116.
113. Orphanides, G., Wu, W.H., Lane, W.S., Hampsey, M., and Reinberg, D. (1999) The chromatin-specific transcription elongation factor FACT comprises human SPT16 and SSRP1 proteins. Nature 400, 284–288.
114. Formosa, T., Eriksson, P., Wittmeyer, J., Ginn, J., Yu, Y., and Stillman, D.J. (2001) Spt16-Pob3 and the HMG protein Nhp6 combine to form the nucleosome-binding factor SPN. EMBO J. 20, 3506–3517.

115. Brewster, N.K., Johnston, G.C., and Singer, R.A. (2001) A bipartite yeast SSRP1 analog comprised of Pob3 and Nhp6 proteins modulates transcription. Mol. Cell. Biol. 21, 3491–3502.
116. Muchardt, C., Reyes, J.C., Bourachot, B., Leguoy, E., and Yaniv, M. (1996) The hbrm and BRG-1 proteins, components of the human SNF/SWI complex, are phosphorylated and excluded from the condensed chromosomes during mitosis. EMBO J. 15, 3394–3402.
117. Sif, S., Stukenberg, P.T., Kirschner, M.W., and Kingston, R.E. (1998) Mitotic inactivation of a human SWI/SNF chromatin remodeling complex. Genes Dev. 12, 2842–2851.
118. Phelan, M.L., Schnitzler, G.R., and Kingston, R.E. (2000) Octamer transfer and creation of stably remodeled nucleosomes by human SWI-SNF and its isolated ATPases. Mol. Cell. Biol. 20, 6380–6389.
119. Havas, K., Flaus, A., Phelan, M., Kingston, R., Wade, P.A., Lilley, D.M.J., and Owen-Hughes, T. (2000) Generation of superhelical torsion by ATP-dependent chromatin remodeling activities. Cell 103, 1133–1142.
120. Corona, D.F., Längst, G., Clapier, C.R., Bonte, E.J., Ferrari, S., Tamkun, J.W., and Becker, P.B. (1999) ISWI is an ATP-dependent nucleosome remodeling factor. Mol. Cell 3, 239–245.
121. Wang, H.-B. and Zhang, Y. (2001) Mi2, an auto-antigen for dermatomyositis, is an ATP-dependent nucleosome remodeling factor. Nucleic Acids Res. 29, 2517–2521.
122. Dhalluin, C., Carlson, J.E., Zeng, L., He, C., Aggarwal, A.K., and Zhou, M.M. (1999) Structure and ligand of a histone acetyltransferase bromodomain. Nature 399, 491–496.
123. Jacobson, R.H., Ladurner, A.G., King, D.S., and Tjian, R. (2000) Structure and function of a human TAFII250 double bromodomain module. Science 288, 1422–1425.
124. Owen, D.J., Ornaghi, P., Yang, J.-C., Lowe, N., Evans, P.R., Ballario, P., Filetici, P., and Travers, A.A. (2000) The structural basis for the recognition of acetylated histone H4 by the bromodomain of histone acetyltransferase Gcn5p. EMBO J. 19, 6141–6149.
125. Nicolas, R.H. and Goodwin, G.H. (1996) Molecular cloning of polybromo, a nuclear protein containing multiple domains including five bromodomains, a truncated HMG-box, and two repeats of a novel domain. Gene 175, 233–240.
126. Horikawa, I. and Barrett, J.C. (2002) cDNA cloning of the human polybromo-1 gene on chromosome 3p21. DNA Seq. 13, 211–215.
127. Humphrey, G.W., Wang, Y., Russanova, V.R., Hirai, T., Qin, J., Nakatani, Y., and Howard, B.H. (2001) Stable histone deacetylase complexes distinguished by the presence of SANT domain proteins CoREST/kiaa0071 and Mta-L1. J. Biol. Chem. 276, 6817–6824.
128. Kuzmichev, A., Nishioka, K., Erdjument-Bromage, H., Tempst, P., and Reinberg, D. (2002) Histone methyltransferase activity associated with a human multiprotein complex containing the Enhancer of Zeste protein. Genes Dev. 16, 2893–2905.
129. Müller, J., Hart, C.M., Francis, N.J., Vargas, M.L., Sengupta, A., Wild, B., Miller, E.L., O'Connor, M.B., Kingston, R.E., and Simon, J.A. (2002) Histone methyltransferase activity of a Drosophila Polycomb group repressor complex. Cell 111, 197–208.
130. Czermin, B., Melfi, R., McCabe, D., Seitz, V., Imhof, A., and Pirrotta, V. (2002) Drosophila enhancer of Zeste/ESC complexes have a histone H3 methyltransferase activity that marks chromosomal Polycomb sites. Cell 111, 185–196.
131. Boyer, L.A., Langer, M.R., Crowley, K.A., Tan, S., Denu, J.M., and Peterson, C.L. (2002) Essential role for the SANT domain in the functioning of multiple chromatin remodeling enzymes. Mol. Cell 10, 935–942.
132. Thomas, J.O. and Travers, A.A. (2001) HMG1 and 2, and related 'architectural' DNA binding proteins. Trends Biochem.Sci. 26, 167–174.
133. Reeves, R. (2001) Molecular biology of HMGA proteins: hubs of nuclear function. Gene 277, 63–81.
134. Wang, W., Chi, T., Xue, Y., Zhou, S., Kuo, A., and Crabtree, G.R. (1998) Architectural DNA binding by a high-mobility-group/kinesin-like subunit in mammalian SWI/SNF-related complexes. Proc. Natl. Acad. Sci. USA 95, 492–498.
135. Papoulas, O., Daubresse, G., Armstrong, J.A., Jin, J., Scott, M.P., and Tamkun, J. (2001) The HMG-domain protein BAP111 is important for the function of the BRM chromatin-remodeling complex *in vivo*. Proc. Natl. Acad. Sci. USA 98, 5728–5733.

136. Chi, T.H., Wan, M., Zhao, K., Taniuchi, I., Chen, L., Littman, D.R., and Crabtree, G.R. (2002) Reciprocal regulation of CD4/CD8 expression by SWI/SNF-like BAF complexes. Nature 418, 195–199.
137. Peterson, C.L. (1996) Multiple SWItches to turn on chromatin? Curr. Opin. Genet. Dev. 6, 171–175.
138. Bourachot, B., Yaniv, M., and Muchardt, C. (1999) The activity of mammalian brm/SNF2α is dependent on a high-mobility-group protein I/Y-like DNA binding domain. Mol. Cell. Biol. 19, 3931–3939.
139. Bouazoune, K., Mitterweger, A., Längst, G., Imhof, A., Akhtar, A., Becker, P.B., and Brehm, A. (2002) The dMi-2 chromodomains are DNA binding modules important for ATP-dependent nucleosome mobilization. EMBO J. 21, 2430–2440.
140. Akhtar, A., Zink, D., and Becker, P.B. (2000) Chromodomains are protein-RNA interaction modules. Nature 407, 405–409.
141. Kortschak, R.D., Tucker, P.W., and Saint, R. (2000) ARID proteins come in from the desert. Trends Biochem. Sci. 25, 294–299.
142. Damelin, M., Simon, I., Moym, T.I., Wilson, B., Komili, S., Tempst, P., Roth, F.P., Young, R.A., Cairns, B.R., and Silver, P.A. (2002) The genome-wide localization of Rsc9, a component of the RSC chromatin-remodeling complex, changes in response to stress. Mol. Cell 9, 563–573.
143. O'Hara, P.J., Horowitz, H., Eichinger, G., and Young, E.T. (1988) The yeast ADR6 gene encodes homopolymeric amino acid sequences and a potential metal-binding domain. Nucleic Acids Res. 16, 10153–10169.
144. Dallas, P.B., Pacchione, S., Wilsker, D., Bowrin, V., Kobayashi, R., and Moran, E. (2000) The human SWI-SNF complex protein p270 is an ARID family member with non-sequence-specific DNA binding activity. Mol. Cell. Biol. 20, 3137–3146.
145. Treisman, J.E., Luk, A., Rubin, G.M., and Heberlein, U. (1997) eyelid antagonizes wingless signaling during Drosophila development and has homology to the Bright family of DNA-binding proteins. Genes Dev. 11, 1949–1962.
146. Vázquez, M., Moore, L., and Kennison, J.A. (1999) The trithorax group gene osa encodes an ARID-domain protein that genetically interacts with the brahma chromatin-remodeling factor to regulate transcription. Development 126, 733–742.
147. Collins, R.T. and Treisman, J.E. (2000) Osa-containing Brahma chromatin remodeling complexes are required for the repression of wingless target genes. Genes Dev. 14, 3140–3152.
148. Collins, R.T., Furukawa, T., Tanese, N., and Treisman, J.E. (1999) Osa associates with the Brahma chromatin remodeling complex and promotes the activation of some target genes. EMBO J. 18, 7029–7040.
149. Zhao, K., Wang, W., Rando, O.J., Xue, Y., Swiderek, K., Kuo, A., and Crabtree, G.R. (1998) Rapid and phosphoinositol-dependent binding of the SWI/SNF-like BAF complex to chromatin after T lymphocyte receptor signaling. Cell 95, 625–636.
150. Cairns, B.R., Erdjument-Bromage, H., Tempst, P., Winston, F., and Kornberg, R.D. (1998) Two actin-related proteins are shared functional components of the chromatin-remodeling complexes RSC and SWI/SNF. Mol. Cell 2, 639–651.
151. Galarneau, L., Nourani, A., Boudreault, A.A., Zhang, Y., Heliot, L., Allard, S., Savard, J., Lane, W.S., Stillman, D.J., and Côté, J. (2000) Multiple links between the NuA4 histone acetyltransferase complex and epigenetic control of transcription. Mol. Cell 5, 927–937.
152. Ikura, T., Ogryzko, V.V., Grigoriev, M., Groisman, R., Wang, J., Horikoshi, M., Scully, R., Qin, J., and Nakatani, Y. (2000) Involvement of the TIP60 histone acetylase complex in DNA repair and apoptosis. Cell 102, 463–743.
153. Saha, A., Wittmeyer, J., and Cairns, B.R. (2002) Chromatin remodeling by RSC involves ATP-dependent DNA translocation. Genes Dev. 16, 2120–2134.
154. Narlikar, G.J., Phelan, M.L., and Kingston, R.E. (2001) Generation and interconversion of multiple distinct nucleosomal states as a mechanism for catalyzing chromatin fluidity. Mol. Cell 8, 1219–1230.
155. Hamiche, A., Kang, J.G., Dennis, C., Xiao, H., and Wu, C. (2001) Histone tails modulate nucleosome mobility and regulate ATP-dependent nucleosome sliding by NURF. Proc. Natl. Acad. Sci. USA 98, 14316–14321.

156. Clapier, C.R., Nightingale, K.P., and Becker, P.B. (2002) A critical epitope for substrate recognition by the nucleosome remodeling ATPase ISWI. Nucleic Acids Res. 30, 649–655.
157. Verreault, A., Kaufman, P.D., Kobayashi, R., and Stillman, B. (1996) Nucleosome assembly by a complex of CAF-1 and acetylated histones H3/H4. Cell 87, 95–104.
158. Owen-Hughes, T., Utley, R.T., Côté, J., Peterson, C.L., and Workman, J.L. (1996) Persistent site-specific remodelling of a nucleosome array by transient action of SWI/SNF complex. Science 273, 513–516.
159. Lorch, Y., Zhang, M., and Kornberg, R.D. (1999) Histone octamer transfer by a chromatin-remodeling complex. Cell 96, 389–392.
160. Côté, J., Peterson, C.L., and Workman, J.L. (1996) Perturbation of nucleosome core structure by the SWI/SNF complex persists after its detachment, enhancing subsequent transcription factor binding. Proc. Natl. Acad. Sci. USA 95, 4947–4952.
161. Imbalzano, A.N., Schnitzler, G.R., and Kingston, R.E. (1996) Nucleosome disruption by human SWI/SNF is maintained in the absence of continued ATP hydrolysis. J. Biol. Chem. 271, 20726–20733.
162. Schnitzler, G., Sif, S., and Kingston, R.E. (1998) Human SWI/SNF interconverts a nucleosome between its base state and a stable remodeled state. Cell 94, 17–27.
163. Lorch, Y., Cairns, B.R., Zhang, M., and Kornberg, R.D. (1998) Activated RSC-nucleosome complex and persistently altered form of the nucleosome. Cell 94, 29–34.
164. Lorch, Y., Zhang, M., and Kornberg, R.D. (2001) RSC unravels the nucleosome. Mol. Cell 7, 89–95.
165. Schnitzler, G.R., Cheung, C.L., Hafner, J.H., Saurin, A.J., Kingston, R.E., and Lieber, C.M. (2001) Direct imaging of human SWI/SNF-remodeled mono- and polynucleosomes by atomic force microscopy employing carbon nanotube tips. Mol. Cell. Biol. 21, 8504–8511.
166. Kassabov, S.R., Zhang, B., Persinger, J., and Bartholomew, B. (2003) SWI/SNF unwraps, slides, and rewraps the nucleosome. Mol. Cell 11, 391–403.
167. Guyon, J.R., Narlikar, G.J., Sullivan, E.K., and Kingston, R.E. (2001) Stability of a human SWI-SNF remodeled nucleosomal array. Mol. Cell. Biol. 21, 1132–1144.
168. Aoyagi, S., Narlikar, G., Zheng, C., Sif, S., Kingston, R.E., and Hayes, J.J. (2002) Nucleosome remodeling by the human SWI/SNF complex requires transient global disruption of histone-DNA interactions. Mol. Cell. Biol. 22, 3653–3662.
169. Sengupta, S.M., VanKanegan, M., Persinger, J., Logie, C., Cairns, B.R., Peterson, C.L., and Bartholomew, B. (2001) The interactions of yeast SWI/SNF and RSC with the nucleosome before and after chromatin remodeling. J. Biol. Chem. 276, 12636–12644.
170. Bazett-Jones, D.P., Côté, J., Landel, C.C., Peterson, C.L., and Workman, J.L. (1999) The SWI/SNF complex creates loop domains in DNA and polynucleosome arrays and can disrupt DNA-histone contacts within these domains. Mol. Cell. Biol. 19, 1470–1478.
171. Aalfs, J.D., Narlikar, G.J., and Kingston, R.E. (2001) Functional differences between the human ATP-dependent nucleosome remodeling proteins BRG1 and SNF2H. J. Biol. Chem. 276, 34270–34278.
172. Kassabov, S.R., Henry, N.M., Zofall, M., Tsukiyama, T., and Bartholomew, B. (2002) High-resolution mapping of changes in histone-DNA contacts of nucleosomes remodeled by ISW2. Mol. Cell. Biol. 22, 7524–7534.
173. Längst, G. and Becker, P.B. (2001) ISWI induces nucleosome sliding on nicked DNA. Mol. Cell 8, 1085–1092.
174. Flaus, A. and Owen-Hughes, T. (2003) Mechanisms for nucleosome mobilisation. Biopolymers 68, 563–578.
175. Polach, K.J. and Widom, J. (1995) Mechanism of protein access to specific DNA sequences in chromatin a dynamic equilibration model for gene regulation. J. Mol. Biol. 254, 130–149.
176. Widom, J. (2001) Role of DNA sequence in nucleosome stability and dynamics. Q. Rev. Biophys. 34, 269–324.
177. Liu, L.F. and Wang, J.C. Supercoiling of the DNA template during transcription. Proc. Natl. Acad. Sci. USA 84, 7024–7027.

178. Baer, B.W. and Rhodes, D. (1983) Eukaryotic RNA polymerase II binds to nucleosome cores from transcribed genes. Nature 301, 482–488.
179. Brower-Toland, B.D., Smith, C.L., Yeh, R.C., Lis, J.T., Peterson, C.L., and Wang, M.D. (2002) Mechanical disruption of individual nucleosomes reveals a reversible multistage release of DNA. Proc. Natl. Acad. Sci. USA 99, 1960–1965.
179a. Belotserkovskaya, R., Oh, S., Bondarenko, V.A., Orphanides, G., Studitsky, V.M., and Reinberg, D. (2003) FACT facilitates transcription-dependent nucleosome alteration. Science 301, 1890–1893.
180. Davey, C.A., Sargent, D.F., Luger, K., Maeder, A.W., and Richmond, T.J. (2002) Solvent mediated interactions in the structure of the nucleosome core particle at 1.9 Å resolution. J. Mol. Biol. 319, 1097–1113.
181. Low, C.M., Drew, H.R., and Waring, M.J. (1986) Echinomycin and distamycin induce rotation of nucleosome core DNA. Nucleic Acids Res. 14, 6785–6801.
182. Suto, R.K., Edayathumangalam, R.S., White, C.L., Melander, C., Gottesfeld, J.M., Dervan, P.B., and Luger, K. (2003) Crystal structures of nucleosome core particles in complex with minor groove DNA-binding ligands. J. Mol. Biol. 326, 371–380.
183. Singleton, M.R. and Wigley, D.B. (2002) Modularity and specialization in superfamily 1 and 2 helicases. J. Bacteriol. 184, 1819–1826.
184. Walker, J.E., Saraste, M., Runswick, M.J., and Gay, N.J. (1982) Distantly related sequences in the $\tilde{\alpha}$ and β-subunits of ATP synthase, myosin, kinases and other ATP-requiring enzymes and a common nucleotide binding fold. EMBO J. 1, 945–951.
185. Dillingham, M.S., Soultanas, P., Wiley, P., Webb, M.R., and Wigley, D.B. (2001) Defining the roles of individual residues in the single-stranded DNA binding site of PcrA helicase. Proc. Natl. Acad. Sci. USA 98, 8381–8387.
186. Whitehouse, I., Stockdale, C., Flaus, A., Szczelkun, M.D., and Owen-Hughes, T. (2003) Evidence for DNA translocation by the ISWI chromatin-remodeling enzyme. Mol. Cell. Biol. 23, 1935–1945.
187. Bianco, P.R., Brewer, L.R., Corzett, M., Balhorn, R., Yeh, Y., Kowalczykowski, S.C., and Baskin, R.J. (2001) Processive translocation and DNA unwinding by individual RecBCD enzyme molecules. Nature 409, 374–378.
188. Smith, D.E., Tans, S.J., Smith, S.B., Grimes, S., Anderson, D.L., and Bustamante, C. (2001) The bacteriophage straight ϕ29 portal motor can package DNA against a large internal force. Nature 413, 748–752.
189. Fyodorov, D.V. and Kadonaga, J.T. (2002) Dynamics of ATP-dependent chromatin assembly by ACF. Nature 418, 897–900.
190. Lohmann, T.M. and Bjornson, K.P. (1996) Mechanisms of helicase-catalyzed DNA unwinding. Annu. Rev. Biochem. 65, 169–214.
191. Guasch, A., Pous, J., Ibarra, B., Gomis-Ruth, F.X., Valpuesta, J.M., Sousa, N., Carrascosa, J.L., and Coll, M. (2002) Detailed architecture of a DNA translocating machine: the high-resolution structure of the bacteriophage ϕ29 connector particle. J. Mol. Biol. 315, 663–676.
192. Rodríguez, A.C. and Stock, D. (2002) Crystal structure of reverse gyrase: insights into the positive supercoiling of DNA. EMBO J. 21, 418–426.
193. Rodríguez, A.C. (2002) Studies of a positive supercoiling machine. Nucleotide hydrolysis and a multifunctional latch in the mechanism of reverse gyrase. J. Biol. Chem. 277, 29865–29873.
194. Gavin, I., Horn, P.J., and Peterson, C.L. (2001) SWI/SNF chromatin remodeling requires changes in DNA topology. Mol. Cell 7, 97–104.
195. Travers, A.A. (1992) The reprogramming of transcriptional competence. Cell 69, 573–575.
196. Velankar, S.S., Soultanas, P., Dillingham, M.S., Subramanya, H.S., and Wigley, D.B. (1999) Crystal structures of complexes of PcrA DNA helicase with a DNA substrate indicate an inch-worm mechanism. Cell 97, 75–84.
197. Asturias, F.J., Chung, W.H., Kornberg, R.D., and Lorch, Y. (2002) Structural analysis of the RSC chromatin-remodeling complex. Proc. Natl. Acad. Sci. USA 99, 13477–13480.
198. Smith, C.L., Horowitz-Scherer, R., Flanagan, J.F., Woodcock, C.L., and Peterson, C.L. (2003) Structural analysis of the yeast SWI/SNF chromatin remodeling complex. Nat. Struct. Biol. 10, 141–145.

199. Travers, A.A. and Thomas, J.O. Chromosomal HMG-box proteins. This volume.
200. Bonaldi, T., Längst, G., Strohner, R., Becker, P.B., and Bianchi, M.E. (2002) The DNA chaperone HMGB1 facilitates ACF/CHRAC-dependent nucleosome sliding. EMBO J. 21, 6865–6873.
201. Moreira, J.M.A. and Holmberg, S. (2000) Chromatin-mediated transcriptional regulation by the yeast architectural factors NHP6A and NHP6B. EMBO J. 19, 6804–6813.
201a. Ragab, A. and Travers, A. (2003) HMG-D and histone H1 after the local accessibility of nucleosomal DNA. Nucleic Acids Res., 7083–7089.
202. Nightingale, K., Dimitrov, S., Reeves, R., and Wolffe, A.P. (1996) Evidence for a shared structural role for HMG1 and linker histones B4 and H1 in organizing chromatin. EMBO J. 15, 548–561.
203. Ner, S.S., Blank, T., Pérez-Parallé, M.L., Grigliatti, T.A., Becker, P.B., and Travers, A.A. (2001) HMG-D and histone H1 interplay during chromatin assembly and early embryogenesis. J. Biol. Chem. 276, 37569–37576.
204. Prelich, G. and Winston, F. (1993) Mutations that suppress the deletion of an upstream activating sequence in yeast: involvement of a protein kinase and histone H3 in repressing transcription in vivo. Genetics 135, 665–676.
205. Santisteban, M.S., Arents, G., Moudrianakis, E.N., and Smith, M.M. (1997) Histone octamer function *in vivo*: mutations in the dimer-tetramer interfaces disrupt both gene activation and repression. EMBO J. 16, 2493–2506.
206. Fleming, A.B. and Pennings, S. (2001) Antagonistic remodeling by Swi-Snf and Tup1-Ssn6 of an extensive chromatin region forms the background for FLO1 gene regulation. EMBO J. 20, 5219–5231.
207. Kruger, W., Peterson, C.L., Sil, A., Coburn, C., Arents, G., Moudrianakis, E.N., and Herskowitz, I. (1995) Amino acid substitutions in the structured domains of histones H3 and H4 partially relieve the requirement of the yeast SWI/SNF complex for transcription. Genes Dev. 9, 2770–2779.
208. Recht, J. and Osley, M.A. (1999) Mutations in both the structured domain and N-terminus of histone H2B bypass the requirement for Swi-Snf in yeast. EMBO J. 18, 229–240.
209. Wechser, M.A., Kladde, M.P., Alfieri, J.A., and Peterson, C.L. (1997) Effects of Sin-versions of histone H4 on yeast chromatin structure and function. EMBO J. 16, 2086–2095.
209a. Muthurajan, U.M., Bao, Y., Forsberg, J., Edayathumangalam, R.S., Dyer, P.H., White, C.L., and Luger, K. (2004) Crystallographic and biochemical studies of nucleosome core particles containing histone sin mutants. EMBO J. 23(2), in press.
210. Kurumizaka, H. and Wolffe, A.P. (1997) Sin mutations of histone H3: influence on nucleosome core structure and function. Mol. Cell. Biol. 17, 6953–6969.
210a. Flaus, A., Rencurel, C., Ferreira, H., Wiechens, N., and Owen-Hughes, T. (2004) Sin mutations after inherent nucleosome mobility. EMBO J. 23(2), in press.
211. Horn, P.J., Crowley, K.A., Carruthers, L.M., Hansen, J.C., and Peterson, C.L. (2002) The SIN domain of the histone octamer is essential for intramolecular folding of nucleosomal arrays. Nat. Struct. Biol. 9, 167–171.
212. Schirmer, T. and Evans, P.R. (1990) Structural basis of the allosteric behaviour of phosphofructokinase. Nature 343, 140–145.
213. Clarkson, M.J., Wells, J.R., Gibson, F., Saint, R., and Tremethick, D.J. (1999) Regions of variant histone His2AvD required for *Drosophila* development. Nature 399, 694–697.
214. Suto, R.K., Clarkson, M.J., Tremethick, D.J., and Luger, K. (2000) Crystal structure of a nucleosome core particle containing the variant histone H2A.Z. Nat. Struct. Biol. 7, 1121–1124.
215. Santisteban, M.S., Kalashnikova, T., and Smith, M.M. (2000) Histone H2A.Z regulates transcription and is partially redundant with nucleosome remodeling complexes. Cell 103, 411–422.
216. Deleted.
217. Kang, J.G., Hamiche, A., and Wu, C. (2002) GAL4 directs nucleosome sliding induced by NURF. EMBO J. 21, 1406–1413.
218. Flaus, A. and Owen-Hughes, T. (2003) Dynamic properties of nucleosomes during thermal and ATP-driven mobilisation. Biopolymers 68, 563–578.
219. Lomvardas, S. and Thanos, D. (2002) Modifying gene expression programs by altering core promoter architecture. Cell 110, 261–271.

220. Buttinelli, M., Di Mauro, E., and Negri, R. (1993) Multiple nucleosome positioning with unique rotational setting for the *Saccharomyces cerevisiae* 5S rRNA gene in vitro and in vivo. Proc. Natl. Acad. Sci. USA 90, 9315–9319.
221. Di Mauro, E., Verdone, L., Chiappini, B., and Caserta, M. (2002) *In vivo* changes of nucleosome positioning in the pretranscription state. J. Biol. Chem. 277, 7002–7009.
222. Wu, L. and Winston, F. (1997) Evidence that Snf-Swi controls chromatin structure over both the TATA and UAS regions of the SUC2 promoter in *Saccharomyces cerevisiae*. Nucleic Acids Res. 25, 4230–4234.
223. Yasui, D., Miyano, M., Cai, S., Varga-Weisz, P., and Kohwi-Shigematsu, T. (2002) SATB1 targets chromatin remodelling to regulate genes over long distances. Nature 419, 641–645.
224. Sudarsanam, P., Iyer, V.R., Brown, P.O., and Winston, F. (2000) Whole-genome expression analysis of snf/swi mutants of *Saccharomyces cerevisiae*. Proc. Natl. Acad. Sci. USA 97, 3364–3369.
225. Holstege, F.C., Jennings, E.G., Wyrick, J.J., Lee, T.I., Hengartner, C.J., Green, M.R., Golub, T.R., Lander, E.S., and Young, R.A. (1998) Dissecting the regulatory circuitry of a eukaryotic genome. Cell 95, 717–728.
226. Flanagan, J.F. and Peterson, C.L. (1999) A role for the yeast SWI/SNF complex in DNA replication. Nucleic Acids Res. 27, 2022–2028.
227. Davie, J.K. and Kane, C.M. (2000) Genetic interactions between TFIIS and the Swi-Snf chromatin-remodeling complex. Mol. Cell. Biol. 20, 5960–5973.
228. Hara, R. and Sancar, A. (2002) The SWI/SNF chromatin-remodeling factor stimulates repair by human excision nuclease in the mononucleosome core particle. Mol. Cell. Biol. 22, 6779–6787.
229. Gaillard, H., Fitzgerald, D.J., Smith, C.L., Peterson, C.L., Richmond, T.J., and Thoma, F. (2003) Chromatin remodeling activities act on UV-damaged nucleosomes and modulate DNA damage accessibility to photolyase. J. Biol. Chem. 278, 17655–17663.
230. Moreira, J.M. and Holmberg, S. (1999) Transcriptional repression of the yeast *CHA1* gene requires the chromatin-remodeling complex RSC. EMBO J. 18, 2836–2844.
231. Koyama, H., Itoh, M., Miyahara, K., and Tsuchiya, E. (2002) Abundance of the RSC nucleosome-remodeling complex is important for the cells to tolerate DNA damage in *Saccharomyces cerevisiae*. FEBS Lett. 531, 215–221.
232. Ruhf, M.L., Braun, A., Papoulas, O., Tamkun, J.W., Randsholt, N., and Meister, M. (2001) The *domino* gene of *Drosophila* encodes novel members of the SWI2/SNF2 family of DNA-dependent ATPases, which contribute to the silencing of homeotic genes. Development 128, 1429–1441.
233. Deuring, R., Fanti, L., Armstrong, J.A., Sarte, M., Papoulas, O., Prestel, M., Daubresse, G., Verardo, M., Moseley, S.L., Berloco, M., Tsukiyama, T., Wu, C., Pimpinelli, S., and Tamkun, J.W. (2000) The ISWI chromatin-remodeling protein is required for gene expression and the maintenance of higher order chromatin structure in vivo. Mol. Cell 5, 355–365.
234. Badenhorst, P., Voas, M., Rebay, I., and Wu, C. (2002) Biological functions of the ISWI chromatin remodeling complex NURF. Genes Dev. 16, 3186–3198.
235. Goldmark, J.P., Fazzio, T.G., Estep, P.W., Church, G.M., and Tsukiyama, T. (2000) The Isw2 chromatin remodeling complex represses early meiotic genes upon recruitment by Ume6p. Cell 103, 423–433.
236. Fazzio T.G., Kooperberg, C., Goldmark, J.P., Neal, C., Basom, R., Delrow, J., and Tsukiyama, T. (2001) Widespread collaboration of Isw2 and Sin3-Rpd3 chromatin remodeling complexes in transcriptional repression. Mol. Cell. Biol. 21, 6450-6460.
237. Collins, N., Poot, R.A., Kukimoto, I., Garcia-Jimenez, C., Dellaire, G., and Varga-Weisz, P.D. (2002) An ACF1-ISWI chromatin-remodeling complex is required for DNA replication through heterochromatin. Nat. Genet. 32, 627–632.
238. Cosma, M.P., Tanaka, T., and Nasmyth, K. (1999) Ordered recruitment of transcription and chromatin remodeling factors to a cell cycle- and developmentally regulated promoter. Cell 97, 299–311.
239. Krebs, J.E., Fry, C.J., Samuels, M.L., and Peterson, C.L. (2000) Global role for chromatin remodeling enzymes in mitotic gene expression. Cell 102, 587–598.

240. Agalioti, T., Lomvardas, S., Parekh, B., Yie, J., Maniatis, T., and Thanos, D. (2000) Ordered recruitment of chromatin modifying and general transcription factors to the IFN-β promoter. Cell 103, 667–678.
241. Hassan, A.H., Neely, K.E., and Workman, J.L. (2001) Histone acetyltransferase complexes stabilize swi/snf binding to promoter nucleosomes. Cell 104, 817–827.
242. Sharma, V.M., Li, B., and Reese, J.C. (2003) SWI/SNF-dependent chromatin remodeling of RNR3 requires TAF(II)s and the general transcription machinery. Genes Dev. 17, 502–515.
243. Natarajan, K., Jackson, B.M., Zhou, H., Winston, F., and Hinnebusch, A.G. (1999) Transcriptional activation by Gcn4p involves independent interactions with the SWI/SNF complex and the SRB/mediator. Mol. Cell 4, 657–664.
244. Neely, K.E., Hassan, A.H., Wallberg, A.E., Steger, D.J., Cairns, B.R., Wright, A.P., and Workman, J.L. (1999) Activation domain-mediated targeting of the SWI/SNF complex to promoters stimulates transcription from nucleosome arrays. Mol. Cell 4, 649–655.
245. Yudkovsky, N., Logie, C., Hahn, S., and Peterson, C.L. (1999) Recruitment of the SWI/SNF chromatin remodeling complex by transcriptional activators. Genes Dev. 13, 2369–2374.
246. Fryer, C.J. and Archer, T.K. (1998) Chromatin remodelling by the glucocorticoid receptor requires the BRG1 complex. Nature 393, 88–91.
247. Wallberg, A.E., Neely, K.E., Hassan, A.H., Gustafsson, J.A., Workman, J.L., and Wright, A.P. (2000) Recruitment of the SWI-SNF chromatin remodeling complex as a mechanism of gene activation by the glucocorticoid receptor tau1 activation domain. Mol. Cell. Biol. 20, 2004–2013.
248. Liu, R., Liu, H., Chen, X., Kirby, M., Brown, P.O., and Zhao, K. (2001) Regulation of CSF1 promoter by the SWI/SNF-like BAF complex. Cell 106, 309–318.
249. Kadam, S. and Emerson, B.M. (2003) Transcriptional specificity of human SWI/SNF BRG1 and BRM chromatin remodeling complexes. Mol. Cell 11, 377–389.
250. Barker, N., Hurlstone, A., Musisi, H., Miles, A., Bienz, M., and Clevers, H. (2001) The chromatin remodelling factor Brg-1 interacts with β-catenin to promote target gene activation. EMBO J. 20, 4935–4943.
251. Schultz, D.C., Friedman, J.R., and Rauscher, F.J., 3rd. (2001) Targeting histone deacetylase complexes via KRAB-zinc finger proteins: the PHD and bromodomains of KAP-1 form a cooperative unit that recruits a novel isoform of the Mi-2α subunit of NuRD, Genes Dev. 15, 428–443.
252. Ng, H.-H., Robert, F., Young, R.A., and Struhl, K. (2002) Genome-wide location and regulated recruitment of the RSC nucleosome-remodeling complex. Genes Dev. 16, 806–819.
253. Boeger, H., Griesenbeck, J., Strattan, J.S., and Kornberg, R.D. (2003) Nucleosomes unfold completely at a transcriptionally active promoter. Mol. Cell 11, 1587–1598.
254. Reinke, H. and Hörz, W. (2003) Histones are first hyperacetylated and then lose contact with the activated PHO5 promoter. Mol. Cell 11, 1599–1607.
255. Simic, R., Lindstrom, D.L., Tran, H.G., Roinick, K.L., Costa, P.J., Johnson, A.D., Hartzog, G.A., and Arndt, K.M. (2003) Chromatin remodeling protein Chd1 interacts with transcription elongation factors and localizes to transcribed genes. EMBO J. 22, 1846–1856.
256. Squazzo, S.L., Costa, P.J., Lindstrom, D.L., Kumer, K.E., Simic, R., Jennings, J.L., Link, A.J., Arndt, K.M., and Hartzog, G.A. (2002) The Paf1 complex physically and functionally associates with transcription elongation factors *in vivo*. EMBO J. 21, 1764–1774.
257. Winkler, G.S., Kristjuhan, A., Erdjument-Bromage, H., Tempst, P., and Svejstrup, J.Q. (2002) Elongator is a histone H3 and H4 acetyltransferase important for normal histone acetylation levels *in vivo*. Proc. Natl. Acad. Sci. USA 99, 3517–3522.
258. Ng, H.-H., Robert, F., Young, R.A., and Struhl, K. (2003) Targeted recruitment of Set1 histone methylase by elongating Pol II provides a localized mark and memory of recent transcriptional activity. Mol. Cell 11, 709–719.
259. Schaft, D., Roguev, A., Kotovic, K.M., Shevchenko, A., Sarov, M., Neugebauer, K.M., and Stewart, A.F. (2003) The histone 3 lysine 36 methyltransferase, SET2, is involved in transcriptional elongation. Nucleic Acids Res. 31, 2475–2482.

260. Corey, L.L., Weirich, C.S., Benjamin, I.J., and Kingston, R.E. (2003) Localized recruitment of a chromatin-remodeling activity by an activator in vivo drives transcriptional elongation. Genes Dev. 17, 1392–1401.
261. Shen, X., Xiao, H., Ranallo, R., Wu, W.H., and Wu, C. (2003) Modulation of ATP-dependent chromatin-remodeling complexes by inositol polyphosphates. Science 299, 112–114.
262a. Gozani, O., Karuman, P., Jones, D.R., Ivanov, D., Cha, J., Lugovskoy, A.A., Baird, C.L., Zhu, H., Field, S.J., Lessnick, S.L., Villasenor, J., Mehrotra, B., Chen, J., Rao, V.R., Brugge, J.S., Ferguson, C.G., Payrastre, B., Myszka, D.G., Cantley, L.C., Wagner, G., Divecha, N., Prestwich, G.D., and Yuan J. (2003) The PHD finger of the chromatin-associated protein ING2 functions as a nuclear phosphoinositide receptor. Cell 114, 99–111.
262b. Rando, O.J., Zhao, K., Janmey, P., and Crabtree, G.R. (2002) Phosphatidylinositol-dependent actin filament binding by the SWI/SNF-like BAF chromatin remodeling complex. PNAS 99, 2824–9.
263. Mizuguchi, G., Shen, X., Landry, J., Wu, W.H., Sen, S., and Wu, C. (2003) ATP-driven exchange of histone H2AZ variant catalysed by SWR1 chromatin remodeling complex. Science, epub 27, Nov. 2003.
264. Sternberg, P.W., Stern, M.J., Clark, I., and Herskowitz, I. (1987) Activation of the yeast HO gene by release from multiple negative controls. Cell 48, 567–577.
265. Peterson, C.L. and Workman, J.L. (2000) Promoter targeting and chromatin remodeling by the SWI/SNF complex. Curr. Opin. Genet. Dev. 10, 187–192.
266. Cosma, M.P., Tanaka, T., and Nasmyth, K. (1999) Ordered recruitment of transcription and chromatin remodeling factors to a cell cycle- and developmentally regulated promoter. Cell 97, 299–311.
267. Gregory, P.D., Schmid, A., Zavari, M., Munsterkotter, M., and Hörz, W. (1999) Chromatin remodelling at the PHO8 promoter requires SWI-SNF and SAGA at a step subsequent to activator binding. EMBO J. 18, 6407–6414.
268. Krebs, J.E., Kuo, M.H., Allis, C.D., and Peterson, C.L. (1999) Cell cycle-regulated histone acetylation required for expression of the yeast HO gene. Genes Dev. 13, 1412–1421.
269. Hsu, J.M., Huang, J., Meluh, P.B., and Laurent, B.C. (2003) The yeast RSC chromatin-remodeling complex is required for kinetochore function in chromosome segregation. Mol. Cell. Biol. 23, 3202–3215.
270. Verbsky, M.L. and Richards, E.J. (2001) Chromatin remodeling in plants. Curr. Opin. Plant Biol. 4, 494–500.
271. Sarnowski, T.J., Swiezewski, S., Pawlikowska, K., Kaczanowski, S., and Jerzmanowski, A. (2002) AtSWI3B, an Arabidopsis homolog of SWI3, a core subunit of yeast Swi/Snf chromatin remodeling complex, interacts with FCA, a regulator of flowering time. Nucleic Acids Res. 30, 3412–3421.
272. Sawa, H., Kouike, H., and Okano, H. (2000) Components of the SWI/SNF complex are required for asymmetric cell division in C. elegans. Mol. Cell 6, 617–624.
273. Armstrong, J.A., Papoulas, O., Daubresse, G., Sperling, A.S., Lis, J.T., Scott, M.P., and Tamkun, J.W. (2002) The *Drosophila* BRM complex facilitates global transcription by RNA polymerase II. EMBO J. 21, 5245–5254.
274. Gregg, R.G., Willer, G.B., Fadool, J.M., Dowling, J.E., and Link, B.A. (2003) Positional cloning of the young mutation identifies an essential role for the Brahma chromatin remodeling complex in mediating retinal cell differentiation. Proc. Natl. Acad. Sci. USA 100, 6535–6540.
275. Klochendler-Yeivin, A., Fiette, L., Barra, J., Muchardt, C., Babinet, C., and Yaniv, M. (2000) The murine SNF5/INI1 chromatin remodeling factor is essential for embryonic development and tumor suppression. EMBO Rep. 1, 500–506.
276. Roberts, C.W., Galusha, S.A., McMenamin, M.E., Fletcher, C.D., and Orkin, S.H. (2000) Haploinsufficiency of Snf5 (integrase interactor 1) pre-disposes to malignant rhabdoid tumors in mice. Proc. Natl. Acad. Sci. USA 97, 13796–13800.
277. Bultman, S., Gebuhr, T., Yee, D., La Mantia, C., Nicholson, J., Gilliam, A., Randazzo, F., Metzger, D., Chambon, P., Crabtree, G., and Magnuson, T. (2000) A Brg1 null mutation in the mouse reveals functional differences among mammalian SWI/SNF complexes. Mol. Cell 6, 1287–1295.

278. Reyes, J.C., Barra, J., Muchardt, C., Camus, A., Babinet, C., and Yaniv, M. (1998) Altered control of cellular proliferation in the absence of mammalian brahma (SNF2α). EMBO J. 17, 6979–6991.
279. Bochar, D.A., Wang, L., Beniya, H., Kinev, A., Xue, Y., Lane, W.S., Wang, W., Kashanchi, F., and Shiekhattar, R. (2000) BRCA1 is associated with a human SWI/SNF-related complex: linking chromatin remodeling to breast cancer. Cell 102, 257–265.
280. Nie, Z., Yan, Z., Chen, E.H., Sechi, S., Ling, C., Zhou, S., Xue, Y., Yang, D., Murray, D., Kanakubo, E., Cleary, M.L., and Wang, W. (2003) Novel SWI/SNF chromatin-remodeling complexes contain a mixed-lineage leukemia chromosomal translocation partner. Mol. Cell. Biol. 23, 2942–2952.
281. DiRenzo, J., Shang, Y., Phelan, M., Sif, S., Myers, M., Kingston, R., and Brown, M. (2000) BRG-1 is recruited to estrogen-responsive promoters and cooperates with factors involved in histone acetylation. Mol. Cell. Biol. 20, 7541–7549.
282. Lomvardas, S. and Thanos, D. (2001) Nucleosome sliding via TBP DNA binding in vivo. Cell 106, 685–696.
283. de la Serna, I.L., Carlson, K.A., and Imbalzano, A.N. (2001) Mammalian SWI/SNF complexes promote MyoD-mediated muscle differentiation. Nat. Genet. 27, 187–190.
284. Huang, M., Qian, F., Hu, Y., Ang, C., Li, Z., and Wen, Z. (2002) Chromatin-remodelling factor BRG1 selectively activates a subset of interferon-alpha-inducible genes. Nat. Cell Biol. 4, 774–781.
285. Zhang, H.S., Gavin, M., Dahiya, A., Postigo, A.A., Ma, D., Luo, R.X., Harbour, J.W., and Dean, D.C. (2000) Exit from G1 and S phase of the cell cycle is regulated by repressor complexes containing HDAC-Rb-hSWI/SNF and Rb-hSWI/SNF. Cell 101, 79–89.
286. Murphy, D.J., Hardy, S., and Engel, D.A. (1999) Human SWI-SNF component BRG1 represses transcription of the c-fos gene. Mol. Cell. Biol. 19, 2724–2733.
287. Versteege, I., Sevenet, N., Lange, J., Rousseau-Merck, M.F., Ambros, P., Handgretinger, R., Aurias, A., and Delattre, O. (1998) Truncating mutations of hSNF5/INI1 in aggressive paediatric cancer. Nature 394, 203–206.
288. Armstrong, J.A., Bieker, J.J., and Emerson, B.M. (1998) A SWI/SNF-related chromatin remodeling complex, E-RC1, is required for tissue-specific transcriptional regulation by EKLF in vitro. Cell 95, 93–104.
289. Moreau, J.L., Lee, M., Mahachi, N., Vary, J., Mellor, J., Tsukiyama, T., and Goding, C.R. (2003) Regulated displacement of TBP from the PHO8 promoter in vivo requires Cbf1 and the Isw1 chromatin remodeling complex. Mol. Cell 11, 1609–1620.
290. Kikyo, N., Wade, P.A., Guschin, D., Ge, H., and Wolffe, A.P. (2000) Active remodeling of somatic nuclei in egg cytoplasm by the nucleosomal ATPase ISWI. Science 289, 2360–2363.
291. LeRoy, G., Orphanides, G., Lane, W.S., and Reinberg, D. (1998) Requirement of RSF and FACT for transcription of chromatin templates in vitro. Science 282, 1900–1904.
292. Strohner, R., Nemeth, A., Jansa, P., Hofmann-Rohrer, U., Santoro, R., Längst, G., and Grummt, I. (2001) NoRC—a novel member of mammalian ISWI-containing chromatin remodeling machines. EMBO J. 20, 4892–4900.
293. Santoro, R., Li, J., and Grummt, I. (2002) The nucleolar remodeling complex NoRC mediates heterochromatin formation and silencing of ribosomal gene transcription. Nat. Genet. 32, 393–396.
294. LeRoy, G., Loyola, A., Lane, W.S., and Reinberg, D. (2000) Purification and characterization of a human factor that assembles and remodels chromatin. J. Biol. Chem. 275, 14787–14790.
295. Tate, P., Lee, M., Tweedie, S., Skarnes, W.C., and Bickmore, W.A. (1998) Capturing novel mouse genes encoding chromosomal and other nuclear proteins. J. Cell Sci. 111, 2575–2585.
296. Bochar, D.A., Savard, J., Wang, W., Lafleur, D.W., Moore, P., Côté, J., and Shiekhattar, R. (2000) A family of chromatin remodeling factors related to Williams syndrome transcription factor. Proc. Natl. Acad. Sci. USA 97, 1038–1043.
297. Bozhenok, L., Wade, P.A., and Varga-Weisz, P. (2002) WSTF-ISWI chromatin remodeling complex targets heterochromatic replication foci. EMBO J. 21, 2231–2241.

298. MacCallum, D.E., Losada, A., Kobayashi, R., and Hirano, T. (2002) ISWI remodeling complexes in Xenopus egg extracts: identification as major chromosomal components that are regulated by INCENP-aurora B. Mol. Biol. Cell 13, 25–39.
299. Hakimi, M.A., Bochar, D.A., Schmiesing, J.A., Dong, Y., Barak, O.G., Speicher, D.W., Yokomori, K., and Shiekhattar, R. (2002) A chromatin remodelling complex that loads cohesin onto human chromosomes. Nature 418, 994–998.
300. Yoo, E.J., Jin, Y.H., Jang, Y.K., Bjerling, P., Tabish, M., Hong, S.H., Ekwall, K., and Park, S.D. (2000) Fission yeast hrp1, a chromodomain ATPase, is required for proper chromosome segregation and its overexpression interferes with chromatin condensation. Nucleic Acids Res. 28, 2004–2011.
301. Ogas, J., Cheng, J.C., Sung, Z.R., and Somerville, C. (1997) Cellular differentiation regulated by gibberellin in the Arabidopsis thaliana pickle mutant. Science 277, 91–94.
302. Ogas, J., Kaufmann, S., Henderson, J., and Somerville, C. (1999) PICKLE is a CHD3 chromatin-remodeling factor that regulates the transition from embryonic to vegetative development in Arabidopsis. Proc. Natl. Acad. Sci. USA 96, 13839–13844.
303. Eshed, Y., Baum, S.F., and Bowman, J.L. (1999) Distinct mechanisms promote polarity establishment in carpels of Arabidopsis. Cell 99, 199–209.
304. Solari, F. and Ahringer, J. (2000) NURD-complex genes antagonise Ras-induced vulval development in *Caenorhabditis elegans*. Curr. Bio. 10, 223–226.
305. Unhavaithaya, Y., Shin, T.H., Miliaras, N., Lee, J., Oyama, T., and Mello, C.C. (2002) MEP-1 and a homolog of the NURD complex component Mi-2 act together to maintain germline-soma distinctions in C. elegans. Cell 111, 991–1002.
306. Khattak, S., Lee, B.R., Cho, S.H., Ahnn, J., and Spoerel, N.A. (2002) Genetic characterization of Drosophila Mi-2 ATPase. Gene 293, 107–114.
307. Kim, J., Sif, S., Jones, B., Jackson, A., Koipally, J., Heller, E., Winandy, S., Viel, A., Sawyer, A., Ikeda, T., Kingston, R., and Georgopoulos, K. (1999) Ikaros DNA-binding proteins direct formation of chromatin remodeling complexes in lymphocytes. Immunity 10, 345–355.
308. Schmidt, D.R. and Schreiber, S.L. (1999) Molecular association between ATR and two components of the nucleosome remodeling and deacetylating complex, HDAC2 and CHD4. Biochemistry 38, 14711–14717.
309. Fujita, N., Jaye, D.L., Kajita, M., Geigerman, C., Moreno, C.S., and Wade, P.A. (2003) MTA3, a Mi-2/NuRD complex subunit, regulates an invasive growth pathway in breast cancer. Cell 113, 207–219.
310. Darst, R.P., Wang, D., and Auble, D.T. (2001) MOT1-catalyzed TBP-DNA disruption: uncoupling DNA conformational change and role of upstream DNA. EMBO J. 20, 2028–2040.
311. Andrau, J.C., Van Oevelen, C.J., Van Teeffelen, H.A., Weil, P.A., Holstege, F.C., and Timmers, H.T. (2002) Mot1p is essential for TBP recruitment to selected promoters during in vivo gene activation. EMBO J. 21, 5173–5183.
312. Chicca, J.J., 2nd, Auble, D.T., and Pugh, B.F. (1998) Cloning and biochemical characterization of TAF-172, a human homolog of yeast Mot1, Mol. Cell. Biol. 18, 1701–1710.
313. Amedeo, P., Habu, Y., Afsar, K., Scheid, O.M., and Paszkowski, J. (2000) Disruption of the plant gene MOM releases transcriptional silencing of methylated genes. Nature 405, 203–206.
314. Jeddeloh, J.A., Bender, J., and Richards, E.J. (1998) The DNA methylation locus DDM1 is required for maintenance of gene silencing in Arabidopsis. Genes Dev. 12, 1714–1725.
315. Gendrel, A.V., Lippman, Z., Yordan, C., Colot, V., and Martienssen, R.A. (2002) Dependence of heterochromatic histone H3 methylation patterns on the Arabidopsis gene DDM1. Science 297, 1871–1873.
316. Dennis, K., Fan, T., Geiman, T., Yan, Q., and Muegge, K. (2001) Lsh, a member of the SNF2 family, is required for genome-wide methylation. Genes Dev. 15, 2940–2944.
317. Lee, D.W., Zhang, K., Ning, Z.Q., Raabe, E.H., Tintner, S., Wieland, R., Wilkins, B.J., Kim, J.M., Blough, R.I., and Arceci, R.J. (2000) Proliferation-associated SNF2-like gene (PASG): a SNF2 family member altered in leukaemia. Cancer Res. 60, 3612–3622.

318. Gibbons, R.J., McDowell, T.L., Raman, S., O'Rourke, D.M., Garrick, D., Ayyub, H., and Higgs, D.R. (2000) Mutations in ATRX, encoding a SWI/SNF-like protein, cause diverse changes in the pattern of DNA methylation. Nat. Genet. 24, 368–371.
319. van Gool, A.J., Verhage, R., Swagemakers, S.M., van de Putte, P., Brouwer, J., Troelstra, C., Bootsma, D., and Hoeijmakers, J.H. (1994) RAD26, the functional *S. cerevisiae* homolog of the Cockayne syndrome B gene ERCC6. EMBO J. 13, 5361–5369.
320. Woudstra, E.C., Gilbert, C., Fellows, J., Jansen, L., Brouwer, J., Erdjument-Bromage, H., Tempst, P., and Svejstrup, J.Q. (2002) A Rad26-Def1 complex coordinates repair and RNA pol II proteolysis in response to DNA damage. Nature 415, 929–933.
321. Troelstra, C., van Gool, A., de Wit, J., Vermeulen, W., Bootsma, D., and Hoeijmakers, J.H. (1992) ERCC6, a member of a subfamily of putative helicases, is involved in Cockayne's syndrome and preferential repair of active genes. Cell 71, 939–953.
322. van Gool, A.J., Citterio, E., Rademakers, S., van Os, R., Vermeulen, W., Constantinou, A., Egly, J.M., Bootsma, D., and Hoeijmakers, J.H. (1997) The Cockayne syndrome B protein, involved in transcription-coupled DNA repair, resides in an RNA polymerase II-containing complex. EMBO J. 16, 5955–5965.
323. Verhage, R.A., van Gool, A.J., de Groot, N., Hoeijmakers, J.H., van de Putte, P., and Brouwer, J. (1996) Double mutants of *Saccharomyces cerevisiae* with alterations in global genome and transcription-coupled repair. Mol. Cell. Biol. 16, 496–502.
324. Reed, S.H., You, Z., and Friedberg, E.C. (1998) The yeast RAD7 and RAD16 genes are required for postincision events during nucleotide excision repair. In vitro and in vivo studies with rad7 and rad16 mutants and purification of a Rad7/Rad16-containing protein complex. J. Biol. Chem. 273, 29481–29488.
325. Ulrich, H.D. and Jentsch, S. (2000) Two RING finger proteins mediate cooperation between ubiquitin-conjugating enzymes in DNA repair. EMBO J. 19, 3388–3397.
326. Emery, H.S., Schild, D., Kellogg, D.E., and Mortimer, R.K. (1991) Sequence of RAD54, a Saccharomyces cerevisiae gene involved in recombination and repair. Gene 104, 103–106.
327. Van Komen, S., Petukhova, G., Sigurdsson, S., Stratton, S., and Sung, P. (2000) Superhelicity-driven homologous DNA pairing by yeast recombination factors Rad51 and Rad54. Mol. Cell 6, 563–572.
328. Petukhova, G., Stratton, S., and Sung, P. (1998) Catalysis of homologous DNA pairing by yeast Rad51 and Rad54 proteins. Nature 393, 91–94.
329. Jaskelioff, M., Van Komen, S., Krebs, J.E., Sung, P., and Peterson, C.L. (2003) Rad54p is a chromatin remodeling enzyme required for heteroduplex DNA joint formation with chromatin. J. Biol. Chem. 278, 9212–9218.
330. Alexeev, A., Mazin, A., and Kowalczykowski, S.C. (2003) Rad54 protein possesses chromatin-remodeling activity stimulated by the Rad51-ssDNA nucleoprotein filament. Nat. Struct. Biol. 10, 182–186.
331. Alexiadis, V. and Kadonaga, J.T. (2002) Strand pairing by Rad54 and Rad51 is enhanced by chromatin. Genes Dev. 16, 2767–2771.
332. Sukhodolets, M.V., Cabrera, J.E., Zhi, H., and Jin, D.J. (2001) RapA, a bacterial homolog of SWI2/SNF2, stimulates RNA polymerase recycling in transcription. Genes Dev. 15, 3330–3341.
333. Caruthers, J.M. and McKay, D.B. (2002) Helicase structure and mechanism. Curr. Opin. Struct. Biol. 12, 123–133.

J. Zlatanova and S.H. Leuba (Eds.) *Chromatin Structure and Dynamics: State-of-the-Art*

DOI: 10.1016/S0167-7306(03)39017-9

CHAPTER 17

What happens to nucleosomes during transcription?

Vaughn Jackson

Department of Biochemistry, Medical College of Wisconsin, Milwaukee, WI 53226, USA. Tel.: 414-456-8776; Fax: 414-456-6510; E-mail: jacksonv@mcw.edu

1. Introduction

What happens to nucleosomes during transcription has been a question that has been investigated from a number of different experimental approaches and has been extensively reviewed [1–6]. Investigations designed to evaluate the nucleosomal state of active genes in cells have generally observed that these structures are in a more open, generally disrupted state. Yet a number of *in vitro* experiments have suggested that polymerases can transcribe through nucleosomes with limited disruption. At the same time when the conditions of transcription are different from those of the previous studies, an altered structure is observed and in some instances a complete displacement of histones from DNA. There does not appear to be a simple answer to this question. Since the transcription conditions that are used in *in vitro* experimentation substantially define what answers are obtained, perhaps it is of value to review more closely what is known about the *in vivo* state of nucleosomes and then determine whether the *in vitro* studies have accurately reflected that condition.

2. In vivo studies of transcription on nucleosomes

2.1. Nuclease studies

One of the very early research tools that were used to study the nucleosomal state of active genes were the nucleases, DNase I and Micrococcal nuclease. With the development of protocols for the isolation of nuclei from cells, it was possible to add these reagents to probe the accessibility of DNA. DNase I makes single nicks in double stranded DNA and when the DNA is associated with histones within the nucleosome, the DNA is extensively protected. Those nicks that are observed are found to occur only after extensive digestion and are limited to the outside surface of the DNA in 10 base increments [7,8]. Weintraub and Groudine in 1976 [9] first used this nuclease and observed that when nuclei from chicken erythrocytes were treated with DNase I, the active β-globin gene was preferentially solubilized. This selective solubilization was not observed in nuclei of fibroblast and brain in which this gene was inactive. The rapid solubilization (enhanced cleavage) was interpreted as indicating the presence of an altered nucleosome structure. Since this early study a number of genes have been characterized in this way with the general conclusion that genes with greater frequency of transcription tend to be more sensitive to the action of DNase I (reviewed in Refs. [10–12]). Micrococcal nuclease was also found to be useful in this analysis. This nuclease cleaves both double and single stranded DNA. It does not readily cleave the DNA of the nucleosome, but does show preferential cleavage between them which results in the well characterized 200 bp DNA ladder that is seen by a partial digest of chromatin [13,14]. Bellard *et al.* [15] observed that when nuclei from oviduct were treated with this enzyme, the characteristic DNA ladder of the active ovalbumin gene was very diffuse. Other active genes such as the *Drosophila* heat shock [16] and yeast galactokinase gene [17] produced a similar diffuse DNA ladder. In contrast the nucleosomal state of the active β-globin gene of chick erythrocyte produced a well-defined ladder, although the multiples of 200 bp were shifted in size by 100 bp, which could be interpreted as indicating that the edges of the nucleosome encompassing the regions associated with histones H2A, H2B were now increasingly accessible to the nuclease [18]. Using this nuclease, nucleosome structure has also been found on the highly active ribosomal genes of a variety of species, yet as before there is a much enhanced sensitivity to cleavage which again suggests a more open, extended nucleosomal conformation [19–22].

The question remains as to whether the entire length of the active gene is in this perturbed state. With the development of a method referred to as indirect labeling, it has been possible to map the entire length of genes with these nucleases [23,24]. In this method nuclei are first exposed to a partial digest with the nucleases, the DNA isolated and treated with a restriction nuclease, which cleaves the DNA at low frequency for the gene of interest. After the DNA is electrophoretically separated and transferred to a paper matrix, which immobilizes the DNA, the DNA is then hybridized with a DNA probe which is specific to the restriction site. Since the probe hybridizes to the end of the DNA fragment, the final length will be defined by the cleavage site on the other end, which was produced when the nuclei were exposed to either DNase I or Micrococcal nuclease. This method has provided substantial information regarding active genes and again the results indicate that perturbation occurs throughout the transcribed gene and is dependent on the relative frequency of transcription [23–28]. Of particular note is the work of Lee and Garrard [29] in which they observed a preference for cleavage in the center of the nucleosome on the 3′ end of the HSP70 gene, a complete nucleosome splitting. In all of these studies it is generally difficult to define whether changes occur in nucleosome structure just prior to transcription, as a direct result of transcription or as a long-term result of transcription. The rather rapid change in nucleosome structure as a consequence of induction in the heat shock genes is strongly suggestive of a direct transcription involvement [29]. Cavalli and Thoma [30] have shown that a rapid rearrangement of nucleosomes is observed upon the rapid activation of a strong GAL promoter in yeast. This rearrangement is also brought to a normal nucleosomal state in an equally rapid process when transcription is repressed. These results taken together strongly suggest that the actual process of transcription causes much of the alteration in distribution and structure of the nucleosomes.

The question remains as to whether the absolute number of nucleosomes changes as a function of transcription. Pederson and Morse [31] have attempted to address this question by determining the linking number (number of negative coils held by the histones) that is observed *in vivo* in the presence and absence of transcription on a yeast plasmid carrying the HSP26 gene. They observed a stable maintenance of the linking number. Their conclusion from these studies was that if nucleosome perturbation had occurred, the perturbation was not sufficient to change the linking number or if it was, the same number of perturbed nucleosomes must have reformed normal structure within 1 to 5 min. De Bernardin *et al.* [32] also addressed this question on transcriptionally active SV40 minichromosomes. In this latter experiment the minichromosomes were treated with psoralen to produce intra-strand crosslinks between the DNA strands on all regions of the DNA except where nucleosomes were present. Nucleosomes prevent psoralen crosslinking. By examining the number of single-stranded bubbles (nucleosomal sites) on the SV40 DNA using electron microscopy, they observed no change in the number or relative distribution of nucleosomes as a result of transcription. In this latter study it is unclear whether perturbed nucleosomes also protect DNA from psoralen crosslinking. Nevertheless, there is an indication that for the transcription of most genes, the number of nucleosomes remains unchanged. A specific example of a gene in which this conclusion may not be valid are the ribosomal genes. Electron micrographic analyses of chromatin spreads from the embryos of the milkbug *Oncopeltus fasciatus* show the presence of extended regions on these genes in which the "beads on a string" appearance which is characteristic of nucleosome structure is absent [33]. Applying the same psoralen treatment described above, Sogo *et al.* [34,35] have shown that for the ribosomal genes of both *Dictyostelium discoideum* or Friend leukemic cells, crosslinking is much more extensive on those genes compared to non-transcribed regions. Thus the diffuse nuclease digestion pattern, which was described in the previous section for the ribosomal genes may not only indicate a more open nucleosome conformation but also a significant displacement of histones due to the extensive transcription that is characteristic of these genes.

2.2. Histones of active genes

Because these nucleases preferentially solubilize the DNA of active genes, methods have been developed to examine the proteins associated with the DNA. These methods take advantage of two main characteristics of the chromatin from active genes, its solubility in 100 mM NaCl and its solubility in moderate concentrations of Mg^{2+} (2–5 mM). What has been found is that all four of the core histones H2A, H2B, H3, and H4 are present in these solubilized fractions [36–38]. Histone H1 is depleted. The core histones are also found to be enriched in the highly acetylated forms, which suggest a potential role for acetylation in transcription. The combination of high levels of acetylation and depletion of H1 is thought to be the source of the enhanced solubility in these fractionation procedures [39,40]. These characteristics may also have relevance to the accessibility of DNA for initiation and subsequent elongation of transcription. *In vivo* formaldehyde crosslinking and subsequent isolation of active genes based on density differences caused by the associated RNA transcript have confirmed the presence of all the histones; again as before in a highly acetylated state [41]. It still remains undefined, however how depleted these genes are of histone H1. Immunocytochemistry of the Balbiani rings indicates no change in H1 content between repressed and activated states of the gene [42]. Kamaka and Thomas [43] observed a depletion of H1 when using a UV crosslinking protocol that was coupled with immunoprecipitation of H1-associated DNA. Mirzabekov's laboratory [44,45] has developed an alternative protocol for assessing histone content on active genes in nuclei. In this protocol the nuclei are treated with dimethyl sulfate and then, in consecutive order the DNA is depurinated and exposed to $NaBH_4$ to complete the crosslink. With this approach those histones, which are crosslinked to DNA can be electrophoretically separated from those that do not. The associated DNA can then be probed for sequence content. By also including a step just prior to crosslinking which involves a protease treatment, they were able to differentiate crosslinks with the N-terminal tails (no protease treatment) from crosslinks with the globular regions (with protease treatment) of the histones. When the crosslinking was done with the N-terminal tails, they observed no difference in the quantity of histones that were present in the coding regions of either an active or inactive HSP70 gene. When the crosslinking was through the globular domains alone, H1 was found depleted, H2A, H2B to an intermediate extent and H3, H4 least of all. In contrast, the promoter region was found to be totally depleted of histones with or without protease treatment. Taken as a whole, whatever perturbation is occurring it does not result in dramatic losses in specific histone subtypes from the transcribed DNA, but perhaps a loss of interactions, as demonstrated by the variable crosslinking that is observed with the globular domains.

Except for histone H4, each of the other histone types are found in different isoforms and are called histone variants. The chapter by Pehrson (Chapter 8) will provide a more in-depth discussion of these forms. It is worth pointing out that with regard to transcription through nucleosomes, some of these variants are expressed in a replication-independent process and are found in active gene fractions that have been prepared using the nuclease-sensitive solubilization procedures described above. Of particular note are two minor histone variants, H2A.Z and H3.3. Both are expressed throughout the cell cycle and incorporated into the nucleosomes of active genes ([38,39], see reviews [46,47]). For example, both *Tetrahymena*, H2A.Z (termed *Tetrahymena* hv1) and an H3.3-like histone (hv2) are preferentially present in the active macronucleus and are expressed in the micronucleus just prior to the time when this nucleus becomes transcriptionally active [48,49]. Suto *et al.* [50] have determined the crystal structure of a nucleosome containing H2A.Z and observed subtle disruption both in the interphase between H2A.Z and H2B as well as at the interphase with the H3, H4 tetramer. The implication from such a replication-independent deposition is that these histones are replacing existing histones that have been displaced from the nucleosomes possibly during transcription. Disruption of both histone–histone and histone–DNA interactions would be required for such an exchange to occur. Physical studies with H2A.Z-containing nucleosomes have indicated that higher order structure is substantially minimized when this variant is present [51]. It remains to be determined whether either of these variants actually facilitate the transcription process or serve as markers for interaction with other proteins that do.

Another metabolic modification that has been correlated with an involvement in transcription is ubiquitination. In particular, ubiquitinated-H2B (uH2B) is highly enriched in the nuclease-solubilized fraction of chick erythrocyte, which contains the transcriptionally active globin gene [37,38]. Treatment of breast cancer cells with inhibitors of transcription (actinomycin D or 5,6-dichloro-1-B-D-ribofuranosylbenzimidazole, DRB) results in a disappearance of uH2B from these fractions [52]. This selective localization to active genes, however, does not appear to be universal. Active immunoglobulin genes are depleted in this modification [53]. *In vitro* studies that were designed to examine the effects of having either uH2A or uH2B in the nucleosome have indicated that there was no increase in susceptibility to DNase I cleavage [54]. The lysines that are modified (lys 120 for H2B and lys 119 for H2A) are located in the nucleosome structure where nucleosome disruption would not necessarily be required to facilitate conjugation to ubiquitin [55]. Therefore, it is generally conceded that this modification probably does not change the stability of the nucleosome. It has recently been found that ubiquitination of H2B has a role in facilitating the methylation of specific lysines in H3 (lysines 4 and 79) which in turn enhances telomeric silencing in yeast [56,57]. As a result it has been proposed that the ubiquitin modification serves as a marker for the repression rather than the activation of genes. Considerable more research will be required before it will be possible to define how this unusual modification is used to mark gene activity either positively or negatively.

In histone H3 of many eucaryotes, there is a single cysteine at position 110. Histones H2A, H2B, H4, and H1 do not contain a cysteine. From the nucleosome structure [55,58,59], the cysteine from each H3 in the H3, H4 tetramer are within a few angstroms of each other buried within the octameric complex of core histones. It is known that sulfhydryl-modifying reagents are not able to access these cysteines in nucleosomes that have been isolated either from nuclei or prepared by reconstitution of DNA with histones by a variety of methods [60,61]. It was then of interest to note that for the active ribosomal genes of *Physarum polycephalum*, the bulky sulfhydryl reagent idoacetamidofluorescein was able to modify these cysteines [62]. When using ^{3}H iodoacetate, these investigators were also able to show that highly acetylated H3 was preferentially modified [63]. Chan *et al.* [64] have observed that the salt soluble fraction (enriched in active genes) from nuclease-treated, chick erythrocyte nuclei were highly enriched in nucleosomes that contained H3's which were highly accessible to a Hg-containing, SH-modifying group. Allfrey's laboratory has further characterized the accessibility of the cysteines by the use of Hg-affinity columns and has obtained extensive enrichment for active genes [65]. Of particular interest is one study in which the cysteine of H3 in the nucleosomes of the c-fos and c-myc genes was reversibly unavailable when transcription was inhibited by exposure of the cells to α-amanitin, a RNA synthesis inhibitor [66]. In summary, these observations indicate that the exposure of the cysteine is correlated with the active process of transcription. Substantial nucleosomal perturbation would be required. Electron micrograph analyses of these nucleosomes show a highly perturbed "U" type structure [67].

2.3. *In vivo nucleosomal dynamics*

From the previous description nucleosomal perturbation appears to be a common mechanism for facilitating transcription. With the observation that the core histones including H1 remain present on these genes, there is a sense that the perturbation is insufficient to cause displacement of histones. These approaches have not identified, however, whether histones are exchanging (releasing and re-associating) during the transcription process. What was needed was an *in vivo* approach to address this question. One approach that was used was to radiolabel newly synthesized histones and allow deposition to occur within the nucleus. During the radiolabeling, the cells were also exposed to the heavy base analog iododeoxyuridine, which incorporates into the newly replicated DNA. The increase in density of the newly replicated DNA allows one to separate newly replicated chromatin from unreplicated chromatin. To preserve protein–DNA interactions during the fractionation procedure, the cells were exposed to formaldehyde at the completion of the pulse and after crosslinking for 2 hr, the cells were disrupted and the chromatin isolated. This chromatin was then broken into smaller fragments by sonication and applied to a CsCl gradient, which separates based on density. After fractionation and reversal of the formaldehyde crosslinks, the quantity and type of labeled histone associated with the replicated DNA was determined. The enrichment for replicated DNA in these experiments was known to be 20-fold and when the level of enrichment was determined for the newly synthesized histones, new H3, H4 were also enriched 20-fold [68]. New H2A, H2B were only 3-fold enriched. New histone H1 showed no enrichment. The new H3, H4 were selectively depositing on replicated DNA whereas a substantial level of new H2A, H2B were not. This result meant that old H2A, H2B were substituting for the new H2A, H2B, which in turn were being assembled into unreplicated chromatin. An exchange was occurring for H2A, H2B, which was not for H3, H4. For H1, the exchange between new and old H1 was so extensive that no preference for new H1 deposition on the replicated DNA was detected. What regions of the chromatin were providing the source of the old H2A, H2B? In additional experiments, it was observed that if this pulse was chased up to the next replication cycle, the labeled H3, H4 continued to remain on that same newly replicated DNA and the 3-fold enrichment of H2A, H2B also remained unchanged [69].

If there were to be a general phenomenon in which old H2A, H2B exchanged throughout the chromatin, one might have expect this 3-fold enrichment to rapidly decrease during the chase. It was an initial indication that perhaps a limited region in the genome was responsible for much of this exchange. Subsequently, it was observed that when mature chick erythrocytes (no DNA synthesis) were radiolabeled to examine the deposition sites for the low level of histones that were synthesized in these cells, the new H2A, H2B were preferentially associated with nucleosomes enriched in active genes, as isolated by the nuclease-sensitive solubility methods described above [70,71]. Because of the very high enrichment of new H3, H4 for the replication fork, there must be very limited exchange of those histones with old H3, H4 in other regions of the chromatin and implies that irrespective of nuclear processes, there is minimal release of those histones.

An additional approach, that was used to study nucleosomal dynamics, was to label cells with both radioactive and ^{15}N, ^{13}C, ^{2}H dense amino acids. After a short pulse, the nuclei were isolated and exposed to formaldehyde at pH 9. In these conditions the crosslink is primarily between the histones in the octamer and not with DNA. The crosslinked octamers were then isolated and applied to a $CsSO_4$ gradient to separate the octamers based on the percentage of density-labeled histones that were present in them. The formaldehyde crosslinks were then reversed and what was observed was that the octamers containing labeled H3, H4 had a density, which was 45% of what would have been expected if the octamer had contained 100% density-labeled histones. This result indicated that the labeled H3, H4 were organized as a homogenous tetramer (all four histones density-labeled) and primarily crosslinked to two unlabeled H2A, H2B dimers. The labeled H2A, H2B were found to be in a crosslinked octamer with a density of 24% (2 of 8 proteins density-labeled) which was an indication that the labeled H2A, H2B were as a homogenous dimer (one H2A and one H2B density-labeled) interacting with one unlabeled H2A, H2B dimer and one unlabeled H3, H4 tetramer [72]. This type of distribution was to be expected given the level of exchange of H2A, H2B that was observed in the previous experiment in which the subtypes of newly synthesized histone were characterized at the replication fork [68]. When this experiment was repeated except now in the presence of the transcription inhibitor, actinomycin D, the density of the octamer containing the labeled H2A, H2B increased dramatically [73]. The labeled H3, H4 were now found to be significantly associated with the labeled H2A, H2B with a substantial increase in overall density. These results were interpreted as indicating that the newly synthesized H2A, H2B were now preferentially associated with the new H3, H4 on the replicated DNA. The ultimate conclusion from this study was that when transcription is limited, H2A, H2B exchange in the nucleosomes of those genes is also limited. In theory, the release of H2A, H2B might be expected to enhance transcription as potentially half of the core histones and their contact with DNA would be temporarily displaced leaving only the H3, H4 tetramer to circumvent. Reassociation of an H2A, H2B (new or old) to the H3, H4 after polymerase passage would allow the nucleosome to reform and maintain the normal complement of nucleosomes. These procedures were also able to detect a low background level of H2A, H2B exchange which was independent of both transcription and replication. It would appear that nucleosomes are in a dynamic state in which nuclear processes other than transcription and replication are able to shift the equilibrium towards a basal level of nucleosomal disruption and exchange of H2A, H2B.

A third approach that was recently applied to address nucleosomal dynamics in the cell was to tag histones H3, H4, and H2B with GFP (green fluorescent protein) and then stably maintain them in transformed Hela cells. In order to address the relative dynamics of histone diffusion in the nucleus, a procedure referred to as FRAP (Fluorescent Recovery After Photobleaching) was applied. In this innovative technique, a small region of the nucleus in the cell is bleached with high intensity light and then the rate in which the GFP-labeled protein repopulates the bleached region is measured. Kimura and Cook [74] observed that there were two kinetically definable diffusion rates for GFP-H2B, $\sim 40\%$ diffused very slowly and 3% very rapidly. The rapid diffusion of the 3% would not occur if the RNA synthesis inhibitor, DRB was present. In contrast GFP-H3 or GFP-H4 were minimally diffusible in the presence and absence of this drug. Their conclusion was that H2B was exchanging as a result of active transcription and H3, H4 were not.

Therefore, in evaluating transcription through nucleosomes with *in vitro* protocols, we must be aware that the *in vivo* experimentation does indicate that a general unfolding of the nucleosome can occur to allow this exchange. This unfolding may result in a general small depletion of all four histones under conditions of high frequency transcription (i.e., the ribosomal genes), but the preferential condition is an unfolding that does not result in loss of nucleosomes. The observation that H2A, H2B are preferentially exchanging and not H3, H4 is of importance. The H3, H4 tetramer may serve as the central complex of proteins upon which nucleosomal disassembly and assembly can be rapidly done. With regard to histone H1, Mistelli *et al.* [75] have applied the FRAP methodology to examine this histone's mobility and have observed extensive exchange in both euchromatin and heterochromatin. This result is an indication that when present on active genes, the mobility of H1 may also be a pre-requisite for effective transcription through nucleosomes.

3. *In vitro studies of transcription on nucleosomes*

3.1. *The eucaryotic polymerases*

There have been many studies that have examined transcription on chromatin templates *in vitro*. The general sense is that transcription does occur through the nucleosomes, but there is substantial premature termination and repression. An example of earlier experiments in this regard is the work of Wasylyk and Chambon, 1979 [76] in which histones were reconstituted on SV40 form 1 DNA and subsequently transcribed with calf thymus RNA polymerases I and II. They observed that extensive repression was relieved by increasing the ammonium sulfate concentration from 40 mM to 300 mM and concluded that a major reason for the increased transcription was the neutralization of charge on the basic N-terminal regions of the histones by the higher salt concentrations. Since this earlier work, a number of investigators have extended those studies with the eucaryotic polymerases. With regard to RNA polymerase III the general consensus is that transcription is much reduced with a tendency for the polymerase to terminate at natural pause sites in the DNA sequence that is in the nucleosome [77–79]. In studies which involve transcription through tandem arrays of nucleosomes, chromatin folding from nucleosome–nucleosome interactions was also found to be an additional factor in the frequency of termination [77,78]. Since generally polymerase III transcription *in vivo* is primarily on short genes of roughly 200 bp in length, transcription on tandem arrays may not be directly relevant to this polymerase. However, Izban and Luse [80] have reported similar observations with RNA polymerase II, a polymerase designed to transcribe great distances. In addition they have observed that even in the absence of H2A, H2B, enhanced termination at pause sites occurs with H3, H4 alone [81]. As originally observed by Wasylyk and Chambon [76], efficient transcription by these polymerases requires elevated salt concentrations of 300 mM. To get around this problem, the procaryotic RNA polymerases (T7 and SP6) have been used extensively to study transcription through nucleosomes. These polymerases exhibit considerable processivity through nucleosomes at salt concentrations of 100 mM NaCl or lower. Even here the pause sites that are directed by the DNA sequence appear to be the preferred sites of premature termination when single nucleosomes were studied [82]. When nucleosomes in tandem array were transcribed, pausing between the nucleosomes was also seen which again emphasizes the importance of nucleosome–nucleosome interactions in defining transcriptional efficiency [83]. Given that these characteristics are common for all polymerases, it provides the rationale for applying the procaryotic polymerases to an analysis of what happens to the nucleosomes during processive transcription through them. An additional added feature of these procaryotic polymerases is their efficient and quantitative initiation at well defined promoters. These proteins are relatively easy to purify to homogeneity as they are single polypeptides and not multi-subunit complexes. As a result the transcriptional conditions can be more clearly defined and the ultimate results from the experimentation interpreted with less ambiguity.

3.2. The procaryotic polymerases

Using the unique characteristics of the procaryotic polymerases, a number of investigators have attempted to define what happens to the histones during transcription. In the mid 1980s two different laboratories did a set of important experiments. Lorch *et al.* [84] observed that during transcription with SP6 RNA polymerase of a DNA that contained an SP6 promoter, flanked by DNA sufficient to hold one nucleosome, the histones of that nucleosome were displaced, as observed by an altered mobility of the DNA after gel electrophoresis. Subsequently, Losa and Brown [85] applied a similar approach except that in this instance they used the nucleosome positioning sequence from the *Xenopus borealis* 5S RNA gene to position their nucleosome. They observed no change in the nucleosome content nor did they observe a change in the nucleosome position on the fragment during transcription. Thus began a point of controversy with respect to what is happening to the histones. Subsequently, Prunell's laboratory [86] repeated these latter studies and in this instance included a competitor DNA. In this instance they observed an exchange of histones to the competitor. The conclusion from these studies was that if sufficient competitor DNA is present it is possible to detect histone release. There is probably more to it than just the quantity of competitor. The differences in these results represent a more fundamental problem that has continued to plague *in vitro* transcription studies. The question remains as to what ionic strength appropriately describes the condition in the cell. For all three of these studies, the ionic strength was rather low. In the Losa and Brown study, the conditions were 14 mM NH_4Cl, 10 mM Hepes, 7 mM $MgCl_2$, and 6 mM spermidine. For the Prunell study the conditions were 10 mM NaCl, 40 mM Tris, 3 mM $MgCl_2$, and 1 mM spermidine. The higher level of Mg^{2+} and spermidine in the former study [85], when coupled with the lower overall ionic strength, causes extensive binding of these polycations to the nucleosome. The overall effect of these conditions on nucleosomes can be more clearly seen when tandem arrays of nucleosomes are present. Aggregation and insolubility is nearly complete (unpublished observations). Gallego *et al.* [87] have also examined transcription on short templates and in this instance used T7 RNA polymerase. When a competitor DNA was present, the nucleosomal template was seen to convert to naked DNA, as assayed by gel mobility. As might be expected in this latter study, the ionic strength conditions were closer to physiological, i.e., 50 mM NaCl, 40 mM Tris, 1.7–8 mM $MgCl_2$, and 0.5–1 mM spermidine. The overall conclusion from these earlier studies is that histones can be displaced from DNA during transcription, although the overall frequency of this displacement remained unknown.

3.3. The "spooling" model

Clark, Studitsky, and Felsenfeld [88–92] have extensively evaluated transcription on single nucleosomes utilizing SP6 RNA polymerase. Based on their data they have developed a model which is shown in Fig. 1A. This model, referred to as the "spooling" model, proposes that when transcription occurs into the first 25 bp of the nucleosome, the polymerase pauses (step 2). DNA behind the polymerase binds the octamer surface, which further inhibits the action of the polymerase due to steric hinderance of the polymerase's rotation (step 3). For transcription to continue the DNA behind the polymerase must dissociate to allow the polymerase to continue to transcribe an additional 35 bp (step 4). At this point sufficient surface of the octamer is now available to permit substantial levels of

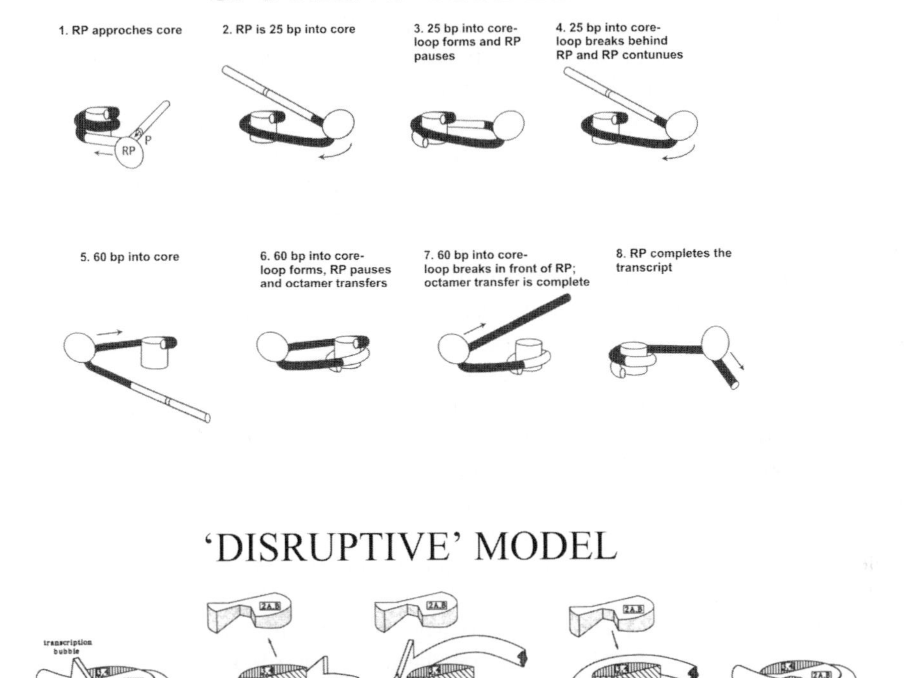

Fig. 1. Two models to describe the process of transcription through nucleosomes. The "spooling" model is taken from Studitsky *et al.* [89]. The RP (RNA polymerase) is shown to cause octamer displacement from the DNA that is being transcribed. The octamer is transferred to the DNA that was previously transcribed which occurs in a series of eight steps. The "disruptive" model is taken from van Holde *et al.* [3]. The octamer is shown to be disrupted in a series of steps (A–E) in which the two H2A, H2B dimers are displaced by the RNA polymerase and subsequently shown to reassociate after the polymerase has passed.

binding of DNA behind the polymerase (steps 5 and 6). The octamer can now fully transfer to this DNA, which permits the polymerase to transcribe the remaining 85 bp of DNA without the bound histones (steps 7 and 8). Data in support of this model are at several levels. In the first instance these investigators determined the transcriptional pause sites on a 227 bp template. This particular sequence very effectively positions the nucleosome away from the flanking SP6 promoter and provides a means to accurately measure length of transcript relative to the promoter start site [90]. Additional data were obtained by a characterization of the transcribed template. The octamer was now found to protect sequences encompassing the promoter, which is an indication that the histones have transferred behind the transcribing polymerase [88,89]. Electron micrographic analysis also showed the repositioning of the nucleosome on the transcribed template [91]. By examining the size of the transcribed template using gel electrophoresis, they also concluded that the octamer remained intact throughout the transfer. This phenomenon was not dependent on a procaryotic polymerase for they also observed that RNA polymerase III repositions a nucleosome in a similar manner [92]. The ability of histones in the nucleosome to slide along DNA is a characteristic that has been extensively studied and is well known [93]. A positive aspect to this model is that after the polymerase has transcribed the 60 bp into the nucleosome, the subsequent looping of the retrograde DNA to the octamer allows the polymerase to transcribe the remaining 85 bp of DNA in the 145 bp nucleosome without interference by associated histones. This model would be consistent with the *in vivo* studies, which indicated that transcription causes minimal perturbation of the histones in the nucleosome. It should be noted that these observations do not preclude the transfer of the octamer to other polyanions. These investigators also observed that when an excess of competitor DNA was present, 50% of the octamer transferred to the competitor [90].

Subsequently, Peng and Jackson [94] examined transcription on multi-nucleosomal templates and used a protocol that would allow direct detection of the histones that were transferred to the competitor DNA. At an ionic strength of 100 mM NaCl, 40 mM Hepes, 4 mM $MgCl_2$, and 1 mM EDTA, they observed efficient transcription with T7 RNA polymerase on a 8.9 kb plasmid containing up to an average of 32 nucleosomes. The transcription was done in the presence of a eucaryotic topoisomerase I (topo I) and as a result of transcription, the linking number of the plasmid was seen to dramatically decrease. This inability to maintain negative coils was interpreted as indicating a loss of histones. Subsequent experimentation indicated that the histones were transferring to the nascent RNA and not to the competitor DNA. Because of the heterogeneous size of the transcripts, it was not directly possible to separate RNA from template DNA on sucrose gradients. Those histones that were bound to the RNA could be determined by destroying the RNA with RNAase A treatment, while in the presence of excess competitor DNA (885 bp). The histones that transferred to this DNA could then be separated from the template DNA by sedimentation on a sucrose gradient. By using a high competitor/template ratio, it was then possible to determine the quantity of histones that were displaced from the template. An SDS PAGE analysis of the histones showed that equimolar quantities of H3, H4, H2A, and H2B were released at a frequency of transcription in which 1 in 4 nucleosomes were disrupted. With that frequency it would appear that this polymerase prefers to transcribe through nucleosomes rather than disrupt them. It was also determined that for the RNA to be an effective competitor the size of the nascent transcript had to be greater than 200 bases. Therefore, in regard to the previous studies in which templates consisting of a single nucleosome were used, a short DNA template produces equally short RNA transcripts. These transcripts would not be effective competitors and minimal displacement would be observed. It is necessary to transcribe in the presence of an excess of competitor

DNA, since proximity effects demand that the DNA or nascent RNA closest to the histones at the point of disruption will be the polyanion for which those histones will preferentially reassociate. Ten Heggeler-Bordier *et al.* [95] have verified these observations. They used immuno-electron microscopy to determine what happens to histones after transcription with T7 RNA polymerase of a multi-nucleosomal template and also observed transfer to the nascent RNA. In contrast, Kirov *et al.* [96] have reported that no histones displace during transcription with this polymerase. However, as described above, transcriptional efficiency and ultimately histone displacement is not efficient in very low ionic strength conditions.

When considering the results of the Felsenfeld laboratory as well as these latter studies, one might conclude that the core histones transfer to the RNA as an octameric complex. O'Neill *et al.* [97] have shown that T7 RNA polymerase can transcribe through crosslinked octamers rather efficiently. However, Feng *et al.* [98] have shown that at physiological ionic strength a facilitated transfer between two polyanions would be required to prevent the H2A, H2B from being displaced from H3, H4. It is well known that the dimer–tetramer interphase is not stable at 100 mM NaCl in the absence of DNA [99]. It is also known that even at this ionic strength, H2A, H2B will rapidly transfer from an RNA or DNA and associate with H3, H4, when those H3, H4 are bound to a separate DNA or RNA [100]. Therefore it is unclear whether the transfer of all four histones occurs as intact octamers, which may be possible because of the proximity of the RNA transcript or sequentially through a stepwise process whereby H2A, H2B and H3, H4 are separately displaced and subsequently recombine.

3.4. The "disruptive" model

The model of Fig. 1B is taken from a review by van Holde *et al.* [3] which I refer to as the "disruptive" model. In this model the polymerase causes conditions (step A) which promote not only the displacement of the entry site H2A, H2B dimer from DNA, but also from the H3, H4 tetramer (step B). As a result of this disruption, the polymerase is free to transcribe through the tetramer alone without a general displacement from its associated DNA (step C). The H2A, H2B dimer is now free to reassociate to the vacated entry site (step D) to re-establish contacts with both the DNA and the H3, H4 tetramer. As transcription proceeds into the exit site H2A, H2B dimer, these proteins are now displaced from both the DNA and the H3, H4 tetramer in a similar manner as the entry site H2A, H2B dimer (step E). A positive feature with regard to this model is that by displacement of H2A, H2B, the polymerase is able to transcribe the DNA with half the histones displaced prior to transcription. Therefore both models, "spooling" and "disruptive", describe mechanisms which would favorably enhance the process of transcription. Support for the disruptive model comes from the substantial *in vivo* information which suggests that nucleosomes undergo substantial disruption during transcription, as was described in the previous section. Of particular note are those observations which indicate that a discrete population of H2A, H2B is highly mobile in the presence of active transcription [74]. Baer and Rhodes [101] observed that RNA polymerase II preferentially binds nucleosomes deficient in one copy of H2A, H2B and Gonzalez and Palacian [102] have observed that transcription with this polymerase is more efficient on nucleosomes deficient in H2A, H2B. What is unclear regarding this model is the mechanism (step A) which would promote the displacement of the entry site H2A, H2B dimer. The polymerase may directly disrupt the H2A, H2B interactions with DNA and with H3, H4 or perhaps indirectly by the production of topological stress. A number of studies have been done to evaluate this potential indirect effect.

3.5. Transcription-induced topological effects

The twin-domain model for generating both positive and negative stresses during polymerase action was described by Liu and Wang in 1987 [103] and has gained general acceptance based on a large number of *in vitro* and *in vivo* studies demonstrating these stresses in both procaryotes [104–109] and eucaryotes [110–113]. The model is based on the premise that RNA polymerase is restricted from rotation around the DNA helix due to the viscous drag of the RNA transcript. Since 10–15 bp of the DNA helix is maintained in an open state at the site of transcription, the DNA is forced to rotate. This rotation results in overwinding in advance of the polymerase (one positive coil per 10.5 bp transcribed) and a similar underwinding of the helix in the wake of the polymerase (one negative coil per 10.5 bp transcribed). This positive stress, which has the capacity to form a right-handed superhelix could have a disruptive influence on the left-handed supercoil of the nucleosome and therefore indirectly cause its disruption. Lee and Garrard [114] examined the effect of positive stress on chromatin *in vivo* using a yeast strain that is Δtopl–top2ts, but is also expressing *E.coli* topo I. *E coli* topo I relaxes exclusively negatively coiled DNA and is not active on positively coiled DNA. This strain also carries the 2μ plasmid. At the permissive temperature of 24 °C, the plasmid has a normal number of negative coils, consistent with the presence of a normal complement of nucleosomes and a functional topo II. *E. coli* topo I activity is masked by the extensive activity of the topo II which relaxes both negative and positive coils equally well. However, at the non-permissive temperature of 37 °C in which the topo II is inactive, the activity of *E. coli* topo I predominates and causes a preferential relaxation of the negative coils that are induced by the RNA polymerase. As a result these investigators observed high levels of positive coils on the plasmid. They also observed that the nucleosomes of the rarely transcribed REP2 locus on this plasmid were highly sensitive to DNase I. This increased sensitivity was not due to loss of nucleosomes. When this same region was digested with micrococcal nuclease, they observed a DNA ladder similar to what was observed in the permissive condition. Further confirmation that a normal number of nucleosomes was still present on this plasmid came from experiments in which the isolated minichromosome was treated with eucaryotic topo I. This topoisomerase relaxes both positive and negative stresses equally well and therefore in this experiment removes the excess positive stress from the minichromosome. The number of negative coils that were held by the nucleosomes was found to be the same as was observed in the permissive condition. The conclusion from this study is that positive stress does not remove nucleosomes, it alters them sufficient enough to create substantial nuclease sensitivity. This *in vivo* experiment was the first indication that the presence of positive stress could be a factor in causing perturbation within the nucleosome. Positive stress has also been proposed to be the cause of the nucleosome-splitting observed in the 3′ end of the *Drosophilia* HSP70 gene that was described in the earlier section [29].

Using *in vitro* experiments, Peng and Jackson [115] have further characterized the factors that facilitate the maintenance of transcription-induced stress in chromatin. They transcribed a 6 kb closed circular plasmid with T7 RNA polymerase in the presence of *E. coli* topo I. The plasmid had been reconstituted with an increasing number of nucleosomes. They observed that with as little as 3 nucleosomes in a plasmid capable of holding 30 nucleosomes, the general viscosity of the DNA was increased sufficiently that transcription-induced stresses were significantly maintained in the plasmid. Because of the bulky nature of the nucleosome, there is a reduced rate of DNA rotation, which limits the translational flux of the transcriptionally-induced positive and negative stresses around the circular DNA. *E.coli* topo I is then able to relax the accumulating negative stress, causing a large accumulation of transcription-induced positive stress on the plasmid. In the absence of histones the negative and positive stresses rapidly neutralize themselves on the circular DNA. *E.coli* topo I is not able to detect and remove negative stress and no accumulating positive stress is observed. It was also observed that histones H3, H4 and not H2A, H2B were the critical proteins that facilitated the presence of the transcription-induced stresses. This observation is another indication of the importance of H3, H4 as the central core of proteins that define the assembly [116–118] and potential disassembly of the nucleosome. The overall conclusion from this study is that the presence of nucleosomes greatly enhances the maintenance of transcription-induced stresses; stresses that would be available to influence the structural integrity of those nucleosomes.

Thus when we consider the models of Fig. 1 in which DNA is represented as a tube, we need to be cognizant that DNA is not a tube, but a right-handed helix, undergoing substantial topological stresses. In theory one positive and one negative coil are produced for every 10.5 bp of DNA (one turn of the helix) that is transcribed by an RNA polymerase which is larger in size than that of the histones in the nucleosome. Therefore, extensive disruption of histone–DNA and histone–histone interactions with the resultant spooling and/or disruption of the nucleosome might be anticipated.

3.6. Histone chaperones that release H2A, H2B

To further evaluate the effect of positive stress on nucleosome structure, Clark and Felsenfeld [119,120] reconstituted histones onto either positively or negatively coiled DNA and compared the nucleosomal state using gel mobility assays, chemical crosslinking, sedimentation and CD analysis. They were not able to differentiate any difference in the nucleosomes between the two stresses. They did observe, however, that in mixing experiments histones preferentially bound the negatively coiled DNA. Jackson [121] also addressed this question and observed an altered nucleosome structure on the positively coiled DNA. His conclusion was based on the extended time that was required to remove the positive coils from the reconstituted complex when exposed to a eucaryotic topo I. If nucleosomes were present and each one was holding a negative coil, the unrestrained positive coils should have been rapidly removed. The rate of removal was found to occur over a 30 min period at 23 °C as compared to negatively coiled DNA, for which that reconstitute established the expected number of restrained negative coils within 15 s, the shortest time that was measured. To confirm the altered state, formaldehyde crosslinking was also applied in order to preserve the state of the nucleosome prior to the topo I treatment. Additional data are shown in Fig. 2 which illustrate the difference in stability of nucleosomes when present on positively or negatively coiled DNA and also show the importance of histone chaperones. In this experiment the histones H3, H4, H2A, and H2B were reconstituted on these DNAs and subsequently exposed to NAP1 (nucleosome assembly protein 1), a histone chaperone found in relative abundance in the nucleus and frequently a part of chromatin assembly complexes [122–124]. NAP1 has a preference for binding histones H2A, H2B but it will also assemble H3, H4 on DNA [125]. After an incubation at 35 °C for 5 min, the sample was sedimented on a 5–20% sucrose gradient and both the distribution of DNA and protein was determined. As shown in Fig. 2A ~45% of the H2A, H2B on the positively coiled DNA have been extensively displaced while the H3, H4 continue to remain bound to the DNA. For the negatively coiled DNA (Fig. 2B), no detectable H2A, H2B were displaced. Therefore under circumstances in which a chaperone is available to interact with H2A, H2B, these proteins are readily removable when positive stress is present.

Orphanides *et al.* [127] have isolated a subunit complex termed FACT (facilitates chromatin transcription) that has been shown to interact with nucleosomes and in particular H2A, H2B. Of particular interest is the observation that FACT is unable to facilitate RNA polymerase II transcription through nucleosomes when crosslinked octamers are present [128]. When FACT is present, a 30 min incubation at 30 °C results in the formation of a 390 base transcript whereas in its absence no transcript is observed. Because of the extended time that is required to observe a transcript as compared to transcripts from a naked template, the action of the FACT complex is insufficient to remove the highly inhibitory conditions of the nucleosome brought about by what is assumed to be the remaining H3, H4. Salt concentration above physiological are still required [80]. Kireeva *et al.* [129] have examined transcription on a positioned nucleosome with RNA polymerase II. They also found that transcription was inhibited except under conditions of 300 mM salt. In those conditions they observed transcription through the nucleosome without the retrograde displacement that was previously observed for SP6 polymerase [88–90] and RNA polymerase III [92]. They also observed that the

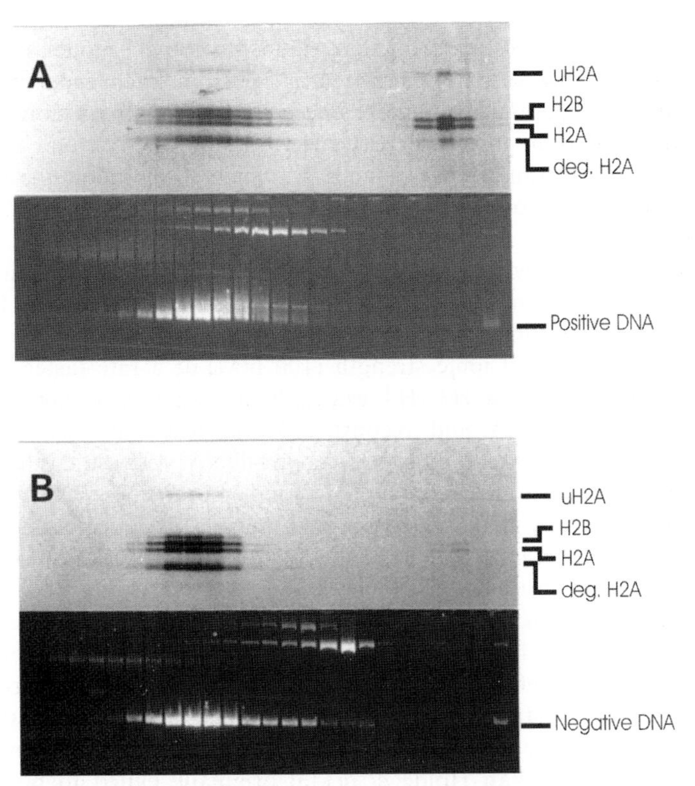

Fig. 2. NAPI facilitates H2A, H2B release from nucleosomes that are on positively coiled DNA (A) but not negatively coiled DNA (B). The positively coiled DNA (6.0 kb) with a superhelical density of +0.05 and negatively coiled DNA (6.0 kb) with a superhelical density of –0.05 were reconstituted with ^{3}H lysine, ^{3}H arginine-labeled histones H3, H4, H2A, H2B by NaCl dialysis from 2.0 M to 1.2 M to 0.6 M to 0.1 M NaCl over a 14 h period. The samples were incubated with NAP1 at 35 °C for 5 min and applied to a 5–20% sucrose/100 mM NaCl/40 mM Tris, pH 7.8 gradient. After sedimentation at 200,000 × g for 5 h, fractions were collected and the distribution of DNA (bottom panel) was determined on agarose gel and the distribution of protein (top panel) on SDS-PAGE followed by fluorography. These data are unpublished observations (V. Levchenko and V. Jackson). The deg-H2A is degraded H2A in which a 15 amino acid peptide of the "C" terminal has been proteolytically removed. When H2A, H2B is no longer present in a nucleosome, the "C" terminal region is sensitive to proteolysis [126] from a protease which is a minor contaminate in the NAP1 preparation.

nucleosomal template changed to a mobility on gel electrophoresis which was consistent with the loss of a H2A, H2B dimer. To test for this possibility they added back exogenous H2A, H2B and were able to produce the original mobility of the nucleosomal template. This displacement of H2A, H2B in the higher salt condition may simulate the action of FACT. It is not likely that the 300 mM NaCl concentration is responsible for the lack of retrograde positioning of the nucleosome for Walters and Studitsky [130] have applied those same conditions to the SP6 polymerase and continued to observe the same repositioning. It may be that what is being observed in the RNA polymerase II process are mechanisms which are best represented by the "disruptive" model such that the H3, H4 tetramer is able to maintain its relative position even under the influence of a large RNA polymerase II complex. These studies have been done with small linear templates that do not preserve transcription-induced stress. It would be of interest to establish transcription conditions which would allow the presence of such stresses in an environment in which a chaperone (FACT or NAP1) were also included.

The Jackson laboratory has studied the binding of H2A, H2B and H3, H4 separately on positively and negatively coiled DNA. They found that H2A, H2B have a high preference for negatively coiled DNA and will transfer from positive to negative at physiological ionic strength (100 mM) at a rate faster than 5 s at 4°C [131]. In contrast when H3, H4 were added to a mix of highly positively and negatively coiled DNA and reconstituted by NaCl dialysis, the H3, H4 maintained a preference for the positively coiled DNA [132,100]. When both the H2A, H2B and H3, H4 were reconstituted together onto these DNAs, a 20 fold preference for negatively coiled DNA was observed [100], consistent with the earlier observations of Clark and Felsenfeld ([119], see also Ref. [133]). In fact not only do the core histones have a preference for negatively coiled DNA, in assembly experiments it has been observed that the rate of nucleosome formation is virtually instantaneous for this DNA [134]. Step A in the "dispersive" model of Fig. 1B could now be defined as the formation of transcription-induced positive stress that functions to stabilize the H3, H4 tetramer on DNA and at the same time cause the displacement of H2A, H2B (Step B), particularly when facilitated by a histone chaperone. Van Holde *et al.* [3] originally eluted to positive stress as being a potential factor based on the earlier experiments of Pfaffle *et al.* [135], which were designed to examine the effects of these induced stresses on nucleosomes. The additional experimentation since that time has now provided more credibility to the function of these induced stresses. The preference of H2A, H2B to bind negatively coiled DNA also gives a clearer understanding of Step D in which H2A, H2B is shown to reassociate with H3, H4 and DNA after polymerase passage. The negative stress in the wake of the polymerase would be a powerful factor to cause a rather rapid reformation of those interactions.

3.7. The chiral transition of H3, H4

The preference of H3, H4 to bind positively coiled DNA [100,136] was ascribed to a decrease in the helical pitch brought about by the overwinding of the DNA helix. Eight arginines in the globular regions of the H3, H4 tetramer form salt bridges with the DNA phosphates in the minor groove [59]. Prunell's laboratory subsequently evaluated the state of the tetramer on a small 359 bp miniplasmid. They have proposed that the tetramer undergoes a "chiral" transition in which there is a re-orientation around the H3–H3 interphase of the tetramer [136]. Instead of a left-handed pitch on the surface of the H3, H4 tetramer, a right-handed pitch is now present which could occur only if there was either a full 360° rotation around that interphase or a complete disruption and re-formation of that interphase. Positively coiled miniplasmids were found to facilitate this transition. A chemical modification of the cysteine 110 in H3 with 5,5′-dithiobis-2-nitrobenzoic acid (DTNB) was also found to hold the tetramer in this right-handed state [136]. Chemical modifications with either N-ethylmaleimide (NEM) or monobromobimane (mBrB) prevented this transition and held the H3,H4 tetramer in the left-handed state [137]. It was also found that when in the right-handed conformation, H3, H4 would not efficiently form nucleosomes with H2A, H2B [138]. If these observations are applied to the "disruptive" transcription mechanism of Fig. 1B, one would predict that the chiral transition of the H3, H4 tetramer could not occur in a nucleosome unless both the entry and exit site H2A, H2B dimers were first displaced from the tetramer and the positively coiled DNA. Only then would a 360° rotation or disruption of the H3–H3 interphase be possible. This conclusion is based on the known contact sites between H2A, H2B and H3, H4 in the nucleosome [55,58,59]. It is unclear at this time whether both this chiral transition and/or stabilization of interactions in the minor groove are responsible for the preference of H3, H4 for positively coiled DNA. That such a preference exists is of importance for it provides a potential mechanism for ensuring that H3, H4 is minimally displaced during transcription; a condition that is observed *in vivo*. As described in the previous section, the cysteine 110 of H3 becomes accessible to chemical reagents as a result of transcription *in vivo* [62] and active genes can be isolated on Hg columns because of this accessibility [65,66]. It is tempting to speculate that in the process of changing from the left to right-handed state, the cysteines might become accessible and is a partial explanation for the "U" type structures that have been seen for fractions isolated by those procedures [67].

3.8. Histone acetylation

Even with the removal of H2A, H2B, transcription of templates containing H3, H4 continues to be highly inhibitory [81,139]. This leads us back to the original observations of Wasylyk and Chambon [76] in which they suggested that the N-terminal tails were responsible for the transcriptional repression that could be relieved when 300 mM ammonium sulfate was present. *In vitro* transcription studies have indicated that removal of the N-terminal tails by either trypsin or clostripain treatment of the histones results in a less repressive transcription [139,140]. Chemical acetylation of the histones, which is expected to in part simulate the removal of the N-terminal tails, does increase transcription efficiency [141]. However, studies with naturally occurring acetylated histones have resulted in mixed reviews regarding their effect. Protacio *et al.* [140] observed that acetylation of H3, H4 and not H2A, H2B greatly enhanced T7 RNA polymerase transcription of a small nucleosomal template. Yet Mathias *et al.* [142] and Roberge *et al.* [143] did not observe any effect on transcription when using RNA polymerases I, II, and III respectively on a multi-nucleosomal template. It would appear that the effects of this modification are more apparent on single nucleosomes rather than on multi-nucleosomes. Kim *et al.* [144] have recently purified the human Elongator complex [145] which binds RNA polymerase II and has a histone acetyl transferase activity (HAT) with specificity for H3 and less so for H4. When present along with the FACT complex and acetyl CoA (for acetylation by the HAT), transcription through a small nucleosomal template was enhanced. Inhibition was still substantial compared to naked DNA, but it is an indication that the acetylation process partly relieved the salt dependence of transcription by the polymerase. What other factors are missing that would reduce the remaining inhibition remains to be determined. On a follow up note, an interesting observation by Sivolob *et al.* [146] is that the H3 tails are an inhibitory factor in preventing the transition of the H3, H4 tetramer from the left to right-handed state. Acetylation was found to facilitate this transition. Morales and Richard-Foy [147] have observed a similar dependence of this transition on acetylation. This observation suggests that it may be more than correlative that highly acetylated histones are part of nucleosomal complexes that bind Hg columns [65,66]. Perhaps when transcription protocols are able to establish relevant transcription-induced stresses, it will be possible to test whether the chiral transition as supported by positive stress and acetylation may be yet another additional factor that facilitates transcription through the H3, H4 tetramer when on multi-nucleosomal templates. The chapters by Ausio and Abbott (Chapter 10) and Turner (Chapter 11) will provide a more detailed description of the potential function of histone acetylation. For a more extensive review on protein factors, which facilitate transcription through nucleosomes by RNA polymerase II, see the review by Orphanides and Reinberg [148].

3.9. Histone H1

As indicated from the studies *in vivo*, histone Hl is present on active genes although perhaps at a somewhat depleted state because of displacement during transcription [149]. *In vitro* studies have been done to determine what happens to nucleosomes if H1 is present during transcription. The results of those studies have shown that nothing happens to the nucleosomes. H1 causes complete repression of initiation and elongation [150,151]. Because H1 facilitates the formation of higher-order structure, transcription is almost totally inhibited irrespective of whether a procaryotic or eucaryotic polymerase is used [152–154]. How then is this effect of H1 alleviated *in vivo*? A major factor appears to be the HMG14/17 proteins, which are enriched in the nuclease-solubilized fractions that contain active genes [155]. By immuno-selection techniques these proteins also appear to be associated with gene regions that are not only actively transcribed, but also have the potential for transcription [156]. Incorporation of these proteins into nascent nucleosomes has been shown to induce an extended chromatin conformation [157]. Ding *et al.* [158] have also shown that during *in vitro* transcription with a Hela nuclear extract and a SV40 minichromosome template containing H1, the addition of HMG14 suppressed the H1-mediated repression. Thus non-histone proteins such as the HMGs may contribute significantly to transcription of multi-nucleosomal templates when H1 is present. The chapters by Jerzmanowski (Chapter 4) and West and Bustin (Chapter 6) will provide a more detailed picture of the role of these proteins in this and other nuclear functions.

4. Summary

The studies which have examined the state of the nucleosome within the cell have generally shown that under conditions of active transcription the nucleosomes are frequently in a disrupted state in which there appears to be an exchange of H2A, H2B and probably histone H1. The exchange of histones H3, H4 is limited. These histones appear to serve as the core set of proteins that facilitate reassembly of the nucleosome by the reassociation of H2A, H2B. Therefore the disrupted state is generally not a state in which histones are missing. It is a state in which interactions between histones and DNA as well as between the histones themselves are being transiently displaced through the action of the polymerase. Histone modifications such as acetylation and ubiquitination as well as the incorporation of histone variants may result in structural changes that destabilize the nucleosome either directly or perhaps indirectly by acting as markers for interaction with other proteins. The *in vitro* studies have indicated the potential involvement of two additional processes that could facilitate nucleosome disruption, transcriptionally-induced positive stress and the action of histone chaperones. The chaperones facilitate the release of H2A, H2B, particularly under the conditions of positive stress. This same positive stress may also alter the H3–H3 interphase in the H3, H4 tetramer which could have profound effects in facilitating transcription through the tetramer as well as ensuring stability of the tetramer on the transcribed DNA. The *in vitro* studies indicate that all four histones at a frequency of 1 in 4 nucleosomes can be displaced from a nucleosome and transferred to the nascent RNA during transcription. There must be *in vivo* mechanisms that limit the displacement of H3, H4, particularly, when one considers the highly concentrated state of the polyanionic chromatin [159] and transcribed RNA in the nucleus of a cell. The *in vitro* studies have also indicated that the different RNA polymerases may not transcribe through nucleosomes in the same way. A "spooling" mechanism is observed for SP6 RNA polymerase and RNA polymerase III in which histones are retrogradingly repositioned behind the transcribing polymerase. For RNA polymerase II a "dispersive" mechanism is observed in which H2A, H2B is displaced but not H3, H4. These observations suggest that the remarkable structure that is the nucleosome has the potential to be transcribed in more than one way and is an indication that both mechanisms may have relevance for particular transcription conditions in the nucleus of a cell. With regard to our question at the beginning of our discussion, the *in vitro* studies do appear to recapitulate the *in vivo* conditions and have in fact contributed significantly to our understanding of what happens to nucleosomes during transcription.

References

1. Thoma, F. (1991) Trends Genet. 7, 175–177.
2. Kornberg, R.D. and Lorch, Y. (1991) Cell 67, 833–836.
3. van Holde, K.E., Lohr, D.E., and Robert, C. (1992) J. Biol. Chem. 267, 2837–2840.
4. Kornberg, R.D. and Lorch, Y. (1992) Annu. Rev. Cell Biol. 8, 563–587.
5. Adams, C.C. and Workman, J.L. (1993) Cell 72, 305–308.
6. Felsenfeld, G., Clark, D., and Studitsky, V. (2000) Biophys. Chem. 86, 231–237.
7. Noll, M. (1974) Nucleic Acids Res. 1, 1573–1578.
8. Lutter, L.C. (1978) J. Mol. Biol. 124, 391–420.
9. Weintraub, H. and Groudine, M. (1976) Science 193, 848–856.
10. Elgin, S.C.R. (1981) Cell 27, 413–415.
11. Gross, D.S. and Garrard, W.T. (1988) Annu. Rev. Biochem. 57, 159–197.
12. Reeves, R (1984) Biochem. et Biophys. Acta 782, 343–393.
13. Noll, M. (1974) Nature 251, 249–251.
14. Noll, M. and Kornberg, R.D. (1977) 109, 393–404.
15. Bellard, M., Gannon, F., and Chambon, P. (1977) In: Chromatin, Vol. XLII. Cold Spring Harbor Symp. Quant. Biol., Cold Spring Harbor, NY, pp. 700–791.
16. Wu, C., Wong, Y.C., and Elgin, S.C.R. (1979) Cell 16, 807–814.
17. Lohr, D. (1983) Nucleic Acids Res. 11, 6755–6773.
18. Sun, Y.L., Yuan, A.X., Bellard, M., and Chambon, P. (1986) EMBO J. 5, 293–300.
19. Reeves, R. (1978) Biochemistry 17, 4908–4916.
20. Colavito-Shepansky, M. and Gorovsky, M.A. (1983) J. Biol. Chem. 258, 5944–5954.
21. Davies, A.H., Reudehaber, T.L., and Garrard, W.T. (1983) J. Mol. Biol. 167, 133–155.
22. Johnson, E.M., Allfrey, V.G., Bradbury, E.M., and Matthews, H.R. (1978) Proc. Natl. Acad. Sci. USA 75, 1116–1120.
23. Wu, C. (1980) Nature 286, 854–860.
24. Nedospasov, S.A. and Georgiev, G.P. (1980) Biochem. Biophys. Res. Commun. 92, 532–539.
25. Samal, B. and Worcel, A. (1981) Cell 23, 401–409.
26. Udvardy, A. and Schedl, P. (1984) J. Mol. Biol. 172, 385–403.
27. Cartwright, I.L. and Elgin, S.C.R. (1986) Mol. Cell. Biol. 6, 779–791.
28. Benezra, R., Cantor, C.R., and Axel, R. (1986) Cell 44, 697–704.
29. Lee, M.S. and Garrard, W.T. (1991) EMBO J. 10, 607–615.
30. Cavalli, G. and Thoma, T. (1993) EMBO J. 12, 4603–4613.
31. Pederson, D.S. and Morse, R.H. (1990) EMBO J. 9, 1873–1881.
32. De Bernardin, W., Koller, T., and Sogo, J.M. (1986) J. Mol. Biol. 191, 469–482.
33. Foe, V.E. (1977) In: Chromatin, Vol. XLII. Cold Spring Harbor Symp. Quant. Biol., Cold Spring Harbor, NY, pp. 723–740.
34. Sogo, J.M., Ness, P.J., Widmer, R.M., Parish, R.W., and Koller, T. (1984) J. Mol. Biol. 178, 897–928.
35. Conconi, A., Widmer, R.M., Koller, T., and Sogo, J.M. (1989) Cell 57, 753–761.
36. Sanders, S. (1978) J. Cell Biol. 79, 97–109.
37. Ridsdale, J.D. and Davie, J.R. (1987) Nucleic Acids Res. 15, 1081–1096.
38. Nickel, B.E., Allis, C.D., and Davie, J.R. (1989) Biochemistry 28, 958–963.
39. Perry, M. and Chalkley, R. (1981) J. Biol. Chem. 256, 3313–3318.
40. Ridsdale, J.A., Hendzel, M.J., Delcuve, G.P., and Davie, J.R. (1990) J. Biol. Chem. 265, 5150–5156.
41. Ip, Y.T., Jackson, V., Meier, J., and Chalkley, R. (1986) J. Biol. Chem. 263, 14044–14052.
42. Ericsson, C., Grossbach, U., Bjorkroth, B., and Daneholt, B. (1990) Cell 60, 73–83.
43. Kamakaka, R.T. and Thomas, J.O. (1990) EMBO J. 9, 3997–4006.
44. Karpov, V.L., Preobrazhenskaya, O.V., and Mirzabekov, A.D. (1984) Cell 36, 423–431.
45. Nacheva, G.A., Guschin, D.Y., Preobrazhenskaya, O.V., Karpov, V.L., Ebralidse, K.K., and Mirzabekov, A.D. (1989) Cell 58, 27–36.
46. Ausio, J. and Abbott, D.W. (2002) Biochemistry 41, 5945–5949.

47. Ahmed, K. and Henikoff, S. (2002) Proc. Natl. Acad. Sci. USA 99, 16477–16484.
48. Stargell, M.S., Bowen, J., Dadd, C.A., Dedon, P.C., Davis, M., Cook, R.G., Allis, C.D., and Gorovsky, M.A. (1993) Genes Dev. 7, 2641–2651.
49. Allis, C.D. and Wiggins, J.C. (1984) Dev. Biol. 101, 282–294.
50. Suto, R.K., Clarkson, J.J., Tremethick, D.J., and Luger, K. (2000) Nat. Struct. Biol. 7, 1121–1124.
51. Fan, J.Y., Gordon, F., Luger, K., Hansen, J.C., and Tremethick, D.J. (2002) Nat. Struct. Biol. 9, 172–176.
52. Davie, J.R. and Murphy, L.C. (1990) Biochemistry 29, 4752–4757.
53. Huang, S.Y, Barnard, M.B., Xu, M., Matsui, S., Rose, S.M., and Garrard, W.T. (1986) Proc. Natl. Acad. Sci. USA 83, 3738–3742.
54. Davies, N. and Lindsey, G. (1994) Biochim. Biophys. Acta 1218, 187–193.
55. Luger, K., Mader, A.W., Richmond, R.K., Sargent, D.F., and Richmond, T.J. (1997) Nature 389, 251–260.
56. Sun, Z.W. and Allis, C.D. (2002) Nature 418, 104–108.
57. Ng, H.H., Xu, R.M., Zhang, Y., and Struhl, K. (2002) J. Biol. Chem. 277, 34655–34667.
58. Arents, G., Burlingame, R.W., Wang, B., Love, W.E., and Moudrianakis, E.N. (1991) Proc. Natl. Acad. Sci. USA 88, 10148–10152.
59. Harp, J.M., Hanson, B.L., Timm, D.E., and Bunick, G.J. (2000) Acta Crystallogr. D Biol. Crystallogr. 56, 1513–1534.
60. Wong, N.T.N. and Candido, E.P.M. (1978) J. Biol. Chem. 253, 8263–8268.
61. Feinstein, D.L. and Moudrianakis, E.N. (1986) Biochemistry 25, 8409–8418.
62. Prior, C.P, Cantor, C.R., Johnson, E.M., Littau, V.C., and Allfrey, V.G. (1983) Cell 34, 1033–1042.
63. Walker, J., Chen, T.A., Sterner, R., Berger, M., Winston, F., and Allfrey, V.G. (1990) J. Biol. Chem. 265, 5736–5746.
64. Chan, S., Attisano, L., and Lewis, P.N. (1988) J. Biol. Chem. 263, 15643–15651.
65. Johnson, E.M., Sterner, R., and Allfrey, V.G. (1987) J. Biol. Chem. 262, 6943–6946.
66. Chen, T.W., Sterner, R., Cozzolino, A., and Allfrey, V.G. (1990) J. Mol. Biol. 212, 481–493.
67. Bazett-Jones, D.P., Mendez, E., Czarnota, G.J., Ottensmeyer, F.P., and Allfrey, V.G. (1996) Nucleic Acids Res. 24, 321–329.
68. Jackson, V. and Chalkley, R. (1985) Biochemistry 24, 6921–6930.
69. Jackson, V. and Chalkley, R. (1981) J. Biol. Chem. 256, 5095–5103.
70. Hendzel, M.J. and Davie, J.R. (1990) Biochem. J. 271, 67–73.
71. Perry, C.A., Dadd, C.A., Allis, C.D., and Annunziato, A.T. (1993) Biochemistry 32, 13605–13614.
72. Jackson, V. (1987) Biochemistry 26, 2315–2324.
73. Jackson, V. (1990) Biochemistry 29, 719–731.
74. Kimura, H. and Cook, P.R. (2001) J. Cell Biol. 153, 1341–1353.
75. Misteli, T., Gunjan, A., Hock, R., Bustin, M., and Brown, D.T. (2000) Nature 408, 877–881.
76. Wasylyk, B. and Chambon, P. (1979) Eur. J. Biochem. 98, 317–327.
77. Morse, R.H. (1989) EMBO J. 8, 2343–2351.
78. Felts, S.J., Weil, P.A., and Chalkley, R. (1990) Mol. Cell. Biol. 10, 2390–2401.
79. Hansen, J.C. and Wolffe, A.P. (1992) Biochemistry 31, 7977–7988.
80. Izban, M.G. and Luse, D.S. (1991) Genes Dev. 5, 683–696.
81. Chang, C.-H. and Luse, D.S. (1997) J. Biol. Chem. 272, 23427–23434.
82. Protacio, R.U. and Widom, J. (1996) J. Mol. Biol. 256, 458–472.
83. O'Neill, T.E., Roberge, M., and Bradbury, E.M. (1992) J. Mol. Biol. 223, 67–78.
84. Lorch, Y., LaPointe, J.W., and Kornberg, R.D. (1987) Cell 49, 203–210.
85. Losa, R. and Brown, D.D. (1987) Cell 50, 801–808.
86. O'Donohue, M.-F., Duband-Goulet, I., Hamiche, A., and Prunell, A. (1994) Nucleic Acids Res. 22, 937–945.
87. Gallego, F., Fernandez-Busquets, X., and Daban, J.-R. (1995) Biochemistry 34, 6711–6719.
88. Clark, D.J. and Felsenfeld, G. (1992) Cell 71, 11–22.
89. Studitsky, V.M., Clark, D.J., and Felsenfeld, G. (1994) Cell 76, 371–382.

90. Studitsky, V.M., Clark, D.J., and Felsenfeld, G. (1995) Cell 83, 19–27.
91. Bednar, J., Studitsky, V.M., Grigoryev, S.A., Felsenfeld, G., and Woodcock, C.L. (1999) Mol. Cell 4, 377–386.
92. Studitsky, V.M., Kassavetis, G.A., Geiduschek, E.P., and Felsenfeld, G. (1997) Science 278, 1960–1963.
93. Meersseman, G., Pennings, S., and Bradbury, E.M. (1992) EMBO J. 11, 2951–2959.
94. Peng, H.F. and Jackson, V. (1997) Biochemistry 36, 12371–12382.
95. ten Heggeler-Bordier, B., Muller, S., Monestier, M., and Wahli, W. (2000) J. Mol. Biol. 299, 853–858.
96. Kirov, N., Tsaneva, I., Einbinder, E., and Tsanev, R. (1992) EMBO J. 11, 1941–1947.
97. O'Neill, T.E., Smith, J.G., and Bradbury, E.M. (1993) Proc. Natl. Acad. Sci. USA 90, 6203–6207.
98. Feng, H.-P., Scherl, D.S., and Widom, J. (1993) Biochemistry 32, 7824–7831.
99. Godfrey, J.E., Eickbush, T.H., and Moudrianakis, E.N. (1980) Biochemistry 19, 1339–1346.
100. Jackson, S., Brooks, W., and Jackson, V. (1994) Biochemistry 33, 5392–5403.
101. Baer, R.W. and Rhodes, D. (1983) Nature 301, 482–488.
102. Gonzalez, P.J. and Palacian, E. (1989) J. Biol. Chem. 264, 18457–18462.
103. Liu, L.F. and Wang, J.C. (1987) Proc. Natl. Acad. Sci. USA 84, 7024–7027.
104. Wu, H.Y., Shyy, S., Wang, J.C., and Liu, L.F. (1988) Cell 53, 433–440.
105. Tsao, Y.P., Wu, H.Y., and Liu, L.F. (1989) Cell 56, 111–118.
106. Lodge, J.K., Kazic, T., and Berg, D.E. (1989) J. Bacteriol. 171, 2181–2187.
107. Ostrander, E.A., Benedetti, P., and Wang, J.C. (1990) Science 249, 1261–1265.
108. Rahmouni, A.R. and Wells, R.D. (1992) J. Mol. Biol. 223, 131–144.
109. Cook, D.N., Ma, D., Pon, N.G., and Hearst, J.E. (1992) Proc. Natl. Acad. Sci. USA 89, 10603–10607.
110. Brill, S.J. and Sternglanz, R. (1988) Cell 54, 403–411.
111. Giaever, G.N. and Wang, J.C. (1988) Cell 55, 849–856.
112. Ljungman, M. and Hanawalt, P.C. (1992) Proc. Natl. Acad. Sci. USA 89, 6055–6059.
113. Kramer, P.R. and Sinden, R.R. (1997) Biochemistry 36, 3151–3158.
114. Lee, M.-S. and Garrard, W.T. (1991) Proc. Natl. Acad. Sci. USA 88, 9675–9679.
115. Peng, H.F. and Jackson, V. (2000) J. Biol. Chem. 275, 657–668.
116. Gruss, C. and Sogo, J.M. (1992) Bioessays 14, 1–8.
117. Smith, S. and Stillman, B. (1991) EMBO J. 10, 971–980.
118. Dong, F. and van Holde, K.E. (1991) Proc. Natl. Acad. Sci. USA 88, 10596–10600.
119. Clark, D.J. and Felsenfeld, G. (1991) EMBO J. 10, 387–395.
120. Clark, D.J., Ghirlando, R., Felsenfeld, G., and Eisenberg, H. (1993) J. Mol. Biol. 234, 297–301.
121. Jackson, V. (1993) Biochemistry 32, 5901–5912.
122. Ishmi, Y. and Kikuchi, A (1991) J. Biol. Chem. 266, 7025–7029.
123. Workman, J.L. and Kingston, R.E. (1998) Ann. Rev. Biochem. 67, 545–579.
124. Ito, T., Tyler, J.K., and Kadonaga, J.T. (1997) Genes Cells 2, 593–600.
125. Ishmi, Y., Hirosumi, J., Sato, W., Sugasawa, K., Yokota, S., Hanoaka, F., and Yamada, M. (1984) Eur. J. Biochem. 142, 431–439.
126. Eickbush, T.H., Godfrey, J.E., Elia, M.C., and Moudranakis, E.N. (1988) J. Biol. Chem. 263, 18972–18978.
127. Orphanides, G., LeRoy, G., Chang, C.H., Luse, D.S., and Reinberg, D. (1998) Cell 92, 105–116.
128. Orphanides, G., Wu, W.H., Lane, W.S., Hampsey, M., and Reinberg, D. (1999) Nature 400, 284–288.
129. Kireeva, M.L., Walter, W., Tchernajenko, V., Bondarenko, V., Kashlev, M., and Studitsky, V.M. (2002) Mol. Cell 9, 541–552.
130. Walters, W. and Studitsky, V.M. (2001) J. Biol. Chem. 276, 29104–29110.
131. Brooks, W. and Jackson, V. (1994) J. Biol. Chem. 269, 18155–18166.
132. Jackson, V. (1995) Biochemistry 34, 10607–10619.
133. Patterson, H.-G. and von Holt, C. (1993) J. Mol. Biol. 229, 623–636.
134. Pfaffle, P. and Jackson, V. (1990) J. Biol. Chem. 265, 16821–16829.

135. Pfaffle, P., Gerlach, V., Bunzel, L., and Jackson, V. (1990) J. Biol. Chem. 265, 16830–16840.
136. Hamiche, A., Carot, V., Alilat, M., De Lucia, F., O'Donohue, M.-F., Revet, B., and Prunell, A. (1996) Proc. Natl. Acad. Sci. USA 93, 7588–7593.
137. Hamiche, A. and Richard-Foy, H. (1998) J. Biol. Chem. 273, 9261–9269.
138. Alilat, M., Sivolob, A., Revet, B., and Prunell, A. (1999) J. Mol. Biol. 291, 815–841.
139. Hernandez, F., Lopez-Alarcon, L., Puerta, C., and Palacian, E. (1998) Arch. Biochem. Biophys. 358, 98–103.
140. Protacio, R.U., Li, G., Lowary, P.T., and Widom, J. (2000) Mol. Cell. Biol. 20, 8866–8878.
141. Hernandex, F., Puerta, C., Lopez-Alarcon, L., and Palacian, E. (1995) Biochem. Biophys. Res. Commun. 213, 232–238.
142. Mathias, D.J., Oudet, P., Wasylyk, B., and Chambon, P. (1978) Nucleic Acids Res. 5, 3523–3547.
143. Roberge, M., O'Neill, T.E., and Bradbury, E.M. (1991) FEBS Lett. 288, 215–218.
144. Kim, J.-H., Lane, W.S., and Reinberg, D. (2002) Proc. Natl. Acad. Sci. USA 99, 1241–1246.
145. Winkler, G.S., Kristjuhan, A., Erdjument-Gromage, H., Tempst, P., and Svejstrup, J.O. (2002) Proc. Natl. Acad. Sci. USA 99, 3517–3522.
146. Sivolob, A., De Lucia, F., Alilat, M., and Prunell, A. (2000) J. Mol. Biol. 295, 55–69.
147. Morales, V. and Richard-Foy, H. (2000) Mol. Cell. Biol. 20, 7230–7237.
148. Orphanides, G. and Reinberg, D. (2000) Nature 407, 471–475.
149. Zlatanova, J., Caiafa, P., and van Holde, K.E. (2000) FASEB J. 14, 1697–1704.
150. Croston, G.E., Kerrigan, L.A., Lira, L.M., Marshak, D.R., and Kadonaga, J.T. (1991) Science 251, 643–649.
151. O'Neill, T.E., Meersseman, G., Pennings, S., and Bradbury, E.M. (1995) Nucleic Acids Res. 23, 1075–1082.
152. Allan, J., Cowling, G.J., Harborne, N., Cattini, P., Craigie, R., and Gould, H. (1981) J. Cell Biol. 90, 279–288.
153. Thoma, F., Koller, T., and Klug, A. (1979) J. Cell Biol. 83, 403–427.
154. Leuba, S.H., Zlatanova, J., and van Holde, K. (1993) J. Mol. Biol. 229, 917–929.
155. Weisbrod, S. and Weintraub, H. (1979) Proc. Natl. Acad. Sci. USA 76, 630–635.
156. Druckmann, S., Mendelson, E., Landsman, D., and Bustin, M. (1986) Exp. Cell Res. 166, 486–496.
157. Trieschmann, L., Alfonso, P.J., Crippa, M.P., Wolffe, A.P., and Bustin, M. (1995) EMBO J. 14, 1478–1489.
158. Ding, H.-F., Bustin, M., and Hansen, U. (1997) Mol. Cell Biol. 17, 5843–5855.
159. Louters, L. and Chalkley, R. (1985) Biochemistry 24, 3080–3085.

Subject Index